Engineering Data Analysis with MATLAB®

This book provides a concise overview of a variety of techniques for analyzing statistical, scientific, and financial data, using MATLAB® to integrate several approaches to data analysis and statistics.The chapters offer a broad review of computational data analysis, illustrated with many examples and applications. Topics range from the basics of data and statistical analysis to more advanced subjects such as probability distributions, descriptive and inferential statistics, parametric and non-parametric tests, correlation, and regression analysis. Each chapter combines theoretical concepts with practical MATLAB® applications and includes practice exercises, ensuring a comprehensive understanding of the material.

With coverage of both basic and more complex ideas in applied statistics, the book has broad appeal for undergraduate students up to practicing engineers.

Dr. Tanvir Mustafy is a distinguished figure in the field of structural engineering, renowned for his expertise in complex finite element modeling at the micro level of structures, injury biomechanics, machine learning, and analytic mechanics. With a rich academic and research background, Dr. Mustafy has made significant contributions to the advancement of structural engineering. He holds a master of science (MSc) degree in structural engineering from the University of Alberta, Canada, and a bachelor of science (BSc) degree from the Bangladesh University of Engineering and Technology (BUET) in Dhaka, Bangladesh. Throughout his career, Dr. Tanvir Mustafy has consistently demonstrated a passion for research and a commitment to pushing the boundaries of structural engineering. His work continues to inspire and shape the future of the field, making him a respected authority in the academic and scientific community.

Dr. Tauhid Rahman earned his PhD in environmental engineering from Tohoku University, Japan, in 2009. He did his MSc in environmental engineering, land, and water engineering from KTH, Sweden, and his BSc in civil engineering from Bangladesh University of Engineering and Technology, Bangladesh. He is currently a professor in the CE Department of MIST. His research interests are water quality modeling, land use change detection, climate change, water insecurity, micro-climate effects, etc.

Nafisa Siddiqui is an accomplished educator with a bachelor of science in mathematics from the University of Texas at Austin, USA. She has served as a mathematics teaching assistant at Texas A&M University–Kingsville and the University of Texas at Tyler, demonstrating her dedication to facilitating mathematical learning. Nafisa's commitment to education and her strong academic background make her a valuable asset in the field of mathematics.

Engineering Data Analysis with MATLAB®

Tanvir Mustafy, Tauhid Rahman and Nafisa Siddiqui

CRC Press
Taylor & Francis Group
Boca Raton London New York

CRC Press is an imprint of the
Taylor & Francis Group, an **informa** business

First edition published 2024
by CRC Press
2385 NW Executive Center Drive, Suite 320, Boca Raton FL 33431

and by CRC Press
4 Park Square, Milton Park, Abingdon, Oxon, OX14 4RN

CRC Press is an imprint of Taylor & Francis Group, LLC

© 2024 Tanvir Mustafy, Tauhid Rahman and Nafisa Siddiqui

Library of Congress Cataloging-in-Publication Data

Names: Mustafy, Tanvir, author. | Rahman, Tauhid, author. | Siddiqui, Nafisa, author.
Title: Engineering data analysis with MATLAB® / Tanvir Mustafy, Tauhid Rahman and Nafisa Siddiqui.
Description: First edition. | Boca Raton FL : CRC Press, 2024. | Includes index.
Identifiers: LCCN 2023050857 | ISBN 9781032506586 (hardback) | ISBN 9781032507712 (paperback) | ISBN 9781003399582 (ebook)
Subjects: LCSH: Engineering--Statistical methods--Data processing. | Mathematical statistics--Data processing. | Quantitative research--Data processing. | MATLAB.
Classification: LCC TA340 .M87 2024 | DDC 620.00285--dc23/eng/20240426
LC record available at https://lccn.loc.gov/2023050857

ISBN: 978-1-032-50658-6 (hbk)
ISBN: 978-1-032-50771-2 (pbk)
ISBN: 978-1-003-39958-2 (ebk)

DOI: 10.1201/9781003399582

Typeset in Sabon
by Deanta Global Publishing Services, Chennai, India

Contents

5 Continuous probability distribution   247

6 Descriptive statistics

7 Inferential statistics

Preface

This book, *Engineering Data Analysis with MATLAB: Principles and Applications*, is a comprehensive resource for engineers, scientists, and students. It aims to provide valuable insights and practical knowledge to help you master the art of data analysis, specifically within the MATLAB environment.

In today's data-centric world, the ability to analyze and derive meaningful insights from data is paramount. This book is designed to equip you with the necessary skills and knowledge to excel in data analysis. Whether you are new to the field or an experienced practitioner, our goal is to offer a structured approach to understanding and applying data analysis techniques.

The book covers a wide range of topics from the basics of data and statistical analysis to more advanced subjects like probability distributions, descriptive and inferential statistics, parametric and non-parametric tests, correlation, and regression analysis. Each chapter combines theoretical concepts with practical MATLAB applications, ensuring a comprehensive understanding of the material. Our hope is that this book serves as a valuable companion in your journey through the world of data analysis. We have drawn upon academic and industry experiences to provide you with a resource that will enable you to confidently tackle data analysis challenges in various contexts.

We extend our gratitude to all those who contributed to the creation of this book, including colleagues, students, mentors, and reviewers. Your insights and support have been invaluable in bringing this project to fruition. We invite you to dive into the pages of *Engineering Data Analysis with MATLAB: Principles and Applications* and explore the dynamic field of data analysis. May this book empower you to make informed decisions, solve complex problems, and excel in your endeavors.

Chapter 1

Getting started

Data is being generated every second by numerous organizations around the globe. These companies depend on data to draw conclusions and finally use that data to shape their decision-making process, crucial for their businesses. However, the data being generated in its raw form is usually out of date and is often unstructured, inaccurate, and overly complex. It holds very little value until it is converted to a form such that it can then be analyzed.

To get the most out of this we need to first clean the data, transforming it so that it becomes easier to interpret and analyze. This entire process is data analysis. This sounds like it's extremely important and indeed it is. Undoubtedly data analysis is a priority for every business as it holds the key to unlocking potential profit, and clearly only choosing the right data analysis tool can turn these troves of data into usable information. So overall we could say that data analysis is defined as a process of cleaning and transforming data so that it becomes easier to interpret and analyze to detect trends, discover useful information, and draw conclusions from that information.

To achieve this, we use a number of data analysis tools like Microsoft Excel, Tableau, SAS, Python, R, MATLAB, etc. This book introduces basic procedures and methods of data analysis in MATLAB. MATLAB provides an integrated environment for a wide range of scientific and engineering applications like signal and image processing, control designing, machine learning, and most importantly data analysis. It is widely used by scientists and engineers in a range of industries for its excellent capabilities in performing analyses and optimizations and in deploying mission-critical calculations. Moreover, it has data exploration features such as graphs, descriptive statistics, and visualization which allow users to get valuable insights from the data warehouse.

At the end of a book we shall have a clear understanding of how we can use this software platform to parse through reams of data and derive useful intelligence from it across content areas.

1.1 WHAT IS DATA ANALYSIS?

Data analysis is an extremely sophisticated technique that involves several pre-steps to collect and cleanse large volume of unstructured data. There are different ways in which data analytics can be performed. However, there are five steps in a typical model workflow including:

1. Collecting the data
2. Importing the data into MATLAB
3. Data cleaning and transforming
4. Visualizing the data

DOI: 10.1201/9781003399582-1

5. Analyzing the data
6. Interpreting the results

In the coming sections we shall discuss each of the steps in detail.

1.1.1 Collecting the data

Data collection is the process of extracting and collecting voluminous amounts of data usually in an unstructured form like text, video, audio, XML files, records, or other images that is relevant to the current objective and needs to be analyzed. The data which is to be analyzed must be collected from different valid sources to ensure its accuracy such that the related decisions are valid. This makes defining the required data a prerequisite as it provides a baseline for company improvement. This step may involve any number of tasks to get the data in-hand, such as by doing experiments, conducting surveys or questionnaires, or searching for various sources ranging from organizational databases to the information in web pages. During this process it is necessary to identify the type of data to be analyzed and the methods to be used to collect the data. Most of the data collected is of two types, one of which is known as "qualitative data" which is non-numerical data such as words and sentences and is collected through methods of observations, one-to-one interviews, conducting focus groups, and similar methods. The other type is "quantitative data" which is in numerical forms and can be calculated using different scientific tools and sampling data.

It is widely recognized that Excel is one of the most commonly used tools for data analysis among engineers. It is an extremely powerful analysis tool for engineers due to its ability and versatility. Excel provides a convenient and familiar platform for organizing, analyzing, and storing data in a tabular format. This is why engineers rely on Excel to store their data.

In this book we use Excel to import data into MATLAB. Excel files are widely compatible and can be easily read by MATLAB, making it convenient for engineers to transfer data between the two platforms without the need for conversion.

1.1.2 Importing the data into MATLAB

Even though Excel offers a wide range of spreadsheet functionalities it is not advanced enough to cater to all data analysis needs, especially for complex engineering applications. For more complex analyses, for larger datasets, engineers may need to explore dedicated statistical software or programming languages like MATLAB.

These tools offer more advanced functionalities and are designed specifically for data analysis and modeling. MATLAB provides a comprehensive Data Analysis Toolbox that offers a wide range of built-in functions and tools for data manipulation, exploration, visualization, and modeling. Hence we import the Excel data into MATLAB to access and utilize the data directly in MATLAB for further analysis and processing.

Importing data actually means loading data into our analysis platform from an external file like txt, jpg, clipboard, etc. This won't work or operate without an import function. MATLAB provides multiple import functions to import data from various file formats such as comma-separated values (CSV), Excel, and text files. CSV is a simple file format used for storing tabular data, where each line of the file represents a single row and each value in a row is separated by a comma. In MATLAB, you can use the readtable function to import data from a CSV file. The readtable function reads the CSV file and creates a MATLAB table object, which allows for the easy manipulation and analysis of the data. Here's an example (Table 1.1) of how to import data from a CSV file using the readtable function.

Table 1.1

After importing the data, you can access and manipulate the table using various table functions and indexing operations available in MATLAB.

Another commonly used function for importing tabular data is xlsread even though it is not recommended. Another function is import data which allows users to load various data files from spreadsheet files, delimited text files, and fixed-width text files. Moreover, we can use text files, Excel files, notepad, and also images in different formats. The basic syntax of the import function requires the function along with the name of the file which we are going to use in our program.

```
A = importdata('image.jpg');
image(A)
```

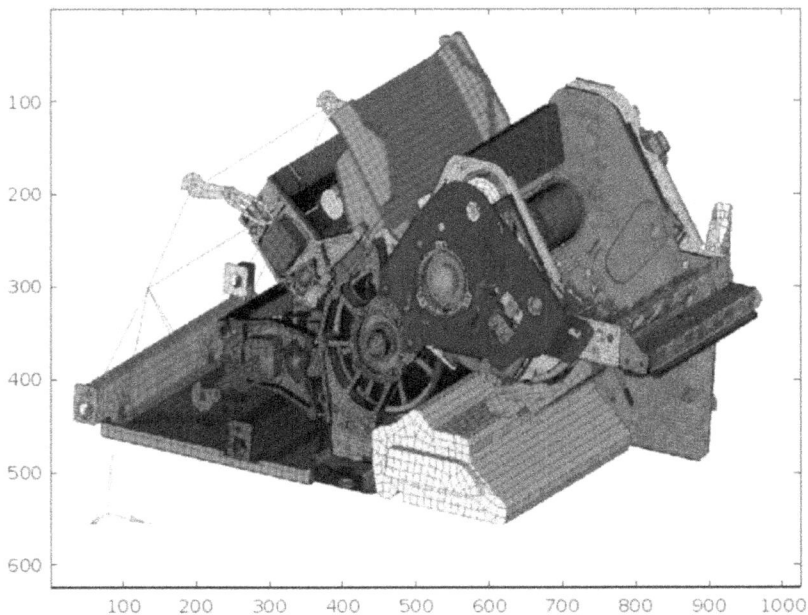

Figure 1.1

1.1.3 Data cleaning and transforming

As mentioned earlier, raw data is seldom usable in its current form. There will be errors within it, like missing or repeated values or even an overly complex structure. Even though the flaws might seem extremely minor, these can actually be quite pernicious: even the tiniest inaccuracies can skew our results and we could end up with a Type I or II error in our conclusion.

Hence it is required that the data collected is first subjected to cleaning and processing to remove or fix those blunders. This step is extremely important before the analysis as the accuracy of our analysis entirely depends on the quality of our data and we do not want the results to be influenced by the wrong points. The primary duties involved in cleaning the data include getting rid of errors, duplicates, extreme outliers, or unwanted data points that are encountered when the data is aggregated from multiple sources. Data transforming/processing could include bringing structure to our data which could help in mapping and maneuvering the data in a simple manner.

Even though the entire process of cleaning and transforming data seems like a challenging task, MATLAB provides numerous functions to make this process faster, more efficient, and reliable so we can focus on our analysis and problem solving. Common data cleaning tasks include:

- **removing missing data**

```
R = rmmissing(data) removes missing entries from an array or table.
Missing values are defined according to the data type of our data:
```

- NaN
- NaT
- <missing>
- <undefined>
- {''}

```
File = readtable("missingdata.csv", VariableNamingRule = "preserve")
R = rmmissing(File)
```

- **removing outliers**

```
B = rmoutliers(data, method) detects and removes outliers from the data
using method specified as one of these values.
```

Table 1.2

Cement	Blast Furna	Water	Coarse Ag	Fine Aggre	Strength
{''}	120.2	162	1040	676	79.99
540	138.4	162	1055	676	61.89
332.5	142.5	228	932	594	40.27
332.5	142.5	228	932	594	NaN
198.6	132.4	192	<missing>	825.5	44.3

Table 1.3

```
R =

  2×6 table

    Cement    Blast Furnace Slag    Water    Coarse Aggregate    Fine Aggregate    Strength
    _____    _____    _____    _____    _____    _____

       540                 138.4      162                1055               676       61.89
     332.5                 142.5      228                 932               594       40.27
```

Method	Description
"median"	Outliers are defined as elements more than three scaled MAD from the median. The scaled MAD is defined as c*median(abs(A − median(A))), where c = −1/(sqrt(2)*erfcinv(3/2)).
"mean"	Outliers are defined as elements more than three standard deviations from the mean. This method is faster but less robust than "median".
"quartiles"	Outliers are defined as elements more than 1.5 interquartile ranges above the upper quartile (75%) or below the lower quartile (25%). This method is useful when the data in A is not normally distributed.
"grubbs"	Outliers are detected using Grubbs' test for outliers, which removes one outlier per iteration based on hypothesis testing. This method assumes that the data in A is normally distributed.
"gesd"	Outliers are detected using the generalized extreme Studentized deviate test for outliers. This iterative method is similar to "grubbs" but can perform better when there are multiple outliers masking each other.

Example 1

```
Data = readtable("outlierdata.csv", VariableNamingRule = "preserve");
R = rmoutliers(Data);
```

- filling missing data

```
F = fillmissing(Data, method) fills missing entries using the method
specified as one of these values.
```

Table 1.4

Material A	Material B	Material C
705	1019	974.4
720	1033.5	975.84
74	1041.1	976.32
740	1054.9	977.04
755	1063.1	978.24
765	10.79	980.64
770	1109.3	980.88
770	1128.8	9820.32
785	1130.2	985.2
800	1140.3	986.4

Table 1.5

```
R =

    7×3 table

        Materia A      Material B      Material C

        _____       _____        _____

          705            1019            974.4
          720            1033.5          975.84
          740            1054.9          977.04
          755            1063.1          978.24
          770            1109.3          980.88
          785            1130.2          985.2
          800            1140.3          986.4
```

'previous'	Previous nonmissing value
'next'	Next nonmissing value
'nearest'	Nearest nonmissing value
'linear'	Linear interpolation of neighboring, nonmissing values (numeric, duration, and datetime data types only)
'spline'	Piecewise cubic spline interpolation (numeric, duration, and datetime data types only)
'pchip'	Shape-preserving piecewise cubic spline interpolation (numeric, duration, and datetime data types only)
'makima'	Modified Akima cubic Hermite interpolation (numeric, duration, and datetime data types only)

Example 1:

```
file = readtable("missingdata2.csv", VariableNamingRule = "preserve")
```

Now if we look at Table 1.6 we will notice a missing value in all of the columns except for the third one. Now which filling method do you think we should use in this example?

Let's look at the first column. Notice that the column of temperature data linearly increases by 2; therefore the missing value should be 38 + 2 = 40. The second column is a similar case. Since the second column of temperature is in Kelvin, each temperature value in this column is equal to temperature (°C) + 273. Since the second column depends on the values of the first column and is added to a constant, the values in the second column also increase proportionately by 2. Therefore, the missing value in this column should be 305 + 2 = 307. The best filling method in this case would be 'linear'.

Now let's look at the fourth column of resistance values. Clearly the values are decreasing but not linearly. Therefore, we can only approximate the values. In such a situation it is

Table 1.6

filename =

16×6 table

Temperature(°C)	Temperature(°K)	1/T (/K)	Resistance (k?)	ln R(?)	Current(mA)
20	293	0.003413	57.5	4.0518	0.078902
22	295	0.0033898	50.9	3.9299	0.11375
24	297	0.003367	46.6	3.8416	0.14823
26	299	0.0033445	41.8	NaN	0.20538
28	301	0.0033223	37.2	3.6163	0.29138
30	303	0.0033003	33.3	3.5056	NaN
32	305	0.0032787	29.8	3.3945	0.56682
34	NaN	0.0032573	27.6	3.3178	0.71345
36	309	0.0032362	25.4	3.2347	0.91536
38	311	0.0032154	23.1	3.1398	1.2169
NaN	313	0.0031949	21.5	3.0681	1.5093
42	315	0.0031746	19.7	2.9806	1.962
44	317	0.0031546	18.2	2.9014	2.4882
46	319	0.0031348	NaN	2.8332	3.0531
48	321	0.0031153	15.8	2.76	3.8029
50	323	0.003096	16.3	2.7912	3.4636

a good idea to visualize the data which would give a rough approximation of the missing value.

As we can see there is a gap in observation 14 which is the missing value. Now notice the decreasing curve, which suggests that the resistance should be between 20 and 15 so around 17.

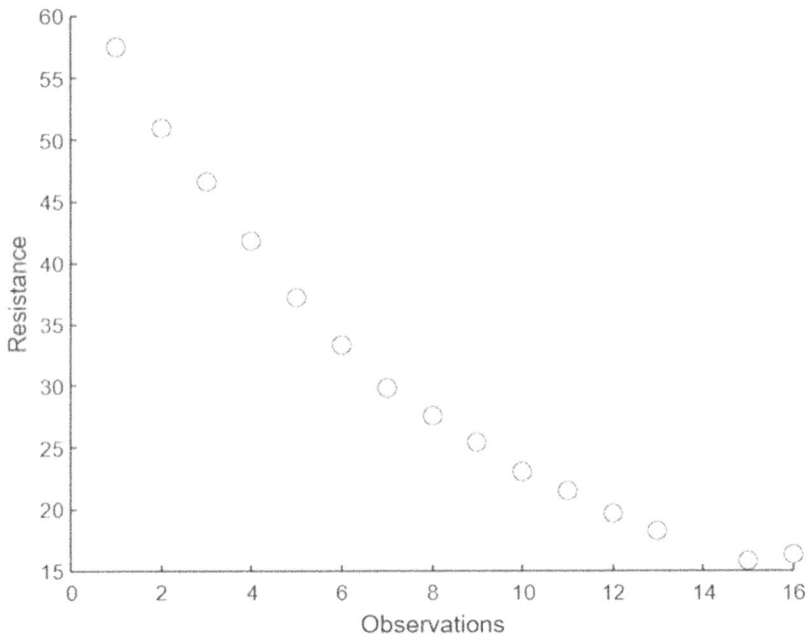

Figure 1.2

```
fillmissing(filename, "linear")
fillmissing(filename, "spline")
```

Table 1.7

ans =

16×6 table

Temperature(°C)	Temperature(°K)	1/T(/K)	Resistance(kohms)	lnR	Current(mA
20	293	0.003413	57.5	4.0518	0.078902
22	295	0.0033898	50.9	3.9299	0.11375
24	297	0.003367	46.6	3.8416	0.14823
26	299	0.0033445	41.8	3.729	0.20538
28	301	0.0033223	37.2	3.6163	0.29138
30	303	0.0033003	33.3	3.5056	0.4291
32	305	0.0032787	29.8	3.3945	0.56682
34	307	0.0032573	27.6	3.3178	0.71345
36	309	0.0032362	25.4	3.2347	0.91536
38	311	0.0032154	23.1	3.1398	1.2169
40	313	0.0031949	21.5	3.0681	1.5093
42	315	0.0031746	19.7	2.9806	1.962
44	317	0.0031546	18.2	2.9014	2.4882
46	319	0.0031348	17	2.8332	3.0531
48	321	0.0031153	15.8	2.76	3.8029
50	323	0.003096	16.3	2.7912	3.4636

Table 1.8

ans =

16×6 table

Temperature(°C)	Temperature(°K)	1/T(/K)	Resistance(kohms)	lnR	Current(mA)
20	293	0.003413	57.5	4.0518	0.078902
22	295	0.0033898	50.9	3.9299	0.11375
24	297	0.003367	46.6	3.8416	0.14823
26	299	0.0033445	41.8	3.7336	0.20538
28	301	0.0033223	37.2	3.6163	0.29138
30	303	0.0033003	33.3	3.5056	0.41985
32	305	0.0032787	29.8	3.3945	0.56682
34	307	0.0032573	27.6	3.3178	0.71345
36	309	0.0032362	25.4	3.2347	0.91536
38	311	0.0032154	23.1	3.1398	1.2169
40	313	0.0031949	21.5	3.0681	1.5093
42	315	0.0031746	19.7	2.9806	1.962
44	317	0.0031546	18.2	2.9014	2.4882
46	319	0.0031348	16.719	2.8332	3.0531
48	321	0.0031153	15.8	2.76	3.8029
50	323	0.003096	16.3	2.7912	3.4636

1.1.4 Visualizing the data

After we are done cleaning and transforming the imported data into MATLAB, it is a good idea to plot the data so that we can explore its features. This step in the data analysis process involves creating interactive visualizations by selecting the most appropriate charts and graphs.

1.1.5 Analyzing the data

Now that we have cleaned, processed, and organized the data, it is ready for the analysis. Using various data analysis tools and techniques we can look for hidden patterns and relationships and find insights and predictions. Basic analysis methods and graphs can be used, as well as more advanced methods like clustering, principal component analysis, or other dimension reduction methods. (In this book we will be focusing on the statistical methods of data analysis in MATLAB and how it can be used to interpret the data and draw meaningful insights from it.)

1.1.6 Visualizing the results

The goal of this step is to produce outputs from the conclusions in the previous step that will help in communicating and sharing insights with the people concerned more clearly and efficiently. The results of the data analysis are to be reported in a format as required by the users to support their decisions and further action.

1.2 WHAT IS STATISTICAL ANALYSIS?

Statistical analysis is a specific subset of data analysis that focuses on applying various statistical methods and techniques to analyze and interpret data. Now, understanding the data type is essential for selecting appropriate statistical methods, visualizations, and data preprocessing techniques.

1.2.1 Types of data

Data can be categorized in two primary ways: by the nature of the values it represents (categorical or numerical) and by the number of variables or dimensions involved in the analysis (univariate, bivariate, and multivariate).

We will look into the two types of data—categorical and numerical—in the next chapter, but for now let's focus on the data based on the number of variables which can be classified as univariate, bivariate, and multivariate.

1.2.1.1 Univariate data

This data consists of observations based on one single characteristic or attribute. For example, a chemical engineer in a manufacturing plant monitors the temperature inside a chemical reactor during a chemical reaction; the univariate variable in this example is the temperature measurements.

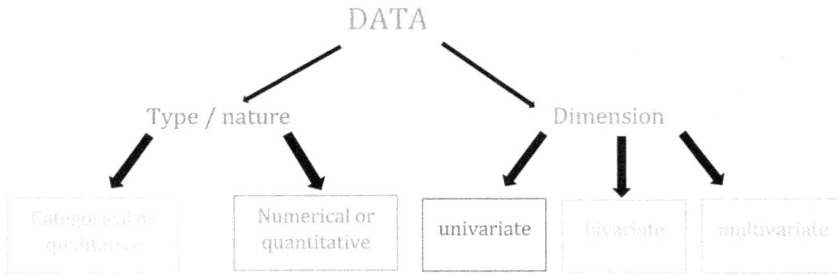

Figure 1.3

1.2.1.2 Bivariate data

A bivariate dataset involves two variables or dimensions such that each data point in the first variable has a corresponding data point in the second variable. This pairing implies that there is a relationship or association between the two variables, and changes in one variable can influence or be influenced by changes in the other variable. Therefore, in a bivariate dataset, one variable is typically considered the independent variable, and the other is considered the dependent variable. For example, the same chemical engineer above now wants to understand how the heat treatment of steel affects its hardness. Therefore, in this case, the independent variable is the temperature at which the steel is subjected to the heat treatment process, and the dependent variable is the hardness of the steel after heat treatment.

1.2.1.2.1 Further examples

Explanatory/predictor variable	Response variable
Compressive strength	Intrinsic permeability
Roadway surface temperature	Pavement deflection
Oil viscosity	Wear volume
Engine displacement	Miles per gallon
Effective life of a cutting tool	Cutting speed

1.2.1.2.2 Multivariate data

Finally, the multivariate data as one would guess involves multiple variables. While bivariate data has two variables, typically one independent and one dependent, multivariate data has three or more variables with multiple independent and dependent variables. For example, an engineer wants to determine what the life of a cutting tool depends on. He uses factors like the cutting speed, feed rate, tool geometry, tool material, cutting fluid, etc., given in Figure 1.4.

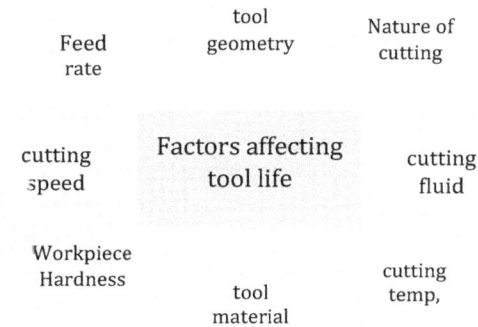

tool
geometry

Feed
rate

Nature of
cutting

cutting
speed

Factors affecting
tool life

cutting
fluid

Workpiece
Hardness

tool
material

cutting
temp,

Figure 1.4

1.2.2 Types of analysis

Based on these three data types in Section 1.2.1, there are three main types of analysis:

- Univariate analysis
- Bivariate analysis
- Multivariate analysis

Let's discuss each of these types of analysis.

1.2.2.1 Univariate analysis

Univariate analysis is the simplest of the three types of analyses, where the data we are analyzing contains only one variable. For example an engineer is working on testing the tensile strength of a new material for a structural application. He has collected a dataset containing the values of the tensile strength of 30 samples of the material. The data here is the tensile strength which is a single variable. The main objective of the univariate analysis is to describe the data and find patterns that exist within it.

212.9, 215.7, 217.6, 220.4, 221.2, 223.3, 223.6, 223.8,
225.5, 226.4, 229.6, 232.1, 233.2, 234.5, 238.4,
238.5, 239.4, 240.7, 241.1, 244.8, 245.3,
247.7, 251.6, 253.3, 256.6, 257.9,
260.2, 262.1, 263.3, 263.6

Now what kind of analysis can we do with this dataset?

When dealing with a dataset of a single variable, finding a single value that summarizes the central position of the data is a common and fundamental step in data analysis. This single value is often referred to as a measure of central tendency. The most commonly used measures of central tendency include the mean, the median, and the mode.

Each measure of central tendency (mean, median, mode) has its own strengths and is appropriate in different situations based on the characteristics of the data which we shall be discussing it in just a second!

Knowing only the central value doesn't give us a complete picture of the data. We need to understand how spread out or dispersed the data points are around the central value. This measure is called the measure of dispersion. It tells us how much individual data points deviate from the central value. The most commonly used measures of dispersion include the range, mean deviation, quartile deviation, standard deviation, and variance.

We previously mentioned that the choice of which measure of central tendency (e.g., mean, median, mode) to use depends on the specific characteristics of the data we are working with. When we refer to the "characteristics of the data", we are talking about the specific properties or features of the dataset. These characteristics can be determined using both graphical and non-graphical methods. Graphical methods are often preferred because they are quick and intuitive, providing a visual representation of the data's properties. On the other hand non-graphical methods provide quantifiable measures that can be used for more precise characterization. Two commonly used non-graphical measures are skewness and kurtosis (Chapter 6). These two measurements are typically discussed in relation to the normal distribution as a point of reference. The normal distribution is commonly used as a benchmark because of certain characteristics (Chapter 5).

The three measurements above are used to summarize or describe the data. This branch of statistics that only focuses on summarizing and describing data is called descriptive statistics.

One should also take notice that the data given above is only a sample and not the entire population. So, descriptive statistics are calculated based on a sample.

Now let's take a step back and think about the objective of this sample data. Why have we collected these 30 samples? What was the end goal? One thing we all need to learn is that the quantitative data is of no use without interpretation. Understanding why people collect samples is considered the fundamental aspect of statistical analysis. In statistics samples serve as the basis for making inferences about the entire population from which the sample was taken. This implies that the data our engineer collected is just a subset of the entire population. Now one may ask then why not collect the population if we want to make statements about the population. Think realistically!

Collecting data from an entire population, even though it seems the only right way to make statements about that population, is often impractical or infeasible because we clearly do not know how big that population is. Think about our example for a second. The population of tensile strength measurements for a material could be infinite because tensile strength can have an infinite number of potential measurements. Therefore, it is impossible to measure or observe every single individual in the population. This branch of statistics that draws conclusions about the population from sample data is called inferential statistics.

Now how do we do that? Remember, when conducting statistical analysis, the ultimate goal is often to make inferences or draw conclusions about a larger population based on information obtained from a representative sample. So, to perform inferential statistics effectively, we need certain characteristics and information about both the sample and the population. These characteristics of samples and populations are described by numbers called statistics and parameters.

Our example of tensile strength perfectly illustrates why inferential statistics is essential. Tensile strength is significant in materials used for the construction of building structures like bridges, where safety and reliability are of paramount importance. So engineers and manufacturers need to ensure that a new material consistently meets certain strength requirements. To ensure the long-term reliability of a bridge or any structure, it's crucial to use materials with consistent and known population parameters. If we only focused on the sample statistic (the mean tensile strength of a few samples), we might miss variations which could lead to unexpected failures or degradation over time, potentially risking people's safety.

It is important to note that the parameter is unknown because it would be impossible to collect information from every member of the population. Therefore, we rely on sample statistics to estimate and draw conclusions about these unknown population parameters.

Remember that parameters and statistics are numerical values that summarize any measurable characteristic of a population and the sample. There is a range of possible attributes that we can evaluate, which give rise to various types of parameters and statistics. Some of the most commonly used parameters and parameters are means, medians, standard deviations, proportions, and correlations (Chapter 10).

The process of making decisions about the population involves the estimation **of population parameters** using sample statistics which in turn is calculated from the data collected in a sample. Once we have estimated population parameters and obtained sample statistics, the next step is hypothesis testing where we actually make the decisions.

These steps are complex areas of study in statistics and considered extensive because they involve a multitude of techniques, methods, and underlying assumptions which we shall discuss in Chapter 7. For now we can conclude that statistical inference is divided into two major areas: parameter estimation and hypothesis testing.

In conclusion, we can categorize univariate analysis into two broad categories: descriptive and inferential statistics, which we shall cover in the first half of the textbook.

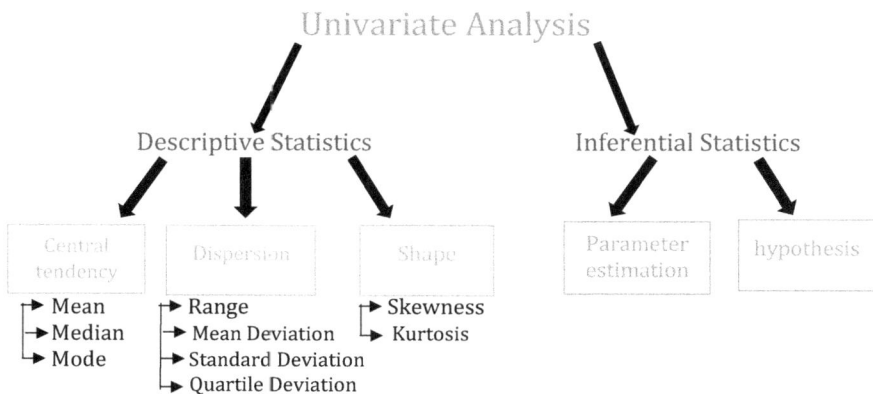

Univariate Analysis

Descriptive Statistics

Inferential Statistics

| Central tendency | Dispersion | Shape | Parameter estimation | hypothesis |

- Mean
- Median
- Mode

- Range
- Mean Deviation
- Standard Deviation
- Quartile Deviation

- Skewness
- Kurtosis

Figure 1.5

1.2.2.2 Bivariate analysis

As the name itself suggests, this analysis deals with data containing two variables. Bivariate analysis involves exploring cause-and-effect relationships between the two variables—how changes in one variable might influence or explain changes in another variable. Now the two variables could either be dependent or completely independent of each other. Imagine an engineer is working in a manufacturing company that produces metal components. He is interested in understanding the relationship between temperature and the thermal expansion coefficient (α) of a particular metal material. He collects the data and gets the results below.

T(K)	100	200	300	400	500	600	700	800	900	1000
$\alpha(T)[10^{-6}/K]$	0.12	0.83	1.25	1.56	2.25	2.76	3.04	3.28	3.81	4.37

Looking at the table we can immediately conclude that even though they are not proportional they are correlated—as the temperature increases, the thermal expansion coefficient also increases. To get more reliable results these variables are plotted on X and Y axes on the graph.

The nature of the dependent and independent variables is a critical consideration when conducting bivariate analysis. This consideration helps determine the appropriate statistical techniques, the type of analysis to perform, and the interpretation of the results.

We have already discussed that the types of data can be either numerical or categorical, and since bivariate analysis always involves examining the relationship between two variables, there can be three types of scenarios:

- Categorical vs. categorical
- Categorical vs. numerical
- Numerical vs. numerical

Understanding the different scenarios in bivariate analysis is crucial because the approach and techniques you use can vary significantly depending on the types of variables involved.

Bivariate analysis, like univariate analysis, consists of two main components: the descriptive component and the inferential component. Let's take a closer look at each of these components.

Unlike bivariate data, univariate data analyzes only one variable; therefore descriptive statistics is used to describe and understand the characteristics and distribution of that variable, whereas bivariate data analysis involves the examination of two variables simultaneously to understand the relationship or association between them. Hence, the primary goal of this analysis is to uncover patterns, trends, and the strength of this relationship. Therefore the descriptive component is only concerned with the visualization of the relationship using graphs like scatterplots, boxplots, and numerical tabulations (crosstabulation, etc.).

One should always remember that visualization tools alone are never enough in confirming that a relationship exists between two variables. It is extremely crucial to pair visualization with inferential statistical methods to confirm, quantify, and draw reliable conclusions about relationships between variables. Therefore, both components, visualization and inference, are essential and complementary elements in the process of bivariate analysis. As stated previously, when conducting a bivariate hypothesis test, it's crucial to consider the nature of both the dependent and independent variables. We shall discuss the data types for both categorical and numerical variables, along with the corresponding hypothesis tests for each data type.

Table 1.9

Experiment	ΔT (°C)	Thermal conductivity (W/m·K)	Area (m²)	Heat transfer coefficient (W/ m²·K)
1	5	0.15	0.2	1.5
2	10	0.18	0.3	2.0
3	15	0.22	0.4	2.75
4	20	0.25	0.5	3.0
5	25	0.28	0.6	3.5
6	30	0.32	0.7	4.0
7	35	0.35	0.8	4.375
8	40	0.38	0.9	4.75
9	45	0.42	1.0	5.25
10	50	0.45	1.1	5.5

1.2.2.3 Multivariate analysis

Finally, multivariate analysis is when we are comparing more than two variables.

Suppose one wants to understand the factors that influence the heat transfer coefficient. An engineer collects data on the following variables:

1. **The temperature** difference between the media
2. **The thermal conductivity** of the material in the heat exchanger
3. **The area** over which the energy is transferred

1.3 STATISTICAL ANALYSIS IN MATLAB

Within each of these types of analysis, there are both graphical and non-graphical methods. Data visualization is the practice of using graphical methods, such as charts, graphs, plots, and other visual representations, to explore, analyze, and communicate data effectively. It complements non-graphical (statistical) methods, by providing a visual context for the data, making patterns and insights more accessible. Data visualization is a crucial step in the analysis process, regardless of the type of data we are working with, since it helps us to better understand the patterns, trends, and relationships and draw meaningful conclusions.

MATLAB is a powerful tool for both non-graphical and graphical methods of data analysis. We are using MATLAB since it provides a rich set of numerous statistical functions to analyze data of one or multiple variables. It offers a wide range of capabilities for both aspects, making it a versatile choice for researchers and analysts. Let's explore some of the visual tools available in MATLAB for different types of data analysis.

1.3.1 Data visualization in univariate analysis

As previously stated, data can be categorical or quantitative. Various types of graphs can be used to understand data. The standard types of graphs include:

1) **Frequency distribution table:** frequency distribution tables can be used for both categorical and numerical data. A frequency distribution table is a table that counts the number of times each value (ungrouped data), range of values (grouped data), or category occurs in a dataset. In Table 1.10, we use a frequency table to analyze the distribution of flange face types, while in Table 1.11, we use a frequency table to analyze the distribution of bolt lengths to ensure they meet design specifications.

Table 1.10

Types		Frequency
{'F.lat'	}	10
{'Raised'	}	14
{'Tongue'	}	17
{'Grooved'	}	9
{'Spigot'	}	12
{'Recess'	}	18
{'Ring_Recess'	}	22
{'O-Ring_Groove'}		15

Table 1.11

Bolt_Length	Frequency
110	8
112	12
116	10
120	7
125	9
128	13
130	15
132	5
136	8
142	10

2) **Bar chart:** bar charts represent data using rectangular bars, where the height of each bar corresponds to the frequency of each category in a dataset. In Figure 1.6 we use bar charts to represent the Monthly Energy Consumption in a Factory. The bar chart In Figure 1.7 shows the Young's modulus of different systems with varied a/b ratio.

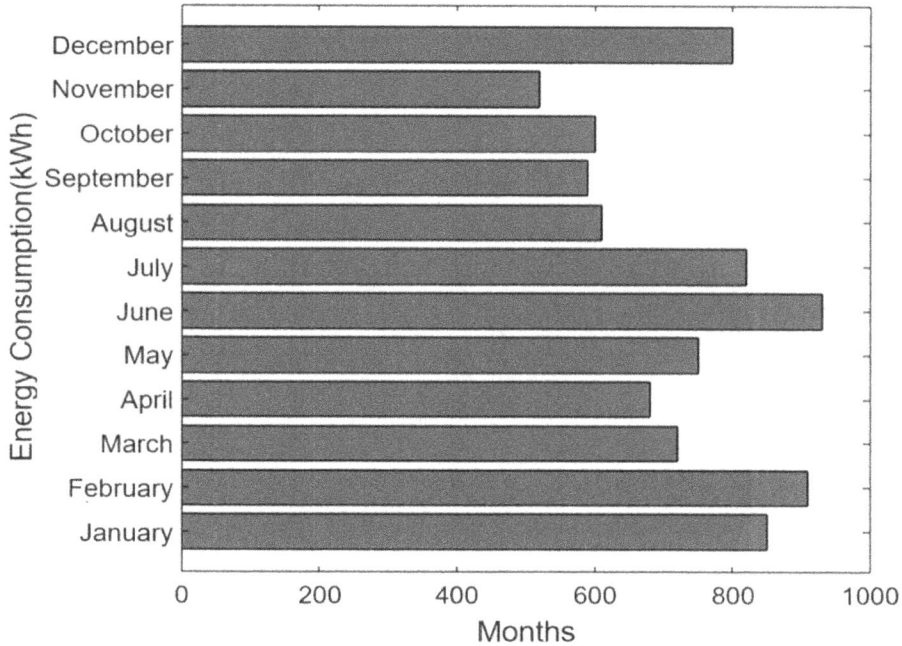

Figure 1.6

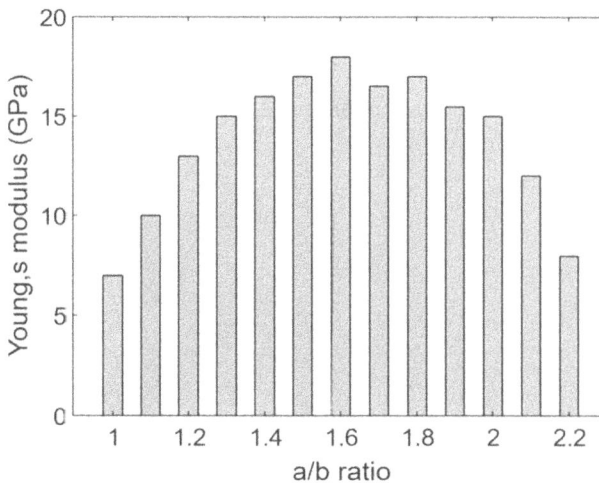

Figure 1.7

3) **Pie chart**: pie charts are similar to bar charts but instead of rectangular bars, data is displayed in a circular format, with each "slice" of the pie representing a proportion or percentage of the whole dataset that a specific category occupies. In Figure 1.8 we used the Pie chart showing the percentage of publications at www.sciencedirect.com on different features of the processing of AISI 316L via the DED process. There is an example of a simple pie chart in Figure 1.10 that engineers might use to represent

the distribution of energy sources in a power grid. Each sector of the pie corresponds to a different energy source, showing its percentage contribution to the total energy generation

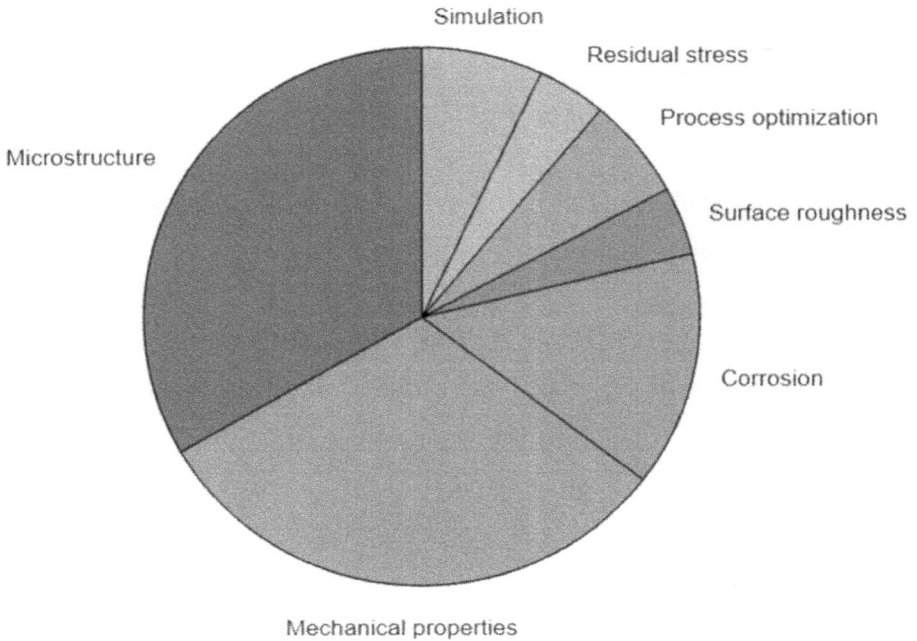

Figure 1.8

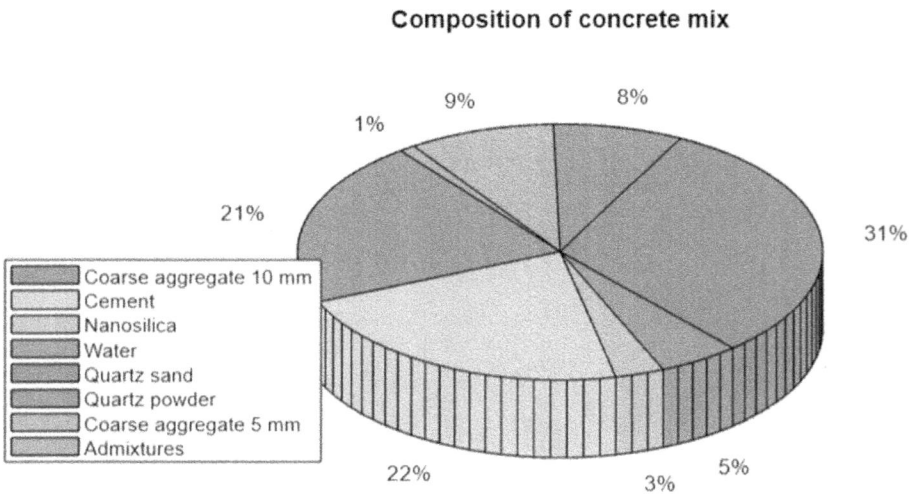

Figure 1.9

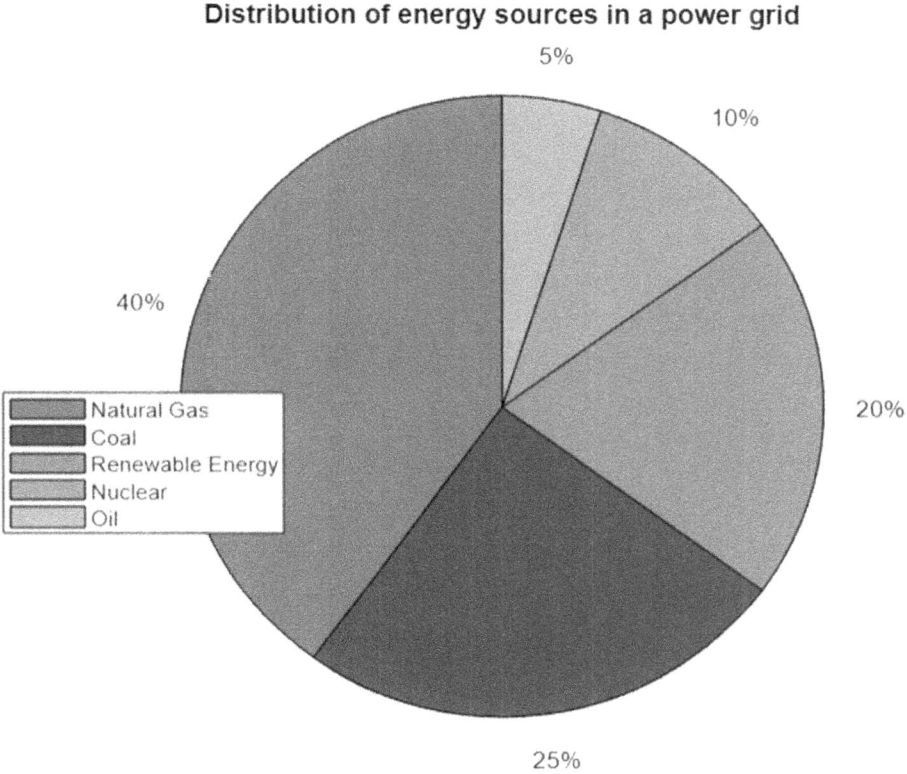

Distribution of energy sources in a power grid

5%

10%

40%

20%

Natural Gas
Coal
Renewable Energy
Nuclear
Oil

25%

Figure 1.10

4) **Histograms:** even though both bar graphs and histograms make use of rectangular bars, they are used with different types of data. While bar graphs are used to represent categorical data or discrete data, histograms represent numerical or continuous data. Another fundamental difference between histograms and bar graphs is that there are gaps between bars in a bar graph but in the histogram, the bars are adjacent to each other. For example, the histogram in Figure 1.5 displays the number of component failures with time in hours.

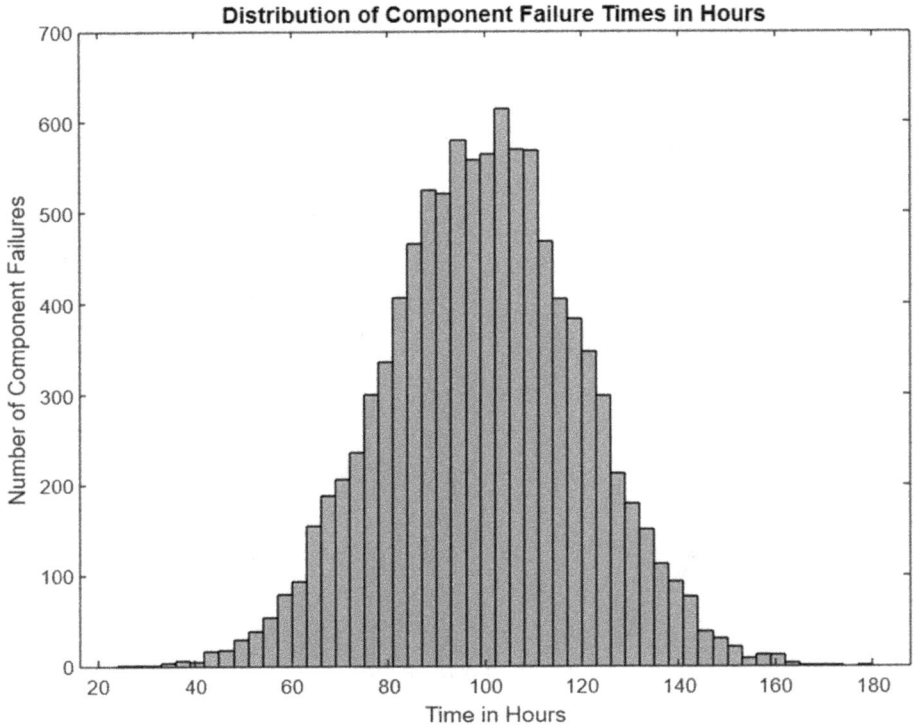

Figure 1.11

5) **Density curve**: while a histogram shows the frequency of values in each range, a density plot shows the proportion of those values in that range. Another notable contrast is that a histogram is made up of bars that touch each other, but a density plot is a smooth curve that shows the distribution of the data in a more continuous way, for example in Figure 1.6.

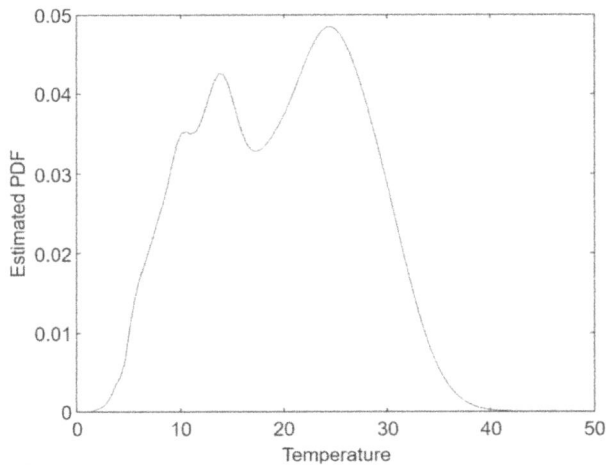

Figure 1.12

6) **Boxplot:** a boxplot provides a summary of the data's distribution including the median, first, and third quartiles, the minimum and maximum values, and also the potential outliers. See Figure 1.7.

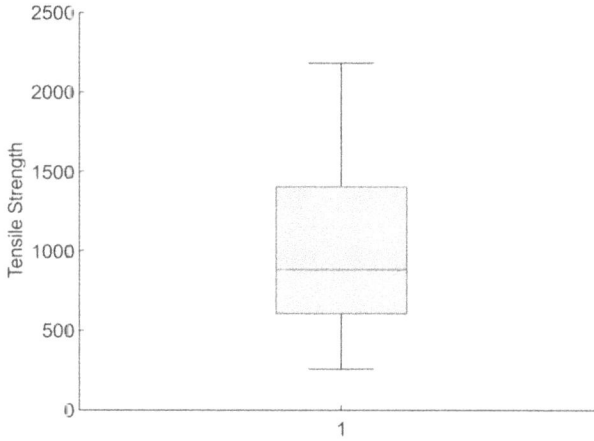

Figure 1.13

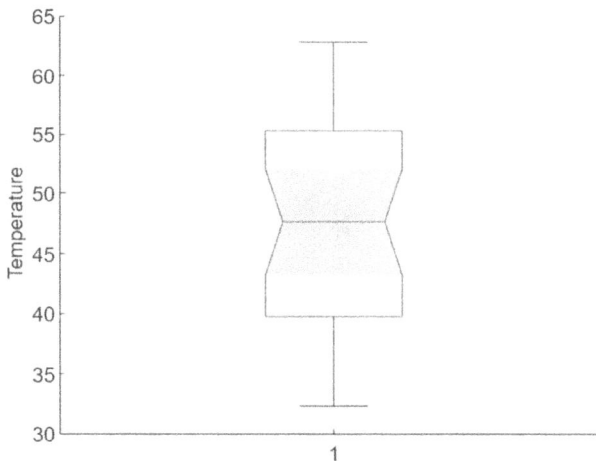

Figure 1.14

Table 1.12

Contingency Table:

	Defective	Non-Defective	Total
Capacitor	23	7	60
Diode	15	5	40
Inductor	12	18	60
Resistor	15	10	50
Transistor	11	14	390

The distribution of a single quantitative variable is typically plotted with a histogram or kernel density plot.

1.3.2 Data visualization in bivariate analysis

1.3.2.1 Categorical vs. categorical

To study the relationship between two categorical variables, a contingency table (also called a crosstab or two-way table) or a comparative bar graph is typically used. A contingency table is a special type of frequency distribution table, which displays frequencies for combinations of two categorical variables.

1.3.2.2 Categorical vs. quantitative

When plotting the relationship between a categorical variable and a quantitative variable, a large number of graph types are available, for example the bar chart we have seen in Figure 1.3 or the pie chart in Figure 1.4. We have already seen boxplots and kernel density plots in the univariate analysis in Figure 1.7. However, we can draw boxplots for different categories.

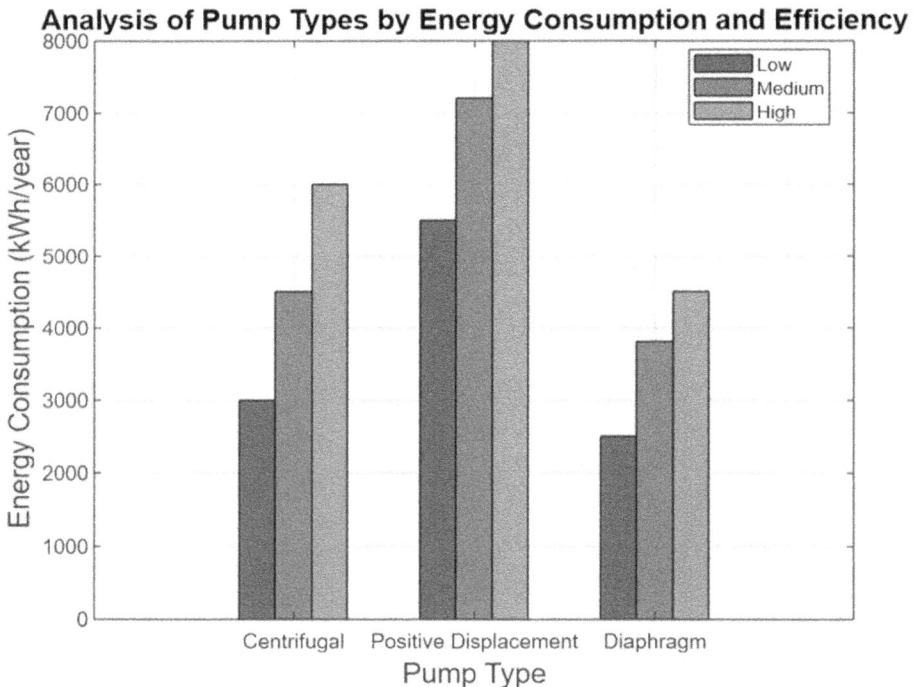

Figure 1.15

1.3.2.3 Quantitative vs. quantitative

In this kind of analysis, both variables of the bivariate data, which includes a dependent and an independent variable, have a numerical value. Let's discuss some of the statistical plots for visualizing bivariate data in MATLAB, using the Statistics and Machine Learning Toolbox™.

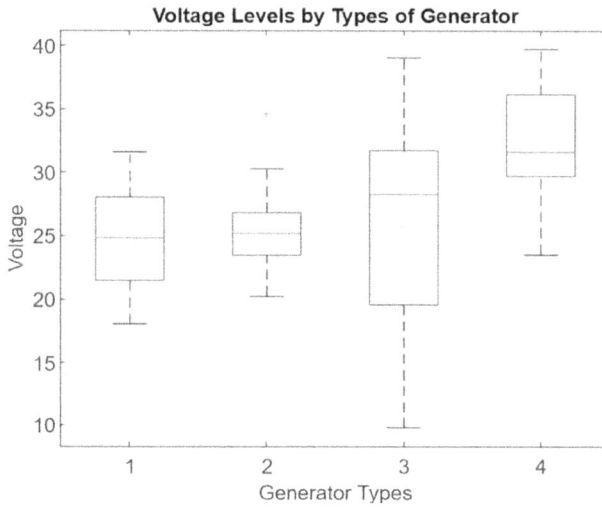

Figure 1.16

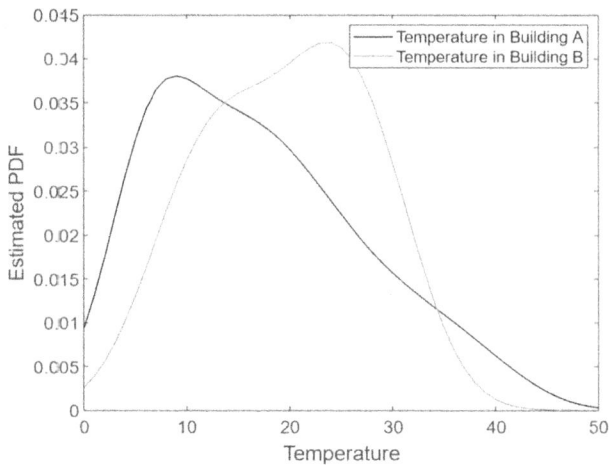

Figure 1.17

1) **Scatterplot:** A scatter plot is one of the most powerful visualization tool for exploring the relationship between two continuous variables in bivariate data. Scatterplots are particularly useful for assessing the association or "correlation" between two variables. The pattern of points in the plot can suggest whether there is a positive, negative, or no correlation between the variables. Scatterplots are also extremely helpful for checking assumptions of statistical models. For example, linear regression assumes a linear relationship between variables, and this can be visually assessed through a scatterplot. Below there are an extensive array of scatterplots, each showcasing distinct patterns and relationships between two variables.

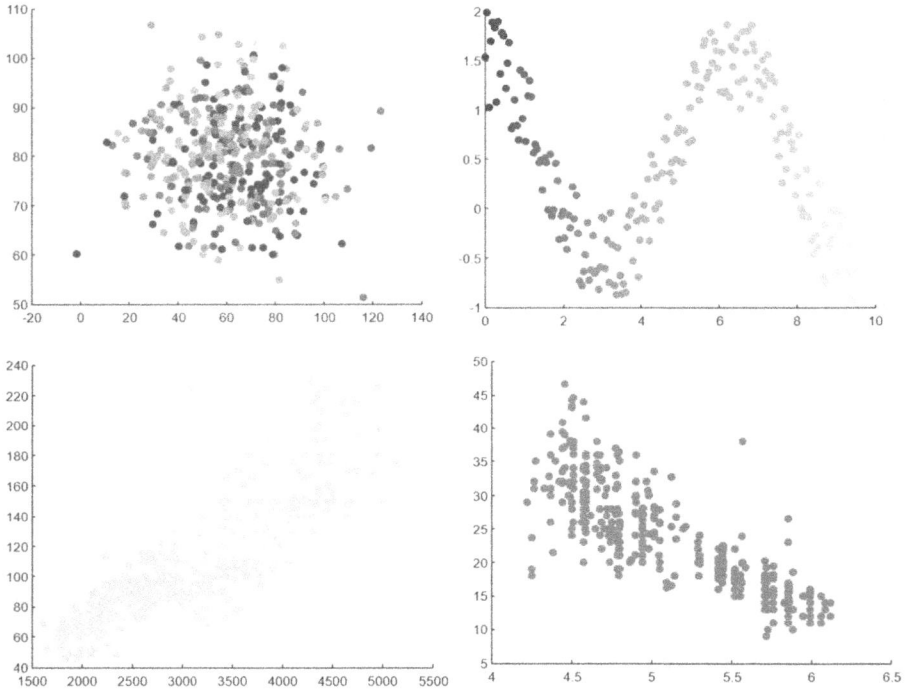

Figure 1.18

2) **Bivariate histogram plot:** Three-dimensional histograms serve as a visualization tool for cross tabulations involving two variables. Essentially, they join two individual (univariate) histograms, allowing for an exploration of the frequencies at the intersection of values from both analyzed variables. Typically, in this graphical representation, a 3D bar is constructed for each "cell" in the cross tabulation table, with the bar's height indicating the frequency of values within that specific cell. This method provides a comprehensive view of the bivariate distribution by accommodating different categorization approaches for each of the two variables under examination.

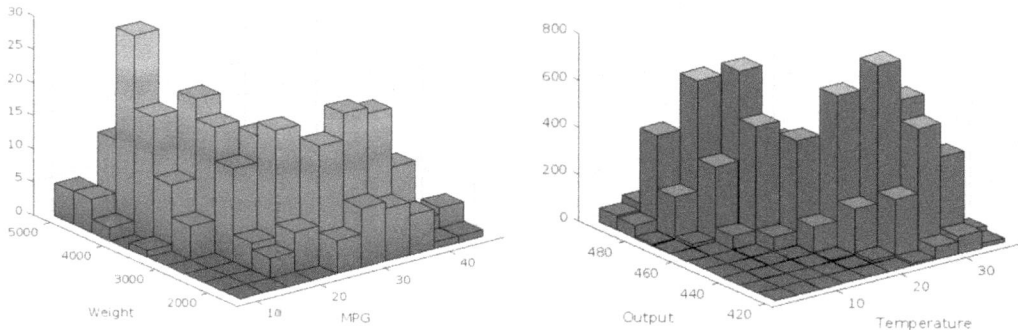

Figure 1.19

1.3.3   Data Visualization in Multivariate Analysis

Previously we saw how to visualize both univariate and bivariate data using various statistical plots. Such data are easy to visualize using scatter plots, bivariate histograms, boxplots, etc. However, we have seen that multivariate data can involve a large number of variables, making direct visualization more difficult. In this section we shall look at a dataset and explore some of the ways to visualize these high-dimensional data in MATLAB.

We will use a dataset that contains various measured variables for 398 automobiles. We'll illustrate multivariate visualization using the values for fuel efficiency (in miles per gallon, MPG), acceleration, engine displacement, weight, and horsepower. We will use the number of cylinders to group observations.

1) **Scatterplot:** As discussed earlier, multivariate data has multiple independent variables say so when dealing with multiple independent variables and a dependent variable, a scatterplot matrix allows for the simultaneous examination of relationships between all pairs of variables on a single page in a matrix format. Each scatter plot in the matrix visualizes the relationship between a pair of variables, allowing many relationships to be explored in one chart. If we have k independent variables and one dependent variable, the scatterplot matrix will have (k+1) rows and (k+1) columns. The diagonal of the matrix typically contains univariate plots for each variable, illustrating their individual distributions.

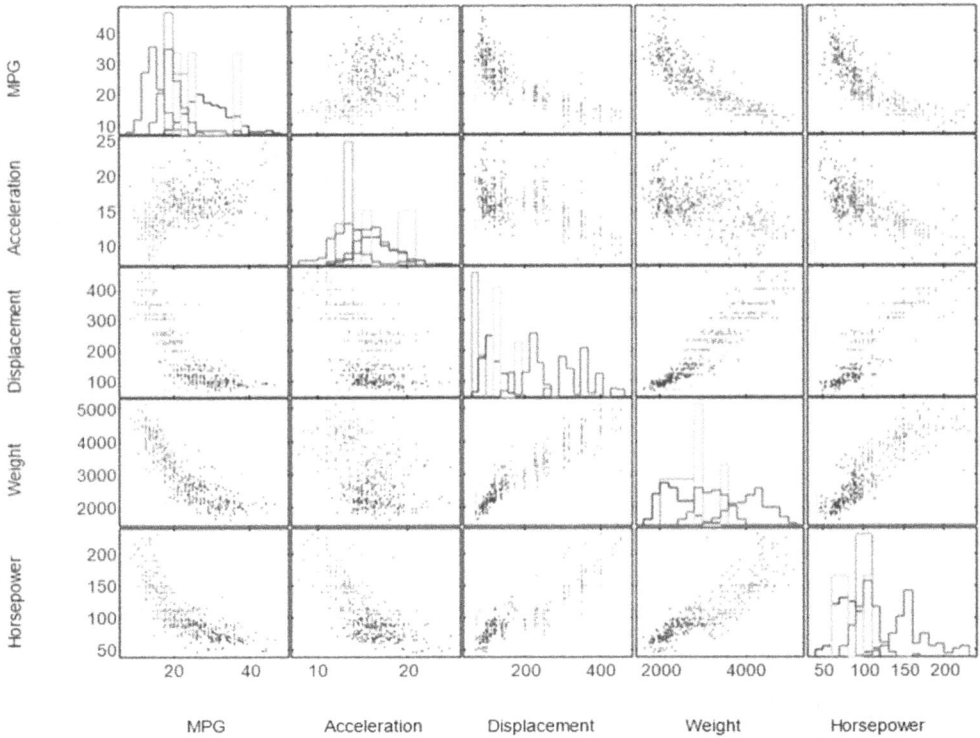

Figure 1.20

2) **Parallel Coordinates Plots:** The scatter plot matrix only displays bivariate relationships. However, there are other alternatives that display all the variables together,
allowing you to investigate higher-dimensional relationships among variables. The
most straight-forward multivariate plot is the parallel coordinates plot. In this plot,
the coordinate axes are all laid out horizontally, instead of using orthogonal axes as
in the usual Cartesian graph. Each observation is represented in the plot as a series of
connected line segments. For example, we can make a plot of all the cars with 4, 6, or
8 cylinders, and color observations by group.

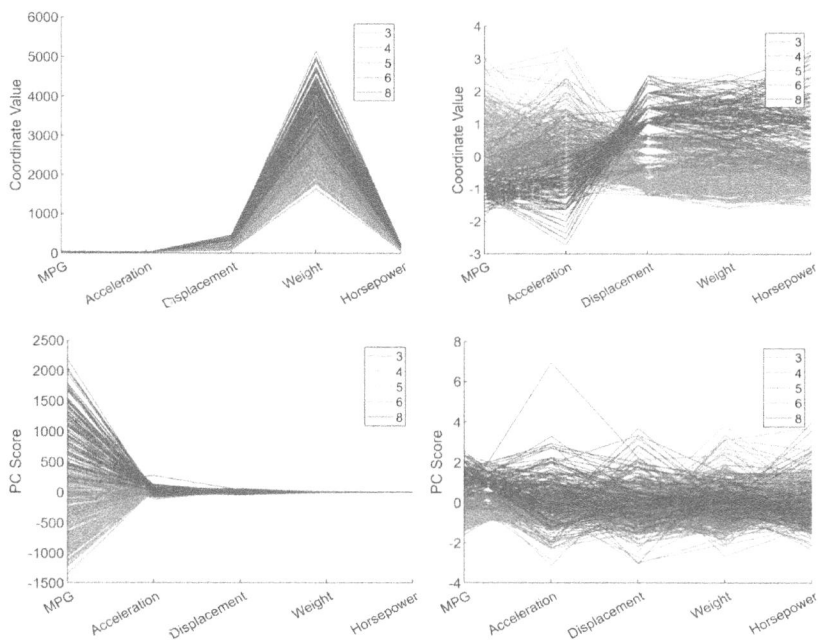

Figure 1.21

3) **Andrews Plots:** Another similar type of multivariate visualization is the Andrews plot. This plot represents each observation as a smooth function over the interval [0,1].

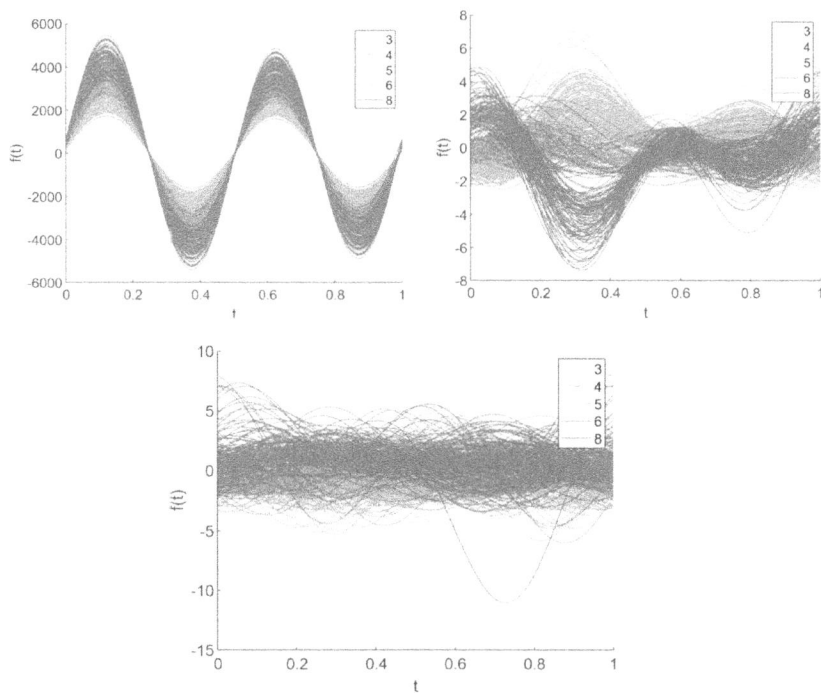

Figure 1.22

This textbook comprehensively covers both graphical and non-graphical methods of analysis involved in univariate, bivariate, and multivariate analysis. It provides hands-on experience for statistical data analysis with MATLAB to reinforce the concepts and help readers gain proficiency in using MATLAB for statistical data analysis. Each of the chapters in the textbook starts with a theoretical explanation of the concepts allowing the readers to develop a solid foundation and understanding of the statistical principles and techniques. This theoretical knowledge serves as a basis for the practical application of those concepts using MATLAB. Therefore, after presenting the theoretical concepts, the textbook then provides step-by-step instructions, code snippets, and examples to demonstrate the practical application of those concepts using the software. In the next chapter, we are going to see univariate, bivariate, and multivariate analysis in data visualization using MATLAB.

Chapter 2

Data types and visualization

So far, we have learned that statistics is a field that involves collecting, organizing, summarizing, analyzing, and finally making inferences or drawing conclusions from data. Notice that the definition sounds like it follows a systematic pattern or set of steps to make sense of data effectively. First we collect the data, then we organize and summarize the data, and finally we make some estimations or draw conclusions from this data. These steps provide a structured framework for conducting analyses of vast amounts of data. Our book aims to cover all of these steps in detail, not just theoretically but also practically. The good news is this book is divided into three parts, where each part has chapters corresponding to each step of the analysis.

In this chapter we shall study the process of collecting data for different kinds of analysis, learn about the different data types, and finally utilize MATLAB's visualization tools for exploring and presenting data.

2.1 TYPES OF DATA

According to the definition, the first step of this analysis is the collection of data. But before we even get to collect the data, it is necessary to first understand what kind of data we are going to collect. Think about it this way, if we want to do grocery shopping, don't we need a list? How else are we supposed to know why and what exactly we are going to purchase? Similarly, to analyze any dataset, we first need to know what kind of data we're working with. This is because certain statistical methods can only be used with certain data types in order to make accurate conclusions or get meaningful insights. Therefore, knowing the types of data you are dealing with enables you to analyze the data correctly and get accurate results. It is basically like a road map for properly conducting a statistical investigation!

Now in statistics there are only two types of data: qualitative and quantitative data. However, these types are divided into two and four categories of data respectively.

2.1.1 Qualitative/categorical variables

As the word suggests, these kinds of data are non-numeric, that is, they cannot be directly counted or measured using numbers. Instead, they are arranged into categories that can be based on gender, university, college major, states, countries, or anything that does not have a number directly associated with it. Since these variables fall into categories, quantitative data can be referred to as categorical data. Even though these variables are mostly categorized in terms of words or natural language specifications they can sometimes hold numerical values, but they are strictly descriptive. For examples we can have categories based on

DOI: 10.1201/9781003399582-2

years like 1900, 1920, … 2020. Understand that even if these are numbers, such values are not mathematically meaningful; they have no numerical meaning.

As mentioned earlier, qualitative data can be subdivided into two parts.

2.1.1.1 Nominal

Nominal data is a kind of qualitative data that aids in the labeling of variables into different categories without offering any intrinsic order. That is, it cannot be placed in any ranks (from highest or lowest). For example, the category gender is nominal because there is no ranking in the levels. Different types of cutting tool materials are another example because there is no order among "high-speed steel (HSS)", "tungsten carbide", "ceramics", "cubic boron nitride (CBN)", and "diamond". Also notice that each of these mentioned categories has two or more levels, i.e. for gender, we have "male" and "female", and for tool materials we have five. Say we're collecting data on the number of students enrolling in Stanford by major. We might use a numbering system to denote different majors: say, 1 to represent architecture, 2 to represent business, 3 for computer science, 4 for political science, 5 for health science, and so on. Notice that although we are using numbers instead of words to label each category these numbers do not indicate any kind of hierarchy (e.g. health as represented by the number 5 does not imply that it is worth more or is better than architecture represented by the number 1 and vice versa).

Some other examples of nominal data in engineering include:

- **Material types** (e.g. metals, concrete, steel, wood, ceramics, textile)
- **Different types of power plants** (e.g. thermal, nuclear, hydropower, solar)
- **Switch types** (e.g. toggle, pushbutton, selector, joystick, limit)
- **Types of bolts** (e.g. anchor, carriage, eye, hanger, hex, lag)
- **Different types of gears** (e.g. spur, helical, herringbone, bevel, hypoid)
- **Temperature measuring instruments** (e.g. thermometer, thermostat, thermistor, thermopile)

Notice that none of the categories represent level with a meaningful order of magnitudes.

2.1.1.2 Ordinal

Just like nominal variables, ordinal variables too can be divided into categories but **with a logical order implied in the levels**. They are placed in order according to their position (for example, first, second, third, etc.). For instance, in the case of material strength levels, you can rank them from "low strength" to "high strength" based on their relative strength; this variable would be ordinal because there is a clear order in the levels. Here the three levels, namely "high strength", "moderate strength", and "low strength", are ranked from the most positive (high strength), to the middle response (moderate strength), to the least positive (low strength). Other examples can include:

- Education level categorized into "high school", "bachelor's degree", "master's degree", and "PhD", which have a clear order based on the level of education achieved.
- Classification of temperature for a hazardous area classification (HAC) system, categorized into "T1", "T2", "T3", "T4", "T5", and "T6" which have a clear order based on the level of maximum permissible surface temperature.

2.1.2 Qualitative/categorical variables

In opposition to qualitative data, a **quantitative** variable **or numerical variable** is a variable that reflects a notion of **magnitude**, if the values it can take are **numbers**. A quantitative variable **consists of values that represent counts or measurements of a certain quantity**, for instance, age, height, number of cigarettes smoked, etc. Quantitative variables are divided into **discrete** and **continuous**.

2.1.2.1 Discrete and continuous

Discrete variables are variables that can take on only a finite (or countably infinite) number of outcomes. In contrast, continuous variables are values that can fall anywhere corresponding to points on a line segment. Some examples are weight and temperature, voltage, flowrate, etc. The table provides a general summary of the differences between discrete and continuous variables.

2.1.2.2 Interval and ratio

Notice the two examples of continuous data—temperature and say volume. We can measure temperatures below 0 degrees Celsius, such as –10 degrees, but it is not possible to have negative volume. The zero point on a volume scale (e.g., liters or cubic meters) represents the complete absence of substance or volume.

This difference is an example of ratio and interval.

Now let's look back at the example of the material strength levels, in ordinal data. Even though we can rank the levels, we cannot place a "value" on them; we cannot say that "moderate strength" is twice as positive as "low strength or "low strength" is thrice as negative as "high strength". An interval scale possesses all the characteristics of ordinal data. It allows us to group variables into categories and order them, but unlike ordinal data, interval data also allows us to assess the degree of difference between any two values.

An interval scale is a numeric scale that has equal distances between adjacent values. These distances are called "intervals". To put it another way, the differences between points on the scale are equivalent. An example of this type of data is the temperature in degrees Fahrenheit. The difference between 80°F and 75°F is the same as the difference between 60°F and 55°F.

The other important feature we saw at the beginning is that an interval scale always lacks what's known as a "true zero". Simply stated, on an interval scale zero does not denote the complete absence of something. By default, this means that zero on an interval scale is simply another variable. For instance, 0°F does not mean the complete absence of temperature. The lack of a true zero in interval scales makes it impossible to multiply or divide scores on interval scales. However other, arithmetic operations—addition and subtraction—can be performed on interval data. Due to its quantitative character, almost all statistical analysis is applicable on interval data. For example 30°C is not twice as hot as 15°C. Similarly, –5°F is not half as cold as –10°F.

Table 2.1

Type of variable	What does the data represent?	Examples
Discrete variables	Counts of individual items or values.	• No. of students in a class
Continuous variables	Measurements of continuous or non-finite values.	• Distance • Volume • Current

Ratio-scaled data looks a lot like interval-scaled data. The only difference between an interval scale and a ratio scale is the former's ability to dip below zero. In contrast to interval data, the zero point has a special meaning in ratio-scaled data: it indicates the absence of whatever property is being measured (for example, volume). Due to this there can be no negative numerical value in ratio data.

Another significant difference is the ability to perform all arithmetic operations including multiplication and division, which was not possible in interval data.

In ratio data, the difference between 1 and 2 is the same as the difference between 3 and 4, but also here 4 is twice as much as 2. This comparison is impossible in interval data.

Examples of ratio variables include:

- Enzyme activity, dose amount, reaction rate, flow rate, concentration, pulse, weight, length, temperature in Kelvin (0.0 Kelvin really does mean "no heat"), survival time.

Both interval and ratio scales are measurement scales that are typically associated with continuous data.

2.1.3 Types of data in statistical analysis

A central concept in statistics is a variable's level of measurement. In brief, it helps us to interpret and manipulate data in the right way. Now that we can differentiate between the data types it should be obvious that we need to visualize and analyze numerical data differently than categorical data; otherwise it would result in a wrong analysis. In other words, distinguishing between data types help you decide which visualization tool and statistical technique to use for which analysis.

Let's explore the data types typically associated with different types of statistical analysis along with examples for each.

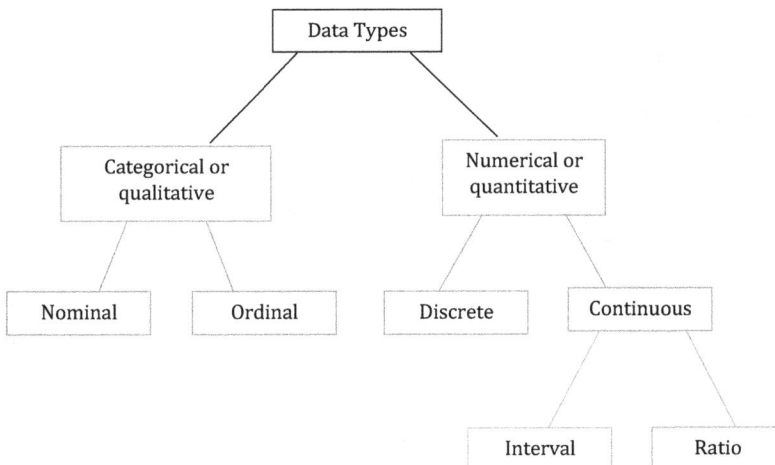

Figure 2.1

2.1.3.1 Univariate analysis

Since univariate analysis focuses on examining and describing a single variable at a time, a univariate variable can be either categorical or numerical.

1. Categorical: as previously stated, categorical univariate data involves observations that are non-numerical and can be placed into distinct categories or groups.
   i. Nominal: these categories do not have any order.

   For example, let's imagine the data is about all the employees in a company. Each of the categories, "job title", "department", "business unit", "gender", and "ethnicity", represents a separate categorical variable.

   The first category, "job title", could include sr. manager, technical architect, director, computer systems manager, etc. For the second category, we have IT, finance, sales, etc. For the third we have research and development, manufacturing, specialty products, and corporate. For gender we have two options, and finally for ethnicity we have four groups, Black, Asian, Caucasian, and Latino.

Table 2.2

Full Name	Job Title	Department	Business Unit	Gender	Ethnicity
Emily Davies	Sr. Manger	IT	Research & Development	Female	Black
Theodore Dinh	Technical Architect	IT	Manufacturing	Male	Asian
Luna Sanders	Director	Finance	Speciality Products	Female	Caucasian
Penelope Jordan	Comp. System Manager	IT	Manufacturing	Female	Caucasian
Austin Vo	Sr. Analyst	Finance	Manufacturing	Male	Asian
Joshua Gupta	Account Representation	Sales	Corporate	Male	Asian
Ruby Barnes	Manager	IT	Corporate	Female	Caucasian
Luke Martin	Analyst	Finance	Manufacturing	Male	Black
Easton Bailey	Manager	Accounting	Manufacturing	Male	Caucasian
Madeline Walk	Sr. Analyst	Finance	Speciality Products	Female	Caucasian
Savannah Ali	Sr. Analyst	Human Resource	Manufacturing	Female	Asian
Camila Rogers	Controls Engineer	Engineering	Speciality Products	Female	Caucasian
Eli Jones	Manager	Human Resource	Manufacturing	Male	Caucasian
Everleigh Ng	Sr. Manager	Finance	Research & Development	Female	Asian
Robert Yang	Sr. Analyst	Accounting	Speciality Products	Male	Asian
Isabella Xi	Vice President	Marketing	Research & Development	Female	Asian
Bella Powell	Director	Finance	Research & Development	Female	Black
Camila Silva	Sr. Manager	Marketing	Speciality Products	Male	Latino
David Barnes	Director	IT	Corporate	Male	Caucasian
Adam Dang	Director	Sales	Research & Development	Male	Asian
Elias Alvarado	Sr. Manager	IT	Manufacturing	Male	Latino
Eva Rivera	Director	Sales	Manufacturing	Female	Latino
Logan Rivera	Director	IT	Research & Development	Male	Latino
Leonardo Dixo	Analyst	Sales	Speciality Products	Male	Caucasian
Mateo Her	Vice President	Sales	Speciality Products	Male	Asian
Jose Henderson	Director	Human Resource	Speciality Products	Male	Black

Unfortunately, there's no way to arrange these in a hierarchy; hence the data is nominal.

(Data taken from https://www.thespreadsheetguru.com/sample-data/.)

We have already seen that categorical data can take numerical values even though these values do not carry any inherent numerical meaning or significance. Let's look at the example below.

(Data taken from https://community.tableau.com/s/question/0D54T00000CWeX8SAL/sample-superstore-sales-excelxls.)

Table 2.3

Row ID	Order ID	Order Date	Ship Date	Customer ID	Customer Name	Postal Code	Product ID
1	2016152156	11/8/2016	11/11/2016	12520	Claire Gute	42420	101798
2	2016152156	11/8/2016	11/11/2016	12520	Claire Gute	42420	100454
3	2016138688	6/12/2016	6/16/2016	13045	Darrin Van Huff	90036	100240
4	2015108966	10/11/2015	10/18/2015	20335	Sean O'Donnell	33311	100577
5	2015108966	10/11/2015	10/18/2015	20335	Sean O'Donnell	33311	100760
6	2014115812	6/9/2014	6/14/2014	11710	Brosina Hoffman	90032	101487
7	2014115812	6/9/2014	6/14/2014	11710	Brosina Hoffman	90032	102833
8	2014115812	6/9/2014	6/14/2014	11710	Brosina Hoffman	90032	102275
9	2014115812	6/9/2014	6/14/2014	11710	Brosina Hoffman	90032	103910
10	2014115812	6/9/2014	6/14/2014	11710	Brosina Hoffman	90032	102892
11	2014115812	6/9/2014	6/14/2014	11710	Brosina Hoffman	90032	101539
12	2014115812	6/9/2014	6/14/2014	11710	Brosina Hoffman	90032	102033
13	2017114412	4/15/2017	4/20/2017	10480	Andrew Allen	28027	102365
14	2016161389	12/5/2016	12/10/2016	15070	Irene Maddox	98103	103656
15	2015118983	11/22/2015	11/26/2015	14815	Harold Pawlan	76106	102311
16	2015118983	11/22/2015	11/26/2015	14815	Harold Pawlan	76106	100756
17	2014105893	11/11/2014	11/18/2014	19075	Pete Kriz	53711	104186
18	2014167164	5/13/2014	5/15/2014	10270	Alejandro Grove	84084	100107
19	2014143336	8/27/2014	9/1/2014	21925	Zuschuss Donatelli	94109	103056
20	2014143336	8/27/2014	9/1/2014	21925	Zuschuss Donatelli	94109	101949
21	2014143336	8/27/2014	9/1/2014	21925	Zuschuss Donatelli	94109	102215
22	2016137330	12/9/2016	12/13/2016	16585	Ken Black	68025	100246
23	2016137330	12/9/2016	12/13/2016	16585	Ken Black	68025	101492

Here the columns Order ID, Customer ID, Postal code, and Product ID are all numerical but are treated as categorical variables.

ii. Ordinal: these categories have an order. For example for the education levels of engineers, categories like "associate degree", "bachelor's degree", "master's degree", and "PhD" represent an ordinal scale of increasing education. Also notice the major column is a nominal variable.

2. Numeric: numerical univariate data involves observations that are numerical.

i. Discrete: for example, counting the number of specific components in an assembly, such as bolts, screws, or electronic components, or the number of times a certain failure mode has occurred in a system. Let's look at the data below. From our discussion we can already guess that the region and item categories are nominal since region represents different geographical regions and the item column categorizes items into distinct groups. Therefore, these columns don't have a hierarchical or natural order. However the unit column representing the number of items sold is a discrete variable.

(Data taken from https://www.contextures.com/xlsampledata01.html#data.)

ii. Continuous: unlike discrete data, which consists of distinct, separate values, continuous data can take on an infinite number of values within a given interval. For example the values of tensile strength, temperature, voltage, etc. We already saw an example of a dataset of tensile strength values in Chapter 1.

2.1.3.2 Bivariate analysis

In bivariate analysis, where we are examining the relationship or association between two variables, we typically work with different combinations of data types for the two variables involved. Bivariate data can have the following combination of data types.

2.1.3.2.1 Numerical and numerical

Here, both the dependent and independent variables of the bivariate data have a numerical value. We already saw an example in the previous section studying the relationship between temperature and the thermal expansion coefficient (α) of a particular metal material.

Table 2.4

EEID	Name	Birth Date	Gender	Major	Level	Annual Salary
E04239	Alison Mitchell	3/28/1995	Female	-	Associate	$67,742
E02387	Alan Smith	2/5/2001	Male	-	Associate	$41,604
E02832	David Jordan	3/1/2002	Male	-	Associate	$54,913
E02572	Gary Chan	9/2/1999	Male	Chemical Eng.	Bachelor	$93,099
E03344	David Keith	1/19/2000	Male	Chemical Eng.	Bachelor	$109,851
E02071	Dylan Henderson	5/17/1997	Male	Civil Eng.	Bachelor	$90,172
E04533	Ivy Cavill	9/25/2002	Female	Comp. Science	Bachelor	$93,527
E00163	Emily davis	9/10/1992	Female	Comp. Science	Bachelor	$85,837
E04116	Penelope Jordan	1/10/1996	Female	Economics	Bachelor	$186,503
E03680	Luke Martin	4/1/1999	Male	Economics	Bachelor	$86,140
E03838	John Lennon	5/16/1997	Male	Electrical Eng.	Bachelor	$87,203
E00549	Alex Brown	7/1/2000	Male	Electrical Eng.	Bachelor	$94,270
E01639	Eva Lee	10/2/1995	Female	Biomedical Eng.	Masters	$95,409
E04625	Ruby Barnes	3/23/1994	Female	Biomedical Eng.	Masters	$96,331
E00591	Kevin Robinson	7/21/1996	Male	Comp. Science	Masters	$97,333
E04105	Ben Jones	4/6/1994	Male	Electrical Eng.	Masters	$119,975
E04332	Harry Smith	1/17/1995	Male	Electrical Eng.	Masters	$97,336
E04732	Madeline Walker	2/15/1993	Female	Mechanical Eng.	Masters	$101,703
E01550	James Brown	8/7/1996	Male	Mechanical Eng.	Masters	$109,746
E03496	Tyler Olson	4/4/1996	Male	Mechanical Eng.	Masters	$97,078
E00530	Shay Johnson	2/7/1984	Female	Civil Eng.	PhD	$115,086
E00671	Carson Lu	10/12/1988	Male	Civil Eng.	PhD	$129,998
E00884	Luna Sanders	11/5/1986	Female	Comp. Science	PhD	$134,828
E00644	Ryan Roberts	11/24/1985	Male	Comp. Science	PhD	$120,994
E03484	Isabella Yang	1/2/1983	Female	Electrical Eng.	PhD	$132,787

Table 2.5

Order Date	Region	Rep	Item	Units	UnitCost	Total
1/6/2020	East	Jones	Pencil	95	1.99	189.05
1/23/2020	Central	Kivell	Binder	50	19.99	999.5
2/9/2020	Central	Jardine	Pencil	36	4.99	179.64
2/26/2020	Central	Gill	Pen	27	19.99	539.73
3/15/2020	West	Sorvino	Pencil	56	2.99	167.44
4/1/2020	East	Jones	Binder	60	4.99	299.4
4/18/2020	Central	Andrews	Pencil	75	1.99	149.25
5/5/2020	Central	Jardine	Pencil	90	4.99	449.1
5/22/2020	West	Thompson	Pencil	32	1.99	63.68
6/8/2020	East	Jones	Binder	60	8.99	539.4
6/25/2020	Central	Morgan	Pencil	90	4.99	449.1
7/12/2020	East	Howard	Binder	29	1.99	57.71
7/29/2020	East	Parent	Binder	81	19.99	1619.19
8/15/2020	East	Jones	Pencil	35	4.99	174.65
9/1/2020	Central	Smith	Desk	2	125	250
9/18/2020	East	Jones	Pen Set	16	15.99	255.84
10/5/2020	Central	Morgan	Binder	28	8.99	251.72
10/22/2020	East	Jones	Pen	64	8.99	575.36
11/8/2020	East	Parent	Pen	15	19.99	299.85
11/25/2020	Central	Kivell	Pen Set	96	4.99	479.04
12/12/2020	Central	Smith	Pencil	67	1.29	86.43
12/29/2020	East	Parent	Pen Set	74	15.99	1183.26

Another example of numerical independent and dependent variables is **temperature and pressure in a gas tank**. The results are presented in the following bivariate data table.

Temperature	174.3	225.1	273.8	320.2	372.7	424.6	479.4	522.5	578.9	625.3
Pressure	36.2	46.4	56.8	67.3	77.6	88.1	99.4	107.8	113.2	132.6

2.1.3.2.2 Categorical and categorical

Both variables in bivariate data are categorical. For example, we want to study the relationship between vehicle safety ratings and drivers' injury severity.

	Drivers' injury severity				
Vehicle safety ratings	Minimal	Minor	Severe	Fatal	Total
5	25	32	15	05	77
4	20	30	20	09	79
3	28	33	21	11	93
2	35	28	18	15	96
1	38	33	26	19	116

2.1.3.2.3 Numerical and categorical

This is when one of the variables is numerical, and the other is categorical. For example we want to study the relationship between the yield strength (MPa) (numerical dependent variable) and the type of material (categorical independent variable). We collect data on different material samples and their yield strength values given below.

Material	Yield strength (MPa)
A	38
AX	46
AY	69
B	25
BX	72
BY	53
C	79
CX	81
CY	87
ABC	95

2.1.3.3 Multivariate analysis

For multivariate data, let's look back at our example where both the dependent variable (heat transfer coefficient) and all three independent variables (**temperature, thermal conductivity, area) are numerical.** Consider another example where an automobile manufacturer wants to assess the factors influencing a vehicle's fuel efficiency (miles per gallon, or mpg) and suspects that these factors include the "number of cylinders", "displacement", "horsepower", "weight", "acceleration", "model year", and "origin".

Among these parameters, "displacement", "horsepower", "weight", and "acceleration" are pure numerical variables and "model year" and "origin" are categorical. However, the number of cylinders can be considered either numerical or categorical depending on the context of the analysis and the specific research goals.

Both numerical and categorical data have different characteristics and require different approaches for analysis, which we shall learn in this textbook. We will explore these classifications in more detail in Chapter 2.

2.2 COLLECTING THE DATA

In the previous section we first learned the types of data so that we can decide what type of insights we require. Now that we know the data types, we get out and collect the data, but surprisingly this is not always the data that we are interested in! Wait! What? So before we delve into the topic we need to get familiar with some terms and concepts.

Population: generally, population refers to the people who live in a particular area at a specific time or a group of people with a common characteristic.

However, in statistical terms, population is defined in a slightly different way. It refers to data in your study of interest. Now this data is usually so large that it is simply impossible to collect. So what do we do now? What else do we collect? Now comes the second important terminology—sample. A sample is a smaller but statistically significant portion drawn from the population that has the characteristics of the larger population. So instead of the population we collect a sample of the population which is smaller, more manageable, and less time consuming and also more realistic.

Now here are several ways to obtain samples from a population.

Simple random sampling: this is selecting samples through the use of a random number generator or some means of random selection which implies that each member of a population has an equal chance of being selected for the sample. Even though it might seem like the simplest and most convenient method to obtain samples, it is not always perfect and should not always be used. This is because the randomness of the selection does not guarantee that the data collected is reflective of the entire population.

Systematic random sampling: in systematic sampling we first arrange or sort the target population into some order and then we choose a random starting point and select every nth element from then onward to be in the sample. A simple example would be to select every fifth name from the list of students to come to the board and solve a math problem (an "every fifth" sample, also referred to as "sampling with a skip of 5").

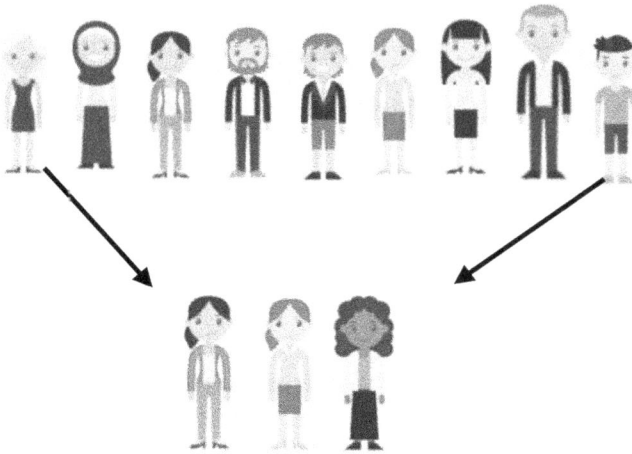

Figure 2.2

Stratified random sampling: in this sampling we split a population into k number of groups and then we randomly select some members from each group to be in the sample.

Figure 2.3

For example, let's say we want to sample the students pursuing their bachelor's degree to see how many hours of sleep they get every week, and we want the sample to be representative of all grade levels including their majors (freshmen, sophomore, junior, and senior from every major). It would make sense to divide the students into their distinct grade levels and then into their individual majors.

2.2.1 Cluster sample

Cluster sampling divides the population into groups, or clusters. Some of these clusters are randomly selected. Then, all of the individuals in the chosen cluster are selected to be in the sample. This process is often used because it can be cheaper and more time-efficient.

For example, while surveying households within a city, we might choose to select 100 city blocks and then interview every household within the selected blocks, rather than interviewing random households spread out over the entire city.

2.3 DATA VISUALIZATION AND GRAPHICAL REPRESENTATION OF A VARIABLE

Now that we have collected our data using one of the sampling techniques, what do we do next? According to the definition of statistics we now begin the process of analyzing, but the question is where or how do we start? Understand that the sample we collected really does not give out anything. Often the data we collect is unstructured inhibiting us from seeing the main features or characteristics of the data and making it impossible to analyze.

Moreover, realize that even though we collect a sample which is a much **smaller, manageable version of a larger group**, it still can be huge. Therefore, it's necessary to have a clear idea of what the collected sample represents, which is best achieved by giving it visual context; the term we use is data visualization.

Data visualization refers to the use of charts and graphs to visually display, analyze, and identify trends, outliers, and patterns and interpret the ever-increasing data flow to predict and make decisions. It not only provides a powerful means of summarizing and presenting data, but also provides the shape and pattern of the data, which is critical in data analysis and decision making. In a nutshell, it is necessary for our analysis.

Now there are dozens of different ways to represent statistical data graphically. Choosing the most appropriate chart depends on a variety of different factors—the nature of the data and the type of analysis.

Let's look at all of the different types of graphical representations commonly used for the three types of data analysis.

2.3.1 Univariate analysis

2.3.1.1 Frequency distribution table

A frequency distribution table is often the first type of visualization tool used in univariate analysis because it describes **the number of observations for each possible value of a variable** whether it's numerical or categorical. Here the word "frequency" tells us **the number of times** a data value occurs. It can be thought of as a count or tally of observations for each unique value or category recorded and displayed in a tabular format called a **frequency distribution table**.

As well as just displaying the actual counts, it can also present the percentage or proportion of observations in each category or interval. In the latter instance, the table is called a **relative frequency distribution table**. This is useful for comparing the relative importance of different categories, especially when dealing with datasets of varying sizes.

A frequency distribution table summarizes all the data mostly under two columns (or three for a **relative frequency distribution table**). However, we could also add another column which is usually optional.

The first column lists all the outcomes of a sample space. As stated earlier, the outcomes can be categorical or numerical. The second column includes the tally marks of each outcome and is typically optional. The third column lists the frequency of each outcome. In the case of **relative frequency distribution tables**, the fourth column includes the percentage or proportion of observations in each outcome or category relative to the total number of observations.

As mentioned earlier, frequency distribution tables can be used to summarize and analyze both categorical and numerical data. Let's now discuss the frequency distribution for each type of data.

2.3.1.1.1 Frequency distribution table for qualitative data

In a frequency distribution table for categorical data, the table would list the categories along with the corresponding frequencies or counts. Examples in engineering could include the type of material used, the classification of defects, or the failure modes of a system.

2.3.1.1.2 Example 2.1

Let's look at the example in Table 2.2; we want to create a frequency distribution table for each of the categories, "job title", "department", "business unit", "gender", and "ethnicity". Here's how the frequency distribution table might look.

Variable	Category	Tally marks	Frequency/count
Job title	Sr. manager	⊥⊥⊦	5
	Technical architect	|	1
	Director	⊥⊦⊦ |	6
	Computer systems manager	|	1
	Sr. analyst	|||	3
	Account representative	|	1
	Manager	|||	3
	Analyst	||	2
	Controls engineer	|	1
	Vice president	||	2
Department	IT	⊥⊦⊦ ||	7
	Finance	⊥⊦⊦ |	6
	Sales	⊥⊦⊦	5
	Accounting	||	2
	Human resources	||	2
	Engineering	||	2
	Marketing	|	1
Business unit	Research and development	⊥⊦⊦ |	6
	Manufacturing	⊥⊦⊦||||	9
	Specialty products	⊥⊦⊦ ||	7
	Corporate	|||	3
Gender	Male	⊥⊦⊦ ⊥⊦⊦ |||	13
	Female	⊥⊦⊦ |||| ||	12
Ethnicity	Black	|||	3
	Asian	⊥⊦⊦ ||||	9
	Caucasian	⊥⊦⊦ ||||	9
	Latino	||||	4

2.3.1.1.3 Example 2.2

Let's look at the example in Table 2.4. We want to create a **relative frequency distribution table** for each of the categories, "gender", major", and "level of education".

Variable	Category	Tally marks	Frequency	Percentage
Gender	Male	⊥⊥⊦ ⊥⊥⊦ ⊥⊥⊦	15	$\frac{15}{25} \times 100 = 60\%$
	Female	⊥⊥⊦ ⊥⊥⊦	10	$\frac{10}{25} \times 100 = 40\%$
Major	Biomedical engineering	||	2	$\frac{2}{25} \times 100 = 8\%$
	Chemical engineering	||	2	$\frac{2}{25} \times 100 = 8\%$
	Civil engineering	|||	3	$\frac{3}{25} \times 100 = 12\%$
	Computer science	⊥⊥⊦	5	$\frac{5}{25} \times 100 = 20\%$
	Economics	||	2	$\frac{2}{25} \times 100 = 8\%$
	Electrical engineering	⊥⊥⊦	5	$\frac{5}{25} \times 100 = 20\%$
	Mechanical engineering	|||	3	$\frac{3}{25} \times 100 = 12\%$
	-	|||	3	$\frac{3}{25} \times 100 = 12\%$
Level	Associate	|||	3	$\frac{3}{25} \times 100 = 12\%$
	Bachelor	⊥⊥⊦ ||||	9	$\frac{9}{25} \times 100 = 36\%$
	Master	⊥⊥⊦ |||	8	$\frac{8}{25} \times 100 = 32\%$
	PhD	|||	5	$\frac{5}{25} \times 100 = 20\%$

2.3.1.1.4 Frequency distribution table for numerical data

We have seen that numerical data can be discrete or continuous. Numerical data, whether it's discrete or continuous, can be presented as a single value or defined as a range or an interval of numbers. When numerical data is represented as a single value in a frequency distribution table, it is referred to as an "ungrouped frequency distribution table". On the other hand, when the outcome is represented by intervals or ranges of values in a frequency distribution table, that table is typically called a "grouped frequency distribution table". Therefore, based on the categorization of the numerical data, there are two types of frequency distribution tables:

* Grouped frequency distribution tables
* Ungrouped frequency distribution tables

Let's explore both ungrouped and grouped frequency distribution tables.

2.3.1.1.5 The frequency distribution table for ungrouped data

A frequency distribution table for ungrouped data is a table where each row corresponds to a single distinct value or category and displays the frequency or count of that specific value in the dataset. Discrete data is often presented in ungrouped frequency distribution tables, but it can also be presented in grouped frequency distribution tables when the dataset is extremely large. Let's examine both situations.

2.3.1.1.6 Example 2.3

Consider a dataset that represents the number of design iterations required to optimize a mechanical component. The dataset consists of the following counts of design iterations.

Number of iterations	Tally marks	Frequency/count
5	‖ ‖	4
6	‖	3
7	‖ ‖	4
8	‖‖‖ ‖	6
10	‖ ‖	5
12	‖	3

In this example:

* "Number of iterations" represents the distinct values or categories in the dataset, indicating the number of design iterations required (ranging from 5 to 12 iterations).
* "Frequency" indicates the count or frequency of cases where that specific number of design iterations was needed. For instance, five cases required four design iterations, eight cases required six iterations, and so on

2.3.1.1.7 The frequency distribution table for grouped data

Notice that there are only six distinct numerical categories, and for a small number of distinct numerical categories, like six, presenting ungrouped data can provide a clear and detailed view of the data distribution. However, when dealing with a larger number of categories, such as 50 or more, using grouped data becomes more practical and efficient. If we create a frequency distribution table for all 50 or more observations, then it will form a large table. Remember the whole point of the frequency distribution table is to present a high volume of data in a more precise usable form. Grouping the data into intervals or class intervals serves the purpose of condensing a high volume of data into a more concise and usable form, making it easier to analyze and draw meaningful insights from the data.

As mentioned earlier, grouped frequency distribution tables can be applied to both discrete and continuous data. So let's take a look at examples of grouped frequency distribution tables for both discrete and continuous data.

2.3.1.1.8 Example 2.4

The time taken in hours for the failure of 75 specimens of a metal subjected to fatigue failure tests is as shown.

$$18, \quad 20, \quad 28, \quad 20, \quad 15, \quad 25, \quad 36, \quad 29, \quad 31, \quad 35, \quad 27,$$
$$15, \quad 27, \quad 13, \quad 34, \quad 18, \quad 26, \quad 08, \quad 21, \quad 29, \quad 14,$$
$$23, \quad 16, \quad 23, \quad 11, \quad 17, \quad 06, \quad 33, \quad 28, \quad 14, \quad 24, \quad 15,$$
$$29, \quad 26, \quad 49, \quad 22, \quad 19, \quad 36, \quad 22, \quad 37, \quad 22, \quad 19$$
$$30, \quad 15, \quad 25, \quad 36, \quad 40, \quad 25, \quad 38, \quad 27, \quad 09, \quad 31, \quad 42$$
$$17, \quad 11, \quad 22, \quad 19, \quad 23, \quad 37, \quad 26, \quad 18, \quad 08, \quad 41, \quad 10$$
$$24, \quad 13, \quad 42, \quad 05, \quad 20, \quad 18, \quad 29, \quad 34, \quad 45, \quad 37, \quad 26$$

Time taken (hours)	Tally marks	Frequency/count
1–10	⊥⊦⊦ l	6
11–20	⊥⊦⊦ ⊥⊦⊦ ⊥⊦⊦ ⊥⊦⊦ l ll	23
21–30	⊥⊦⊦ ⊥⊦⊦ ⊥⊦⊦ ⊥⊦⊦ ⊥⊦⊦ ll	27
31–40	⊥⊦⊦ ⊥⊦⊦ llll	14
41–50	⊥⊦⊦	5

2.3.1.1.9 Example 2.5

Let's bring back our continuous dataset of tensile strength:

$$212.9, \quad 215.7, \quad 217.6, \quad 220.4, \quad 221.2, \quad 223.3, \quad 223.6, \quad 223.8,$$
$$225.5, \quad 226.4, \quad 229.6, \quad 232.1, \quad 233.2, \quad 234.5, \quad 238.4,$$
$$238.5, \quad 239.4, \quad 240.7, \quad 241.1, \quad 244.8, \quad 245.3,$$
$$247.7, \quad 251.6, \quad 253.3, \quad 256.6, \quad 257.9,$$
$$260.2, \quad 262.1, \quad 263.3, \quad 263.6$$

Tensile strength	Tally marks	Frequency/count
210–220	|||	3
220–230	⊥⊦⊦⊦ |||	8
230–240	⊥⊦⊦⊦		6
240–250	⊥⊦⊦⊦	5
250–260	||||	4
260–270	||||	4

Let's look at the two examples above. If we look carefully we will notice that the two frequency distribution tables are a little different from each other. The first table has upper limits that are not repeated as lower limits in the preceding interval, whereas the upper limits in the second table are. These two types of frequency distribution are called inclusive and exclusive frequency distribution. Each type has its own advantages and use cases, depending on the context and data at hand, so let's discuss each of them.

2.3.1.1.10 Inclusive frequency distribution

The upper limit of one class interval does not repeat itself as the lower limit of the next class interval. In this distribution all the values falling within the range from the lower limit up to the upper limit are included in the class interval. For example, the range for 1–10 hours means that all the hours from 1 to 10 are included, for 11–20 hours all the hours from 11 to 20 are included, and so on.

Notice that there is a gap between the upper limit of one class interval and the lower limit of the next class interval. For example, our class intervals are 1–10, 11–20, 21–30, 31–40, and so on. In this case, the gap between the upper limit of one class interval and the lower limit of the next class interval is 1, since it does not include all the values between 20 and 21 excluded. In fact, this gap could typically range anywhere from 0.1 to 1.0. So, the gap between class intervals only contains decimal/fractional values, and therefore it is difficult to perform statistical analysis with inclusive series. In those cases, the inclusive series is converted into an exclusive series. For the same reason, we typically use an inclusive series only if the value is discrete/an integer number and not in decimal form.

2.3.1.1.11 Exclusive frequency distribution

In this kind of frequency distribution, the upper limit of one class interval is repeated as the lower limit of the next class interval. This means the same value appears twice, but as each value needs to belong to only one class interval, the frequencies corresponding to the specific class interval do not include the value of its upper limit. For example, for the class interval 210–220, 220 would belong to the preceding interval (220–230), and 230 would belong to the next interval, and so on. Unlike in inclusive series, an exclusive series does not have any gap between the class intervals which means it maintains the continuity of the data and also ensures there is no overlap between the intervals since it does not include the upper limit of the interval. This is typically why this approach is used for both discrete and continuous data.

Therefore, we conclude that there are two methods of classifying data according to class intervals:

Exclusive method	Inclusive method
The upper limit of the previous class intervals repeats in the lower limit of the next class interval.	The upper limit of the previous class intervals does not repeat in the lower limit of the next class interval.
We include only the value of the lower limit and do not include the value of the upper limit in the distribution table. The upper limit of a class interval is counted in the next immediate class.	Both limits of a class interval are counted in the same class.
The upper limit of a class interval and the lower limit of the next class are the same.	The upper limit of a class interval and the lower limit of the next class are different. The difference is generally of one.

2.3.1.1.12 Conversion of inclusive series into exclusive series

Not surprisingly, it's more common and preferred to work with exclusive class intervals when analyzing a frequency distribution. Therefore, whenever an inclusive class interval (where both limits are included) is encountered, it's often necessary to convert it to an exclusive class interval (where the upper limit is excluded) to facilitate proper analysis and visualization. This conversion is done by using class boundaries. In a frequency distribution, **class boundaries** are the values that separate the classes.

We use the following steps to calculate the class boundaries in an inclusive data frequency distribution:

1. Calculate the difference between the upper limit of one class interval and the lower limit of the next class interval.
2. Divide the result by two.
3. Subtract the result from the lower limit and add the result to the upper limit for each class.

Let's look at Example 2.4:

(i) First, we find the difference between the upper limit of one class interval and the lower limit of the succeeding class interval. For example, if we look at the first row of the table, the upper limit of the class interval $1-10$ is 10. The lower limit of the next class interval $11-20$ is 11. The difference is

$$11-10 = 1$$

(ii) Secondly, half of that difference is added to the upper limit of a class interval and half is subtracted from the lower limit of the class interval. Since the difference found in step 1 is 1, half of that difference will be 0.5. Therefore we add 0.5 to the upper limit and subtract 0.5 from the lower limit for each class interval.

Following the steps our table will now become:

Time taken (hours)	Tally marks	Frequency/count
0.5–10.5	ⅬⅬⅬⅬ |	6
10.5–20.5	ⅬⅬⅬⅬ ⅬⅬⅬⅬ ⅬⅬⅬⅬ ⅬⅬⅬⅬ | ||	23
20.5–30.5	ⅬⅬⅬⅬ ⅬⅬⅬⅬ ⅬⅬⅬⅬ ⅬⅬⅬⅬ ⅬⅬⅬⅬ ||	27
30.5–40.5	ⅬⅬⅬⅬ ⅬⅬⅬⅬ ||||	14
40.5–50.5	ⅬⅬⅬⅬ	5

2.3.1.1.13 MATLAB

To display a frequency table of the data we use the function

```
tabulate(Data)
```

which displays a frequency table of the data displaying categories, their frequencies, and the percentages of each category.

Let's do Example 2.3.

```
Data = categorical(x);
tabulate(Data)
```

Table 2.6

Value	Count	Percent
5	4	16.00%
6	3	12.00%
7	4	16.00%
8	6	24.00%
10	5	20.00%
12	3	12.00%

We can also modify the frequency table by the following syntax

```
Data = categorical(x);
t = cell2table(tabulate(Data),'VariableNames', {'Iteration','Count',
                                     'Percentage'});
t.Iteration = categorical(t.Iteration)
```

Let's do Example 2.2.

```
Data = readtable("Employees_Info.csv", VariableNamingRule = "preserve");
Gender = Data.Gender;
Major = Data.Major;
```

```
Level = Data.Level;
Data = categorical(Major);
t = cell2table(tabulate(Data),'VariableNames',{'Major','Count','Perce
nt'});
t.Major = categorical(t.Major)
```

Table 2.7

Iteration	Count	Percentage
5	4	16
6	3	12
7	4	16
8	6	24
10	5	20
12	3	12

Table 2.8

```
t =

  8x3 table
```

Major	Count	Percent
-	3	12
Biomedical Engineering	2	8
Chemical Engineering	2	8
Civil Engineering	3	12
Computer Science	5	20
Economics	2	8
Electrical Engineering	5	20
Mechanical Engineering	3	12

What about intervals?

Unfortunately, MATLAB does not have a built-in function for displaying grouped frequency distribution tables directly in tabular form.

MATLAB's default visualization tool for grouped frequency distributions is the histogram. Histograms are a common and effective way to visually represent the distribution of data in grouped intervals.

However, MATLAB's default functions for creating histograms does not always generate precise grouped frequency distribution tables. While it can provide a graphical overview of how data is distributed in intervals (bins), the primary focus is on visualization rather than tabular output; therefore we have to modify the function to create a more accurate and reliable interval frequency distribution.

2.3.1.2 Bar chart

Bar graphs have rectangular bars whose height represents the frequency counts of values for the different levels of a categorical variable. The independent variable of a bar chart is generally categorical which can be thought of as labels like the months in the following example or days of the week, states or countries, years like 2000, 2001,....2020, or grades, as discussed in Section 3.3.1. In contrast, the dependent variable will be numeric in nature. Its values determine the height of each bar which is proportionate to the numerical value of that particular category. So basically, bar graphs display large sets of data by categorizing them based on their numerical values.

For example, in a survey, the satisfaction levels of customers were recorded for different products. The data collected suggests that there might be variations in the satisfaction levels across the products. If we create a bar graph to visualize the satisfaction levels of the products, our graph would look something like Figure 2.4.

In an engineering survey, the preferences for different programming languages were collected from a group of engineers. The data collected includes the number of engineers who prefer each programming language. Create a bar graph to visualize the preferences of programming languages (Figure 2.5).

2.3.1.2.1 MATLAB

The main function for creating a bar graph in MATLAB is called bar. The main syntax for the function is

```
bar(x, y, BarWidth, style, 'FaceColor', , 'EdgeColor', , 'LineWidth', ,
'LineStyle', )
```

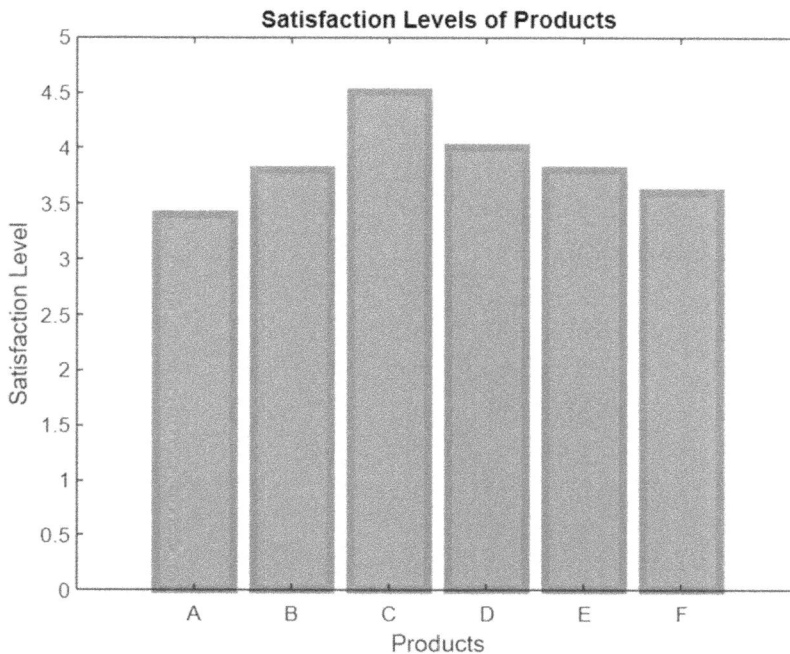

Figure 2.4

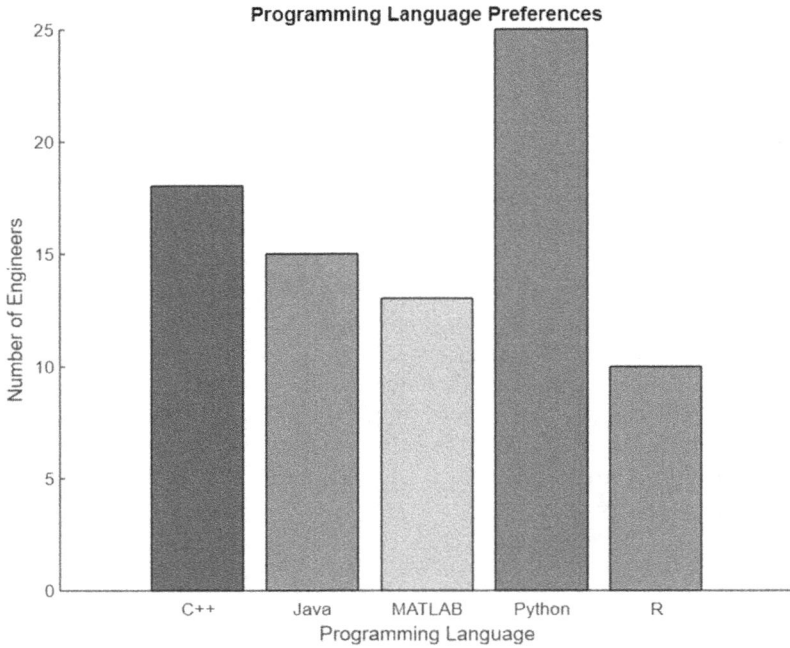

Figure 2.5

BarWidth: bar width, specified as a fraction of the total space available for each bar. The default of 0.8 means the bar width is 80% of the space from the previous bar to the next bar, with 10% of that space on each side. If the width is 1, then the bars within a group touch one another. Let's explore the difference between specifying a bar width of 0.8 and 1.0 in a bar graph in Figure 2.6.

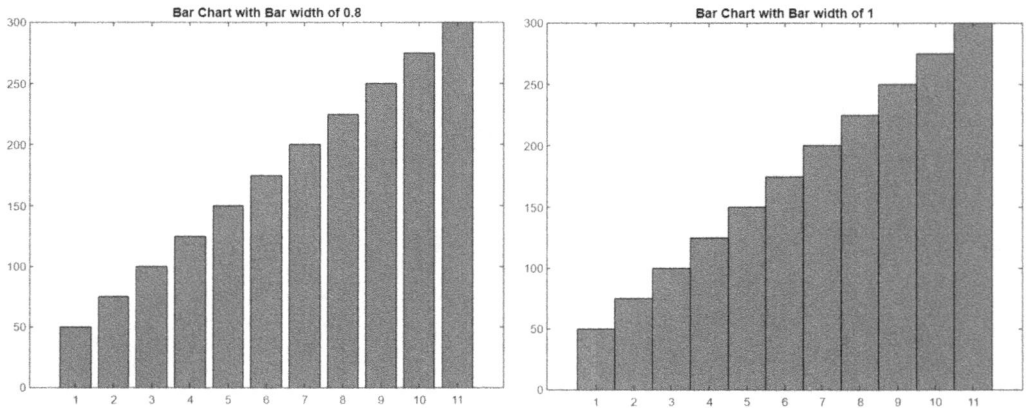

Figure 2.6

2.3.1.2.2 Style

2.3.1.2.2.1 'GROUPED'(DEFAULT)

Displays multiple bars side by side, centered around their corresponding x value, allowing for easy comparison between different categories or groups. Each category or group is represented by a separate set of bars, and the bars within each group are grouped together based on a shared attribute or characteristic.

The graphs we have seen and discussed so far were plotted on the basis of only one categorical variable. But guess what, MATLAB allows us to plot numeric values for two or more levels of the same category instead of only one.

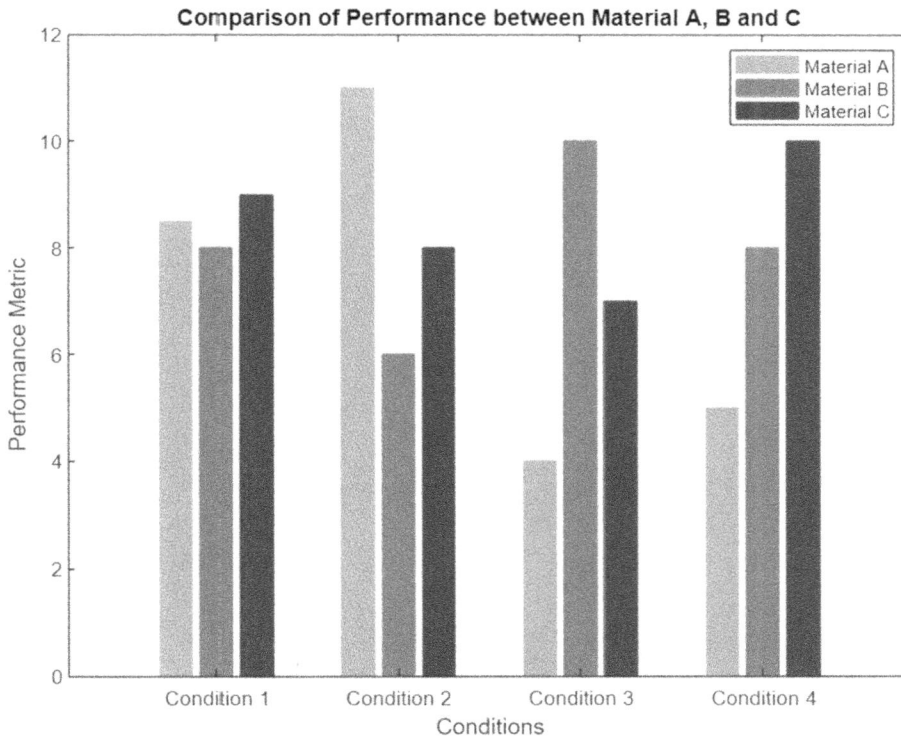

Figure 2.7

2.3.1.2.2.2 'STACKED'

Like grouped, stacked also presents multiple data in a single graph except it displays all the parameters in a single bar. They are stacked on top of each other rather than being placed side by side. So, for example, if instead of grouped our data was stacked then it looks like (Figures 2.8 and 2.9).

Each bar represents a category or group, and the height of the bar is divided into segments to represent different subcategories or components. This shows the composition of each level of a category you're comparing. The total height of each stacked bar represents the total value or sum of the subcategories. However, the problem with stacked bars is that it can become harder to read when representing a lot of information.

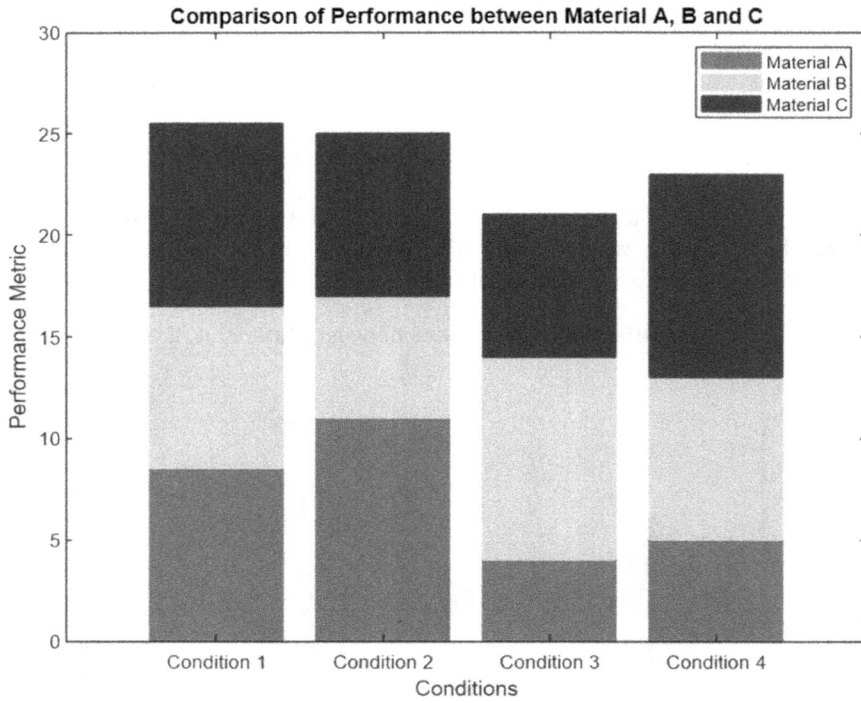

Figure 2.8

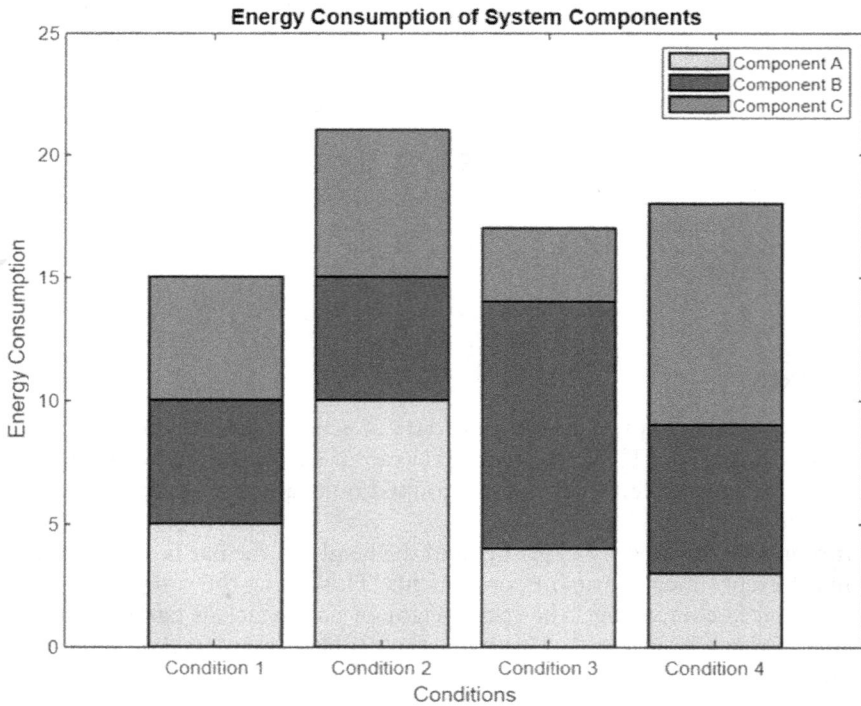

Figure 2.9

2.3.1.2.3 LINESTYLE

Line style of bar outlines, specified as one of the line styles in this table.

Line style	Description	Resulting line
"-"(default)	Solid line	_____
"–"	Dashed line	_ _ _ _ _ _
":"	Dotted line	....
"-."	Dash-dotted line	_ . _ . _ . _ . _
"none"	No line	No line

Let's now look at the codes that displayed all these graphs.

2.3.1.2.4 Code for Figure 2.4

```
products = {'A', 'B', 'C', 'D', 'E', 'F'};
satisfactionLevels = [3.4, 3.8, 4.5, 4.0, 3.8, 3.6];
% Create bar graph
figure;
bar(satisfactionLevels,'FaceColor',[0.900 0.4250 0.0980],
                   'EdgeColor',[0.980 0.0280 0.0840],'LineWidth',3.5);
title('Satisfaction Levels of Products');
xlabel('Products');
ylabel('Satisfaction Level');
xticks(1:length(products));
xticklabels(products);
ylim([0 5]);
```

2.3.1.2.5 Code for Figure 2.5

```
languages = {'C++', 'Java', 'MATLAB', 'Python', 'R'};
Count = [18, 15, 13, 25, 10];

figure;
hold on;

% Define a color for each bar
colors = lines(length(Count));

for i = 1:length(Count)
  bar(i, Count(i), 'FaceColor', colors(i,:));
end

hold off;

% Add labels and title
title('Programming Language Preferences');
xlabel('Programming Language');
ylabel('Number of Engineers');

% Set x-axis labels
xticks(1:length(languages));
xticklabels(languages);
```

2.3.1.2.6 Code for Figure 2.7

```
Conditions = categorical({'Condition 1', 'Condition 2', 'Condition 3',
                          'Condition 4'});
Conditions = reordercats(Conditions,{'Condition 1', 'Condition 2',
                          'Condition 3',
                                       'Condition 4'});
Performance = [8.5 8 9; 11 6 8; 4 10 7 ; 5 8 10];
b = bar(Conditions, Performance, 'grouped')
colors = [1 0 0; 0 1 0; 0 0 1];
for i = 1:numel(b)
  b(i).FaceColor = colors(i,:);
  b(i).EdgeColor = colors(i,:);
end
xlabel("Conditions")
ylabel("Performance Metric")
title("Comparison of Performance between Material A, B and C")
legend("Material A", "Material B", "Material C")
xlabel("Conditions")
ylabel("Performance Metric")
title("Comparison of Performance between Material A, B and C")
legend("Material A", "Material B", "Material C")
```

2.3.1.2.7 Code for Figure 2.9

```
Conditions = categorical({'Condition 1', 'Condition 2', 'Condition 3',
'Condition 4'});
Conditions = reordercats(Conditions,{'Condition 1', 'Condition 2',
'Condition 3',
                               'Condition 4'});
Consumption = [5 5 5; 10 5 6; 4 10 3; 3 6 9];
b = bar(Conditions, Consumption, 'stacked', LineWidth = 1.2)
colors = [1 0.25 0.5; 0.25 1 0.50; 0.50 0.25 1];
for i = 1:numel(b)
  b(i).FaceColor = colors(i,:);
end

xlabel("Conditions")
ylabel("Energy Consumption")
title("Energy Consumption of System Components")
legend("Component A", "Component B", "Component C")

features = {'Durability', 'Performance', 'Ease of Use', 'Design'};
% Define satisfaction levels (random data)
satisfactionLevels = [4.2, 4.5, 3.8, 4.1];

% Create bar graph
figure;
bar(satisfactionLevels, 'FaceColor',[0 0.8 0.8],'EdgeColor','black','Line
Width', 1.2)
title('Satisfaction Levels for Different Product Features');
xlabel('Product Features');
ylabel('Satisfaction Level');
set(gca, 'XTickLabel', features); % Set x-axis tick labels

% Add data labels to the bars
```

```
text(1:length(satisfactionLevels), satisfactionLevels,
num2str(satisfactionLevels'),
        'HorizontalAlignment','center', 'VerticalAlignment','bottom');

% Adjust the y-axis limits to provide space for the data labels
ylim([0, max(satisfactionLevels) + 0.5]);
```

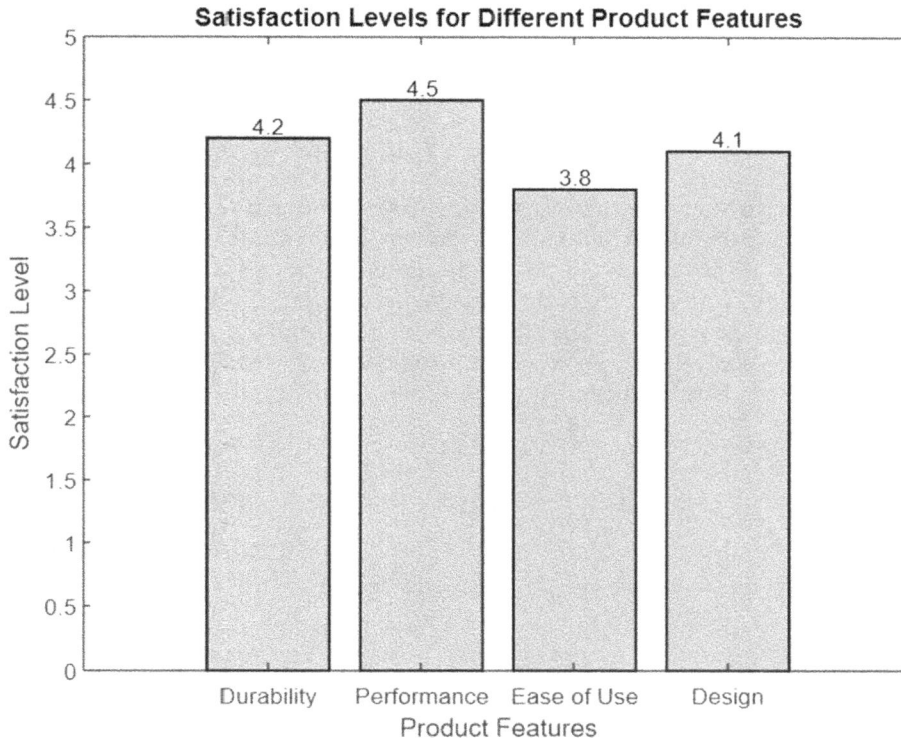

Figure 2.10

2.3.1.3 Pie chart

A pie chart is a circle that is divided into several segments or sectors. We can also think of it as slices of a pizza but they are not always necessarily equally divided. Each slice represents a proportion or percentage of observations of a certain category. Pie charts are often used in business. Examples include showing which products bring in the most revenue or different employment sectors with a mathematics degree or even the number of **"Yes" responses compared to "No" responses on a survey.** Pie charts can be helpful for showing the relationship of parts to the whole.

Unlike bar charts, pie charts fail to display multiple dimensions of a certain category. Like bar charts, pie charts are not used for quantitative data. Instead, pie charts are used for qualitative data.

However unlike bar charts, pie charts fail to display multiple dimensions of a certain category.

2.3.1.3.1 MATLAB

The main function for creating a pie chart in MATLAB is called pie.
1) *pie()* draws a pie chart using the data in X. Each slice of the pie chart represents an element in X.
   - If sum(X) ≤ 1, then the values in X directly specify the areas of the pie slices. pie draws only a partial pie if sum(X) < 1.
   - If sum(X) > 1, then pie normalizes the values by X/sum(X) to determine the area of each slice of the pie.
   - If X is of data type categorical, the slices correspond to categories. The area of each slice is the number of elements in the category divided by the number of elements in X.

2.3.1.3.2 Example 1

Let's say we want to visualize the relative popularity or usage of different programming languages within a particular context. For instance, let's consider a scenario where a survey was conducted to determine the programming languages used by a group of software developers. The data collected from the survey could be represented in a pie chart. Each programming language would be represented by a slice of the pie, with the size of the slice proportional to the percentage who reported using that particular language.
%percentage using that language

```
languages = {'Java', 'Python', 'MATLAB', 'PHP', 'JavaScript', 'C++',
'Other'}
X = [0.299 0.235 0.191 0.073 0.062 0.059 0.002];
figure
pie(percentage, languages)
title("Most popular programming languages among a group of software
developers")
```

Now notice that
$0.299 + 0.235 + 0.191 + 0.073 + 0.062 + 0.059 + 0.002 = 0.921 < 1$

Most popular programming languages among a group of software developers

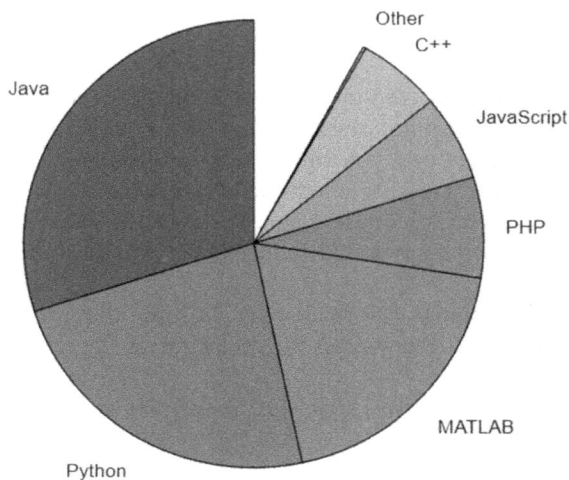

Figure 2.11

The pie chart allows viewers to quickly grasp the relative popularity of different programming languages at a glance. Larger slices indicate greater usage or popularity, while smaller slices represent less usage or popularity. It provides a visual representation of the distribution of programming languages within the surveyed group, enabling easy comparison and understanding of the overall landscape of programming language usage.

2.3.1.3.3 Example 2

In an engineering manufacturing process, the defects of a product were classified into different categories. The data collected includes the number of defects in each category. Let's create a pie chart to visualize the distribution of defects in the manufacturing process.

```
defectCounts = [25, 15, 10, 5];
% Create pie chart
figure;
pie(defectCounts,'%.3f%%')
title('Distribution cf Defects in Manufacturing Process')
% Add legends
legend({'Category A', 'Category B', 'Category C', 'Category D'})
```

$25 + 15 + 10 + 5 = 55 > 1$ so

$$\text{Area for Category A} = \frac{25}{55} \times 100 = 45.45\%$$

$$\text{Area for Category B} = \frac{15}{55} \times 100 = 27.27\%$$

$$\text{Area for Category C} = \frac{10}{55} \times 100 = 18.18\%$$

$$\text{Area for Category D} = \frac{5}{55} \times 100 = 9.09\%$$

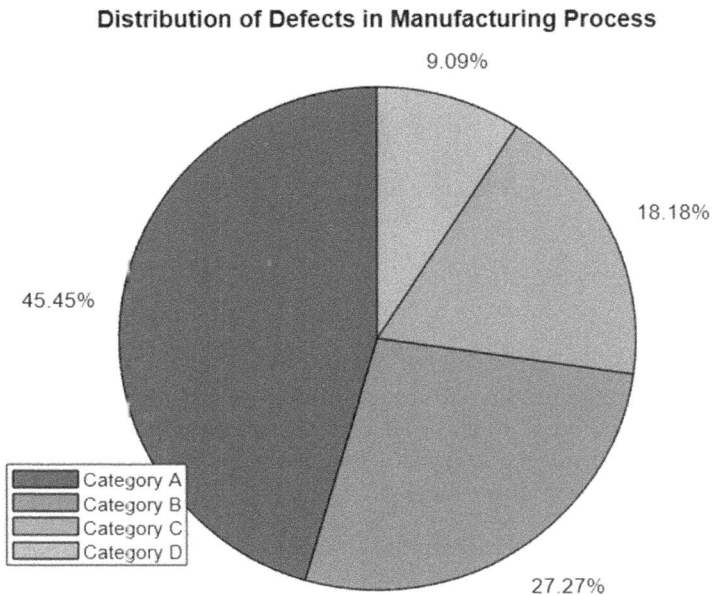

Figure 2.12

pie(X, explode)

offsets slices from the pie. explode is a vector or matrix of zeros and nonzeros that correspond to X. The pie function offsets slices for the nonzero elements only in explode.

If X is of data type categorical, then explode can be a vector of zeros and nonzeros corresponding to categories, or a cell array of the names of categories to offset.

2.3.1.3.4 Example

```
defectCounts = [25, 15, 10, 5];
explode = [1, 0, 0, 0];
% Create pie chart
figure
pie(defectCounts,explode, '%.2f%%')
title("Distribution of Defects in Manufacturing Process")
legend({'Category A', 'Category B', 'Category C', 'Category D'})
```

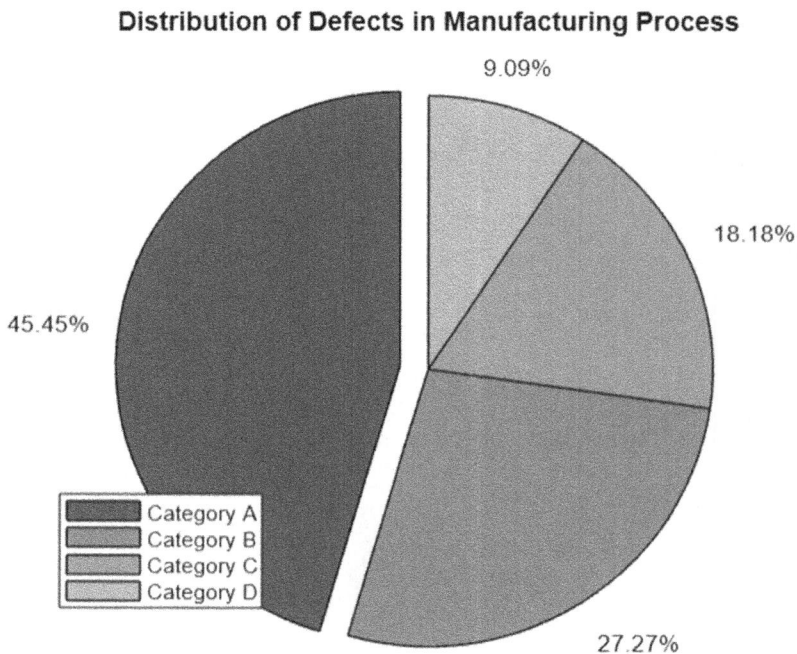

Figure 2.13

2.3.1.3.5 Example

In the field of energy management or sustainability, let's say an engineer wants to use a pie chart to showcase the breakdown of energy consumption in an industrial facility. Each energy source would be displayed as a slice, demonstrating the relative contribution of each component to the total energy consumption.

```
Source = categorical({'Electricity', 'Natural Gas', 'Petroleum', 'Biomass',
                      'Coal', 'Geothermal', 'Hydroelectric'});
Percentage = [53.03 36.36 7.44 1.23 0.77 0.17 0.01];
```

```
pie(Source, Percentage)
title("Energy Consumption for Commercial Sector")
```

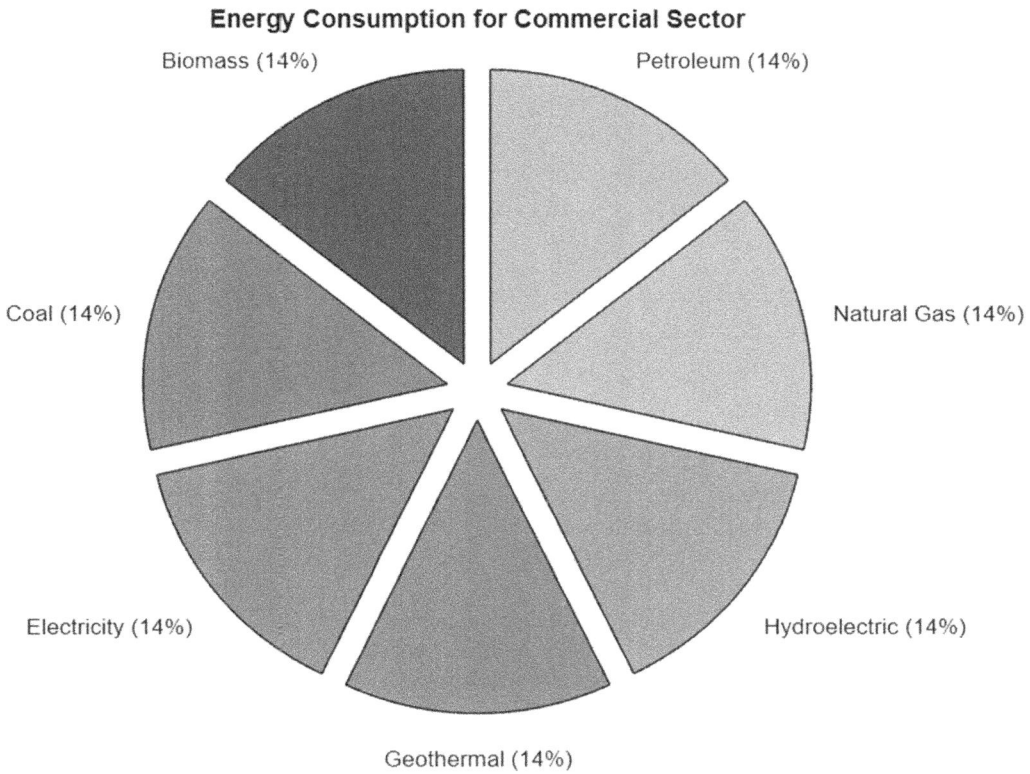

Energy Consumption for Commercial Sector

Biomass (14%)   Petroleum (14%)

Coal (14%)   Natural Gas (14%)

Electricity (14%)   Hydroelectric (14%)

Geothermal (14%)

Figure 2.14

2.3.1.3.6 Example

An engineer wants to use a pie graph to represent the composition of two different materials to understand the materials' properties and behavior.

```
Composition = {'Iron','Chromium','Nickel','Carbon','Silicon','Mangane
se'};
MaterialA = [47.12 21.53 13.46 5.88 8.62 3.39];
MaterialB = [38.27 20.41 16.64 11.80 9.73 3.15];
t = tiledlayout(1,2,'TileSpacing','loose');
% Create pie charts
ax1 = nexttile;
pie(ax1, MaterialA, '%.2f%%')
title('Material A')
ax2 = nexttile;
pie(ax2, MaterialB, '%.2f%%')
title('Material B')
% Create legend
lgd = legend(Composition);
```

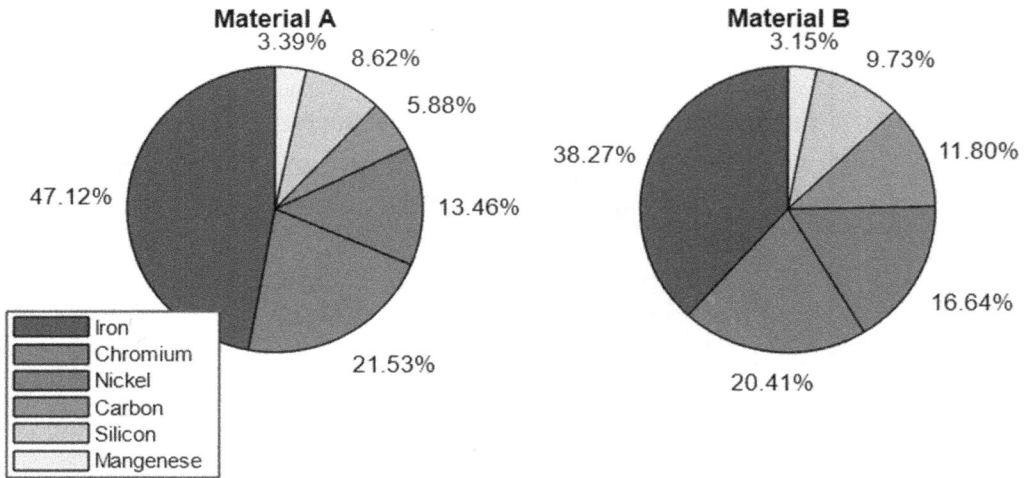

Material A

3.39%   8.62%

5.88%

47.12%   13.46%

Iron
Chromium
Nickel
Carbon
Silicon
Mangenese

21.53%

Material B

3.15%   9.73%

38.27%   11.80%

16.64%

20.41%

Figure 2.15

2.3.1.4 Histogram

A histogram is similar to a bar graph, but it is based on the frequency of numerical values rather than their actual values. The data is organized into intervals and the bars represent the frequency of the values in that range. That is, the height of each bar counts how many values of the data fall into that particular range. Another important distinction is the presence of the gaps between the bars. While bar graphs have gaps between each bar, histograms do not. This makes sense since the data in histograms is continuous and not discrete.

Remember the distinction:

Bar graph	*Histogram*
Discrete data	Continuous data
Each data point into a "category", based on an attribute. So each rectangular bar signifies a category and the height of that bar **represents the frequency or the number of observations in that category**.	Each data point into a "bin", based on a range of values for each bin. Here every bar is grouped into ranges. The heights of the bars are determined by the number of data values that fall in that range.
Can be space between the bars.	There is no space between the bars.

Histogram graphs are classified into different types based on the distribution of the data. The shape of the frequency distribution in a particular type of histogram can offer a better understanding of the variations.

2.3.1.4.1 MATLAB

The main function for creating a histogram in MATLAB is called histogram.

So far, we have learned that bar charts are typically used for displaying categorical data, while histograms are commonly used for visualizing the distribution of continuous data. However, MATLAB allows us to use the histogram function for both types of data. But it is important to understand that even though we can use the histogram function for both

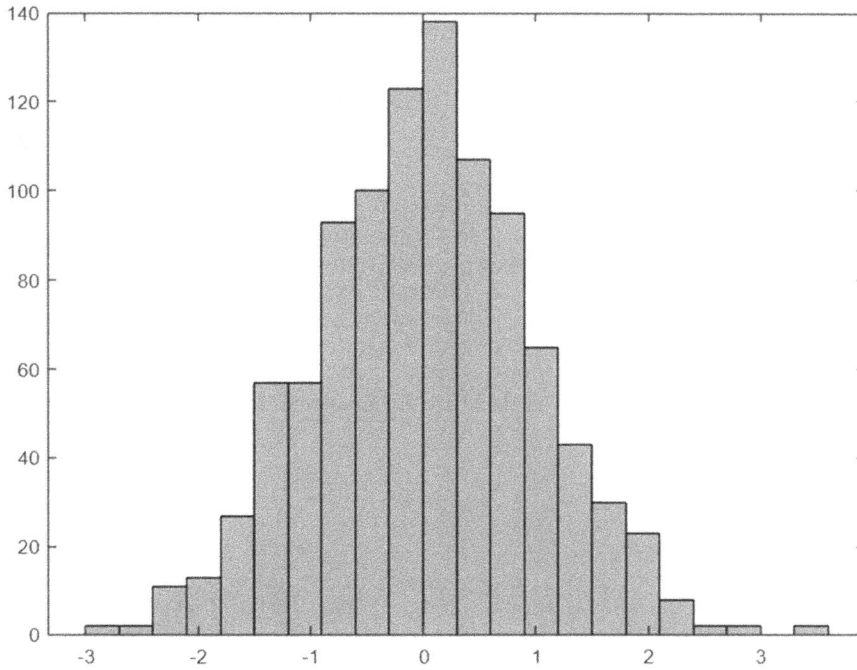

Figure 2.16

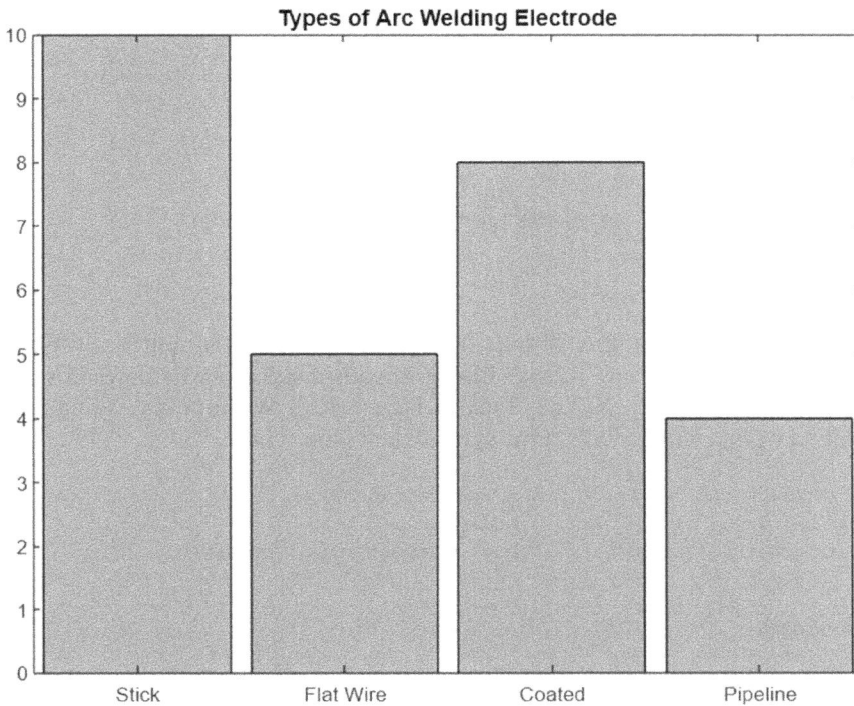

Figure 2.17

continuous and categorical data in MATLAB, it's recommended to use bar for categorical data and histogram when working with continuous data.

The input arguments for the histogram function in MATLAB can vary depending on whether we are working with continuous or categorical data.

```
histogram(x, nbins, BinEdges, BinCounts, 'BinLimits', [bmin,bmax],
'BinMethod', 'integers', 'BinWidth', , 'DisplayStyle','', 'EdgeAlpha',
'EdgeColor','', FaceAlpha', , 'FaceColor', '', LineStyle, "",
LineWidth, , 'Normalization','', 'Orientation', '')
histogram('Categories', Categories, 'BinCounts', BarWidth, DisplayOrder,
'DisplayStyle','', 'EdgeAlpha', , 'EdgeColor', 'r', FaceAlpha',
'FaceColor', '' , LineStyle, "", LineWidth, 'NumDisplayBins',
'Orientation', '')
```

The histogram plot in Figure 2.17 and Figure 2.18 is the default behavior of the histogram function.

2.3.1.4.2 Code for Figure 2.17

```
x = randn(1000,1);
h = histogram(x)
```

So let's see the default values of each argument.

2.3.1.4.3 Code for Figure 2.18

```
A = [1 1 3 2 1 2 2 4 1 3 3 1 2 2 3 1 3 1 4 1 4 4 3 1 3 3 1];
C = categorical(A, [1 2 3 4], {'Stick', 'Flat Wire', 'Coated' ,
'Pipeline'})
h = histogram(C)
title("Types of Arc Welding Electrode")
```

Let's first discuss each of the arguments for the continuous data.

2.3.1.4.4 Nbins

This specifies the desired number of bins for the histogram. The input is always a positive integer, and if we do not specify it, then histogram automatically calculates how many bins to use based on the values in X. Let's look at Figure 2.17. Without specifying it we get 22 bars, but let's say we want 15 bins. Our syntax then would be

```
x = randn(1000, 1);        x = randn(1000, 1);
nbins = 15;                nbins = 30;
h = histogram(x, nbins)    h = histogram(x, nbins)
```

2.3.1.4.5 BinEdges

This parameter is used to specify the edges of the bins in a histogram. It allows us to define custom binning intervals for our data. It can be an array that specifies the edges of the bins. Without specifying it we get binEdges in a sequence starting from −3.00 and ending at 3.60 with an increment of 0.30. Let's now change the binEdges.

```
    BinCounts: [2 2 11 13 27 57 57 93 100 123 138 107 95 65 43 30 23 8 2 2 0 2]
BinCountsMode: 'auto'
    BinEdges: [-3.0000 -2.7000 -2.4000 -2.1000 -1.8000 -1.5000 -1.2000 -0.9000 -
    BinLimits: [-3.0000 3.6000]
BinLimitsMode: 'auto'
    BinMethod: 'auto'
    BinWidth: 0.3000
  DisplayName: 'x'
 DisplayStyle: 'ba~'
    EdgeAlpha: 1
    EdgeColor: [0 3 0]
    FaceAlpha: 0.6300
    FaceColor: 'auto'
    LineStyle: '-'
    LineWidth: 0.5000
Normalization: 'count'
      NumBins: 22
  Orientation: 'vertical'
```

Figure 2.18

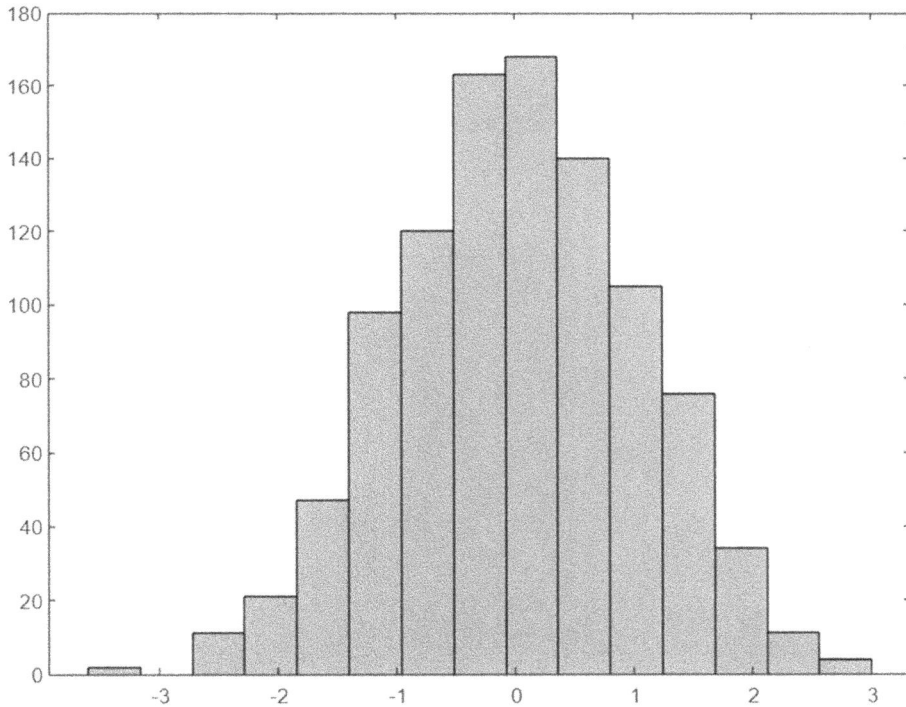

Figure 2.19

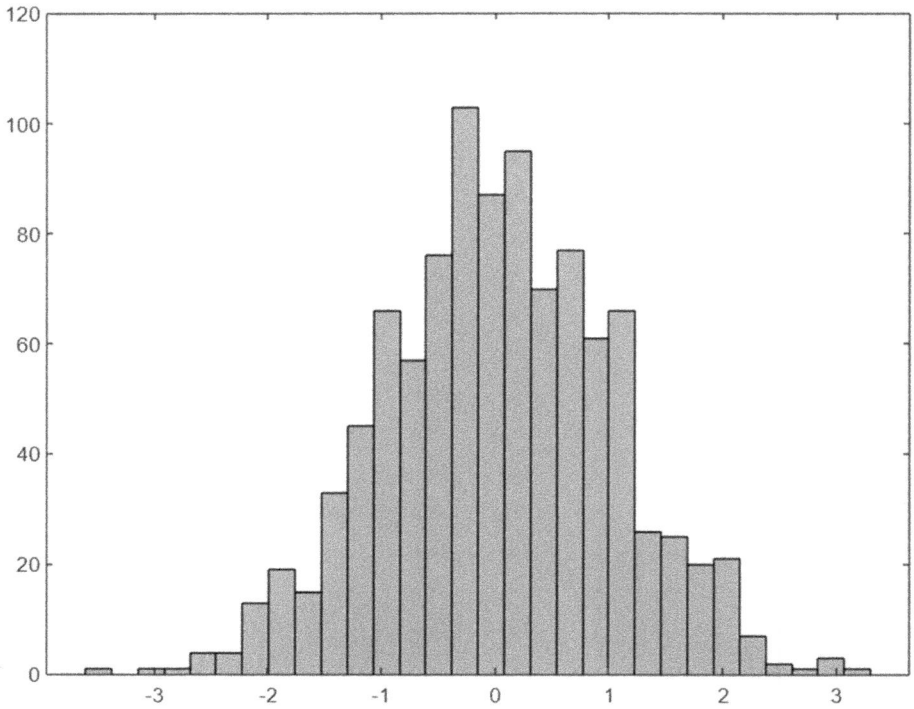

Figure 2.20

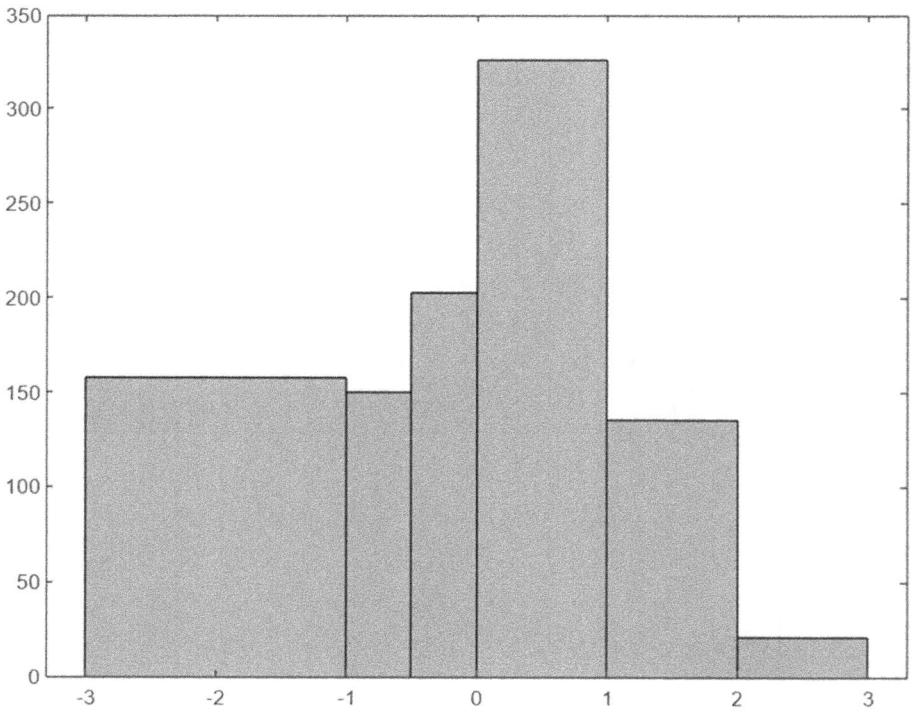

Figure 2.21

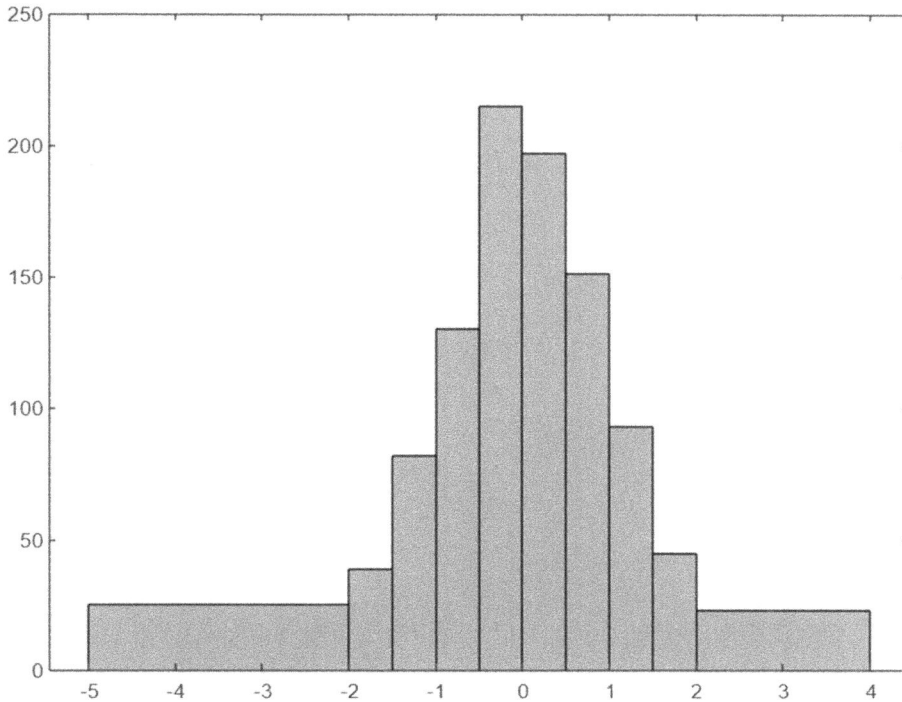

Figure 2.22

2.3.1.4.6 Example

```
data = randn(1000, 1); % Random data    data = randn(1000, 1); % Random data
binEdges = [-3, -1, -0.5, 0, 1, 2, 3];   binEdges = [-5 -2:0.5:2 4];
histogram(data, binEdges);               histogram(data, binEdges);
```

The resulting histogram in Figure 2.22 will have bins with ranges [–3, –1), [–1, –0.5), [–0.5, 0), [0, 1), [1, 2), [2,3)].

2.3.1.4.7 BinCounts

```
histogram('BinEdges', [-3, -1, -0.5, 0, 1, 2, 3],'BinCounts', [20 12 09
16 15 28]);
```

2.3.1.4.8 BinLimits

BinLimits is specified as a two-element vector, [min, max]. This option plots a histogram using the values in the input array, X, that fall between min and max inclusive.

2.3.1.4.9 BinMethod

Binning algorithm, specified as one of the values in this table.

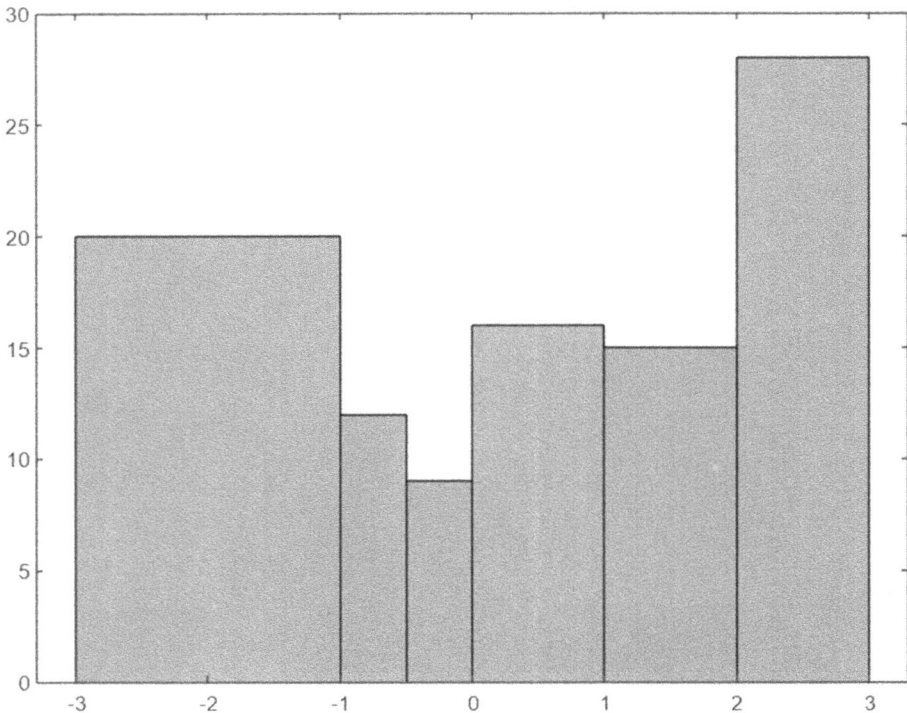

Figure 2.23

Value	Description
'auto'	The default 'auto' algorithm chooses a bin width to cover the data range and reveal the shape of the underlying distribution.
'scott'	Scott's normal reference rule is a widely used method for determining the optimal bin width (bin size) when constructing histograms for data that is assumed to be approximately normally distributed. Bin width is given by $$h = \frac{3.5\sigma}{\sqrt[3]{n}}$$ where σ = sample standard deviation
'fd'	If we have more diverse or non-normally distributed data, Freedman–Diaconis' choice is typically used as it is less sensitive to outliers in the data. Bin width is given by: $$h = 2 * \frac{IQR}{\sqrt[3]{n}}$$ where IQR = is the interquartile range of x
'integers'	The integer rule is useful with integer data, as it creates a bin for each integer. It uses a bin width of 1 and places bin edges halfway between integers.
'sturges'	Sturges' rule is popular due to its simplicity. It chooses the number of bins to be $ceil(1 + log_2(n))$.
'sqrt'	It chooses the number of bins to be $ceil\left(\sqrt{n}\right)$

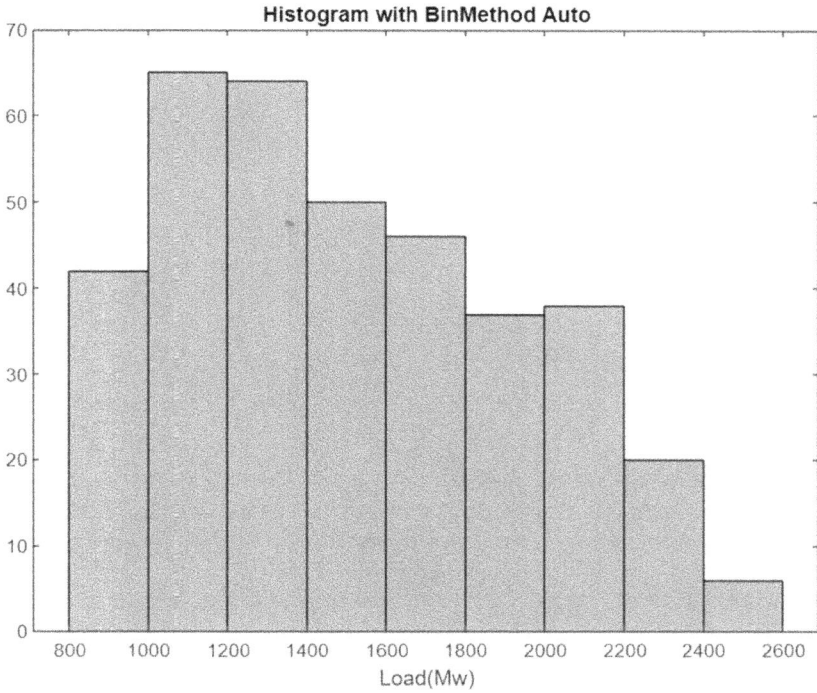

Figure 2.24

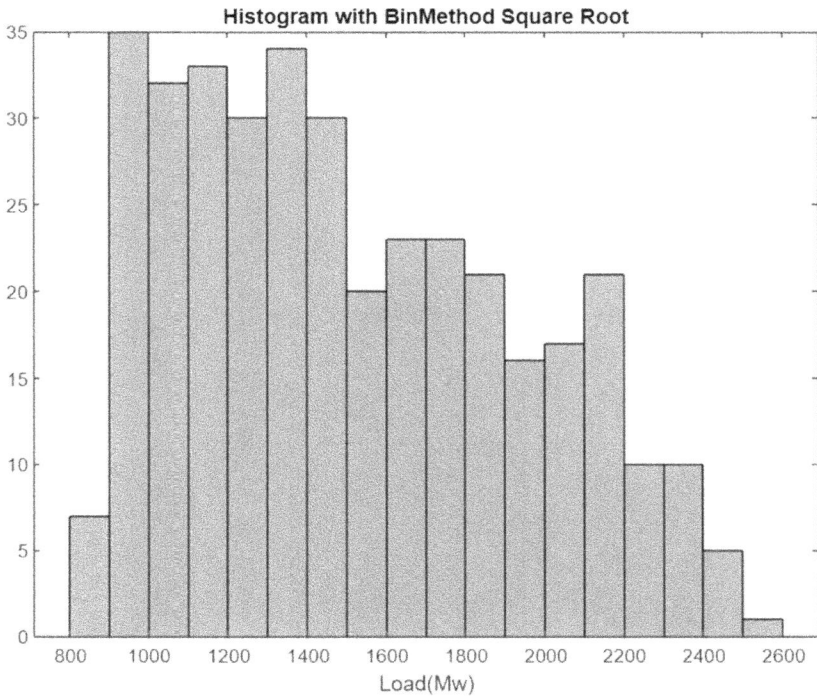

Figure 2.25

2.3.1.4.10 Code for Figure 2.25

```
Data = readtable("Load.csv", VariableNamingRule = "preserve")
Load = Data.Load;
histogram(Load)
title("Histogram with BinMethod Auto")
xlabel("Load(Mw)")
```

2.3.1.4.11 Code for Figure 2.26

```
histogram(Load, 'BinMethod','sqrt')
title("Histogram with BinMethod Square Root")
```

2.3.1.4.12 BinWidth

This assigns the width of each class interval. Let's create the histogram from Example 2.4. If we look at Figure 2.27 we will observe that without specifying the BinWidth our histogram uses a bin width of 1. Since the bin limits are from 5 to 50 and all the values are integers, a bin width of 5 seems a reasonable choice. See Figure 2.28.

```
histogram(Time)                          histogram(Time, 'BinWidth', 5)
```

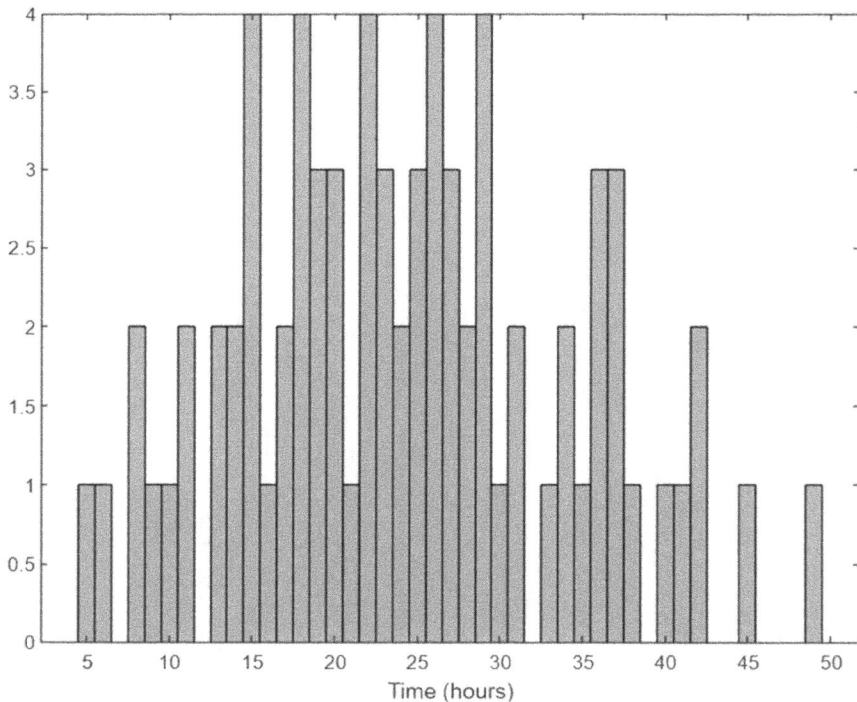

Figure 2.26

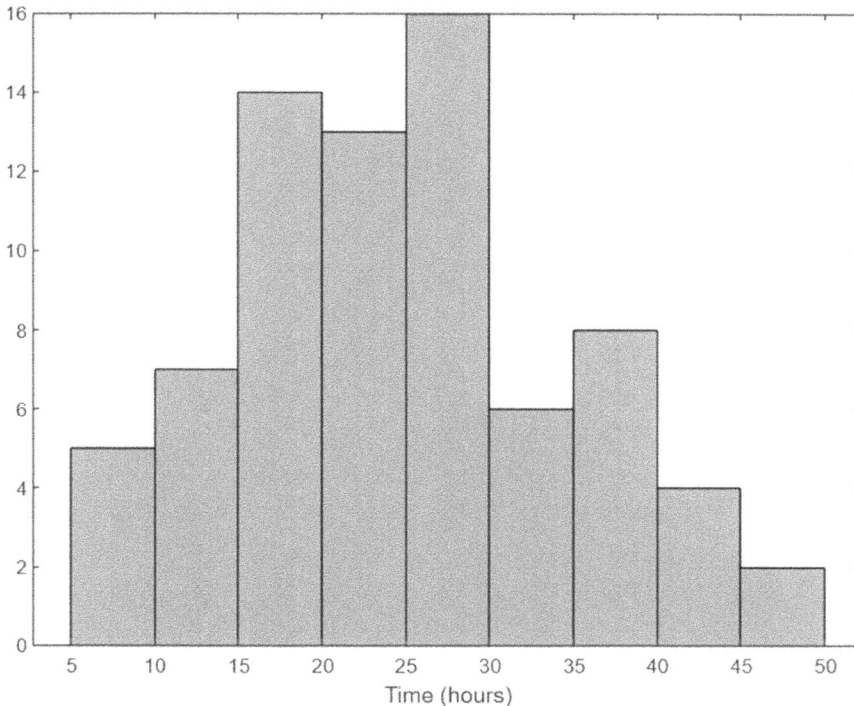

Figure 2.27

2.3.1.4.13 DisplayStyle

This property is used to control the appearance of the bars in a histogram.
'bar' (default): Displays filled bars.
'stairs': Displays an outline of the bars using stair-step lines.
All the histograms so far have been displayed in bar. Let's create the same graphs as in Figures 2.27 and 2.28 in stairs now.

```
histogram(Time,'DisplayStyle','stairs')    histogram(Time,'BinWidth',
                                                5,'DisplayStyle','stairs')
```

2.3.1.4.14 EdgeAlpha

This property is used to control the transparency of the edges between the bars. It specifies the opacity of the edges on a scale from 0 to 1, where 0 is fully transparent (invisible) and 1 is fully opaque (solid). All the histograms displayed earlier have an EdgeAlpha of 1 that is fully opaque.

```
histogram(Load, 'EdgeAlpha', 0.25)    histogram(Load, 'EdgeAlpha', 0)
```

2.3.1.4.15 FaceAlpha

This property is used to control the transparency of the bars. It specifies the opacity of the bar on a scale from 0 to 1, where 0 is fully transparent (invisible) and 1 is fully opaque (solid). Again all the histograms displayed earlier have an FaceAlpha of 1 that is fully opaque.

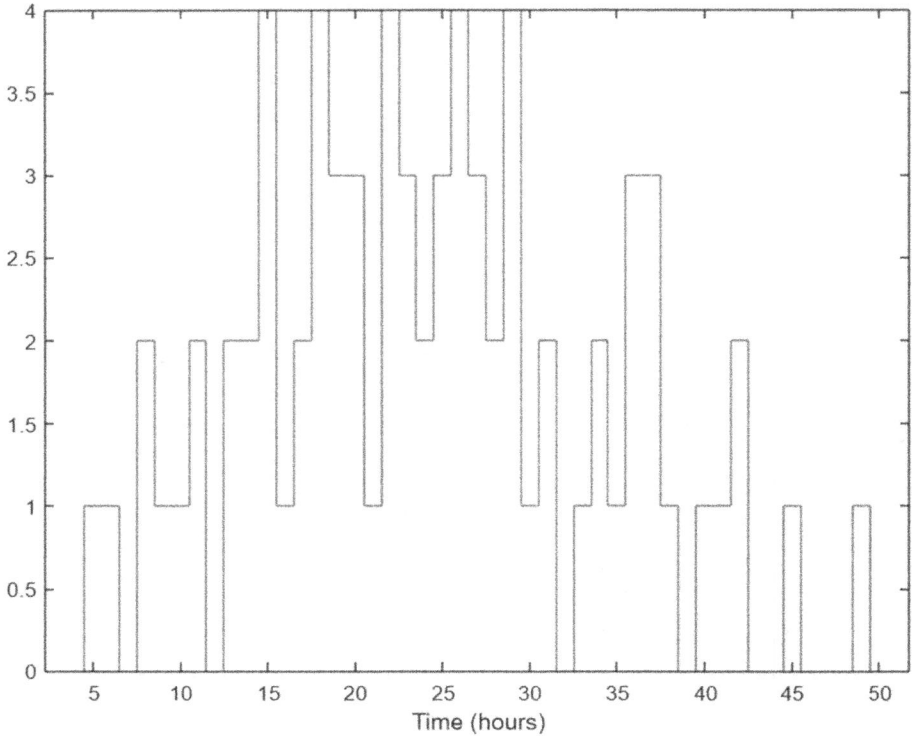

Figure 2.28

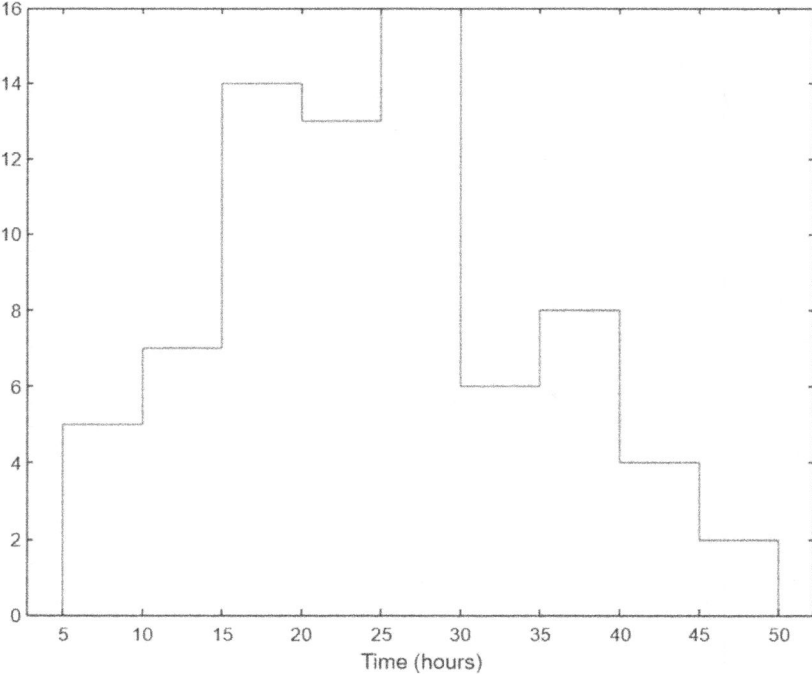

Figure 2.29

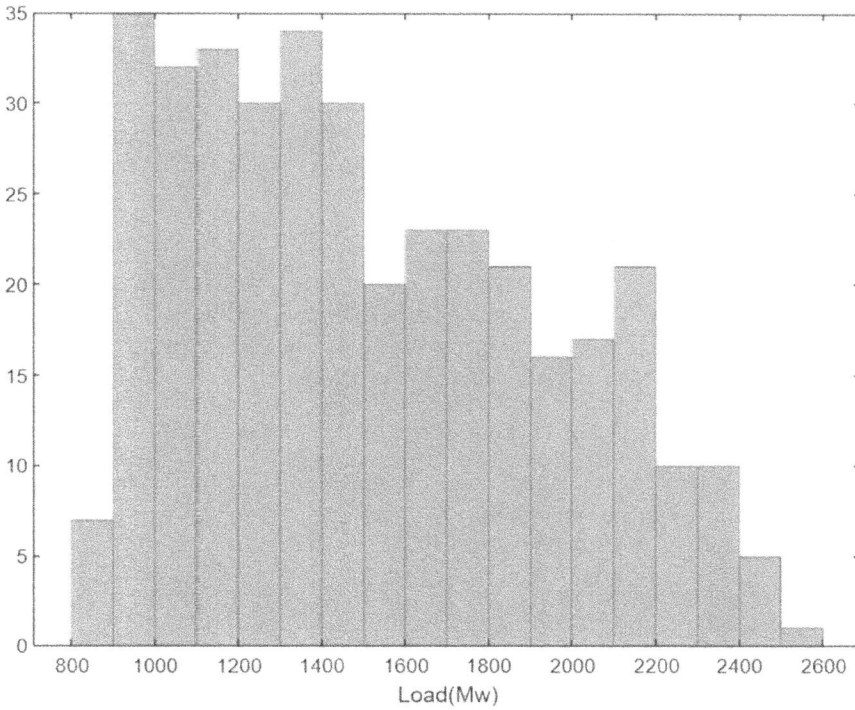

Figure 2.30

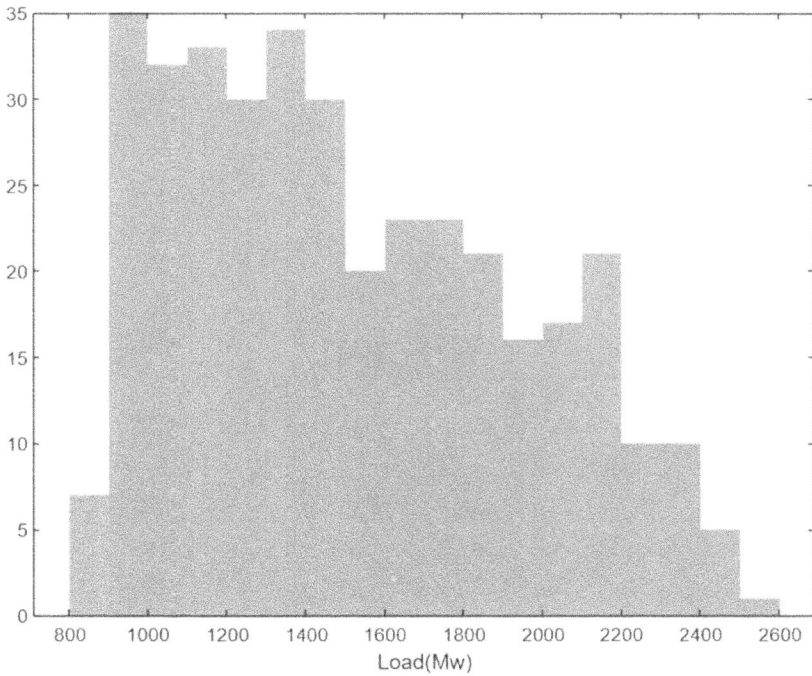

Figure 2.31

```
histogram('Categories',{'Unfilled section', 'Scale pits', 'Cold shut',
                        'Die shift', 'grain growth', 'Flakes',
                        'Surface cracking', 'Incomplete forging',
                        'Residual stress in forging',
                        'Incomplete filling of dies',
                        'Cracking at the flash', 'Internal cracks'},
            'BinCounts',[25 18 20 25 15 30 21 16 09 12 24 11])
```

2.3.1.4.16 'BarWidth'

The input is specified as a scalar value in the range [0,1]. The default value is 0.9, which means that the bar width is 90% of the space from the previous bar to the next bar, with 5% of that space on each side. If we set this property to 1, then adjacent bars touch. This only applies when we have categorical data.

```
histogram('Categories',{'Unfilled section', 'Scale pits', 'Cold shut',
                        'Die shift', 'grain growth', 'Flakes',
                        'Surface cracking', 'Incomplete forging',
                        'Residual stress in forging',
                        'Incomplete filling of dies',
                        'Cracking at the flash', 'Internal cracks'},
            'BinCounts',[25 18 20 25 15 30 21 16 09 12 24 11],
            'BarWidth',0.85)
```

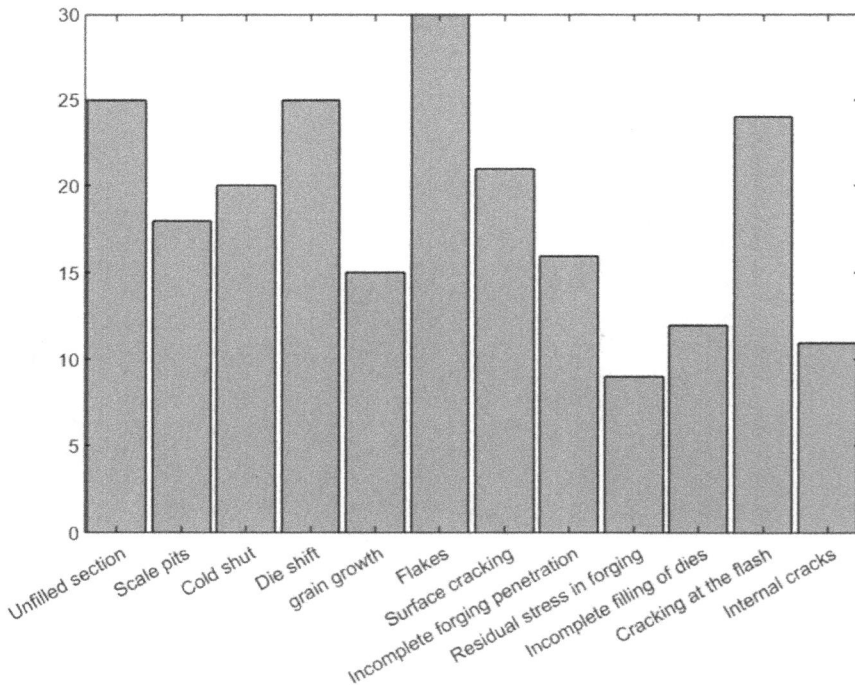

Figure 2.32

2.3.1.4.17 'DisplayOrder'

This option only applies to histograms of categorical data.

This property is used to specify the order in which the elements of a chart or plot are displayed.

'data': the default 'data' value uses the category order in the input data.

'ascend': the histogram displays with increasing bar heights.

'descend': the histogram displays with decreasing bar heights.

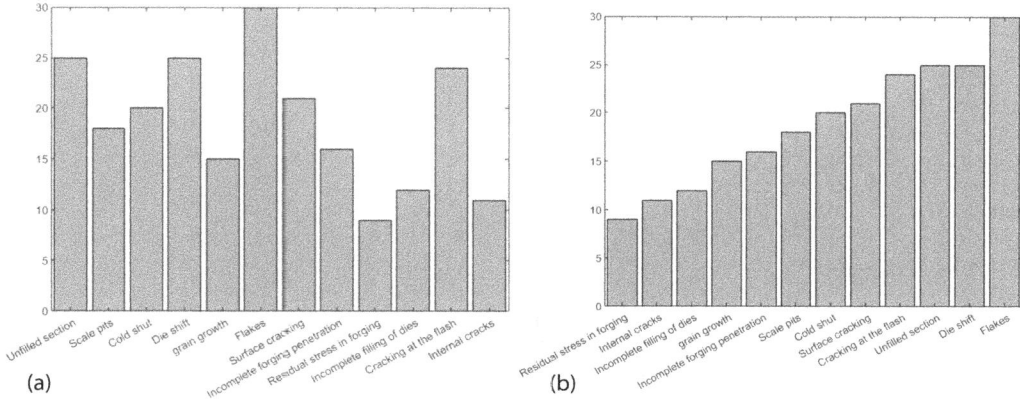

(a)   (b)

Figure 2.33

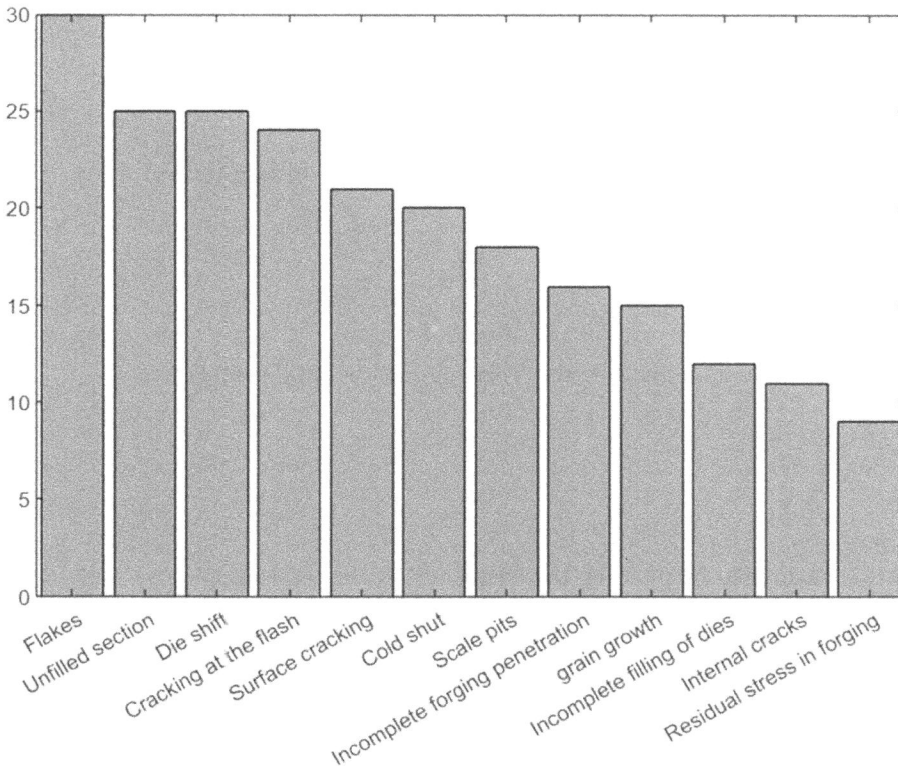

Figure 2.34

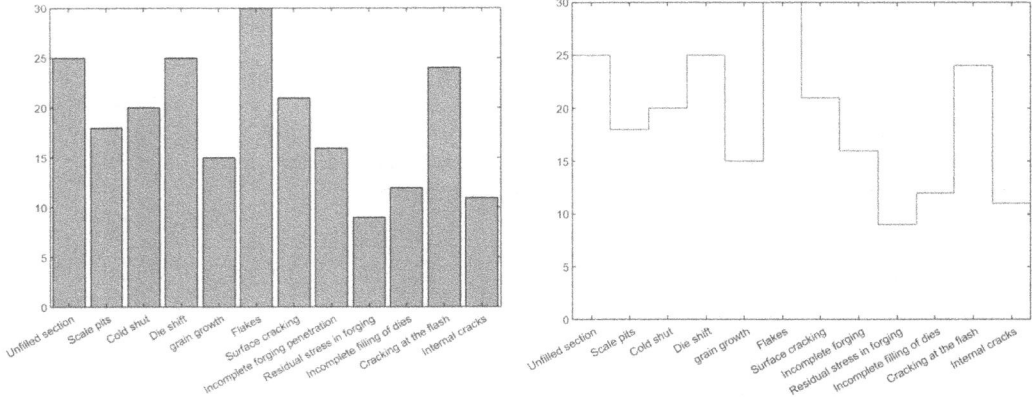

Figure 2.35

2.3.1.4.18 'FaceAlpha'

This property is used to control the transparency of the bars. It specifies the opacity of the bar on a scale from 0 to 1, where 0 is fully transparent (invisible) and 1 is fully opaque (solid).

2.3.1.4.19 'Normalization'

1. 'count' (default)
   This counts the frequency or number of occurrences of each observation.

2. 'countdensity'
   This is calculated by dividing the frequency by the width of the bin so

$$cd_i = \frac{f_i}{w_i}$$

3. 'probability'
   This is calculated by dividing the frequency by the total sample size so

$$p_i = \frac{f_i}{N}$$

4. 'pdf'
   This is calculated by dividing the frequency by the total sample size and width of the bin so

$$pdf_i = \frac{f_i}{N \times w_i}$$

5. 'cumcount'
   This is the cumulative count where each bin value is the cumulative number of observations in that bin and all previous bins.

$$cc_i = \sum_{j=1}^{i} c_j$$

6. 'cdf'

$$cdf_i = \sum_{j=1}^{i} \frac{c_j}{N}$$

2.3.1.4.20 Normal distribution

A histogram chart is said to be of the *normal distribution* if it is bell-shaped. It is a symmetrical arrangement of a dataset that has only one peak in the distribution with exactly similar data distribution on both sides.

2.3.1.4.21 Example

Let's say we want to create a histogram to display the distribution of tensile strength of our sample data. This can help us understand the material's characteristics and select the appropriate material for specific applications.

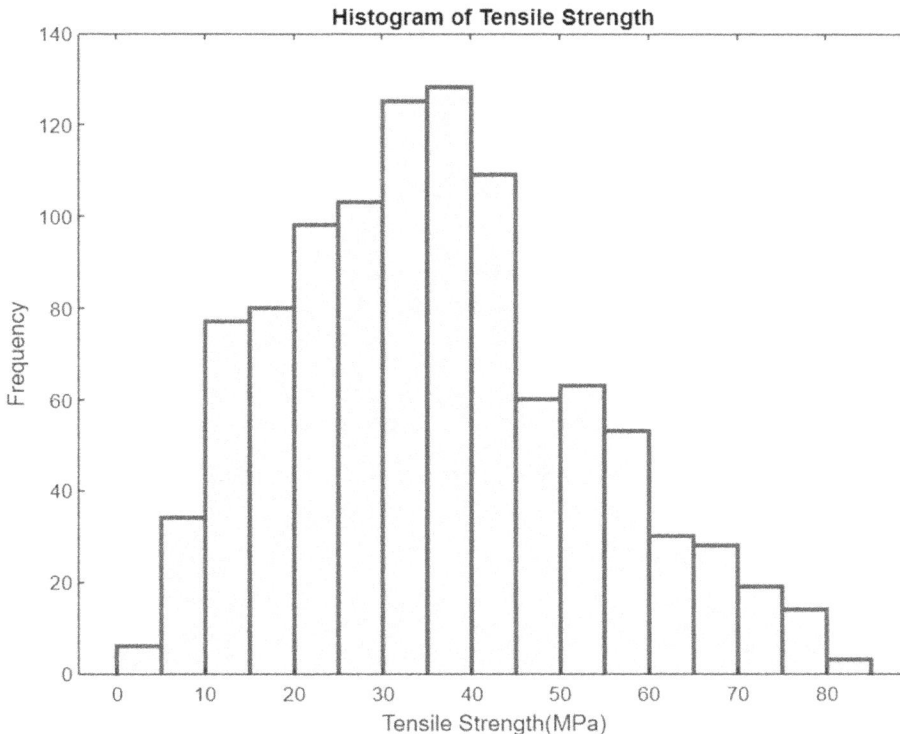

Figure 2.36

```
data = readtable("Strength_data.csv", VariableNamingRule="preserve");
Strength = data.Strength;
h = histogram(Strength, 'EdgeColor','b', 'FaceColor', 'y','BinWidth', 5,
              'LineWidth', 2)
xlabel("Tensile Strength(MPa)")
ylabel("Frequency")
title("Histogram of Tensile Strength")
```

2.3.1.4.22 Bimodal distribution

In this kind of histogram, there are two peaks visible rather than just one.

```
data = readtable("mpg vs horsepower.csv", VariableNamingRule="preserve");
MPG = data.mpg;
h = histogram(MPG, 'EdgeColor','r', 'FaceColor', 'c','BinWidth',
              2, 'LineWidth', 2.5)
xlabel("MPG")
ylabel("Number of vehicles")
title("Histogram of vehicle MPG")
```

Our histogram exhibits two distinct peaks or modes in the distribution of vehicle MPG. This indicates the presence of two dominant subpopulations or groups of vehicles with different weight ranges.

```
data = readtable("Energyconsumption.csv", VariableNamingRule
="preserve");
Power = data.Output;
h = histogram(Power, 'EdgeColor','k', 'FaceColor', [0.85 0 0.18],
              'BinWidth', 2, 'LineWidth', 2, 'Normalization',
              'probability')
xlabel("Power Output")
ylabel("Probability")
title("Histogram")
```

2.3.1.4.23 Skewed distribution

There are two types of skewed distributions: right-skewed distribution and left-skewed distribution. As the name suggests, a distribution which has a larger number of bins to the right side of the average is said to be right-skewed, whereas a distribution with a larger number of bins to the left side of the average is said to be left-skewed.

```
data = readtable("auto-mpg.csv", VariableNamingRule="preserve");
Weight = data.weight;
h = histogram(Weight,'EdgeColor','k','FaceColor', [0.4660 0.6740 0.1880],
              'BinWidth', 100, 'LineWidth', 1.5)
xlabel("Weight(kg)")
ylabel("Number of vehicles")
title("Histogram of vehicle weight")
```

Now if we analyze the shape of the graph, it appears to be left-skewed. This indicates that the majority of vehicles in our dataset have weights that are lower than the average or median weight.

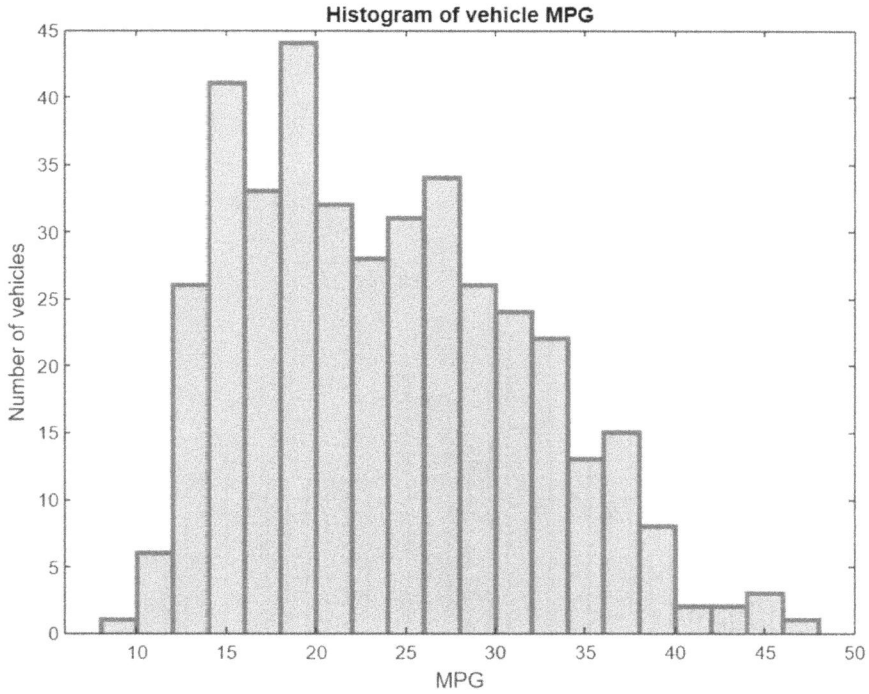

Figure 2.37

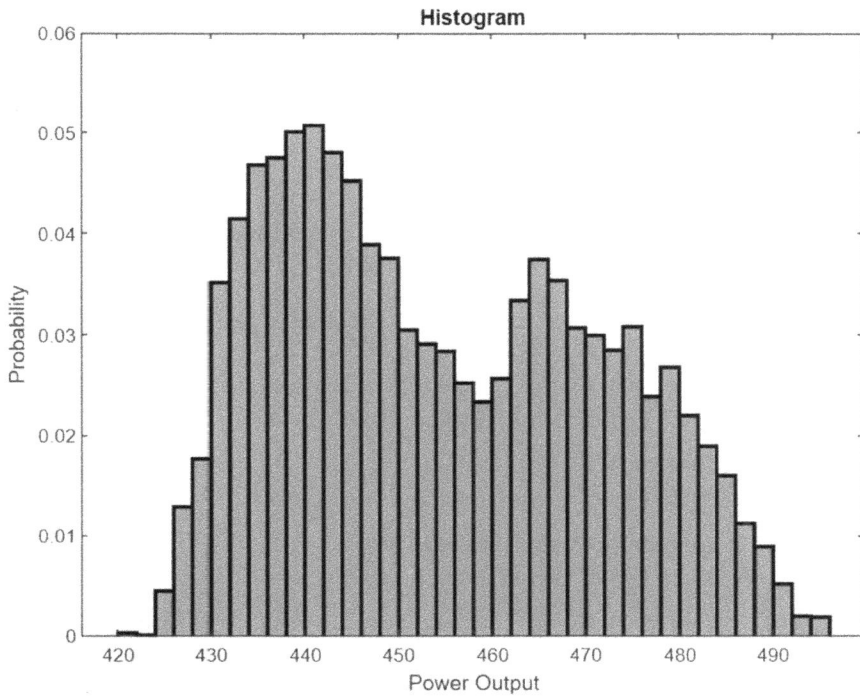

Figure 2.38

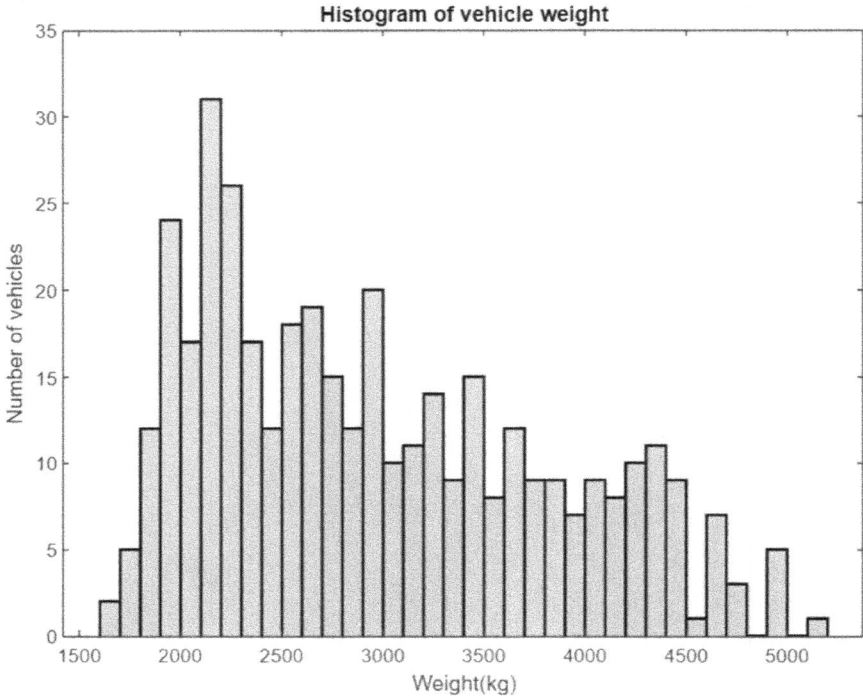

Figure 2.39

2.3.1.4.24 Random distribution

This type of distribution lacks a particular pattern and produces several peaks.

Now sometimes we want to compare two datasets together using a histogram.

2.3.1.4.25 Example

Let's say we have data on tensile strengths for two different materials and now we want to plot a histogram for each of them in the same figure.

```
data = readtable("2 Strengths.csv", VariableNamingRule = "preserve");
Material_A = data.A;
Material_B = data.B;
h2 = histogram(Material_A);
hold on;
h3 = histogram(Material_B);
xlabel("Tensile Strength(MPa)")
ylabel("Frequency")
title("Tensile Strength of Two Materials")
legend('Material A', 'Material B');
```

You may notice that Material B has more data points, making it difficult to compare the two. Moreover the bin width of Material A is wider, making it difficult to compare.

Now to deal with different sample sizes we shall normalize the histograms. Normalizing the histogram means we divide the data in each set by its own volume, converting the

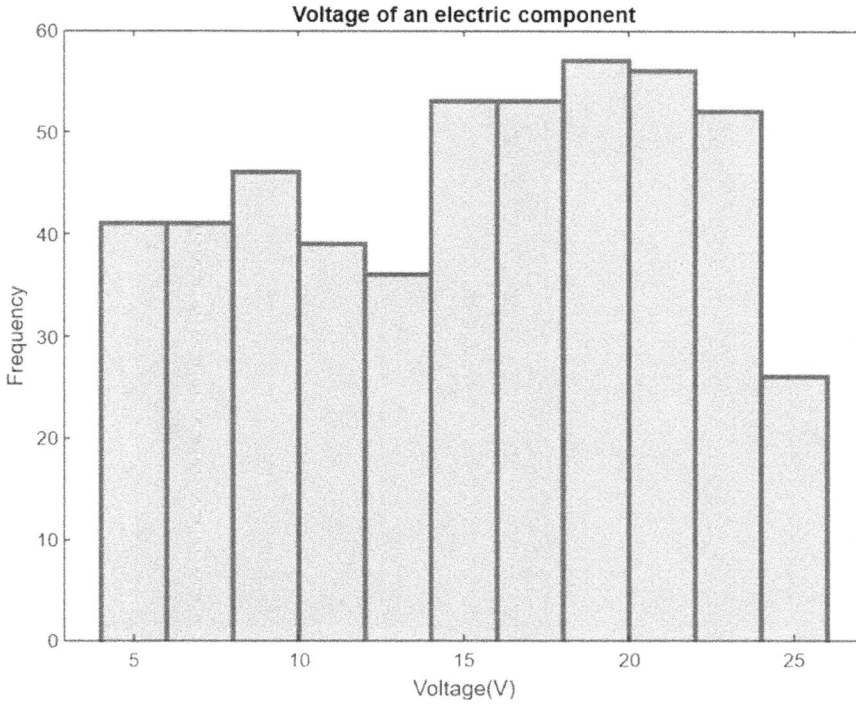

Figure 2.40

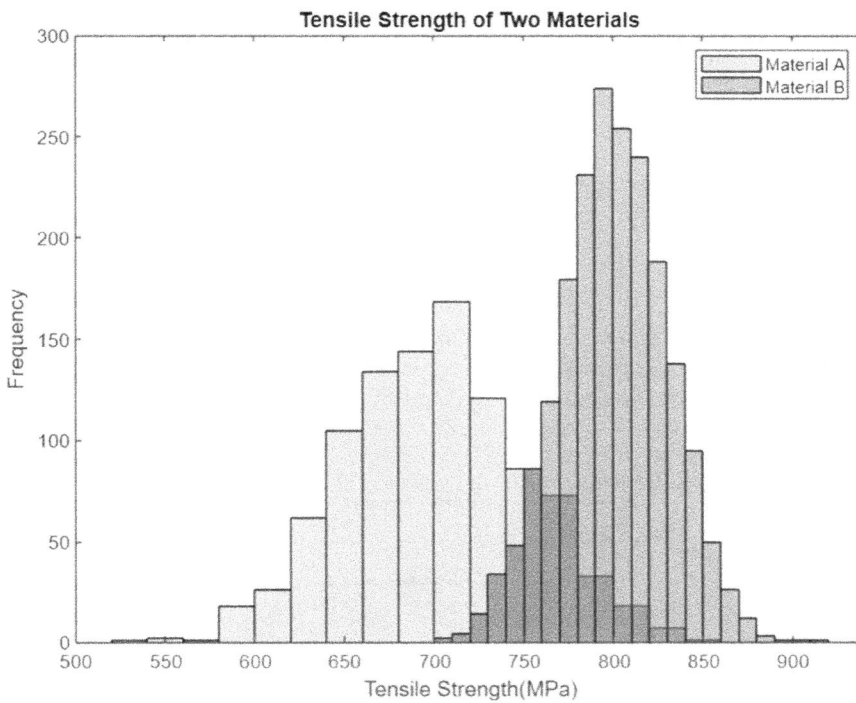

Figure 2.41

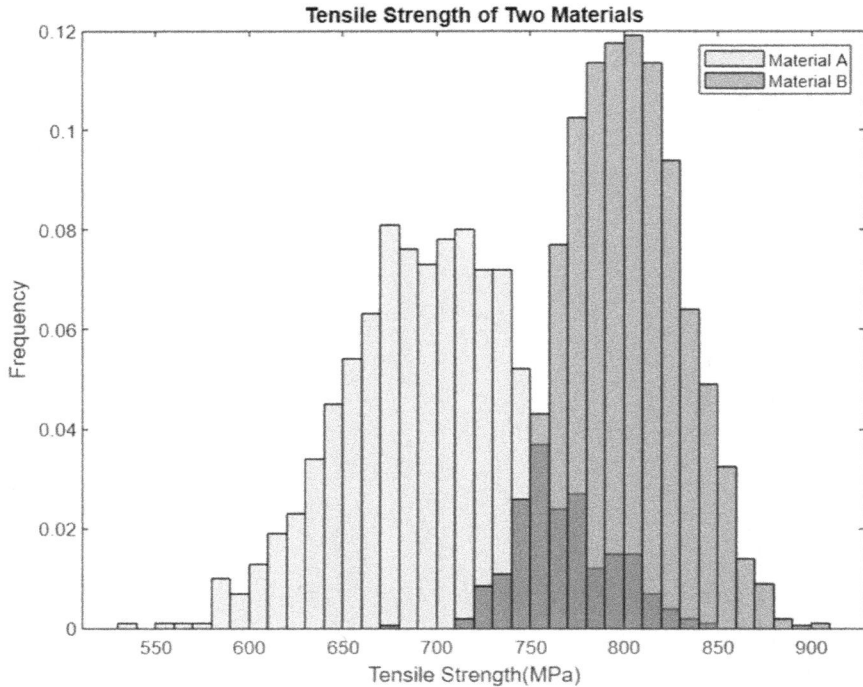

Figure 2.42

frequency to percentage so that all of the bar heights add to 1. By doing this, two datasets can be accurately compared to each other. To eliminate the differences in the bin widths we use a uniform bin width.

```
h1.Normalization = 'probability';
h1.BinWidth = 10;
h2.Normalization = 'probability';
h2.BinWidth = 10;
```

2.3.1.5 Boxplot

Another visual representation of dataset is the boxplot. A box plot displays the five-number summary of a dataset: (going from ascending order) the minimum value, the first quartile, the median, the third quartile, and the maximum value. The following diagram shows a typical boxplot of an arbitrary dataset.

Figure 2.43

Lets make some observations from the figure

The first thing we observe is a rectangular box and whiskers that sprout from the two ends of the box.

Let's talk about the box first

The box encloses the middle half of a dataset- that is around 50 **percent of the data fall inside the box**. Here the quartiles (first and third) mark the ends of the box.

The first/lower quartile is where Twenty-five percent of scores fall below it and thus, 75% of data are above it.

The third/upper quartile is where Seventy-five percent of the scores fall below it and therefor 25% of data are above it.

The length of the box represents the interquartile range which is the difference between the upper/third and the lower/first quartile (More in Chapter)

Notice that there is A line drawn at the middle of the box presenting the sample median.

The second/median quartile is where Half the data values are greater and half are lower than the value.

The median or second quartile does not necessarily have to be between the first and third quartiles, it can be on one, or the other, or even both depending on the dataset.

Lets look at the whiskers now

The whisker that extends from the ends of the box represents the scores that fall outside of the middle 50% . The "whiskers" extend from the ends of the box until they reach the sample maximum and minimum. Therefore The smallest(Minimum) and largest(Maximum) data values are labelled at the endpoints of the whiskers.

The box plot are useful as they give a good, quick picture of a dataset. They provide a visual summary of the data enabling researchers to quickly identify mean values, the dispersion of the data set, and signs of skewness.(More about them in Chapter)

The length of the whiskers of the box in its whiskers and the position of the line in the box also tells us whether the sample is symmetric or skewed, either to the right or left.

When the median is in the middle of the box, and the whiskers are about the same length on both ends of the box, then the distribution is symmetric. Lets draw a boxplot from Figure 2.40. We have seen from the histogram that the data is symmetric now lets create a boxplot for the same MPG values and check where our median lies.

As we can see the median lies right in the middle of the box therefore the data indeed is symmetric.

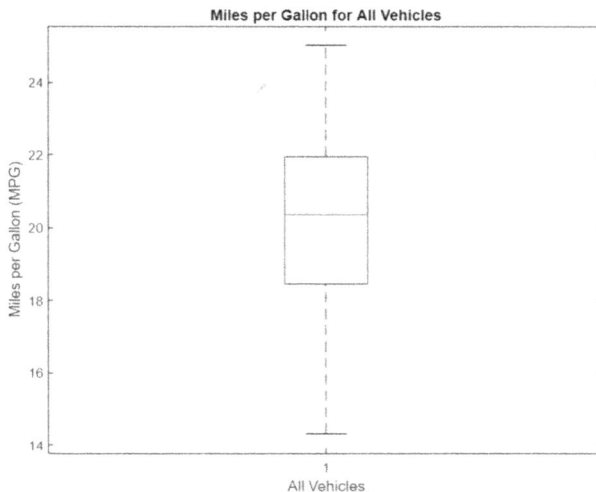

Figure 2.44

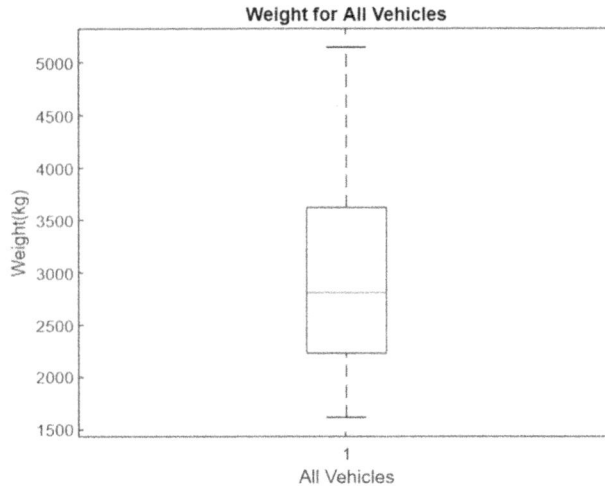

Figure 2.45

In a normal distribution, the mean and the median are the same number while the mean and median in a skewed distribution become different numbers: While symmetrical distribution has whiskers of equal length and the median sits exactly on the middle, a skewed distribution entails one tail being longer than the other.

When the median is closer to the bottom of the box, and if the whisker is shorter on the lower end of the box, then the distribution is positively skewed (skewed right). A right-skewed distribution will have the mean to the right of the median. Lets use our Figure from 2.42. When the median is closer to the top of the box, and if the whisker is shorter on the upper end of the box, then the distribution is negatively skewed (skewed left). A left-skewed, negative distribution will have the mean to the left of the median

Kurtosis is a measure of whether the data are heavy-tailed or light-tailed relative to a normal distribution.

Recall that In histogram If the data are symmetric, they have about the same shape on either side of the middle. In other words, if you fold the histogram in half, it looks about the same on both sides. In a symmetric distribution, the mean and median are nearly the same, and the two whiskers has almost the same length. The box length gives an indication of the sample variability and the line across the box shows where the sample is centered.

For a symmetric distribution, long whiskers, relative to the box length, can betray a heavy tailed population and short whiskers, a short tailed population. So, provided the number of points in the sample is not too small, the boxplot also gives us some idea of the "shape" of the sample, and by implication, the shape of the population from which it was drawn. This is all important when considering appropriate analyses of the data.

Although boxplots can be drawn in any orientation, most statistical packages seem to produce them vertically by default, as shown on the right, rather than horizontally.

MATLAB

```
boxchart(xgroupdata, ydata, 'PlotStyle', ' ', 'BoxStyle', ' ',
'MedianStyle', ' ')
```

`'PlotStyle'`

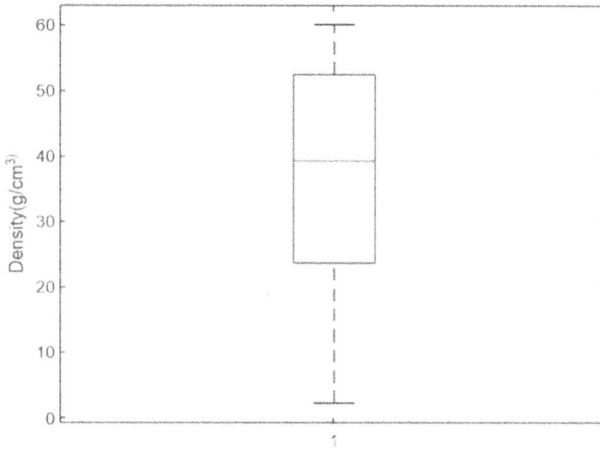

Figure 2.46

Plot style, specified as one of the following.
- 'traditional' (default)
- 'compact'

   In Figure 2.60, 2.61 and 2.63 we saw a boxplot of the plot style 'traditional'. Lets now look at the boxplot of the plot style 'compact'

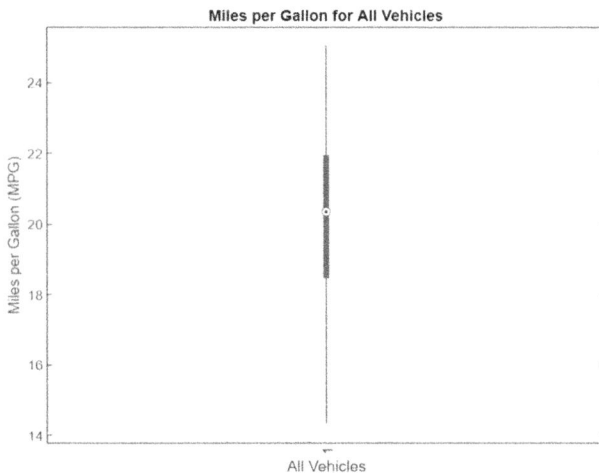

Figure 2.47

Weight for All Vehicles

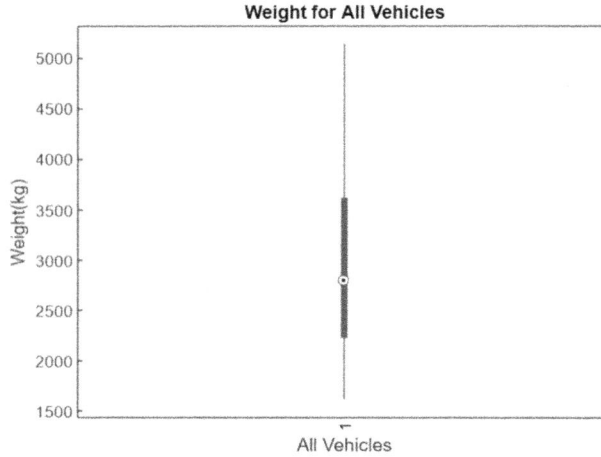

Figure 2.48

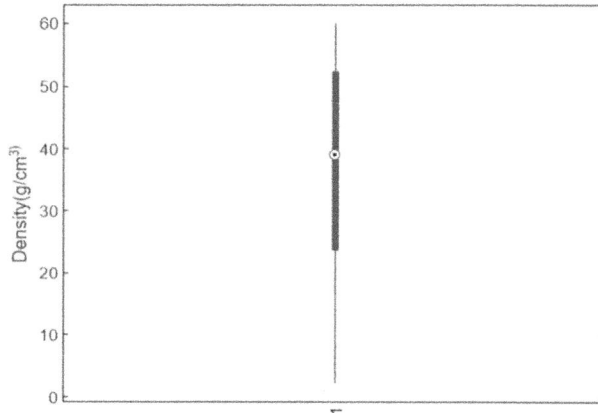

Figure 2.49

`'BoxStyle'`

Box style, specified as one of the following.

```
'outline'- This is the default if 'PlotStyle' is 'traditional'.
'filled'-This is the default if 'PlotStyle' is 'compact'.
```

`'MedianStyle'`

Median style, specified as one of the following.

```
'line'-This is the default when 'PlotStyle' is 'traditional'.
'target'-This is the default when 'PlotStyle' is 'compact'.
```

Example

```
data = readtable("500_Height_Weight.csv", VariableNamingRule =
"preserve");
Height = data.Height;
Weight = data.Weight;

tiledlayout(1,2)

% Left axes
ax1 = nexttile;
boxchart(ax1,Height, 'BoxFaceColor','red')
ylabel(ax1,'Height(cm)')
title(ax1, "Height of 500 People")

% Right axes
ax2 = nexttile;
boxchart(ax2,Weight, 'BoxFaceColor','blue')
ylabel(ax2,'Weight(kg)')
title(ax2,'Weight of 500 People')
```

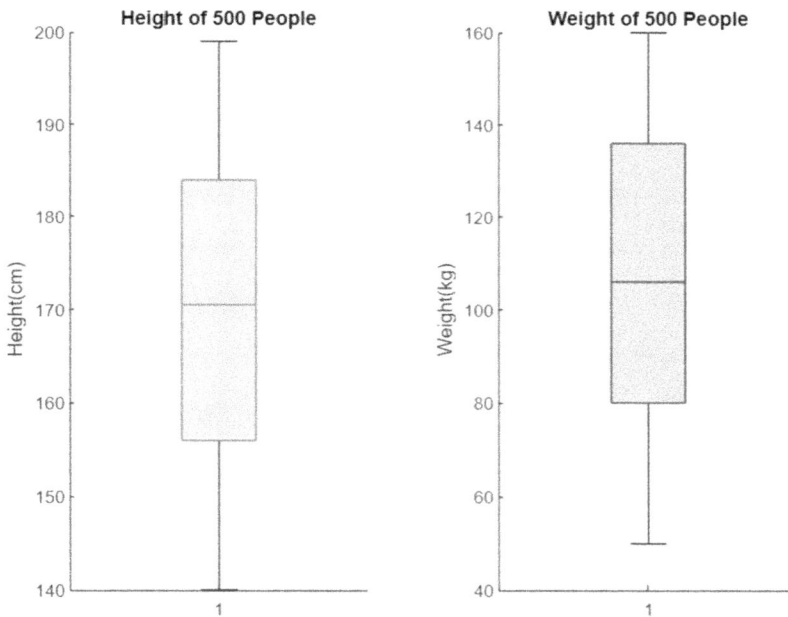

Figure 2.50

Here we display a side-by-side pair of box charts using the `tiledlayout` and `nexttile` functions. The first boxplot represents the distribution of heights while the second one displays the weights of 500 people. The function `boxplot` creates a visual representation of the data, but does not return numeric values which we discussed.

The dataset used to draw the boxplot above includes both genders and now we want to create box plots to display both height and weight data separately for both males and females, creating two sets box plots, one for each gender

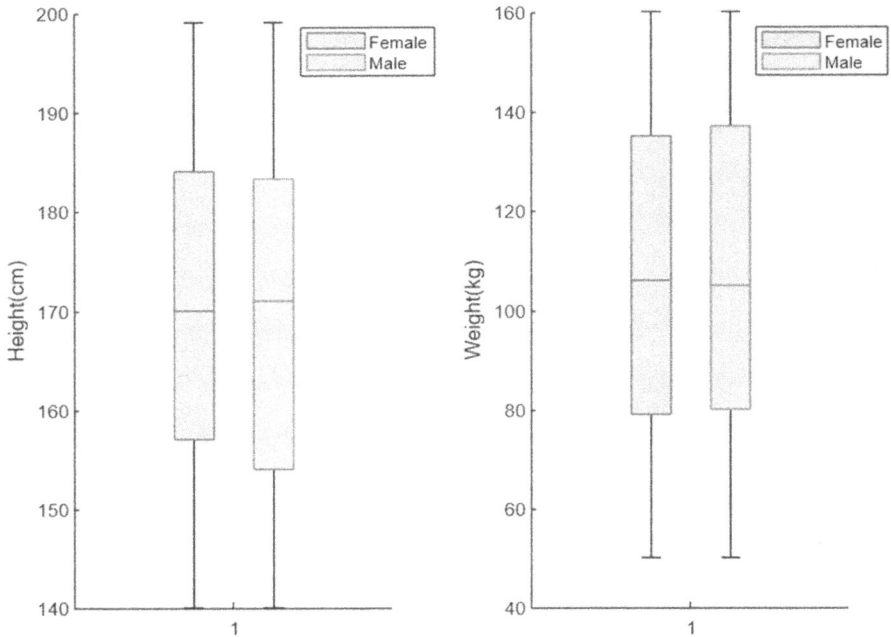

Figure 2.51

```
data = readtable('500_Person_Gender_Height_Weight_Index.csv');
Height = data.Height;
Weight = data.Weight;
Gender = data.Gender;

tiledlayout(1,2)

% Left axes
ax1 = nexttile;
boxchart(ax1,Height,'GroupByColor',Gender)
ylabel(ax1,'Height(cm)')
legend

% Right axes
ax2 = nexttile;
boxchart(ax2,Weight,'GroupByColor',Gender)
ylabel(ax2,'Weight(kg)')
legend
```

In the first set of axes, we display two box charts of Height values, one for the gender male and the other for females. Then in the second set of axes we do the same for the Weight.

Example

Lets say we have a dataset of tensile strength measurements for three different materials: Material A, Material B, and Material C.

```
data = readtable('3_Materials_Strengths.csv');

% Extract the data for each material
materialA = data.A;
materialB = data.B;
materialC = data.C;

colorA = [1 0 0];
colorB = [0 0.4740 0];
colorC = [0 0 1];

% Create a figure
figure;

% Create the boxplots for each material with custom colors
boxplot([materialA, materialB, materialC], 'Labels', {'A', 'B', 'C'},
        'Colors', [colorA; colorB; colorC]);

% Set labels and title
xlabel('Materials');
ylabel('Strength');
title('Strength Comparison');
```

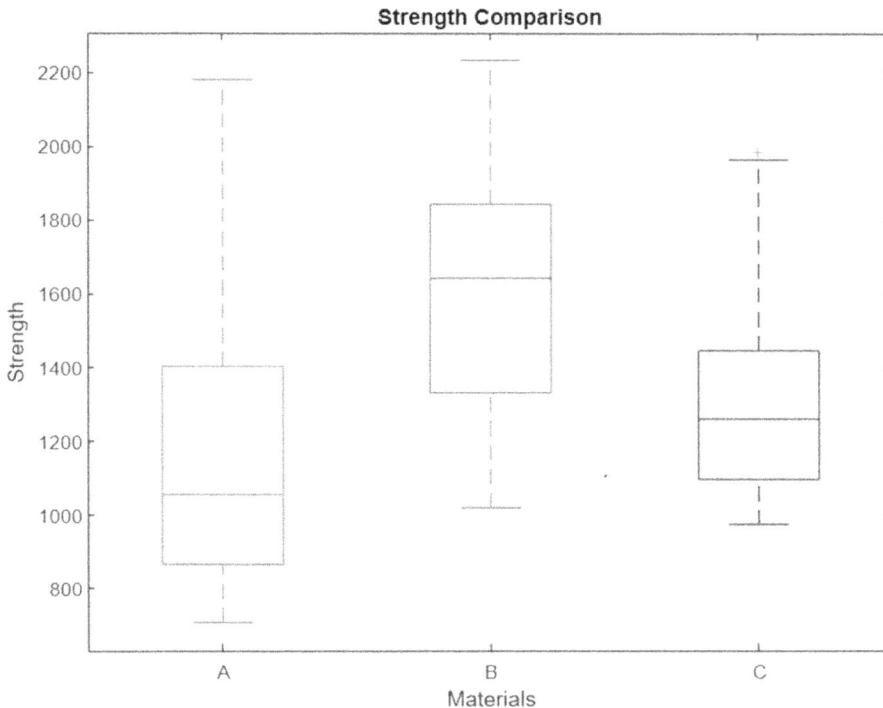

Figure 2.52

Lets analyze each of the data individually. Take a look at the boxplot A and boxplot C representing Material A and Material C. The top whisker is longer than the bottom whisker, indicating an asymmetric distribution of the data-more specifically a positively skewed data.

Example

```
data = readtable('auto-mpg.csv', VariableNamingRule = 'preserve');
MPG = data.mpg;
Cylinder = data.cylinders;
colors = ['r', 'g', 'b', 'k', 'm'];

% Create a figure
figure;

% Create the boxplots for each material with custom colors
boxplot(MPG, Cylinder, 'Colors', colors);

% Set labels and title
xlabel('Number of cylinders')
ylabel('Fuel efficiency');
title('Mileage Data');
```

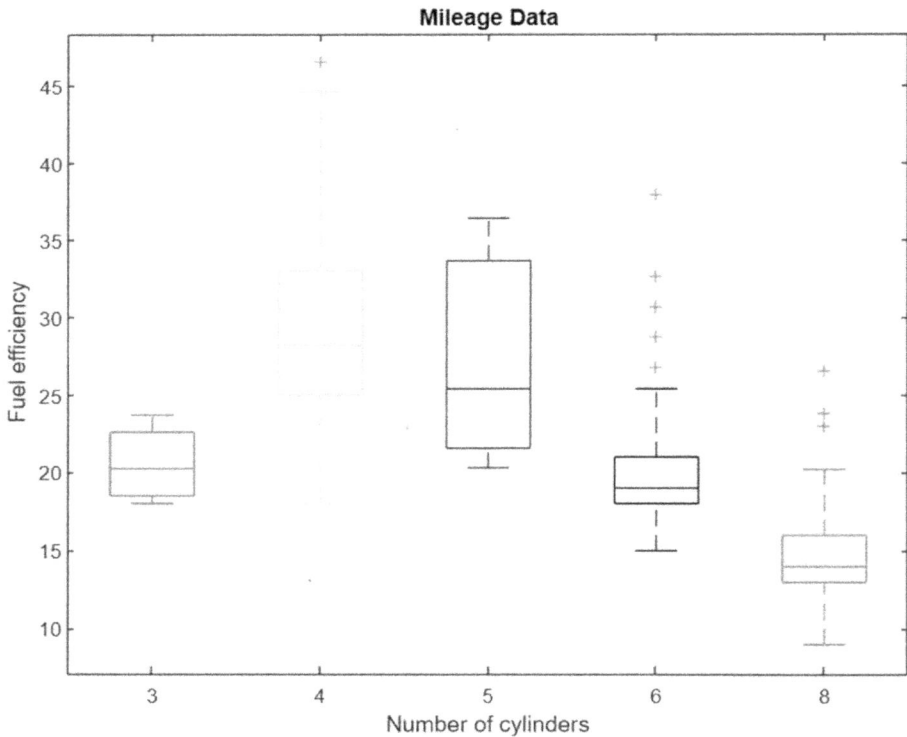

Figure 2.53

2.3.2 Scatterplot

Multivariate analysis is a type of data visualization that graphs pairs of numerical data, with one variable plotted on the horizontal axis and the other on the vertical axis. Each data point is then plotted at the intersection of the corresponding values on the two axes but not joined. It is the type of plot that is used when trying to determine whether the two variables plotted are related. In other words, it is considered the simplest way to study the correlation between these two variables (more in Chapter 10). The pattern of the points on the graph determines the strength of an association between two variables. If there is a certain pattern in the plot, the variables are associated; if not then they are not. Generally, there are three types of relationships that can be observed in a scatterplot:

- Positive correlation: the points on the scatterplot roughly follow an upward trend, indicating that as one variable increases, the other variable also tends to increase.

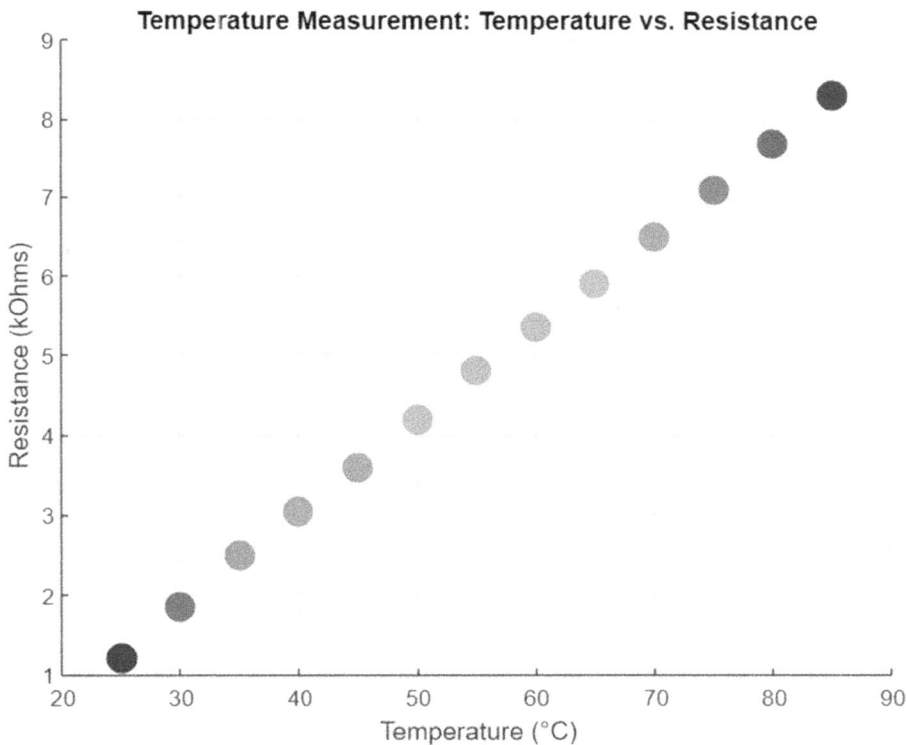

Figure 2.54

For example, you are working on a project to investigate the relationship between the temperature (x) and the corresponding resistance of a temperature sensor (y) in an engineering system. You have collected a set of data points and want to visualize the linear positive relationship between the resistance and temperature using a scatterplot.

The above graph Figure 2.54 is what a perfectly positive correlation scatterplot looks like. It represents a strong linear relationship between resistance and temperature where, as the temperature increases, the resistance of the temperature sensor increases consistently. As we can see, the data points are clustered tightly around a rising line that slopes upwards from left to right. The correlation coefficient (r) for a perfectly positive correlation would be +1.

- Negative correlation: the points on the scatterplot roughly follow a downward trend, suggesting that as one variable increases, the other variable tends to decrease.

For example, we are conducting a study to analyze the fuel efficiency of different vehicles in relation to their speed. We have collected data on the speed (x) in kilometers per hour and the corresponding fuel consumption rate (y) in liters per kilometer for a set of vehicles. We want to visualize the linear negative relationship between speed and fuel efficiency using a scatterplot.

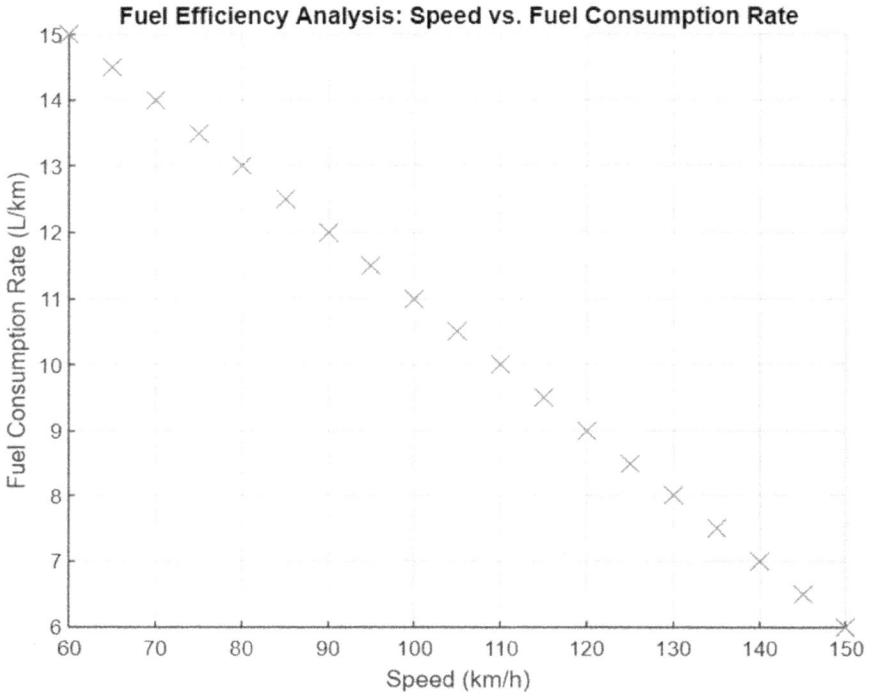

Figure 2.55

Figure 2.55 is another perfect but negatively correlated scatterplot, as higher speeds are generally associated with increased fuel consumption

- No correlation: the points on the scatterplot do not exhibit a clear trend or pattern, so neither a straight line nor a nonlinear pattern, indicating that there is little to no relationship between the variables. Therefore, when one variable changes, it does not influence the other variable.

In engineering, an example of a "no correlation" scenario with temperature could be the study of the hardness of a material as a function of temperature. In almost all cases, the hardness of a material does not exhibit a clear correlation or relationship with temperature.

For instance, let's consider a study where the hardness of a metal alloy is measured at various temperatures ranging from 30°C to 65°C. However, due to the complex microstructural changes, phase transformations, or other factors, the hardness of the alloy may not show a significant correlation with temperature.

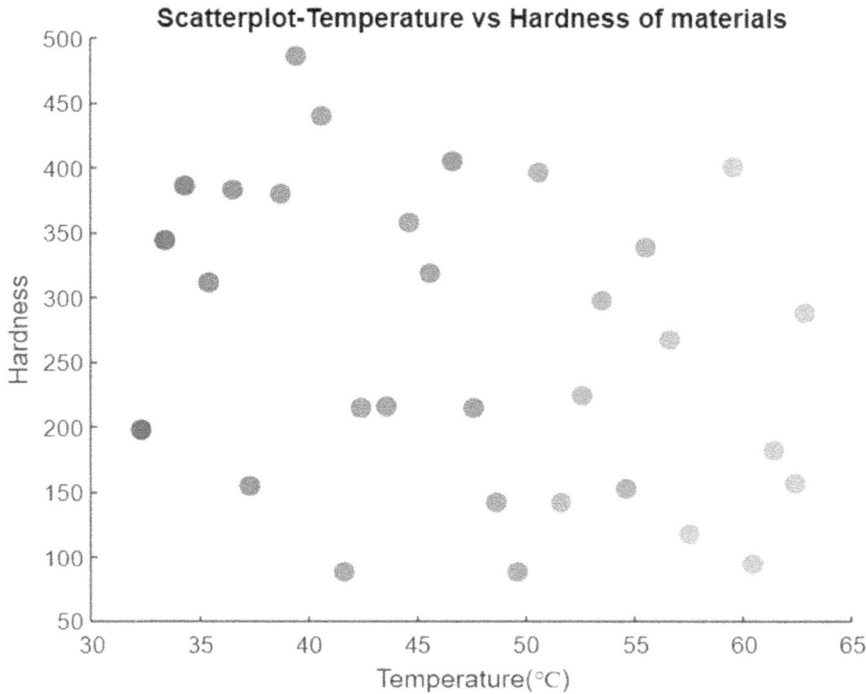

Figure 2.56

In Figure 2.56, the scatterplot displays the temperature values on the x-axis and the hardness values on the y-axis. As we can see, the points are randomly scattered without any noticeable trend or pattern, indicating a lack of correlation between temperature and hardness. This "no correlation" scenario suggests that changes in temperature do not have a significant impact on the hardness of the metal alloy.

A scatterplot does not just reveal the strength of the relationship between two variables but also the type of relationship between them. If the data points on a scatterplot have somewhat a straight-line pattern, we say the relationship is linear. However, in many cases, the relationship between two variables does not follow a straight line but instead exhibits a curved or nonlinear pattern. Such a relationship is said to be nonlinear.

Let's look at some examples of nonlinear relationships between variables in the real world.

- Quadratic relationships: when plotted on a scatterplot, this relationship exhibits a "U" shape indicating that the relationship between the variables being plotted follows a quadratic pattern. As the independent variable x increases or decreases, the response variable y also changes, but not in a constant or proportional manner. Instead, the relationship exhibits a curvature, suggesting that the rate of change is not constant throughout the range of values. This type of relationship is commonly observed when analyzing data involving variables that have a squared or higher-order relationship.

2.3.2.1 Example 1

We are analyzing the voltage-current characteristics of a diode in an electrical circuit. To study the behavior of the diode, we have collected data on the applied voltage (x) across the diode and the corresponding current (y) flowing through it. We suspect that the relationship between voltage and current is nonlinear and want to visualize it using a scatterplot.

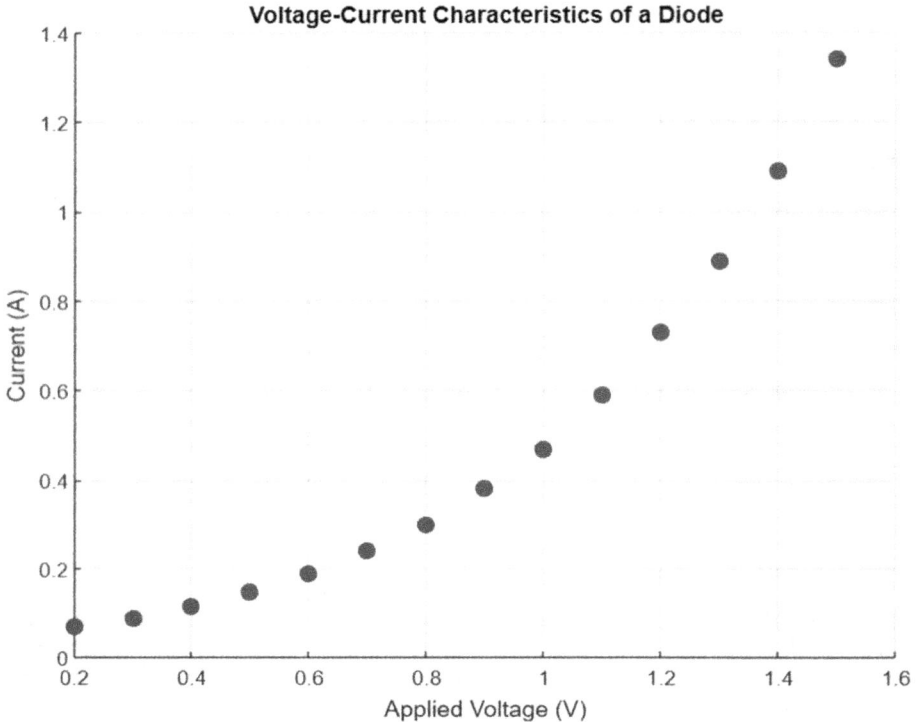

Figure 2.57

2.3.2.2 Example 2

You are investigating the trajectory of a projectile launched into the air. To analyze the projectile's motion, you have collected data on the horizontal distance (x) traveled by the projectile and the corresponding vertical height (y) above the launch point. You suspect that the relationship between distance and height follows a quadratic pattern and want to visualize it using a scatterplot.

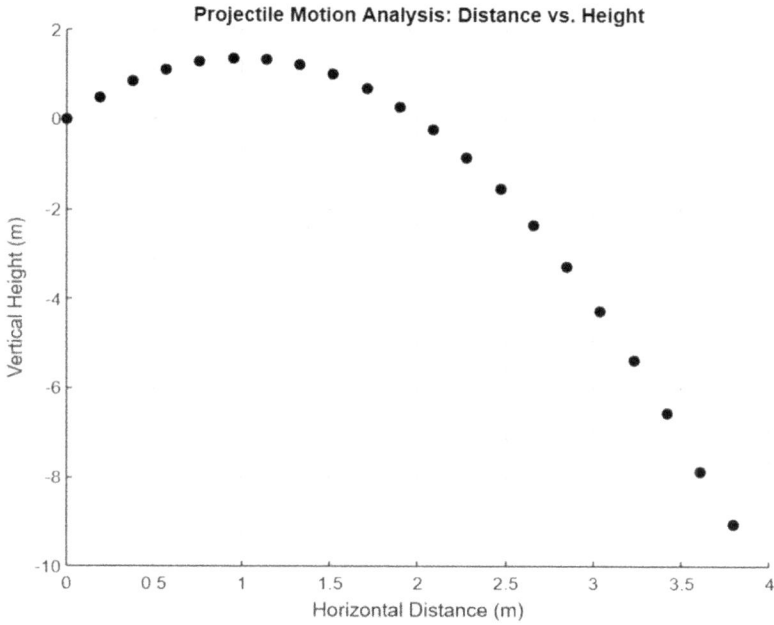

Figure 2.58

- Cubic relationships: when plotted on a scatterplot, this kind of relationship typically has two distinct curves indicating that the relationship between the variables being plotted follows a cubic pattern.

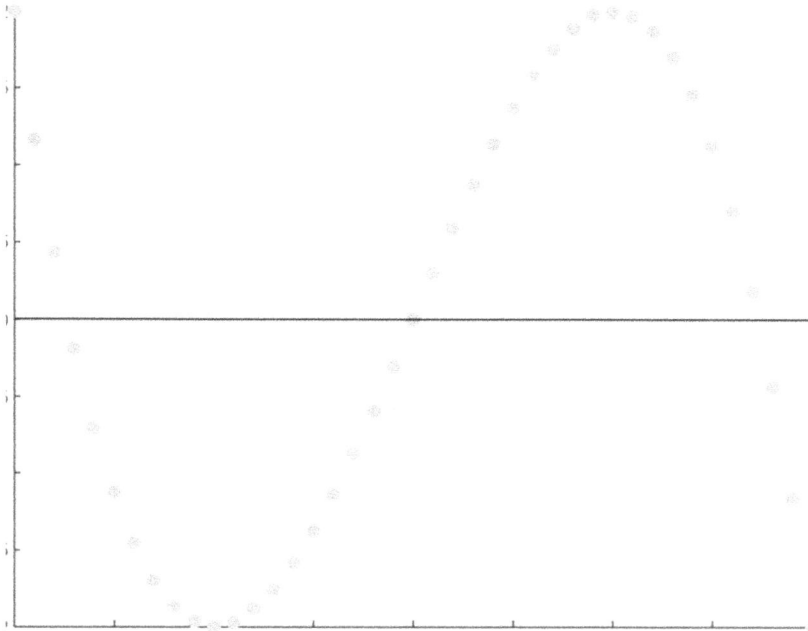

Figure 2.59

- Exponential relationships: when plotted on a scatterplot, this kind of relationship displays a curved pattern forming an exponential curve. An increasing pattern indicates that the relationship between the variables being plotted follows an exponential growth while a decreasing curve represents an exponential decay. In an exponential relationship, as the independent variable x increases, the y variable responds at an accelerating increasing or decreasing rate. In a scatterplot, an exponential relationship is typically visualized as a curve that starts gradually but then can either steepen or completely flatten out, depending on whether it represents exponential growth or decay.

2.3.2.3 Example

Suppose an engineer wants to determine the percentage of light transmission through a glass pane to optimize natural lighting within a building. To determine this percentage of transmission he collects a set of data points and wants to visualize the relationship between the number of panes used and the corresponding percentage of light transmitted through the panes using a scatterplot (Figure 2.50).

Exponential functions are important to circuit analysis because they're solutions to many problems in which a circuit contains resistors, capacitors, and inductors. In all engineering applications, understanding the relationship between the decay time constant and amplitude allows engineers to characterize system dynamics, assess stability, optimize performance, and design control strategies.

In mechanical engineering, an example of an exponentially increasing scatterplot can be observed in the relationship between applied stress and material deformation in a ductile material, as described by the stress-strain curve (Figure 2.52).

- Logarithmic relationships: when plotted on a scatterplot, this relationship displays a curved pattern forming a logarithmic curve. Similar to exponential relationships, an increasing pattern indicates that the relationship between the variables being plotted follows a logarithm growth while a decreasing curve represents a decay pattern.

Periodic relationships: when plotted on a scatterplot, this relationship is characterized by a repeating pattern of data points. It indicates that the relationship between the variables being plotted follows a cyclic or oscillating pattern. The points may form a sinusoidal curve, such as a sine or cosine wave, that repeats at a regular interval.

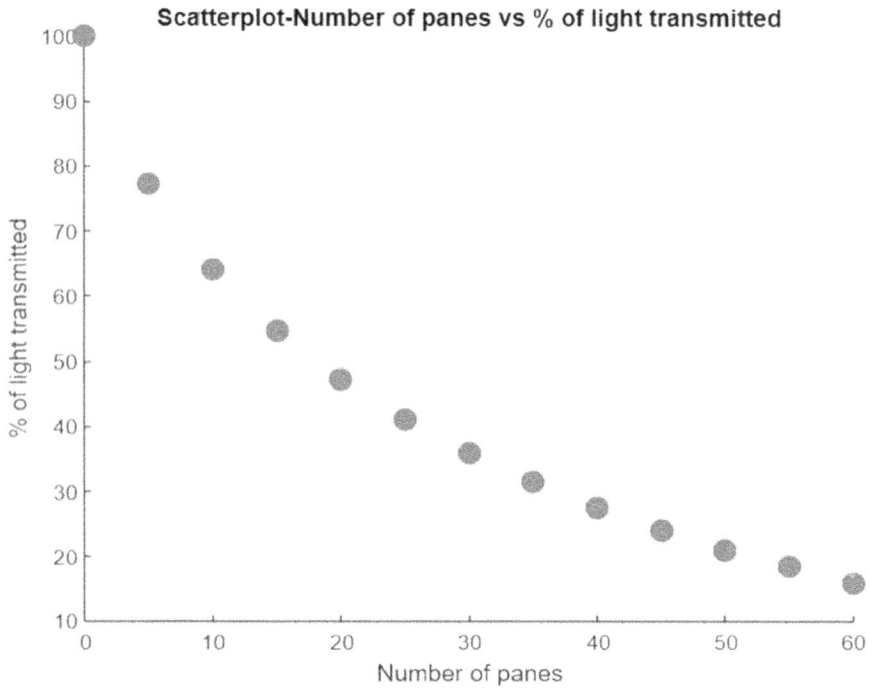

Figure 2.60

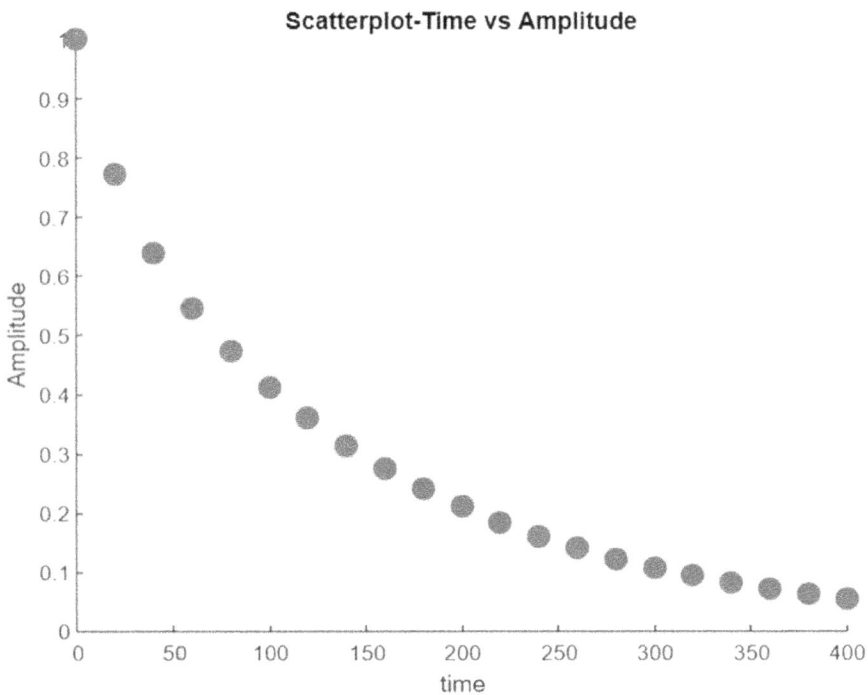

Figure 2.61

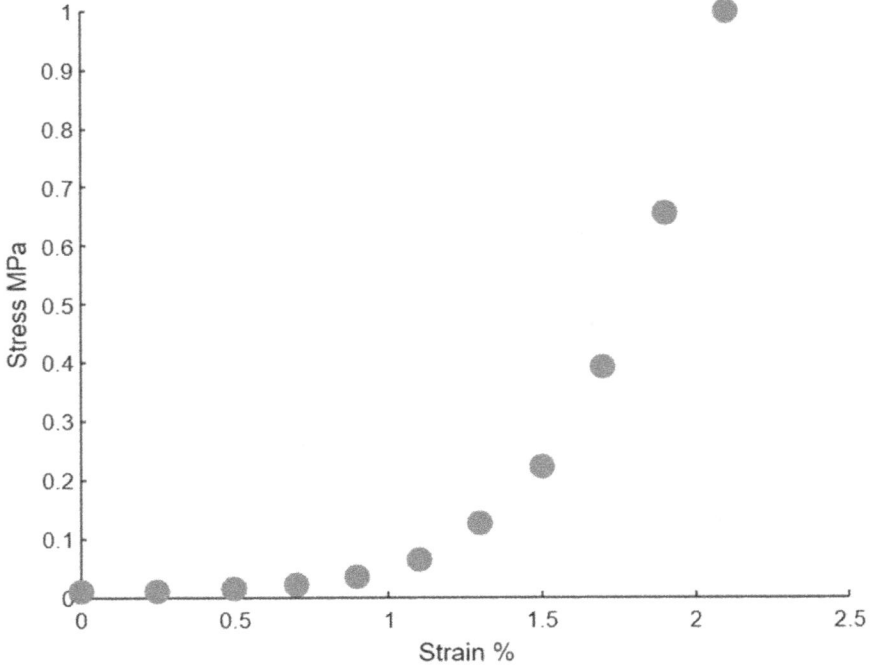

Figure 2.62

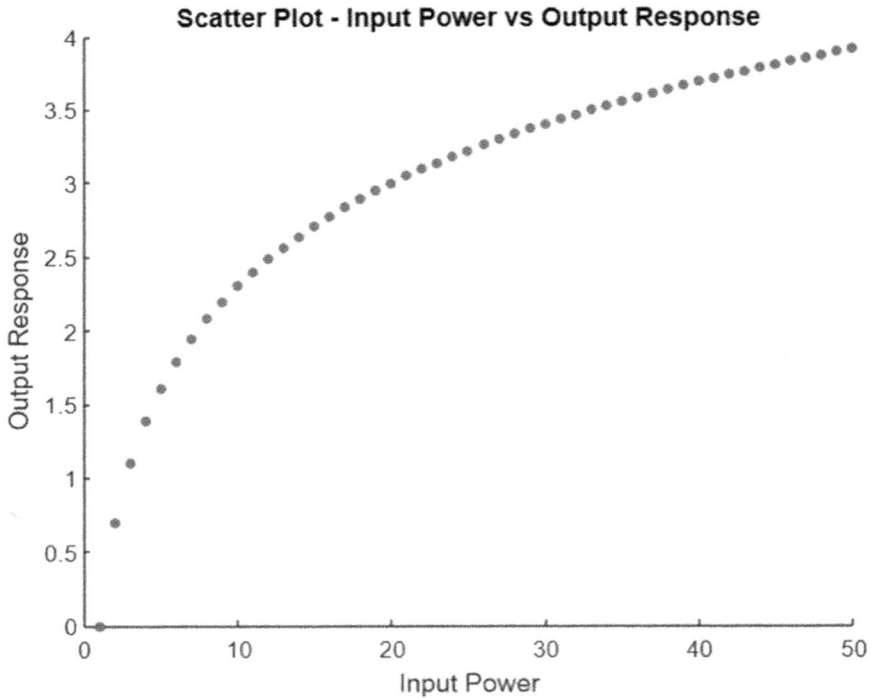

Figure 2.63

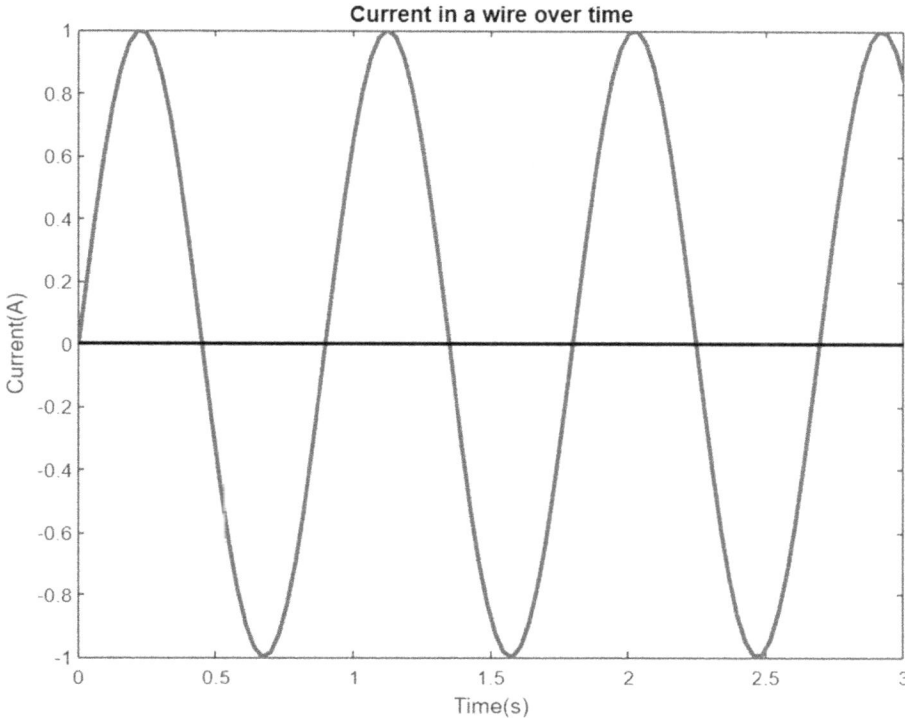

Figure 2.64

To create a scatterplot in MATLAB we use the function scatter. The syntax of the function is:

```
scatter(x, y, size, 'Marker','', 'MarkerFaceColor','', 'MarkerEdgeColor',
'');
```

Let's discuss each of the input arguments.

x and *y* coordinates

We can use our scatterplot for both one and multiple sets of data.

For a single set of data, we specify x and y as any combination of row or column vectors of the same length.

Now in some cases one can choose to plot multiple sets of data in a single scatterplot to compare relationships or identify correlations or associates. In that case the syntax would be:

scatter($[x_1; x_2;\ldots\ldots x_n], y_1; y_2;\ldots\ldots y_n]$)

However, adding multiple sets of data (let's say more than two) to a scatterplot can potentially overcrowd the plot, making it difficult to interpret the individual relationships. So, therefore, it is essential to ensure that the plot remains clear and can be distinguished. The most important note is to ensure that the variables being plotted are comparable or have compatible scales. If the datasets have significantly different ranges or units of measurement, it might be more appropriate to plot them on separate scatterplots for better visualization.

The next argument, size, specifies the size of the markers at the locations specified by the *x* and *y*.

The 'MarkerSize' allows us to define the size of the markers as any positive value, typically ranging from one to a few hundred pixels. In MATLAB, the size property of markers in a scatterplot can be specified either as a scalar value, a vector, or a matrix. An empty array specifies the default size of 36 points.

When we specify the size as a scalar value, we set a single size for all markers in the scatterplot.

```
Resistance = [1.2, 2.5, 3.6, 4.8, 5.9, 7.1, 8.3];
Temperature = [25, 35, 45, 55, 65, 75, 85];
size = 200

% Create scatter plot
scatter(Temperature, Resistance, size)

% Add labels and title
xlabel('Temperature (°C)')
ylabel('Resistance (kOhms)')
title('Temperature Measurement: Temperature vs. Resistance)

% Add grid
grid on;
```

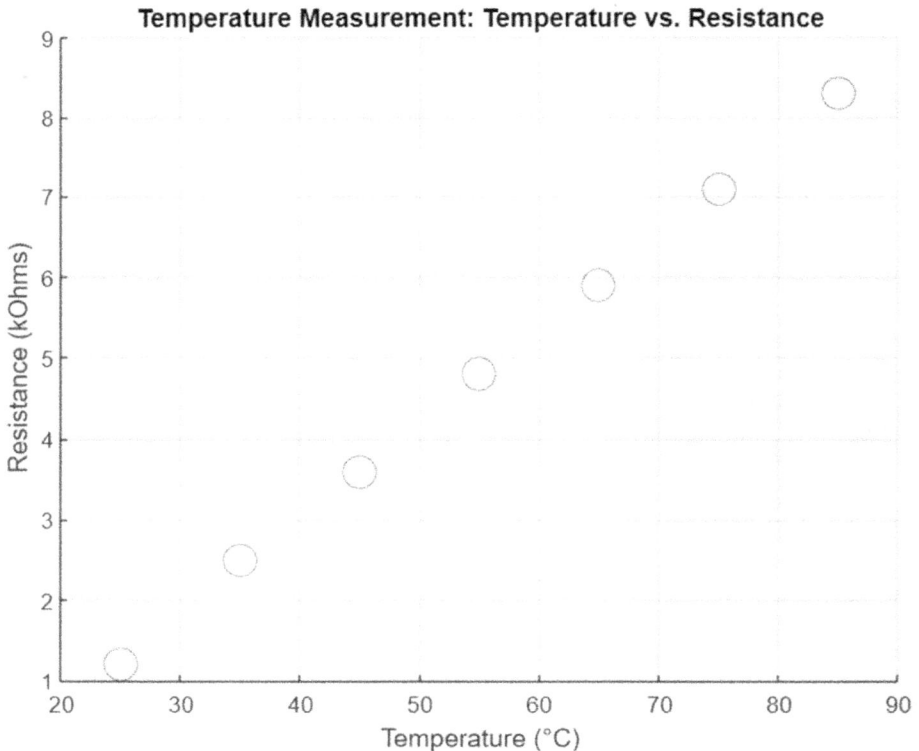

Figure 2.65

When presenting as a vector we are basically assigning different sizes to individual markers. Remember the length of the vector should have the same length as x and y.

```
Size = [40 80 120 160 200 240 280];
% Create scatter plot
scatter(Temperature, Resistance, size)

% Add labels and title
xlabel('Temperature (°C)')
ylabel('Resistance (kOhms)')
title('Temperature Measurement: Temperature vs. Resistance)

% Add grid
grid on;
```

The other way of doing it is using the function linspace($x_1, x_2,$ length(x)).

This function generates a row vector of x elements evenly spaced points between x_1 and x_2. The resistance values in our dataset range from 1 to 9 and we therefore have:

```
size = linspace(1, 9, length(Resistance))
scatter(Temperature, Resistance, size * 50)
```

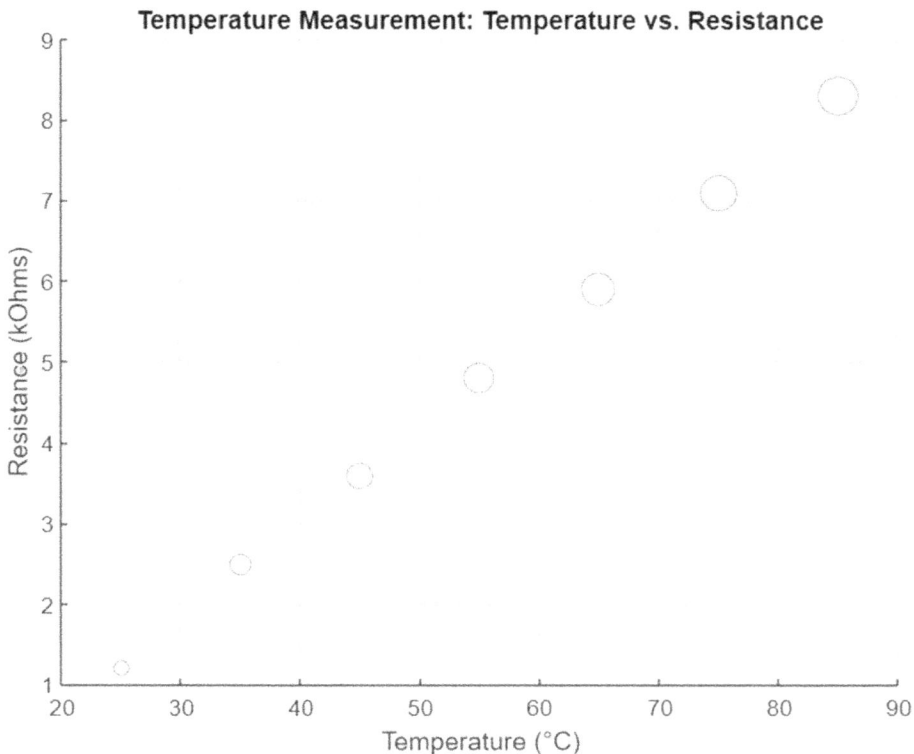

Figure 2.66

As stated earlier, the size property of markers in a scatterplot can be represented as a matrix. The matrix should have the same size as the data points used in the scatterplot. Each element of the matrix corresponds to the size of the marker at the corresponding data point.

The next argument is the 'Marker' property which specifies the type of marker. Here are some commonly used marker types:

"o"	Circle
"+"	Plus sign
"*"	Asterisk
"."	Point
"x"	Cross
"_"	Horizontal line
"|"	Vertical line
"square"	Square
"diamond"	Diamond
"^"	Upward-pointing triangle
"v"	Downward-pointing triangle
">"	Right-pointing triangle
"<"	Left-pointing triangle
"pentagram"	Pentagram
"hexagram"	Hexagram

To specify the marker type in a scatterplot, we first call out the parameter 'Marker' and then specify the type using any one of the above. In MATLAB for scatterplots the default marker type is a circle, "o".

Next is the MarkerEdgeColor which is used to specify the color of the marker edges or outlines.

Then finally we have the MarkerFaceColor property which is used to specify the fill color of markers. Here are some commonly used marker colors:

Color name	Short name	Appearance	Color name	Short name	Appearance
"red"	"r"		"magenta"	"m"	
"green"	"g"		"yellow"	"y"	
"blue"	"b"		"black"	"k"	
"cyan"	"c"		"white"	"w"	

In addition to the above colors we can use RGB triplets and hexadecimal color codes to specify the marker color. In MATLAB, we can specify colors using RGB triplet values, which represent the red, green, and blue components of a color. Each component is a value between 0 and 1. Here are some examples of using RGB triplet colors in MATLAB:

[1 0 0] represents pure red.
[0 1 0] represents pure green.
[0 0 1] represents pure blue.
[1 1 0] represents yellow.

[1 0 1] represents magenta.
[0 1 1] represents cyan.
[0 0 0] represents black.
[1 1 1] represents white.

We can customize the intensity of each RGB component to create a wide range of colors. Now corresponding to the RGB triplets we have hexadecimal color codes to specify colors using the "#RRGGBB" format, where "RR", "GG", and "BB" represent the hexadecimal values for the red, green, and blue color components, respectively.

Just like the 'MarkerSize', the MarkerEdgeColor property can only be expressed as a scalar value, a vector, or a matrix. Unfortunately, using a vector or matrix to specify the MarkerFaceColor property is not supported in MATLAB and we will see why that is the case in a bit.

As discussed earlier, when we provide a scalar value, it will be applied to all markers, resulting in the same edge and filled color for all markers in the plot. For example,

```
scatter(Resistance, Temperature, size, 'MarkerFaceColor', 'yellow',
                                 'MarkerEdgeColor', 'black')
Let's specify the color using an RGB triplet.


scatter(Resistance, Temperature, size, 'MarkerFaceColor', [1 1 0],
                                 'MarkerEdgeColor', [0 0 0])
```

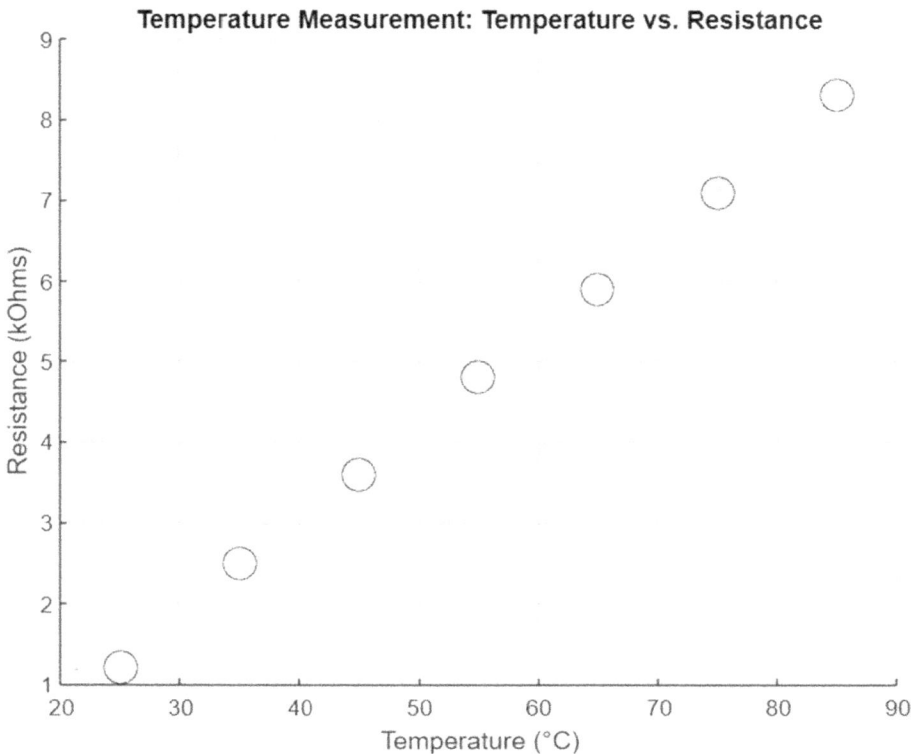

Figure 2.67

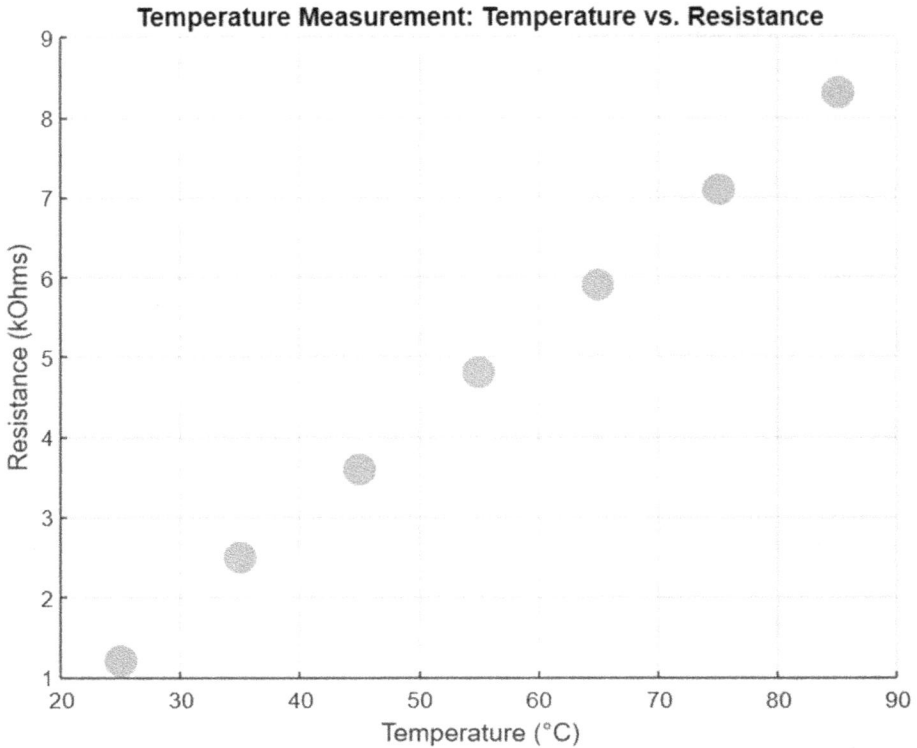

Figure 2.68

The above code sets the marker face color and marker edge color of all markers in the scatterplot to yellow and black.

Let's customize the colors.

```
scatter(Resistance, Temperature, size, 'MarkerFaceColor', [1 0.56 0.75],
                                  'MarkerEdgeColor', 'none')
```

Notice we set the value of MarkerEdgeColor to none which means we want to remove the marker edges, making them invisible. This is useful when we want to show only the filled markers without any outlines.

In contrast, if we provide a vector or matrix of color values, it allows us to assign different edge colors to individual markers. Each element of the vector or matrix corresponds to the edge color of a specific marker. To assign different colors to each point we use a colormap. The syntax of the function is:
```
colormap(target, map)
```

2.3.2.4 Colormap name

The following table lists the predefined colormaps.

Map	Color scale	Map	Color scale
parula		bone	
turbo		copper	
hsv		pink	
hot		sky	
cool		jet	
spring		lines	
summer		colorcube	
autumn		prism	
winter		flag	
gray		white	

```
edgeColor = linspace(1, 9, length(x));
scatter(x, y, size, edgeColor, "filled")
colormap(gca, "parula")
```

Notice how the above function does not include MarkerEdgeColor or MarkerFaceColor.

2.3.3 Line graphs

A line graph—also known as a line plot or a line chart—is a graph that uses straight lines to connect individual data points. Choosing a line chart when ordering and joining the data points by their x-axis values highlights meaningful changes in the vertical y-axis variable. Analysts use these charts in many settings to display different types of information.

Line charts are similar to scatterplots except that they connect the data points with lines. For example:

Now to create the line graphs in MATLAB we use the following functions

plot:	2-D line plot
plot3:	3-D point or line plot

2.3.3.1 Plotting 2-D data

The main function for creating a 2-D plot is called plot. When the function plot is called, it opens a new figure window. The main syntax for plot is:

$plot(x_1, y_1, x_2, y_2, \ldots x_n,$ 'linestyle_marker_color')

This function plots multiple pairs of x- and y- coordinates on the same set of axes. The next argument specifies the line style, marker, and color. We have already discussed the types of markers and colors we use in scatterplot. Let's now walk through the types of line styles we can use in line graphs.

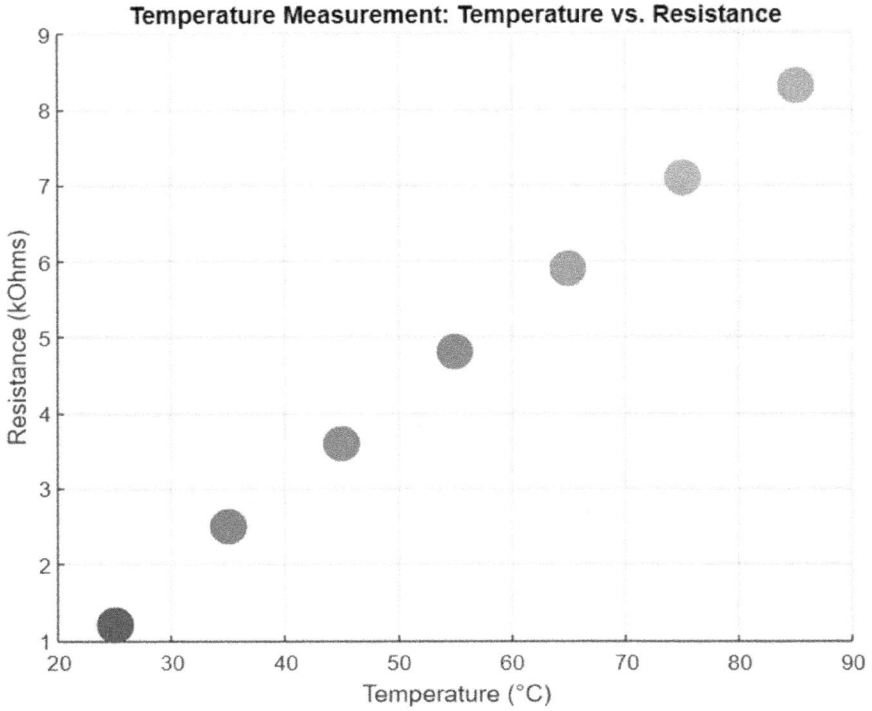

Figure 2.69

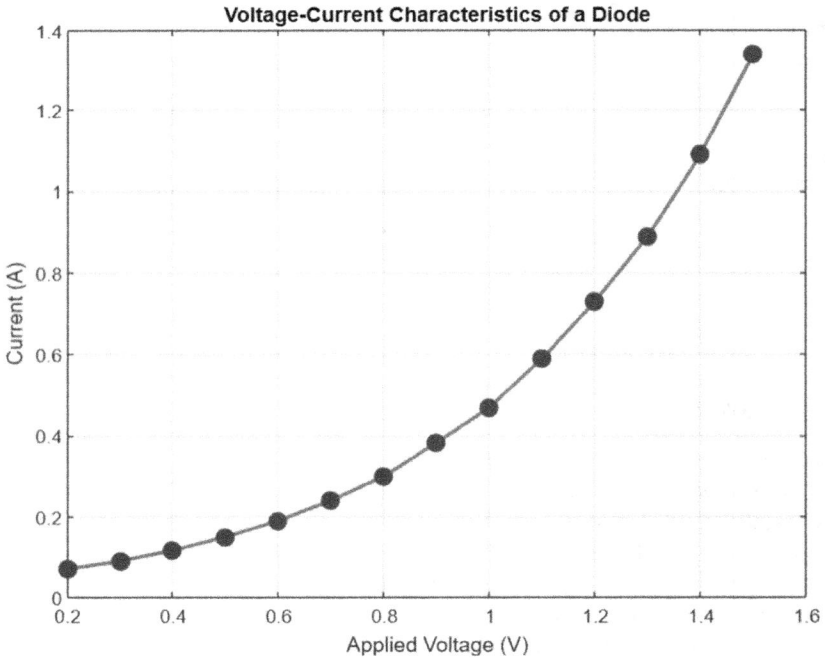

Figure 2.70

Line style	Description	Resulting line
"-"	Solid line	————————
"--"	Dashed line	– – – – – –
":"	Dotted line	....
"-."	Dash-dotted line	– · – · – · – · –

Note that the symbols can appear in any order. It is not necessary to specify all three characteristics—line style, marker, and color—all the time. For example, if we do not specify the line style and specify the marker, then the plot shows only the marker and no line.

2.3.3.2 Example

1. $plot(x, y, ":")$: creates a plot with a dotted line with the default color.
2. $plot(x, y, 'g')$: creates a plot with green markers.
3. $plot(x, y, 'o')$ If you specify a marker symbol and do not specify a line style, then plot displays only the markers with no line connecting them, hence creating a plot with only circle markers. $plot(x, y, '-o')-$: creates a plot with a solid line with circle markers.

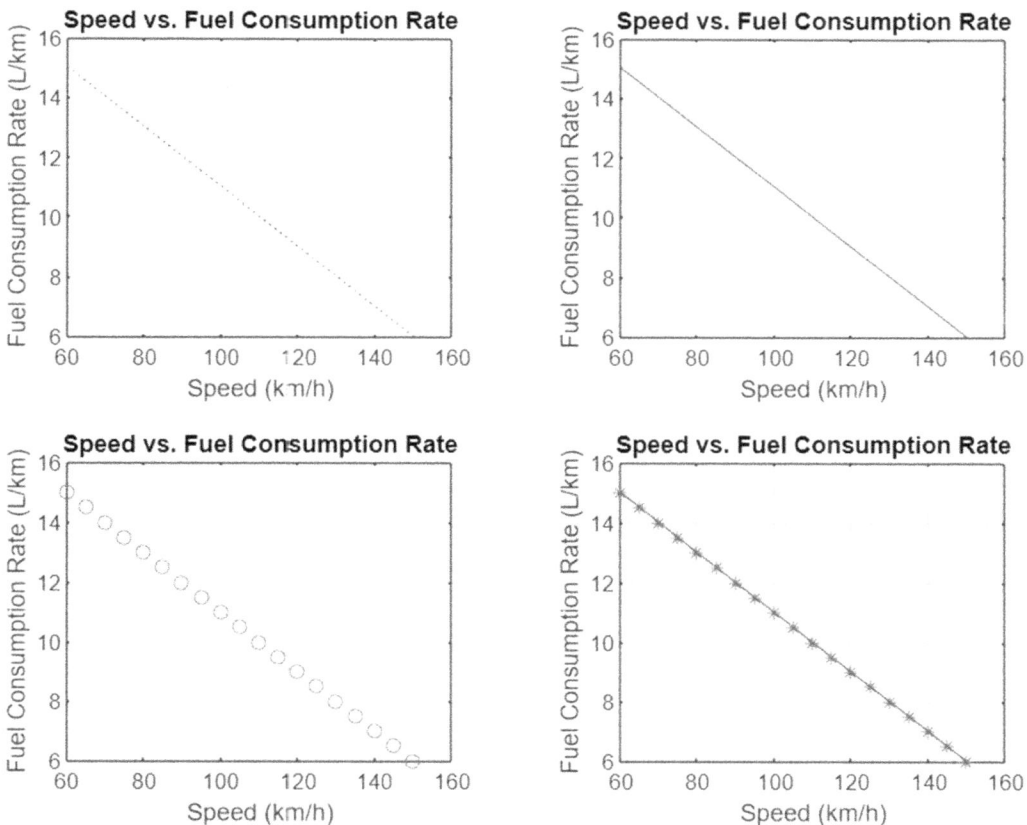

Figure 2.71

Customized examples:

```
%'r-' creates a plot with a dashed red line of width 2 and circular blue
markers of size 8.
plot(Voltage, Current, 'r-', 'LineWidth', 2, 'Marker', 'o',
   'MarkerEdgeColor', 'b', 'MarkerFaceColor', 'b', 'MarkerSize', 8);
%'g-.' creates a plot with a dash-dotted green line of width 3 and
circular black markers of size 6.
plot(Horizontal, Vertical, 'g-.', 'LineWidth', 3, 'Marker', 'o',
'MarkerEdgeColor', 'k', 'MarkerFaceColor', 'k', 'MarkerSize', 6);
plot(Time, Amplitude, 'r-', 'LineWidth', 3, 'Marker', 'o',
   'MarkerEdgeColor', 'g','MarkerFaceColor', 'g', 'MarkerSize', 8);
xlabel("time")
ylabel("Amplitude")
title("Line Graph-Time vs Amplitude")
plot(Panes, Perc,'y-', 'LineWidth', 3, 'Marker', 'o', 'MarkerEdgeColor',
[0.4940 0.1840 0.5560], 'MarkerFaceColor', [0.4940 0.1840 0.5560],
                  'MarkerSize', 8);
xlabel("Number of panes")
ylabel("% of light transmitted")
title("Line Graph-Number of panes vs % of light transmitted")
plot(x, y,'c-', 'LineWidth', 3, 'Marker', 'o', 'MarkerEdgeColor',
      'red', 'MarkerFaceColor', 'red', 'MarkerSize', 8);
xlabel("Strain %")
ylabel("Stress MPa")
```

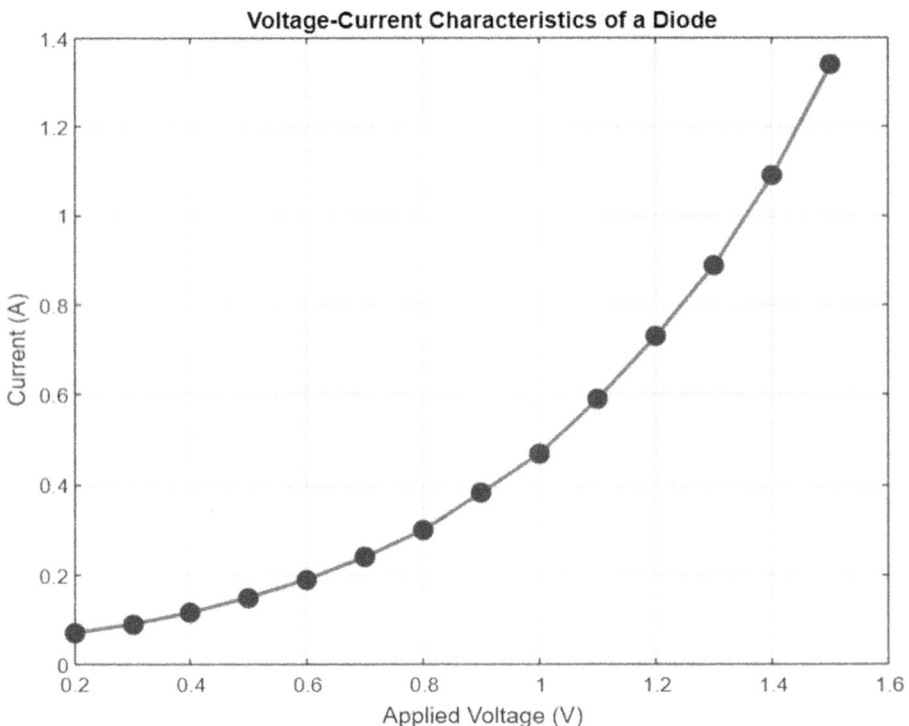

Figure 2.72

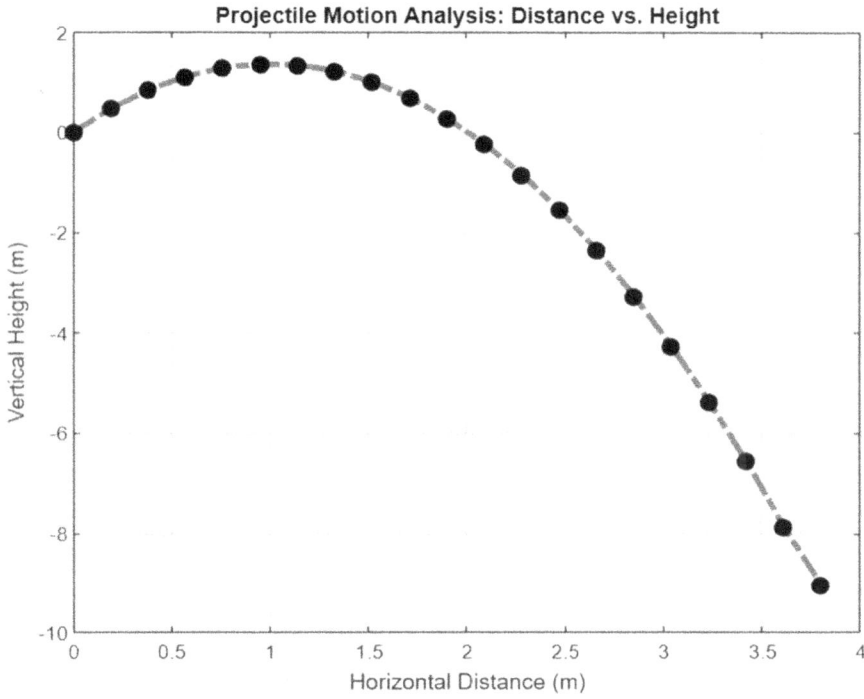

Figure 2.73

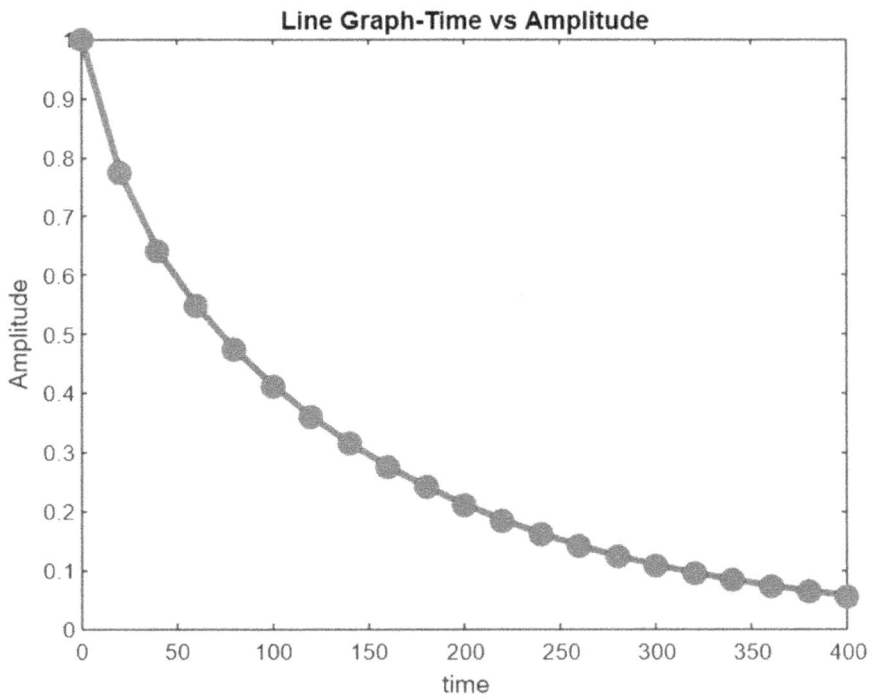

Figure 2.74

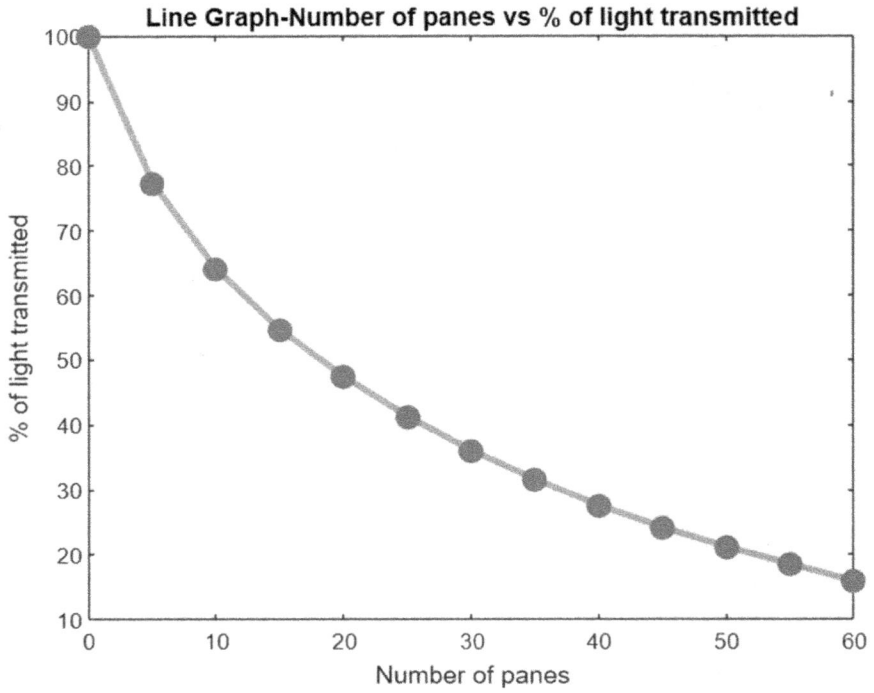

Figure 2.74

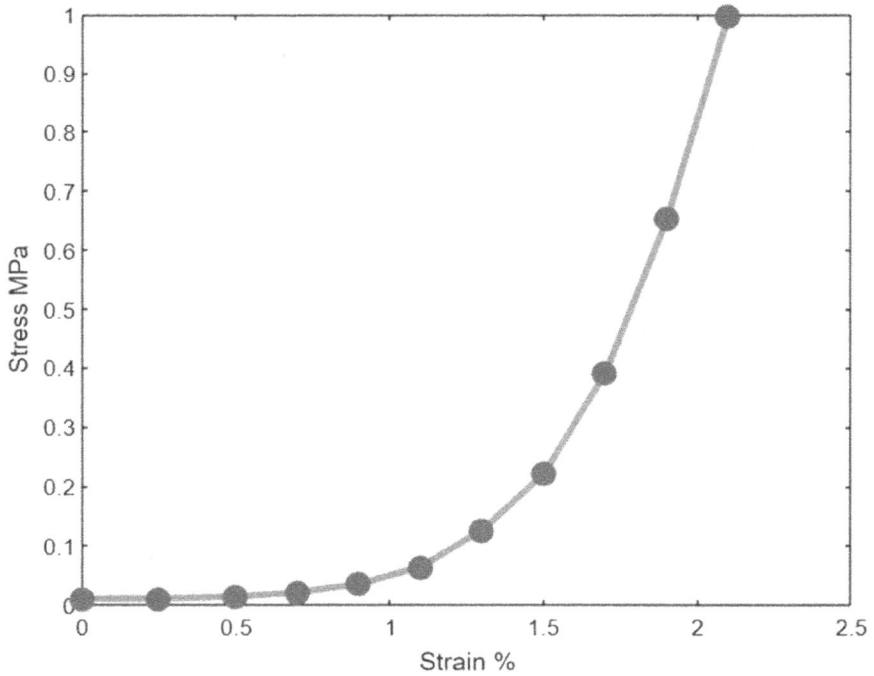

Figure 2.76

Showing a change over time for a measure is one of the significant advantages of using line graphs in engineering. They are commonly used to visualize trends, patterns, and changes over a continuous period.

One of the most common examples of line graphs displaying data over time is damping. Damping is a phenomenon that reduces the amplitude of oscillations or vibrations in a system over time.

```
amplitude = 1;      % Initial amplitude of the oscillation
damping_ratio = 0.1;  % Damping ratio
natural_frequency = 2; % Natural frequency of the oscillation
time = 0:0.1:10;    % Time values
% Calculate the amplitude of the damped oscillation
damped_amplitude = amplitude * exp(-damping_ratio * natural_frequency *
time) .* cos(natural_frequency * sqrt(1 - damping_ratio^2) * time);
% Plotting the damping graph
plot(time, damped_amplitude, 'b-', 'LineWidth', 2);
yLine = 0; % Y-coordinate for the horizontal line
xLine = [0, 10]; % X-coordinates for the line (spanning the entire
x-axis)
line(xLine, [yLine, yLine], 'Color', 'k', 'LineWidth', 2.5) % Drawing the
line in red ('r') with a dashed line style ('--')
% Adding labels to x-axis and y-axis
xlabel('Time');
ylabel('Amplitude');
% Adding a title to the graph
title('Damping Graph');
```

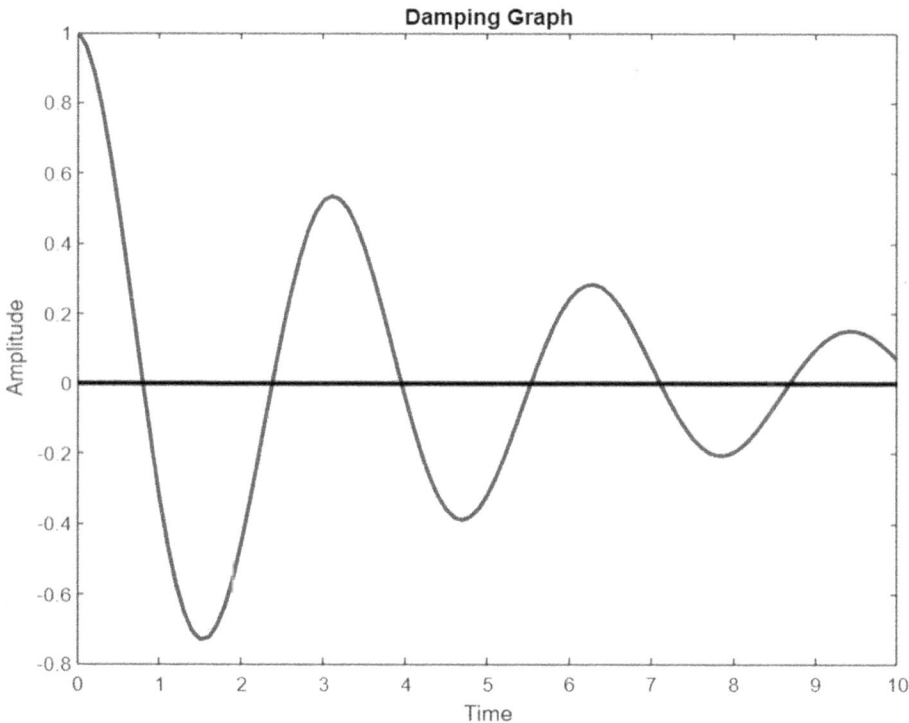

Figure 2.77

Analyzing the graph, we see the decay of the amplitude over time. As the system dissipates energy due to damping forces, the amplitude gradually decreases. The rate of decay depends on the damping ratio.

2.3.3.3 Multiple line graphs

Ohm's law states that the current (I) flowing through a conductor is directly proportional to the voltage (V) applied across it, given a constant resistance (R). Mathematically, Ohm's law can be expressed as V = IR. So we shall create a line graph representing Ohm's law for different resistors, by varying the voltage across each resistor and measuring the resulting current.

```
x = [0 1 2 3 4 5 6 7 8 9 10] ;
y = x;
plot(x, y, 'r', 'LineWidth', 1.5)
hold on
y1 = 0:0.5:5;
plot(x, y1, 'g', 'LineWidth', 1.5)
hold on
y2 = 0:0.25:2.5;
plot(x, y2, 'b', 'LineWidth', 1.5)
hold on
y3 = 0:0.10:1;
plot(x, y3, 'k', 'LineWidth', 1.5)
hold on
y4 = 0:0.05:0.5;
```

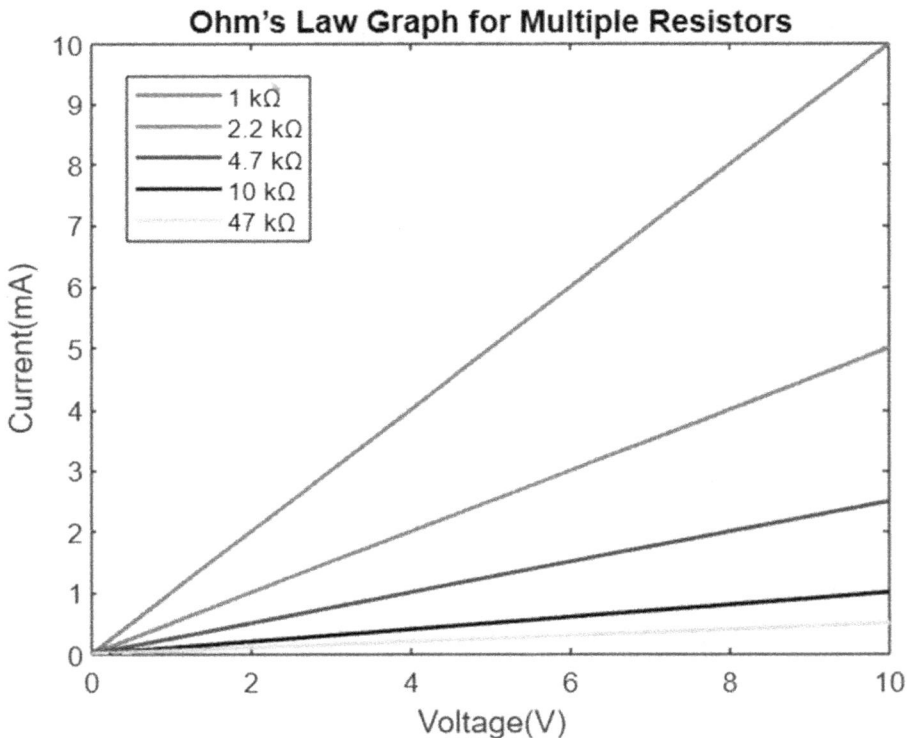

Figure 2.78

```
plot(x, y4, 'm', 'LineWidth', 1.5)
hold off
xlabel('Voltage(V)')
ylabel('Current(mA)')
title('Ohm's Law Graph for Multiple Resistors')
legend('1 kΩ','2.2 kΩ','4.7 kΩ','10 kΩ','47 kΩ')
```

The resulting line graph in Figure 2.78 demonstrates how the current varies with different resistances according to Ohm's law. As resistance increases, the current decreases proportionally, as indicated by the negative slope of the lines connecting the data points.

We can also plot our line graphs in 3-D using the syntax:

$$plot(x, y, z)$$

$$plot(x_1, y_1, z_1, x_2, y_2, z_2)$$

$$plot(x_1, y_1, z_1, x_2, y_2, z_2 \ldots x_n, y_n, z_n)$$

2.3.4 Histogram

2.3.5 Boxplot

Another visual representation of a dataset is the boxplot. A box plot displays a five-number summary of a dataset: (in ascending order) the minimum value, the first quartile, the median, the third quartile, and the maximum value. The following diagram shows a typical boxplot of an arbitrary dataset.

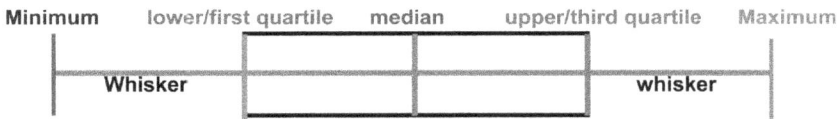

Figure 2.79

Let's make some observations from the figure.

The first thing we observe is a rectangular box and whiskers that sprout from the two ends of the box.

Let's talk about the box first.

The box encloses the middle half of a dataset, that is, around 50% **of the data fall inside the box**. Here the quartiles (first and third) mark the ends of the box.

Twenty-five percent of scores fall below the first/lower quartile, and thus, 75% of data are above it.

Seventy-five percent of the scores fall below the third/upper quartile, and therefore 25% of data are above it.

The length of the box represents the interquartile range which is the difference between the upper/third and the lower/first quartile.

Notice that there is a line drawn in the middle of the box presenting the sample median.

Half of the data values are greater and half are lower than the second/median quartile.

The median or second quartile does not necessarily have to be between the first and third quartiles; it can be on one, or the other, or even both depending on the dataset.

Let's look at the whiskers now.

The whisker that extends from the ends of the box represents the scores that fall outside of the middle 50% . The "whiskers" extend from the ends of the box until they reach the sample maximum and minimum. Therefore the smallest (minimum) and largest (maximum) data values are labeled at the endpoints of the whiskers.

Boxplots are useful as they give a good, quick picture of a dataset. They provide a visual summary of the data, enabling researchers to quickly identify mean values, the dispersion of the dataset, and signs of skewness.

The length of the whiskers of the box and the position of the line in the box also tell us whether the sample is symmetric or skewed, either to the right or left.

When the median is in the middle of the box, and the whiskers are about the same length on both ends of the box, then the distribution is symmetric.

In a normal distribution, the mean and the median are the same number, while the mean and median in a skewed distribution are different numbers: while a symmetrical distribution has whiskers of equal length and the median sits exactly in the middle, a skewed distribution means one tail is longer than the other.

When the median is closer to the bottom of the box, and if the whisker is shorter on the lower end of the box, then the distribution is positively skewed (skewed right). A right-skewed distribution will have the mean to the right of the median.

When the median is closer to the top of the box, and if the whisker is shorter on the upper end of the box, then the distribution is negatively skewed (skewed left). A left-skewed, negative distribution will have the mean to the left of the median.

Kurtosis is a measure of whether the data is heavy-tailed or light-tailed relative to a normal distribution.

Histograms and boxplots are very similar in that they both provide insight into the shape of the distribution and help determine if the distribution is symmetric or skewed.

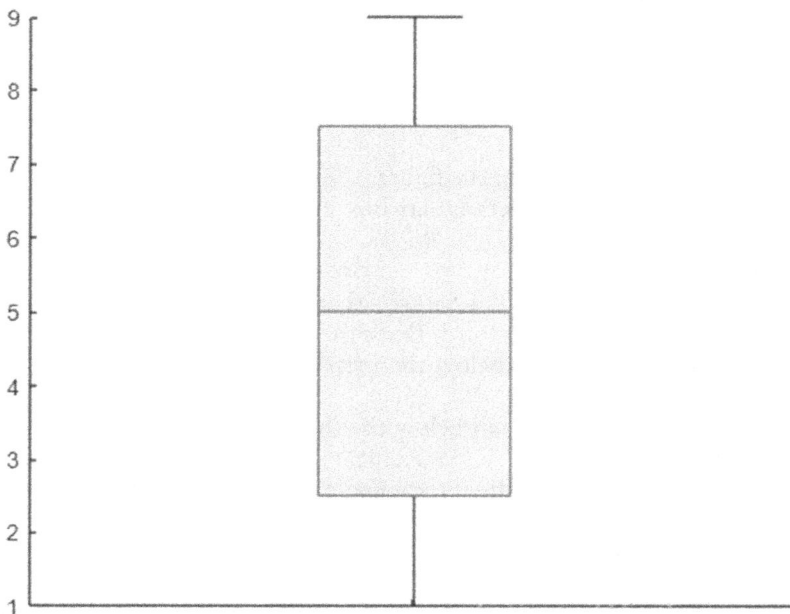

Figure 2.80

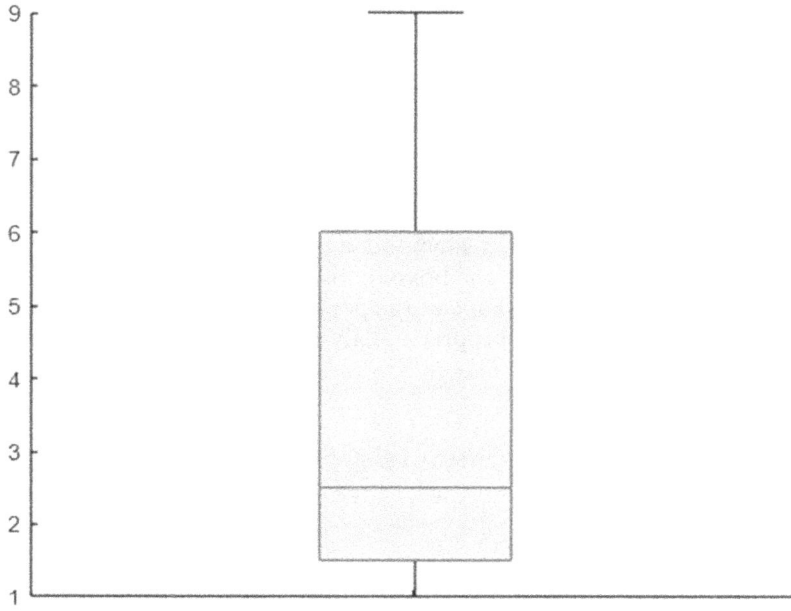

Figure 2.81

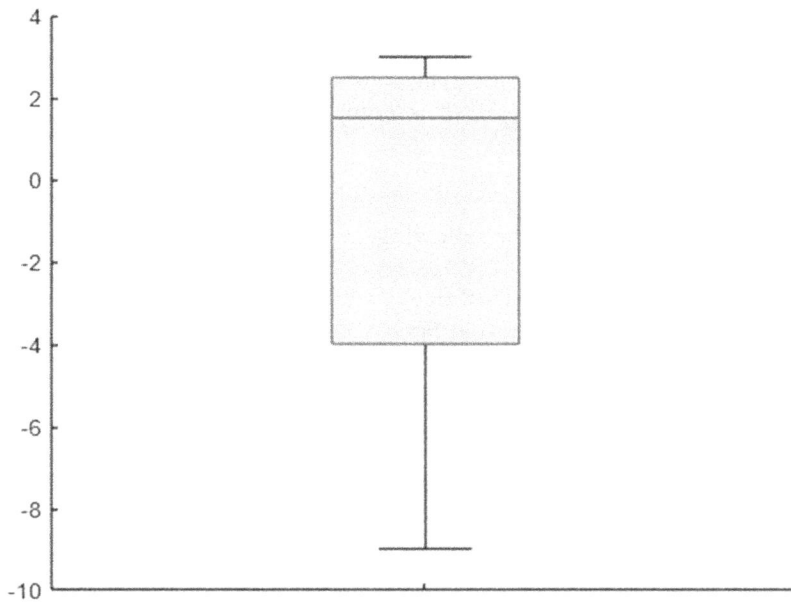

Figure 2.82

Recall that in a histogram, if the data are *symmetric*, they have about the same shape on either side of the middle. In other words, if you fold the histogram in half, it looks about the same on both sides.

In a symmetric distribution, the mean and median are nearly the same, and the two whiskers are almost the same length.

The box length gives an indication of the sample variability, and the line across the box shows where the sample is centered.

For a symmetric distribution, long whiskers, relative to the box length, can betray a heavy-tailed population and short whiskers, a short-tailed population. So, provided the number of points in the sample is not too small, the boxplot also gives us some idea of the "shape" of the sample and, by implication, the shape of the population from which it was drawn. This is all important when considering appropriate analyses of the data.

2.3.5.1 MATLAB

```
boxchart(xgroupdata, ydata, 'PlotStyle', ' ', 'BoxStyle', ' ',
'MedianStyle', ' ', )
```

2.3.5.2 'PlotStyle'

Plot style is specified as one of the following.

- 'traditional' (default)
- 'compact'

2.3.5.3 'BoxStyle'

Box style is specified as one of the following.

- 'outline': this is the default if 'PlotStyle' is 'traditional'.
- 'filled': this is the default if 'PlotStyle' is 'compact'.

2.3.5.4 'MedianStyle'

Median style is specified as one of the following.

- 'line': this is the default when 'PlotStyle' is 'traditional'.
- 'target': this is the default when 'PlotStyle' is 'compact'.

2.3.5.5 *Example*

```
data = readtable("500_Height_Weight.csv", VariableNamingRule="preserve");
Height = data.Height;
Weight = data.Weight;
tiledlayout(1,2)
% Left axes
ax1 = nexttile;
boxchart(ax1,Height, 'BoxFaceColor','red')
ylabel(ax1,'Height(cm)')
title(ax1, "Height of 500 People")
% Right axes
ax2 = nexttile;
boxchart(ax2,Weight, 'BoxFaceColor','blue')
ylabel(ax2,'Weight(kg)')
title(ax2,'Weight of 500 People')
```

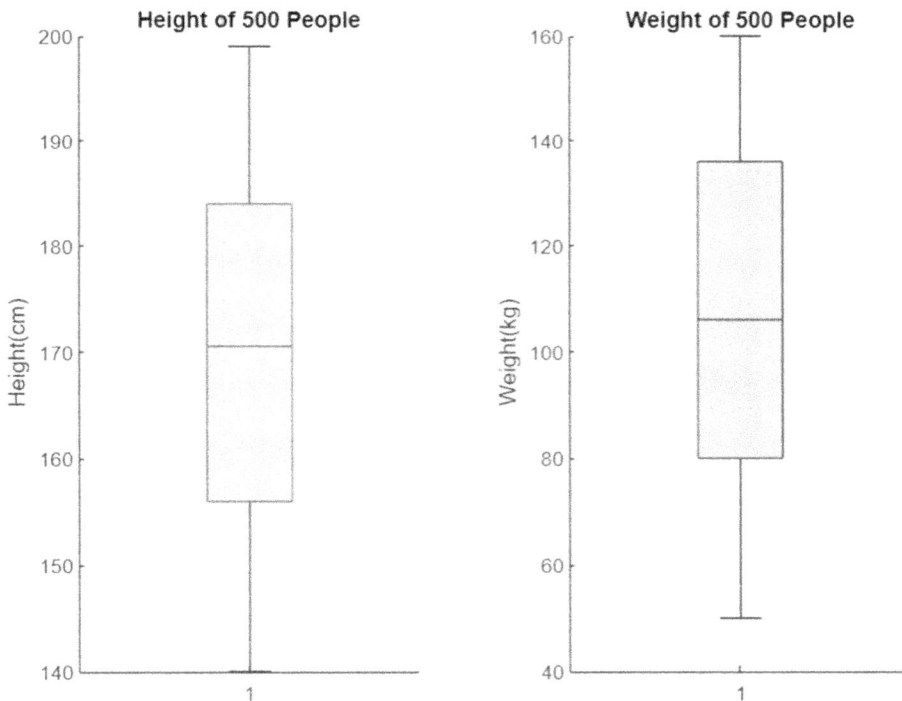

Figure 2.83

Here we display a side-by-side pair of box charts using the tiledlayout and nexttile functions. The first boxplot represents the distribution of heights while the second one displays the weights of 500 people. The function boxplot creates a visual representation of the data but does not return numeric values which we discussed. To calculate the relevant summary statistics for the sample data, we use the following functions:

- min(data): find the minimum value in the sample data.
- max(data): find the maximum value in the sample data.

- median(data): find the median value in the sample data.
- quantile(data, [0.25 0.75]): find the quantile values in the sample data.
- iqr: find the interquartile range in the sample data.
- grpstats: calculate summary statistics for the sample data, organized by group.

Hence our statistics for the sample data is:

```
Minimum =

   140


Maximum =

   199


Median =

   170.5000


Quantile =

   156    184


IQR =

   28
```

Similarly if we estimate the statistics for the weight:

```
Minimum =

   50


Maximum =

   160


Median =

   106


Quantile =

   80    136


IQR =

   56
```

Let's look at the data real quick:

```
data =

  500×4 table

      Gender        Height      Weight      Index
    _____      _____      _____      _____

    {'Male'  }       174          96          4
    {'Male'  }       189          87          2
    {'Female'}       185         110          4
    {'Female'}       195         104          3
    {'Male'  }       149          61          3
    {'Male'  }       189         104          3
    {'Male'  }       147          92          5
    {'Male'  }       154         111          5

        :            :           :           :

    {'Female'}       198          50          0
    {'Female'}       170          53          1
    {'Male'  }       152          98          5
    {'Female'}       150         153          5
    {'Female'}       184         121          4
    {'Female'}       141         136          5
    {'Male'  }       150          95          5
    {'Male'  }       173         131          5

        Display all 500 rows.
```

```
data = readtable('500_Person_Gender_Height_Weight_Index.csv');
Height = data.Height;
Weight = data.Weight;
Gender = data.Gender;
tiledlayout(1,2)
```

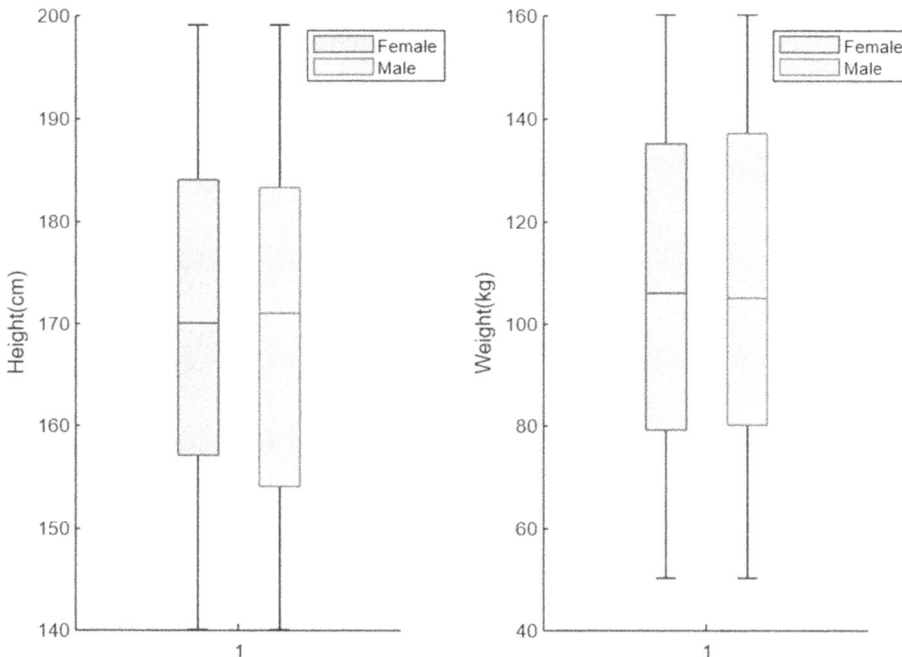

Figure 2.84

```
% Left axes
ax1 = nexttile;
boxchart(ax1,Height,'GroupByColor',Gender)
ylabel(ax1,'Height(cm)')
legend
% Right axes
ax2 = nexttile;
boxchart(ax2,Weight,'GroupByColor',Gender)
ylabel(ax2,'Weight(kg)')
legend
```

In the first set of axes, we display two box charts of height values, one for the gender male and the other for females. Then in the second set of axes we do the same for the weight.

2.3.5.6 Example

Let's say we have a dataset of tensile strength measurements for three different materials: Material A, Material B, and Material C.

data =

302×3 table

A	B	C
705	1019	974.4
720	1033.5	975.84
740	1041.1	976.32
740	1054.9	977.04
755	1063.1	978.24
765	1079	980.64
770	1109.3	980.88
770	1128.8	982.32
:	:	:
NaN	2149.7	1855.2
NaN	2156.6	1891.2
NaN	2177.2	1903.2
NaN	2197.9	1905.6
NaN	2232.4	1919.8
NaN	NaN	1924.8
NaN	NaN	1962
NaN	NaN	1982.4

Display all 302 rows.

Here N/A represents an empty cell which indicates that the sample sizes of the three materials are not equal.

```
data = readtable('3_Materials_Strengths.csv');
% Extract the data for each material
materialA = data.A;
materialB = data.B;
materialC = data.C;
colorA = [1 0 0];
colorB = [0 0.4740 0];
colorC = [0 0 1];
% Create a figure
figure;
% Create the boxplots for each material with custom colors
boxplot([materialA, materialB, materialC], 'Labels', {'A', 'B', 'C'},
    'Colors', [colorA; colorB; colorC]);
% Set labels and title
xlabel('Materials');
ylabel('Strength');
title('Strength Comparison');
```

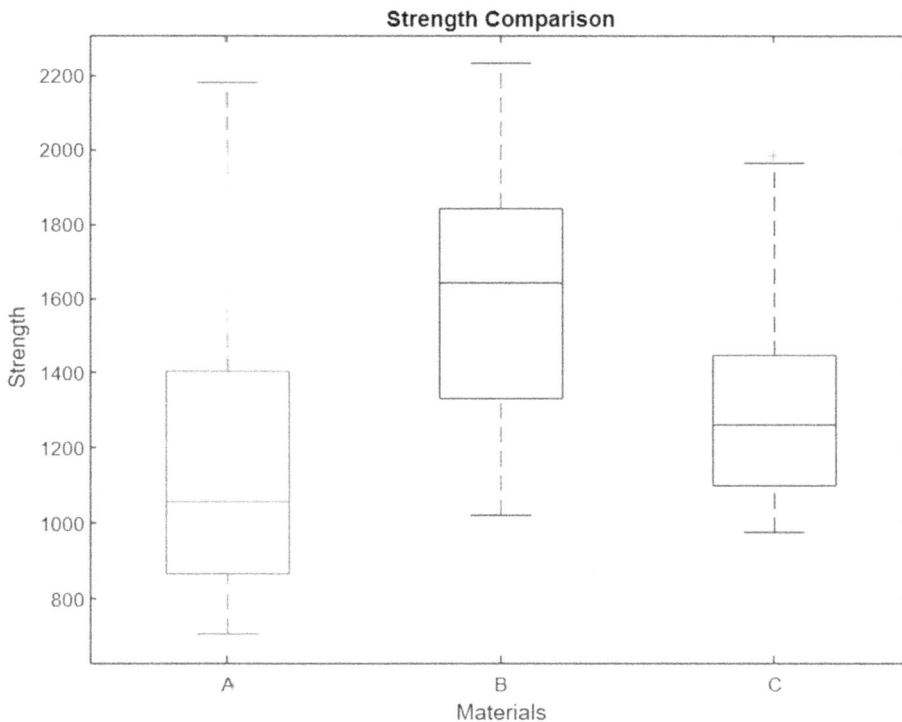

Figure 2.85

Let's analyze each of the data individually. Take a look at boxplot A and boxplot C representing Material A and Material C. The top whisker is longer than the bottom whisker, indicating an asymmetric distribution of the data—more specifically positively skewed data. If we draw a histogram we see

```
histogram(materialA, 'BinLimits',[700,2200],'BinWidth',200,
    'FaceColor',[0.3010 0.7450 0.9330], 'LineWidth',2)
xlabel("Tensile Strength")
```

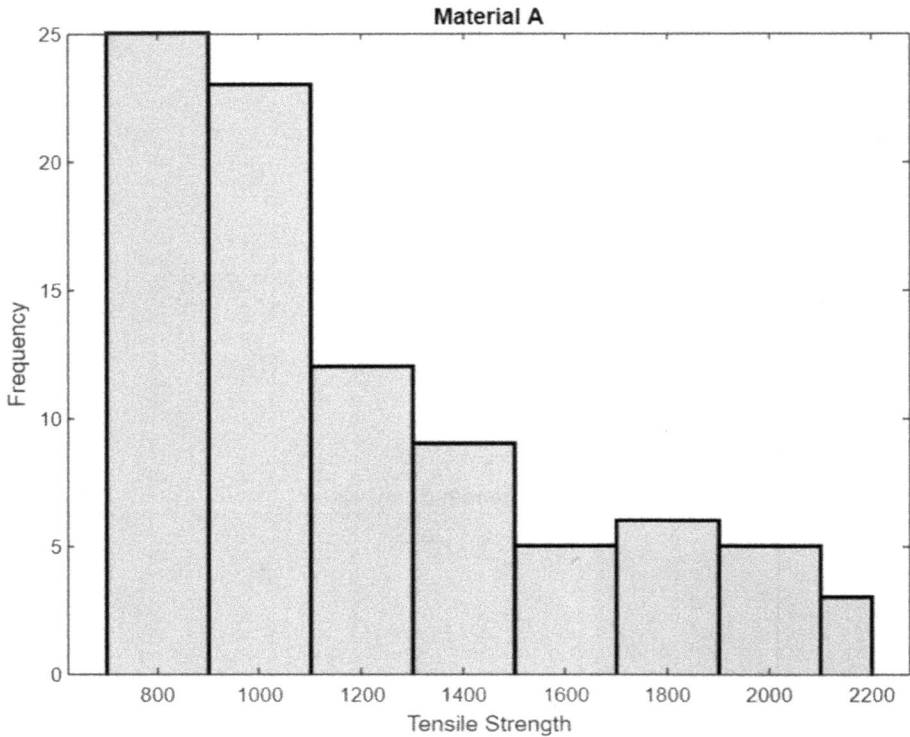

Figure 2.86

```
ylabel("Frequency")
title("Material A")
```

2.3.5.7 Example

```
data = readtable('auto-mpg.csv', VariableNamingRule='preserve');
MPG = data.mpg;
Cylinder = data.cylinders;
colors = ['r', 'g', 'b', 'k', 'm'];
% Create a figure
figure;
% Create the boxplots for each material with custom colors
boxplot(MPG, Cylinder, 'Colors', colors);
% Set labels and title
xlabel('Number of cylinders')
ylabel('Fuel efficiency');
title('Mileage Data');
```

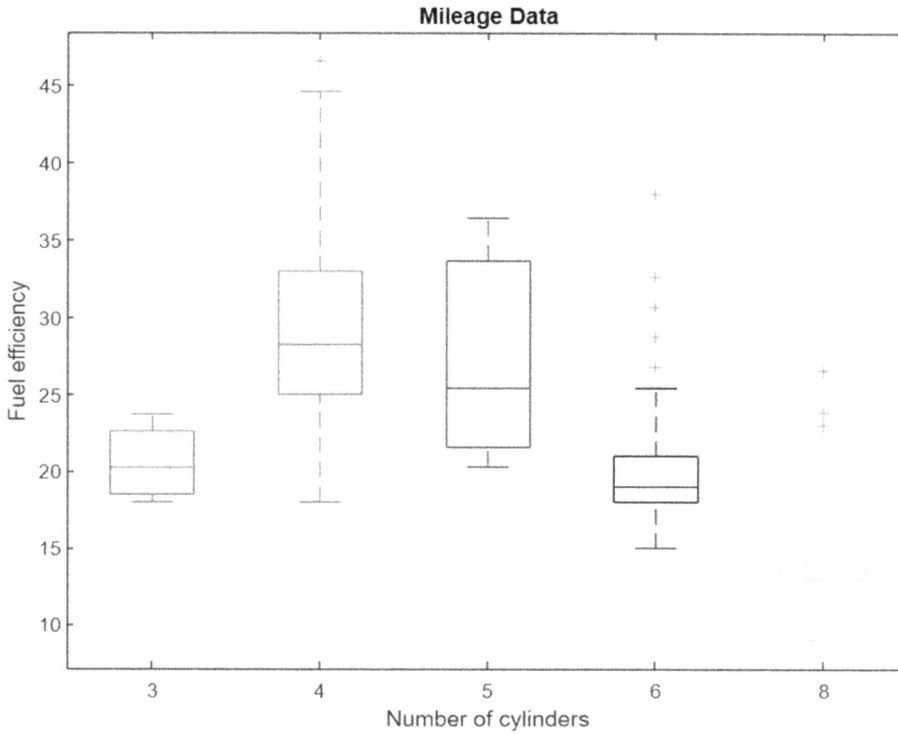

Figure 2.87

In this chapter we discussed the seven most commonly used techniques for graphical representation in MATLAB. However these are just a few examples of graphical representation techniques in MATLAB. Depending on your specific data and requirements, there may be other specialized functions or visualization methods available in MATLAB.

Chapter 3

Random variable and probability distribution

Recall the definition of statistics—the four-step process that encompasses the field. Now, so far we've discussed the first two steps:

Producing data—types of data we can encounter and how they are obtained.
Data visualization—tools that help us get a first feel for the data by exposing their features using graphs and numbers.

Also remember that the way statistics works is that we use a sample data relevant to the problem at hand, to learn about the larger population from which it was drawn. Ideally, the sample should be random so that it represents the population well. Now when we use any of the four techniques we learned in the previous chapter to collect our sample data, we assume that there is no inherent bias in this sampling procedure. This means that the selection of a sample from a population is purely based on the principle of randomization, that is, random selection or chance where one random sample may give a fairly accurate representation of the population in research, whereas another might be "off" purely because of chance.

So regardless of the methods used one can never be 100% sure that the sample represents the population completely. Moreover, just by looking at the samples it is almost impossible to determine how much the sample data differs from the population. This uncertainty is where probability comes into the picture. We use probability to quantify how much we expect random samples to vary. This gives us a way to draw conclusions about the population in the face of the uncertainty that is generated by the use of a random sample.

This proves that probability is a fundamental concept used extensively in statistical and data analysis. Let's discuss the areas where the probability is applied in statistical and data analysis:

- Probability distribution: a probability distribution describes the likelihood of different outcomes in a random variable (more in this chapter).
- Probability mass function (PMF): for discrete random variables, the probability mass function assigns probabilities to each possible outcome. It gives the probability of observing a specific value of the random variable.
- Probability density function (PDF): for continuous random variables, the probability density function represents the relative likelihood of different values. Unlike the probability mass function, it gives the density rather than the probability at a specific value.
- Cumulative distribution function (CDF): the cumulative distribution function gives the probability that a random variable takes on a value less than or equal to a given value.
- Sampling and estimation: probability is used to sample from populations and estimate population parameters based on sample data. Techniques such as random sampling and probability-based sampling designs are used to ensure representative and unbiased samples.

DOI: 10.1201/9781003399582-3

- Hypothesis testing: probability plays a crucial role in hypothesis testing. It allows statisticians to assess the likelihood of observing certain data under a specific hypothesis and make decisions about the validity of the hypothesis.

In this chapter we will introduce the concept of probability and how it's used in probability distribution and describe the functions and discuss how to estimate the parameters of the various distributions in MATLAB using those functions.

3.1 WHAT IS PROBABILITY?

Probability can be defined as the possibility of an outcome from all possible events. When we think of probability, the first example we think of is the coin flip problem. When we flip a coin in the air, what is the possibility of getting a head? We all know that it is 1/2. But how and why?

The answer to this question is based on the number of all possible outcomes and of the outcome that we want which is heads. Here the possibility is either heads or tails; hence we have only two possible outcomes to this, and out of those two we are looking for heads, so there is only one outcome we want from the two. So, the probability of a head is 1/2.

As mentioned, the probability is based on the number of all possible outcomes and of the outcome that we want. This favorable outcome is called an event. An event is a subset of the sample space which is a set of all possible outcomes of an experiment. The sample space and event are denoted by the symbols S and E where $E \subseteq S$. So for example for rolling a die, our sample space, $S = \{1,2,3,4,5,6\}$, so:

subset E_1 could be $\{1,3,5\}$ which represents the set of odd numbers
or $\{2,4,6\}$ which represents the set of even numbers.

So whatever the E_1 is we can see the number of events is $n(E_1) = 3$.

Now that we are familiar with the two most important probability terms, we now can conclude that probability is the ratio of the number of times an outcome occurs compared to all the possible outcomes or probability $P(E) = \dfrac{\text{number of ways the desired outcome can occur}}{\text{number of all the possible outcomes i.e sample space}}$
$= \dfrac{n(E)}{n(S)}$.

The probability of an event always lies between 0 and 1 and hence is always positive. Therefore, it can be defined as the percentage of a particular event. Zero indicates the impossibility of an event, in other words, it can never occur. On the other hand, 1 indicates the certainty of a particular event; it will surely happen, for example, the probability of the sun rising from the east.

Now how can we apply the concept of probability in engineering? Let's look at the following situations:

- A system is composed of three components and only operates when all these components function. The probability of failure for each component is 0.21, 0.34, and 0.42. What is the probability that the entire system fails?
  What is the probability that at least one of the components fails?
- In a mechanical assembly process, there are two critical characteristics that need to be within specified limits:
  Event A: surface roughness within a specified range

Event B: friction coefficient within a specified range

Given:

The probability of the surface roughness is within the specified range: P(A) = 0.68.

The probability of the friction coefficient given that the surface roughness is within the specified range is P(B)=0.85.

What is the probability of both surface roughness and the friction coefficient being within specified ranges?

- In a manufacturing process, the diameter of a component is known to follow a normal distribution with a mean of 10 cm and a standard deviation of 0.2 cm. What is the probability that a randomly selected component has a diameter greater than 10.5 cm?
- In a quality control process, the weight of a product is measured, and it is known to follow a normal distribution with a mean of 500 grams and a standard deviation of 10 grams. What is the probability that a randomly selected product has a weight between 490 grams and 510 grams?
- A company is conducting a reliability study on a new product. The product has been tested for 500 hours, and 10 failures were observed. Based on this data, estimate the failure rate of the product using a Poisson distribution.
- A manufacturer produces light bulbs, and the lifetimes of the bulbs follow an exponential distribution with a mean lifetime of 1000 hours. What is the probability that a bulb will last more than 1500 hours?
- An engineering team is conducting a Monte Carlo simulation to estimate the reliability of a bridge under different loading conditions. Random samples of loads are generated from a known probability distribution, and the bridge's response is evaluated. What is the probability that the bridge's deflection exceeds a certain threshold value?
- An automotive company is analyzing the failure times of a specific engine part. The failure times follow a Weibull distribution. Determine the parameters of the Weibull distribution using the observed failure times and estimate the probability of failure within a certain time interval.
- A manufacturing process produces cylindrical rods, and the length of the rods follows a uniform distribution between 10 cm and 15 cm. What is the probability that a randomly selected rod has a length between 12 cm and 13 cm?

In the first problem we see three events E_1, E_2 and E_3 where

E_1 = failure of first component and $P(E_1) = 0.21$

E_2 = failure of second component and $P(E_2) = 0.34$

E_3 = failure of third component and $P(E_3) = 0.42$

None of these events are dependent on each other—the outcome of one event is unaffected by the outcome of the others. Hence these events are called independent events whose occurrence has no effect on the occurrence or non-occurrence of another event. To put it another way, the likelihood of occurrence of one event does not affect the likelihood of occurrence of the other event. Flipping a coin twice is an example of independent events. We know that the probability of the coin landing on heads on the first flip is 1/2 and since the second event is independent of the first then the probability of it landing on heads on the second flip is also 1/2. This is because the probabilities of the two outcomes do not depend on each other. If we now flip the coin let's say 500 times the probability of getting heads in every single flip will still be 1/2. This means the probability of an event will remain the same no matter how many times the same experiment is done. The probability for independent events is computed as the product of the probabilities of all the independent events, so the probability of getting two heads is: $(1/2)(1/2) = 1/4$.

So therefore when two or more events are independent, then

$$P(A \cap B \cap ...Z) = P(A)P(B).....P(Z)$$

Mathematically, if events E_1, E_2, and E_3 are independent, the probability of all three events occurring can be calculated by multiplying their individual probabilities:

$$P(E_1 \text{ and } E_2 \text{ and } E_3) = P(E_1) * P(E_2) * P(E_3)$$

So going back to the same question, let's analyze the situations where the entire system fails.

It has already been mentioned that a defect in any one of the components can result in the failure of the entire system. So in situations like:

The failure of only the first component: $P(E_1) \ *(1 - P(E_2)) * (1 - P(E_3))$

$$= 0.21 * (1 - 0.34) * (1 - 0.42) = 0.080388$$

The failure of only the second component: $(1 - P(E_1)) \ * P(E_2) * (1 - P(E_3))$

$$= (1 - 0.21) * 0.34 * (1 - 0.42) = 0.155788$$

The failure of only the third component: $(1 - P(E_1)) \ * (1 - P(E_2)) * P(E_3)$

$$= (1 - 0.21) * (1 - 0.34) * 0.42 = 0.218988$$

The failure of components 1 and 2: $P(E_1) \ * P(E_2) * (1 - P(E_3))$

$$= 0.21 * 0.34 * (1 - 0.42) = 0.041412$$

The failure of components 1 and 3: $P(E_1) \ *(1 - P(E_2)) * P(E_3)$

$$= 0.21 * (1 - 0.34) * 0.42 = 0.058212$$

The failure of components 2 and 3: $(1 - P(E_1)) \ * P(E_2) * P(E_3)$

$$= (1 - 0.21) * 0.34 * 0.42 = 0.112812$$

The failure of all components: $P(E_1)) \ * P(E_2) * P(E_3)$

$$= 0.21 * 0.34 * 0.42 = 0.029988$$

To calculate the probability that the entire system fails we sum up all the probabilities of the occurrences that can cause the system to fail; hence we get

$$= 0.080388 + 0.155788 + 0.218988 + 0.041412 + 0.058212 + 0.112812 + 0.029988$$

$$= 0.697588$$

Here the idea of the sum of probabilities is typically used in the context of mutually exclusive events. When events are mutually exclusive, it means that they cannot occur simultaneously. Think of the flipping of a coin. Let's say event A is the coin landing on heads and B is landing on tails. In a single flip of a coin, the coin can either land on heads or on tails. It cannot land on both. So we could say that their intersection A ∩ B is null or impossible, making it zero. Since the two events have nothing in common, they are called mutually exclusive events. Therefore, if two or more events are **mutually exclusive**, then the probability that they occur at the same time is as follows.

In the case of two events:

$$P(A \cup B) = P(A) + P(B) - P(A \cap B) = P(A) + P(B) - 0 = P(A) + P(B)$$

In the case of three events:

$$P(A \cup B \cup C) = P(A) + P(B) + P(C) - P(A \cap B) - P(A \cap C) - P(B \cap C) + P(A \cap B \cap C)$$

In our example we have a total of seven events which have nothing in common therefore to calculate the probability that the entire system fails, any one of the events occurring is the sum of their individual probabilities.

Let's look at the second example.

Our examples say that the probability of the friction coefficient given that the surface roughness is within the specified range is P(B)=0.85. What does this statement mean? It basically states that the probability of the friction coefficient if and only if the surface roughness is within the specified range is 0.85. This is an example of a dependent event since the probability of the friction coefficient is affected by the probability of the surface roughness. Realize that our probability of the friction coefficient would have been different if the surface roughness were within the specified range. Such events are represented by

P(friction coefficient | surface roughness is within the specified range) = $P(B \mid A)$

Mathematically dependent events are given by

$$P(A \cap B) = P(A) \times P(B \mid A)$$

We want to calculate the probability of both surface roughness and friction coefficient being within the specified ranges. This means

$$P(\text{surface roughness} \cap \text{friction coefficient}) = P(A \cap B)$$

We are given $P(B \mid A) = 0.85$ and P(A) = 0.68. Hence

$$P(A \cap B) = 0.68 \times 0.85 = 0.578$$

Here $P(B \mid A)$ represents the conditional probability of event B given that event A has occurred. In other words, it calculates the probability of event B happening, given that event A has already occurred. Here the vertical line stands for "GIVEN THAT". Therefore the conditional probability is given by

$$P(B \mid A) = \frac{P(A \cap B)}{P(A)} \tag{3.1}$$

Similarly we can define $P(A \mid B)$ as

$$P(A \mid B) = \frac{P(A \cap B)}{P(B)} \tag{3.2}$$

Rearranging Equations 3.1 and 3.2 we can get

$$P(A \cap B) = P(B).P(A \mid B) = P(A).P(B \mid A) \tag{3.3}$$

Then we have mutually **inclusive events** which *can* occur at the same time. For example, let E_1 be the event that a die lands on an even number and let E_2 be the event that a die lands on a number less than four. From the given information we can define the sample space for the two events as $E_1 = \{2,4,6\}$ and $E_2 = \{1,2,3\}$. Notice that there is an overlap between the two sample spaces. They both include 2. Thus, events E_1 and E_2 are mutually inclusive because they can both occur at the same time. It's possible for the die to land on a number that is even *and* is less than four.

Therefore, we can conclude that if the events are **mutually inclusive**, then the probability that they occur simultaneously will be some number greater than zero. Hence

$$P(A \cup B) = P(A) + P(B) - P(A \cap B)$$

where $P(A \cap B) > 0$

Similarly we can also calculate the probability for three **mutually inclusive** events A, B, and C:

$$P(A \cup B \cup C) = P(A) + P(B) + P(C) - P(A \cap B) - P(A \cap C) - P(B \cap C) + P(A \cap B \cap C)$$

3.1.1 Independent events

Independent events are events where the occurrence or non-occurrence of one event does not affect the probability of the other event. In other words, the outcome of one event has no influence on the likelihood of the other event.

3.1.2 Dependent events

Dependent events are events where the occurrence or non-occurrence of one event affect the probability of the other event. In other words, the outcome of one event provides information about the likelihood of the other event. The probability of the second event is conditional on the outcome of the first event.

3.1.3 Mutually exclusive events

Mutually exclusive events are events that cannot occur together. If one event happens, then the other event cannot occur simultaneously. The events have no common elements or outcomes.

3.1.4 Mutually inclusive events

Mutually inclusive events are events that can occur together. It means that the events have some elements or outcomes in common, and they can happen simultaneously.
Some of the properties of conditional probability include:

$$P(A \mid B \cap C) = \frac{P(A \cap B \mid C)}{P(B \mid C)}$$

$$P(A \cup B \mid C) = P(A \mid C) + P(B \mid C) - P(A \cap B \mid C)$$

$$P(\sim A \mid B) = 1 - P(A \mid B)$$

Two events and are termed as independent if $P(A \mid B) = P(A)$

Bayes' theorem is also known as the extended version of conditional probability which states when a sample is a disjoint union of events, and event A overlaps this disjoint union, then the probability that one of the disjoint partitioned events is true given A is true is:

$$P(A \mid B) = \frac{P(A \cap B)}{P(B)} \quad \text{from Equation 3.1}$$

$$= \frac{P(A) \cdot P(B \mid A)}{P(B)} \quad \text{from Equation 3.3}$$

Equivalently,

$$P(A_i \mid B) = \frac{P(A_i) \cdot P(B \mid A_i)}{P(B)}$$

Similarly $P(B \mid A) = \dfrac{P(A \cap B)}{P(A)}$ from Equation 3.2

$$= \frac{P(B) \cdot P(A \mid B)}{P(A)} \quad \text{from Equation 3.3}$$

$$P(B_i \mid A) = \frac{P(B_i) \cdot P(A \mid B_i)}{P(A)}$$

3.1.5 Law of total probability

Suppose the sample space Ω is divided into three disjoint events B1, B2, B3. Then for any event A:

$$P(A) = P(A \cap B_1) + P(A \cap B_2) + P(A \cap B_3)$$

$$\text{from eq } (3) = P(B_1)P(A \mid B_1) + P(B_2)P(A \mid B_2) + P(B_3)P(A \mid B_3)$$

therefore $P(A) = \Sigma P(B_k) P(A \mid B_k)$

Similarly we can say $P(B) = \Sigma P(A_k) P(B \mid A_k)$

The top equation says "if A is divided into three pieces then P (A) is the sum of the probabilities of the pieces". The bottom Equation 3.3 is called the law of total probability. It is just a rewriting of the top equation using the multiplication rule.

3.1.6 Using both the first Bayes' theorem and the law of total probability

$$P(A_k \mid B) = \frac{P(A_i) \cdot P(B \mid A_i)}{P(B)} = \frac{P(A_i) \cdot P(B \mid A_i)}{\Sigma P(A_k) P(B \mid A_k)}$$

Similarly,

$$P(B_k \mid A) = \frac{P(B_i) \cdot P(A \mid B_i)}{P(A)} = \frac{P(B_i) \cdot P(A \mid B_i)}{\Sigma P(B_k) P(A \mid B_k)}$$

3.2 RANDOM VARIABLE

Now that we have a grip on the concept of probability in the previous section, here in this section we will introduce a topic related to probability.

3.2.1 What is a random variable?

Notice that we used probability to calculate the outcome of a certain event. Now a random variable is a variable whose possible values depend on the outcomes of a certain random experiment. Not clear?

Let's look at an example. Say we have a coin and we assign the number of heads as our random variable. Now if we toss it let's say five times then the possible numbers of heads we can get in these five tosses are 0, 1, 2, 3, 4, and 5. This implies that a random variable can take on different real values. This really makes it clear why it is called "random" since this experiment may result in any one of the six values of heads.

In an engineering context:

- Number of defects in a manufactured electronic chip.
- Number of software bugs in a program.
- Number of components in a system that exceed a certain temperature threshold.
- Number of network outages in a telecommunications system.
- Temperature of a chemical reactor.
- Pressure within a gas pipeline.
- Velocity of fluid flow in a channel.
- Length of a steel rod.
- Power output of a wind turbine.

Since there could be any values, a random variable is represented using uppercase letters. More commonly, X is used to denote a random variable.

Consequently $P(X = x)$ represents the probability of the random variable X taking on any particular value x.

3.2.2 Types of random variables

Random variables are classified into discrete and continuous variables. The main difference between the two categories is the type of possible values that each variable can take.

A discrete random variable can have an exact value—for example the number of defects, the number of software bugs, the number of components.

On the other hand, a continuous random variable will lie within a specific range—for example temperature, pressure, velocity, length. A continuous random variable can take on any value within a specified range or interval, including decimal values and an infinite number of possible values in between.

Let's discuss each of the variables in detail.

3.2.2.1 Discrete random variables

A **discrete random variable** is one which may take on only a countable or finite number of distinct values like 0, 1, 2, 50, 1000, etc. Since discrete.... $P(X = x)$ represents the probability of the random variable X taking on any particular value x.

Properties:

- The probability of each value is between 0 and 1, $0 \leq P(X = x) \leq 1$.
- Sum of all probabilities $\sum P(x) = 1$.
- Examples:
  - Number of failed inspections in a quality control process.
  - Number of errors in a communication channel.
  - Number of power outages in a residential area.
  - Number of pumps operating in a water distribution system.
  - Number of units produced in a manufacturing batch.
  - Number of times a machine undergoes maintenance.

3.2.2.1.1 A discrete uniform distribution?

A discrete uniform distribution is when all the distinct random variables have the exact same probability values, so everything is constant or just a number. In a uniform probability distribution, all random variables have the same or uniform probability; thus, it is referred to as a discrete uniform distribution. For example, when we roll a die all of the possible outcomes (1–6) have an equal chance of occurring. Therefore, the probability of any one number on the die is the same or uniform.

3.2.2.2 Continuous random variables

A continuous variable on the other hand can take on an infinite number of possible values like 0.03, 1.2374553, etc. While a discrete variable, as mentioned earlier, only takes on countable whole numbers, a continuous variable can include values with infinite decimal places. Like discrete variables, continuous variables also possess certain properties:

- Probability of **any one particular outcome** is 0.
- The area under the curve (i.e. the indefinite integral) is 1.
- Examples include height, weight, and temperature.

An easy way to make the distinction between a discrete and a continuous variable is that discrete variables are usually whole numbers with no decimals. Continuous variables on the other hand frequently take the form of decimals so could be any real numbers. For instance, the number of people in a classroom is a discrete variable because it's always a whole number, while a person's height would be continuous since it can go up to any decimal place.

3.2.2.3 MATLAB

3.2.2.3.1 Generate random variables in MATLAB?

The base MATLAB package has two functions for generating the two variables.

  1) **The function** *rand* generates uniformly distributed random variables in the interval (0 1):
     ```
     r = rand;
     ```
     Function with no arguments returns a pseudorandom value drawn from the standard uniform distribution on the open interval (0,1):
     ```
     r = rand(n);
     ```
     This returns an n-by-n matrix of uniformly distributed random numbers between 0 and 1.

3.2.2.3.2 Example

$$r = rand(5)$$

```
r =

    0.8147    0.0975    0.1576    0.1419    0.6557
    0.9058    0.2785    0.9706    0.4218    0.0357
    0.1270    0.5469    0.9572    0.9157    0.8491
    0.9134    0.9575    0.4854    0.7922    0.9340
    0.6324    0.9649    0.8003    0.9595    0.6787
```

To create a graphical representation of the rand function in MATLAB, we can generate a histogram of the random numbers it produces.

```
random_numbers = rand(50);
histogram(random_numbers, 'EdgeColor', 'k', 'FaceColor', [0.9290
                   0.6940 0.1250], 'LineWidth', 2);
% Add labels and title to the plot
xlabel('Random Numbers');
ylabel('Frequency');
title('Histogram of Random Numbers generated by rand');
  r = rand(m, n);
```

This returns an m-by-n matrix containing independent pseudorandom values drawn from the standard uniform distribution on the open interval (0,1).

3.2.2.3.3 Example

```
r = rand(3, 2)

                 r =

              0.7577    0.6555
              0.7431    0.1712
              0.3922    0.7060
```

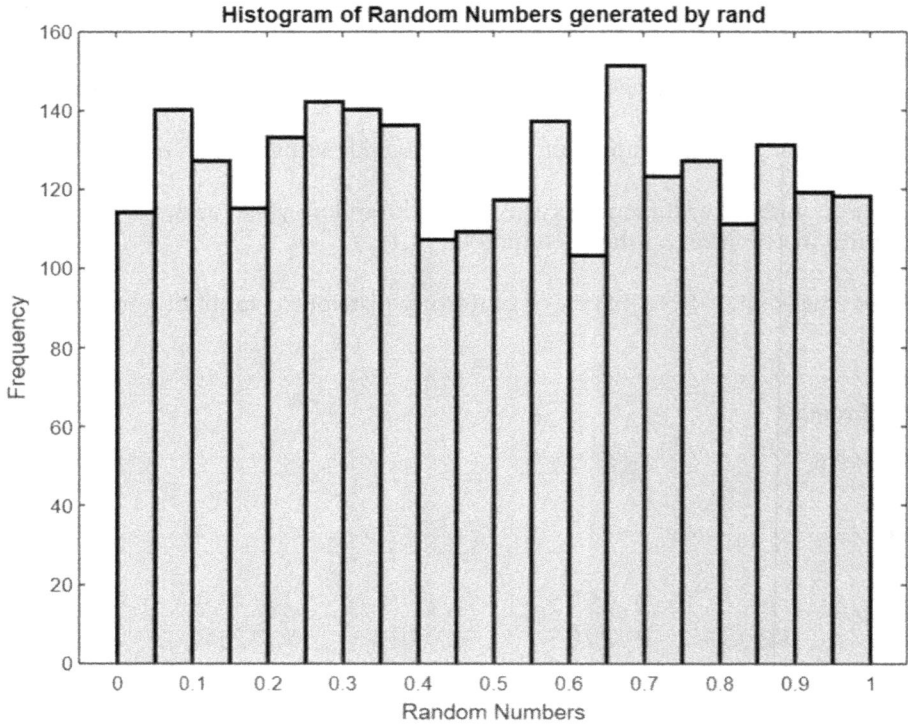

Figure 3.1

```
random_numbers = rand(1000,1);
histogram(random_numbers, 'EdgeColor','k','FaceColor',
                    [0.6350 0.0780 0.3840], 'LineWidth', 2);
% Add labels and title to the plot
xlabel('Random Numbers');
ylabel('Frequency');
title('Histogram of Random Numbers generated by rand');
r = rand([z, m, n])
```

This returns z number of m-by-n array of random numbers.

3.2.2.3.4 Example

$$r = rand\left(\left[3,2,3\right]\right)$$

This creates a 3-by-2-by-3 array of random numbers.

```
r = a + (b - a) * rand(m , n)
```

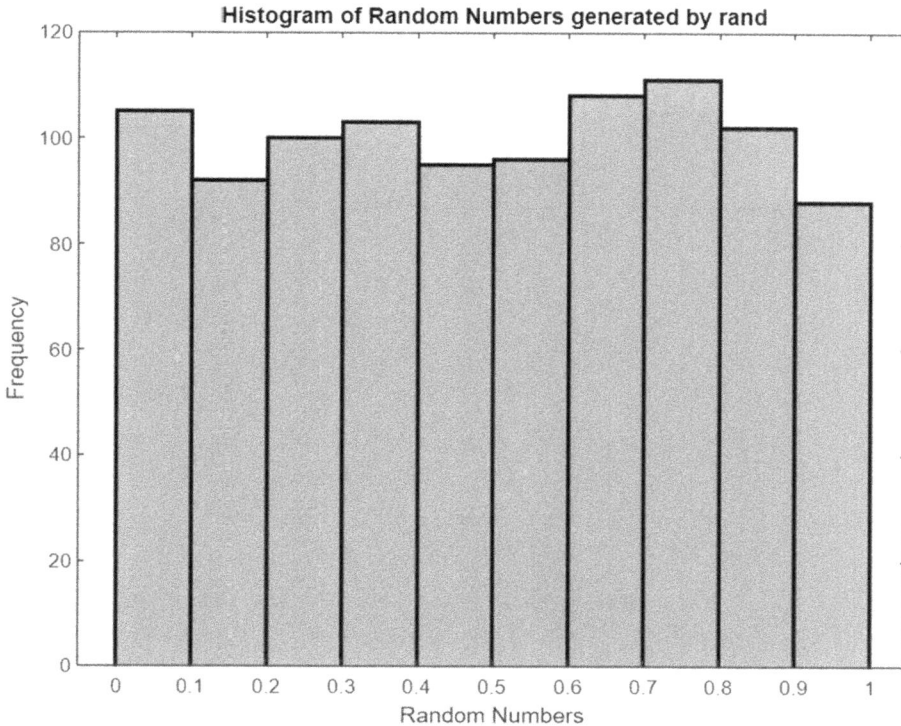

Figure 3.2

3.2.2.3.5 Example

Generate a 10 -by- 1 column vector of uniformly distributed numbers in the interval (−5,5).

```
r = -5 + (5 + 5) * rand(1000, 1)
histogram(r, 'EdgeColor',"blue", 'FaceColor', [0 0.6780 0.7840],
          'LineWidth', 2)
% Add labels and title to the plot
xlabel('Random Numbers');
ylabel('Frequency');
title('Histogram of Random Numbers generated by rand');
```

2) There is another function *randi* which also generates uniformly distributed random variables but instead of any real numbers it only returns integers. The function generates numbers between the interval 1 and *max* where the *max* indicates the largest integer in the interval.

```
r = randi(max)
```

This returns a pseudorandom scalar integer between 1 and max.

```
r = randi(max, n)
```

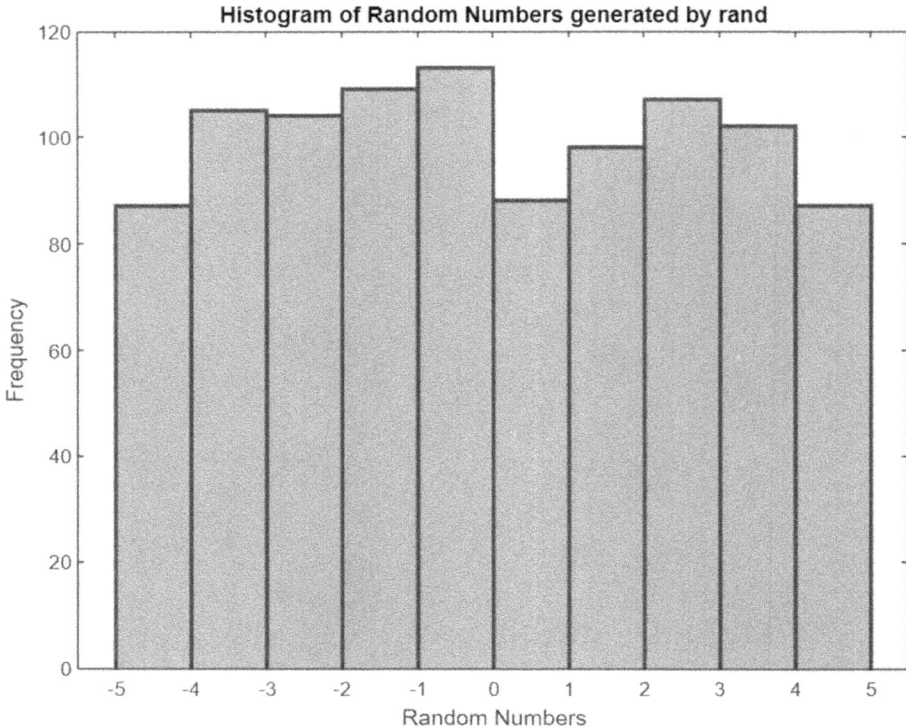

Figure 3.3

This returns an n-by-n matrix of pseudorandom integers drawn from the discrete uniform distribution on the interval [1, max].

```
r = randi(max, m, n)
```

This returns an m-by-n matrix of pseudorandom integers drawn from the discrete uniform distribution on the interval [1, max].

3.2.2.3.6 Example

```
r = randi(15, 1, 10)
```

This generate a row vector of 10 integer numbers from 1 to 15.

```
r =

     13    8    2    6   14    3   12   15    3   14

r = randi([min, max], m, n)
```

All the functions mentioned above generate integers drawn from the discrete uniform distribution on the interval [1, *max*]. Now if we want to change the starting point of our interval, we can simply specify the point we want our interval to start along with the endpoint, max, as arguments in square brackets, and the rest of the syntax remains the same as the above syntaxes.

3.2.2.3.7 Example

```
r = randi([-5, 5], 10, 1)
```

This generates a column vector of 10 integer numbers from −5 to 5.

```
r =

    -2
     2
    -5
    -2
     4
    -1
    -4
     0
    -1
     3
```

We have already discussed the two main functions which generate random numbers for the two variables—discrete and continuous. However, there are different types of discrete and continuous distributions which we will start talking about in the next section and in the upcoming chapters. Now to generate a variable of that particular distribution we will use a root word specifying the distribution along with the letters **rnd** attached to it (there will be more discussion on this in the upcoming chapters). The syntax of the function is

$$r = rootrnd\left(parameter_1, parameter_2, ...parameter_k, m, n\right)$$

Here the **root** is the type of distribution we want with its parameters as its arguments which returns an **m**-by-**n** matrix of uniformly distributed random variables of a specific distribution. Again we will talk more about different types of distributions and their parameters in upcoming sections and chapters.

3.3 PROBABILITY DISTRIBUTIONS

3.3.1 What is a probability distribution?

The probability distribution has a strong relationship with random variables as its definition is based on the term. Hence it was necessary to first get familiar with the term which will now, in turn, make it easier to understand the definition.

So the probability distribution is the statistical function that describes the likelihood of obtaining all possible values that a random variable can take. In other words it tells us how likely it is for the random variable X to take on different values of x. It lists all the possible values and likelihoods that a random variable X can take within a given range. In our previous example we tossed a coin five times and assigned the number of heads as our random variable. Therefore, the probability distribution for the number of heads in five tosses is:

X = no. of heads	0	1	2	3	4	5
$P(X)$	$P(X=0)$	$P(X=1)$	$P(X=2)$	$P(X=3)$	$P(X=4)$	$P(X=5)$

As we can see we used a table which represents a possible outcome (number of heads) that can occur in five coin tosses. In fact probability distribution could be defined as the table or equations showing respective probabilities of different possible outcomes of a defined event or scenario. When we describe the values in the range of a random variable in terms of the probability of their occurrence, we are essentially talking about the probability distribution of the random variable. In other words, the probability distribution of a random variable can be determined by calculating the probability of occurrence of every value in the range of the random variable.

Now keep in mind that probability distributions are theoretical since obtaining infinitely large samples in practice is impossible. They are idealized frequency distributions intended to represent the population from which the sample was taken.

In statistics we use a capital letter to represent a random variable—usually the letter X —and a lower-case letter to represent different values of the random variable X. In this book particularly:

- X represents the random variable 'X'
- $P(X)$ represents the probability of 'X'
- $P(X=x)$ refers to the probability that the random variable X is equal to a particular value, denoted by 'x'. For example, $P(X=2)$ refers to the probability that the random variable X is equal to 1.

Now you must be wondering what the probability of each outcome is. This is determined by the type of probability distribution which depends on the nature of the random variable and the underlying process being modeled. There are two different types of probability distribution depending on the two types of (random) variable we have discussed earlier.

3.3.2 Types of probability distribution

3.3.2.1 Discrete probability distribution

A discrete probability distribution is a probability distribution that represents the probabilities of a discrete random variable. Recall that a discrete random variable is one that can take on a countable set of distinct values. Now if you look back at our coin example the possible values for the number of heads can range from zero (no heads) to the total number of tosses (e.g., five heads in five tosses) which implies that our random variable (number of heads) can only take on a countable set of distinct values. Hence it is a discrete random variable which is clearly represented by discrete probability distribution.

Since, probability in general, by definition, must add up to 1, the summation of all the possible outcomes must sum to 1. Therefore

$$P(X=0)+P(X=1)+P(X=2)+P(X=3)+P(X=4)+P(X=5)=1$$

3.3.2.2 Continuous probability distribution

Unlike discrete variables, the probability of a continuous random variable taking on a specific value is always zero. This is because the number of possible values for a continuous random variable is uncountably infinite, and the probability of any single value is infinitesimally small. Therefore, instead of assigning probabilities to individual values, the probability for a continuous random variable is defined over intervals or ranges of values.

Suppose the tire pressure in an automobile is usually between 0 and 32 psi. The pressure can take any value in the range of 0 to 32 psi. Another example could be the flow rate of a fluid; therefore it is only possible to calculate the probability of a continuous variable within a certain interval, like $P(X \leq k)$ or $P(a \leq X \leq b)$.

3.3.3 Calculating the probabilities in a probability distribution

3.3.3.1 Probability mass function and cumulative density function

There are two main functions associated with a discrete random variable. These are

   i. Probability mass function (PMF)
  ii. Probability distribution function/cumulative distribution function (CDF).

3.3.3.1.1 Probability Mass Function (PMF)

The PMF can be defined as a function denoted by $f(x)$ that is directly used to give the probability of a discrete random variable, X, being exactly equal to some value, x. The formula is given as follows:

$$f(x) = P(X = x)$$

Each probability mass function satisfies the following conditions:

  (i) $f(x) \geq 0$
     This property implies that $f(x)$ is never negative.
  (ii) $\Sigma f(x) = 1$ i.e. for all x
     This property concludes that the sum of $f(x)$ over the entire sample space will always be 1.

3.3.3.1.2 Cumulative density function (CDF)

While the PMF provides the probability of a particular outcome, the cumulative distribution function provides the probability of seeing a result less than or equal to a particular value of the random variable, that is, the probability when X is less than or equal to x, for every value x.

$$F(x) = P(X \leq x)$$

For a discrete random variable, the cumulative distribution function is found by summing up the probabilities.

3.3.3.2 Probability density function and cumulative density function

As we have seen earlier, the probability distribution of a continuous variable is a little different than that of a discrete variable because of the infinite values that a continuous variable could assume, making the probability of the variable equal to one specific value, zero. In other words, continuous random variables are only concerned with a value in a particular interval and not only that particular value since in a continuous variable only value ranges can have a non-zero probability.

Just like discrete variables there are two main functions associated with such a continuous random variable. These are the

   i. Probability density function (PDF)
   ii. Probability distribution function/cumulative distribution function (CDF)

3.3.3.2.1 The probability density function (PDF)

For discrete random variables, we use the probability mass function which is analogous to the probability density function. Like the PMF of a discrete variable, the PDF of a continuous variable is also a function that can be denoted by $f(x)$. However the difference between PDF and PMF is that PDF cannot calculate the probability of X taking on the exact value, x. This is again because the probability of the continuous variable equaling one specific value is always zero. So when working with *continuous variables*, we always compute the probability of X that lie within a certain interval, $P(a \leq X \leq b)$. Moreover, unlike PMF, PDF cannot be directly used. Instead the function needs to be integrated, which means we need to find the area underneath the curve of the probability density function. Thus, the PDF is given by

$$\int_a^b f(x)\,dx = P(a \leq x \leq b)$$

Thus every continuous random variable has a PDF, instead of a PMF, that defines the relative likelihood that a random variable X has a particular value. Why do we need this new construct? We already said that $P(X = a) = 0$ for any value of a, and so a "PMF" for a continuous random variable would equal 0 for any input and be useless. It wouldn't satisfy the constraint that the sum of the probabilities is 1 (assuming we could even sum over

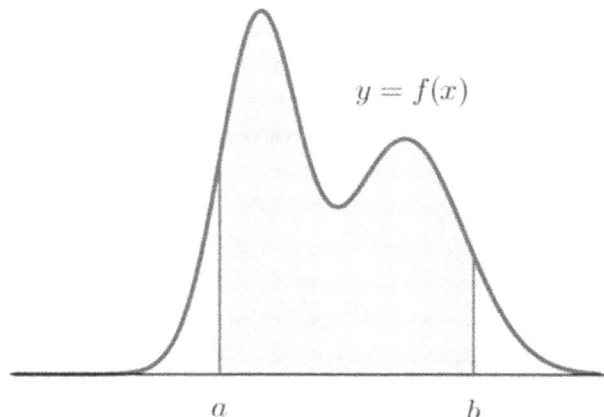

Figure 3.4

uncountably many values; we can't). Instead, we have the idea of a probability density function where the x-axis has values in the random variable's range (usually an interval), and the y-axis has the probability density (not mass).

Some of the properties of a PDF include:

- The PDF is non-negative for all possible values, i.e. $f(x) \geq 0$, for all x.
- The area between the density curve and horizontal x-axis is equal to 1, i.e.

$$\int_{-\infty}^{\infty} f(x) dx = 1$$

Because the area of a line segment is always 0, the definition of the probability distribution of a continuous random variable implies that for any value of X, say a, the probability that X assumes the exact value a is 0. This property implies that whether or not the endpoints of an interval are included makes no difference concerning the probability of the interval. So we can conclude that

$$P(a \leq X \leq b) = P(a < X \leq b) = P(a \leq X < b) = P(a < X < b) = \int_{a}^{b} f(x) dx$$

3.3.3.2.2 The cumulative distribution function (CDF)

The cumulative distribution function (CDF) of a continuous random variable X is defined in exactly the same way as the CDF of a discrete random variable.

$$F(x) = P(X \leq x)$$

But how do we calculate the probability using CDF?

$$F(x) = P(X \leq x) = \int_{-\infty}^{x} f(x) dx \text{ where } f(x) \text{ is } pdf \text{ of } X \tag{3.3}$$

Now we have seen how we can find the probability of a continuous variable using the PDF of the variable, but how do we find it using the CDF?

$$\int_{-\infty}^{b} f(x) dx = \int_{-\infty}^{a} f(x) dx + \int_{a}^{b} f(x) dx$$

$$\int_{a}^{b} f(x) dx = \int_{-\infty}^{b} f(x) dx - \int_{-\infty}^{a} f(x) dx$$

From Equations 3.1 and 3.2, $F(b) - F(a) = P(a \leq x \leq b)$ \hfill (3.4)

There are three "types" of probability calculations that we will deal with for continuous variables. Notice that in all of the probability calculations we are always integrating the PDF, just with different points.

1) When calculating $P(X \le k)$

Since $F(x) = P(X \le x)$

Then we can write $P(X \le k) = \int_{-\infty}^{k} f(x)\,dx$.

This "tells us" that the probability that the continuous random variable X is less than or equal to some value k equals the area enclosed by the probability density function and the horizontal axis, between $-\infty$ and k.

2) When calculating $P(a \le X \le b)$

$$P(a \le X \le b) = \int_{a}^{b} f(x)\,dx$$

3) When calculating $P(X \ge k)$

$$P(X \ge k) = \int_{k}^{+\infty} f(x)\,dx$$

3.3.3.2.3 Connection between the CDF and PDF of a continuous variable

We denoted the PDF in both discrete and continuous variable as $f(x)$ and the CDF is given by $F(x)$

From Equations 3.1 and 3.3, $P(a \le x \le b) = \int_{a}^{b} f(x)\,dx = F(b) - F(a)$

$$f(x) = \frac{dF(x)}{dx} = F'(x)$$

$$F(x) = \int_{-\infty}^{x} f(x)\,dx$$

Here we can see that the CDF corresponds to the integral of the PDF, and the PDF corresponds to the derivative of the CDF.

3.3.4 Mean and variance of a probability distribution

3.3.4.1 Discrete probability distribution

3.3.4.1.1 Mean

The mean of a discrete probability distribution gives the weighted average of all possible values of the discrete random variable. It is also known as the expected value. The formula for the mean of a discrete random variable is given as follows:

$$E[X] = \sum x P(X = x)$$

3.3.4.1.2 Variance

The discrete probability distribution variance gives the dispersion of the distribution about the mean. It can be defined as the average of the squared differences of the distribution from the mean, μ. The formula is given below:

$$Var[X] = \sum (x - \mu)^2 P(X = x)$$

3.3.4.2 *Continuous probability distribution*

3.3.4.2.1 *Mean*

If $f(x)$ is the probability density function of the random variable X, then the mean is given by the following formula:

$$E[X] = \mu = \int_{-\infty}^{\infty} x f(x) dx$$

3.3.4.2.2 *Variance*

The variance is given by:

$$Var[X] = \int_{-\infty}^{\infty} (x - \mu)^2 f(x).dx$$

3.3.4.2.3 *Example Problems*

1) A part selected for testing is equally likely to have been produced on any one of five cutting tools.

Cutting Tool	1	2	3	4	5
Probability	0.20	0.20	0.20	0.20	0.20

   a) What is the probability that the part is from tool 4?
   b) What is the probability that the part is from tool 3 or tool 5?
   c) What is the probability that the part is not from tool 1?

Solutions

   a) $P(X = 4) = 0.20$
   b) $P(3 \le X \le 5) = P(X = 3) + P(X = 4) + P(X = 5) = 0.20 + 0.20 + 0.20 = 0.60$
   c) $1 - P(X = 1) = 1 - 0.20 = 0.80$

2) The number of days it takes to fix the defective parts produced by a machine and the probability that it will take that number of days are in the table is given in the following table.

No of Days	1	2	3	4	5	6	7	8	9	10
Probability	0.20	0.12	0.09	0.05	0.04	0.17	0.13	0.06	0.11	0.03

   a) Find the mean number of days to fix defects.
   b) Find the variance for the number of days to fix defects.
   c) Find the standard deviation for the number of days to fix defects.
   d) Find probability that it will take 2 days to fix the defect.
   e) Find probability that it will take at least 6 days to fix the defect.
   f) Find probability that it will take at most 3 days to fix the defect.
   g) Find probability that it will take more than 8 days to fix the defect.

Solutions

   a) $E[X] = (1 \times 0.20) + (2 \times 0.12) + \dots (9 \times 0.11) + (10 \times 0.03) = 4.81$
   b) $Var[X] = ((1 - 4.81)^2 \times 0.20) + ((2 - 4.81)^2 \times 0.12) + \dots ((10 - 4.81)^2 \times 0.03) = 8.3939$
   c) $SD = \sqrt{8.3939} = 2.8972$

d) $P(X = 2) = 0.12$
e) $P(X \geq 6) = P(X = 6) + P(X = 7) + \ldots P(X = 10) = 0.17 + 0.13 + \ldots 0.03 = 0.05$
f) $P(X \leq 3) = P(X = 1) + P(X = 2) + P(X = 3) + 0.20 + 0.12 + 0.09 = 0.41$
g) $P(X > 8) = P(X = 9) + P(X = 10) = 0.11 + 0.03 = 0.14$

3. Let X be a random variable with PDF given by

$$f(x) = \begin{cases} \dfrac{1}{2}x^2, & 0 \leq x \leq 1 \\ 0, & \text{otherwise} \end{cases}$$

0, otherwise

Find the cumulative distribution function $(CDF), F(x)$.

We know $F(x) = \displaystyle\int_{-\infty}^{x} f(x)\,dx$.

When x is in the interval $(-\infty, 0)$ then $F(x) = \displaystyle\int_{-\infty}^{x} f(x)\,dx = \int_{-\infty}^{x} 0 \, dx = 0$

When x is in the interval $[0,1]$ then $F(x) = \displaystyle\int_{-\infty}^{x} f(x)\,dx = \int_{-\infty}^{0} f(x)\,dx + \int_{0}^{x} f(x)\,dx$

$$= \int_{-\infty}^{0} 0 \, dx + \int_{0}^{x} \frac{1}{2}x^2 \, dx$$

$$= 0 + \frac{1}{2}\left[\frac{x^3}{3}\right]$$

When x is in the interval $[1,\infty)$ then $F(x) = \displaystyle\int_{-\infty}^{x} f(x)\,dx$

$$= \int_{-\infty}^{0} f(x)\,dx + \int_{0}^{1} f(x)\,dx + \int_{1}^{x} f(x)\,dx$$

$$= \int_{-\infty}^{0} 0 \, dx + \int_{0}^{1} \frac{1}{2}x^2 \, dx + \int_{1}^{x} 0 \, dx$$

$$= 0 + \frac{1}{2}\left[\frac{1}{3}\right] + 0 = \frac{1}{6}$$

Hence our cumulative distribution function $(CDF), F(x)$

$$F(x) = \begin{cases} 0 & 0 < x \\ \dfrac{x^3}{6}, & 0 \leq x \leq 1 \\ \dfrac{1}{6}; & x > 1 \end{cases}$$

3.3.4.2.4 Example

Let X be a random variable with PDF given by

$$f(x) = \begin{cases} cx^2, & |x| \leq 1 \\ 0, & \text{otherwise} \end{cases}$$

a. Find the constant c.
b. Find $E[X]$ and $Var(X)$.
c. Find $P(X \geq \frac{1}{2})$.

To find c, we can use $\int_{-\infty}^{\infty} f(x)dx = 1$

$$1 = \int_{-1}^{1} cx^2 dx + 0$$

$$= c\frac{x^3}{3}\Big| = c\left[\frac{1}{3} - \frac{-1}{3}\right] = c\left[\frac{2}{3}\right]$$

$$\frac{\frac{1}{2}}{\frac{3}{3}} = c = \frac{3}{2}$$

b) $E[X] = \int_{-1}^{1} x.\left(\frac{3}{2}x^2\right)dx = \frac{3}{2}\frac{x^4}{4}\Big| = 0$

$Var[X] = \int_{-1}^{1} (x-0)^2 \frac{3}{2}x^2.dx = \int_{-1}^{1}\frac{3}{2}x^4.dx = \frac{3}{2}\frac{x^5}{5}\Big| = \frac{3}{10} - \left(-\frac{3}{10}\right) = \frac{6}{10}$

c) $P(x) = \int_{\frac{1}{2}}^{1}\frac{3}{2}x^2 dx = \frac{7}{16}$

	Discrete	Continuous
Variable	Discrete variables take on distinct, fixed and countable values.	Continuous variables take on any value within a range, and the number of possible values within that range is infinite.
Outcomes	Have countable outcomes and we can assign a probability to each of the outcomes	
Defined by	Probability mass punction (PMF) cumulative distribution function	Probability density function (PDF) Cumulative distribution function
Calculating probability	PMF denoted by $f(x)$ can be directly used to calculate the probability of the exact outcome $$P(X = x) = f(x)$$ A discrete experiment can only have countable outcomes and we can assign a probability to each of the outcomes	PDF also denoted by $f(x)$ needs to be integrated to calculate the probability of a range/interval $$P(a \leq X \leq b) = \int_{a}^{b} f(x)dx$$ Because there are infinite values that a continuous variable could take on, the probability that a continuous random variable will assume a particular value is zero $$P(X = x) = 0$$

3.4 TYPES OF DISCRETE AND CONTINUOUS PROBABILITY DISTRIBUTION

Now let's go back to our coin problem. We still have not estimated the probabilities of each outcome. The only information we have is that it is a discrete random variable and to calculate the probability of each outcome we use the PMF associated with that distribution. Now the question is what is the PMF of this particular example?

If we look back at the PMF, PDF, and CDF we are not given a formula to calculate the probabilities. This is because the formulas for these functions vary depending on the specific probability distribution being considered.

In Figure 3.5 is the list of different types of discrete and continuous distributions we are going to talk about in the next two chapters.

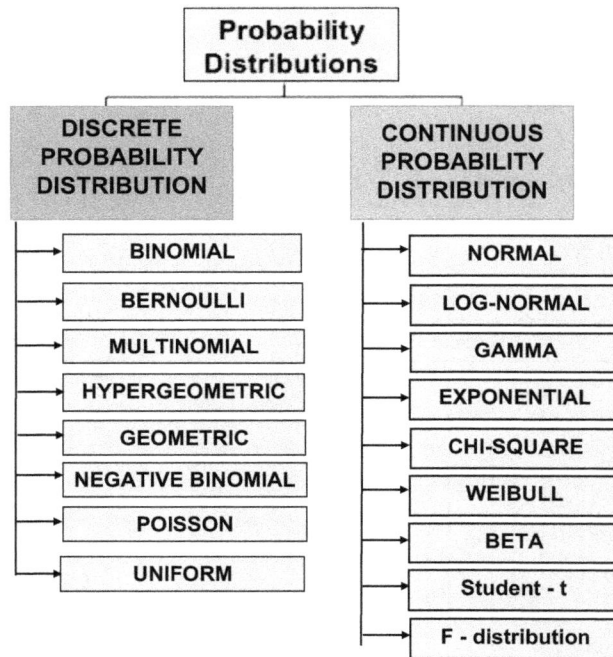

Figure 3.5

3.5 PROBABILITY DISTRIBUTIONS IN MATLAB

Now that we have talked so much about the probability distributions, its types, and differences and shown examples theoretically, how are we going to do all that in MATLAB? Now we will talk about some of the probability distributions available in MATLAB and illustrate the functions to estimate the probabilities of those distributions.

Before jumping into the calculations of probability density functions and cumulative distribution functions of a given dataset, there are certain terms we need to get familiar with.

3.5.1 Probability distribution object

So far we have seen that probability distributions can be categorized as discrete or continuous. We have discussed the different types of discrete and continuous probability distribution. Each of these distributions has specific parameters that govern certain aspects of the distribution. The parameters determine the characteristics of those distributions, for example the shape and probabilities.

Before we can calculate the probabilities in MATLAB using the probability functions, we first create a probability distribution object. This is because these probability distribution objects provide a simple way to work with data and distributions.

We are going to learn two ways to generate a probability distribution object in MATLAB:

1) The first function we use to do the task is makedist. The syntax of the function is:

```
Pd = makedist ('Distname')
```

This function creates a probability distribution object for the distribution Distname, using the default parameter values.

3.5.1.1 Example

```
Pd = makedist ('Binomial)
```

```
pd =

    BinomialDistribution

    Binomial distribution
        N =    1
        p = 0.5
```

We created a binomial distribution object using the default parameter values, which correspond to the parameters of the standard binomial distribution.

```
Pd = makedist ('Normal')
```

```
pd =

    NormalDistribution

    Normal distribution
          mu = 0
       sigma = 1
```

We created a normal distribution object using the default parameter values, which correspond to the parameters of the standard normal distribution.

In contrast to the first one we can also create a probability distribution object for the distribution *distname* along with its parameter values specified

$$pd = makedist('Distname', parameter_1, Value, parameter_2, Value, ...parameter_k, Value).$$

3.5.1.2 Example

Create a binomial distribution object with parameter values $n = 30$ and $p = 0.25$.

$$pd = makedist('Binomial','n',30,'p',0.25)$$

```
pd =

  BinomialDistribution

  Binomial distribution
    N =    30
    p = 0.25
```

Create a normal distribution object with parameter values $mu = 75$ and $sigma = 10$.

$$pd = makedist('Normal','mu',75,'sigma',10)$$

```
pd =

  NormalDistribution

  Normal distribution
     mu = 75
  sigma = 10
```

2) The other function we are using to build our object is *fitdist* which is only used when you already have a dataset and you want to fit the distribution specified by *Distname* to the data.

$$pd = fitdist(x,'Distname')$$

3.5.1.3 Example

```
data = readtable("Temperature.csv", VariableNamingRule = "preserve")
Temperature = data.Temperature;
pd = fitdist(Temperature, 'Normal')
```

```
pd =

  NormalDistribution

  Normal distribution
     mu = 47.6634   [44.2451, 51.0816]
  sigma =  9.3191   [7.447, 12.4566]
```

$$pd = fitdist(x, 'Distname', parameter_1, Value, parameter_2, Value,....parameter_k, Value)$$

3.5.1.4 Example

Going back to our coin example:

```
Heads = 0:1:5;
x = Heads'
pd = fitdist(x,'Bincmial','NTrials',5)
```

```
                    pd =

              BinomialDistribution

              Binomial distribution
                 N =   5
                 p = 0.5   [0.31297, 0.68703]
```

3.5.2 Calculating the probabilities

Now that we are done creating the probability distribution object, the next step will actually be calculating the probabilities of each distribution. In MATLAB we can actually perform these calculations in three different ways which are discussed below. Whatever way you choose to evaluate the probability density function and cumulative distribution function, we always add *pdf* and *cdf* to the base word

$$y = pdf\left('Distname',\ x,\ parameterValue_1,\ parameterValue_2,\ldots\ldots parameterValue_k\right)$$

3.5.2.1 Example

```
% Define the range of x values for the PDF plot
x = linspace(-3, 3, 100);

% Parameters for the first normal distribution
mu1 = 0;
sigma1 = 1;
% Compute PDF values for the first normal distribution
y1 = pdf('Normal', x, mu1, sigma1);

% Parameters for the second normal distribution
mu2 = 0;
sigma2 = 0.75;
% Compute PDF values for the second normal distribution
y2 = pdf('Normal', x, mu2, sigma2);

% Parameters for the third normal distribution
mu3 = 0;
sigma3 = 0.5;
% Compute PDF values for the second normal distribution
y3 = pdf('Normal', x, mu3, sigma3);
```

```
% Parameters for the fourth normal distribution
mu4 = 0;
sigma4 = 0.25;
% Compute PDF values for the second normal distribution
y4 = pdf('Normal', x, mu4, sigma4);

figure
plot(x, y1, 'r', x, y2, 'k', x, y3, 'g', x, y4, 'b', 'LineWidth',2)
title('Multiple PDFs');
xlabel('x');
ylabel('PDF');
legend('PDF 1', 'PDF 2', 'PDF 3', 'PDF 4');
```

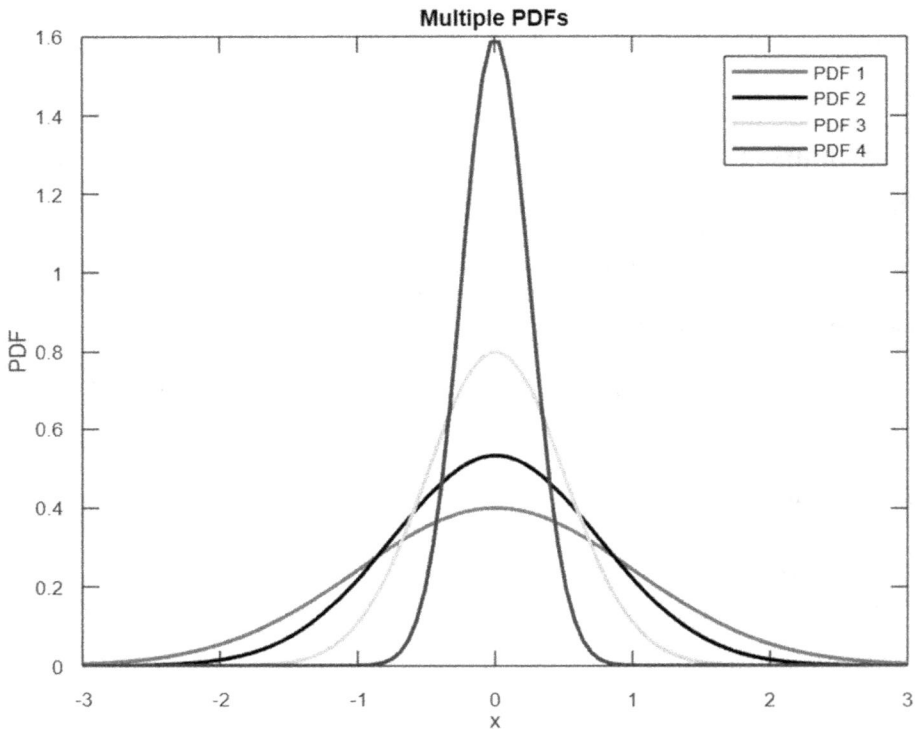

Figure 3.6

Notice that:

$$y = pdf\,(pd, x)$$

This returns the pdf of the probability distribution object pd, evaluated at the values in x.

3.5.2.2 Example

Now we can finally calculate the probabilities of each outcome when flipping the coin:

```
Heads = 0:1:5;
x = Heads';
pd = fitdist(x,'Binomial','NTrials',5);
y = pdf(pd,x)
```

$$y =$$

$$
\begin{aligned}
&0.0312\\
&0.1562\\
&0.3125\\
&0.3125\\
&0.1562\\
&0.0312
\end{aligned}
$$

Hence our probability distribution table will now look like:

X = no. Of heads	0	1	2	3	4	5
P(X)	0.0312	0.1562	0.3125	0.3125	0.1562	0.0312

3.5.2.3 Example

```
data = readtable("Energy consumption.csv", VariableNamingRule="prese
rve");
Temperature = data.Temperature;
pd = fitdist(Temperature, 'Normal');

x = linspace(min(Temperature), max(Temperature), 100);
y = pdf(pd,x);

plot(x, y, "r", 'LineWidth', 2)
title('Normal Distribution PDF');
xlabel('Temperature');
ylabel('PDF');
```

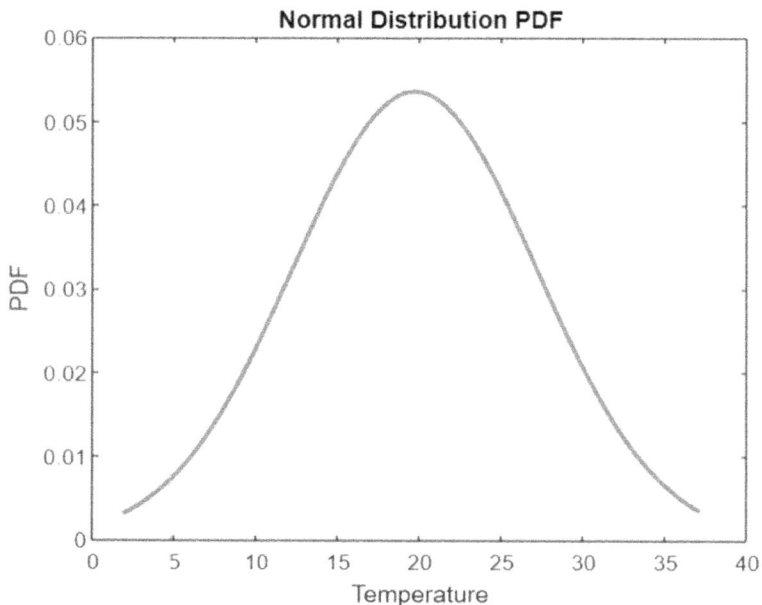

Figure 3.7

Notice that if we choose this function, we do not really have to include the name of the parameters. Instead we are only specifying the values of the parameters of the distributions.

Another function:

$$y = RootWordpdf\left(x, \; parameterValue_1, \; parameterValue_2, \ldots \ldots parameterValue_k\right)$$

The second function makes use of the root word of different distributions to run the pdf operation.

```
% Define the range of x values for the PDF plot
x = linspace(-3, 3, 100);

% Parameters for the first normal distribution
mu1 = 0;
sigma1 = 1;
% Compute PDF values for the first normal distribution
y1 = normpdf(x, mu1, sigma1);

% Parameters for the second normal distribution
mu2 = 0.25;
sigma2 = 0.75;
% Compute PDF values for the second normal distribution
y2 = normpdf(x, mu2, sigma2);

% Parameters for the third normal distribution
mu3 = 0.5;
sigma3 = 0.5;
% Compute PDF values for the second normal distribution
y3 = normpdf(x, mu3, sigma3);

% Parameters for the fourth normal distribution
mu4 = 0.75;
sigma4 = 0.25;
% Compute PDF values for the second normal distribution
y4 = normpdf(x, mu4, sigma4);

figure
plot(x, y1, 'r', x, y2, 'k', x, y3, 'g', x, y4, 'b', 'LineWidth',2)
title('Multiple PDFs');
xlabel('x');
ylabel('PDF');
legend('PDF 1', 'PDF 2', 'PDF 3', 'PDF 4');
```

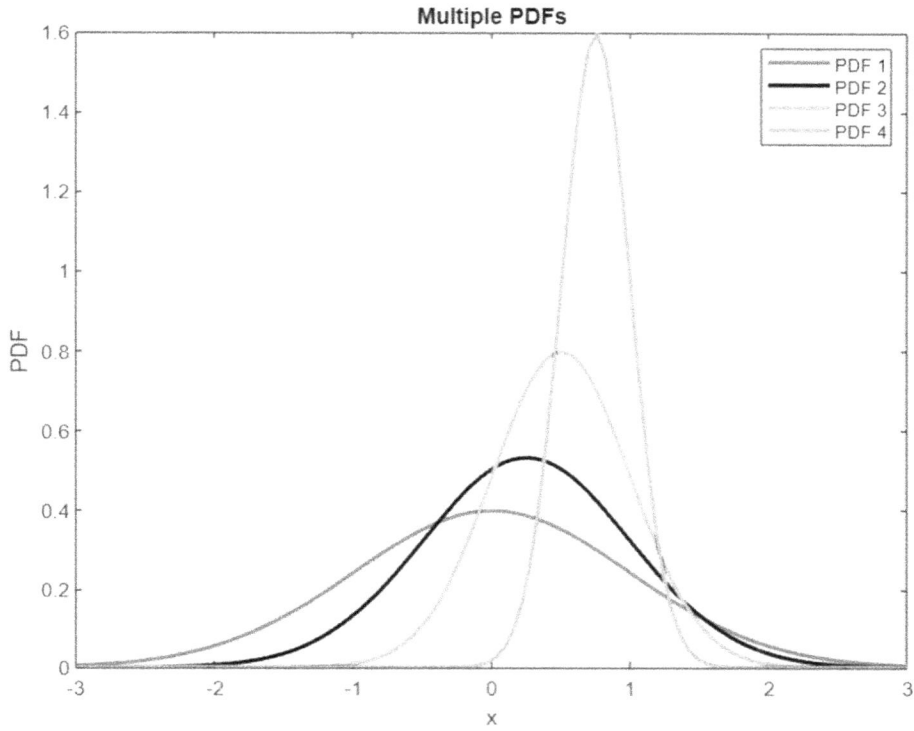

Figure 3.8

Below I have provided a table of some of the distributions from both discrete and continuous variables along with their root words and parameters.

3.5.2.4 Discrete distribution

Distribution	Root Word	Parameters	Example
Binomial	*bino*	1) *n* = *number of trials* 2) *p* = *probability of success*	*'Binomial','n',30,'p',0.25*
Bernoulli	*bino*	*p* = *probability of success*	*'Binomial','n',1,'p',0.25*
Multinomial	*mn*	*Probabilities* = *Outcome probabilities*	*'Multinomial','Probabilities',* $\begin{bmatrix} 1 & 1 & 1 \\ 2 & 3 & 6 \end{bmatrix}$
Hypergeometric	*hyge*	1) *M* = *size of the population.* 2) *K* = *no. of items with the desired characteristic in the population* 3) *n* = *number of samples drawn.*	*'Hypergeometric','M',10,'K',6, 'n',5*
Geometric	*geo*	*p* = *probability of success*	*'Geometric','p',0.3*
Negative Binomial	*nbin*	1) *R* = *number of successes* 2) *P* = *probability of success*	*'NegativeBinomial','R',5,'P',0.1*
Poisson	*poiss*	*lamda*(λ) = *expected value*	*'Poisson','lamda',5*
Uniform	*unid*	*N*	

3.5.2.5 Continuous distribution

Distribution	Root Word	Parameters	Example
Normal	norm	1) $mu(\mu) = mean$	$'Normal','mu',75,'sigma',10$
		2) $sigma(\sigma) = standard\ deviation$	
Exponential'	exp	$mu(\mu) = mean = rate$	$'Exponential','mu',2$
'Lognormal'	logn	1) $mu(\mu) = mean$	$'Lognormal','mu',log(500),$
		2) $sigma(\sigma) = standard\ deviation$	$'sigma',1$
'Beta'	beta	1) $a = first\ shape\ parameter$	$'Beta','a',2,'b',4$
		2) $b = second\ shape\ parameter$	
'Gamma'	gam	1) $a = shape\ parameter$	$'Gamma','a',2,'b',4$
		2) $b = scale\ parameter$	
'Weibull'	wbl	1) $A = scale\ parameter$	$'Weibull','A',2,'B',5$
or 'wbl'		2) $B = shape\ parameter$	
		$nu(\nu) - degrees\ of\ freedom$	
		$nu(\nu) - degrees\ of\ freedom$	
Uniform	unif	1) $a = lower\ endpoint$	$'Uniform','Lower',-4,$
		2) $b = upper\ endpoint$	$'Upper',2$

The second column shows the root words that MATLAB uses to specify a particular distribution. For example, the root word for a binomial distribution is *bino* . To evaluate the probability density function, we add *pdf* to the root word: *binopdf* . Likewise, we can also add *cdf* to the root word, such as *binocdf* which will calculate the cumulative density function of the binomial distribution. Similarly, there are other functions which can be added to the root word to estimate that function.

Functions	What that function calculates
pdf	*Probability density function*
cdf	*Cumulative density function*
rnd	*Generates random variables*
stat	*Mean and variance*
fit	*Estimate parameters*

The third column lists the parameters of each of the distributions that are needed as arguments for some of the functions we have mentioned. Recall from our discussion that we have a **probability mass function** instead of a density function when we work with discrete variables. But here we are only talking about pdf ; this is because the acronym pdf is used to refer to both the probability mass function and the probability density function.

Let's do some examples to calculate the PDF of a certain distribution using the three different ways and check if all three of them give out the same value.

1) In one day, a quality inspector tests 100 electric generators; 4% of the generators showed defects.

a) Compute the probability that the inspector will find no defective boards on any given day.

```
DefectiveBoard = 0
Using Method 1
y = pdf('Binomial', DefectiveBoard, 100, 0.04)

Method 2
y = binopdf(DefectiveBoard, 100, 0.04)

Method 3
pd = makedist('Binomial', 'N', 100, 'p', 0.04)
y = pdf(pd, DefectiveBoard)
```

Regardless of the function we use, our output is

```
y =

0.0169
```

This proves that we can calculate the PDF of a discrete variable using any of the syntaxes.

b) Finding the probability distribution

```
DefectiveBoard = 0:1:10;
x = DefectiveBoard';

Using Method 1
y = pdf('Binomial', DefectiveBoard, 100, 0.04);
figure
bar(x, y, 1)
xlabel('Observation')
ylabel('Probability')
title("Probability Distribution of Defective Boards")

Using Method 2
y = binopdf(DefectiveBoard, 100, 0.04)
figure
bar(x, y, 1)
xlabel('Observation')
ylabel('Probability')
title("Probability Distribution of Defective Boards")

Using Method 3
pd = makedist('Binomial', 'N', 100, 'p', 0.04)
y = pdf(pd, DefectiveBoard)
```

```
figure
bar(x, y, 1)
xlabel('Observation')
ylabel('Probability')
title("Probability Distribution of Defective Boards")
```

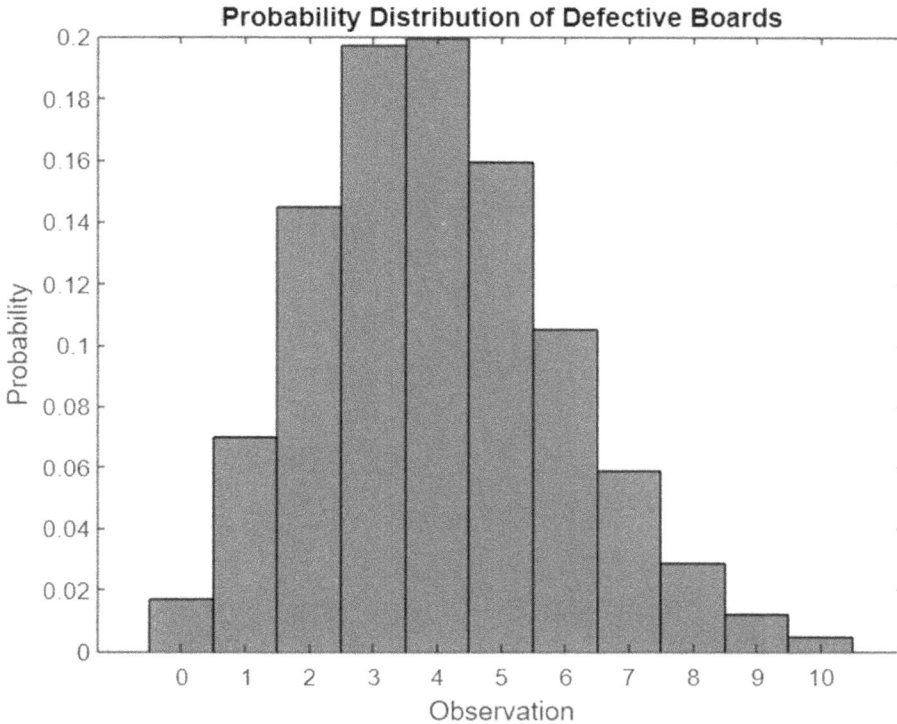

Figure 3.9

Just by changing the PDF to CDF we can actually use the same syntax to calculate the cumulative density function of a distribution.

2) Consider a telecommunications network. Say you want to analyze the number of calls received in a specific hour which follows a Poisson distribution. To collect the data, you record the number of calls received during 15 randomly chosen hours. Here's the dataset:

Hour	1	2	3	4	5	6	7	8	9	10	11	12	13	14	15
Calls	12	8	10	9	11	13	10	7	11	15	12	8	10	11	

What is the probability that ten calls are received in one particular hour?

If we look at the table we will see that the Poisson distribution is characterized by a single parameter, lambda (λ), which represents the average rate of occurrence of events in a fixed interval of time or space. Therefore

```
Hour = 1:15;
Calls = [12 8 10 9      11 13 10 7 11 15 12      8 10 11 7]
lambda = sum(Calls)/numel(Hour)
x = 10
```

Using Method 1
```
probability = pdf('Poisson', x, lambda)
```

Method 2
```
probability = poisspdf(x, lambda)
```

Method 3
```
pd = makedist('Poisson', 'lambda', lambda)
y = pdf(pd, x)
```

```
                        probability =

                          0.1247
```

Let's show both the PDF and the CDF for a Poisson distribution with the calculated lambda.

```
X = min(Calls):1:max(Calls)
y = poisspdf(x, lambda);
% Get the values of the CDF
z = poisscdf(x, lambda);

% Construct the plots for PDF
subplot(2,1,1)
plot(x ,y,'b'o'' 'LineWid'h', 1.5)
ylabe'('Poisson P'F')
titl'('Poisson Distribution' )

% Construct the plots for CDF
subplot(2,1,2)
stairs(x,z)
ylabel('Poisson CDF')
xlabel('X')
```

```
OUTPUT
```

Let's do an example for continuous distribution.

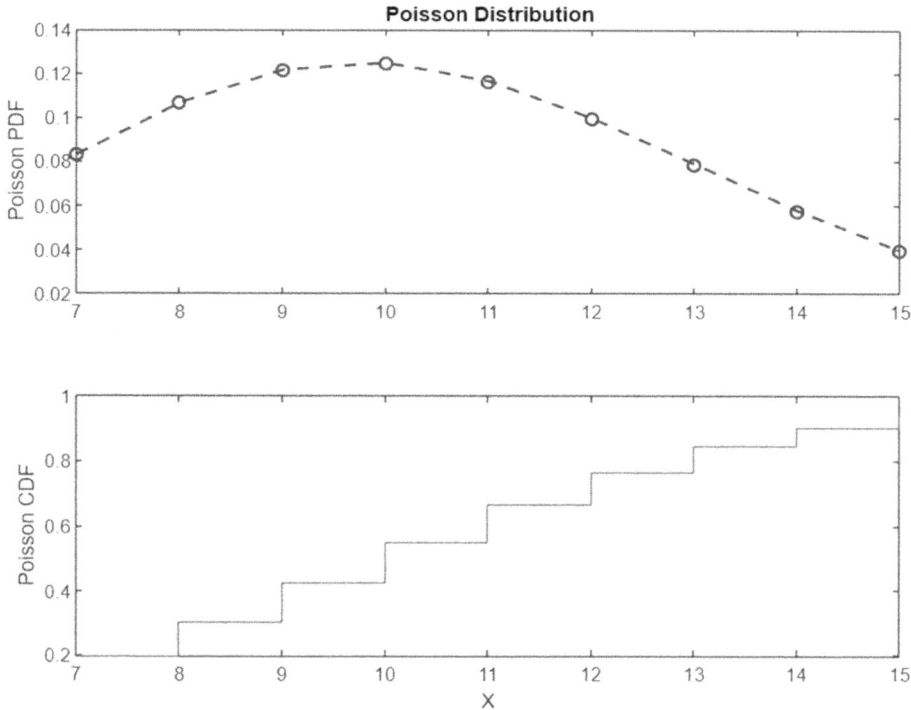

Figure 3.10

3.5.2.6 Example

Compute the PDFs of the normal distribution with several mean and standard deviation parameters. Let's say we have four datasets of bolt diameters produced by a machine. Each dataset represents a different batch of bolts. We calculate the PDF for each dataset and then plot the PDFs for all four datasets on the same graph.

```
data = readtable("Bolts.csv", VariableNamingRule="preserve");
Batch1 = data.Bolts1
Batch2 = data.Bolts2
Batch3 = data.Bolts3
Batch4 = data.Bolts4

[mean1, std1] = normfit(Batch1);
[mean2, std2] = normfit(Batch2);
[mean3, std3] = normfit(Batch3);
[mean4, std4] = normfit(Batch4);

x = 0:0.01:7;
y1 = gampdf(x, mean1, std1);
y2 = gampdf(x, mean2, std2);
y3 = gampdf(x, mean3, std3);
y4 = gampdf(x, mean4, std4);
plot(x, y1, 'LineWidth', 2)
hold on
```

```
plot(x, y2,'LineWidth', 2)
plot(x, y3, 'LineWidth', 2)
plot(x, y4, 'LineWidth', 2)
hold off
ylabel("Normal PDF")
title("Multiple Normal PDF")
legend('Batch 1', 'Batch 2','Batch 3','Batch 4')
```

3.5.2.7 Example

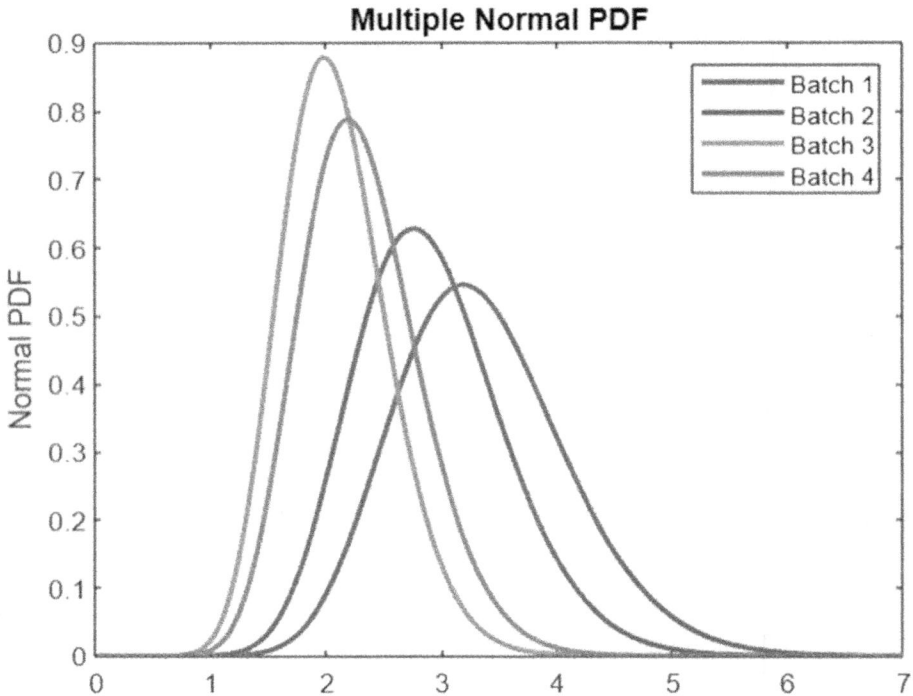

Figure 3.11

A group of engineers are studying the particle size distribution of a granular material by recording their size.

```
data = readtable("Granular size.csv", VariableNamingRule="preserve");
size = data.Size;
% Estimate the parameters of the lognormal distribution
params = lognfit(size);
mu = params(1);
sigma = params(2);

x = min(size):0.5:max(size)
% Calculate the PDF values for the dataset
pdf = lognpdf(x, mu, sigma);

% Construct the plots for PDF
```

```
plot(x ,pdf,'m', 'LineWidth', 1.5)
ylabel('Lognormal PDF')
title('Lognormal Distribution' )
```

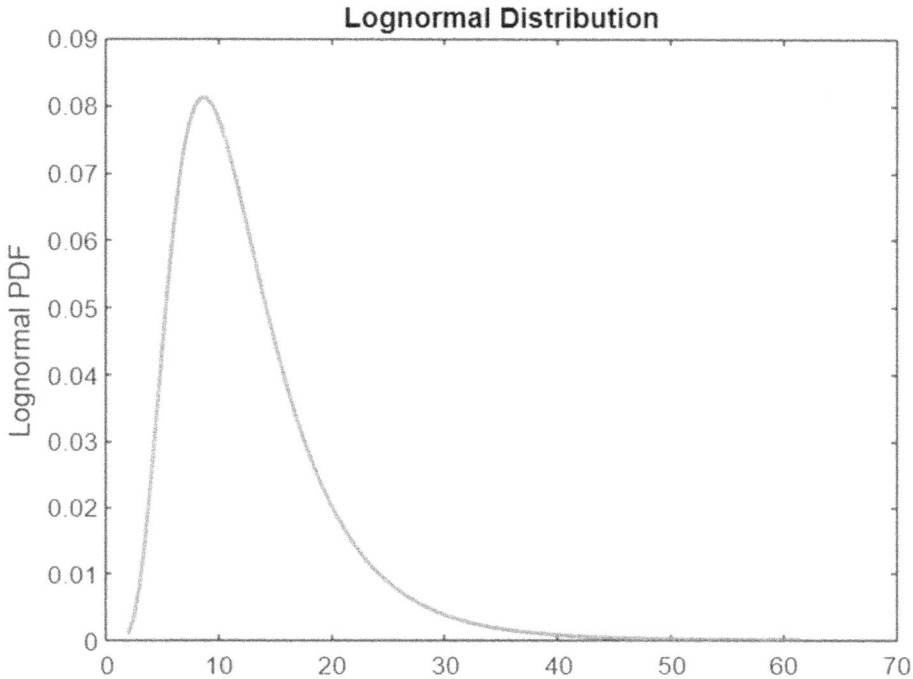

Figure 3.12

3.5.2.8 Example

Suppose you are an engineer working in the automotive industry, in the field of material strength analysis. You are interested in modeling the tensile strength of a particular type of steel alloy used in the manufacturing of car parts.

```
data = readtable("Tensile Strength.csv", VariableNamingRule="preserve");
strength = data.Tensile_Strength;

%converting strength MPa to GPa so that it falls in the range[0,1]
x = strength / 1000

%To ensure that the filtered data retains the decimal values, you can
convert the data to a floating-point type (e.g., double) before
performing the filtering.
values = double(x);

% Filter the data to keep only values less than 1
filtered_data = values(values < 1)

% Estimate the parameters of the lognormal distribution
params = betafit(filtered_data);
```

```matlab
FirstShape = params(1);
SecondShape = params(2);

x = 0:0.001:1
% Calculate the PDF values for the dataset
pdf = betapdf(x, FirstShape, SecondShape);

% Construct the plots for PDF
plot(x, pdf, 'Color', [0.856 0.3567 0.05], 'LineWidth', 1.5)
ylabel('Beta PDF')
title('Beta Distribution')
```

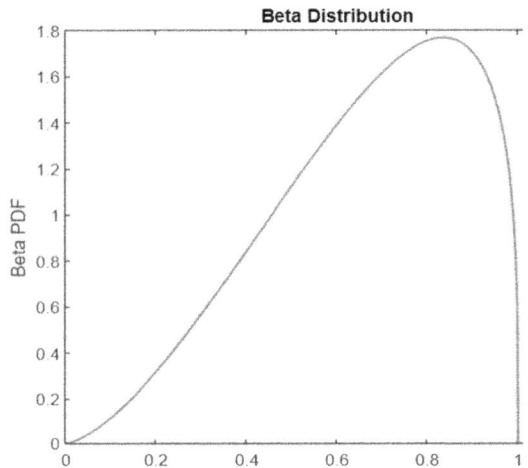

Figure 3.13

3.5.2.9 Example

In engineering, it is often important to analyze rainfall data for various applications, such as designing stormwater drainage systems or assessing flood risks. The gamma distribution can be used to model the intensity or amount of rainfall within a specific time period. Let's say we have a dataset that represents the hourly rainfall intensities (in millimeters) recorded over a period of time. We want to analyze this data using the gamma distribution.

```matlab
data = readtable("Rainfall Intensity.csv", VariableNamingRule="preserve");
Intensity = data.Intensity;

params = gamfit(filtered_data);
Shape = params(1);
Scale = params(2);

x = min(Intensity):2:max(Intensity);
% Calculate the PDF values for the dataset
pdf = gampdf(x, Shape, Scale)

% Construct the plots for PDF
plot(x, pdf,'Color', [0.856 0.2567 0.25], 'LineWidth', 1.5)
ylabel('Beta PDF')
title('Beta Distribution' )
```

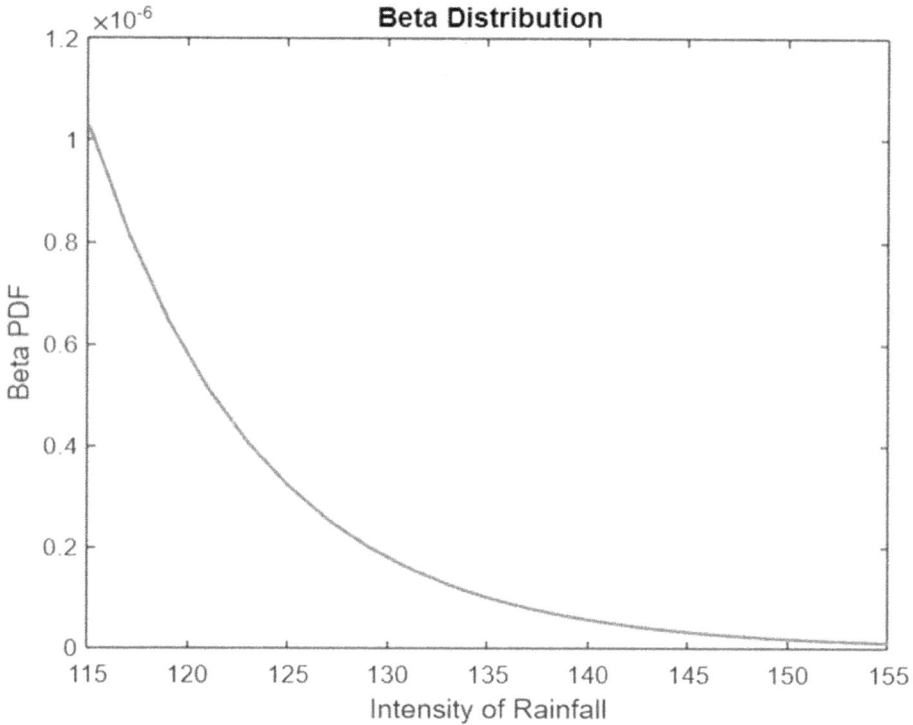

Figure 3.14

3.5.2.10 Examples

```
% Parameters for Normal distribution
mu = 0;
sigma = 1;
% Parameters for exponential distribution
lambda = 0.5;

%Parameter for Gamma distribution
a1 = 2;
b1 = 1;

%Parameter for Beta distribution
a2= 2;
b2 = 2;

%Parameter for Weibull distribution
a3 = 3;
b3 = 5;

% Generate a range of x-axis values
x = linspace(0, 5, 100); % Adjust the range as needed

% Calculate the PDF values for each distribution
pdf_normal = normpdf(x, mu, sigma);
pdf_exponential = exppdf(x, lambda);
```

```
pdf_gamma = gampdf(x, a1, b1);
pdf_beta = betapdf(x, a2, b2);
pdf_Weibull = wblpdf(x, a3, b3);

% Plot the PDFs of each distribution on the same graph
plot(x, pdf_normal, 'b-', 'LineWidth', 2)
hold on;
plot(x, pdf_exponential, 'r--', 'LineWidth', 2)
plot(x, pdf_gamma, 'g:', 'LineWidth', 2)
plot(x, pdf_beta, 'y-', 'LineWidth', 2)
plot(x, pdf_Weibull, 'k-.', 'LineWidth', 2)

% Add labels and legends
xlabel('X');
ylabel('Probability Density');
title('Multiple Continuous Distributions');
legend('Normal', 'Exponential', 'Gamma', 'Beta', 'Weibull');
```

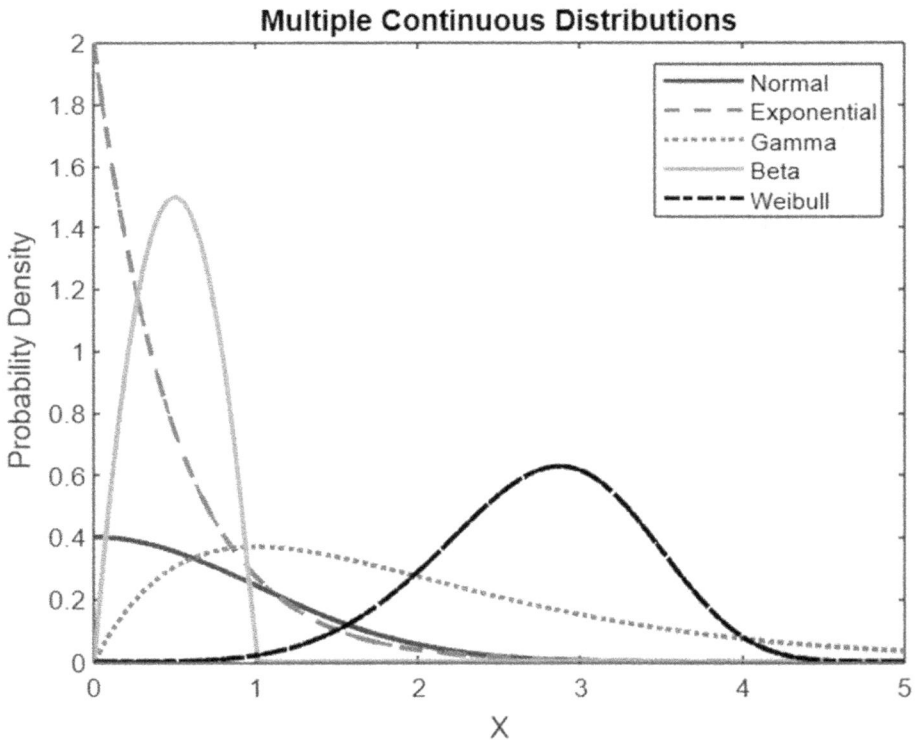

Figure 3.15

2) Compute the pdfs of the exponential distribution with several mean parameters.

Consider a complex mechanical system composed of five components: A, B, C, D, and E. Each component has a different failure rate and is known to have an exponential distribution with the average amount of time given below:

Component A = 132 hours
Component B = 225 hours
Component C = 450 hours

Let's graph the distribution.

```
x = 0:0.001:0.02;
ComponentA = 132;
ComponentB = 225;
ComponentC = 450;

rateA = 1 / ComponentA
rateB = 1 / ComponentB
rateC = 1 / ComponentC

yA = exppdf(x,rateA);
yB = exppdf(x,rateB);
yC = exppdf(x,rateC);

plot(x, yA, 'LineWidth', 2)
hold on
plot(x, yB, 'LineWidth', 2)
plot(x, yC, 'LineWidth', 2)
hold off
xlabel('Observation')
ylabel('Probability Density')
legend('Component A', 'Component B', 'Component C')
```

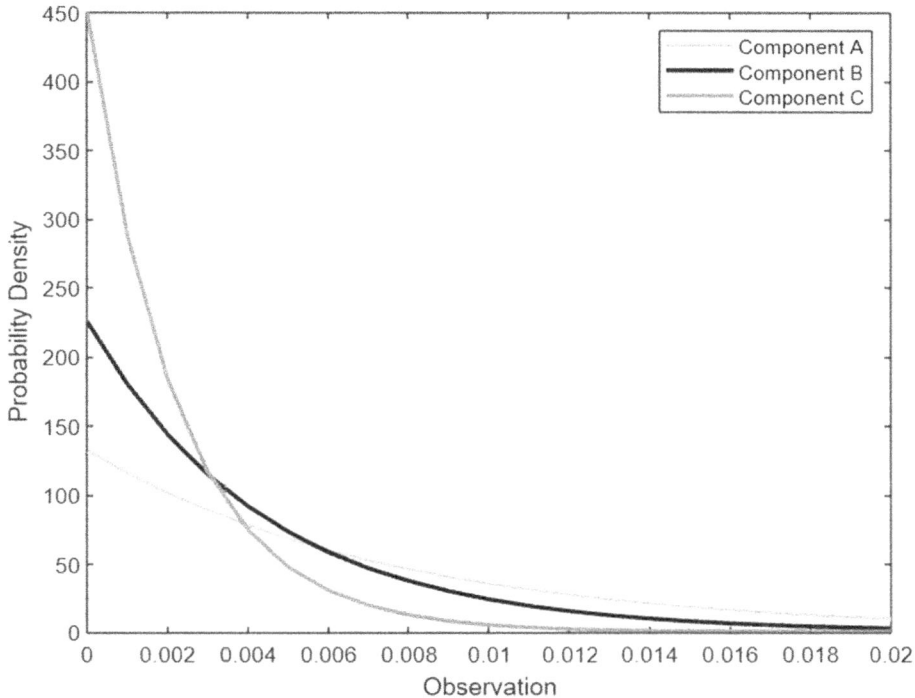

Figure 3.16

3.6 SUMMARY

In this chapter, we first introduced the concept of random variables and the types of probability distribution of those variables theoretically. Then we went through some of the functions that MATLAB offers to deal with those probability distributions.

In the next two chapters we will get into more detail on the different types of discrete and continuous probability distributions and do a lot more examples theoretically and on MATLAB.

Chapter 4

Discrete probability distribution

In the previous chapter we introduced the topic of probability distribution and discussed the two types of distributions based on the type of data generated by the experiments.

We learned how to estimate the probability distribution for both discrete and continuous variables in MATLAB. In this chapter we will learn about different types of discrete probability distributions commonly used in the engineering field. There are many discrete probability distributions that are used in different scenarios. Here, we will discuss some of the most important ones in great detail, including their properties and conditions along with some examples both in theory and in MATLAB.

4.1 TYPES OF DISCRETE DISTRIBUTION

Recall that discrete distribution in statistics is a probability distribution that calculates the likelihood of a particular discrete, finite outcome. In simple words, the discrete probability distribution helps us to find the chances of the occurrence of a certain event expressed in terms of positive, non-decimal, or whole numbers as opposed to a continuous distribution. These probability distributions are represented by using a mathematical function called the probability mass function (PMF). A PMF equation looks like this: $P(X = x)$. That just means "the probability that X takes on some value x" which could be interpreted as the probability of some individual event happening. Now probabilities for different situations will be different and therefore the PMFs will be different.

In this chapter, we introduce some of the most commonly occurring PMFs that describe the shapes of their associated distributions. We are going to use the rules and techniques we learned from the previous chapter to calculate the probabilities of different events occurring. Now in this and in the next chapter we will follow two paths to compute the PDF, CDF, mean, and variance of different distributions. In the first path we will solve and understand a problem theoretically. The example will help us get more familiar with discrete probability distributions and, when given a problem, understand which PMF of which distribution is appropriate. Then in the next path we will learn how to use pre-existing functions available in MATLAB to evaluate the same problem and then practice some more after. The goal of the chapter is to realize that MATLAB includes robust codes that can solve many of the problems that face us on a daily basis.

MATLAB provides seven discrete distributions that we can use for statistical analysis. These are:

1) Binomial distribution
2) Bernoulli distribution

DOI: 10.1201/9781003399582-4

3) Multinomial distribution
4) Hypergeometric distribution
5) Geometric distribution
6) Negative binomial distribution
7) Poisson distribution
8) Uniform distribution (discrete)

4.2 BINOMIAL DISTRIBUTION

Binomial distribution arises from experiments in which there is a fixed number of repeated trials which is denoted by n, with only two outcomes, and only one of the outcomes is counted. One outcome is usually referred to as a success and the other as a failure depending on the situation. It could be any experiment such as yes/no, a defective/not defective item, heads/tails when you toss a coin, etc. So any kind of event where there are only two outcomes will have a binomial distribution.

The following conditions need to be satisfied for the experiment to have a binomial distribution:

1. Fixed number of n trials.
2. Each trial is independent.
3. Only two outcomes are possible (success and failure).
4. The probability of success (p) for each trial is constant.

If all of these conditions are met, then the random variable X can be considered to follow a binomial distribution.

Binomial distribution is useful in calculating the probabilities of the following situations:

- Getting two heads when you toss a coin ten times.
  Here the number of successes in this case heads is two and the number of trials $n = 10$.
- Getting 30 voltage fluctuations in a series of 50 power supply units.
- Getting 25 equipment failures in a set of 100 machines.
- Getting 12 defective circuit boards in a production batch of 25.

There are two parameters n and p used in a binomial distribution.

Parameter	Description
n	Number of trials. It could be any positive integer.
p	Probability of success in a single trial.

4.2.1 Calculations in theory

If a *random variable* X follows a binomial distribution, then the probability of getting k number of successes in n number of trials is given by

$$P(k:n,p) = {}^{n}C_{k}\ p^{k}\left(1-p\right)^{n-k} = \binom{n}{k}p^{k}\left(1-p\right)^{n-k} = \frac{n!}{k!(n-k)!}p^{k}\left(1-p\right)^{n-k}$$

The mean of the binomial distribution is $\mu = np$.
The variance of the binomial distribution is $\sigma^2 = np(1-p)$.

In the previous chapter we talked about the probability distribution for discrete variables. Like any other discrete distribution, the binomial probability distribution summarizes the likelihood that a value will take one of two independent values under a given set of parameters or assumptions. So for example if we toss a coin 20 times the probability distribution of the number of heads during these 20 tosses with $p=0.5$ can be shown by.

Binomial Probability Distribution

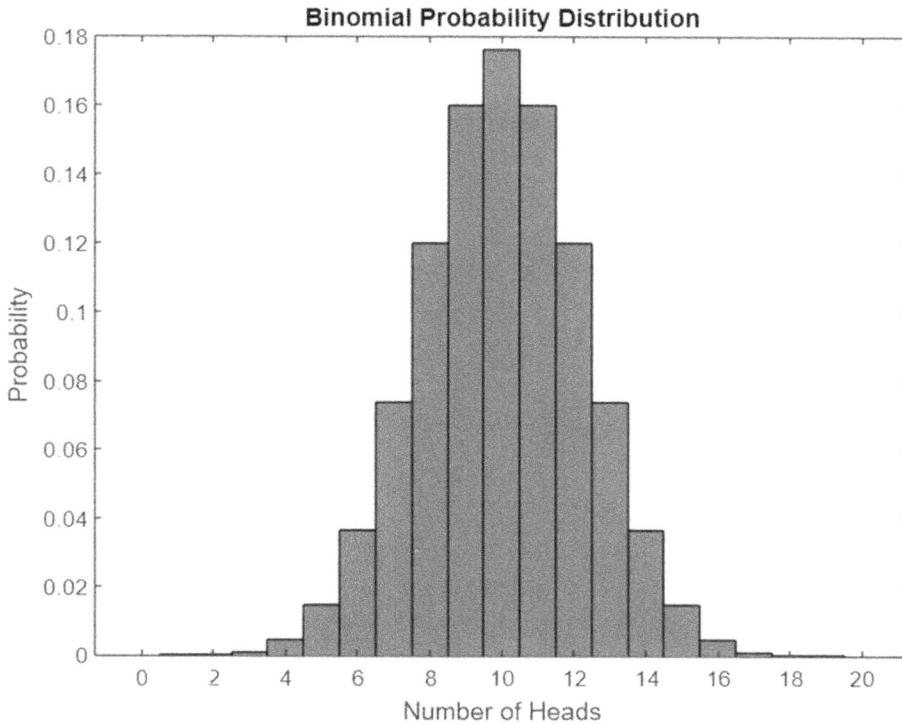

Figure 4.1

As we can see the bar graph in Figure 4.1 exhibits a symmetric shape, resembling a bell curve.

Note that the probability of getting ten heads is higher than the probability of any other number of successes or heads. The probability smoothly declines as we approach greater numbers of heads and greater numbers of tails.

Now let's perform the same experiment and instead of 20 we now repeat the experiment 50 times and draw a similar histogram with all of the possible heads that we can get in 50 tosses and the probability of getting that many heads.

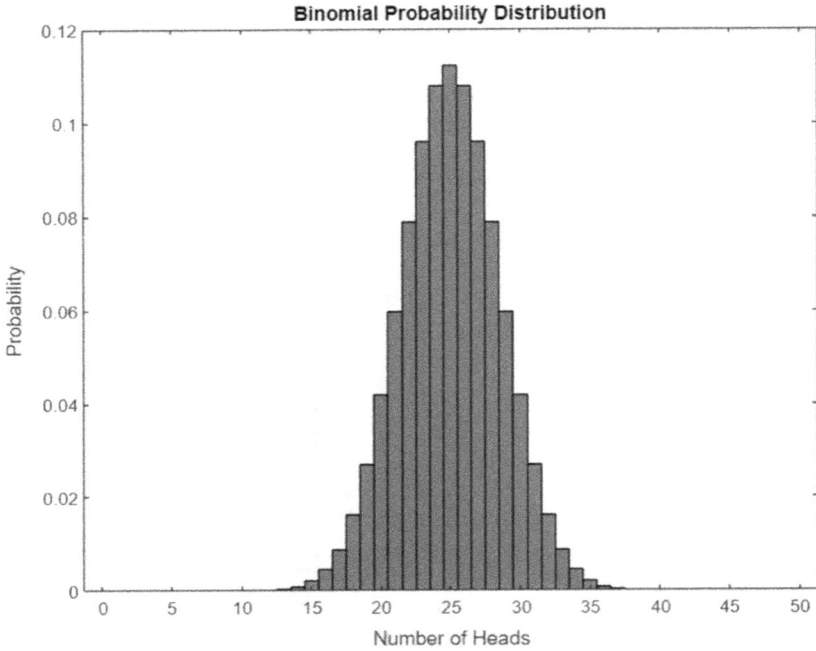

Figure 4.2

Notice how the shape of the curve in Figure 4.2 for 50 trials is much smoother than the curve for 20 trials. Now let's go a little further and now toss the coin 50 more times.

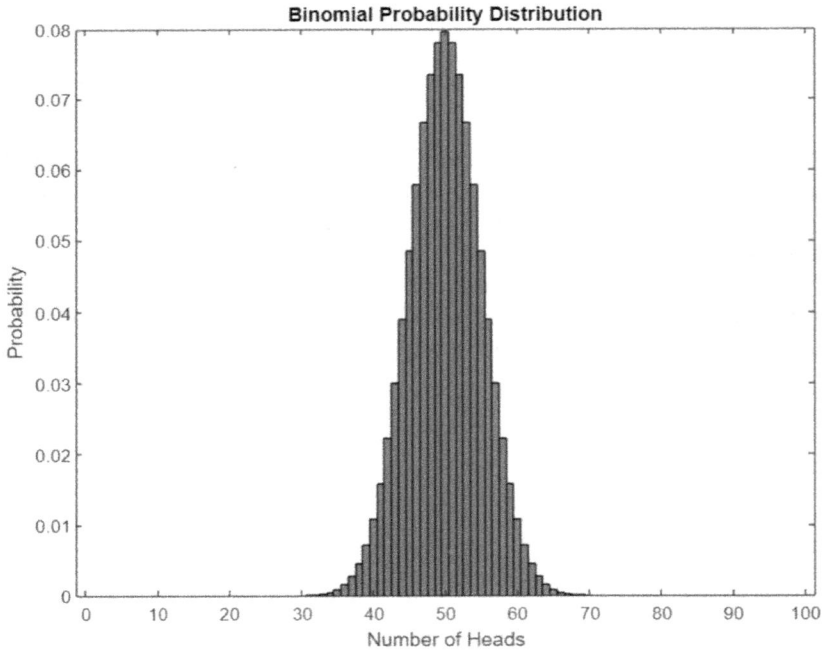

Figure 4.3

Notice that as we keep increasing the number of trials keeping the probability $p=0.5$ the same throughout, the shape of the graph in Figure 4.3 gets more smooth and the vertical bars form a **symmetrical bell shape** which is also known as a "normal curve" or "bell curve" (we will talk about this in the next chapter).

We'll examine the shape of the distribution with this biased coin in Figure 4.4, where it is tossed 20 times, and in Figure 4.5, where it is tossed 100 times.

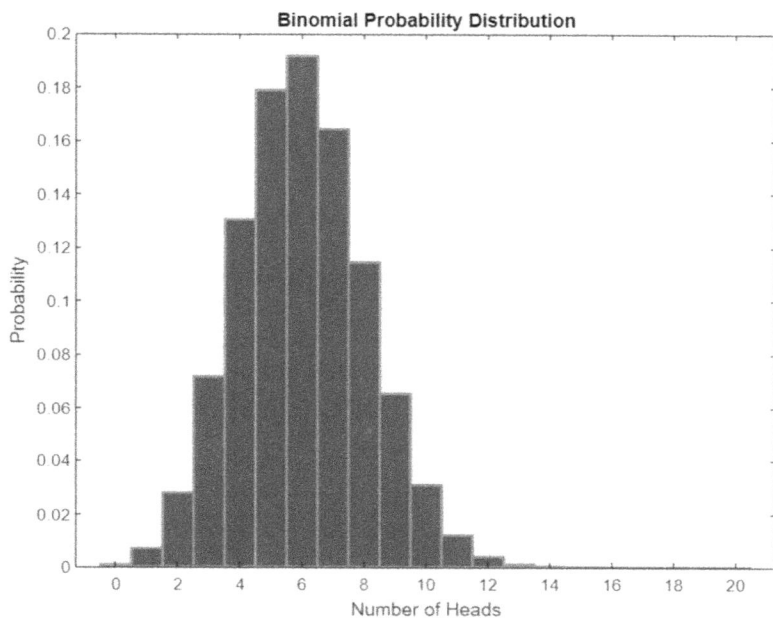

Figure 4.4

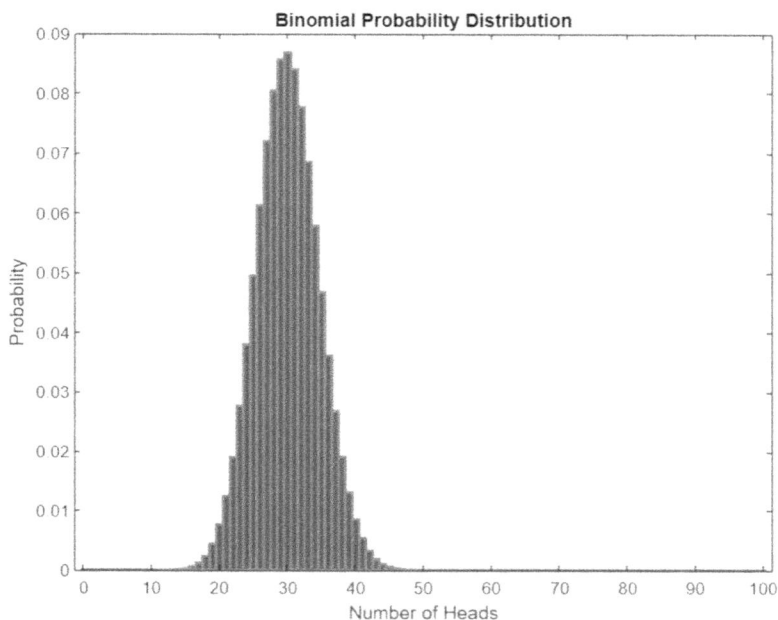

Figure 4.5

Even though the histogram still maintains a similar bell curve shape, we can see it is no longer centered around the middle and instead is shifted towards the lower value. This suggests that the bias of the coin is influencing the outcomes.

This implies that the binomial can also be asymmetrical which is called a skewed distribution.

Whenever $p=0.5$, the binomial distribution will be symmetrical, regardless of how large or small the value of n.

However, when $p \neq 0.5$ and the sample size n is not large enough, then the distribution will be skewed.

If $p < 0.5$, the distribution will be positive or right skewed.
If $p > 0.5$, the distribution will be negative or left skewed.

The closer p is to 0.5 and the larger the number of observations in the sample, n, the more symmetrical the distribution will be.

In engineering, the binomial distribution is often used in quality control to model the number of defective items in a sample. It helps in determining the probability of having a certain number of defects in a given batch or production run.

It is also used to assess the reliability and failure rates of systems or components. It can be employed to estimate the probability of a certain number of failures occurring within a specified time period.

4.2.2 Example

Let's now learn to estimate the probability distribution of the above example both when $p = 0.5$ and $p = 0.3$.

1) If the experiment is tossing a coin 20 times, what is the probability of:
   a) Probability distribution
   b) Getting at most one head.
   c) Getting more than four heads.
   d) Getting less than ten heads.
   e) Getting at least twelve heads
   f) Getting between two and five heads
      I. Inclusive
      II. Exclusive
   g) Find the mean, variance, and deviation.

4.2.3 Solution

a) The probability distribution

$$P(X=0) = \frac{20!}{0!(20-0)!} 0.50^0 (1-0.5)^{20-0} = 0.50^{20}$$

$$P(X=1) = \frac{20!}{1!(20-1)!} 0.50^1 (1-0.5)^{20-1} = 0.00001$$

$$P(X=2) = \frac{20!}{2!(20-2)!} 0.50^2 (1-0.5)^{20-2} = 0.00018$$

$$P(X=3) = \frac{20!}{3!(20-3)!} 0.50^3 (1-0.5)^{20-3} = 0.00108$$

$$P(X=4) = \frac{20!}{4!(20-4)!} 0.50^4 (1-0.5)^{20-4} = 0.00462$$

$$P(X=5) = \frac{20!}{5!(20-5)!} 0.50^5 (1-0.5)^{20-5} = 0.01478$$

$$P(X=6) = \frac{20!}{6!(20-6)!} 0.50^6 (1-0.5)^{20-6} = 0.03696$$

$$P(X=7) = \frac{20!}{5!(20-7)!} 0.50^7 (1-0.5)^{20-7} = 0.07392$$

$$P(X=8) = \frac{20!}{5!(20-8)!} 0.50^8 (1-0.5)^{20-8} = 0.12013$$

$$P(X=9) = \frac{20!}{5!(20-9)!} 0.50^9 (1-0.5)^{20-9} = 0.16017$$

$$P(X=10) = \frac{20!}{10!(20-10)!} 0.50^{10} (1-0.5)^{20-10} = 0.17619$$

$$P(X=11) = \frac{20!}{11!(20-11)!} 0.50^{11} (1-0.5)^{20-11} = 0.16017$$

$$P(X=12) = \frac{20!}{12!(20-12)!} 0.50^{12} (1-0.5)^{20-12} = 0.12013$$

$$P(X=13) = \frac{20!}{13!(20-13)!} 0.50^{13} (1-0.5)^{20-13} = 0.07392$$

$$P(X=14) = \frac{20!}{14!(20-14)!} 0.50^{14} (1-0.5)^{20-14} = 0.03696$$

$$P(X=15) = \frac{20!}{15!(20-15)!} 0.50^{15} (1-0.5)^{20-15} = 0.01478$$

$$P(X=16) = \frac{20!}{16!(20-16)!} 0.50^{16} (1-0.5)^{20-16} = 0.00462$$

$$P(X=17) = \frac{20!}{17!(20-17)!} 0.50^{17} (1-0.5)^{20-17} = 0.00108$$

$$P(X=18) = \frac{20!}{18!(20-18)!} 0.50^{18} (1-0.5)^{20-18} = 0.00018$$

$$P(X=19) = \frac{20!}{19!(20-19)!} 0.50^{19} (1-0.5)^{20-19} = 0.00001$$

$$P(X = 20) = \frac{20!}{20!(20-20)!} 0.50^{20}(1-0.5)^{20-20} = 0.50^{20}$$

Let's now discuss the probabilities. Notice that the distribution of possible outcomes starts with fewer successes (or more failures) on the edges and increases toward the center which is given by np, where n is the number of trials and p is the probability of success in a single trial. In this case our n=20 and

p=0.5; hence our center would be $20*0.5 = 10$. As we move away from this center value, the likelihood of having more or fewer successes decreases symmetrically on both sides.

b) $P(X \le 1) = P(X = 0) + P(X = 1)$

$$= \frac{20!}{0!(20-0)!} 0.50^0 (1-0.5)^{20-0} + \frac{20!}{1!(20-1)!} 0.50^1 (1-0.5)^{20-1}$$

$$= 0.00002$$

c) $P(X > 4) = P(X = 5) + P(X = 6)...... + P(X = 19) + P(X = 20)$

$$= 0.99409$$

Or

$$= 1 - \left[P(X = 0 + P(X = 1) + P(X = 2) + P(X = 3) + P(X = 4) \right] = 1 - P(X \le 4)$$

$$= 1 - \left[0.50^{20} + 0.00001 + 0.00018 + 0.00108 + 0.00462055206 \right] = 0.99410$$

d) $P(X < 10) = P(X \le 9) = P(X = 0) + P(X = 1) + ... + P(X = 8) + P(X = 9)$

$$= 0.50^{20} + 0.00001 + ... + 0.12013435363 + 0.16018 = 0.41187$$

e) $P(X \ge 12) = P(X = 12) + (X = 13) + ...P(X = 19) + P(X = 20)$

$$= 1 - P(X \le 11) = 1 - \left[P(X = 0) + P(X = 1) + ...P(X = 10) + P(X = 11) \right]$$

$$= 1 - \left(0.50^{20} + 0.00001 + .. + 0.1762 + 0.16018 \right) = 0.2517$$

f) Getting between two and five heads

I. Inclusive

When we say inclusive of 2 and 5, then the probability we seek is the probability of getting exactly 2 heads out of 20 tosses, plus the probability of getting exactly 3, plus exactly 4, plus exactly 5.

$$P(2 \le X \le 5) = P(X = 2) + P(X = 3) + P(X = 4) + P(X = 5)$$

$$= 0.00018 + 0.00108 + 0.00462 + 0.01478 = 0.02066$$

II. Exclusive

On the other hand, if we mean exclusive, then the probability we seek is the probability of exactly 3 heads out of 20 tosses plus the probability of exactly 4 heads out of 20 tosses.

$$P(2 < X < 5) = P(X = 3) + P(X = 4) = 0.00108 + 0.00462 = 0.0057$$

g) Mean $= np = 20(0.5) = 10$

Variance $= np(1-p) = 20(0.5)(1-0.50) = 5$

Standard deviation $= \sqrt{5}$

The probabilities when we have a biased coin:

a. The probability distribution

$$P(X = 0) = \frac{20!}{0!(20-0)!} 0.30^0 (1-0.3)^{20-0} = 0.30^{20}$$

$$P(X = 1) = \frac{20!}{1!(20-1)!} 0.30^1 (1-0.3)^{20-1} = 0.00683$$

$$P(X = 2) = \frac{20!}{2!(20-2)!} 0.30^2 (1-0.3)^{20-2} = 0.02784$$

$$P(X = 3) = \frac{20!}{3!(20-3)!} 0.30^3 (1-0.3)^{20-3} = 0.0716$$

$$P(X = 4) = \frac{20!}{4!(20-4)!} 0.30^4 (1-0.3)^{20-4} = 0.13042$$

$$P(X = 5) = \frac{20!}{5!(20-5)!} 0.30^5 (1-0.3)^{20-5} = 0.17886$$

$$P(X = 6) = \frac{20!}{6!(20-6)!} 0.30^6 (1-0.3)^{20-6} = 0.03696$$

$$P(X = 7) = \frac{20!}{5!(20-7)!} 0.50^7 (1-0.5)^{20-7} = 0.07392$$

$$P(X = 8) = \frac{20!}{5!(20-8)!} 0.50^8 (1-0.5)^{20-8} = 0.12013$$

$$P(X = 9) = \frac{20!}{5!(20-9)!} 0.50^9 (1-0.5)^{20-9} = 0.16017$$

$$P(X = 10) = \frac{20'}{10!(20-10)!} 0.50^{10} (1-0.5)^{20-10} = 0.17619$$

$$P(X=11) = \frac{20!}{11!(20-11)!} 0.50^{11} (1-0.5)^{20-11} = 0.16017$$

$$P(X=12) = \frac{20!}{12!(20-12)!} 0.50^{12} (1-0.5)^{20-12} = 0.12013$$

$$P(X=13) = \frac{20!}{13!(20-13)!} 0.50^{13} (1-0.5)^{20-13} = 0.07392$$

$$P(X=14) = \frac{20!}{14!(20-14)!} 0.50^{14} (1-0.5)^{20-14} = 0.03696$$

$$P(X=15) = \frac{20!}{15!(20-15)!} 0.50^{15} (1-0.5)^{20-15} = 0.01478$$

$$P(X=16) = \frac{20!}{16!(20-16)!} 0.50^{16} (1-0.5)^{20-16} = 0.00462$$

$$P(X=17) = \frac{20!}{17!(20-17)!} 0.50^{17} (1-0.5)^{20-17} = 0.00108$$

$$P(X=18) = \frac{20!}{18!(20-18)!} 0.30^{18} (1-0.3)^{20-18} = 0.0000000361$$

$$P(X=19) = \frac{20!}{19!(20-19)!} 0.30^{19} (1-0.3)^{20-19} = 0.00000000163$$

$$P(X=20) = \frac{20!}{20!(20-20)!} 0.30^{20} (1-0.3)^{20-20} = 0.30^{20}$$

b. $P(X \le 1) = P(X=0) + P(X=1)$

$$= \frac{20!}{0!(20-0)!} 0.30^0 (1-0.3)^{20-0} + \frac{20!}{1!(20-1)!} 0.30^1 (1-0.3)^{20-1}$$

0.00763

c. $P(X>4) = P(X=5) + P(X=6)\ldots\ldots + P(X=19) + P(X=20)$

Or

$$= 1 - \left[P(X=0 + P(X=1) + P(X=2) + P(X=3) + P(X=4) \right] = 1 - P(X \le 4)$$

$$= 1 - \left[0.70^{20} + 0.00684 + 0.027846 + 0.071604 + 0.130421 \right] = 0.7625$$

d. $P(X<10) = P(X \le 9) = 0.70^{20} + 0.00684 + \ldots + 0.1144 + 0.06537 = 0.952$

e. $P(X \geq 12) = 1 - P(X \leq 11) = 1 - \left(0.70^{20} + 0.00684 + \ldots 0.03082 + 0.01201\right) = 0.005$

f. Getting between two and five heads
 I. Inclusive

$$P(2 \leq X \leq 5) = P(X = 2) + P(X = 3) + P(X = 4) + P(X = 5)$$

$$= 0.02784 + 0.0716 + 0.13042 + 0.17886 = 0.40872$$

 II. Exclusive

$$P(2 < X < 5) = P(X = 3) + P(X = 4) = 0.0716 + 0.13042 = 0.20202$$

g. Mean $= np = 20(0.3) = 6$
 Variance $= np(1 - p) = 20(0.3)(1 - 0.30) = 4.2$
 Standard deviation $= \sqrt{4.2} = 2.05$

4.2.3.1 Inverse binomial

The cumulative binomial distribution lets us obtain the probability of observing less than or equal to x successes in n trials, with the probability p of success on a single trial. The inverse binomial calculates the inverse of the cumulative binomial distribution—for a given number of independent trials, it estimates the smallest value of s (the number of successes) for which the cumulative binomial distribution is greater than or equal to a supplied probability. Mathematically:

$$P(X \leq s) \geq Y$$

Let's take the same problem and find the minimum number of fair coin tosses needed so that the probability of getting at least five heads is at least 90%.
 We wish to find x such that

$$P(X \geq 5) \geq 0.90$$

$$1 - P(X \leq 4) = 1 - \left(P(X = 0) + P(X = 1) + P(X = 2) + P(X = 3) + P(X = 4)\right) \geq 0.90$$

$$= P(X \leq 4) = P(X = 0) + P(X = 1) + P(X = 2) + P(X = 3) + P(X = 4) \leq 0.1$$

Our cumulative probability is 0.10, the probability of success is 0.5, number of successes = 4.

$$\frac{s!}{0!(s-0)!}0.50^{s} + \frac{s!}{1!(s-1)!}0.50^{s} + \frac{s!}{2!(s-2)!}0.5^{s} + \frac{s!}{3!(s-3)!}0.50^{s} + \frac{s!}{4!(s-4)!}0.50^{s} \leq 0.1$$

$$0.50^{s} + \left(s \times 0.50^{s}\right) + \left(\frac{s*(s-1)}{2} \times 0.50^{s}\right) + \left(\frac{s*(s-1)*(s-2)}{6} \times 0.50^{s}\right)$$

$$+ \frac{s*(s-1)*(s-2)*(s-3)}{24}0.50^{s} \leq 0.1$$

This gives $x = 13.7331$.

We know x must be an integer.

$$P(X=13) = 0.50^{13} + \left(13 \times 0.50^{13}\right) + \left(\frac{13*(13-1)}{2} \times 0.50^{13}\right) + \left(\frac{13*(13-1)*(13-2)}{6} \times 0.50^{13}\right)$$

$$+ \left(\frac{13*(13-1)*(13-2)*(13-3)}{24} \times 0.50^{13}\right) = 0.13342$$

$$P(X=14) = 0.50^{14} + \left(14 \times 0.50^{14}\right) + \left(\frac{14*(14-1)}{2} \times 0.50^{14}\right) + \left(\frac{14*(14-1)*(14-2)}{6} \times 0.50^{14}\right)$$

$$+ \left(\frac{14*(14-1)*(14-2)*(14-3)}{24} \times 0.50^{14}\right) = 0.08978$$

Hence the minimum number of trials required so that the probability of getting at least 5 heads is at least 90% is 14.

g. We are given $P(X \geq 1) > 0.99$

$1 - P(X < 1) > 0.99$

$-P(X < 1) > -1 + 0.99$

$-P(X < 1) > -0.01$

By rules of inequality we flip the signs

$P(X < 1) < 0.01$

$P(X \leq 0) < 0.01$

Our cumulative probability is 0.01, the probability of success is 0.5, number of success = 0

$$\frac{s!}{0!(s-0)!} 0.50^s < 0.01$$

$0.50^s < 0.01$

This gives s = 6.644

We know s must be an integer

$P(X=6) = 0.50^6 = 0.015625$

Clearly this value is not less than or equal to 0.01

$P(X=7) = 0.50^7 = 0.0078125 < 0.01$

Hence the minimum number of trials required so that the probability of getting at least 1 head is more than 99% is 7

4.2.4 Calculations in MATLAB

4.2.4.1 Create a binomial distribution object

4.2.4.1.1 Default parameters

```
pd = makedist('Binomial')

pd =

    BinomialDistribution

    Binomial distribution
        N =   1
        p = 0.5
```

4.2.4.1.2 Specified parameters

```
pd = makedist('Binomial', 'n', 20, 'p', 0.05)

pd =

    BinomialDistribution

    Binomial distribution
        N =    20
        p = 0.05
```

4.2.4.2 Fitting the binomial distribution to sample data

The fitting of the binomial distribution to sample data involves estimating the parameters of the binomial distribution that best describe the observed data. This process allows you to find the most suitable values for the number of trials (n) and the probability of success (p) in the binomial distribution.

4.2.4.2.1 Default parameters

```
pd = fitdist(x, 'Binomial')
```

4.2.4.2.2 Specified parameters

```
pd = fitdist(x, 'Binomial', 'NTrials', ___, 'p', ____)
```

4.2.4.3 Use distribution-specific functions

4.2.4.3.1 binofit

This returns a maximum likelihood estimate of the probability of success and $100(1-\textbf{alpha})\%$ confidence intervals in a given binomial trial based on the number of successes, x, observed in n independent trials.

```
[phat, pci] = binofit(x, n, alpha)
```

4.2.4.3.2 Example

A manufacturing process produces electronic components, and a quality control engineer is interested in estimating the proportion of defective components. A sample of 75 components is selected, and 17 of them are found to be defective.

```
[phat, pci] = binofit(58, 75)
```

```
phat =

    0.7733


pci =

    0.6621    0.8621
```

4.2.4.3.3 binornd

This generates random numbers from the binomial distribution specified by the number of trials n and the probability of success for each trial. The syntax is:

```
binornd(n, p, size)
```

4.2.4.3.4 Example

Suppose an engineering team is conducting a reliability analysis of a specific component. Based on some assumptions, they estimate that the probability of a component failure during a certain time period is 0.025. They want to simulate the number of component failures that could occur in a sample of 100 components during this time period.

```
N = 100; % Sample size (number of components)
p = 0.025; % Probability of failure
s = 10
% Generate 10 random numbers representing component failures
failures = binornd(n, p, [1, s])
```

```
failures =

    1    0    4    1    1    3    2    2    2    3
```

4.2.4.3.5 Example

Now instead of a specific component the same engineering team is conducting a reliability analysis of multiple batches of components. For each batch, they want to simulate the number of defective components based on the known probabilities of defects. They have the following information for five batches:

```
% Number of components in each batch
n = [100 75 56 82 91];
% Probabilities of defects for each batch
p = [0.034 0.056 0.082 0.15 0.012];
% Generate random numbers representing defective components for each
batch
defective_counts = binornd(n, p);
```

$$defective\_counts =$$

$$2 \quad 6 \quad 8 \quad 14 \quad 1$$

4.2.4.3.6 binopdf

This function computes the binomial probability density function at each of the values in x using the corresponding number of trials in n and the probability of success for each trial in p. In other words, it finds the probability that **exactly x** successes occur during n trials where the probability of success on a given trial is equal to p. The syntax is:
```
binopdf(x, n, p)
```

4.2.4.3.7 Example

Suppose the same engineering team now want to assess the probability of a specific number of defective components in a batch of 82. As in the previous example the probability of a component being defective is 0.025 and we want to calculate the probability of having exactly:

a) Five defective components.
b) k defective components in the batch, where k can vary from 0 to 20.

```
x = 5;
n = 82;
p = 0.025;
binopdf(x, n, p)
```

$$ans =$$

$$0.0379$$

```
x = 0:1:20;
n = 82;
p = 0.025;
y = binopdf(x, n, p)
bar(x, y , 1, 'EdgeColor', [0 0 0] , 'FaceColor', [0.5 0.25 0.5],
   'LineWidth',2)
```

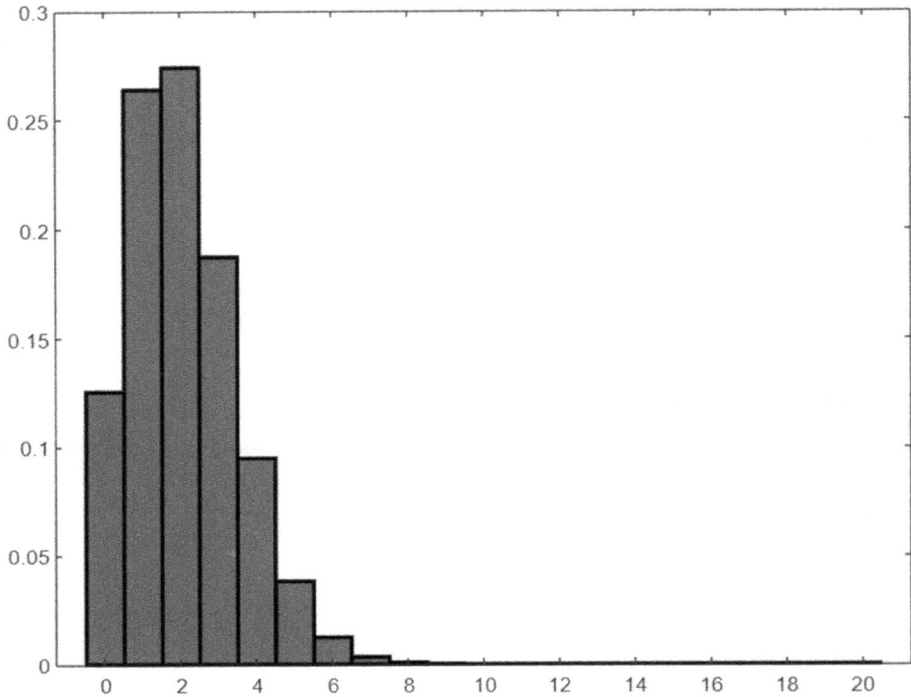

Figure 4.6

4.2.4.3.8 Binocdf

4.2.4.3.8.1 BINOCDF(X, N, P)

This function computes a binomial cumulative distribution function at each of the values of x using the corresponding number of trials, n, and the probability of success for each trial, p. In other words, it calculates the probability that the number of successes is less than, or equal to, x where the number of trials is given as n and the probability of success for each trial is p.

4.2.4.3.8.2 BINOCDF(X, N, P, 'UPPER')

This returns the complement of the binomial cumulative distribution function for each value of x.

4.2.4.3.9 Example

Suppose a manufacturer wants to test the reliability of their air conditioning units. They have a batch of 75 units, and based on historical data and engineering analysis, they estimate that each unit has a 23% chance of failing the performance test. Find the probability that:

a) Twenty or fewer units failed to complete the test.

```
X = 20;
n = 75;
p = 0.23;
binocdf(x, n, p)
```
$$ans =$$
$$0.8152$$

b) Fewer than ten units failed to complete the test.

$$P(X < 10) = P(X \le 9)$$

```
x = 10;
n = 75;
p = 0.23;
binocdf(x - 1, n, p)
```
$$ans =$$
$$0.0125$$

c) More than 15 units failed to complete the test.

$$P(X > 15) = 1 - P(X \le 15)$$

```
x = 15;
n = 75;
p = 0.23;
1 - binocdf(x, n, p)
binocdf(x, n, p, 'upper')
```
$$ans =$$
$$0.6774$$

$$ans =$$
$$0.6774$$

d) Thirty or more units failed to complete the test.

$$P(X \ge 30) = 1 - P(X \le 29)$$

```
x = 30;
n = 75;
p = 0.23;
1 - binocdf(x - 1, n, p)
```

```
binocdf(x - 1, n, p, 'upper')
```

ans =

7.4566e-04

ans =

7.4566e-04

4.2.4.3.10 Binoinv

This returns the smallest integer x such that the probability of obtaining x successes after N trials—each of which has a probability of success of p—is greater than or equal to Y. We can think of Y as the probability of observing x successes in N independent trials where p is the probability of success in each trial. Each X is a positive integer less than or equal to N.
```
X = binoinv(Y, N, p)
```

4.2.4.3.11 Example

```
Y = 0.90
N = 14
p = 0.5
X = binoinv(1 - Y, N, p)
```

ans =

5

4.2.4.3.12 Binostat

This returns the mean and variance for the binomial distribution with parameters specified by the number of trials, n, and probability of success for each trial, p.

```
[M, V] = binostat(n, p)
```

1) Going back to the problem but this time using MATLAB functions:
   a) Probability distribution of the fair coin.

```
x = 0:20;
n = 20;
p = 0.5;
y = binopdf(x, n, p)
bar(x, y, 1)
title("PMF for a Fair coin")
```

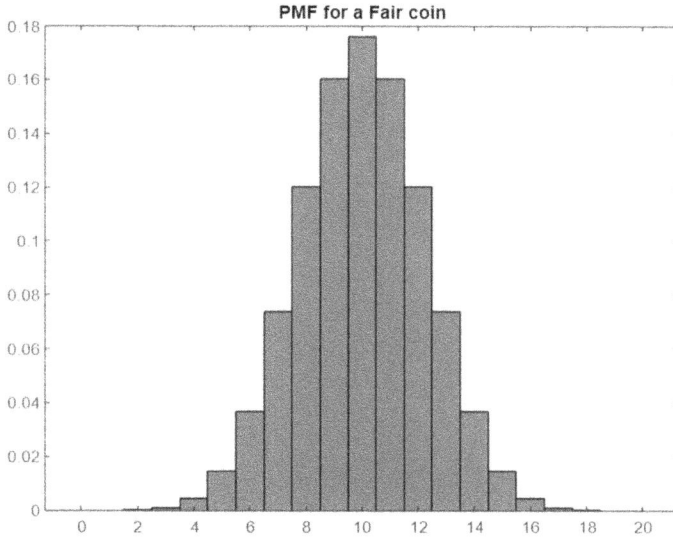

Figure 4.7

b) Getting at most one head.

```
x = 1;
n = 20;
p = 0.5;
y = binocdf(x, n, p);
```

```
y =

    2.0027e-05
```

c) Getting more than four heads.

```
1        p = 0.5;
2        n = 20;
3        x = 4;
4        binocdf(x, n, p, 'upper');
```

```
Command Window
>> Binomial

ans =

    0.9941
```

d) Getting less than ten heads.

$$P(X \geq 2) = 1 - P(X < 2) = 1 - P(X \leq 1)$$

```
1        p = 0.5;
2        n = 20;
3        x = 10;
4        binocdf(x - 1, n, p);
```

```
Command Window
>> Binomial

ans =

    0.4119
```

e) Getting at least 12 heads

```
1    p = 0.5;
2    n = 20;
3    x = 12;
4    binocdf(x - 1, n, p, 'upper');
```

```
Command Window
>> Binomial

ans =

    0.2517
```

f) Getting between two and five heads.

I. Inclusive

$$P(2 \leq X \leq 5) = P(X = 2) + P(X = 3) + P(X = 4) + P(X = 5)$$

$$= P(X \leq 5) - P(X < 2) = P(X \leq 5) - P(X \leq 1)$$

```
x1 = 5;
x2 = 2;
n = 20;
p = 0.5;
y = binocdf(x1, n, p) - binocdf(x2 - 1, n, p);
                    y =

                    0.0207
```

II. Exclusive

$$P(2 < X < 5) = P(X = 3) + P(X = 4) = P(X < 5) - P(X < 2)$$

$$= P(X < 5) - P(X \leq 2) = P(X \leq 4) - P(X \leq 2)$$

```
x1 = 5;
x2 = 2;
n = 20;
p = 0.5;
y = binocdf(x1 - 1, n, p) - binocdf(x2, n, p);
                    y =

                    0.0057
```

g) Find mean, variance, and deviation.

```
n = 20;
p = 0.5;
[Mean, Variance] = binostat(n, p);
                Mean =

                    10


                Variance =
```

5

For the biased coin:

a) Probability distribution of the biased coin.

```
x = 0:20;
n = 20;
p = 0.3;
y = binopdf(x, n, p)
bar(x, y, 1)
title("PMF for a Biased coin")
```

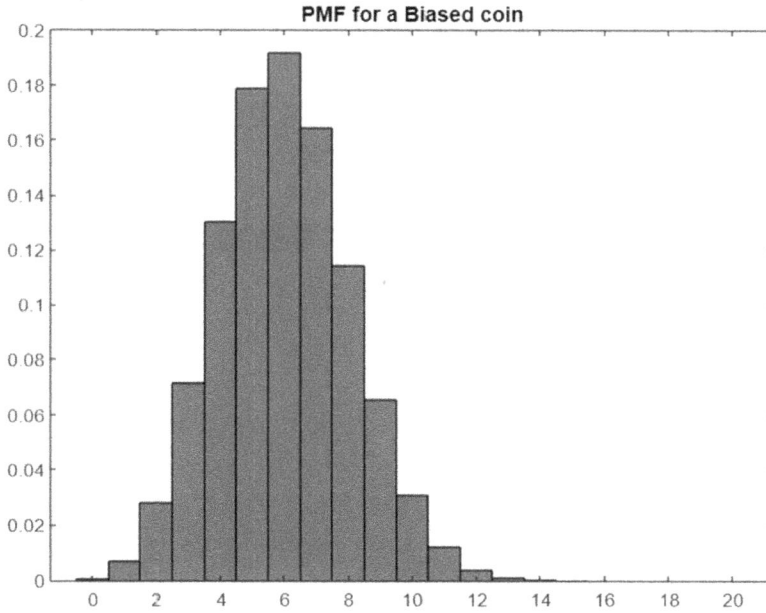

Figure 4.8

b) Getting at most one head.

```
x = 1;
n = 20;
p = 0.3;
y = binocdf(x, n, p);
```

y =

0.0076

c) Getting more than four heads.

```
1    p = 0.3;
2    n = 20;
3    x = 4;
4    binocdf(x, n, p, 'upper');
```

Command Window

```
>> Binomial

ans =

0.7625
```

d) Getting less than ten heads.

$$P(X \geq 2) = 1 - P(X < 2) = 1 - P(X \leq 1)$$

```
1    p = 0.3;
2    n = 20;
3    x = 10;
4    binocdf(x - 1, n, p);
```

Command Window

```
>> Binomial

ans =

    0.9520
```

e) Getting at least 12 heads

```
1    p = 0.3;
2    n = 20;
3    x = 12;
4    binocdf(x - 1, n, p, 'upper');
```

Command Window

```
>> Binomial

ans =

    0.0051
```

f) Getting between two and five heads.
   I. Inclusive

$$P(2 \leq X \leq 5) = P(X = 2) + P(X = 3) + P(X = 4) + P(X = 5)$$

$$= P(X \leq 5) - P(X < 2) = P(X \leq 5) - P(X \leq 1)$$

```
x1 = 5;
x2 = 2;
n = 20;
p = 0.3;
y = binocdf(x1, n, p) - binocdf(x2 - 1, n, p);
                    y =

                        0.4087
```

II. Exclusive

$$P(2 < X < 5) = P(X = 3) + P(X = 4) = P(X < 5) - P(X < 2)$$

$$= P(X < 5) - P(X \leq 2) = P(X \leq 4) - P(X \leq 2)$$

```
x1 = 5;
x2 = 2;
n = 20;
p = 0.3;
y = binocdf(x1 - 1, n, p) - binocdf(x2, n, p);
                    y =

                        0.2020
```

g) Find mean, variance, and deviation.

```
n = 20;
p = 0.3;
[Mean, Variance] = binostat(n, p);
                          Mean =

                            6


                          Variance =

                            4.2000
```

4.2.4.3.13 Example

The FE Exam is a comprehensive exam that tests fundamental knowledge in engineering principles. It is often the first step toward becoming a licensed professional engineer (PE). This examination consists of 110 multi-choice questions, each with 4 suggested answers of which only 1 is correct. Tom is completely unprepared for the exam and just guesses all of the answers to the 110 questions randomly.

a) Find and plot the probability distribution showing the probabilities associated with each number of correct answers.

```
x = 0:1:110;
n = 110;
p = 0.25;
y = binopdf(x, n, p)
bar(x, y , 1, 'EdgeColor', [0 0 0] , 'FaceColor', [0.75 0.75 0.75])
xlabel("Number of correct answers")
ylabel("Probability")
```

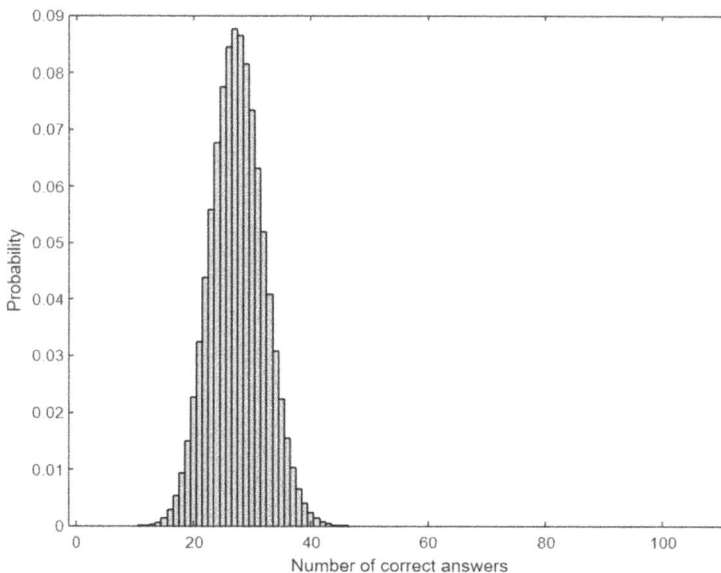

Figure 4.9

b) In order to pass the test he needs to answer 60 or more questions correctly. What is the probability that Tom will pass the exam?

$$P(X \geq 60) = P(X = 60) + P(X = 61) + ... P(X = 110) = 1 - P(X \leq 59)$$

```
1        pass = 59;
2        n = 110;
3        p = 0.25;
4        1 - binocdf(pass, n, p)
5
```

Command Window

```
>> untitled002

ans =

   3.6613e-11
```

c) What is the probability that Tom will get at least one of the questions correct?

$$P(X \geq 1) = 1 - P(X = 0)$$

```
1        n = 110;
2        p = 0.25;
3        1 - binopdf(0, n, p)
4
5
```

Command Window

```
>> untitled002

ans =

   1.0000
```

d) Find the expected value of the number of correct answers.

```
1        n = 110;
2        p = 0.25;
3        Mean = binostat(n, p)
4
```

Command Window

```
>> untitled002

Mean =

   27.5000
```

4.2.4.3.14 Example

We have a batch of 100 bolts and the probability of a bolt being defective is 0.30.

a) Calculate the probability distribution of 50 defective bolts and plot the distribution.
b) Estimate the probability that the sample has less than or equal to 25 defective bolts?
c) Estimate the probability that the sample has less than or equal to 25 defective bolts?
d) What is the probability that the sample has more than 30 defective bolts?

```
x = 0:1:50;
n = 100;
p = 0.30;
y = binopdf(x, n, p)
bar(x, y , 1, 'EdgeColor', [0 0 0] , 'FaceColor', [0.25 0.25 0.5])
xlabel("Number of defective bolts")
ylabel("Probability")
title("Probability distribution of 50 defective bolts")
x = 25;
```

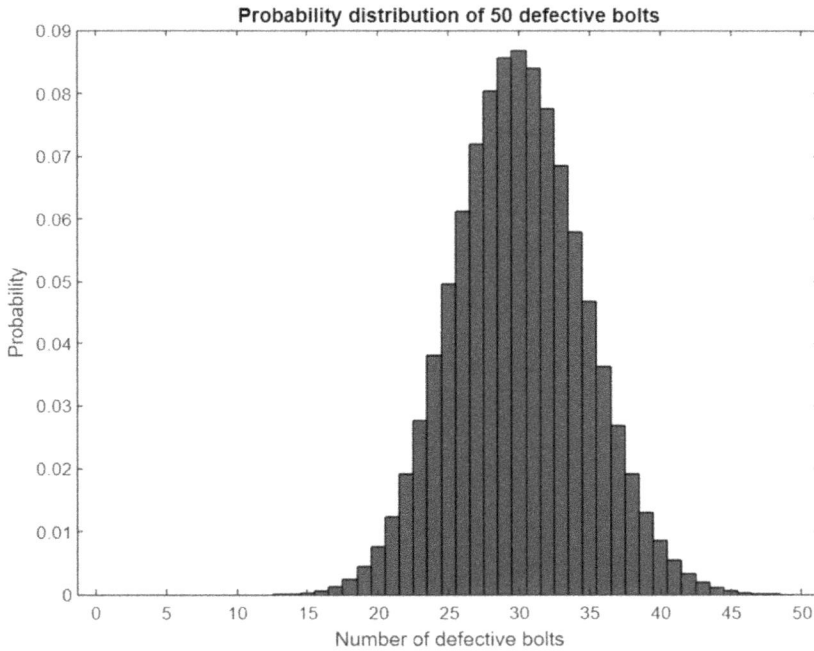

Figure 4.10

```
n = 100;
p = 0.30;
y = binocdf(x, n, p);
```
$$y =$$
$$0.1631$$
```
x = 20;
n = 100;
p = 0.30;
y = binocdf(x-1, n, p);
```
$$y =$$
$$0.0089$$
```
x = 30;
n = 100;
p = 0.30;
y = binocdf(x, n, p, 'upper');
```

$$y =$$

$$0.4509$$

```
x = 40;
n = 100;
p = 0.30;
y = binocdf((x-1, n, p, 'upper');
```
$$y =$$

$$0.0210$$

4.2.4.3.15 Example

An oil production company is considering drilling five oil wells to produce crude oil. The probability of success for each well is 0.33, independent of the results for any other well. The cost of drilling an oil well is $150,000. Each well that is successful will be worth $500,000.

a) What is the probability that one or more wells will be successful?

$$P(X \geq 1) = P(X = 1) + P(X = 2) + P(X = 3) + P(X = 4) + P(X = 5) = 1 - P(X = 0)$$

```
n = 5
p = 0.33
x = 0
1 - binopdf(x, n, p)
```
$$ans =$$

$$0.8650$$

b) What is the expected number of successes?

```
Mean = binostat(n, p)
```
$$Mean =$$

$$1.6500$$

c) What is the expected gain?

```
(Mean * 500000) - (n * 150000)
```
$$x =$$

$$7.5000e+04$$

d) What will be the gain if only one well is successful?

```
(1 * 500000) - (n * 150000)
```
$$x =$$

$$-250000$$

1) According to *Industrial & Engineering Chemistry Research*, approximately 32% of all pipework failures in chemical plants are caused by pipeline corrosion.
   a) What is the probability that out of the next 25 pipework failures at least 12 are due to pipeline corrosion?

$$P(X \geq 12) = P(X = 12) + P(X = 13) + \ldots P(X = 23) + P(X = 24) + P(X = 25)$$

$$= 1 - P(X \leq 11)$$

```
1       n = 25;
2       p = 0.32;
3       x = 11;
4       1 - binocdf(x, n, p)
```

Command Window

```
>> untitled001

ans =

    0.0698
```

this is equivalent to binocdf(x, n, p, 'upper')

b) What is the probability that no more than 5 out of 20 such failures are due to pipe-line corrosion?

$$(X \leq 5) =$$

```
1       n = 20;
2       p = 0.32;
3       x = 5;
4       binocdf(x, n, p)
```

Command Window

```
>> untitled001

ans =

    0.3426
```

c) Suppose, for a particular plant, that, out of the random sample of 50 such failures, exactly 15 are pipeline corrosion. Do you feel that the 32% figure stated above applies to this plant? Comment.

```
1       n = 50;
2       p = 0.32;
3       x = 15;
4       binopdf(x, n, p)
```

Command Window

```
>> untitled001

ans =

    0.1168
```

Since the probability is significant we can conclude that the 32% figure stated above applies to this plant.

4.3 BERNOULLI DISTRIBUTION

A Bernoulli distribution is a special type of binomial distribution where, unlike a binomial distribution, only a single trial is conducted to find the probability of an outcome. Hence number of trials in a Bernoulli distribution is always one. Other than that it has exact same properties as the binomial distribution, with two outcomes—success and failure. The coin toss example is perhaps the easiest way to explain the Bernoulli distribution. When we flip a coin one time the probability that it lands on heads is 1/2. In this case, the random variable X follows a Bernoulli distribution since the coin has been flipped just once. **Now, if we flip a coin multiple times, then the sum of the Bernoulli random variables will follow a binomial distribution.** For example, suppose we flip a coin ten times and we want to know the probability of obtaining heads k times. We would say that the random variable X follows a binomial distribution.

Since the Bernoulli distribution is a special case of the binomial distribution where a single trial is conducted so n would be 1 therefore the only parameter of Bernoulli is the probability of success.

4.3.1 Calculations in theory

If a random variable X follows a bernoulli distribution, then the probability of getting a success in the trial is given by:

$$P(X = x) = p^x(1-p)^{1-x} = \begin{cases} p; & x = 1 \\ 1-p; & x = 0 \end{cases}$$

Recall that the mean and variance of a binomial distribution is given by $E[X]=np$ and $Var = Np(1-p)$. Since we know the number of trials $n = 1$, we can conclude that:

- The mean or expected value of a Bernoulli distribution is given by $E[X] = 1 \times p = p$.
- The variance of a Bernoulli distribution is given by $Var[X] = 1 \times p \times (1-p) = p \times (1-p)$.

So, for our coin example the graphical representation of our probability distribution is

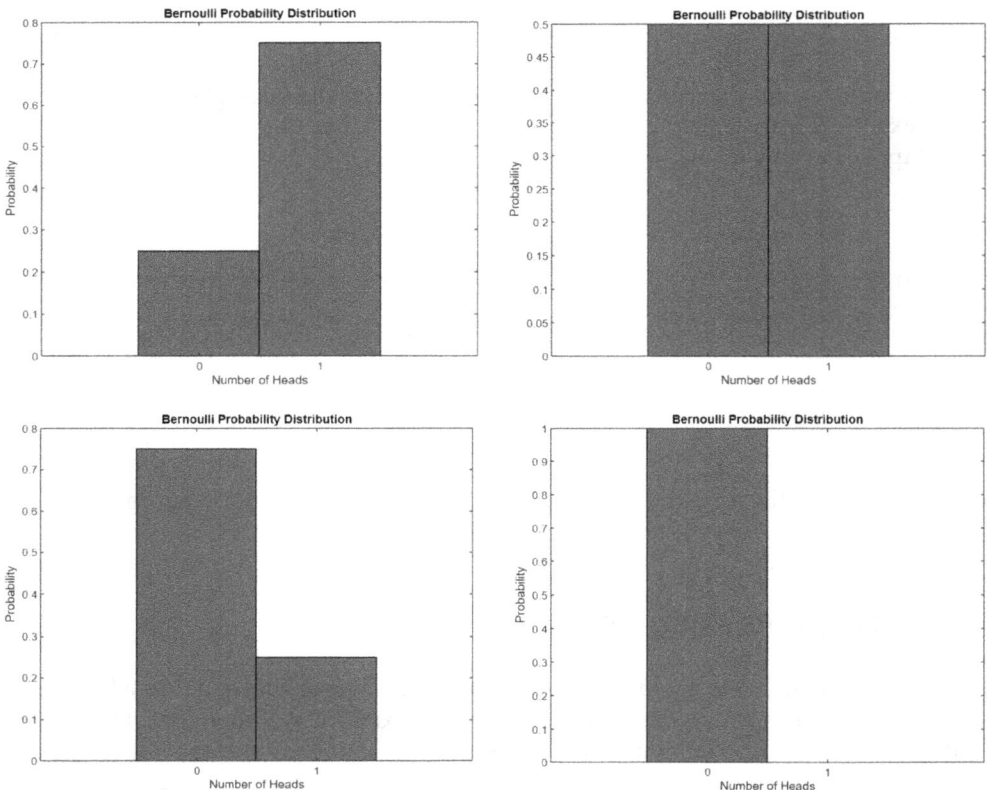

Figure 4.11

Let's consider the scenario of rolling a fair six-sided die. Unlike a coin toss, which has only two possible outcomes (heads or tails) rolling a die results in one of six possible outcomes: 1, 2, 3, 4, 5, or 6. Consequently, each outcome of a die roll cannot be modeled by a Bernoulli random variable, which is defined as the probability distribution of a random variable that takes on only two possible values. However, we can still analyze specific events that can be considered in a binary context. For example, we might be interested in whether the outcome is 1 or not, whether it is 6 or not, or whether the result is an odd number or not. These events can be framed as Bernoulli trials:

- The event of rolling a 1 (success) versus not rolling a 1 (failure).
- The event of rolling a 6 (success) versus not rolling a 6 (failure).
- The event of rolling an odd number (success) versus not rolling an odd number (failure).

4.3.2 Calculations in MATLAB

Unfortunately, MATLAB does not have a specific function for Bernoulli. This is because the Bernoulli distribution can be considered a binomial distribution with the only exception of the number of trials, which is just one for Bernoulli. Therefore, in MATLAB, to calculate the probability distribution for Bernoulli we will use a binomial distribution but assign the number of trials parameter n to 1. Other than that everything else stays the same.

binopdf	binopdf(x,1,p) computes the Bernoulli probability density function at each of the values in x in a single trial n and probability of success for the trial, p.
binocdf	binocdf(x,1,p) computes a binomial cumulative distribution function using the same parameters.
binostat	binostat(1,p) returns the mean and variance for the Bernoulli distribution with parameters specified by probability of success for the trial, p.
binornd	binornd(1,p) generates random numbers from the binomial distribution specified by the number of trials n and the probability of success for each trial p.

4.4 MULTINOMIAL DISTRIBUTION

Recall that in a binomial distribution we perform an experiment n number of times and can have only two outcomes, success or failure, and the binomial random variable is the number of times we obtain one of the two outcomes.

The multinomial distribution is a generalization of the *binomial distribution*. While the binomial distribution gives the probability of x number of "successes" in n independent trials of a two-outcome process, the multinomial distribution gives the probability of each combination of outcomes in n independent trials of a k-outcome process. This is also sometimes referred to as categorical distribution as each possible outcome is treated as a separate category.

Consider the scenario of playing a game of chess n number of times. In each game you can have one of three outcomes: either you win, lose, or the game is a draw. A multinomial distribution determines the combined probability of each outcome in any one trial given by the fixed probabilities p1, p2, and p3. In fact it can calculate the probability of any number of outcomes k.

There is a single parameter p used in a multinomial distribution.

Parameter	Description
p	Probabilities of each outcome

4.4.1 Calculations in theory

If a *random variable X* follows a multinomial distribution, then the probability of getting k number of outcomes is given by

$$P\left(X_1 = x_1, X_2 = x_2, \ldots X_k = x_k\right) = \frac{n!}{n_1! \; n_2! \ldots \ldots n_k!} \; p_1^{n_1} \; p_2^{n_2} \ldots \ldots p_k^{n_k}$$

where,

$$\sum_{i=1}^{k} n_i = n, \sum_{i=1}^{k} p_i = 1.$$

The mean of the multinomial distribution is $E[x_i] = np_i$.

The variance of the multinomial distribution is $Var[x_i] = np_i(1 - p_i)$.

4.4.1.1 Example

An engineer is assessing the structural vulnerability of a building over a five-year period. The engineer has historical data that indicates the probabilities of damage occurrences due to:

- A natural disaster (ND) causing 12% of the damage
- Poor maintenance (PM) causing 24% of the damage
- Foundation problems (F) causing 16% of the damage
- Construction defects (CD) causing 27% of the damage
- Water damage (WD) causing 21% of the damage

If there have been a total of 12 damage occurrences in a building over the 5-year time period, find the probability of:

a) Five damage occurrences caused by poor maintenance, three damage occurrences caused by construction defects, and four damage occurrences caused by water in the building.

$$P\left(5\,PM, 3\,CD, 4\,WD\right) = \frac{12!}{5!3!4!}\left(0.24\right)^5\left(0.27\right)^3\left(0.21\right)^4 = 0.00084$$

b) At most three damage occurrences caused by construction defects.

$$P\left(X \leq 3\,CD\right) =$$

$$= \frac{12!}{0!12!}\left(0.27\right)^0\left(1-0.27\right)^{12} + \frac{12!}{1!11!}\left(0.27\right)^1\left(1-0.27\right)^{11} + \frac{12!}{2!10!}\left(0.27\right)^2\left(1-0.27\right)^{10}$$

$$+ \frac{12!}{3!9!}\left(0.27\right)^3\left(1-0.27\right)^9$$

$$= 0.58624$$

c) At least one damage occurrence caused by earthquakes.

$$P\left(X \geq 1\,ND\right) = P\left(X = 1\right) + P\left(X = 2\right) + \ldots \ldots \ldots \ldots \ldots \ldots P\left(X = 12\right)$$

$$= 1 - P\left(X = 0\right) = 1 - \left[\frac{12!}{0!12!}\left(0.12\right)^0\left(1-0.12\right)^{12}\right] = 0.78432$$

d) Find the mean and variance.

$$E[x_i] = 12(0.12)(0.24)(0.16)(0.27)(0.21) = 0.0031352832$$

$$Var[x_i] = 12((0.12)(0.24)(0.16)(0.27)(0.21))(1-(0.12)(0.24)(0.16)(0.27)(0.21)) = 0.00313$$

4.4.2 Calculations in MATLAB

4.4.2.1 Create a multinomial distribution object

4.4.2.1.1 Default parameters

```
pd = makedist('Multinomial')
                    pd =

                MultinomialDistribution

                Probabilities:
                    0.5000      0.5000
```

4.4.2.1.2 Specified parameters

```
pd = makedist('Multinomial','Probabilities',[1/2 1/4 1/6 1/12])
                pd =

                MultinomialDistribution

                Probabilities:
                    0.5000      0.2500      0.1667      0.0833
```

4.4.2.2 Fit the multinomial distribution to sample data

The fitdist function in MATLAB estimates the parameters of the specified distribution to best fit the data. Unfortunately it is not directly applicable for fitting a multinomial distribution to data. Since the parameters of the multinomial distribution are probabilities associated with each category or outcome, we can use the method of maximum likelihood estimation (MLE) or other appropriate techniques specific to multinomial distributions.

4.4.2.3 Use distribution-specific functions

The following functions are the only ones offered:

mnpdf	$mnpdf([n_1,n_2,\ldots n_k], [p_1,p_2,\ldots p_k]$ returns the PDF for the multinomial distribution with k number of probabilities of each combination of outcomes n^i
mnrnd	$mnrnd(n,p)$ returns random values from the multinomial distribution with parameters n and p

Trying to solve the same problem using MATLAB we get:

a) The probability distribution

```
ND = 0.12;
PM = 0.24;
F = 0.16;
CD = 0.27;
WD = 0.21;
x = categorical({'Natural disaster', 'Poor maintenance', 'Foundation problems', 'Construction defects', 'Water Damage'});
x = reordercats(x,{'Natural disaster', 'Poor maintenance','Foundation problems', 'Construction defects', 'Water Damage'});
y = [ND PM F CD WD];
bar(x, y)
xlabel('Occurences')
ylabel('Probability Mass')
title('Mutinomial Distribution')
```

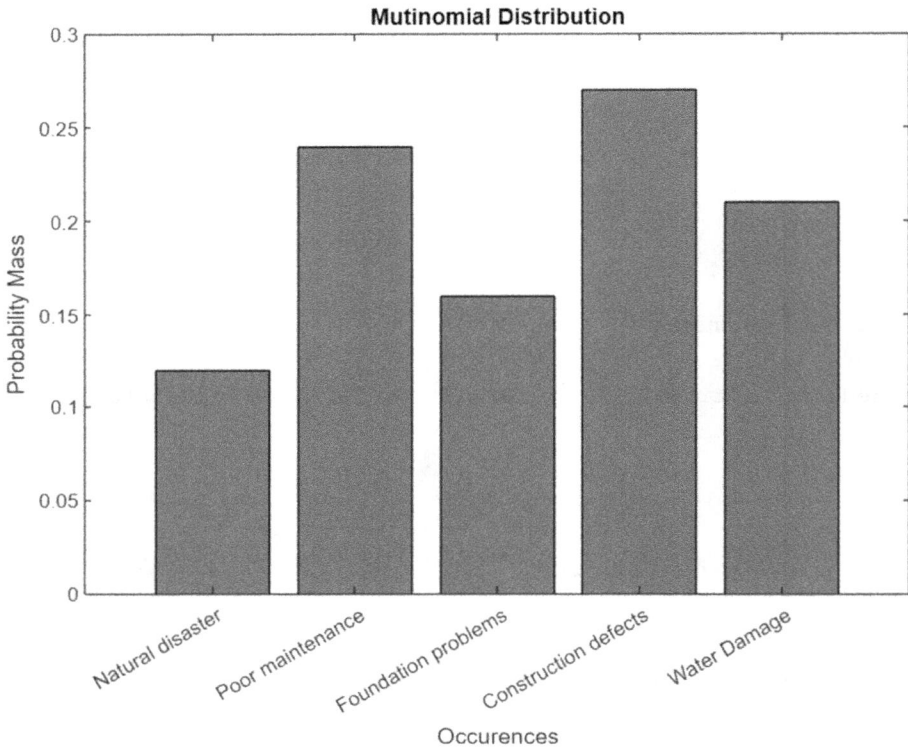

Figure 4.12

```
bar(x, y)
xlabel('Occurences')
ylabel('Probability Mass')
title('Mutinomial Distribution')
```

b) $P(5PM, 3CD, 4WD)$

```
ND = 0.12;
PM = 0.24;
F = 0.16;
CD = 0.27;
WD = 0.21;
mnpdf([0 5 0 3 4],[ND PM F CD WD])
```

                              ans =

                    8.4492e-04

c) $P(X \leq 3CD)$

```
x = 0:3;
y = 12:-1:9;

totalProbability = 0;

for i = 1:numel(x)
    combination = [x(i) y(i)]; % Generate
    probability = mnpdf(combination, [CD (1-CD)]); % Calculate
    totalProbability = totalProbability + probability;
end

disp(totalProbability)
```

                    0.5863

d) $P(X \geq 1ND) = 1 - P(X = 0ND)$

```
1 - mnpdf([0 12],[ND (1-ND)])
```

                              ans =

                    0.7843

4.4.2.4 Example

Two valves are used to control the flow of liquid out of a storage tank (Tank 1) into another storage tank (Tank 2) and controlled by Valves 1 and 2 as seen below. The valves exist in two states: open and closed. Let's first consider all possible combinations of open and closed states for Valve 1 and Valve 2 along with the probability of that configuration having the best flow.

Valve 1	Valve 2	Probability
Open	Open	0.417
Open	Closed	0.325
Closed	Open	0.232
Closed	Closed	0.026

We collected data for 100 experiments with different valve configurations and recorded the number of times the desired flow was achieved for each configuration given below:

Valve 1	Valve 2	Frequency of desired flow
Open	Open	48
Open	Closed	29
Closed	Open	21
Closed	Closed	2

The probability distribution function is

```
OO = 0.417
OC = 0.325
CO = 0.232
CC = 0.026
mnpdf([48 29 21 2],[OO OC CO CC])
                        ans =

                    0.0011
```

4.5 HYPERGEOMETRIC DISTRIBUTION

Recall that if a random variable X has a binomial distribution, the probability is the same for every trial. For a hypergeometric distribution, each trial changes the probability for each subsequent trial because there is no replacement. Hence we could say that the probability of success and failure in a hypergeometric distribution is not fixed unlike in a binomial distribution. The hypergeometric distribution models the total number of successes in a fixed-size sample drawn without replacement from a finite population. Sampling "without replacement" means that once a particular sample is chosen, it is removed from the relevant population for all subsequent selections. Think of a bag with ten marbles of which six are red. Now if you draw a marble from the bag the probability of getting a red marble would be 6/10. Without replacing it, if you choose a second one then the probability of choosing a marble of the same color will be reduced to 5/9. In fact, every time you pick a marble without returning it to the bag the probability of drawing a red marble or a marble of any color is changed. This model can be explained using the hypergeometric distribution. This type of distribution is unlike the binomial distribution where the probability remains constant through the trials.

The **hypergeometric distribution** describes the probability of choosing x objects with a certain characteristic which will be considered a success in n draws without replacement, from a finite population of size M that contains K objects with that characteristic or feature.

As discussed above, the hypergeometric distribution has three parameters that have direct physical interpretations.

- M = the size of the population.
- n = the number of samples drawn from the population.
- K = the number of items with the desired characteristic in the population.

4.5.1 Calculations in theory

If a *random variable* X follows a hypergeometric distribution, then the probability distribution of choosing k objects with a certain feature can be found by the following formula:

$$P(X=x) = \frac{{}^{K}C_x \, {}^{M-K}C_{n-x}}{{}^{M}C_n} = \frac{\binom{K}{x}\binom{M-K}{n-x}}{\binom{M}{n}}$$

In this formula:

- We may select n items from a population of M items in ${}^{M}C_n$ ways.
- We may select x defective items from K defective items in ${}^{K}C_x$ ways.
- We may select $n-x$ non-defective items from $M-K$ non-defective items in ${}^{M-K}C_{n-x}$ ways.

  - The mean of the hypergeometric distribution is *mean* $= \dfrac{nk}{M}$.
  - The variance of the hypergeometric distribution is *var* $= \dfrac{n \times k \times (M-k) \times (M-n)}{M^2(M-1)}$.

Since there are three parameters let's see how the distribution changes as one parameter is altered while keeping the population size and the other parameter constant in the following example.

A helical gear is a type of cylindrical-shaped gear used in industrial equipment. Suppose a mechanical engineer is inspecting a shipment of 100 gears; if he is taking a sample of 50, let's see how the distribution changes as the proportion of "helical" gears changes.

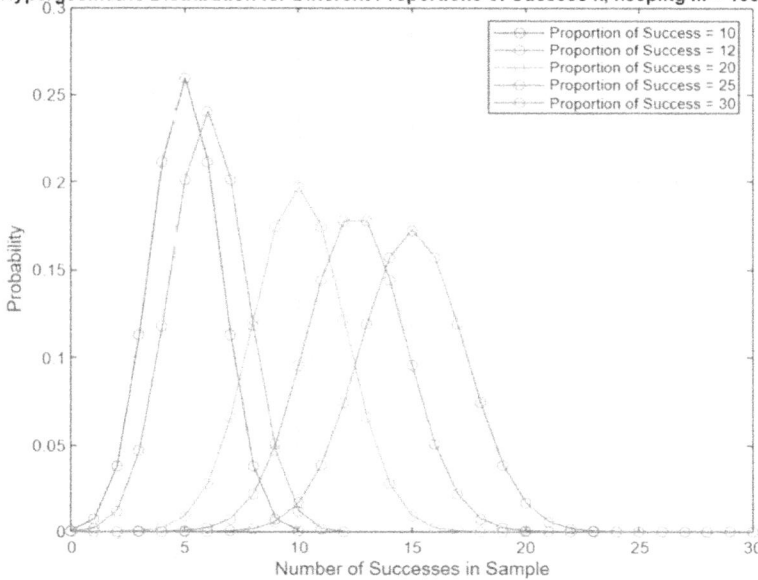

Hypergeometric Distribution for Different Proportions of Success k, keeping M = 100, n = 50

```
M = 100; % Total population size
n = 50;  % Sample size
k_values = [10, 12, 20, 25, 30]; % Different proportion of success in the population.

% Create figure
figure;

% Plot PMF for each proportion of success
legend_labels = cell(length(k_values), 1);
for i = 1:length(k_values)
    k = k_values(i); % Number of success states in the population
    x = 0:min(n, k); % Possible number of successes in the sample
    pmf = hygepdf(x, M, k, n); % Probability mass function
    plot(x, pmf, 'o ', 'DisplayName', ['Proportion of Success = ' num2str(k)]);
    hold on;
    legend_labels{i} = ['Proportion of Success = ' num2str(k)];
end

% Adjust plot properties
title('Hypergeometric Distribution for Different Proportions of Success k, keeping M = 100, n = 50');
xlabel('Number of Successes in Sample');
ylabel('Probability');
legend(legend_labels);
grid on;
hold off;
```

The following graph shows how the distribution changes as the sample size changes, while keeping the population at and proportion of "defective" items constant.

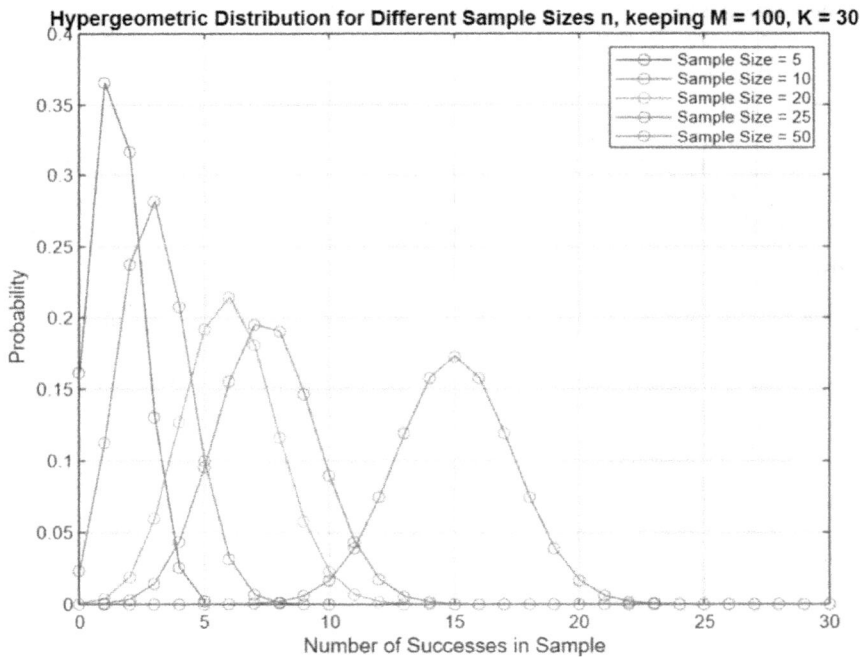

```
M = 100; % Total population size
K = 30;  % Total number of success states in the population
n_values = [5, 10, 20, 25, 50]; % Different sample sizes

% Create figure
figure;

% Plot PMF for each sample size
legend_labels = cell(length(n_values), 1);
for i = 1:length(n_values)
    n = n_values(i); % Sample size
    x = 0:min(n, K); % Possible number of successes in the sample
    pmf = hygepdf(x, N, K, n); % Probability mass function
    plot(x, pmf, 'o', 'DisplayName', ['Sample Size = ' num2str(n)]);
    hold on;
    legend_labels{i} = ['Sample Size = ' num2str(n)];
end

% Adjust plot properties
title('Hypergeometric Distribution for Different Sample Sizes n, keeping M = 100, K = 30');
xlabel('Number of Successes in Sample');
ylabel('Probability');
legend(legend_labels);
grid on;
hold off;
```

4.5.1.1 *Example*

Suppose an engineer is conducting a quality control inspection on a batch of 25 electronic components. The batch contains 6 defective components and 19 non-defective components. The engineer randomly selects a sample of ten components from the batch without replacement. Find:

a) The probability that an equal number of defective and non-defective components will be in the sample.
b) The probability that at most two defective items will be in the sample.
c) The probability that more than 5 non defective items will be in the sample.
d) probability that more than 5 non defective items will be in the sample?.
e) probability that more than 5 non defective items will be in the sample.
f) The expected number of defective items in the sample.
g) The variance of the non defective items in the sample.

a) We select 10 components from $6+19=25$ electronic components in $^{25}C_{10}$ ways; we select 5 defective items from 6 defective items in $^{6}C_{5}$ ways; and we select 5 non-defective items from 19 non-defective items in $^{19}C_{5}$ ways:

$$P\left(X=5 \text{ defective item}\right) = \frac{^{6}C_{5}\,^{19}C_{5}}{^{25}C_{10}} = \frac{\binom{6}{5}\binom{19}{5}}{\binom{25}{10}} = 0.02134$$

$$P\left(X=5 \text{ non defective item}\right) = \frac{^{6}C_{5}\,^{19}C_{5}}{^{25}C_{10}} = \frac{\binom{6}{5}\binom{19}{5}}{\binom{25}{10}} = 0.02134$$

b) Let X = number of defective items in the sample:

$$P(X \leq 2) = P(X=0) + P(X=1) + P(X=2)$$

$$= \frac{\binom{6}{0}\binom{19}{10}}{\binom{25}{10}} + \frac{\binom{6}{1}\binom{19}{9}}{\binom{25}{10}} + \frac{\binom{6}{2}\binom{19}{8}}{\binom{25}{10}} = 0.54466$$

c) Let X = number of non-defective items in the sample:

$$P(X > 5) = P(X=6) + P(X=7) + P(X=8) + P(X=9) + P(X=10)$$

$$= \frac{\binom{6}{4}\binom{19}{6}}{\binom{25}{10}} + \frac{\binom{6}{3}\binom{19}{7}}{\binom{25}{10}} + \frac{\binom{6}{2}\binom{19}{8}}{\binom{25}{10}} + \frac{\binom{6}{1}\binom{19}{9}}{\binom{25}{10}} + \frac{\binom{6}{0}\binom{19}{10}}{\binom{25}{10}} = 0.97747$$

d) Let X = number of non-defective items in the sample:

$$P(X < 4) = P(X=0) + P(X=1) + P(X=2) + P(X=3)$$

$$= \frac{\binom{6}{0}\binom{19}{10}}{\binom{25}{10}} + \frac{\binom{6}{1}\binom{19}{9}}{\binom{25}{10}} + \frac{\binom{6}{2}\binom{19}{8}}{\binom{25}{10}} + \frac{\binom{6}{3}\binom{19}{7}}{\binom{25}{10}} = 0.85296$$

e) Let X = number of non-defective items in the sample:

$$P(X \geq 7) = P(X=7) + P(X=8) + P(X=9) + P(X=10)$$

$$= \frac{\binom{6}{3}\binom{19}{7}}{\binom{25}{10}} + \frac{\binom{6}{2}\binom{19}{8}}{\binom{25}{10}} + \frac{\binom{6}{1}\binom{19}{9}}{\binom{25}{10}} + \frac{\binom{6}{0}\binom{19}{10}}{\binom{25}{10}} = 0.85296$$

4.5.2 Now in MATLAB

4.5.2.1 Create a hypergeometric distribution object

In MATLAB, there is no direct makedist function for creating a hypergeometric distribution.

4.5.2.2 Fitting the hypergeometric distribution to sample data

Just like for a multinomial distribution, MATLAB does not directly support fitting a hypergeometric distribution.

4.5.2.3 Use distribution-specific functions

The functions that MATLAB offers related to the hypergeometric distribution are as follows.

4.5.2.3.1 hygernd

This generates random numbers from the hypergeometric distribution with a corresponding size of the population, M, the number of items with the desired characteristic in the population, K, and the number of samples drawn, n. The syntax is:

```
R = hygernd(M, K, n, [a, b,...])
```

4.5.2.3.2 hygepdf

This computes the hypergeometric PDF at each of the values in r using the corresponding size of the population, M, the number of items with the desired characteristic in the population, K, and the number of samples drawn, n. The syntax is:

```
hygepdf(x, M, K, n)
```

4.5.2.3.3 Example

Suppose we want to determine the probability of selecting 0 to 5 cars with a specific manufacturing defect from a population of 50 cars, where 12 cars have the defect and 38 cars do not have the defect, if we select 15 cars at random.

```
x = 0:1:5;
M = 50;
K = 12;
n = 15;
y = hygepdf(x, M, K, n)
bar(x, y , 1, 'EdgeColor', [0 0 0] , 'FaceColor', [0.7 0.57 0.57],
  'LineWidth', 2)
xlabel("Number of cars")
ylabel("Probability")
```

The hypergeometric distribution graph in Figure 4.13 is helpful because it displays the probability of differing numbers of successes (x) out of the total number of trials (n). In the chart below, the distribution plot finds the likelihood of selecting exactly 0, 1 , 2 , ... up to 5 defective cars in the 15 selections. With this approach, the hypergeometric distribution graph covers the complete range of possible successes up to the total number of trials.

In the chart, each bar represents the probability of selecting a specific number of defective cars over the 15 selections. At a glance, we can see that selecting 3 or 4 defective cars are the most likely outcomes with both having equal probabilities of approximately 0.27.

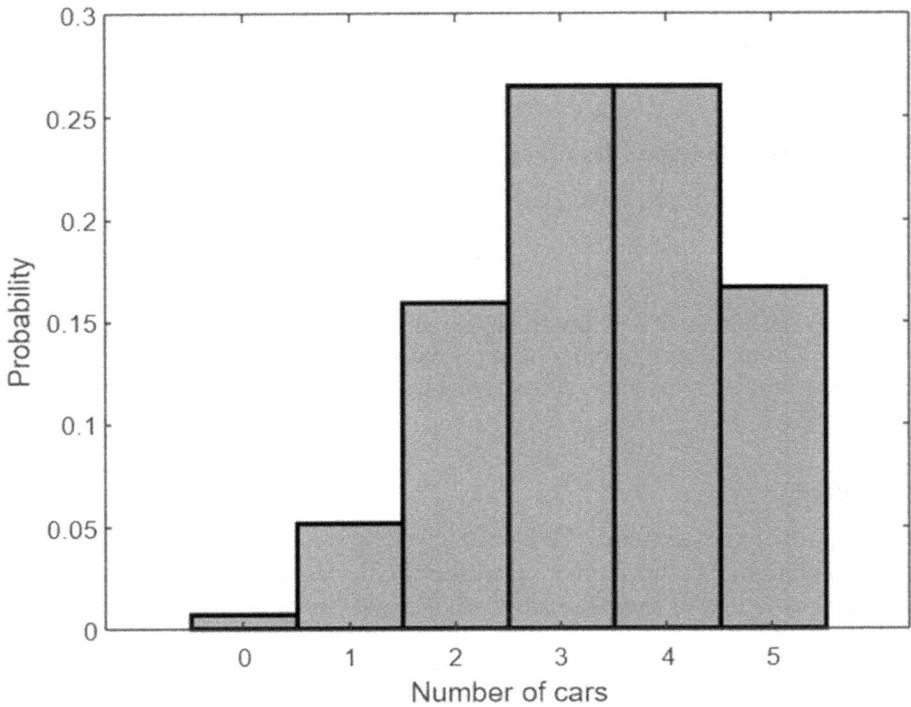

Figure 4.13

4.5.2.3.4 hygecdf

This function computes a binomial cumulative distribution function with the same parameters. The syntax is:

```
hygecdf(x, M, K, n)
```

For example:

1) We want to know the chances of selecting 5 or fewer defective cars in 15 attempts.

```
X = 5;
M = 50;
K = 12;
n = 15;
hygecdf(x, M, K, n)

            ans =

            0.9128
```

2) We want to know the chances of selecting fewer than two defective cars.

$$P(X < 2) = P(X \leq 1)$$

```
1        x = 2;
2        M = 50;
3        K = 12;
4        n = 15;
5        hygecdf(x - 1, M, K, n)
```

Command Window

```
ans =

    0.0584
```

3) We want to know the chances of selecting more than five defective cars.

$$P(X > 5) = 1 - P(X \le 5)$$

```
1        x = 5;
2        M = 50;
3        K = 12;
4        n = 15;
5        1 - hygecdf(x, M, K, n)
6        hygecdf(x, M, K, n, 'upper')
```

Command Window

```
>> untitled002

ans =

    0.8491
```

4) We want to know the chances of selecting eight or more defective cars.

$$P(X \ge 8) = 1 - P(X \le 7)$$

```
1        x = 8;
2        M = 50;
3        K = 12;
4        n = 15;
5        hygecdf(x - 1, M, K, n, 'upper');
6
```

Command Window

```
>> Hypergeometric

ans =

    0.0031
```

4.5.2.3.5 hygeinv

This returns the smallest integer X such that the hypergeometric CDF evaluated at X equals or exceeds P. We can think of P as the probability of observing X defective items in N drawings without replacement from a group of M items where K are defective.

4.5.2.3.6 Example

Suppose you are the quality assurance manager for an aircraft and aerospace component manufacturer. The production line turns out landing gears in batches of 500. You want to sample 50 gears from each batch to see if they have defects. You want to accept 95% of the batches if there are no more than 12 defective gears in the batch. What is the maximum number of defective gears you should allow in your sample of 50?

```
x = hygeinv(0.95, 500, 12, 50)
                        x =

                        3
```

4.5.2.3.7 hygestat

This returns the mean and variance for the hypergeometric distribution with corresponding size of the population, M, number of items with the desired characteristic in the population, K, and number of samples drawn, N. The syntax is:

  [Mean, Variance] = hygestat(M, K, n)
  Now solving the worked example in MATLAB:

  a) What is the probability that an equal number of defective and not defective components will be in the sample?

  $P(X = 5 \text{ defective items})$

```
Defects = 5;
M = 25;
K_Defects = 6;
n = 10;
hygepdf(Defects, M, K_Defects, n)
                              ans =

                              0.0213
```

  $P(X = 5 \text{ non defective items})$

```
Non_Defects = 5;
M = 25;
K_Non_Defects = 19;
n = 10;
hygepdf(Non_Defects, M, K_Non_Defects, n)
```

ans =

0.0213

b) What is the probability that at most two defective items will be in the sample?

$P(X \le 2 \text{ defective items})$

```
Defects = 2;
M = 25;
K_Defects = 6;
n = 10;
hygecdf(Defects, M, K_Defects, n)
```

ans =

0.5447

c) What is the probability that more than five non-defective items will be in the sample?

$P(X \ge 6 \text{ non defective items})$

```
1       Non_Defects = 5;
2       M = 25;
3       K_Non_Defects = 19;
4       n = 10;
5       hygecdf(Non_Defects, M, K_Non_Defects, n, 'upper')
```

Command Window
>> Hypergeometric

ans =

0.9775

d) The probability that less than 4 defective items in the sample?

```
1       Defects = 4;
2       M = 25;
3       K_Defects = 6;
4       n = 10;
5       hygecdf(Defects - 1, M, K_Defects, n);
```

Command Window
>> Hypergeometric

ans =

0.8530

e) the probability that at least 7 non defective items will be in the sample?

```
1        Non_Defects = 7;
2        M = 25;
3        K_Non_Defects = 19;
4        n = 10;
5        hygecdf(Non_Defects - 1, M, K_Non_Defects, n, 'upper')
```

Command Window

```
>> Hypergeometric

ans =

    0.8530
```

f) What is the expected number of defective items and the variance of the non-defective items in the sample?

```
1        M = 25;
2        K_Defects = 6;
3        K_Non_Defects = 19;
4        n = 10;
5
6        [Mean_Defects, Variance_Defects] = hygestat(M, K_Defects, n)
7        [Mean_Non_Defects, Variance_Non_Defects] = hygestat(M, K_Non_Defects, n)
```

Command Window

```
>> untitled002

Mean_Defects =

    2.4000


Variance_Defects =

    1.1400


Mean_Non_Defects =

    7.6000


Variance_Non_Defects =

    1.1400
```

4.5.2.3.8 Example

A portable DVD player requires five batteries to provide sufficient power for extended playback. It can be powered successfully if there are at least three fresh batteries and they have enough charge. There is a collection of 32 AA batteries kept in a toolbox of which 12 batteries are damaged. What is the probability that a selection of five batteries from the toolbox when inserted in the DVD player will power it successfully?

$$P(X \geq 3 \text{ fresh batteries}) = P(X = 3) + P(X = 4) + P(X = 5)$$

$$= \frac{\binom{20}{3}\binom{12}{2}}{\binom{32}{5}} + \frac{\binom{20}{4}\binom{12}{1}}{\binom{32}{5}} + \frac{\binom{20}{5}\binom{12}{0}}{\binom{32}{5}} = 0.73933$$

```
1       M = 32;
2       Defects = 12;
3       K = M - Defects;
4       n = 5;
5       1-hygecdf(2, M, K, n)
6
7
8
9
```

Command Window

```
>> untitled001

ans =

    0.7393
```

Binomial approximation to the hypergeometric distribution

If the population size, M, gets too large in the hypergeometric distribution then we will have problems calculating M! Moreover, when it gets too large, i.e. infinitely large, then it makes absolutely no difference if the experiment is done with or without replacement. Recall that the binomial distribution differs from the hypergeometric in the sense that the binomial experiment is done with replacement and the latter is done without it. So if M in a hypergeometric distribution is so large that replacement does not change anything, then we can actually approximate the hypergeometric distribution by the binomial distribution.

$h(x; M, K, n) \rightarrow b(x; n, p)$ where $p = \dfrac{K}{M}$

For example, if $M = 600$, $n = 16$, $K = 100$, then $h(x; 600, 100, 16) \rightarrow b\left(x; 16, \dfrac{1}{6}\right)$.

x = 0:10

M = 600

$K = 100$

$n = 16$

$y1 = hygepdf\left(x, M, K, n\right);$

$p = \dfrac{K}{M}$

$y2 = binopdf\left(x, n\dfrac{K}{M}\right);$

% Plot the pdf with bars of width 1.

```
x = 0:10;
y = hygepdf(x, 600, 100, 16);
bar(x, y, 1)
title("PMF for Hypergeometric M = 600, K = 100, N = 16")
```

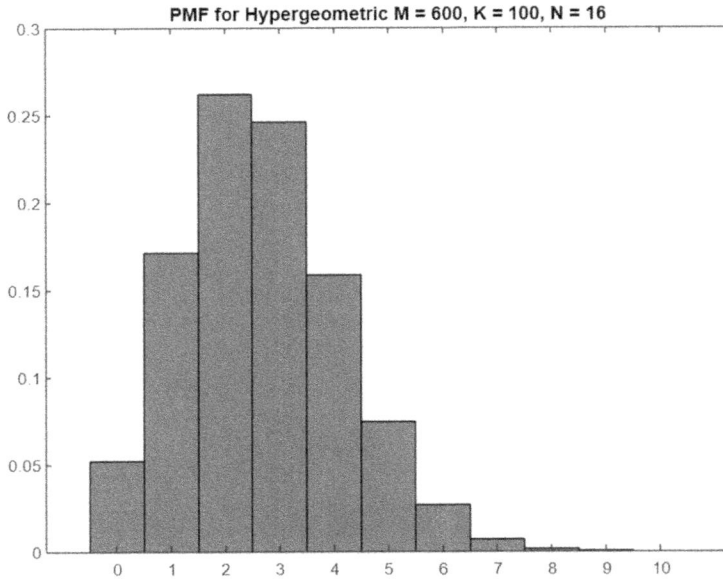

Figure 4.14

```
x = 0:10;
y = binopdf(x, 16, 100/600);
bar(x, y, 1)
title("PMF for Binomial n = 16, p = 100/600")
```

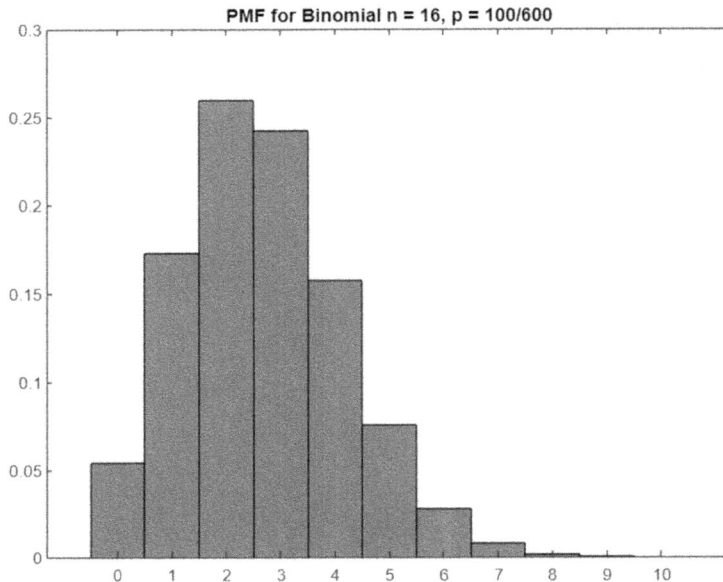

Figure 4.15

4.5.2.3.9 Example

A shipment of 200 standard pump components was ordered for a water supply system out of which 68 were defective. For this consignment to be accepted no more than 5 components in a sample of 30 components selected at random can be defective. What is the probability that the shipment will be accepted? Compare the probability with a binomial distribution.

$$P(X \le 5) = P(X = 0) + P(X = 1) + P(X = 2) + P(X = 3) + P(X = 4) + P(X = 5)$$

```
1       x = 5;
2       M = 200;
3       K = 68;
4       n = 30;
5       hygecdf(x, M, K, n)
6       binopdf(x, n , K/M)
```

Command Window

```
>> untitled001

ans =

    0.0211


ans =

    0.0199
```

As we can see binomial and hypergeometric distributions yield similar results due to the larger population size.

So far we discussed three probability distributions - binomial, multinomial and hypergeometric. One thing to notice is that these distributions are closely related because they all involve a fixed number of trials, n. Deciding which distribution to apply to a given problem depends on two main factors:

1) The number of possible outcomes an experiment can yield.
2) Whether the probability of a particular outcome changes from one trial to the next. As we've observed, the probability only changes when there is no replacement.

4.6 MULTIVARIATE HYPERGEOMETRIC DISTRIBUTION

Just as the binomial distribution can be adjusted to account for multiple random variables (i.e. the multinomial distribution) so too can the hypergeometric distribution account for multiple random variables (multivariate hypergeometric distribution). The multivariate hypergeometric distribution is defined as:

$$P(X_1 = x_1, X_2 = x_2, \ldots X_k = x_k) = \frac{\binom{n_1}{x_1}\binom{n_2}{x_2}\ldots\ldots\binom{n_k}{x_k}}{\binom{N}{X}}$$

Unfortunately there are no separate functions for the multivariate hypergeometric distribution in MATLAB.

TYPE OF DISTRIBUTION	Binomial Distribution	multinomial Distribution	Hyper geometeric Distribution
Number of trials	fixed	fixed	Sample size n is fixed
Dependent or independent	Trials are independent	Trials are independent	Trials are not independent
Number of outcomes	Only 2 Success or failure	k number of outcomes	Only 2 outcomes
Probability of success	Constant throughout	Constant throughout	Changes from trial to trial

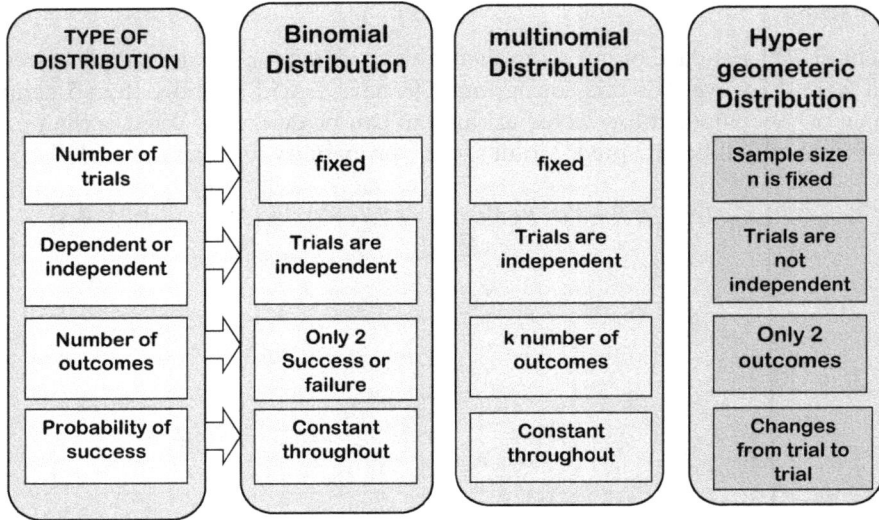

As we can see, the three discussed distributions—binomial, multinomial, and hypergeometric—are closely related in the sense that all of them have a fixed number of trials n. Now which of the three you need to apply for a given problem depends upon:

1) How many outcomes an experiment can yield.
2) Whether the probability of a particular outcome changes from one trial to the next. As we have seen, the probability changes only when there is no replacement.

4.7 GEOMETRIC DISTRIBUTION

The only difference between the binomial random variable and the geometric random variable is the number of trials: the binomial has a fixed number of trials, set in advance, whereas the geometric random variable will conduct as many trials as necessary until the first success.

4.7.1 Calculations in theory

Let's take the example of a coin. We want to calculate the probability that the first head appears on the nth toss.

The probability of getting a tail on the first toss is $\frac{1}{2}$.

The chance of getting a tail for the first time on the second toss is $\frac{1}{2} \times \frac{1}{2} = \frac{1}{4}$.

The chance of getting a tail for the first time on the third toss is $\frac{1}{2} \times \frac{1}{2} \times \frac{1}{2} = \frac{1}{8}$.

The chance of getting a tail for the first time on the fourth toss is $\frac{1}{2} \times \frac{1}{2} \times \frac{1}{2} \times \frac{1}{2} = \frac{1}{16}$.

The chance of getting a tail for the first time on the nth toss is $\left(\frac{1}{2}\right)^n$.

Geometric distribution is used to model the situation where we are interested in finding the probability of a number of failures before the first success or a number of trials (attempts) to get the first success.

Hence there are two different definitions of geometric distributions, one based on the *number of trials (attempts) to get the first success* and the other on the *number of failures before the first success*. The choice of definition is a matter of context.

The former definition counts all trials, including the final, successful one.

In contrast, the latter definition counts only the failures before the success. So if we perform an experiment and it takes n times to get a success then that's equivalent to having $n-1$ failures.

Using the same example of the coin, we want to calculate the probability of getting a tail on the fifth trial. That means we must have heads on every trial before trial 5. Let's picture the situation:

Trial 1   Trial 2   Trial 3   Trial 4   Trial 5
Head   Head   Head   Head   Tail

Let's now interpret the two versions of our definition:

- So to get to our first tail we had to toss it five times. Here the number of trials = 5.
- Using the number of failures form, we don't count the success trial, just the preceding failures. Hence, if we have success on the fifth toss, the fifth trial is excluded; that means we have $5-1=4$ failures or heads.

Regardless of whatever definition you use these two forms of the geometric distribution calculate the same probability. Therefore the geometric random variable is assigned both for the independent trials performed and the consecutive number of failures till the occurrence of the first and only success.

The geometric distribution has only one parameter p which is the probability of success on any trial

Let's use our example to derive the PMF for this distribution.

Since each of our trials is independent, we can simply multiply the probability of each event together, so if we let random variable X denote the number of trials needed to get the first success, then we would have one less failure than we have trials and one success at the end. So since in this example $X = 5$,

$$P(X = 5) = (1-p).(1-p).(1-p).(1-p).p = (1-p)^4 p^1$$

hence the PMF is given by:

$$P(X = n) = p^1 (1-p)^{n-1} \text{ where } n = 1, 2, 3....$$

Now similarly if *random variable X* represents the number of heads before the first tails:

$$P(X = 4) = (1-p).(1-p).(1-p).(1-p).p = (1-p)^4 p^1$$

Therefore, the probability mass function is given by

$P(X = x) = (1-p)^x p^1$ where $= 0, 1, 2, 3....$.

Let's now toss the coin 20 times and see graphically the probabilities of getting a tail on every toss.

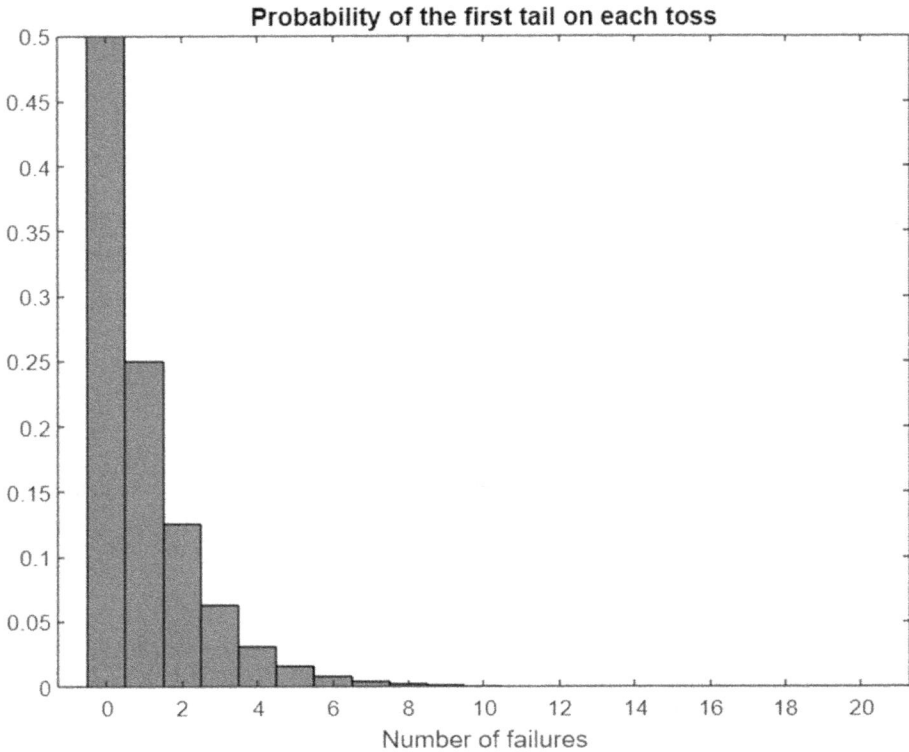

Figure 4.16

$P(X = 0)$ would represent the probability that tails will show up on the first flip and there are therefore zero failures. Similarly, $P(X = 1)$ represents the probability that tails will show up on the second flip and so on. As we can see, the probability that tails will show up on the first flip is 0.50 which implies that the probability is highest in the first trial and it decreases by half with each trial.

What is the probability that we get our first tails on the fifth flip?

Here $n = 5$ because there are five trials, so $P(X = 5) = 0.50^1 (1 - 0.50)^{5-1} = (0.5)^5$.

As mentioned earlier, the success in this example is when the coin lands with tails. What is the probability of observing exactly five heads before tossing a tail?

$$P(X = 5) = (0.5)^1 (1 - 0.50)^5 = (0.5)^6 = 0.015625$$

Another example could be of the tossing of a die. Here the random variable would be the number of failures before getting the first six.

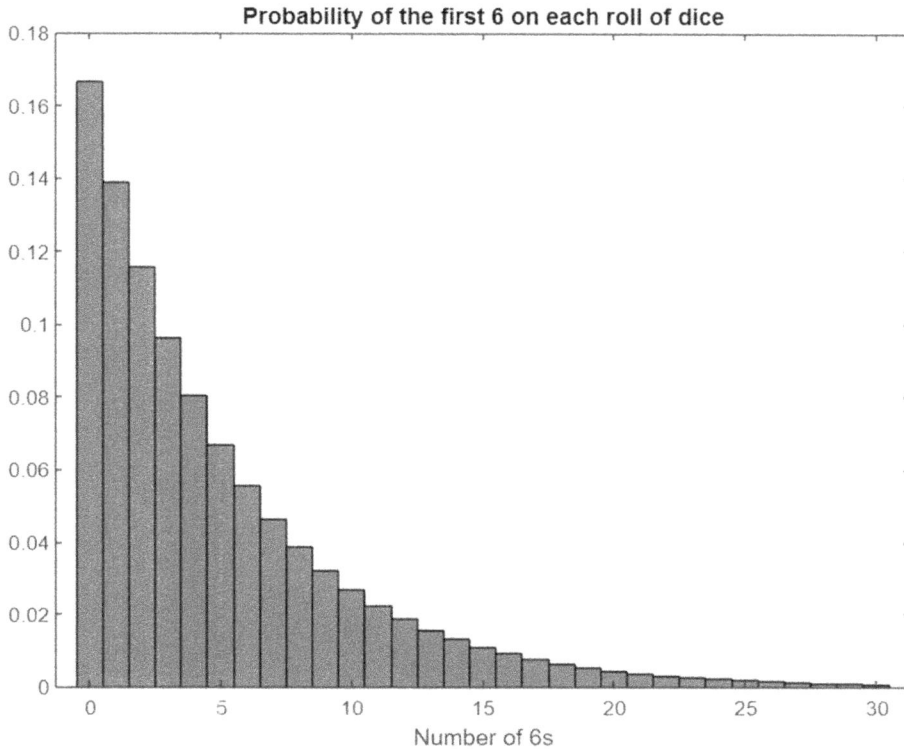

Figure 4.17

Understand that x is equivalent to $n-1$ which represents the number of failures.

As the geometric distribution only needs one success, p has the power 1, p^1. Then the equation is multiplied by the probability of failure $(1-p)$ which is raised by the number of failures, i.e. x or $n-1$.

Therefore the probability of observing exactly five heads before tossing a tail is equivalent to the probability of needing six trials to toss a tail or even the probability of the first tail occurring on the sixth toss. Hence,

$$P(X = 6) = (0.5)^1 (1 - 0.50)^{6-1} = (0.5)^6 = 0.015625$$

4.7.1.1 CDF

Here too we have two variations for the cumulative density function of a geometric distribution.

So again based on the number of trials we have:

$$P(X \le n) = 1 - (1-p)^n$$

When only considering the number of failures we get:

$$P(X \le x) = 1 - (1-p)^{x+1}$$

4.7.1.2 Example

a) Let's determine the probability of observing at most three heads before tossing tails.

$$P(X \leq 3) = 1 - (1 - 0.5)^{3+1} = 1 - (0.5)^4 = 0.9375$$

$$= P(X = 3) + P(X = 2) + P(X = 1) + P(X = 0)$$

$$= 0.5^1 (1 - 0.5)^3 + 0.5^1 (1 - 0.5)^2 + 0.5^1 (1 - 0.5)^1 + 0.5^1 (1 - 0.5)^0$$

$$= 0.5^4 + 0.5^3 + 0.5^2 + 0.5^1 = 0.9375$$

b) Let's determine the probability that it takes fewer than five heads before tossing tails.

$$P(X < 5) = P(X \leq 4) = 1 - (1 - 0.5)^5 = 0.96875$$

$$= P(X = 4) + P(X = 3) + P(X = 2) + P(X = 1) + P(X = 0)$$

$$= 0.5^1 (1 - 0.5)^4 + 0.5^1 (1 - 0.5)^3 + .. + 0.5^1 (1 - 0.5)^0$$

$$= 0.5^5 + 0.5^4 + 0.5^3 + 0.5^2 + 0.5^1 = 0.96875$$

c) Now determine the probability of failing to get a tail within the first ten tosses which is the same as the probability that it takes more than ten tosses to get a tail.

$$P(X > 10) = 1 - P(X \leq 10 \, tosses) = 1 - P(X \leq 10 - 1 \, failures) = 1 - P(X \leq 9 \, failures)$$

$$= 1 - \left(1 - (1 - 0.5)^{9+1}\right) = (0.5)^{10} = 0.0009765625$$

d) Let's determine the probability of observing at least 12 heads before the first tails.

$$P(X \geq 12) = 1 - P(X \leq 11 \, heads)$$

$$= 1 - \left(1 - (1 - 0.5)^{11+1}\right) = (0.5)^{12} = 0.00024$$

The mean of the geometric distribution is mean $= \dfrac{1}{p}$.

The variance of the geometric distribution is var $= \dfrac{1-p}{p^2}$.

4.7.1.3 Example

An engineer is working on a reliability analysis of a system. The system is designed to operate continuously without any failures. The engineer wants to understand the probability of the system operating flawlessly for a certain number of hours before the first failure occurs. Let's say the probability of failure in each hour is 0.015.

a) What is the probability that the system operates flawlessly for exactly 300 hours before the first failure occurs?

$$P(X = 300) = 0.015^1 (1-0.015)^{300} = 0.00016$$

b) What is the probability that the system can operate for 100 or more hours without experiencing a network failure?

$$P(X \geq 100) = 1 - P(Z \leq 99) = 1 - \left(1 - (1-0.015)^{99+1}\right) = (1-0.015)^{100} = 0.2206$$

c) What is the probability that the first network failure occurs after 400 hours? This can be reworded as the probability that it take more than 400 trials to see the first network failure.

$$P(X > 400) = 1 - P(X \leq 400) = 1 - \left(1 - (1-0.015)^{400}\right) = (0.985)^{400} = 0.0023686$$

4.7.1.4 Example

Let's say we have a manufacturing process that produces electronic components, and we are interested in the number of components that need to be tested before encountering a defective one. Suppose the probability of encountering a defective component during testing is 0.06.

a) Find the probability that the first defect is caused by the tenth component tested.
   Now if we visualize it:

   1   2   3   4   5   6   7   8   9   10

Defective component

   As we can see, to get to the target component we had to go through nine components. In this example the success is considered the defective component which we got after going through nine failures(non-defective components). Hence:

$$P(X = 9) = 0.06^1 (1-0.06)^9 = 0.03437$$

b) What is the probability of encountering the first defective component in the first five inspections?
   Maximum number of trials = 5

$$P(X \leq 5) = 1 - (1-0.06)^5 = 0.2660959776$$

c) How many components do we expect to test until one is found to be defective?

$$= \frac{1}{0.06} = 16.66 \text{ so around 16 components}$$

4.7.1.5 Example

Suppose a software application encounters a bug during each execution, and the probability of encountering a bug is 0.034. Using the geometric distribution calculate the probability of encountering the first bug:

a) After seven executions.

$$P(X > 7) = 1 - P(X \leq 7) = 1 - \left[1 - (1 - 0.034)^{7+1} \right] = 0.966^8 = 0.75825$$

b) Within three.

$$P(X \leq 3) = 1 - (1 - 0.034)^3 = 0.098571304$$

c) Exactly on the second execution.

$$P(X = 2) = 0.034^1 (1 - 0.034)^1 = 0.032844$$

d) The expected number of executions until the first bug is:

$$\frac{1}{0.034} = 29.41176$$

4.7.1.6 Inverse geometric probability distribution

We already know that CDF is always equal to the sum of the corresponding PDF:

$$F(x) = P(X \leq x) = \sum_{a \leq x} P(X = a)$$

We have seen that the CDF for the geometric distribution is:

$$P(X \leq n) = 1 - (1 - p)^n$$

Let's say

$$P(X \leq n) = 1 - (1 - p)^n = c$$

$$(1 - p)^n = 1 - c$$

$$\ln(1 - p)^n = \ln(1 - c)$$

$$n \ln(1 - p) = \ln(1 - c)$$

$$n = \frac{\ln(1 - c)}{\ln(1 - p)}$$

Suppose the probability of not transmitting a packet is . To ensure that the probability of failure of the communication system is no more than 25%, what is the maximum number of transmission attempts needed in a row?

$$n = \frac{\ln(1 - 0.25)}{\ln(1 - 0.04)} = 7.04723$$

Since we cannot have a fraction of an attempt, we need to round up to the nearest whole number to ensure that the probability of failure is no more than 20%. Therefore we need a maximum of seven transmission attempts in a row to achieve this level of reliability.

4.7.2 Now in MATLAB

4.7.2.1 Create a geometric distribution object

In MATLAB, there is no direct makedist function for creating a geometric distribution.

4.7.2.2 Fitting the geometric distribution to sample data

MATLAB does not directly support fitting a geometric distribution.

4.7.2.3 Use distribution-specific functions

Even though there are two definitions of the geometric distribution, which we have discussed, MATLAB only models the number of failures for the distribution. So the functions that MATLAB has for working with the distribution are listed below:

geopdf	geopdf(x,p) returns the PDF of the geometric distribution at each value of x using the corresponding probabilities, p.
geocdf	geocdf(x,p) returns the CDF of the geometric distribution, evaluated at each value in x using the corresponding probabilities, p.
geostat	geostat(p) returns the mean and variance of a geometric distribution with the corresponding probability parameter, p.
geornd	geornd(p) generates random numbers from a geometric distribution with probability parameter, p.
geoinv	geoinv(y,p) returns the smallest positive integer x such that the geometric CDF evaluated at x is equal to or exceeds y.

Let's now solve the exercises using MATLAB.
   For example:

1) Determine the probability of observing exactly five heads before tossing tails.

```
1    x = 5;
2    p = 0.5;
3    y = geopdf(x, p);
4
```

Command Window

```
>> Geometric

y =

    0.0156
```

2) Determine the probability that it takes more than ten tosses to get a tail.

```
1        x = 3;
2        p = 0.5;
3        y = geocdf(x,p)
```

Command Window

```
>> untitled001

y =

     0.9375
```

3) Let's determine the probability that it takes fewer than five heads before tossing tails.

```
x = 5;
p = 0.5;
y = geocdf(x - 1, p)
```

4) Let's determine the probability of observing at least 12 heads before the first tail.

```
x = 12;
p = 0.5;
y = geocdf(x - 1, p, 'upper')
```

5) What is the probability that the system operates flawlessly for exactly 300 hours before the first failure given that the probability of failure in each hour is 0.015?

```
1        x = 300;
2        p = 0.015;
3        geopdf(x,p)
```

Command Window

```
>> untitled001

ans =

     1.6105e-04
```

6) What is the probability that the first network failure occurs after 400 hours? This can be reworded as the probability that it take more than 400 trials to see the first network failure.

```
x = 399;
p = 0.015;
y = geocdf(x,p,"upper")
```

```
y =

     0.0024
```

7) What is the probability that the system can operate for 100 or more hours without experiencing a network failure?

   Rephrasing the statement: what is the probability that it takes 100 or more hours to see the first network failure?

```
1        x = 99;
2        p = 0.015;
3        1 - geocdf(x,p)
```

```
>> untitled001

ans =

    0.2206
```

We shall leave rest of the problems as exercises.

4.7.2.4 Example

```
y = 0.25;
p = 0.04;
x = geoinv(y, p);
```

$$X =$$

$$7$$

4.7.2.5 Example

For a production line which has a 21% defective rate, what is the minimum number of inspections that would be necessary for the probability of observing a defective to be more that 80%?

We need to find k so that

$$P(X \leq k) \geq .75$$

Hence:

```
1        y = 0.80;
2        p = 0.21;
3        x = geoinv(y,p)
```

Command Window

```
>> untitled001

x =

    6
```

4.8 NEGATIVE BINOMIAL DISTRIBUTION

Recall that a random variable X has a *geometric distribution* giving the number of trials or the number of failures at which and after which the first success occurs. The negative binomial distribution generalizes the geometric distribution by considering any number of successes.

Just like the geometric distribution, the negative binomial distribution also has two different definitions, one based on the *number of trials (attempts) to get a certain number of successes* and the other on the *number of failures* before achieving a certain number of successes in a series of trials. The choice of the definition again depends on the situation. It is almost the same as a binomial experiment in the sense that unlike the binomial distribution the number of trials is not fixed or is uncertain in a negative binomial experiment.

Using the same example given in the previous section, we would like to know the probability of getting our six thrice. Now just like in the geometric distribution, there are two versions of this distribution. The negative binomial distribution can find both the number of failures that occurred before getting three successes and the number of trials needed to get three successes.

Just like in the geometric distribution, regardless of the definition you use in the negative binomial distribution, these two forms calculate the same probability.

Here the negative binomial random variable models both for the independent trials performed and the number of failures till the occurrence of the desired number of successes.

The negative binomial distribution has two parameters R and p where:

- R :number of successes
- p :probability of success on a given trial

4.8.1 Calculations in theory

If a *random variable* X follows a negative binomial distribution, then the probability of x number of failures before getting R number of successes is given by:

$$P(X = x) = \binom{x+R-1}{R-1} p^R (1-p)^x = \binom{x+R-1}{x} p^R (1-p)^x$$

When X denotes the number of trials needed to get R number of successes then the probability is:

$$P(X = n) = \binom{n-1}{R-1} p^R (1-p)^{n-R}$$

Therefore, a negative binomial variable X can represent both the number of failures before the rth success and the number of trials needed to observe the rth success.

For example, let's say we want to roll five sixes. This means we can get our fifth six in the fifth, sixth, seventh, eighth, etc., die rolls. Now understand that the fifth roll is theoretically our first chance of getting the fifth six, which consequently means we got all five sixes without any failures. Now the probability of that happening realistically is incredibly low which can be seen in the graph below. When failure is one, this means we needed a total of six rolls to get our five successes, when failure is two, a total of seven rolls were needed, and so on.

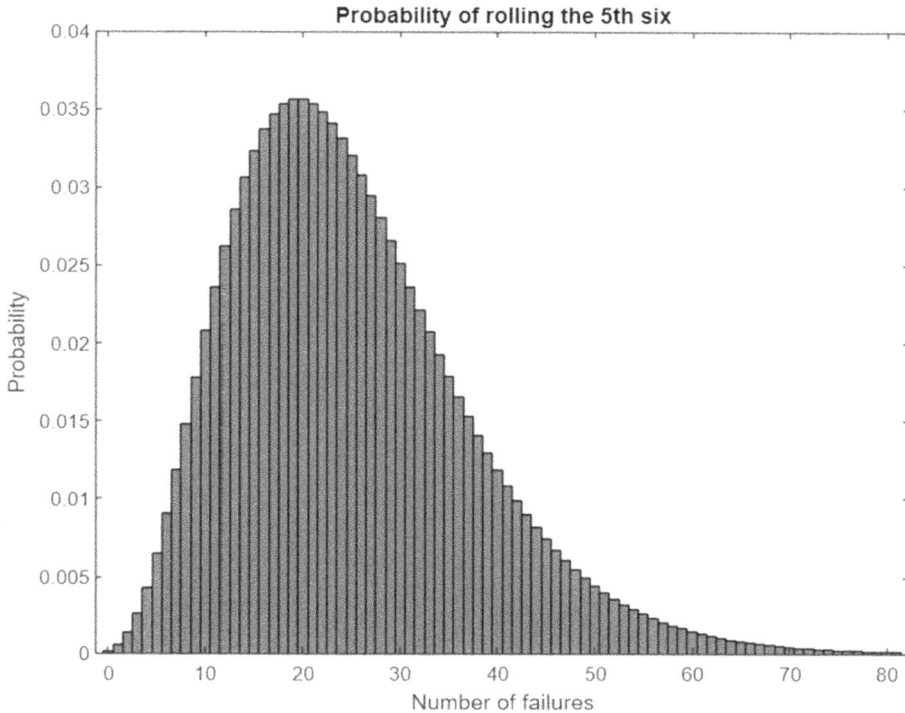

Figure 4.18

The bar graph in Figure 4.18 indicates that the maximum likelihood (0.03563) of roll-ing the fifth six happens after 19 failures; this means it takes around $19 + 5 = 24$ rolls to get all 5 sixes (success). Before 24 rolls (19 failures), our probability of throwing the fifth six increases for each successive roll. After 24 rolls, the likelihood for each roll decreases.

- The mean of the negative binomial distribution is *mean* $= \dfrac{R(1-p)}{p}$.

- The variance of the negative binomial distribution is *var* $= \dfrac{R(1-p)}{p^2}$.

4.8.1.1 Example

The discrete random variable X represents the number of times a biased coin must be tossed until three "heads" have been obtained. Find each of the following:

a) The probability that it took four tosses
b) The probability that it took more than five tosses
c) The probability that it took between six and eight tosses to get that many heads
d) The mean and variance of X

4.8.1.2 Solutions

Note X is the number of trials; hence:

a) $P(X=4) = \binom{4-1}{3-1}(0.50)^3(1-0.50)^{4-3} = \binom{3}{2}(0.50)^3(0.50)^1 = 0.1875$

b) $P(X>5) = 1 - P(X \le 5) = 1 - \left[P(X=3) + P(X=4) + P(X=5)\right]$

$= 1 - \left[\binom{3-1}{3-1}(0.50)^3(1-0.50)^{3-3} + \binom{4-1}{3-1}(0.50)^3(1-0.50)^{4-3} + \binom{5-1}{3-1}(0.50)^3(1-0.50)^{5-3}\right]$

$= 1 - \left[\binom{2}{2}(0.50)^3(0.50)^0 + \binom{3}{2}(0.50)^3(0.50)^1 + \binom{4}{2}(0.50)^3(0.50)^2\right]$

$= 1 - [0.125 + 0.1875 + 0.1875] = 0.50$

c) $P(6 \le X \le 8) =$

- First way $= P(X=6) + P(X=7) + P(X=8)$

$= \binom{6-1}{3-1}0.50^3(1-0.50)^{6-3} + \binom{7-1}{3-1}0.50^3(1-0.50)^{7-3} + \binom{8-1}{3-1}0.50^3(1-0.50)^{8-3}$

$= \binom{5}{2}0.50^3(0.50)^3 + \binom{6}{2}0.50^3(0.50)^4 + \binom{7}{2}0.50^3(0.50)^5$

$= 0.35546875$

- Second way $P(X \le 8) - P(X < 6) = P(X \le 8) - P(X \le 5)$

As we said earlier, regardless of the definition of the negative binomial random variable we choose we will get the same probability. Let's try it out ourselves. So if we change the variable from the number of trials to the number of failures then the number of failures will be:

Number of trials – number of success $= 4 - 3 = 1$

a) $P(X=1) = \binom{1+3-1}{1}0.50^3(1-0.50)^1 = \binom{3}{1}0.50^3(0.50)^1 = 0.1875$

Similarly

b) $P(X>2) = 1 - P(X \le 2) = 1 - \left[P(X=0) + P(X=1) + P(X=2)\right]$

$= 1 - \left[\binom{0+3-1}{0}(0.50)^3(0.50)^0 + \binom{1+3-1}{1}(0.50)^3(0.50)^1 + \binom{2+3-1}{2}(0.50)^3(0.50)^2\right]$

$= 1 - \left[\binom{2}{0}(0.50)^3(0.50)^0 + \binom{3}{1}(0.50)^3(0.50)^1 + \binom{4}{2}(0.50)^3(0.50)^2\right]$

$= 0.50$

c) $P(6 \text{ trials} \le X \le 8 \text{ trials}) = P(6-3 \text{ failures} \le X \le 8-3 \text{ failures})$

$= P(3 \text{ failures} \le X \le 5 \text{ failures})$

- First way $= P(X = 3) + P(X = 4) + P(X = 5)$

$$= \binom{3+3-1}{3}0.50^3(1-0.50)^3 + \binom{4+3-1}{4}0.50^3(1-0.50)^4 + \binom{5+3-1}{5}0.50^3(1-0.50)^5$$

$$= \binom{5}{3}0.50^3(0.50)^3 + \binom{6}{4}0.50^3(0.50)^4 + \binom{7}{5}0.50^3(0.50)^5$$

$$= 0.15625 + 0.1172 + 0.0820 = 0.35545$$

Second way $= P(X \le 5) - P(X < 3) = P(X \le 5) - P(X \le 2)$

d) $E[X] = \dfrac{3(1-0.50)}{0.50} = 3$

$Var[X] = \dfrac{3(1-0.50)}{0.50^2} = 6$

4.8.1.3 Inverse negative binomial probability distribution

The cumulative distribution function can be expressed in terms of the regularized incomplete beta function.

$$F(x) = P(X \le x) = I_p(R, x+1)$$

Where $I_p(R, x+1)$ is the *incomplete beta function*.
Now,

$$I_x(p,q) = \frac{B_x(p,q)}{B(p,q)} = \frac{\displaystyle\int_0^x t^{p-1}(1-t)^{q-1}\, dt}{\displaystyle\int_0^1 t^{p-1}(1-t)^{q-1}\, dt}$$

4.8.1.4 Example

How many times would you need to flip a fair coin to have a 99% probability of having observed ten heads?
So

$$P(X \le x) = 0.99 = I_{0.5}(10,\ x+1)$$

$$= \frac{\int_0^{0.5} t^{10-1}\left(1-t\right)^{(x+1)-1} dt}{\int_0^1 t^{10-1}\left(1-t\right)^{(x+1)-1} dt} = \frac{\int_0^{0.5} t^9 \left(1-t\right)^x dt}{\int_0^1 t^9 \left(1-t\right)^x dt}$$

Clearly this integral is extremely laborious and is more prone to errors. Therefore we leave it to MATLAB or any software library to perform the calculation.

4.8.2 Calculations in MATLAB

4.8.2.1 Create a negative binomial distribution object

4.8.2.1.1 Default parameters

```
pd = makedist('NegativeBinomial')
```

```
pd =

  NegativeBinomialDistribution

  Negative Binomial distribution
    R =   1
    P = 0.5
```

4.8.2.1.2 Specified parameters

```
pd = makedist('NegativeBinomial', 'R', 100, 'p', 0.75)
```

```
pd =

  NegativeBinomialDistribution

  Negative Binomial distribution
    R =   100
    P = 0.75
```

4.8.2.2 Fitting the negative binomial distribution to sample data

4.8.2.2.1 Default parameters

```
pd = fitdist(x, 'NegativeBinomial')
```

4.8.2.2.2 Specified parameters

```
pd = fitdist(x, 'NegativeBinomial', 'R', ___, 'p', ____)
```

4.8.2.3 Use distribution-specific functions

Now how are we going to do this in MATLAB?

Just like in the geometric distribution, MATLAB only models the number of failures, x, before a specified number of successes, r.

nbinpdf	*nbinpdf(x,r,p)* returns the negative binomial PDF at each of the values in x using the corresponding number of successes, r, and probability of success in a single trial, p
nbincdf	*nbincdf(x,r,p)* computes the negative binomial CDF with the same parameters
nbinstat	*nbinstat(r,p)* returns the mean and variance for the negative binomial distribution with the corresponding number of successes, r, and probability of success in a single trial, p
nbinrnd	*nbinpdf(r,p)* generates random numbers from a negative binomial distribution
nbininv	*nbininv (y,r,p)* returns the inverse of the negative binomial CDF with corresponding number of successes, r, and probability of success in a single trial, p

Now solving the problem using MATLAB:

a) The probability distribution
   the distribution graph displays the likelihood of getting the third head after 0, 1, 2, 3... failures.

```
tails = 0:15;
p = 0.5;
heads = 3;
nbinpdf(tails, heads, p)
bar(tails, nbinpdf(tails, heads, p), 1)
xlabel("Number of tails")
ylabel("Probability")
title("Probability of tossing the 3rd head")
```

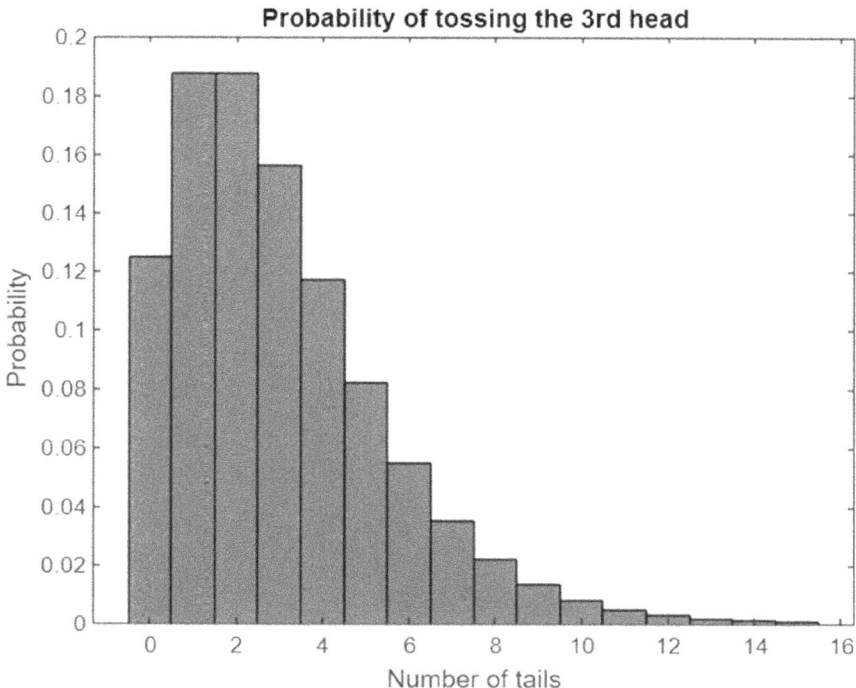

Figure 4.19

As we can see, the probability of getting all three heads is the highest after one failure, that is, after four tosses.

b) The probability that it took four tosses

$$P(X = 4 \text{ trials}) = P(X = 4 - 3 = 1 \text{ failures})$$

```
>> nbinpdf(1,3,0.5)

ans =

    0.1875
```

c) The probability that it took more than five tosses

$$P(X > 5 \text{ trials}) = P(X > 5 - 3 = 2 \text{ failures}) = 1 - P(X \leq 2 \text{ failures})$$

```
>> 1 - nbincdf(2,3,0.5)

ans =

    0.5000
```

d) The probability that it took between six and eight tosses to get that many heads

$$P(6 \text{ trials} \leq X \leq 8 \text{ trials}) = P(6 - 3 \text{ failures} \leq X \leq 8 - 3 \text{ failures})$$

$$= P(3 \text{ failures} \leq X \leq 5 \text{ failures})$$

- First way $= P(X = 3) + P(X = 4) + P(X = 5)$

```
>> nbinpdf(3,3,0.50) + nbinpdf(4,3,0.50) + nbinpdf(5,3,0.50)

ans =

    0.3555
```

- Second way $= P(X \leq 5) - P(X < 3) = P(X \leq 5) - P(X \leq 2)$

```
>> nbincdf(5,3,0.50) - nbincdf(2,3,0.50)

ans =

    0.3555
```

e) The mean and variance of X

```
>> [M,V] = nbinstat(3,0.50)

M =

      3


V =

      6
```

4.8.2.4 Example

```
y = 0.99;
heads = 10;
p = 0.5;
fails = nbininv(y, heads, p);
toss = fails + heads;
```

```
fails =

      23



toss =

      33
```

4.8.2.5 Example

An engineer is responsible for quality control in a manufacturing plant. The engineer wants to investigate the number of inspections needed to find a certain number of defective products. The engineer knows from historical data that the probability of finding a defective product in any given inspection is 0.2. The engineer sets a target of finding ten defective products.
  Find:

a) The probability that it will take 25 inspections to find all 10 defective products.

```
1        Trials = 25;
2        Defects = 10;
3        p = 0.2;
4        nbinpdf(Trials - Defects, Defects, p)
5
```

Command Window

>> untitled001

ans =

    0.0047

b) The probability that it will take at most 15 inspections to find all the defective products.

```
1        Trials = 15;
2        Defects = 10;
3        p = 0.2;
4        nbincdf(Trials - Defects, Defects, p)
5
```

Command Window

>> untitled001

ans =

    1.1323e-04

4.8.2.6 Example

The nitration process of cotton linters plays a crucial role in the production of rocket propellant. This process involves treating the cotton linters with a mixture of nitric acid and sulfuric acid to convert the cellulose fibers into a soluble form. The resulting solution can be further processed and shaped as desired in a later stage with probability 0.74. This shaping process is typically done to form the material into specific shapes or structures that are suitable for their intended application in rocket propellant. What is the probability that exactly 21 lots will be produced in order to obtain the fourth defective lot?

```
1        Lots = 21;
2        Defects = 4;
3        p = 0.74;
4        nbinpdf(Lots - Defects, Defects, p)
5
```

Command Window

>> untitled001

ans =

    3.8760e-08

4.9 POISSON DISTRIBUTION

This is the last discrete distribution we are going to talk about in this chapter.

All of the types of distributions we have discussed so far dealt with outcomes which were either a success or a failure. Even though there were multiple outcomes in multinomial or multivariate hypergeometric distributions these outcomes could still be categorized.

However, in the Poisson distribution, the random variable X is the number of outcomes which are no longer classified as a success or failure during a given interval (of time or space). It describes the probability of obtaining x number of outcomes in a given time or interval taken as λ. In fact, it is the only parameter in the Poisson distribution which is considered the expected number of events in the interval.

4.9.1 Calculations in theory

The probability mass function of the Poisson distribution is: $P(X = x) = \dfrac{(\lambda t)^x}{x!} \cdot e^{-\lambda t}$

Where:
  t is an interval of time or space in which the events occur,
  λ is the mean rate of occurrence per unit time or space.
  The mean of the Poisson distribution is λ.
  The variance of the Poisson distribution is also λ.

The Poisson distribution can be used to model:

1) The number of visitors to a web site per minute
2) The number of arrivals at a car wash in one hour
3) The number of customers entering a McDonald's restaurant per day

4.9.1.1 Example

The number of meteors found by a radar system in any 30-second interval under specified conditions averages 1.54. In a one-minute interval:

a) What is the probability that no meteors are found?
b) What is the probability that fewer than four meteors are seen?
c) What is the probability of finding not more than eight meteors?
d) What is the probability of finding more than three meteors?
e) What is the probability of observing at least five but not more than ten meteors in five minutes of observation?
f) Find variance and standard deviation of the number of meteors in a two-minute interval.

Let's first calculate the rate parameter, λ. To calculate the rate parameter, λ, we need the number of events per interval and interval length.

As given, in a 30-second interval the radar system can expect to find around 1.54 meteors, but our interval length is 1 minute; hence

$$\lambda = \frac{1.54 \; meteor}{0.5 \; minutes} \times 1 \text{ minutes} = 3.08 \text{ meteors / minute}$$

Therefore, our probability mass function is

$$\frac{(3.08 \times 1)^x}{x!} \cdot e^{-(3.08 \times 1)}$$

a) Therefore $P(X = 0) = \dfrac{(3.08 \times 1)^0}{0!} \cdot e^{-(3.08 \times 1)} = e^{-3.08}$

b) $P(X < 4) = P(X = 0) + P(X = 1) + P(X = 2) + P(X = 3)$

$$= \frac{(3.08 \times 1)^0}{0!} \cdot e^{-(3.08 \times 1)} + \frac{(3.08 \times 1)^1}{1!} \cdot e^{-(3.08 \times 1)} + \frac{(3.08 \times 1)^2}{2!} \cdot e^{-(3.08 \times 1)} + \frac{(3.08 \times 1)^3}{3!} \cdot e^{-(3.08 \times 1)} = 0.62931$$

c) $P(X \le 8) = P(X = 0) + P(X = 1) + P(X = 2) + \dots P(X = 6) + P(X = 7) + P(X = 8)$

$$= \frac{(3.08 \times 1)^0}{0!} \cdot e^{-3.08} + \frac{(3.08 \times 1)^1}{1!} \cdot e^{-3.08} + \dots \frac{(3.08 \times 1)^8}{8!} \cdot e^{-3.08}$$

$$= 0.99550$$

d) $P(X > 3) = 1 - P(X \le 3) = 1 - \left[P(X = 0) + P(X = 1) + P(X = 2) + P(X = 3) \right]$

$$= 1 - 0.62931 = 0.37069$$

e) $P(5 \le X \le 10) = P(X = 5) + P(X = 6) + \dots + P(X = 9) + P(X = 10)$

$$= \frac{(3.08 \times 5)^5}{5!} \cdot e^{-(3.08 \times 5)} + \frac{(3.08 \times 5)^6}{6!} \cdot e^{-(3.08 \times 5)} + \dots + \frac{(3.08 \times 5)^{10}}{10!} \cdot e^{-(3.08 \times 5)}$$

$$= 0.09964$$

f) $\lambda = \dfrac{1.54 \; meteor}{0.5 \; minutes} \times 2 \; \text{minutes} = 6.16 \; \text{meteors}$

Mean $= 6.16$, variance $= 6.16$, standard deviation $= \sqrt{6.16}$.

4.9.1.2 Inverse Poisson probability distribution

In a manufacturing plant, machines produce units at an average rate of 20 units per hour. Each technician can handle the production of one unit per hour. If a unit is produced, but no technician is available to handle it, then the unit will be considered defective. Assuming that the production follows a Poisson distribution, what is the minimum number of technicians needed on duty so that defective units are produced at most 10% of the time?

If X is the number of units produced and k is the number of technicians, then k should be set such that $(X > k) \le 0.1$, or equivalently, $P(X \le k) > 0.90$.

$$P(X = 0) = \frac{(20)^0}{0!} \cdot e^{-20} = 2.06 \times 10^{-9}$$

$$P(X = 1) = \frac{(20)^1}{1!} \cdot e^{-20} = 4.12 \times 10^{-8}$$

$$P(X = 2) = \frac{(20)^2}{2!} \cdot e^{-20} = 4.12 \times 10^{-7}$$

$$P(X = 10) = \frac{(20)^{10}}{10!} \cdot e^{-20} = 0.0058$$

$$P(X = 15) = \frac{(20)^{15}}{15!} . e^{-20} = 0.05$$

$$P(X = 20) = \frac{(20)^{20}}{20!} . e^{-20} = 0.0888353$$

$$P(X = 24) = \frac{(20)^{24}}{24!} . e^{-20} = 0.05573$$

$$P(X = 25) = \frac{(20)^{25}}{25!} . e^{-20} = 0.04458765$$

$$P(X = 26) = \frac{(20)^{26}}{26!} . e^{-20} = 0.0343$$

When we add them up, we get around ; therefore the smallest k value is 26.

4.9.2 Calculations in MATLAB

4.9.2.1 Create a Poisson distribution object

4.9.2.1.1 Default parameters

```
pd = makedist('Poisson')
```

```
1        pd = makedist('Poisson')
2
```

Command Window

```
>> untitled002

pd =

    PoissonDistribution

    Poisson distribution
      lambda = 1
```

4.9.2.1.2 Specified parameters

```
1        pd = makedist('Poisson', 'lambda', 3.05)
2
3
```

Command Window

```
>> untitled002

pd =

    PoissonDistribution

    Poisson distribution
      lambda = 3.05
```

4.9.2.2 Fitting the Poisson distribution to sample data

4.9.2.2.1 Default parameters

```
pd = fitdist(x, 'Poisson')
```

4.9.2.2.2 Specified parameters

```
pd = fitdist(x, 'Poisson', 'lambda', ___)
```

4.9.2.3 Use distribution-specific functions

```
[lambdahat,lambdaci] = poissfit(data, alpha)
```

Now how are we going to do it in MATLAB?

poisspdf	*poisspdf(x,lambda)* computes the Poisson probability density function at each of the values of x using the rate parameters in *lambda*.
poisscdf	*poisscdf(x,lambda)* computes the Poisson cumulative distribution function at each of the values of x using the rate parameters in *lambda*.
poisstat	*poisstat(lambda)* returns the mean and variance of the Poisson distribution using parameters in *lambda*.
poissrnd	*poissrnd(lambda)* generates random numbers from the Poisson distribution specified by the rate parameter *lambda*.
poissinv	*poissrnd(p,lambda)* returns the smallest value X such that the Poisson CDF evaluated at X equals or exceeds p, using mean parameters in *lambda*.

Now solving the problem using MATLAB:

a) $P(X = 0)$

```
1    Interval = 0.5;
2    Average = 1.54;
3    TimeInterval = 1;
4    x = 0;
5    lambda = (Average / Interval) * TimeInterval;
6    poisspdf(x, lambda)
7
8
```

> Command Window

```
>> untitled002

ans =

    0.0460
```

b) $P(X < 4) = P(X \le 3)$

```
1        Interval = 0.5;
2        Average = 1.54;
3        TimeInterval = 1;
4        x = 3;
5        lambda = (Average / Interval) * TimeInterval;
6        poisscdf(x, lambda)
7
```

Command Window

```
>> untitled002

ans =

    0.6293
```

c) $P(X \le 8)$

```
1        Interval = 0.5;
2        Average = 1.54;
3        TimeInterval = 1;
4        x = 8;
5        lambda = (Average / Interval) * TimeInterval;
6        poisscdf(x, lambda)
7
```

Command Window

```
>> untitled002

ans =

    0.9955
```

d) $P(X > 3) = 1 - P(X \le 3) = 1 - \left[P(X = 0) + P(X = 1) + P(X = 2) + P(X = 3) \right]$

$$= 1 - 0.62931 = 0.37069$$

```
1        Interval = 0.5;
2        Average = 1.54;
3        TimeInterval = 1;
4        x = 3;
5        lambda = (Average / Interval) * TimeInterval;
6        1 - poisscdf(x, lambda)
7
```

Command Window

```
>> untitled002

ans =

    0 3707
```

$P(5 \le X \le 10)$

```
1        Interval = 0.5;
2        Average = 1.54;
3        TimeInterval = 5;
4        x1 = 10;
5        x2 = 4;
6        lambda = (Average / Interval) * TimeInterval;
7        poisscdf(x1, lambda) - poisscdf(x2, lambda)
8
```

Command Window

```
>> untitled002

ans =

    0.0996
```

f)

```
1        Interval = 0.5;
2        Average = 1.54;
3        TimeInterval = 2;
4        lambda = (Average / Interval) * TimeInterval;
5        [Mean, Variance] = poisstat(lambda)
```

Command Window

```
>> untitled002

Mean =

    6.1600


Variance =

    6.1600
```

4.9.2.4 Example

```
y = 0.90;
lamda = 20;
poissinv(y, lamda);
```

```
                              ans =

                              26
```

4.9.3 Poisson approximation to the binomial distribution

Just like in the hypergeometric distribution, the Poisson distribution too can be used to approximate the binomial distribution. Recall the binomial probability distribution:

$$P(X = x) = \binom{n}{x} p^x (1-p)^{n-x} = \frac{n!}{x!(n-x)!} p^x (1-p)^{n-x}$$

where $x = 0, 1, 2, \ldots, n$

At first glance it might seem like they are not related.

So let's walk through a simple proof showing that the Poisson distribution is really just a special case of the binomial.

Now consider the parameters of the binomial distribution where p is the probability of success and n is the number of trials. Now if we consider a case where:

- The number of trials n is infinitely large
- The probability of success p is negligibly small

Then $\lambda = np$.

Solving for p we get $p = \dfrac{\lambda}{n}$.
Therefore

$$\frac{n!}{x!(n-x)!}\left(\frac{\lambda}{n}\right)^x\left(1-\frac{\lambda}{n}\right)^{a-x}$$

$$= \frac{n!}{(n-x)!}\frac{1}{n^x}\frac{\lambda^x}{x!}\frac{\left(1-\dfrac{\lambda}{n}\right)^n}{\left(1-\dfrac{\lambda}{n}\right)^x}$$

$$= \frac{n(n-1)..(n-x+1)}{n^x}\frac{\lambda^x}{x!}\frac{\left(1-\dfrac{\lambda}{n}\right)^n}{\left(1-\dfrac{\lambda}{n}\right)^x}$$

So as n goes to infinity:

- The first term is cancelled.
- The second term remains as it is.

- The numerator of the third term is $\lim\limits_{n\to\infty}\left(1-\dfrac{\lambda}{n}\right)^n = e^{-\lambda}$.

- The denominator of the third term is $\lim\limits_{n\to\infty}\left(1-\dfrac{\lambda}{n}\right)^x = 1$.

So we are left with

$$1\cdot\frac{\lambda^x}{x!}\cdot\frac{e^{-\lambda}}{1}$$

So, the Poisson distribution is basically the distribution of the number of heads when we toss an infinitely many coins and we expect to see λ heads.

```
n = 50;
p = 0.05;
x = 0:1:10;
y1 = binopdf(x, n, p);
lamda = n*p;
y2 = poisspdf(x, lamda);
figure
bar(x,[y1; y2])
xlabel('Observation')
ylabel('Probability')
title('Binomial and Poisson pdfs of the number of heads')
legend('Binomial Distribution','Poisson Distribution','location','northeast')
```

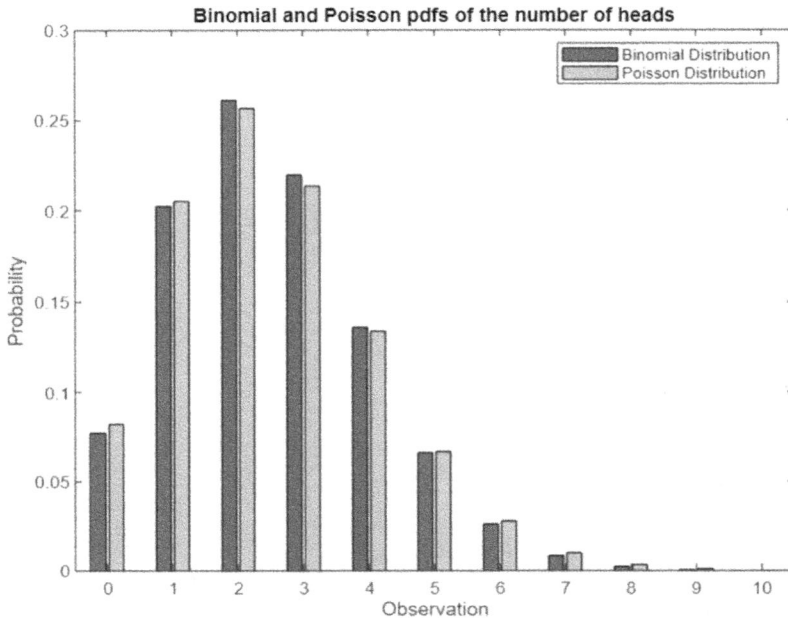

4.9.3.1 Example

In a light bulb manufacturing plant, it is known that 5% of the produced light bulbs are defective under normal circumstances. A sample of 150 bulbs is drawn at random.

  a) What is the probability distribution of the number of defective bulbs?
     What is the probability that:
  b) The sample contains no defective bulbs?
  c) The sample contains more than five defective bulbs?
  d) The sample contains fewer than 3 defective bulbs?

Do this problem both (1) using the binomial distribution, and (2) using the Poisson distribution.

  a) The probability distribution.

```
x = 0:50;
p = 0.05;
n = 50;
yBinomial = binopdf(x, n, p);
```

```
yPoisson = poisspdf(x, )
bar(x, y, 1);
x = [1 2 3];
vals = [2 3 6; 11 23 26];
b = bar(x,vals);
xlabel("Number of Defective Bulbs")
ylabel("Probability")
```

b) The sample contains no defective bulbs.

```
1    x = 0;
2    p = 0.05;
3    n = 50;
4    Binomial = binopdf(x, n, p)
5    Poisson = poisspdf(x, n*p)
6
```

Command Window

```
>> untitled002

Binomial =

    0.0769


Poisson =

    0.0821
```

c) The sample contains more than five defective bulbs.

$$P(X > 10) = 1 - P(X \le 10)$$

```
1    x = 5;
2    n = 50;
3    p = 0.05;
4    lamda = n*p;
5    Binomial = binocdf(x, n, p, "upper");
5    Poisson = poisscdf(x, lamda, "upper");
7
```

Command Window

```
>> approximation

Binomial =

    0.0378


Poisson =

    0.0420
```

d) The sample contains fewer than three defective bulbs.

$$P(X < 20) = P(X \le 19)$$

```
1    x = 19;
2    p = 0.05;
3    n = 50;
4    Binomial = binocdf(x, n, p)
5    Poisson = poisscdf(x, n*p)
6
```

Command Window

```
Binomial =

    1.0000


Poisson =

    1.0000
```

4.10 DISCRETE UNIFORM DISTRIBUTION

A uniform distribution is a type of symmetric *probability distribution* in which all the outcomes have an equal likelihood of occurrence. There are two types of uniform distributions: discrete and continuous. In this chapter we shall discuss the discrete uniform distribution.

Classic examples of experiments where the outcomes are equally likely, resulting in a discrete uniform distribution, are the rolling of a die and the selection of a card from a standard deck. For a fair, six-sided die, the outcome can be any of the numbers 1, 2, 3, 4, 5, or 6 with a 1/6 chance of being rolled and therefore the same probability of occurring. Similarly when we randomly select a card from standard deck of cards all 52 cards have the same probability of being selected.

4.10.1 Calculations in theory

If we let x be a discrete random variable having n values over the interval $[a,b]$, then x has a discrete uniform distribution and its PMF is defined by:

$$P(X = x) = \frac{1}{n}, \ x = 1, 2, 3, \dots . n$$

The mean of the binomial distribution is $\mu = \dfrac{n+1}{2}$.

The variance of the binomial distribution is $\sigma^2 = \dfrac{n^2 - 1}{12}$.

4.10.1.1 Example

Let's say an engineer is designing a random number generator for a simulation study. The generator needs to produce random integers between 1 and 100 (inclusive) with equal probability.

The PMF for the discrete uniform distribution in this example is:

$$P(X = x) = \frac{1}{100}, \ x = 1, 2, 3, \dots . 100$$

a) Find the probability distribution.

$$P(X=1)=\frac{1}{100}$$

$$P(X=2)=\frac{1}{100}$$

$$P(X=3)=\frac{1}{100}$$

.
.
.

$$P(X=100)=\frac{1}{100}$$

b) Find the expected value and the variance.

Using $n=100$ and the above equation:

$$E[X]=\frac{100+1}{2}=50.5$$

$$Var[X]=\frac{100^2-1}{12}=833.25$$

c) What is the probability that the generator produces a number of 25 or less?

$$P(X\le25)=P(X=1)-P(X=2)+..P(X=24)+P(X=25)=\left(\frac{1}{100}\right)*25=0.25$$

d) What is the probability that the generator produces a number of 50 or more?

$$P(X\ge50)=P(X=50)+P(X=51)+..P(X=99)+P(X=100)=\left(\frac{1}{100}\right)*51=0.51$$

e) Find the probability that the generator produces a random integer between 80 and 100.

$$P(80\le X\le100)=P(X\le100)-P(X\le79)=\left(\frac{1}{100}\right)*21=0.21$$

4.10.2 Calculations in MATLAB

MATLAB does not directly support Creating a discrete uniform probability distribution object

4.10.2.1 Fitting the discrete uniform distribution to sample data

MATLAB does not directly support fitting a discrete uniform distribution.

4.10.2.2 Use distribution-specific functions

4.10.2.2.1 unidrnd(n, size)

This generates random numbers from the discrete uniform distribution specified by its maximum value *n* and size indicating the size of the dimension.

4.10.2.2.1.1 EXAMPLE

```
1    n = 100;
2    size = 10;
3    r = unidrnd(n, [1, size])
4
```

```
Command Window
>> untitled002

r =

   40   26   81   44   92   19   27   15   14   87
```

4.10.2.2.2 unidpdf(x, n)

This computes the discrete uniform PDF at each of the values of *x* using the total number of trials, *n*. *x* and *n* can be vectors, matrices, or multidimensional arrays that have the same size.

```
x = 1:5;
y = unidpdf(x, 10);
bar(x, y, 1)
```

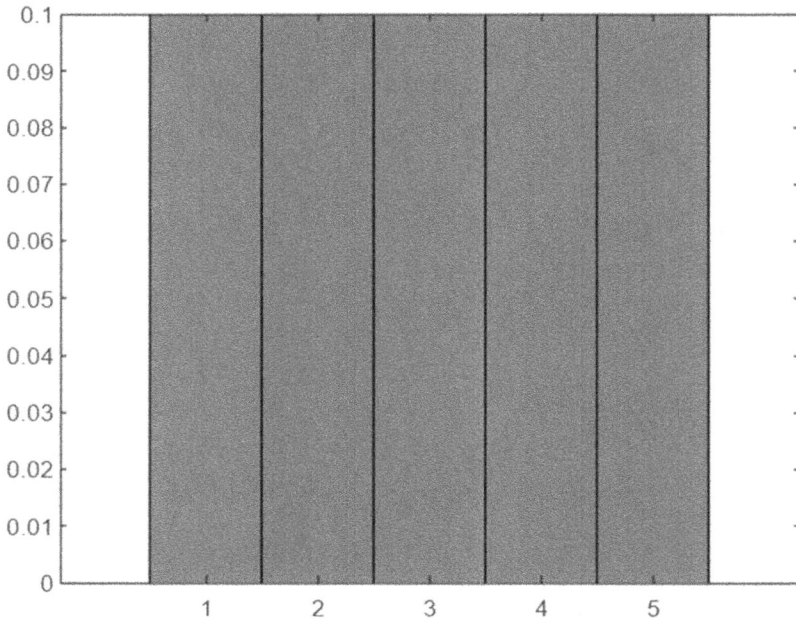

Figure 4.20

4.10.2.2.3 unidcdf(x, n)

This computes the discrete uniform CDF at each of the values of x using the total number of trials, n. x and n can be vectors, matrices, or multidimensional arrays that have the same size.

```
x = 1:5;
y = unidcdf(x, 10);
bar(x, y, 1)
```

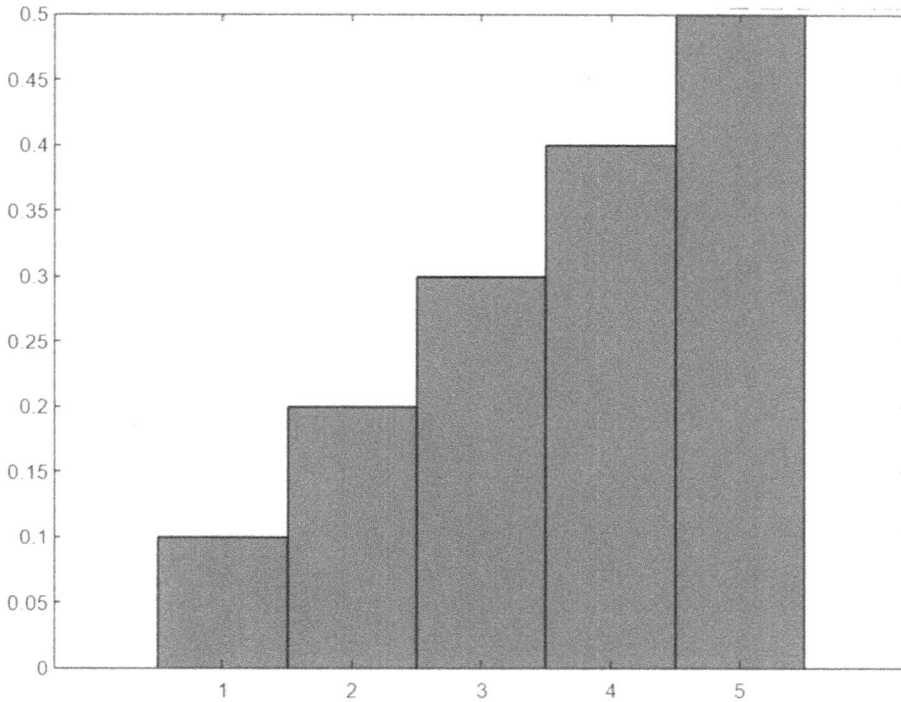

Figure 4.21

4.10.2.2.4 unidinv(p, n)

This returns the smallest positive integer X such that the discrete uniform CDF evaluated at X is equal to or exceeds p. You can think of p as the probability of drawing a number as large as X out of a hat with the numbers 1 through n inside.

4.10.2.2.5 unidstat(n)

This returns the mean and variance of the discrete uniform distribution with minimum value 1 and maximum value n.

4.10.2.3 Solving the problem in MATLAB

1) Solving the problem in MATLAB
   the probability distribution

```
x = 1:1:100;
n = 100;
y = unidpdf(x, n);
bar(x, y, 1)
xlabel("Observation")
ylabel("Probability")
```

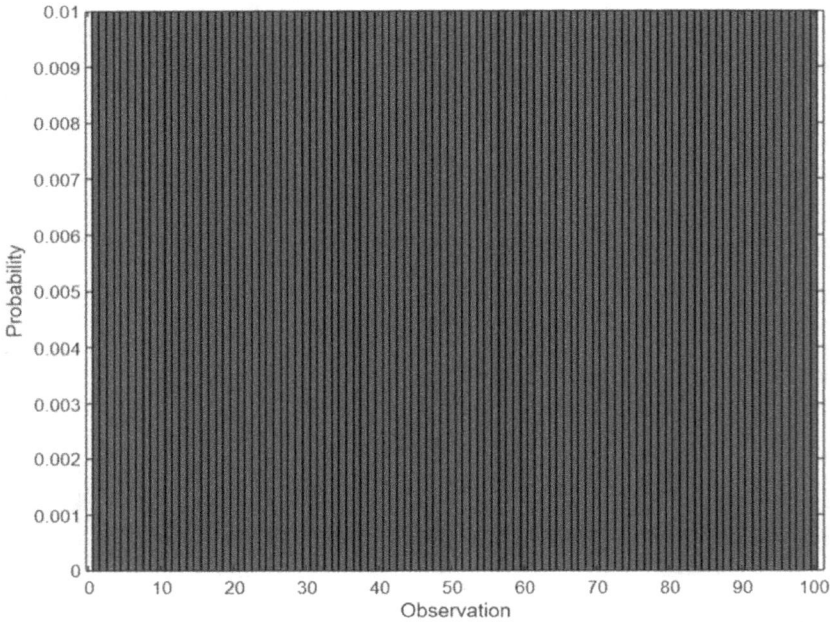

2) Find the expected value and the variance.

```
1       n = 100;
2       [Mean, Variance] = unidstat(n)
3
```

Command Window
```
>> untitled002

Mean =

    50.5000


Variance =

    833.2500
```

3) What is the probability that the generator produces a number of 25 or less?

```
1        x = 25;
2        n = 100;
3        unidcdf(x, n)
4
```

>> Command Window

```
ans =

    0.2500
```

4) What is the probability that the generator produces a number of 50 or more?

```
1        x = 50;
2        n = 100;
3        unidcdf(x - 1, n, 'upper')
4
```

>> Command Window

```
ans =

    0.5100
```

5) Find the probability that the generator produces a random integer between 80 and 100.

```
1        x1 = 80;
2        x2 = 100;
3        n = 100;
4        unidcdf(x2, n) - unidcdf(x1 - 1, n)
```

>> Command Window

```
ans =

    0.2100
```

Distribution name	Parameters	Probability mass function	Mean	Variance
Binomial	n, p	$\binom{n}{x} p^x (1-p)^{n-x}$	np	$np(1-p)$
Bernoulli	p		p	$p(1-p)$
Multinomial		$\dfrac{n!}{n_1! n_2! .. n_k!} p_1^{n_1} p_2^{n_2} \ldots p_k^{n_k}$	np_i	$np_i(1-p_i)$
Hypergeometric	M, K, n	$\dfrac{\binom{K}{x}\binom{--K}{--x}}{\binom{M}{n}}$	$\dfrac{nK}{M}$	$\dfrac{n \times k \times (M-k) \times (M-n)}{M^2(--1)}$

Distribution name	Parameters	Probability mass function	Mean	Variance
Geometric	p	1) $x =$ number of failures $$P(X=x)=p^1(1-p)^x$$ 2) $n =$ number of trials $$P(X=n)=p^1(1-p)^{n-1}$$	$\dfrac{1}{p}$	$\dfrac{1-p}{p^2}$
Negative Binomial	R, p	R, 1) $x =$ number of failures $$P(X=x)=\binom{x+R-1}{R-1}p^R(1-p)^x$$ $$=\binom{x+R-1}{x}p^R(1-p)^x$$ 2) $n =$ number of trials $$P(X=n)=\binom{n-1}{R-1}p^R(1-p)^{-R}$$ $$=\binom{n-1}{x}p^R(1-p)^{-R}$$	$\dfrac{R(1-p)}{p}$	$\dfrac{R(1-p)}{p^2}$
Poisson	λ	$\dfrac{\lambda^x}{x!}\cdot e^{-\lambda}$	λ	λ

In this chapter we talked about types of discrete distribution for calculating the probabilities of different situations. Below is a summary of all distributions along with their parameters and their PMF, mean, and variance.

4.11 SUMMARY

In this chapter, we discussed eight different types of discrete probability distribution along with their properties and a couple of example problems to understand which distribution fits the event. Then we saw the functions that MATLAB offers for different distributions to calculate the probabilities of those events occurring.

In the next chapter we will get into the second type of distribution, the continuous distribution, and will follow the same path of learning to compute the PDF and CDF of those distributions.

Chapter 5

Continuous probability distribution

In Chapter 4 we discussed the discrete probability distribution and its types in detail and worked on a couple of examples for each distribution both theoretically and practically in MATLAB. In this chapter we will now explain the different types of continuous probability distribution just how we did in the previous chapter.

5.1 TYPES OF CONTINUOUS DISTRIBUTION

In Chapter 4 we saw that both discrete and continuous distributions are represented by a mathematical function. For the discrete distribution, the function was called the probability mass function (PMF) and for the continuous distribution it was the probability density function (PDF). We also learned how PDF and PMF differed from each other and how they were similar in the sense that both are used to find the probability of some individual event happening. In Chapter 4 we described the different PMFs applicable in certain situations and how they can be used to calculate probabilities both in theory and in MATLAB.

In this chapter, we will get into a frequently occurring PDF that also describes the shape of its associated distribution. Just like in the previous chapter we are going to use methods from Chapter 4 to calculate the probabilities of different continuous experiments or events occurring.

MATLAB provides multiple continuous distributions that we can use for statistical analysis. However, we shall just discuss the ones that are most commonly used. These are:

1) Normal distribution
2) Lognormal distribution
3) Gamma distribution
4) Exponential distribution
5) Weibull distribution
6) Beta distribution
7) Uniform continuous distribution
8) Rayleigh distribution
9) Extreme value distribution

5.2 NORMAL DISTRIBUTION

A normally distributed random variable is one whose PDF when drawn graphically is a bell curve. The total area under the curve is always 1. A normal distribution is always symmetric because the left and right sides of the distribution perfectly mirror each other.

DOI: 10.1201/9781003399582-5

Every bell curve is symmetric about its mean and lies everywhere above the x-axis, which it approaches asymptotically (arbitrarily closely without touching). The mean, median, and mode of a normal distribution are equal. The median is the mid value, and since the curve is symmetric at the center, the mid value will also be at the center. The mode is the value which occurs the most often, and clearly the highest probability is at the mean value; hence that is also the mode.

5.2.1 Why is this important?

The normal distribution is the most important probability distribution in statistics because many physical and engineering processes exhibit characteristics that can be modeled using the normal distribution. This symmetric distribution fits a wide variety of phenomena, and measurements in engineering.

Another reason why this is of great importance is because many statistical hypothesis tests as well as linear and nonlinear regression rely on assumptions of normality to ensure the validity of results. Engineers use these techniques to analyze experimental data, perform quality control, and make data-driven decisions.

5.2.2 How can I check if my data follows a normal distribution?

MATLAB offers a wide range of tests, both visual and statistical, to check if our dataset meets the assumption of normality. Let's discuss some of those tests.

5.2.2.1 Graphical Tests

If the normal distribution is tested graphically, one looks either at the histogram or even better at the Q-Q plot.

5.2.2.1.1 Q-Q Plot

The most common graphical tool for assessing normality is the Q-Q plot. In these plots, the observed data is plotted against the expected quantiles of a normal distribution. In theory, sampled data from a normal distribution would fall along the dotted line. In reality, even data sampled from a normal distribution, such as the example Q-Q plot below, can exhibit some deviation from the line (Figure 5.1).

Suppose we are working on a project involving the production of resistors, and we want to assess whether the resistance values of the manufactured resistors follow a normal distribution. We have collected resistance measurements for a sample of resistors, and now we want to analyze them using a Q-Q plot.

```
data = readtable("Resistance.csv", VariableNamingRule = "preserve");
Resistance = data.Resistance;
qqplot(Resistance)
ylabel('Resistance Quantiles')
title('QQ Plot of Resistance vs. Standard Normal')
```

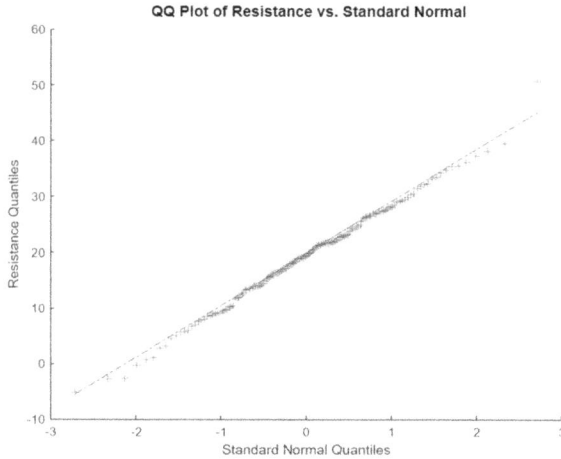

Figure 5.1

Evidently the dataset closely follows a straight line, suggesting that the sample of resistance values has an approximately normal distribution.

The next few examples will show what various Q-Q plots look like when a dataset does not come from a normal distribution.

5.2.2.1.2 Example 1

Miles per gallon (MPG) is a common measurement used in engineering, particularly in the automotive industry, to quantify the fuel efficiency or fuel consumption of vehicles. Now we want to create a Q-Q plot for the MPG of some vehicles.

```
data = readtable("autompg.csv", VariableNamingRule = "preserve");
MPG = data.mpg;
qqplot(MPG)
ylabel('MPG Quantiles')
title('QQ Plot of Miles Per Gallon vs. Standard Normal')
```

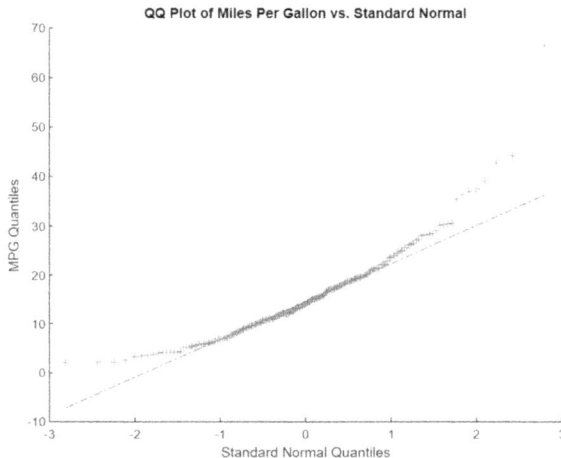

Figure 5.2

Notice that the shape of Figure 5.2 is a little different than Figure 5.1.

At the bottom left of the Q-Q plot, the sample values are greater than the expected normal quantiles, which means the left tail is shortened, whereas at the upper right, the sample values are a bit more positive than expected, which means the right tail is more extended. This is a departure in the direction of a right-skewed distribution. The consequence is a concave up pattern.

5.2.2.1.3 Example 2

Tensile strength is a common measurement used in materials engineering to assess the strength and mechanical properties of materials. Let's say we want to create a Q-Q plot for tensile strength data,

```
data = readtable("Tensile Strength.csv", VariableNamingRule =
"preserve");
Tensile_Strength = data.Tensile_Strength;
qqplot(Tensile_Strength)
ylabel('Tensile Strength Quantiles')
title('QQ Plot of Tensile Strength of Materials vs. Standard Normal')
```

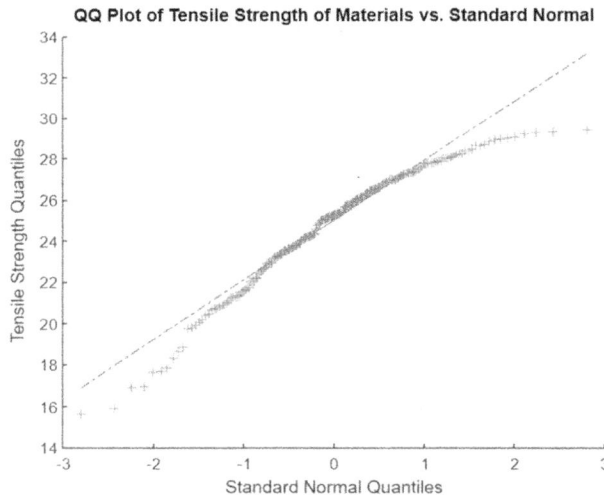

Figure 5.3

Both sample values at the bottom left and the upper right of the Q-Q plot are smaller than expected, which means the left tail is extended and the right tail is shortened. This is a departure in the direction of a left-skewed distribution. The consequence is a concave down pattern.

5.2.2.1.4 Example 3

```
data = readtable("Energy consumption.csv", VariableNamingRule =
"preserve");
Power = data.Output;
qqplot(Power)
ylabel('Power Output Quantiles')
title('QQ Plot of Power Output vs. Standard Normal')
```

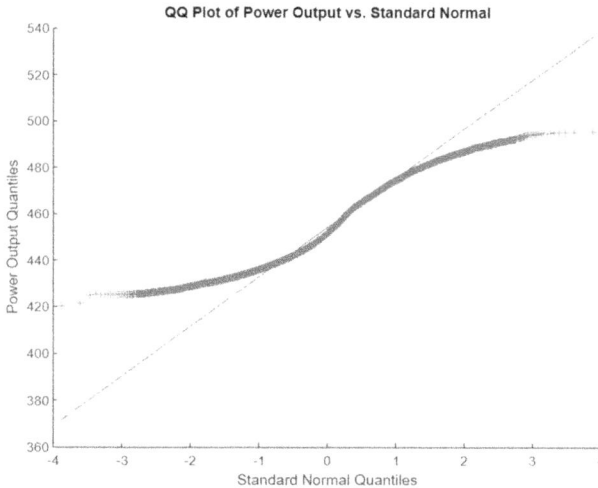

Figure 5.4

This pattern with an "s" shape—a systematic deviation from the straight line, specifically toward the tails of the plot—suggests that the distribution has thick tails. By contrast, a thin-tailed distribution will form a Q-Q plot with very little or negligible deviation at the ends, thus making it a perfect fit for the normal distribution (Figure 5.4).

5.2.2.1.5 Histogram

You may also visually check normality by plotting a frequency distribution, also called a histogram, of the data and visually comparing it to a normal distribution. If you go the histogram route, you plot the normal distribution on the histogram of your data and see if the curve of the normal distribution roughly matches that of the normal distribution curve. Let's now use the same example and check if our conclusion about the individual Q-Q plot is somewhat accurate.

5.2.2.1.6 Example 1

```
data = readtable("Resistance.csv", VariableNamingRule = "preserve");
Resistance = data.Resistance;
histogram(Resistance, 'BinWidth', 5, 'Normalization', 'probability',
         'EdgeColor', [0.6350 0.0780 0.1840],
         'FaceColor', [0.8500 0.3250 0.0980])
ylabel('Probability')
title('Histogram of Resistance')
```

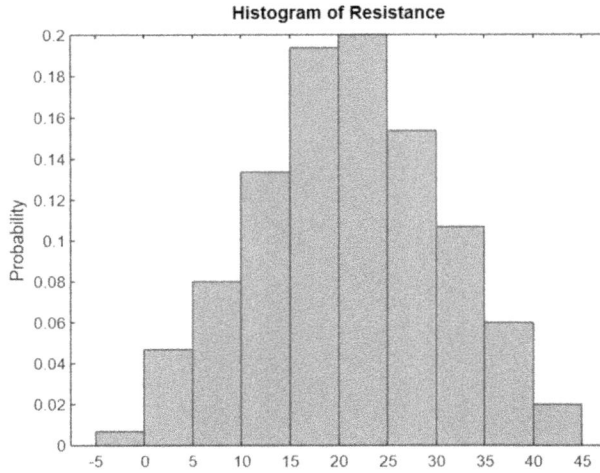

Figure 5.5

5.2.2.1.7 Example 2

```
data = readtable("autompg.csv", VariableNamingRule = "preserve");
MPG = data.mpg;
histogram(MPG, 'BinWidth', 5, 'Normalization', 'probability',
          'EdgeColor', [1 0 1],'FaceColor', [1 0.7740 1])
ylabel('Probability')
title('Histogram of MPG')
```

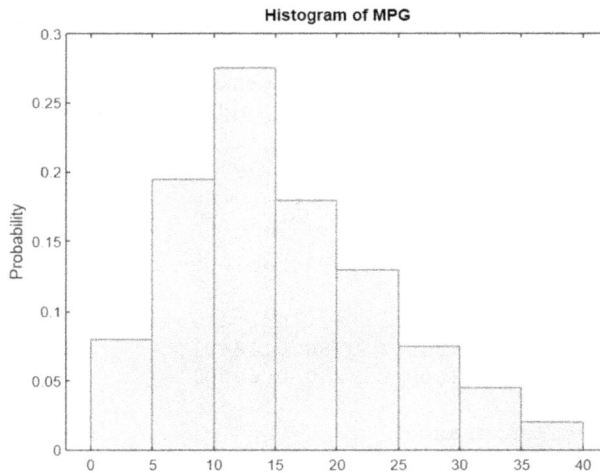

Figure 5.6

Example 3

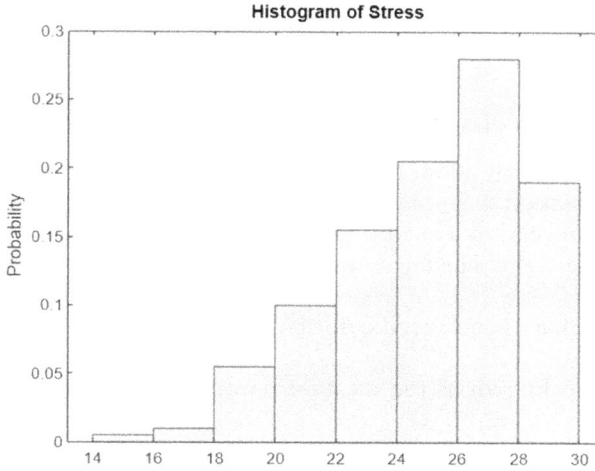
Histogram of Stress

Figure 5.7

Example 4

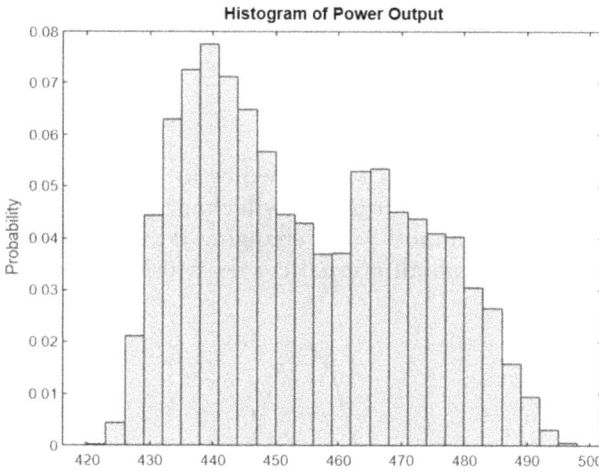
Histogram of Power Output

Figure 5.8

5.2.2.2 *Statistical tests*

Normal distribution can also be tested analytically.

But, to find the distribution of the data, statisticians should not rely only on graphical methods such as histograms or distribution plots. It is also necessary to include the results from the normality tests.

The most common analytical tests to check data for normal distribution are the:

- D'Agostino–Pearson test
- Kolmogorov–Smirnov test

- Shapiro–Wilk test
- Anderson–Darling test

We shall talk about hypothesis tests in Chapter 7.

5.2.2.3 Parameters of the distribution

Now just like any other distribution, the normal distribution too has parameters—the mean μ and the standard deviation σ. Since the area under the curve must always equal 1, a change in the standard deviation causes a change in the shape of the curve; the curve becomes wider or narrower depending on σ. The curve with the larger standard deviation is more spread out/wider.

Hence, we can conclude that a specific normal curve is described by:

- The mean μ, also known as the location parameter, tells us where the center of the curve/peak is.
- The standard deviation σ, the scale parameter, tells us how wide the curve is going to be.

If a variable follows the normal distribution with parameters μ and σ, then it is represented as,

$$X \sim N\left(\mu, \sigma^2\right)$$

5.2.2.4 Shape of the distribution

The shape of any distribution is always determined by its parameters. If that is the case, we can conclude that the shape of our distribution will change as the parameter values change. But, in this case, we absolutely cannot! While changing μ and σ affects the central location and the spread of the distribution all normal curves will have the same bell shape which looks like this (Figure 5.9).

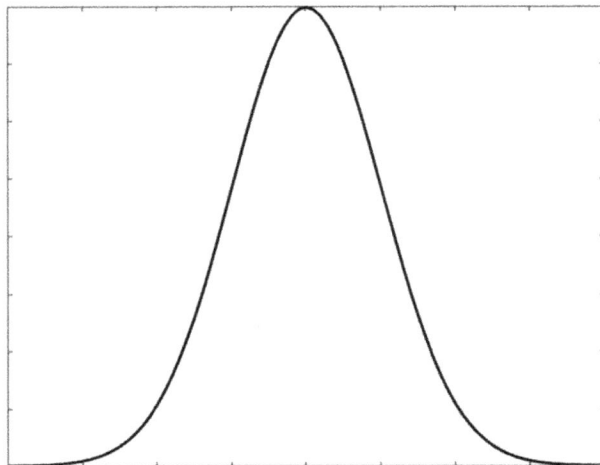

Figure 5.9

However, as mentioned earlier, changing the mean affects the central location of the distribution of a particular dataset. Let's look a plot with varied mean values keeping the standard deviation constant.

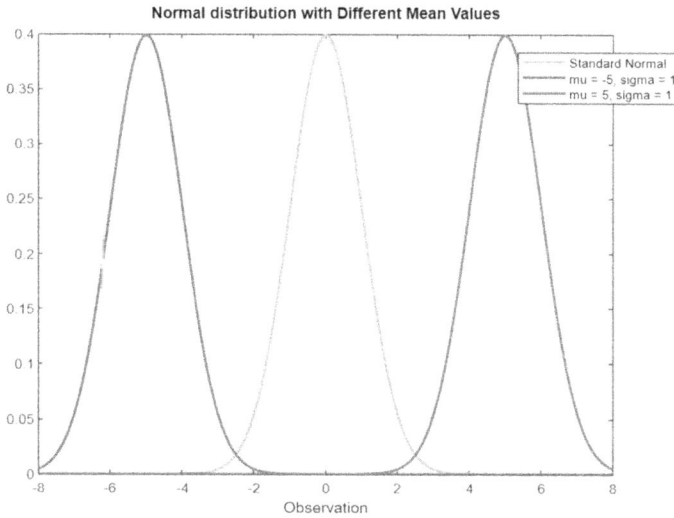

Figure 5.10

As we can see shifting the mean (μ) will change the central location of the distribution.

When μ increases, the distribution shifts to the right, and when μ decreases, it shifts to the left. Notice the shape of the distribution remains symmetric regardless of the mean.

Similarly, let's look at the second plot with varied standard deviation values keeping the mean constant.

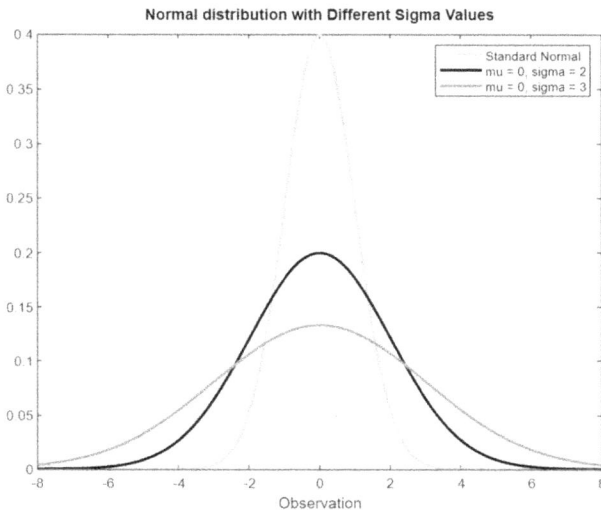

Figure 5.11

Notice in Figure 5.11 that increasing the standard deviation (σ) makes the distribution more spread out, resulting in a wider bell curve, while decreasing the sigma (σ) makes the distribution more concentrated, resulting in a narrower bell curve. As stated earlier, the shape of the distribution remains symmetric regardless of the standard deviation.

Therefore all normal curves, with any value of mean and standard deviation, look the same, and exactly like this. The only differences that might make these distributions look a little different is the center of the distribution and the width of the standard deviation.

5.2.2.5 Calculations in theory

Recall that to determine the probability of a continuous distribution, we solve the integral of the probability density function between two specified points which is basically the area under the PDF curve.

The probability density function of normal distribution $f(x) = \dfrac{1}{\sqrt{2\pi\sigma^2}} e^{-\frac{1}{2}\left(\frac{\mu-x}{\sigma}\right)^2}$

Now if this is the probability density function of the normal distribution, then it would be a pretty troublesome task to integrate! In fact there is no analytical solution for the integral of a normal probability density function. It is one of those integrals which requires numerical methods to find its approximate value. So we use tables to find the values of the integrals. Now for different situations the pair μ and σ will be different, and clearly we cannot build separate tables of probability for each μ and σ. In such cases we do what is called "standardization".

Standardizing is basically "transforming" every normal distribution with different μ and σ into one particular normal distribution, known as the standard normal distribution (also known as the unit normal distribution), which has a mean $\mu = 0$ and a standard deviation $\sigma = 1$. So instead of having an infinite number of area tables for every mean and standard deviation, we have just one called the standard normal tables.

As we have seen, a normal distribution can take on any value as its mean and standard deviation. In the standard normal distribution, the mean and standard deviation are always fixed. We also refer to the standard normal distribution as the Z-distribution because we often speak about this distribution in terms of z-scores. To determine the areas under any normal curve we first transform the data score of our data into a z-score distribution. The z-score, also known as the standard score, allows us to calculate the exact area of any normal distribution and therefore its probabilities using the z or standard normal table.

Standardization also puts different variables on the same scale. Assume you and two of your friends are from three different groups. Three different professors had three different grading scores. Your professor had a grading scale out of 100 so you got a grade of 85; the test has a mean of 75 and a standard deviation of 5. Friend 1's professor had a grading scale out of 500 and he got a grade of 350 with a mean of 200 and a standard deviation of 25. Finally the third professor had a scale of 750 on which Friend 2 got 500 and the mean was 400 with a standard deviation 50. Clearly all three of you scored above average, but who did the best?

Since the grading scale is completely different, the raw datasets are not immediately comparable since they are measured on different scales. Comparisons are meaningless if the scores are not on the same scale.

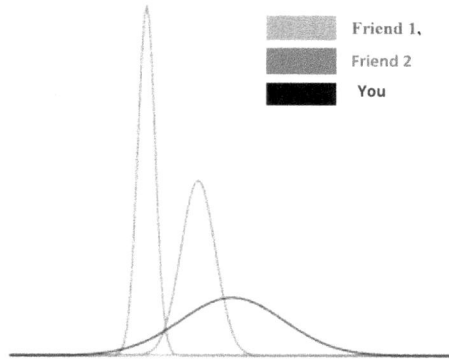

Friend 1,

Friend 2

You

Figure 5.12

So to be able to compare the scores we need to put the three different grading scales onto one scale. Simply stated, to be able to make comparisons, the variables need to be standardized, or put onto the same scale.

So in conclusion converting a normal distribution into the standard normal distribution allows you to calculate the probabilities and compare scores on different distributions with different means and standard deviations.

This entire process of transformation into z-scores can be performed by the following formula:

$$z = \frac{x - \mu}{\sigma}$$

Assume we have a normal distribution with mean μ and standard deviation σ.

We take a data point from this distribution; let's call the value taken x.

A z-score gives you an idea of how far from the mean **a data point is.** But more technically it's a measure of how many standard deviations below or above the population mean a raw score is.

The z-score of x represents the direction and distance from μ to x, with unit σ–.

There is a table which must be used to look up standard normal probabilities. So after we find the z-score we can look up the z-score on the z table and find its corresponding probability.

Notice, the z-score is broken into two parts; the whole number and tenth are looked up along the left side and the hundredth is looked up across the top. The value at the intersection of the row and column is the area under the curve between zero and the z-score looked up. Also note that the table gives the direct value of "less-than" probability $P(X \leq x)$.

But if we have a situation such that we need to calculate:

1) $P(X < x)$ then it is same as calculating $P(X \leq x)$ since $P(X = x) = 0$
2) $P(X > x) = 1 - P(X \leq x)$
3) $P(X \geq x) = 1 - P(X < x) = 1 - P(X \leq x)$ from equation (1)
4) $P(a \leq X \leq b) = P(X \leq b) - P(X < a) = P(X \leq b) - P(X \leq a)$
5) $P(a < X < b) = P(X < b) - P(X \leq a) = P(X \leq b) - P(X \leq a)$

The value of the z-score tells you how many standard deviations you are away from the mean. If a z-score is equal to 0, it is on the mean. If it is positive, then that means the score is above the mean, and a negative score indicates that the score is below the mean. For example, a z-score of -2 means the score is two standard deviations to the left of the mean. The observation is two standard deviations below the mean. A z-score of $+2$ means the score is two standard deviations to the right of the mean. The observation is two standard deviations above the mean.

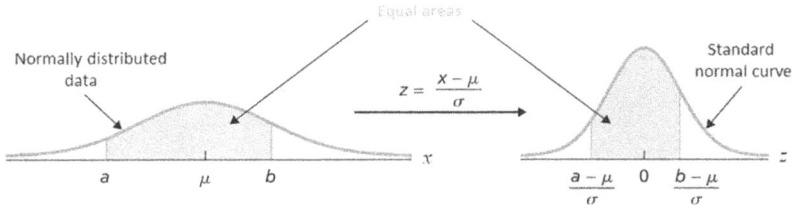

Figure 5.13

Normal distribution is commonly associated with the $68-95-99.7$ rule, or empirical rule, which you can see in the image below (Figure 5.14).

a) 68% of the data is within one standard deviation (σ) of the mean (μ).
b) 95% of the data is within two standard deviations (σ) of the mean (μ).
c) 99.7% of the data is within three standard deviations (σ) of the mean (μ).

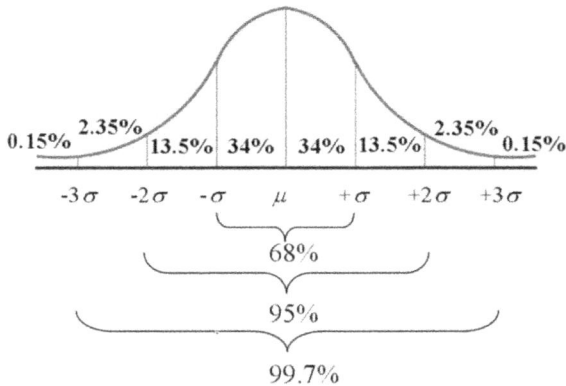

Figure 5.14

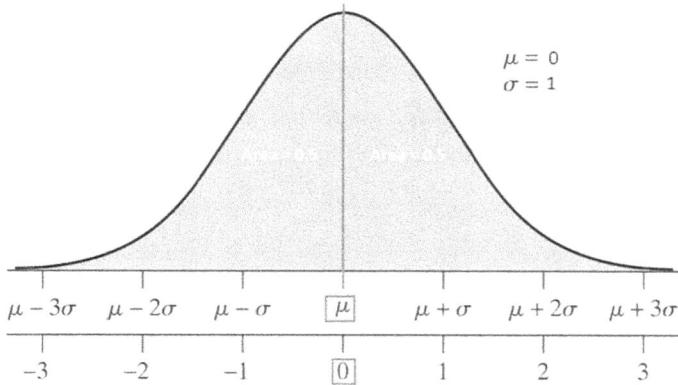

Figure 5.15

5.2.2.6 Examples

The mean score on all 2021 SAT exams was 1060, with a **standard deviation of 217.** Assume the SAT score is normally distributed.

a) Find the probability that a person has a mathematics SAT score of less than 1000.
b) Find the probability that a person has a mathematics SAT score of over 1500.
c) Find the probability that a person has a mathematics SAT score between 1100 and 1200.
d) Find the 90th percentile.
e) How high must an individual score on the SAT in order to score in the highest 5%?

a) $P(x > 1500) = P\left(Z > \dfrac{1500 - 1060}{217}\right) = P(Z > 2.02765) = 1 - P(Z \le 2.02765)$

$= 1 - \sim 0.9788 = \sim 0.9788$

b) $P(x < 1000) =$

$P\left(Z < \dfrac{1000 - 1060}{217}\right) = P(Z < 0.2765) = P(Z \le 0.2765) = \sim 0.609$

$P(1100 < x < 1200)$

c) $= P\left(\dfrac{1100 - 1060}{217} < Z < \dfrac{1200 - 1060}{217}\right) = P(0.184 < Z < 0.645)$

$= P(Z < 0.645) - P(Z \le 0.184) = P(Z \le 0.645) - P(Z \le 0.184)$

$= 0.74055 - 0.57330 = \sim 0.1672$

d) The percentile for a normal distribution is a value that has a specific percentage of the observed data below it. Therefore, if an individual score is in the 90th percentile that means the score is better than 90% of exam takers who took the test therefore we need to find the score k that has 90% of the scores below k

$P(X < ?) = 0.90$

When we go to the table, we find that the value 0.90 is not there exactly, however, the values 0.8997 and 0.9015 are there and correspond to Z values of 1.28 and 1.29, respectively therefore we take the average of the values such that $\dfrac{0.8997+0.9015}{2} \sim 0.9006$ simultaneously we take the average of the corresponding z values and get 1.285 but since our value is slight smaller than 0.9006 hence our z score would be slightly smaller than < 1.285 but for convenience lets take the value

$$P(Z<1.285)=0.90$$

$$Z=1.285=\frac{X-1060}{217}$$

$$X=(217\times1.285)+1060=1338.845$$

e) $P(X>?)=0.05$

If 5% score higher than an individual, then 95% scores lower. In other words,

$$P(X<?)=1-0.05=0.95$$

$$P(X<?)=0.95=P(Z<?)$$

From the table, we find that the value 0.95 is not there exactly, however, the values .94950 and .95053 are there and correspond to Z values of 1.64 and 1.65 respectively therefore we take the average of the values and get 1.645 hence

$$P(Z<1.645)=0.95$$

$$Z=1.645=\frac{X-1060}{217}$$

$$X=(217\times1.645)+1060=1416.965$$

5.2.2.7 Calculations in MATLAB

5.2.2.7.1 Create a normal distribution object

5.2.2.7.1.1 DEFAULT PARAMETERS

```
pd = makedist('Normal')

                        pd =

                 NormalDistribution

             Normal distribution
                    mu = 0
                 sigma = 1
```

5.2.2.7.1.2 SPECIFIED PARAMETERS

```
pd = makedist('Normal', 'mu', 80, 'sigma', 12)
                            pd =

                    NormalDistribution

                    Normal distribution
                        mu = 80
                      sigma = 12
```

5.2.2.7.2 *Fitting the normal distribution to sample data*

5.2.2.7.2.1 DEFAULT PARAMETERS

```
pd = fitdist(x, 'Normal')
```

5.2.2.7.3 *Use distribution-specific functions*

Now that we know all about the normal distribution, how do we make these calculations in MATLAB?

normpdf(x, mu, sigma)	computes the PDF of the normal distribution with mean mu and standard deviation sigma, evaluated at the values in x.
normcdf(x, mu, sigma)	computes the CDF of the normal distribution with mean mu and standard deviation sigma, evaluated at the values in x.
normcdf(x, mu, sigma, 'upper')	returns the complement of the CDF, evaluated at the values in x, using an algorithm that more accurately computes the extreme upper-tail probabilities.
norminv(p, mu, sigma)	returns the inverse of the normal CDF with mean mu and standard deviation sigma, evaluated at the probability values in p.
normstat(mu, sigma)	returns the mean and variance of the normal distribution with the distribution parameters mean mu and standard deviation sigma.
normrnd(mu, sigma, size)	generates an array of normal random numbers, where vector size specifies size(r).

5.2.2.7.4 *Example*

```
x = -5:0.01:5;
PDF = normpdf(x);
CDF = normcdf(x);

tiledlayout(2, 1)
ax1 = nexttile;
plot(ax1, x, PDF)
title(ax1, "Standard normal density")

ax2 = nexttile;
plot(ax2, x, CDF)
title(ax2, "Standard normal cumulative distribution function")
```

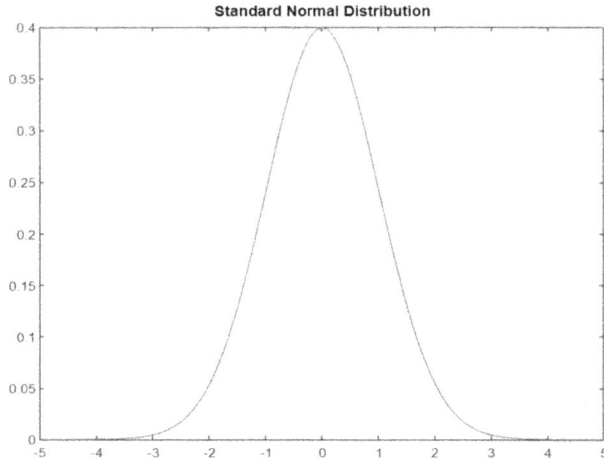

Figure 5.16

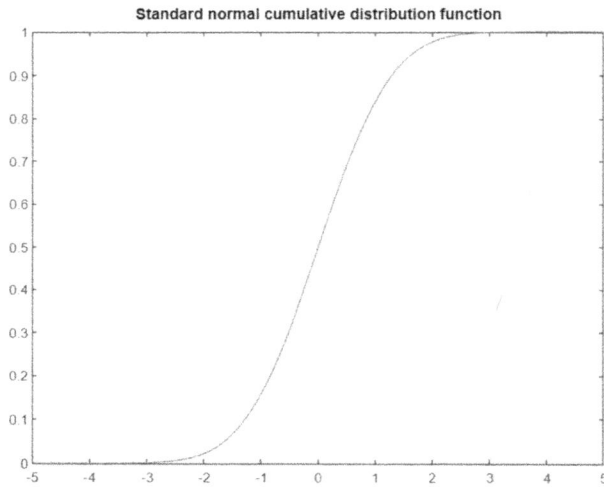

Figure 5.17

This returns the PDF and CDF of the standard normal distribution, evaluated at the values in x.

2) Lets draw multiple normal distribution curve with the same standard mean but different standard deviations in MATLAB.

```
x = -5:0.01:5;

tiledlayout(2,1)
% Top axes
ax1 = nexttile;
plot(ax1, x, normpdf(x), LineWidth = 2)
hold on
plot(ax1, x, normpdf(x, 0, 2),'Color', 'r', LineWidth = 2)
plot(ax1, x, normpdf(x, 0, 0.5), 'Color', 'g', LineWidth = 2)
```

```
hold off
title(ax1, "Normal Distribution with Multiple Sigma")
legend(ax1, 'μ = 0, σ^2 = 1', 'μ = 0, σ^2 = 2', 'μ = 0, σ^2 = 0.5')

ax2 = nexttile;
plot(ax2, x, normcdf(x), LineWidth = 2)
hold on
plot(ax2, x, normcdf(x, 0, 2),'Color', 'r', LineWidth = 2)
plot(ax2, x, normcdf(x, 0, 0.5), 'Color', 'g', LineWidth = 2)
hold off
title(ax2, "Cumuluative Distribution function with Multiple Sigma")
legend(ax2, 'μ = 0, σ^2 = 1', 'μ = 0, σ^2 = 2', 'μ = 0, σ^2 = 0.5')
```

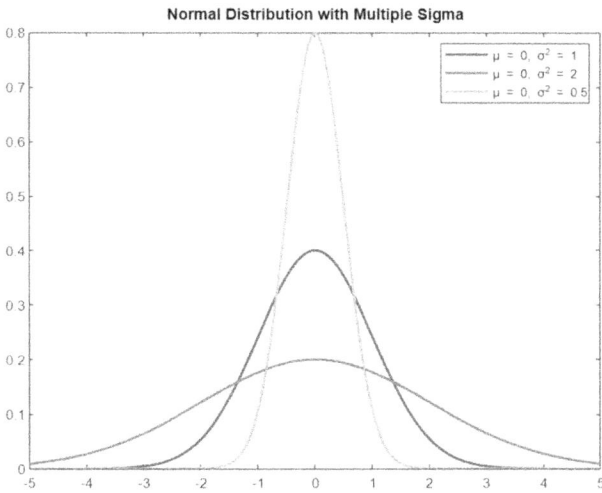

Figure 5.18

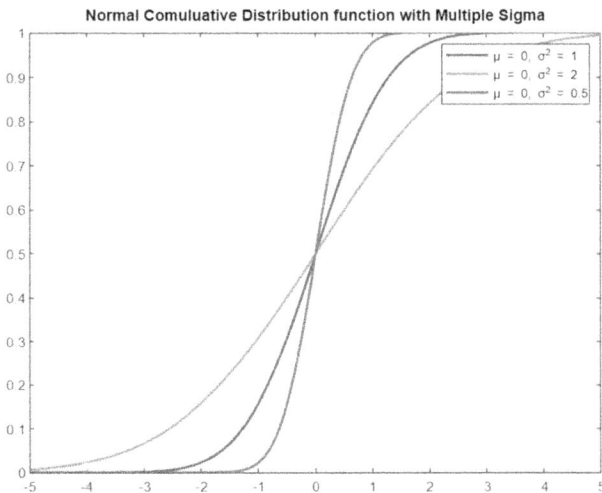

Figure 5.19

As we can see, the histogram of the dataset with a larger standard deviation (red line) has a wider spread. The bars in the histogram are more spread out, indicating a greater dispersion of data points. On the other hand, the histogram of the dataset with a smaller standard deviation (green line) has narrower bars, indicating a tighter clustering of data points around the mean.

```
x = -5:0.01:5;
tiledlayout(2,1)
```

3) Finally draw multiple normal distribution curve with the same standard deviation but different mean values in MATLAB.

```
% Top axes
ax1 = nexttile;
plot(ax1, x, normpdf(x), LineWidth = 2)
hold on
plot(ax1, x, normpdf(x, 1, 1), 'Color', 'm', LineWidth = 2)
plot(ax1, x, normpdf(x, -1, 1), 'Color', [0.9290 0.6940 0.1250], LineWidth = 2)
hold off
title(ax1, "Normal Distribution with Multiple Mean")
legend(ax1, 'μ = 0, σ^2 = 1', 'μ = 1, σ^2 = 1', 'μ = -1, σ^2 = 1')

ax2 = nexttile;
plot(ax2, x, normcdf(x), LineWidth = 2)
hold on
plot(ax2, x, normcdf(x, 1, 1), 'Color', 'm', LineWidth = 2)
plot(ax2, x, normcdf(x, -1, 1), 'Color', [0.9290 0.6940 0.1250], LineWidth = 2)
hold off
title(ax2, "Cumuluative Distribution function with Multiple Mean")
legend(ax2, 'μ = 0, σ^2 = 1', 'μ = 1, σ^2 = 1', 'μ = -1, σ^2 = 1')
```

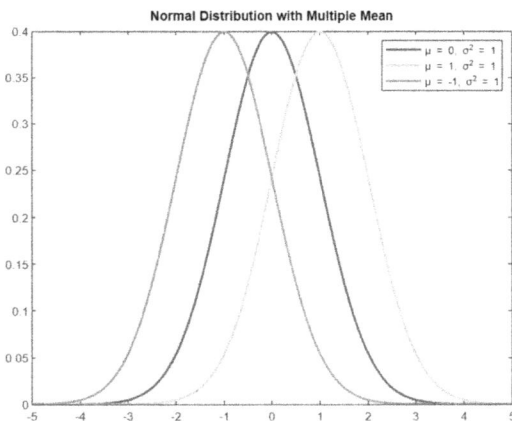

Figure 5.20

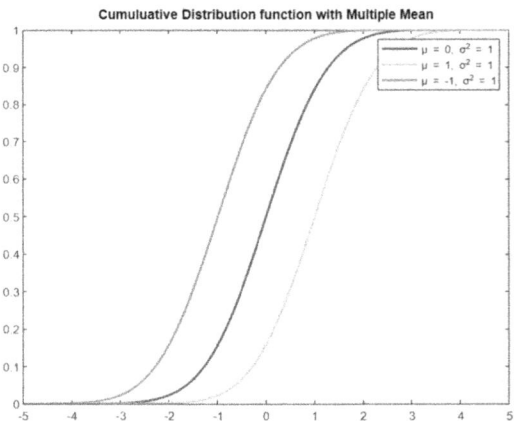

Figure 5.21

In this example, the dataset with the higher mean will have a shift in the entire distribution toward higher values, while the dataset with the lower mean will have a shift toward lower values. However, the spread or width of the histogram bars will be the same for both datasets since they have the same standard deviation.

Solving the same problem using the above functions:

```
mu = 1060;
sigma = 217;

a_less = normcdf(1000, mu, sigma);
b_more = normcdf(1500, mu, sigma, 'upper');
c_between = normcdf(1200, mu, sigma) - normcdf(1100, mu, sigma);
d_percentile = norminv(0.90, mu, sigma);
e_top = norminv(0.95, mu, sigma);
```

```
a_less =

    0.3911


b_more =

    0.0213


c_between =

    0.1675


d_percentile =

    1.3381e+03


e_top =

    1.4169e+03
```

5.2.2.7.5 Practice problems

2) Suppose that we have a dataset concerning the starting salaries of MIT graduates. Find the probability of a randomly selected MIT graduate earning:
    a) Less than $50,000 annually.
    b) More than $120,000 annually.
    c) between $80,000 and $100,000 annually.
    d) Find the 75th percentile for an individual MIT graduate earning.
    e) Find the mean and variance of the income..

The first thing we need to do is test if our dataset is normally distributed.

```
data = readtable("Salary.csv", VariableNamingRule = "preserve");
Salary = data.salary;
histogram(Salary, 'Normalization', 'probability',
'EdgeColor', [0 0.5 0.5], 'FaceColor', [0.4660 0.6740 0.1880],
'LineWidth', 2)
ylabel('Probability')
title('Histogram of Salaries')
```

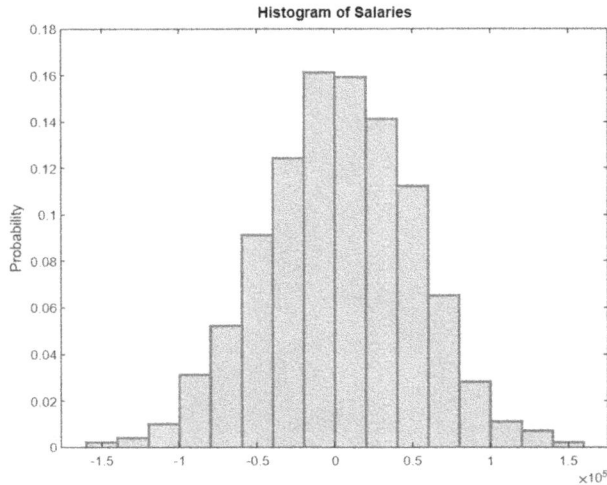

Figure 5.22

As we can see the histogram (Figure 5.23) exhibits a symmetric bell-shaped curve where the majority of values cluster around the mean, with progressively fewer values toward the tails of the distribution. To find the probabilities associated with a normal distribution for a given sample, we first need to estimate the parameters of the normal distribution, mean and sigma. Since we have already tested the normality of the data and confirmed that it follows a normal distribution, we can proceed with estimating our parameters. As seen earlier, we can do this in two ways either use the function fitdist where we need to specify the type of distribution you want to fit (in this case, the normal distribution) or we can use the normfit function, specifically designed for fitting a normal distribution. Now let's perform the calculations.

```
pd = fitdist(Salary, 'Normal');
mu = pd.mu;
sigma = pd.sigma;
x1 = 50000;
x2 = 120000;
x3 = 80000;
x4 = 100000;
less = normcdf(x1, mu, sigma);
more = normcdf(x2, mu, sigma, 'upper');
between = normcdf(x4, mu, sigma) - normcdf(x3, mu, sigma);
percentile = norminv(0.75, mu, sigma);
[Mean, variance] = normstat(mu, sigma);
```

```
pd =

  NormalDistribution

  Normal distribution
       mu = 76012.8   [74407.9, 77617.6]
    sigma = 20463.9   [19390.5, 21664.1]


mu =

   7.6013e+04


sigma =

   2.0464e+04


less =

   0.1018


          more =

              0.0158


          between =

              0.3022


          percentile =

              8.9815e+04


          Mean =

              7.6013e+04


          variance =

              4.1877e+08
```

3) The output voltages for a certain electric circuit is recorded. To calculate the probabilities we need to estimate its parameters so lets use normfit. But before lets check test if our voltage dataset is normally distributed so add that along with the code and the figures.

```
histogram(Voltage, 'Normalization', 'probability', 'EdgeColor', [0.8 0.5 0.5]...
        ,'FaceColor', [0.7 0.3 0.3],'LineWidth', 2)
xlabel("Voltage(V)")
ylabel('Probability')
title('Histogram of Voltages')
```

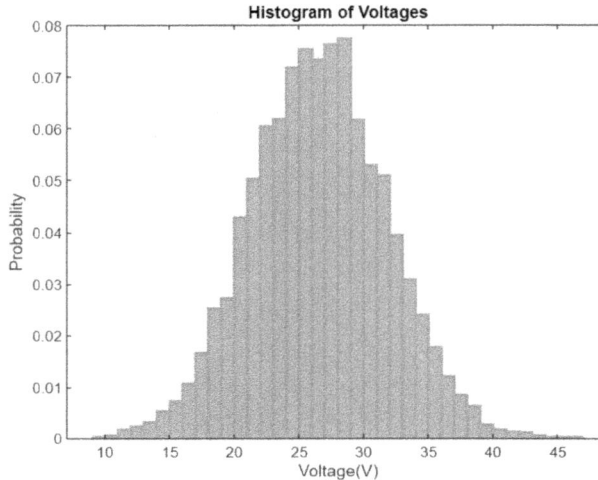

```
[muHat, sigmaHat] = normfit(Voltage);
x1 = 20;
x2 = 30;
x3 = 35;
x4 = 40;
less = normcdf(x1, muHat, sigmaHat);
more = normcdf(x2, muHat, sigmaHat, 'upper');
between = normcdf(x4, muHat, sigmaHat) - normcdf(x3, muHat, sigmaHat);
percentile = norminv(0.85, muHat, sigmaHat);
[Mean, variance] = normstat(muHat, sigmaHat);
```

```
muHat =

   26.4796


sigmaHat =

   5.1773


less =

   0.1054


more =

   0.2483


between =

   0.0454


percentile =

   31.8456


Mean =

   26.4796


variance =

   26.8042
```

5.3 LOGNORMAL DISTRIBUTION

The name itself suggests that this distribution has got to do something with logarithms as well as the normal distribution. In fact this distribution is a direct derivation from the normal distribution. Recall that in a normal distribution, the data is always symmetrically distributed with no skew or outliers. When plotted on a graph, the data follows a bell shape as seen in the previous graph. However, the distribution is not suitable when the data is highly skewed or the data contains outliers. In such a situation, we use this derived normal distribution, the lognormal distribution. In contrast to normally distributed data, lognormally distributed data does not form a symmetric shape but rather slants or skews more toward the right.

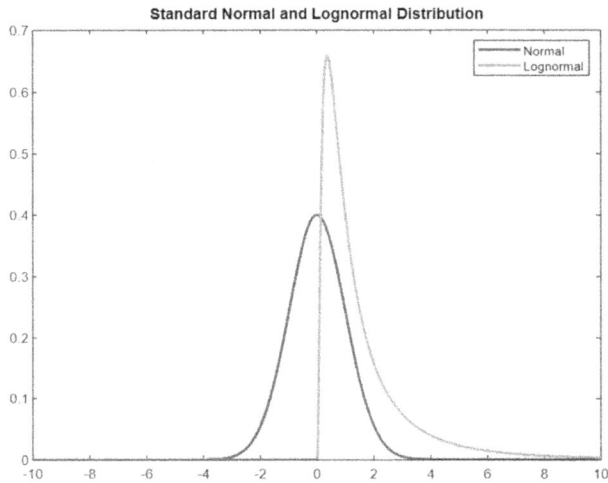

Figure 5.23

Another key difference between the two is that lognormal distributions contain only positive numbers since the log function can only take positive values, whereas the normal distribution can contain negative values.

5.3.1 How can I check if my data follows a lognormal distribution?

As mentioned earlier, instead of having the original raw data normally distributed, the logarithms of this raw data when computed are normally distributed.

Let's look at an example. Say an engineer collects voltage data to analyze and understand the behavior of electrical systems. When we draw a histogram (Figure 5.25) of the raw data we get:

```
data = readtable("Voltage.csv", VariableNamingRule = "preserve");
Voltage = data.Voltage;

tiledlayout(2,1)
ax1 = nexttile;
histogram(ax1, Voltage, 'Normalization', 'probability', 'EdgeColor', [0.57255 0.1412 0.1569], 'FaceColor', [0.67 0.408 0.3412], 'LineWidth', 2)
xlabel(ax1, "Voltage(V)")
title(ax1, "Histogram of Voltages")
ax2 = nexttile;
log_y = log(y)
histogram(ax2, log(Voltage), 'Normalization', 'probability', 'EdgeColor', [0.57255 0.1412 0.1569], 'faceColor', [0.67 0.408 0.3412], 'LineWidth', 2)
xlabel(ax2, "log Voltage(log(V))")
title(ax2, "Histogram of log Voltages")
```

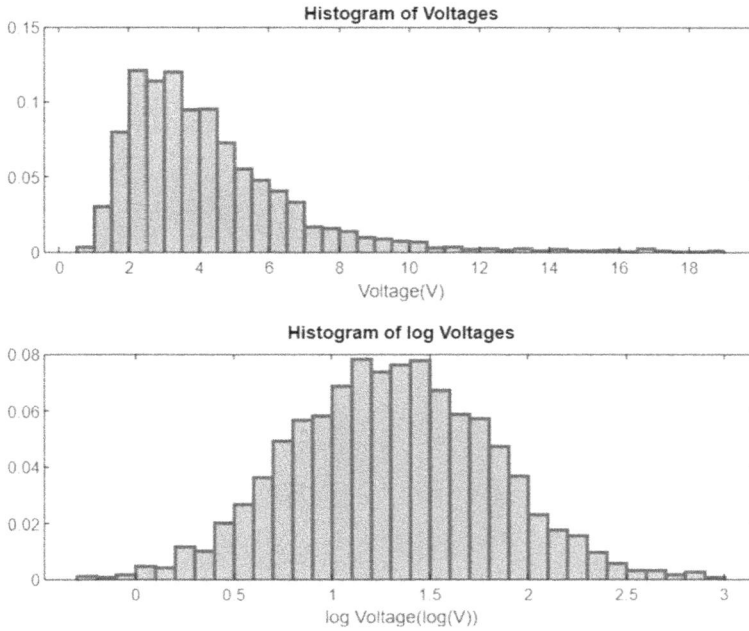

Figure 5.24

As we can see the logarithm transformation of our voltage dataset altered the shape of the distribution. Since the original data is positively skewed, the logarithm transformation made the distribution more symmetric. But what would have happened if our data was not positively skewed?

Let's say another group of engineers collects a dataset of compressor pressure to analyze the performance of compressors. By monitoring the pressure levels at different stages of compression, engineers can assess the efficiency, capacity, and reliability of the compressor.

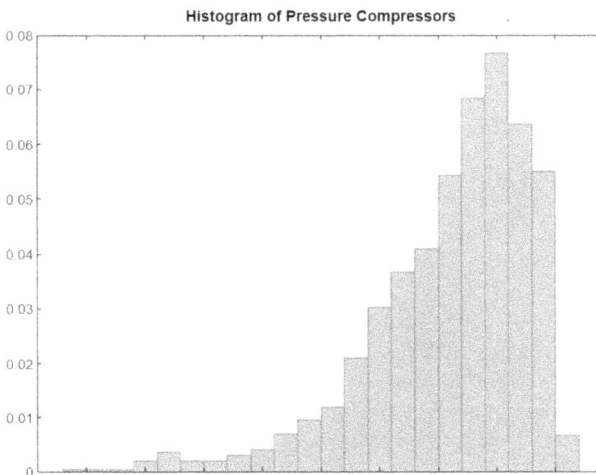

Figure 5.25

On the other hand, if the original data is negatively skewed, the logarithm transformation can make the distribution more skewed.

So how do relate the two variables? If X has a normal distribution then the exponential function of X will have a lognormal distribution.

Similarly, if X has a lognormal distribution then $ln\,X$ is normally distributed.

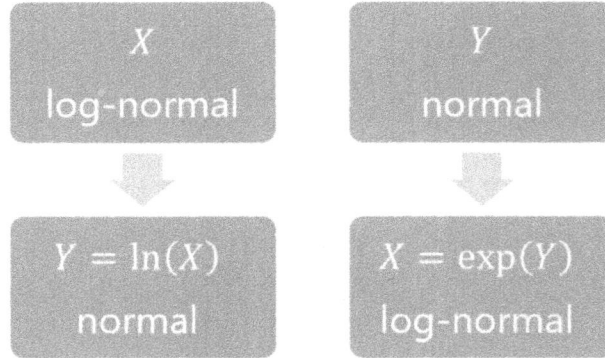

So in conclusion, whenever we take the log of our lognormal data we end up with a normal distribution.

5.3.2 Parameters of the distribution

The above discussion makes it very clear that normal and lognormal are very closely related so it would not be wrong to assume that both of them will have the same parameters—mean μ and standard deviation σ—which are also its location and scale parameters, respectively.

However even though both normal and lognormal have the same parameters they should not be mistaken for each other.

Before the transformation μ and σ are simply two parameters that define our lognormal distribution and not the normal distribution μ and σ.

5.3.3 Shape of the distribution

The lognormal distribution is a right-skewed continuous probability distribution, meaning it has a long tail toward the right.

5.3.4 Calculations in theory

Hence the probability density function of the lognormal distribution with parameters μ and σ is given by

$$f\left(x\right)=\frac{1}{\sqrt{2\pi}\sigma x}e^{-\frac{1}{2}\left(\frac{ln\,x-\mu}{\sigma}\right)^2}$$

As we can see, just like the normal distribution this PDF is equally difficult to work with. Hence we work with the lognormal variable the same way we work with the normal distribution. Therefore, here too we are going to use our z-score distribution to calculate the probabilities.

The cumulative distribution function of a lognormal distribution is

$$F\left(x\right)=P\left(X\leq x\right)=P\left(Z\leq\left(\frac{ln\,x-\mu}{\sigma}\right)\right)$$

Since the normal distribution is symmetrical and non-skewed, all three measures of central tendency are equal. However, this is not the case for lognormal as it is skewed.

- The mean of lognormal distribution $E(X) = e^{\left(\mu + \frac{\sigma^2}{2}\right)}$
- The median of lognormal distribution $E(X) = e^{\mu}$

The above equation is derived by setting the cumulative distribution values equal to 0.5 and solving the resulting equation:

- The mode of lognormal distribution $E(X) = e^{\mu - \sigma^2}$

The above equation is derived by setting the probability distribution function values equal to 0 as the mode denotes the global maximum of distribution.

- The variance of lognormal distribution $Var(X) = (e^{\sigma^2} - 1)e^{(2\mu + \sigma^2)}$

5.3.5 Example

1) The shelf-life (in days) of a certain electronic component is lognormally distributed with $\mu = 1.2$ and $SDc = 0.5$.

   a. Find the value of the density function at $x = 2$.

   $$f(x) = \frac{1}{\sqrt{2\pi}(0.5)(2)} e^{-\frac{1}{2}\left(\frac{\ln 2 - 1.2}{0.5}\right)^2} = 0.23865$$

   b. Find the mean and variance of the lifetime of the electronic component.

   $$E(X) = e^{\left(1.2 + \frac{0.5^2}{2}\right)} = e^{1.325} = 3.76$$

   $$Var(X) = (e^{0.5^2} - 1)e^{(2(1.2) + 0.5^2)} = 4.02010$$

   c. Find the probability that the component works for at most one week.

   $$P(X \le 7) = P\left(Z \le \frac{\ln 7 - 1.2}{0.5}\right) = P(Z \le 1.49) = 0.9319$$

   d. Find the probability that the component works for more than ten days.

   $$(X > 10) = 1 - P(X \le 10) = 1 - P\left(Z \le \frac{\ln 10 - 1.2}{0.5}\right) = 1 - P(Z \le 2.205)$$

   $$= 1 - 0.9862 = 0.0138$$

   e. Find the probability that the component works for between three and five days.

   $$P(3 \le X \le 5) = P(X \le 5) - P(X < 3) = P\left(Z < \frac{\ln 5 - 1.2}{0.5}\right) - P\left(Z < \frac{\ln 3 - 1.2}{0.5}\right)$$

   $$= P\left(Z < \frac{\ln 5 - 1.2}{0.5}\right) - P\left(Z < \frac{\ln 3 - 1.2}{0.5}\right)$$

   $$= P(Z < 0.82) - P(Z < -0.2) = 0.7939 - 0.4207 = 0.3732$$

5.3.6 Calculations in MATLAB

5.3.6.1 Create a lognormal distribution object

5.3.6.1.1 Default parameters

```
pd = makedist('Lognormal')

                       pd =

                  LognormalDistribution

                  Lognormal distribution
                        mu = 0
                     sigma = 1
```

5.3.6.1.2 Specified parameters

```
pd = makedist('Lognormal', 'mu', 20, 'sigma', 5)

                       pd =

                  LognormalDistribution

                  Lognormal distribution
                        mu = 20
                     sigma =  5
```

5.3.6.2 Fitting the lognormal distribution to sample data

5.3.6.2.1 Default parameters

```
pd = fitdist(x, 'Lognormal')
```

5.3.6.3 Use distribution-specific functions

Now that we know all about the normal distribution how do we carry out these calculations in MATLAB?

Now we are going to carry out these calculations in MATLAB using the functions below:

lognpdf(x, mu, sigma)	computes the PDF of the lognormal distribution with mean mu and standard deviation sigma, evaluated at the values in x.
logncdf(x, mu, sigma)	computes the CDF of the lognormal distribution with mean mu and standard deviation sigma, evaluated at the values in x.
logncdf(x, mu, sigma, 'upper')	returns the complement of the CDF, evaluated at the values in x, using an algorithm that more accurately computes the extreme upper-tail probabilities.
logninv(p, mu, sigma)	returns the inverse of the lognormal CDF with mean mu and standard deviation sigma, evaluated at the probability values in p.
lognstat(mu, sigma)	returns the mean and variance of the lognormal distribution with the distribution parameters mean mu and standard deviation sigma.
lognrnd(mu, sigma, size)	generates an array of lognormal random numbers, where vector size specifies size(r).

5.3.6.4 Example

5)   generates an array of lognormal random numbers, where vector size specifies size(r).

```
x = -10:0.01:10;
PDF = lognpdf(x);
plot(x, PDF, 'LineWidth', 2)
title("Standard lognormal density")
```

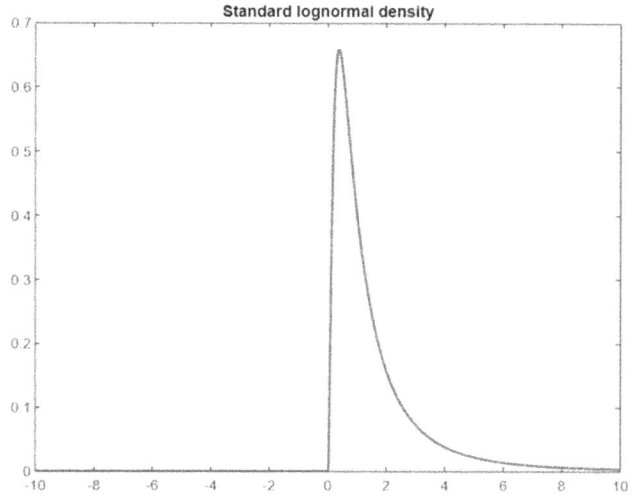

Figure 5.26

```
x = -10:0.01:10;
CDF = logncdf(x);
plot(x, CDF, 'LineWidth', 2)
title("Standard lognormal Cumulative Density Function")
```

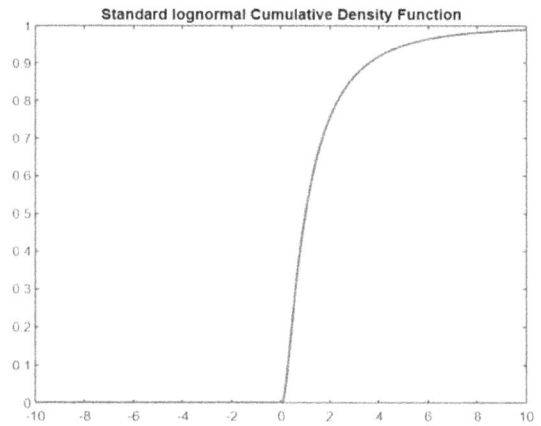

Figure 5.27

2) Lets draw multiple lognormal distribution curve with the same standard mean but different standard deviations in MATLAB.

```
x = 0:0.01:10;

tiledlayout(2,1)
% Top axes
ax1 = nexttile;
plot(ax1, x, lognpdf(x), LineWidth = 2)
hold on
plot(ax1, x, lognpdf(x, 0, 2),'Color', 'k', LineWidth = 2)
plot(ax1, x, lognpdf(x, 0, 0.5), 'Color', 'm', LineWidth = 2)
hold off
title(ax1, "Lognormal Distribution with Multiple Sigma")
legend(ax1, 'μ = 0, σ^2 = 1', 'μ = 0, σ^2 = 2', 'μ = 0, σ^2 = 0.5')

ax2 = nexttile;
plot(ax2, x, logncdf(x), LineWidth = 2)
hold on
plot(ax2, x, logncdf(x, 0, 2),'Color', 'k', LineWidth = 2)
plot(ax2, x, logncdf(x, 0, 0.5), 'Color', 'm', LineWidth = 2)
hold off
title(ax2, "Cumuluative Distribution function with Multiple Sigma")
legend(ax2, 'μ = 0, σ^2 = 1', 'μ = 0, σ^2 = 2', 'μ = 0, σ^2 = 0.5')
```

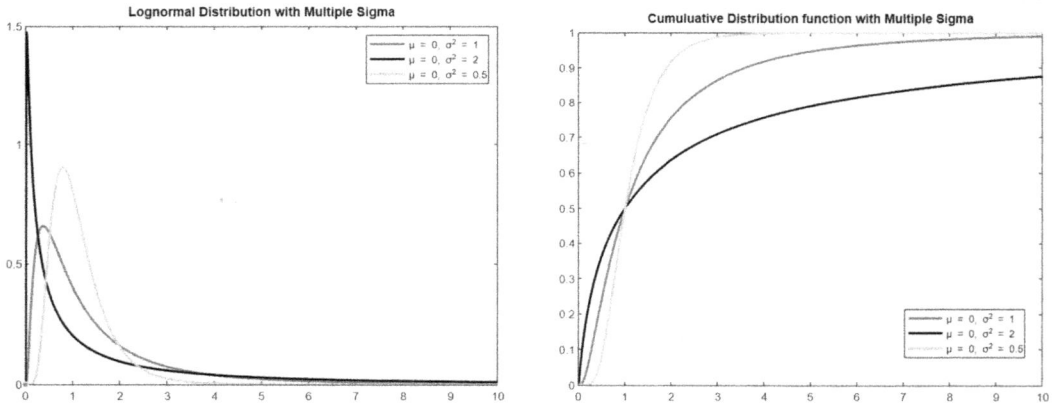

Figure 5.28

3) Finally draw multiple lognormal distribution curve with the same standard deviation but different mean values in MATLAB.

```
x = 0:0.01:10;

tiledlayout(2,1)
% Top axes
ax1 = nexttile;
plot(x, lognpdf(x), LineWidth = 2)
hold on
plot(ax1, x, lognpdf(x, 1, 1),'Color', 'k', LineWidth = 2)
plot(ax1, x, lognpdf(x, -1, 1), 'Color', 'm', LineWidth = 2)
hold off
title(ax1, "Lognormal Distribution with Multiple Mean")
legend(ax1, 'μ = 0, σ^2 = 1', 'μ = 1, σ^2 = 1', 'μ = -1, σ^2 = 1')

ax2 = nexttile;
plot(ax2, x, logncdf(x), LineWidth = 2)
hold on
plot(ax2, x, logncdf(x, 1, 1),'Color', 'k', LineWidth = 2)
plot(ax2, x, logncdf(x, -1, 1), 'Color', 'm', LineWidth = 2)
hold off
title(ax2, "Cumuluative Distribution function with Multiple Mean")
legend(ax2, 'μ = 0, σ^2 = 1', 'μ = 1, σ^2 = 1', 'μ = -1, σ^2 = 1')
```

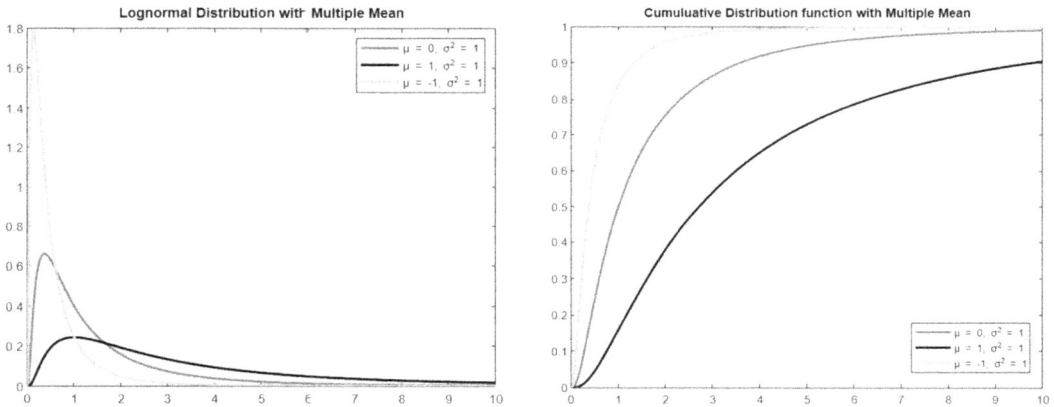

Figure 5.29

Solving the same problem using the above functions:

1) The shelf-life (in days) of a certain electronic component is lognormally distributed with $\mu = 1.2$ and $\sigma = 0.5$.

```
1       mu = 1.2;
2       sigma  = 0.5;
3       x1 = 2;
4       x2 = 7;
5       x3 = 10;
6       x4 = 3;
7       x5 = 5;
8       y = lognpdf(x1, mu, sigma)
9       [mean, Variance] = lognstat(mu, sigma)
10      p1 = logncdf(x2, mu, sigma)
11      p2 = logncdf(x3, mu, sigma, 'upper')
12      p3 = logncdf(x4, mu, sigma, 'upper') - logncdf(x5, mu, sigma, 'upper')
13
14      x = 0:0.01:12;
15      p = logncdf(x, mu, sigma);
16      plot(x, p)
17      grid on
```

```
y =

    0.2387

mean =

    3.7622

Variance =

    4.0201

p1 =

    0.9321

p2 =

    0.0137

p3 =

    0.3739
```

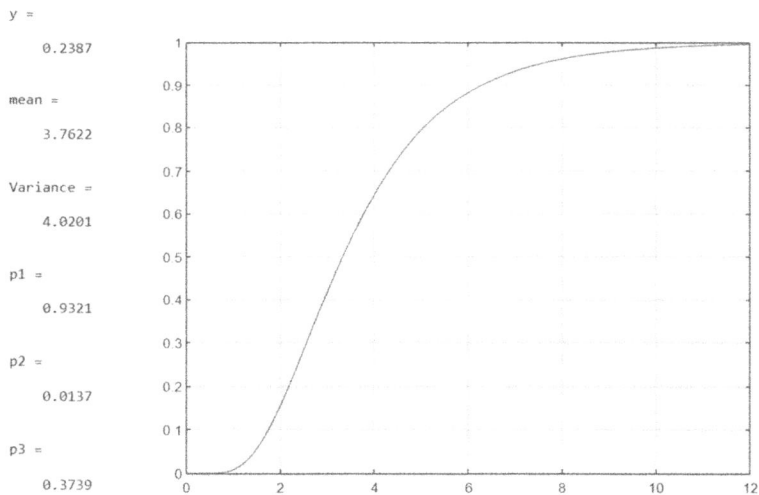

Figure 5.30

5.3.6.5 Example

The temperature of a compressor is a critical parameter in engineering as it impacts performance, equipment reliability, safety, process control, and maintenance.

a) Find the value of the density function at $x = 300$.
b) Find mean and variance of the temperature of a compressor.
c) Find the probability that the temperature exceeds 1000°C.
d) What is the probability that the temperature is less than 500°C?
e) Find the probability that the temperature is between 600 and 1200°C.

```
data = readtable("Temperature logarithm.csv", VariableNamingRule = "preserve");
Temperature = data.Temperature;
pd = fitdist(Temperature, 'Lognormal')
mu = pd.mu
sigma = pd.sigma

tiledlayout(2,1)
ax1 = nexttile;
% Plot a histogram of the generated data
histogram(ax1, Temperature, 'Normalization', 'pdf', 'EdgeColor', [0 0 0], 'FaceColor', [0.22 0.19 0.24], 'LineWidth', 2)
ylabel(ax1, 'PDF')
title(ax1, 'Histogram of Temperature')

ax2 = nexttile;
histogram(ax2, log(Temperature), 'Normalization', 'pdf', 'EdgeColor', [0 0 0], 'FaceColor', [0.22 0.19 0.24], 'LineWidth', 2)
ylabel(ax2, 'PDF')
title(ax2, 'Histogram of log(Temperature)')
```

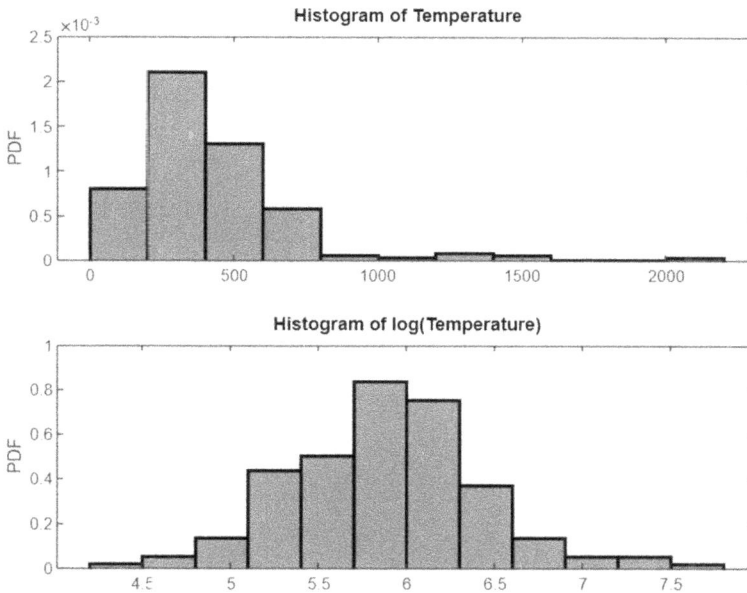

Figure 5.31

Since the data is lognormally distributed we can now make the calculations

```
data = readtable("Temperature", VariableNamingRule = "preserve");
Temperature = data.Temperature;
pd = fitdist(Temperature, 'Lognormal');
mu = pd.mu;
sigma = pd.sigma;
x1 = 300;
x2 = 1000;
x3 = 500;
x4 = 600;
x5 = 1200;
y = lognpdf(x1, mu, sigma);
[mean, Variance] = lognstat(mu, sigma)
p1 = logncdf(x2, mu, sigma);
p2 = logncdf(x3, mu, sigma, 'upper');
p3 = logncdf(x4, mu, sigma, 'upper') - logncdf(x5, mu, sigma, 'upper');
```

```
y =

    0.0023


mean =

    409.3296


Variance =

    5.7736e+04


p1 =

    0.9722


p2 =

    0.2612


p3 =

    0.1526
```

Certain components of motors degrade with time and operating stress. Suppose the lifetime of a motor is recorded.

a) Find mean and variance of the lifetime of the motor.
b) What is the probability that the lifetime is less than 2000 hours?
c) Find the probability that the lifetime exceeds 11000 hours?
d) Find the probability that the lifetime is between 5000 and 10000 hours?

Lets first test if our dataset has the lognormal distribution

```
data = readtable("lifetime of motor.xlsx", VariableNamingRule = "preserve");
Lifetime = data.Lifetime;

tiledlayout(2,1)
ax1 = nexttile;
histogram(ax1, Lifetime, 'Normalization', 'probability', 'EdgeColor', [0.5 0 0.5]...
        ,'FaceColor', [0.6740 0 0.1880], 'LineWidth', 2)
xlabel(ax1, "lifetime(hours)")
ylabel(ax1, 'Probability')
title(ax1, 'Histogram of lifetime of a motor')

ax2 = nexttile;
histogram(ax2, log(Lifetime), 'Normalization', 'probability', 'EdgeColor', [0.5 0 0.5]...
        ,'FaceColor', [0.6740 0 0.1880], 'LineWidth', 2)
xlabel(ax2, "log(hours)")
ylabel(ax2, 'Probability')
title(ax2, 'Histogram of log(lifetime) of a motor')
```

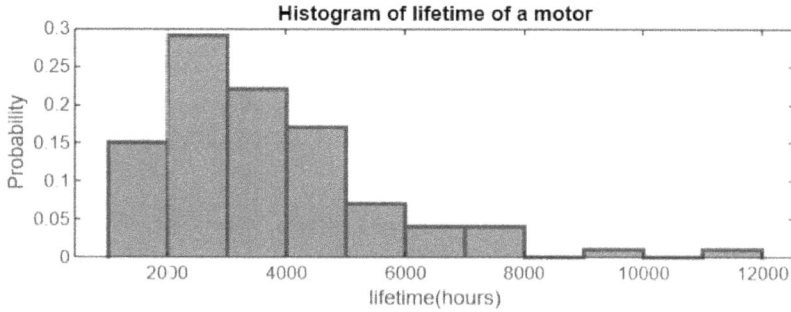

Histogram of lifetime of a motor

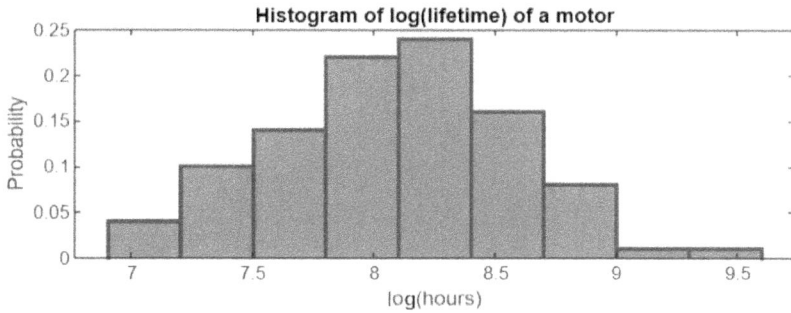

Histogram of log(lifetime) of a motor

```
phat = lognfit(Lifetime);
mu_log = phat(1);
sigma_log = phat(2);
x1 = 2000;
x2 = 11000;
x3 = 5000;
x4 = 10000;
less = logncdf(x1, mu_log, sigma_log);
more = logncdf(x2, mu_log, sigma_log, 'upper');
between = logncdf(x4, mu_log, sigma_log) - logncdf(x3, mu_log, sigma_log);
[Mean, variance] = lognstat(mu_log, sigma_log);
```

```
phat =

    7.9636    0.5025


less =

    0.2352


more =

    0.0038


between =

    0.1288


Mean =

    3.2614e+03


variance =

    3.0558e+06
```

5.4 GAMMA DISTRIBUTION

In statistics, a **gamma distribution** is any one of a family of continuous probability distributions that can be used to model the waiting times between events. A gamma distribution predicts the wait time until a certain number of events occurs randomly at some average rate.

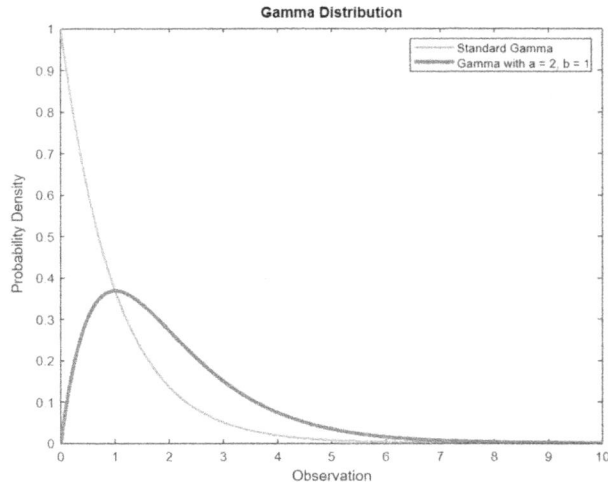

Figure 5.32

```
x = 0:0.01:10;
y1 = gampdf(x, 1, 1);
y2 = gampdf(x, 2, 1);
figure;
plot(x, y1, 'Color', [0.9290 0.1250 0.6940], 'LineWidth', 2)
hold on;
plot(x, y2, 'Color', [0.1250 0.7290 0.6940], 'LineWidth', 3)
hold off
xlabel('Observation')
ylabel('Probability Density')
title("Gamma Distribution")
legend("Standard Gamma", "Gamma with a = 2, b = 1")
```

As we can see, the gamma distribution is similar in appearance to the lognormal distribution. They are both skewed to the right. But are gamma distributions always right- skewed like the lognormal distribution? To find this let's discuss its parameters first.

5.4.1 Parameters of the distribution

The gamma distribution is represented by the two parameters:

- The shape denoted by α
- The rate/scale parameter represented by β

Both parameters are $\alpha, \beta > 0$.

The shape parameter for the gamma distribution specifies the number of events occurring. The shape parameter must be positive, but it does not have to be an integer.

There are two versions of the scale parameter in the gamma distribution. Now β defines the mean wait time between α events. For example, the gamma (α,β) distribution models the time required for α number of events to occur, given the events occur randomly in a Poisson process with a mean time between events equal to *β*. For example, in a highway it was found that there is one accident occurring every four days and we want to evaluate probabilities for the elapsed time of three accidents. In this situation *β* = 4, α = 3.

5.4.2 Shape of the distribution

Now notice that the two distributions we discussed earlier do not have a shape parameter and that's why they had a fixed shape that did not change regardless of the values of their parameters. However, this is not the case for the gamma distribution. Since it has a shape parameter, it represents a family of shapes. As the name suggests, the shape parameter controls the shape of the distribution. The fundamental shapes are characterized by the following values of α:

* When α = 1, we obtain an exponential distribution (discussed in the next section).
* When α = 2.5, the gamma distribution is skewed positively.
* When α = 20, the gamma distribution is now a bell-shaped curve. In fact, the gamma distribution becomes more and more symmetric as the shape parameter α → ∞.

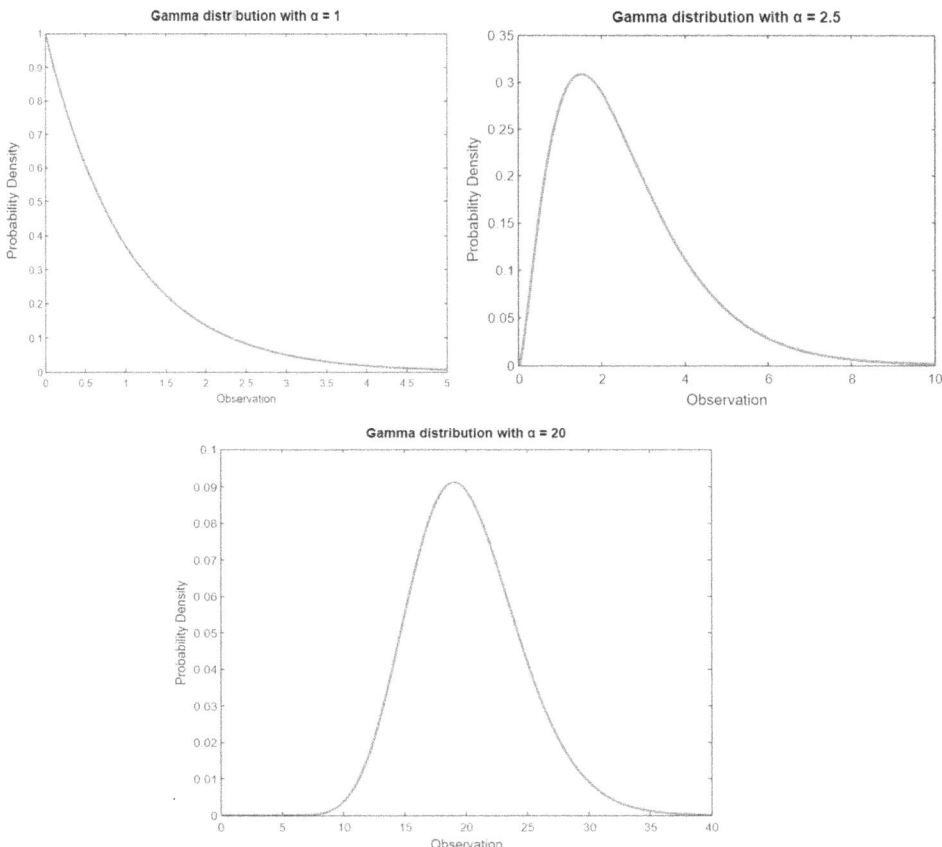

Figure 5.33

The scale parameter controls the height of the distribution's peak. The higher the value of the scale distribution, the more spread it will be, which means it will have a lower peak. However, if the scale parameter is small, the distribution will contract and the peak will be larger.

5.4.3 Compare gamma and normal distribution PDFs

While the gamma and normal distributions are distinct in terms of their characteristics, the gamma distribution closely approximates the normal distribution for a large α with mean $\mu = \alpha\beta$ and variance $= \alpha\beta^2$.

```
x = 250:750;

a = 100;
b = 5;
y_gamma = gampdf(x, a, b);
mu = a*b
sigma = sqrt(a*b^2)
y_normal = normpdf(x,mu,sigma);

plot(x, y_gamma, '-', x, y_normal ,'-.', 'LineWidth', 2)
title('Gamma and Normal pdfs')
xlabel('Observation')
ylabel('Probability Density')
legend('Gamma Distribution', 'Normal Distribution')
```

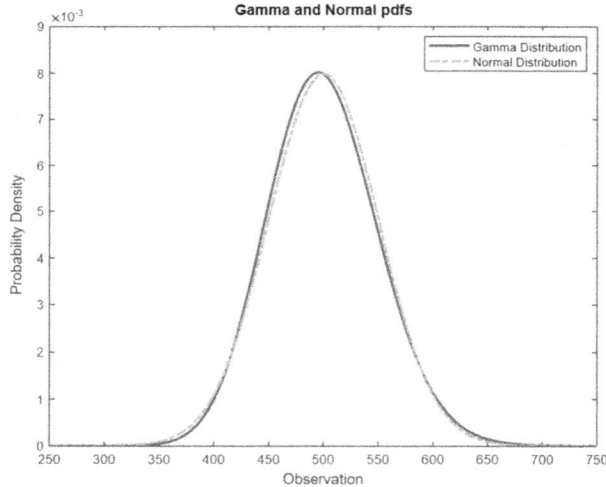

Figure 5.34

In contrast to , lamda (λ) is the mean rate of occurrence during one unit of time in the Poisson distribution. λ and α are reciprocals. To understand why, let's return to our example of one accident occurring every four days on average ($\beta = 4$). An equivalent way to state this is that an average of one-quarter accident occurs during one day ($\lambda = 1 / 4 = 0.25$).

5.4.4 Calculation in theory

Hence the gamma probability density function of the random variable X with shape parameter α and a scale/rate parameter $\beta = \dfrac{1}{\lambda}$ can be defined as:

$$f(x) = \{\frac{1}{\beta^\alpha} \frac{1}{\Gamma(\alpha)} x^{\alpha-1} e^{-\frac{1}{\beta} \cdot x} = \lambda^\alpha \frac{1}{\Gamma(\alpha)} x^{\alpha-1} e^{-\lambda x} \qquad 0 \le x \le \infty$$

- The CDF of the gamma distribution is

$$F(x) = P(X \le k) = \int_c^k f(x)\,dx = \frac{1}{\beta^\alpha} \frac{1}{\Gamma(\alpha)} \int_0^k x^{\alpha-1} e^{-\frac{1}{\beta} \cdot x}\,dx = \lambda^\alpha \frac{1}{\Gamma(\alpha)} \int_0^k x^{\alpha-1} e^{-\lambda x}\,dx$$

- The mean of the gamma distribution is $\alpha\beta = \alpha\dfrac{1}{\lambda}$.

- The variance of the gamma distribution is $\alpha\beta^2 = \alpha\dfrac{1}{\lambda^2}$.

The gamma distribution can be used to model service times, lifetimes of objects, and repair times.

5.4.5 Example

1) The lifetime of a certain piece of equipment is described by a random variable X whose distribution is gamma with parameters $\alpha = 2$ and $\beta = \dfrac{3}{4}$.
   a) Plot the graph of the gamma probability distribution.
   b) Plot the graph of cumulative gamma probabilities.
   c) Find the value of the density function at $x = 5$.

$$\frac{1}{\frac{3}{4}^2} \frac{1}{\Gamma(2)} 5^{(2-1)} e^{-\frac{1}{\frac{3}{4}} \cdot 5} = 0.01131$$

   d) Find the probability that the lifetime of the equipment is at most three units of time.

$$P(X \le 3) = \frac{1}{\frac{3}{4}^2} \frac{1}{\Gamma(2)} \int_0^3 x^{2-1} e^{-\frac{1}{\frac{3}{4}} \cdot x}\,dx = 0.90842$$

   e) Find the probability that the lifetime of the equipment is at least one unit of time.

$$P(X \ge 1) = 1 - P(X < 1) = 1 - \left[\frac{1}{\frac{3}{4}^2} \frac{1}{\Gamma(2)} \int_0^1 x^{2-1} e^{-\frac{1}{\frac{3}{4}} \cdot x}\,dx\right] = 0.61507$$

f) Find the probability that the lifetime of the equipment is less than 2.5 units of time but greater than 1.5 units of time.

$$P(1.5 \le X \le 2.5) = P(X \le 2.5) - P(X \le 1.5)$$

$$= \left[\frac{1}{3^2} \frac{1}{\Gamma(2)} \int_0^{2.5} x^{2-1} e^{-\frac{1}{3} \cdot x} \, dx \right] - \left[\frac{1}{3^2} \frac{1}{\Gamma(2)} \int_0^{1.5} x^{2-1} e^{-\frac{1}{3} \cdot x} \, dx \right] = 0.25141$$

5.4.6 Calculations in **MATLAB**

5.4.6.1 Create a gamma distribution object

5.4.6.1.1 Default parameters

```
pd = makedist('Gamma')
```

```
                  pd =

              GammaDistribution

          Gamma distribution
              a = 1
              b = 1
```

5.4.6.1.2 Specified parameters

```
pd = makedist('Gamma', 'a', 0.25, 'b', 2)
```

```
                  pd =

              GammaDistribution

          Gamma distribution
              a = 0.25
              b =    2
```

5.4.6.2 Fitting the gamma distribution to sample data

5.4.6.2.1 Default parameters

```
pd = fitdist(x, 'Gamma')
```

5.4.6.3 Use distribution-specific functions

Let's carry out these calculations in MATLAB.

gampdf(x, a, b)	computes the PDF of the gamma distribution with the shape parameter a and the scale parameter b, evaluated at the values in x.
gamcdf(x, a, b)	computes the PDF of the gamma distribution with the shape parameter a and the scale parameter b, evaluated at the values in x.
gamcdf(x, a, b, 'upper')	• returns the complement of the CDF, evaluated at the values in x, using an algorithm that more accurately computes the extreme upper-tail probabilities.
gaminv(p, a, b)	returns the inverse of the gamma CDF with the shape parameter a and the scale parameter b, evaluated at the probability values in p.
gamstat(a, b)	returns the mean and variance of the gamma distribution with the shape parameter a and the scale parameter b.
gamrnd(a, b, size)	generates an array of gamma random numbers, where vector size specifies size(r).

Lets draw multiple Gamma distribution curve with the same scale but different shape in MATLAB.

Let's try doing the same problem using the above functions and see for ourselves if these functions give the same answer.

The lifetime of a certain piece of equipment is described by a random variable X whose distribution is gamma with parameters $\alpha = 2$ and $\beta = \frac{3}{4}$.

1. Plot the graph of the gamma probability distribution.

```
a = 2;
b = 0.75;
x = 0:0.01:8;
y_gamma = gampdf(x,a,b);
plot(x, y_gamma, 'LineWidth', 2)
title('Gamma pdfs')
xlabel('Observation')
ylabel('Probability Density')
```

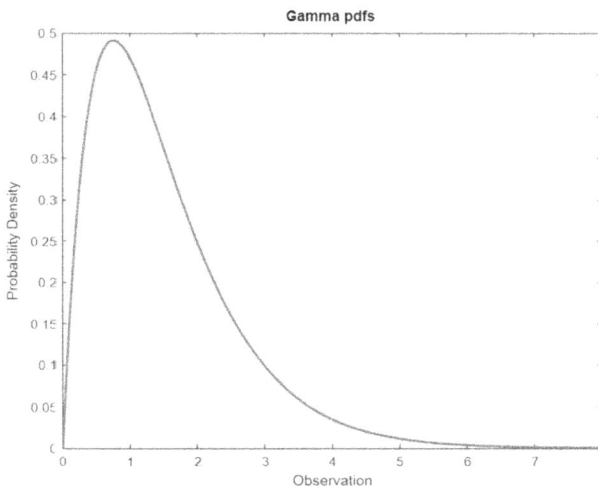

Figure 5.35

2. Plot the graph of cumulative gamma probabilities.

```
A = 2;
b = 0.75;
x = 0:0.01:3;
y_gamma = gamcdf(x,a,b);
plot(x, y_gamma, 'LineWidth', 2)
title('Gamma cdfs')
xlabel('Observation')
ylabel('Probability Density')
```

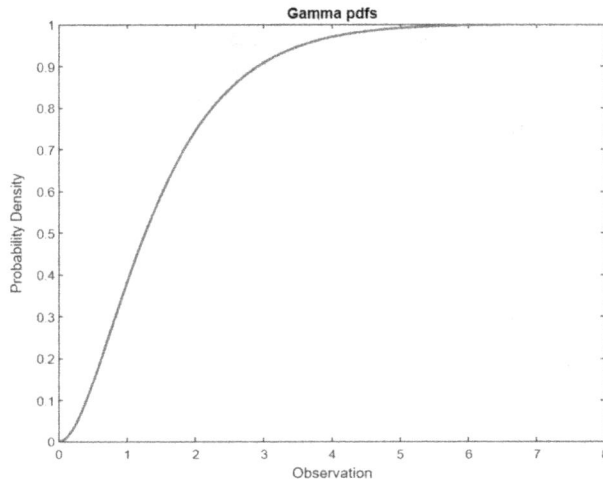

Figure 5.36

3. Find the value of the density function at $x = 5$.
4. Find the probability that the lifetime of the equipment is at most three units of time.
5. Find the probability that the lifetime of the equipment is at least one unit of time.
6. Find the probability that the lifetime of the equipment is less than 2.5 units of time but greater than 1.5 units of time.

Solving the next four problems:

```
1        a = 2;
2        b = 0.75;
3        x1 = 5;
4        x2 = 3;
5        x3 = 1;
6        x4 = 1.5;
7        x5 = 2.5;
8        y = gampdf(x1, a, b);
9        p1 = gamcdf(x2, a, b);
10       p2 = gamcdf(x3, a, b, 'upper');
11       p3 = gamcdf(x4, a, b, 'upper') - gamcdf(x5, a, b, 'upper');
```

Command Window

```
y =

    0.0113


p1 =

    0.9084


p2 =

    0.6151


p3 =

    0.2514
```

In a certain city, the daily consumption of electric power (in millions of kilowatt-hours) can be modeled as a random variable with Gamma distribution. Parameters alpha and beta are to be determined.

a) Estimate the mean daily consumption and variance.
b) Find the probability that the daily consumption is lower than 4.5 millions of kilowatt-hours
c) Find the probability that the daily consumption is higher than 7 millions of kilowatt-hours.
d) What is the probability that the daily consumption is between 2.5 and 5 millions of kilowatt-hours.
e) Find the median of the daily consumption.

```
pd = fitdist(Power, 'Gamma');
Shape = pd.a;
Scale = pd.b;

x1 = 4.5;
x2 = 7;
x3 = 2.5;
x4 = 5;
[Mean, variance] = gamstat(Shape, Scale);
less = gamcdf(x1, Shape, Scale);
more = gamcdf(x2, Shape, Scale, 'upper');
between = gamcdf(x4, Shape, Scale) - gamcdf(x3, Shape, Scale);
median = gaminv(0.5, Shape, Scale);
```

```
pd =

  GammaDistribution

  Gamma distribution
    a = 0.809167   [0.644692, 1.0156]
    b = 5.67884    [4.17659, 7.72143]

Shape =

    0.8092

Scale =

    5.6788
```

Mean =

        4.5951

variance =

        26.0951

less =

        0.6394

more =

        0.2205

between =

        0.2170

median =

        2.8965

5.4.6.4 Practice questions

1. Suppose the time spent (in minutes) by a randomly selected student who uses a terminal connected to a local time-sharing computer facility has a gamma distribution with $\alpha = 5$ and $\beta = 3$.
   a) What is the probability density function for $x = 2$?
   b) Find $P(X < 20)$.
   c) Find $P(60 < X < 40)$.

2. In Westlake, the daily consumption of electric power, in millions of kilowatt hours, is a random variable with X having a gamma distribution with $\alpha = 4$ and $\beta = 2$. What is the probability that the number of kw hours is between 5 and 10?

3. In the annual exam, to solve each problem, Josh takes about half an hour. Find the probability that it will take him somewhere around two to four hours to solve five such problems.

4. Suppose that in a city hospital, an optimal arrival to proper triage of the patient should occur every 15 minutes. What is the probability that the hospital will receive a total of ten patients in less than three hours?

5. Claim amounts on a certain type of policy are modeled as following a gamma distribution with parameters $\alpha = 120$ and $\lambda = 1.20$. Calculate an approximate value for the probability that an individual claim amount exceeds 120.

6. The daily consumption of milk in a city in excess of 20,000 gallons is approximately distributed as a gamma radiation with $\alpha = 2$ and $\lambda = \dfrac{1}{10000}$. The city has a daily stock of 30,000 gallons. What is the probability that the stock is insufficient on a particular day?

7. In a city in Texas, the annual income (in $) is distributed according to gamma distribution with $\alpha = 4$ and $\beta = 10000$. Find $P(X < 26000)$.

The next two distributions are the two special cases of gamma distribution. The chi-square and the exponential distributions are one-parameter distributions that fix one of the two gamma parameters.

5.5 EXPONENTIAL DISTRIBUTION

As seen earlier, when $\alpha = 1$ the gamma distribution becomes an exponential distribution. Recall that the shape value, α, equals the number of events and the gamma distribution models the elapsed time between those events α. In contrast, the exponential distribution predicts the wait time until the first event denoted by $\alpha = 1$. Hence in conclusion the difference between the gamma distribution and exponential distribution is that the **exponential distribution** predicts the wait time until the first event and the **gamma distribution** indicates the wait time until the kth event.

```
x = 0:0.01:14;
y1 = gampdf(x, 1, 1);
y2 = gampdf(x, 2, 2);
y3 = exppdf(x, 2);
figure;
plot(x, y1, 'Color', [0.4940 0.1840 0.5560], 'LineWidth', 2)
hold on;
```

```
plot(x, y2, 'Color', [0.4660 0.6740 0.1880], 'LineWidth', 2)
plot(x, y3, 'Color', [0.1660 0.1740 0.6880], 'LineWidth', 2)
hold off;
xlabel('Observation')
ylabel('Probability Density')
legend('Standard Gamma Distribution', 'Gamma a = 2, b = 2',
                       'Exponential Distribution')
title("Gamma and Exponential PDF")
```

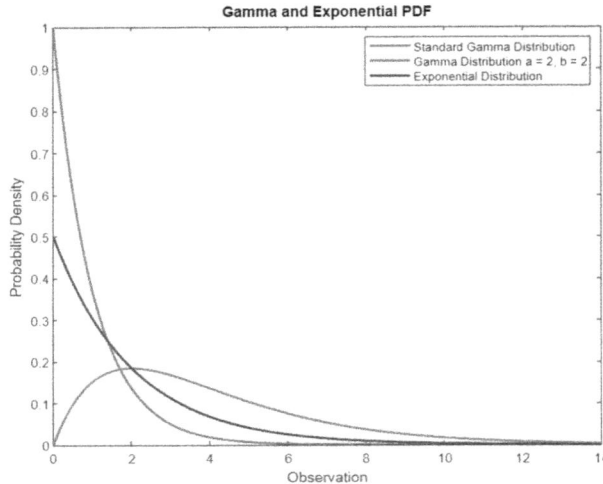

Figure 5.37

The **exponential distribution** is actually helpful in predicting the amount of waiting time until a future event happens.

Generally, the exponential distribution describes the waiting time between Poisson occurrences.

For example,

- The amount of time before an earthquake occurs in a given region
- The waiting time until a customer enters a store
- The time taken before a call center receives the next phone call
- The amount of time until an equipment failure

Please note that this distribution assumes that the average time between events remains constant.

5.5.1 Parameters of the distribution

The distribution has one parameter, λ, which is assumed to be the average rate of arrivals or occurrences of an event in a given time interval.

5.5.2 Shape of the distribution

Notice that our exponential distribution has no shape parameter; therefore it has only one shape. It is always right-skewed and models variables in which small values have relatively high probabilities as compared to the larger values.

5.5.3 Calculation in theory

Recall that the PDF of gamma is

$$f(x) = \{ \frac{1}{\beta^{\alpha}\,\Gamma(\alpha)}\,x^{\alpha-1}e^{-\frac{1}{\beta}\cdot x} = \lambda^{\alpha}\,\frac{1}{\Gamma(\alpha)}\,x^{\alpha-1}e^{-\lambda x} \qquad x > 0$$

$$\text{otherwise}$$

Now if $\alpha = 1$ the gamma distribution then becomes an exponential distribution, so now we get

$$f(x) = \frac{1}{\beta}e^{-\frac{1}{\beta}\cdot x} = \lambda e^{-\lambda x}$$

The CDF of the gamma distribution is

$$F(x) = P(X \le k) = \int_0^k f(x)\,dx = \int_0^k \frac{1}{\beta}e^{-\frac{1}{\beta}\cdot x}\,dx = 1 - e^{-\frac{1}{\beta}k} = 1 - e^{-\lambda k} = \int_0^k \lambda e^{-\lambda x}\,dx$$

- The mean of the exponential distribution is $\beta = \dfrac{1}{\lambda}$.
- The variance of the exponential distribution is $\beta^2 = \dfrac{1}{\lambda^2}$.
- Note: the exponential distribution models the time between events. Time is a continuous variable and the exponential distribution is, correspondingly, a continuous probability distribution. Conversely, the Poisson distribution models the count of events within a fixed amount of time. A count is a discrete variable and the Poisson distribution is a discrete probability distribution.
- The gamma, exponential, and Poisson distributions all model different characteristics of a Poisson process. A Poisson process has independent events occurring at a constant mean rate. All these distributions can use lambda as a parameter, which represents that average rate of occurrence.

5.5.4 Example

The time (in hours) required to repair a machine is an exponential distributed with a mean of two hours. What is the probability that a repair time:

a) Takes exactly six hours?

$$P(X = 6) = \frac{1}{2}e^{-\frac{1}{2}\cdot 6} = \frac{1}{2}e^{-3} = 0.02489$$

b) Takes at most three hours?
   Since the waiting time for the machine to be fixed is two hours, $\alpha = 2$.

$$P(X \le 3) = 1 - e^{-\frac{1}{2}\cdot(3)} = 0.7769$$

c) Exceeds five hours?

$$P(X>5) = 1 - \left[1 - e^{-\frac{1}{2}(5)}\right] = e^{-2.5} = 0.08208$$

d) Takes between two and four hours?e

$$P(2 \leq X \leq 4) = P(X \leq 4) - P(X < 2) = P(X \leq 4) - P(X \leq 2)$$

$$= \left(1 - e^{-\frac{1}{2} \cdot (4)}\right) - (1 - e^{-\frac{1}{2}(2)})$$

$$= e^{-1} - e^{-2} = 0.23254$$

5.5.5 Example

The lifetime of a light bulb of a particular brand is Y hours, where Y can be modeled by an exponential distribution with parameter $\lambda = 0.0225$.

a) Find the mean and variance of the lifetime of a light bulb.
b) Find the probability that the lifetime of a bulb is:
   (i) Precisely 200 hours.
   (ii) Less than 100 hours.
   (iii) More than 120 hours.
   (iv) Between 50 hours and 150 hours.

5.5.6 Solutions

a) Find the mean and variance of the lifetime of a light bulb.

The mean of the exponential distribution is $= \dfrac{1}{0.0225} = 44.44$.

The variance of the exponential distribution is $\beta^2 = \dfrac{1}{0.0225^2} = 1975.30864$.

$$P(X = 200) = (0.0225)e^{-(0.0225)(200)} = 0.00024$$

$$P(X < 100) = 1 - e^{-(0.0225)(100)} = 0.8946$$

$$P(X > 120) = 1 - P(X \leq 120) = 1 - \left(1 - e^{-(0.0225)(120)}\right) = 0.0672$$

$$P(50 \leq X \leq 150) = P(X \leq 150) - P(x \leq 50)$$

$$= \left(1 - e^{-(0.0225)(150)}\right) - \left(1 - e^{-(0.0225)(50)}\right) = 0.29043$$

5.5.7 Calculations in MATLAB

5.5.7.1 Create an exponential distribution object

5.5.7.1.1 Default parameters

```
pd = makedist('Exponential')
```

```
                        pd =

                    ExponentialDistribution

                    Exponential distribution
                        mu = 1
```

5.5.7.1.2 Specified parameters

```
pd = makedist('Exponential', 'mu', 2)
                        pd =

                    ExponentialDistribution

                    Exponential distribution
                        mu = 2
```

5.5.7.2 Fitting the gamma distribution to sample data

5.5.7.2.1 Default parameters

```
pd = fitdist(x, 'Exponential')
```

5.5.7.3 Use distribution-specific functions

Let's carry out these calculations in MATLAB.

exppdf(x, mu)	computes the PDF of the exponential distribution with mean mu, evaluated at the values in x.
expcdf(x, mu)	computes the CDF of the exponential distribution with mean mu, evaluated at the values in x.
expcdf(x, mu, 'upper')	• returns the complement of the CDF, evaluated at the values in x, using an algorithm that more accurately computes the extreme upper-tail probabilities.
expinv(p, mu)	returns the inverse of the exponential CDF with the mean mu, evaluated at the probability values in p.
expstat(mu)	returns the mean and variance of the exponential distribution with the mean parameter mu.
exprnd(mu, size)	generates an array of exponential random numbers, where vector size specifies size(r).

Let's draw multiple exponential distribution curves in MATLAB.

```
x = 0:0.01:2;

lambda1 = 1;
lambda2 = 2;
lambda3 = 3;
lambda4 = 4;
lambda5 = 6;
```

```
y1 = exppdf(x, 1/lambda1);
y2 = exppdf(x, 1/lambda2);
y3 = exppdf(x, 1/lambda3);
y4 = exppdf(x, 1/lambda4);
y5 = exppdf(x, 1/lambda5);

figure;
plot(x, y1, 'Color', 'b', 'LineWidth', 2)
hold on;
plot(x, y2, 'Color', [0.4660 0.6740 0.1880], 'LineWidth', 2)
plot(x, y3, 'Color', 'r', 'LineWidth', 2)
plot(x, y4, 'Color', 'm', 'LineWidth', 2)
plot(x, y5, 'Color', 'k', 'LineWidth', 2)
hold off;
xlabel('Observation')
ylabel('Probability Density')
legend('mu = 1', 'mu = 1/2', 'mu = 1/3', 'mu = 1/4', 'mu = 1/6')
```

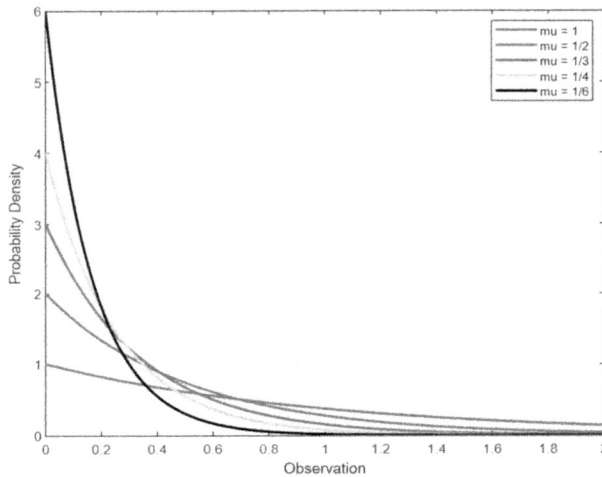

Figure 5.38

Let's go back to the problems.

5.5.7.4 Example 1

a) The probability that a repair takes exactly six hours.
b) The probability that a repair takes at most three hours.
c) The probability that a repair exceeds five hours.
d) The probability that a repair takes between two and four hours.

```
1          mean = 2;
2          x1 = 6;
3          x2 = 3;
4          x3 = 5;
5          x4 = 2;
6          x5 = 4;
7          y1 = exppdf(x1 , mean);
8          y2 = expcdf(x2 , mean);
9          y3 = expcdf(x3 , mean, 'upper');
10         y4 = expcdf(x5 , mean) - expcdf(x4 , mean);
11
```

Command Window

y1 =

   0.0249

y2 =

   0.7769

y3 =

   0.0821

y4 =

   0.2325

On the average, the length of time the computer part lasts is exponentially distributed.

a) What is the probability that a computer part lasts less than 5 years?
b) What is the probability that a computer part lasts more than 7 years?
c) Eighty percent of computer parts last at most how long?
d) What is the probability that a computer part lasts between nine and 11 years?

```
Years = data.Years;
x1 = 5;
x2 = 7;
x3 = 11;
x4 = 9;
pd = fitdist(Years, 'Exponential');
mean = pd.mean;
less = exppdf(5, mean);
more = expcdf(7, mean, 'upper');
y = expinv(0.8, mean);
between = expcdf(11, mean) - expcdf(9, mean);
```

```
                    less =

                      0.0607


                    more =

                      0.4962


                    y =

                      16.0781


                    between =

                      0.0737
```

5.5.7.5 Practice problems

1) The **Strokkur geyser** is considered one of the famous attractions in Iceland and erupts to a height of some 30 meters (98 ft). The mean number of minutes between eruptions for this geyser is around ten minutes. If this geyser has just erupted, what is the probability that we'll have to wait:
   a) Less than five minutes for the next eruption?
   b) Between five and ten minutes?

2) Alaska has experienced more high-magnitude earthquakes more than any other state in the US. On average there is an earthquake of magnitude 8 or greater in Alaska every 13 years. After an earthquake occurs, find the probability that it will take more than 15 years for the next big earthquake to occur.

3) The average lifespan of an Apple desktop computer is five years. The length of time the computer lasts is exponentially distributed. My parents bought an Apple desktop computer for my birthday.
   a) What is the probability that my present will last more than five years?
   b) What is the probability that it will last between 10 and 12 years?

4) Suppose the response time X at a certain on-line computer terminal (the elapsed time between the end of a user's inquiry and the beginning of the system's response to that inquiry) has an exponential distribution with expected response time equal to 5 sec. Find the probability that the response time is between 5 and 10 sec.

5.6 WEIBULL DISTRIBUTION

The Weibull distribution is another distribution that is a variation of the waiting time problem. Recall that the **exponential distribution** models the waiting time until a future event such as a failure happening, keeping the average time interval between those events constant. Unlike the exponential distribution, the Weibull distribution does not restrict the waiting time to be constant. The distribution parameters help us measure whether or not the number of failures is increasing with time, decreasing with time, or remaining constant.

```
x = 0:0.01:10;
y1 = wblpdf(x, 1, 1);
y2 = wblpdf(x, 1, 2);
y3 = wblpdf(x, 2, 1);
figure;
plot(x, y1, 'Color', [0.2630 0.760 0.4410], 'LineWidth', 2)
hold on;
plot(x, y2, 'Color', [0.133 0.4330 0.8450], 'LineWidth', 2)
plot(x, y3, 'Color', [0.5010 0.1450 0.2330], 'LineWidth', 2)
hold off;
xlabel('Observation')
ylabel('PDF')
title("Weibull Distribution")
legend("Standard Weibull","A = 1, B = 2", "A = 2, B = 1")
```

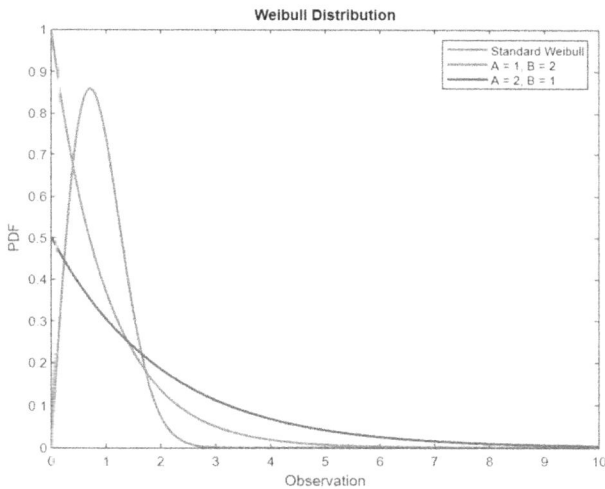

Figure 5.39

5.6.1 Parameters of the distribution

The Weibull distribution has two parameters:

- Scale parameter α
- Shape parameter β

5.6.2 Shape of the distribution

This shape parameter, β, is also known as the slope parameter. This is because the value of β **is equal to the slope of the line in a probability plot**. Depending on the value, this distribution can take on many shapes. To demonstrate it lets look at the following plot.

```
x = 0:0.01:3;
y1 = wblpdf(x, 1, 1);
y2 = wblpdf(x, 1, 0.5);
y3 = wblpdf(x, 1, 2);
y4 = wblpdf(x, 1, 4);
y5 = wblpdf(x, 1, 10);

figure;
plot(x, y1, 'Color', 'r', 'LineWidth', 2)
hold on;
plot(x, y2, 'Color', 'm', 'LineWidth', 2)
plot(x, y3, 'Color', 'y', 'LineWidth', 2)
plot(x, y4, 'Color', 'g', 'LineWidth', 2)
plot(x, y5, 'Color', 'b', 'LineWidth', 2)
hold off;
xlabel('Observation')
ylabel('PDF')
title("Weibull Distribution")
legend("Standard Weibull","B = 0.5", "B = 2", "B = 4", "B = 8")
```

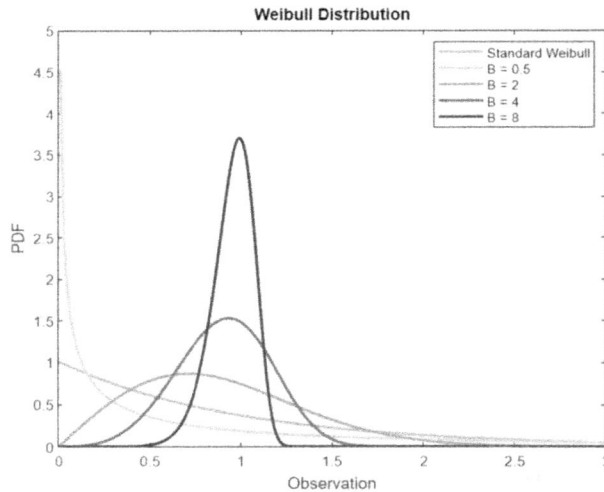

Figure 5.40

As we can see, depending on the value of the scale parameter (β), the Weibull distribution can take the form of various other probability distributions.

- When $\beta < 1$, the Weibull distribution takes on the form of a gamma distribution.
- When $\beta = 1$, the Weibull distribution reduces to the exponential distribution.
- When $\beta = 2$, the PDF is positively skewed so more specifically it is a lognormal distribution.
- When $\beta \approx 4$, the graph looks like a normal distribution, even though there is some deviation.

One of the important characteristics of the shape parameter of the Weibull distribution, beta (β), is its failure rate behavior:

- If $\alpha < 1$, then the failure rate decreases with time.
- If $\alpha > 1$, then the failure rate increases with time.
- If $\alpha = 1$, then the failure rate is constant with time.

Changing α, the scale parameter, does not change the type of shape, but it does stretch out the existing shape if the other two parameters are kept the same.

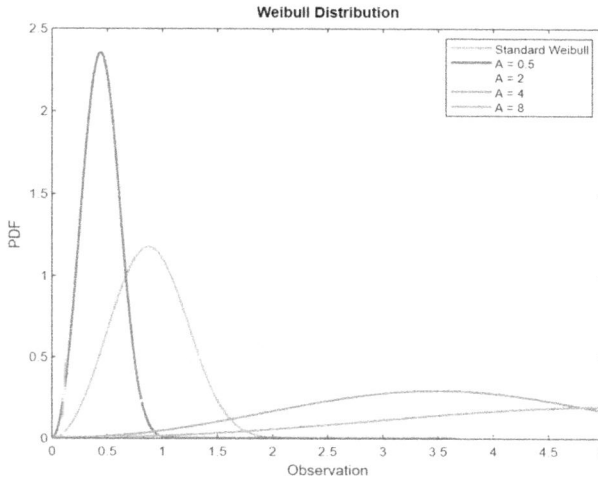

Figure 5.41

```
x = 0:0.01:5;
y1 = wblpdf(x, 1, 3);
y2 = wblpdf(x, 0.5, 3);
y3 = wblpdf(x, 2, 3);
y4 = wblpdf(x, 4, 3);
y5 = wblpdf(x, 6, 3);

figure;
plot(x, y1, 'Color', [0.8500 0.3250 0.0980], 'LineWidth', 2)
hold on;
plot(x, y2, 'Color', [0.4940 0.1840 0.5560], 'LineWidth', 2)
plot(x, y3, 'Color', [0.9290 0.6940 0.1250], 'LineWidth', 2)
plot(x, y4, 'Color', [0.4660 0.6740 0.1880], 'LineWidth', 2)
plot(x, y5, 'Color', [0.3010 0.7450 0.9330], 'LineWidth', 2)
hold off;
xlabel('Observation')
ylabel('PDF')
title("Weibull Distribution")
legend("Standard Weibull","A = 0.5", "A = 2", "A = 4", "A = 8")
```

5.6.3 Calculations in theory

A Weibull distribution has probability density function:

$$f(x) = \frac{\beta}{\alpha}\left(\frac{x}{\alpha}\right)^{\beta-1} e^{-\left(\frac{x}{\alpha}\right)^{\beta}}$$

The CDF is:

$$F(x) = \int_{0}^{k} \frac{\beta}{\alpha}\left(\frac{x}{\alpha}\right)^{\beta-1} e^{-\left(\frac{x}{\alpha}\right)^{\beta}} dx = 1 - e^{-\left(\frac{k}{\alpha}\right)^{\beta}}$$

The mean of the distribution is $\mu = \alpha \Gamma\left(1 + \frac{1}{\beta}\right)$.

The variance of the distribution is $\sigma^2 = \alpha^2 \left\{ \Gamma\left(1 + \frac{2}{\beta}\right) - \left[\Gamma\left(1 + \frac{1}{\beta}\right)\right]^2 \right\}$.

5.6.4 Example

The lifetime X (in hundreds of hours) of an electrical signal amplifier has a Weibull distribution with parameters $\alpha = 10$ and $\beta = 1.50$. Compute the following:

a) Find the value of the density function at $x = 4$.

$$f(x) = \frac{1.50}{10}\left(\frac{4}{10}\right)^{1.50-1} e^{-\left(\frac{4}{10}\right)^{1.5}} = 0.07366$$

b) Find the probability that the lifetime of the electrical signal amplifier is at most eight units of time.

$$P(X \le 8) = \int_{0}^{8} \frac{1.50}{10}\left(\frac{x}{10}\right)^{1.50-1} e^{-\left(\frac{x}{10}\right)^{1.5}} dx = 0.511$$

c) Find the probability that the lifetime of the electrical signal amplifier is at least three units of time.

$$P(X \ge 3) = 1 - P(X \le 3) = 1 - \left[\int_{0}^{3} \frac{1.50}{10}\left(\frac{x}{10}\right)^{1.50-1} e^{-\left(\frac{x}{10}\right)^{1.5}} dx\right] = 1 - 0.1515 = 0.8485$$

d) Find the probability that the lifetime of the equipment is less than 5 but greater than 1.8 units of time.

$$P(1.8 < X < 5) = P(X \le 5) - P(X \le 1.80)$$

$$= \int_{0}^{5} \frac{1.50}{10}\left(\frac{x}{10}\right)^{1.50-1} e^{-\left(\frac{x}{10}\right)^{1.5}} dx - \int_{0}^{1.80} \frac{1.50}{10}\left(\frac{x}{10}\right)^{1.50-1} e^{-\left(\frac{x}{10}\right)^{1.5}} dx$$

$$= 0.29781 - 0.07352 = 0.22429$$

e) Find the mean and variance of the lifetime.

- The mean $= 10\Gamma\left(1+\dfrac{1}{1.5}\right) = 10\Gamma\left(5/3\right) = 9.027452.$

- The variance $= 10^2\left\{\Gamma\left(1+\dfrac{2}{1.5}\right) - \left[\Gamma\left(1+\dfrac{1}{1.5}\right)\right]^2\right\} = 37.57.$

5.6.5 Calculation in MATLAB

5.6.5.1 Create a Weibull distribution object

5.6.5.1.1 Default parameters

```
pd = makedist('Weibull')
                           y =

                           WeibullDistribution

                           Weibull distribution
                              A = 1
                              B = 1
```

5.6.5.1.2 Specified parameters

```
pd = makedist('Weibull', 'A', 5, 'B', 12)
                           pd =

                           WeibullDistribution

                           Weibull distribution
                              A =  5
                              B = 12
```

5.6.5.2 Fitting the normal distribution to sample data

5.6.5.2.1 Default parameters

```
pd = fitdist(x, 'Weibull')
```

5.6.5.3 Use distribution-specific functions

Let's now do these calculations in MATLAB.
  Below are the functions:

wblpdf(x, a, b)	returns the PDF of the Weibull distribution with scale parameter a and shape parameter b, evaluated at the values in x.
wblcdf(x, a, b)	returns the CDF of the Weibull distribution with scale parameter a and shape parameter b, evaluated at the values in x.
wblcdf(x, a, b, 'upper')	returns the complement of the CDF, evaluated at the values in x, using an algorithm that more accurately computes the extreme upper-tail probabilities.
wblinv(p, a, b)	returns the inverse of the Weibull CDF with scale parameter a and shape parameter b, evaluated at the probability values in p.
wblstat(a, b)	returns the mean and variance of the Weibull distribution with scale parameter a and shape parameter b.
wblrnd(a, b, size)	generates an array of Weibull random numbers, where vector size specifies size(r).

Back to the same problem:

1. Find the value of the density function at $x = 4$.
2. Find the probability that the lifetime of the vacuum tube is at most eight units of time.
3. Find the probability that the lifetime of the vacuum tube is at least three units of time.
4. Find the probability that the lifetime of equipment is less than 5 but greater than 1.8 units of time.
5. Plot the graph of the Weibull probability distribution.
6. Plot the graph of cumulative Weibull probabilities.

```
a = 10;
b = 1.50;
x1 = 4;
y1 = wblpdf(x1, a, b);
x2 = 8;
y2 = wblcdf(x2, a, b);
x3 = 3;
y3 = wblcdf(x3, a, b, 'upper');
x4 = 1.8;
x5 = 5;
y4 = wblcdf(x5, a, b) - wblcdf(x4, a, b);
[Mean, Variance] = wblstat(a, b);
```

y1 =

    0.0737

y2 =

    0.5111

y3 =

    0.8485

y4 =

    0.2243

Mean =

    9.0275

Variance =

    37.5690

```
a = 10;
b = 1.50;
x = 0:0.1:40;
y = wblpdf(x, a, b);
plot(x, y, 'Color', 'b', 'LineStyle','-.', 'LineWidth',3)
xlabel("Observation")
ylabel("PDF")
title("Weibull Probability Distribution")
```

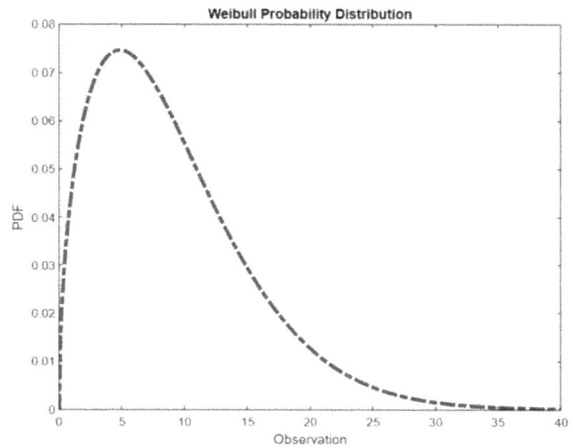

Figure 5.42

```
a = 10;
b = 1.50;
x = 0:0.1:40;
y = wblcdf(x, a, b);
plot(x, y, 'Color', 'b', 'LineStyle','-.', 'LineWidth',3)
xlabel("Observation")
ylabel("CDF")
title("Weibull Cumulative Distribution")
```

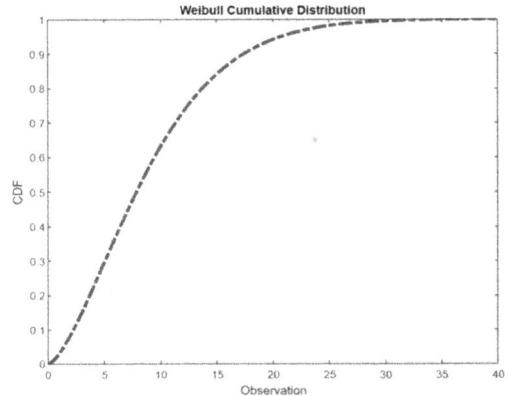

Figure 5.43

The lifetime (in hundreds of hours) of a certain type of vacuum tube has a Weibull distribution where the parameters need to be determined.

a) Find the mean time and the variance until failure
b) Calculate the probability that the tube fails before 6 hours,
c) Calculate the probability that the tube lasts between 1.8 and 6 hours
d) What is the probability that the tube lasts at least 3 hours,

```
Hours = data.Hours;
pd = fitdist(Hours, 'Weibull');
x1 = 6;
x2 = 1.8;
x3 = 3;
scale = pd.A;
shape = pd.B;
[Mean, Variance] = wblstat(scale, shape);
less = wblcdf(x1, scale, shape);
between = wblcdf(x1, scale, shape) - wblcdf(x2, scale, shape);
more = wblcdf(x3, scale, shape, 'upper');
```

```
Mean =

    2.6704


Variance =

    1.9146


less =

    0.9820


between =

    0.6845


more =

    0.3713
```

5.6.5.4 Practice problems

1) Suppose the time to failure, in hours, of a seal in a pump shaft is a Weibull random variable with the parameters $\alpha = 4000$ and $\beta = \frac{1}{5}$.
   a) Find the mean time until failure.
   b) Find the probability that a bearing lasts a least 5000 hours.

2) Assume that the life of a packaged magnetic disk exposed to corrosive gases has a Weibull distribution with $\alpha = 300$ and $\beta = \frac{1}{4}$.
   a) Find the mean time and the variance until failure.
   b) Calculate the probability that the disk lasts at least 600 hours.
   c) Calculate the probability that the disk fails before 500 hours.

3) The time to failure of a very sensitive computer screen follows a Weibull distribution with $\alpha = 1000$ hours and $\beta = 0.6$. What is the probability that the screen will last more than 5000 hours? What is the mean time to failure?

4) If the mean time to failure for a component which follows a Weibull distribution is 1000 hours with a standard deviation of 400 hours, what is the probability that the component will last more than 2000 hours?

5.7 BETA DISTRIBUTION

Until now probabilities have just been numbers in the range 0 to 1. However, if we have uncertainty about our probability, it would make sense to represent our probabilities as random variables. The beta distribution can be used to model events which are constrained to take place within an interval defined by a minimum and maximum value.

The best way to understand it is by modeling the bias of a coin given evidence of the coin flips. Suppose there is a coin that has a probability p of flipping heads and $(1 - p)$ of flipping tails. If we toss it n number of times we will observe some number of heads and tails flips of this coin.

We can first write out the likelihood of an event x, which could be heads or tails; we get

$$\binom{n}{x} p^x \left(1 - p\right)^{n-x}$$

This looks exactly like the binomial distribution from the previous chapter. Also, recall that in the binomial distribution, p was a parameter; however in beta, p will actually be the variable. The beta distribution models a distribution of probabilities. If we don't know the probability of the outcome of an event, we can use the beta distribution to model the distribution of probabilities given the prior information.

5.7.1 Parameters of the distribution

Unlike the gamma and Weibull distributions with a single shape and scale parameter, the beta distribution has two shape parameters, α and β. Both parameters must be positive values.

5.7.2 Shape of the distribution

With two shape parameters, clearly the distribution will take on different shapes depending on the values of the two parameters. Let's look at the plot for different parameter values.

```
x = 0:0.01:1;
y1 = betapdf(x, 1, 1);
y2 = betapdf(x, 0.50, 0.75);
y3 = betapdf(x, 2, 2);
y4 = betapdf(x, 2, 5);
y5 = betapdf(x, 10, 4);
y6 = betapdf(x, 20, 20);
figure;
plot(x, y1, 'Color', [0.8500 0.3250 0.0980], 'LineWidth', 2)
hold on;
plot(x, y2, 'Color', [0.4940 0.1840 0.5560], 'LineWidth', 2)
plot(x, y3, 'Color', [0.9290 0.6940 0.1250], 'LineWidth', 2)
plot(x, y4, 'Color', [0.4660 0.6740 0.1880], 'LineWidth', 2)
plot(x, y5, 'Color', [0 0 0.9330], 'LineWidth', 2)
plot(x, y6, 'Color', [0 0 0], 'LineWidth', 2)
hold off;
xlabel('Observation')
ylabel('PDF')
title("Beta Distribution")
legend("Standard Beta","A = 0.5, B = 0.75", "A = 2, B = 2", "A = 2, B = 5", "A = 10, B = 5", "A = 15, B = 15")
```

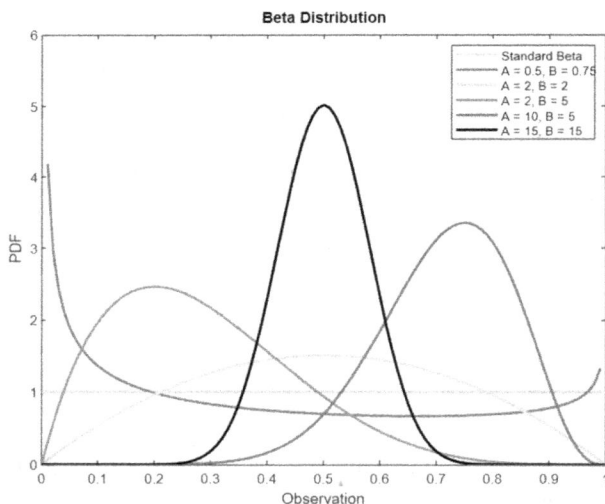

Figure 5.44

What conclusions can be made from this plot? We can conclude:

1) Both shapes equal one: the beta distribution is the uniform distribution.
2) Both shapes are less than one: the distribution is U-shaped.
3) Both shapes are the same and are greater than one: the distribution is symmetric.
4) The first shape is greater than second shape: the distribution is skewed to the left.
5) The first shape is less than second shape: the distribution is skewed to the right.

5.7.3 Calculations in theory

The beta function has the formula.

$$B(x,\alpha,\beta) = \int_0^1 x^{\alpha-1}(1-x)^{\beta-1}\,dx$$

The PDF for a beta $X \sim \text{beta}(\alpha,\beta)$ is:

$$f(x) = \{\frac{1}{B(a,\beta)}x^{\alpha-1}(1-x)^{\beta-1} \quad 0 < x < 1$$

$$0 \qquad\qquad\qquad\qquad\qquad \text{otherwise}$$

where $B(a,\beta) = \dfrac{\Gamma(a)\Gamma(\beta)}{\Gamma(a+\beta)} = \dfrac{(a-1)!(\beta-1)!}{(a+\beta-1)!}$

The CDF for the beta function can be calculated by solving the integral

$$F(x) = P(X \le k) = \int_0^k f(x)\,dx = \int_0^k \frac{1}{B(a,\beta)}x^{\alpha-1}(1-x)^{\beta-1}\,dx$$

The mean of the beta distribution is $\dfrac{a}{a+b}$.

The variance of the beta distribution is $\dfrac{ab}{(a+b)^2(a+b+1)}$.

5.7.4 Example

Suppose that the proportion θ of defective items in a large shipment is unknown. The engineer estimates that the proportion of defective components follows a beta distribution with parameters $\alpha = 4$ and $\beta = 2$.

a) Find the value of the density function at $x = 0.50$.

$$f(x) = \frac{1}{B(4,2)}\, 0.50^{4-1}(1-0.50)^{2-1} = 1.25$$

b) Find the probability that the proportion of defective components is less than 50%.

$$P(X \le 0.50) = \int_0^{0.50} \frac{1}{B(4,2)}x^{4-1}(1-x)^{2-1}\,dx = 0.1875$$

c) Find the probability that the proportion of defective components is more than 80%.

$$P(X \ge 0.80) = 1 - P(X \le 0.80) = 1 - \left[\int_0^{0.80} \frac{1}{B(4,2)}x^{4-1}(1-x)^{2-1}\,dx\right] = 0.26272$$

d) Compute the probability of defective items in the large shipment from 60% to 90%.

$$P(0.60 \le X \le 0.90) = P(X \le 0.90) - P(X \le 0.60)$$

$$= \int_0^{0.90} \frac{1}{B(4,2)}x^{4-1}(1-x)^{2-1}\,dx - \int_0^{0.60} \frac{1}{B(4,2)}x^{4-1}(1-x)^{2-1}\,dx = 0.58158$$

e) Find the mean and variance.

$$\text{Mean} = \frac{4}{4+2} = \frac{4}{6} = \frac{2}{3}$$

$$\text{Variance} = \frac{4*2}{\left(4+2\right)^2\left(4+2+1\right)} = 0.031746$$

5.7.5 Calculations in MATLAB

5.7.5.1 Create a beta distribution object

5.7.5.1.1 Default parameters

```
pd = makedist('Beta')
```

```
pd =

  BetaDistribution

  Beta distribution
    a = 1
    b = 1
```

5.7.5.1.2 Specified parameters

```
pd = makedist('Beta', 'a', 4, 'b', 10)
```

```
pd =

  BetaDistribution

  Beta distribution
    a =  4
    b = 10
```

5.7.5.2 Fitting the beta distribution to sample data

5.7.5.2.1 Default parameters

```
pd = fitdist(x, 'Beta')
```

5.7.5.3 Use distribution-specific functions

betapdf(x, a, b)	computes the beta PDF at each of the values in X using the corresponding parameters in a and b. x, a, and b can be vectors, matrices, or multidimensional arrays that all have the same size. A scalar input is expanded to a constant array with the same dimensions as the other inputs. The parameters in a and b must all be positive, and the values in x must lie on the interval [0, 1].
betacdf(x, a, b)	returns the beta CDF at each of the values in x using the corresponding parameters in a and b.
betacdf(x, a, b, 'upper')	• returns the complement of the CDF, evaluated at the values in x, using an algorithm that more accurately computes the extreme upper-tail probabilities.
betainv(p, a, b)	returns the inverse of the beta CDF with parameters a and b, evaluated at the probability values in p.
betastat(a, b)	with a > 0 and b > 0, returns the mean and variance for the beta distribution with parameters specified by a and b.
betarnd(a, b, [m, n, ..])	generates an m-by-n-by-... array containing random numbers from the beta distribution with parameters a and b. They can each be scalars or arrays of the same size as R.

Using these functions we can now solve problems in MATLAB.
   Solving the same problem:

```
a = 4;
b = 2;
x1 = 0.5;
y1 = betapdf(x1, a, b);
y2 = betacdf(x1, a, b);
x2 = 0.8;
y3 = betacdf(x2, a, b, 'upper');
x3 = 0.6;
x4 = 0.9;
y4 = betacdf(x4, a, b) - betacdf(x3, a, b);
[Mean, Variance] = betastat(a, b);
```

$$y1 =$$

$$1.2500$$

$$y2 =$$

$$0.1875$$

$$y3 =$$

$$0.2627$$

$$y4 =$$

$$0.5816$$

$$Mean =$$

$$0.6667$$

$$Variance =$$

$$0.0317$$

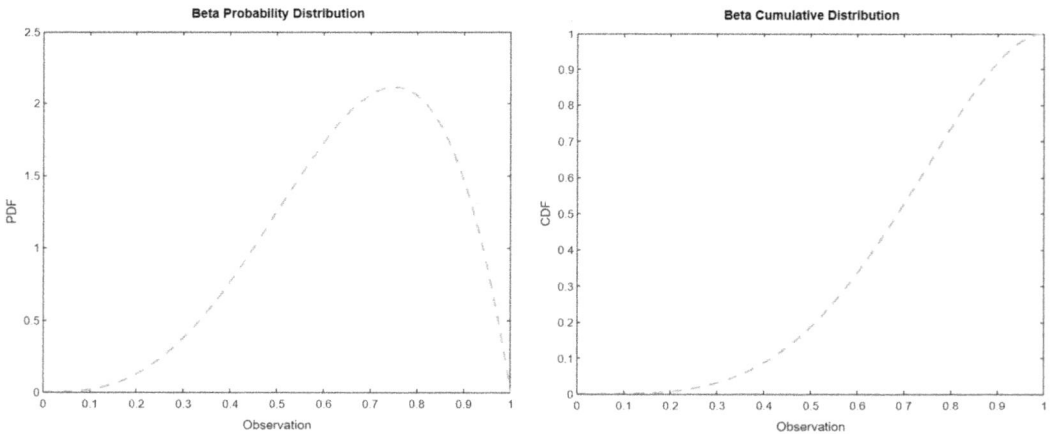

Figure 5.45

5.7.5.4 Calculations in MATLAB

1. An engineer is optimizing a chemical reaction process and wants to model the conversion rate of a reactant. They believe that the conversion rate follows a beta distribution with parameters alpha = 5 and beta = 3. What is the probability that the conversion rate falls between 0.3 and 0.5?

2. In materials engineering, the strength of a material is an important factor. Suppose the tensile strength of a certain material follows a beta distribution with parameters alpha = 5 and beta = 2. Calculate the probability that the tensile strength is greater than 500 MPa.

3. In Travis county, the proportion of highway sections requiring repairs every month is a random variable having the beta distribution with $\alpha = 4$ and $\beta = 6$.
   a) On average, what percentage of the highway sections require repairs in any given year?
   b) Find the probability that at most half of the highway sections will require repairs in any given year.

4. Let's say that we are given a set of student grades for a statistics exam and we find that it is best fit by a beta distribution: $X \sim Beta(\alpha = 7.50 \text{ and } \beta = 2.75)$ What is the probability that a student is below the mean (i.e. expectation)?

5.8 UNIFORM CONTINUOUS DISTRIBUTION

This is a type of probability distribution where every possible outcome has an equal likelihood of happening. With a continuous uniform distribution, just like a discrete uniform distribution, discussed in the previous chapter, every random variable has an equal chance of occurring. However, unlike discrete random variables which have a finite number of outcomes, a continuous random variable can take any real value within a specified range.

5.8.1 Parameters of the distribution

A uniform distribution is defined by two parameters, a and b, where a is the minimum value and b is the maximum value.

5.8.2 Shape of the distribution

Since a uniform distribution holds the same probability for the entire interval, its plot is a rectangle, and hence it is often referred to as a rectangular distribution.

```
x = -1:0.01:6;
y1 = unifpdf(x, 0, 1)
figure
plot(x, y1, 'LineStyle', ':', 'LineWidth', 2, 'Color', 'r')
hold on
y2 = unifpdf(x, -0.75, -0.25)
plot(x, y2, 'LineStyle', '--', 'LineWidth', 2, 'Color', 'k')
y3 = unifpdf(x, -0.5, 1)
plot(x, y3, 'LineStyle', '-.', 'LineWidth', 2, 'Color', 'g')
y4 = unifpdf(x, 2, 5)
plot(x, y4, 'LineStyle', '-', 'LineWidth', 2, 'Color', 'b')
xlabel("Observation")
ylabel("PDF")
title("Uniform Distribution")
legend("Standard Unifrom", "a = 0, b = 1", "a = -0.75, b = -0.25", "a = -0.5, b = 1", "a = 2, b = 5")
```

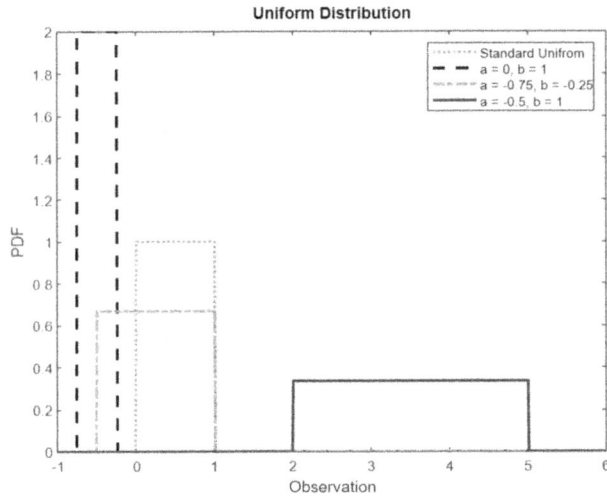

Figure 5.46

In Figure 5.46, we see that the probability density function of a continuous uniform distribution forms a rectangular shape. This occurs because all values within the interval from the lower to the upper limit have an equal probability

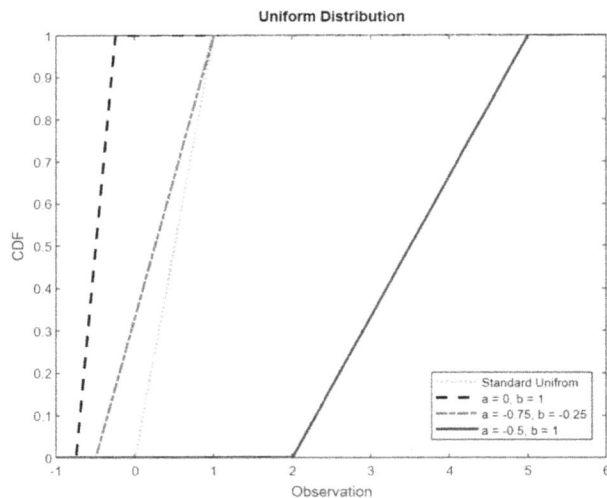

Figure 5.47

Not surprisingly, the two ends of an interval are the two parameters that define the uniform distribution. We use a and b to represent the lower and upper ends of the interval.

A good example of a continuous uniform distribution is an idealized random number generator. This generator produces values that are evenly distributed within the specified range, with no bias toward any particular value. Each value within the range has an equal likelihood of being generated, which aligns with the concept of a continuous uniform distribution.

5.8.3 Calculation in theory

The probability density function and cumulative distribution function for a continuous uniform distribution on the interval are

$$PDF = \begin{cases} \dfrac{1}{b-a}, & a \le x \le b \\ 0, & otherwise \end{cases}$$

$$CDF = \begin{cases} 0, & x < a \\ \dfrac{x-a}{b-a}, & a \le x \le b \\ 1, & x > b \end{cases}$$

$$Mean = \frac{1}{2}(a+b)$$

$$Variance = \frac{1}{12}(b-a)^2$$

5.8.4 Example

Suppose a group of engineers measure the time it takes for a signal to propagate through a communication system. They assume that the time, in seconds, follows a uniform distribution between 0 and 15 seconds, inclusive. Then find the:

1. Probability that it takes exactly 5 seconds.
2. Probability that it takes 20 seconds.
3. Probability that it takes less than 10 seconds.
4. Probability that it takes more than 12 seconds.
5. Probability that it takes more than 12 seconds.
6. Probability that it takes between 5 and 10 seconds
7. Mean, μ, and the standard deviation, σ.

5.8.5 Solution

$$P(X = 5) = \frac{1}{15-0} = 0.0667$$

$$P(X = 20) = 0$$

$$P(X < 10) = \frac{10-0}{15-0} = 0.6667$$

$$P(X > 12) = 1 - P(X \le 12) = 1 - \frac{12-0}{15-0} = 1 - 0.80 = 0.20$$

$$P(5 < X < 10) = \frac{10-0}{15-0} - \frac{5-0}{15-0} = 0.333$$

$$Mean = \frac{1}{2}(0+15) = 7.50$$

$$Variance = \frac{1}{12}(b-a)^2 = 18.75$$

5.8.6 Calculation in MATLAB

5.8.6.1 Create a uniform distribution object

5.8.6.1.1 Default parameters

```
pd = makedist('Uniform')
                        pd =

                    UniformDistribution

                    Uniform distribution
                        Lower = 0
                        Upper = 1
```

5.8.6.1.2 Specified parameters

```
pd = makedist('Uniform', 'Lower', -2, 'Upper', 5)
                        pd =

                    UniformDistribution

                    Uniform distribution
                        Lower = -2
                        Upper =  5
```

5.8.6.2 Use distribution-specific functions

unifpdf(x, a, b)	returns the PDF of the continuous uniform distribution on the interval [a, b], evaluated at the values in x.
unifcdf(x, a, b)	returns the beta CDF at each of the values in x using the corresponding parameters in a and b.
unifcdf(x, a, b, 'upper')	• returns the complement of the CDF, evaluated at the values in x, using an algorithm that more accurately computes the extreme upper-tail probabilities.
unifinv(p, a, b)	returns the inverse of the uniform CDF with parameters a and b, evaluated at the probability values in p.
unifstat(a, b)	returns the mean and variance for the uniform distribution with parameters specified by a and b.
unifrnd(a, b, size)	generates an array of uniform random numbers, where the size vector size specifies size®.

Solving Example 1 using the above functions:

```
Lower = 0;
Upper = 15;
x1 = 5;
y1 = unifpdf(x1, Lower, Upper);
x2 = 20;
y2 = unifpdf(x2, Lower, Upper);
x3 = 10;
y3 = unifcdf(x3, Lower, Upper);
x4 = 12;
y4 = unifcdf(x4, Lower, Upper, 'upper');
x5 = 5;
x6 = 10;
y5 = unifcdf(x6, Lower, Upper) - unifcdf(x5, Lower, Upper);
[Mean, Variance] = unifstat(Lower, Upper);
```

y1 =

0.0667

y2 =

0

y3 =

0.6667

y4 =

0.2000

y5 =

0.3333

Mean =

7.5000

Variance =

18.7500

```
Lower = 0;
Upper = 15;
x = -5:0.01:20;
y = unifpdf(x, Lower, Upper);
plot(x, y, 'Color', [0.6350 0.0780 0.1840], 'LineWidth', 2)
xlabel("Observation")
ylabel("PDF")
title("Uniform Continuous PDF")
```

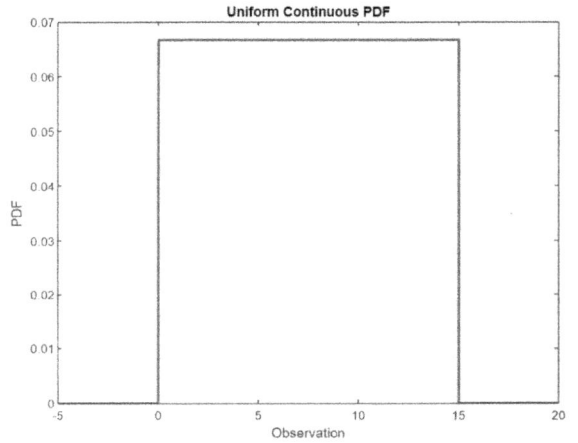

Figure 5.48

```
Lower = 0;
Upper = 15;
x = -5:0.01:20;
y = unifcdf(x, Lower, Upper);
plot(x, y, 'Color', [0.6350 0.0780 0.1840], 'LineWidth', 2)
xlabel("Observation")
ylabel("CDF")
title("Uniform Continuous CDF")
```

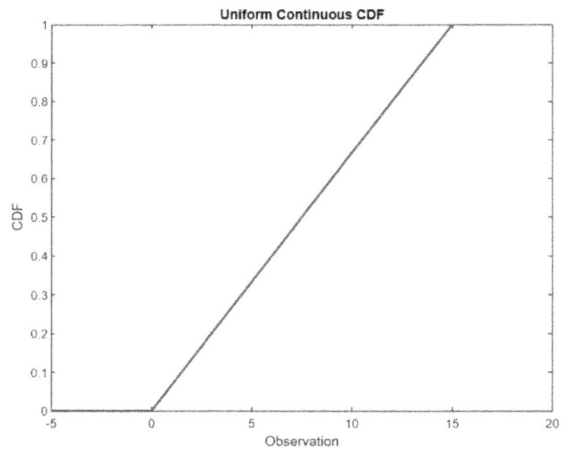

Figure 5.49

Ace Heating and Air Conditioning Service finds that the amount of time it takes to fix a
furnace is uniformly distributed where the parameters need to be estimated.
  a) Find the probability that a randomly selected furnace repair requires more than
      2 hours.
  b) Find the probability that a randomly selected furnace repair requires less than
      3 hours.
  c) Find the 30th percentile of furnace repair times.
  d) The longest 25% of furnace repair times take at least how long? (In other words:
      find the minimum time for the longest 25% of repair times.) What percentile does
      this represent?
  e) Find the mean and standard deviation

```
Hours = data.Hours;
[ahat, bhat] = unifit(Hours) ;
x1 = 2;
x2 = 3;
more = unifcdf(x1, ahat, bhat, 'upper');
less = unifcdf(x2, ahat, bhat);
percentile = unifinv(0.30, ahat, bhat);
toppercentile = unifinv(1-0.25, ahat, bhat);
[Mean, Variance] = unifstat(ahat, bhat);
```

```
more =

    0.8001


less =

    0.6019


percentile =

    2.2491


toppercentile =

    3.3684


Mean =

    2.7466


Variance =

    0.5156
```

5.8.6.3 Practice examples

A heater maintenance and service center finds that the amount of time a licensed plumber needs to complete a process is uniformly distributed between 1.5 and 3 hours.

1. Find the probability that a randomly selected heater repair requires more than two hours.
2. Find the probability that the repair requires less than one hour.
3. Find the mean and standard deviation.

The amount of time a service technician needs to change the oil in a car is uniformly distributed between 15 and 30 minutes. Let X = the time needed to change the oil in a car.

1. Write the random variable X in words. X = _____.
2. Write the distribution.
3. Graph the distribution.
4. Find $P (x > 19)$.

5.9 RAYLEIGH DISTRIBUTION

The **Rayleigh distribution** is a continuous probability distribution used to model random variables that can only take on positive values.

5.9.1 Parameters of the distribution

The Rayleigh distribution is a particular case of a Weibull distribution with shape parameter $\beta = 2$. Therefore it has a single parameter—scale α.

5.9.2 Shape of the distribution

Since the shape parameter is fixed in the Rayleigh distribution, its shape does not change, although it can be scaled.

In Figure 5.5, we observe the Rayleigh distribution with different values of the scale parameter, β. Although the shape of the distribution remains unchanged due to the fixed shape parameter, increasing the scale parameter widens the distribution, spreading it out over a broader range of values.

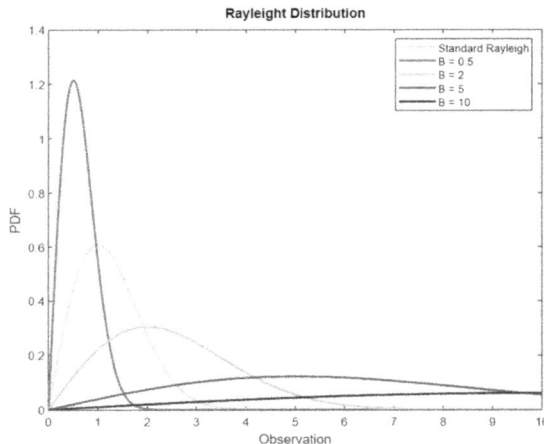

Figure 5.50

As we can see, the larger the value for the scale parameter β, the wider the distribution becomes.

In Figure 5.51, we see a graph illustrating how the cumulative distribution function (CDF) changes as the scale parameter varies

This distribution is widely used for the following:

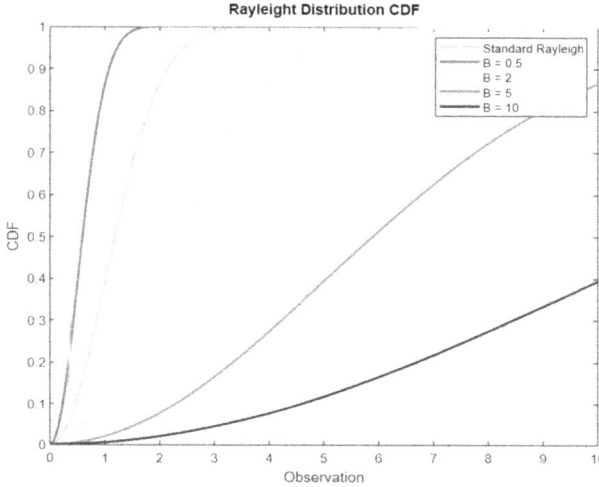

Figure 5.51

- **Communications**—to model multiple paths of densely scattered signals while reaching a receiver.
- **Physical sciences**—to model wind speed, wave heights, sound, or light radiation.
- **Engineering**—to check the lifetime of an object depending upon its age.
- **Medical imaging**—to model noise variance in magnetic resonance imaging.

5.9.3 Calculations in theory

A Rayleigh distribution has probability density function:

$$f(x) = \frac{x}{\alpha^2} e^{-\frac{x^2}{2\alpha^2}}, x > 0$$

The CDF is:

$$F(x) = \int_0^k \frac{x}{\alpha^2} e^{-\frac{x^2}{2\alpha^2}} dx = 1 - e^{-\frac{x^2}{2\alpha^2}}$$

The mean of the distribution is $= \alpha \sqrt{\frac{\pi}{2}}$.

The variance of the distribution is $= \alpha^2 \frac{4 - \pi}{2}$.

4.9.4 Example

Suppose we are studying wind speeds in a particular region, and we collect data over a long period of time. After analyzing the data, we find that the wind speeds follow a Rayleigh distribution with a scale parameter of 10 meters per second (m/s). Find:

a) The probability that the wind speed is exactly 30 m/s.
b) The probability that the wind speed is less than 5 m/s.
c) The probability that the wind speed is greater than 20 m/s.
d) The probability that the wind speed is between 15 and 25 m/s.
e) The mean and variance.

a) $P(X = 30) = \dfrac{30}{10^2} e^{-\frac{30^2}{2(10^2)}} = 0.00333$

b) $P(X < 5) = 1 - e^{-\frac{5^2}{2(10^2)}} = 0.1175$

c) $P(X > 20) = e^{-\frac{20^2}{2(10^2)}} = 0.13533$

d) $P(15 < X < 25) = \left[1 - e^{-\frac{25^2}{2(10^2)}}\right] - \left[1 - e^{-\frac{15^2}{2(10^2)}}\right] = e^{-\frac{15^2}{2(10^2)}} - e^{-\frac{25^2}{2(10^2)}} = 0.2807$

e) mean $= 10\sqrt{\dfrac{\pi}{2}} = 12.5331$

variance $= 10^2 \dfrac{4 - \pi}{2} = 42.92$

5.9.5 Calculations in MATLAB

5.9.5.1 Create a Rayleigh distribution object

5.9.5.1.1 Default parameters

```
pd = makedist('Rayleigh')
```

                         pd =

                  RayleighDistribution

                  Rayleigh distribution
                      B = 1

5.9.5.1.2 Specified parameters

```
pd = makedist('Rayleigh', 'B', 0.5)
```

$$pd =$$

RayleighDistribution

```
Rayleigh distribution
    B = 0.5
```

5.9.5.2 Use distribution-specific functions

raylpdf(x, b)	computes the Rayleigh PDF at each of the values in x using the corresponding scale parameter b.
raylcdf(x, b)	returns the Rayleigh CDF at each of the values in x using the corresponding scale parameter b
raylcdf(x, b, 'upper')	• returns the complement of the CDF, evaluated at the values in x, using an algorithm that more accurately computes the extreme upper-tail probabilities.
raylinv(p, b)	returns the inverse of the Rayleigh CDF with parameter b, evaluated at the probability values in p.
raylstat(b)	returns the mean and variance for the Rayleigh distribution with scale parameter b.
raylrnd(b, size)	generates an array of Rayleigh random numbers, where the size vector size specifies size(r). Using these functions we can now solve problems in MATLAB.

Solving Example I using the above functions:

```
B = 10;
x1 = 30;
y1 = raylpdf(x1, B);
x2 = 5;
y2 = raylcdf(x2, B);
x3 = 20;
y3 = raylcdf(x3, B, 'upper');
x4 = 15;
x5 = 25;
y4 = raylcdf(x5, B) - raylcdf(x4, B);
[Mean, Variance] = raylstat(B);
```

y1 =

    0.0033

y2 =

    0.1175

y3 =

    0.1353

y4 =

    0.2807

Mean =

    12.5331

Variance =

    42.9204

In Figure 5.52, we examine both the probability density function (PDF) and the cumulative distribution function (CDF) for this example

```
B = 10;
x = 0:0.01:40;
y = raylpdf(x, B);
plot(x, y, "LineStyle", "-.", "Color", 'm', 'LineWidth', 2)
xlabel("Observation")
ylabel("PDF")
title("Rayleigh Distribution PDF")
```

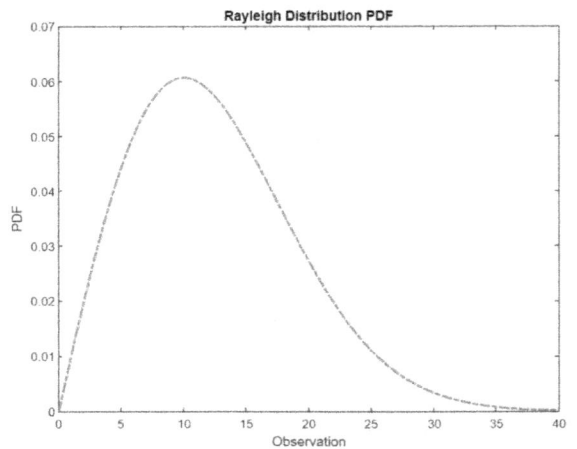

Figure 5.52

```
B = 10;
x = 0:0.01:40;
y = raylcdf(x, B);
plot(x, y, "LineStyle", " .", "Color", 'm', 'Linewidth', 2)
xlabel("Observation")
ylabel("CDF")
title("Rayleigh Distribution CDF")
```

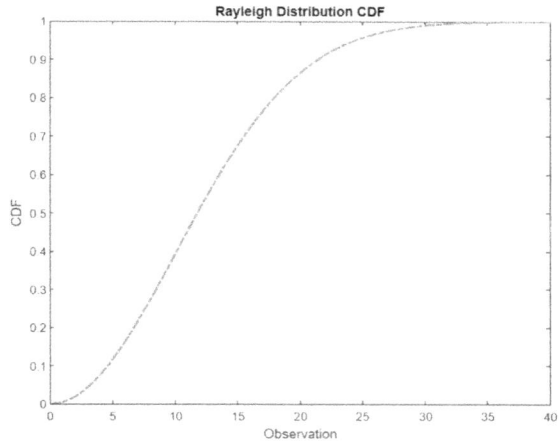

Figure 5.53

2. The scalar miss distances of the AGM Splash 83K considered follows a Rayleigh distribution. Find:

a) The probability that the distance is less than 25 feet.
b) The probability that the distance is greater than 30 feet.
c) The probability that the distance is between 45 and 60 feet.
d) Find the first quartile, median, and third quartile of the distance

```
Distance = data.Distance;

pd = fitdist(Distance, 'Rayleigh') ;
scale = pd.B;
x1 = 25;
x2 = 30;
x3 = 60;
x4 = 45;
less = raylcdf(x1, scale);
more = raylcdf(x2, scale, 'upper');
between = raylcdf(x3, scale) - raylcdf(x4, scale);
quarter = raylinv(0.25, scale);
median = raylinv(0.5, scale);
thirdquarter = raylinv(0.75, scale);
```

```
less =

    0.5237


more =

    0.3436


between =

    0.0765


quarter =

    15.5686


median =

    24.1661


thirdquarter =

    34.1760
```

3. The noise voltage X at the output of a rectifier is found to have a Rayleigh distribution.

a) What is the probability that the noise value is less than 10 V?
b) Calculate the probability that the noise value is more than 40 V.
c) Evaluate the probability that the noise value is between 25 and 50 V.
d) Find the mean and variance of the noise voltage.

```
Output = data.Output;

scale = raylfit(Output);
x1 = 10;
x2 = 40;
x3 = 25;
x4 = 50;
less = raylcdf(x1, scale);
more = raylcdf(x2, scale, 'upper');
between = raylcdf(x4, scale) - raylcdf(x3, scale);
[Mean, Variance] = raylstat(scale);
```

less =

0.1432

more =

0.0844

between =

0.3597

Mean =

22.5452

Variance =

138.8836

5.10 EXTREME VALUE DISTRIBUTION

An **extreme value distribution** is a limiting model for the maximums and minimums of a dataset. It models how large (or small) your data will probably get. It is widely used in various fields, including finance, hydrology, and environmental studies. There are three types of extreme value distributions based on the limiting distribution of the maximum or minimum of a set of random variables.

Type 1: the extreme value distribution type 1 is also referred to as the Gumbel distribution. This is the most commonly discussed one, which by default is known as the extreme value distribution. The extreme value type I distribution has two forms.

- **Minimum:** this is based on the smallest extreme.
- **Maximum:** this is based on the largest extreme.

Before we delve into Type 2 of the extreme value distributions, let's first examine Type 3.

Type 3: The third type of extreme value distribution, Type 3, has already been discussed earlier as the Weibull distribution. However, in this case, it is considered with three parameters.

Recall that in Section 6.5, the Weibull distribution had two parameters:

- Scale parameter
- Shape parameter

In this section, the Weibull distribution is extended by introducing a third parameter, the location parameter. In the two-parameter version of the Weibull distribution, the location parameter is zero.

Type 2: the extreme value distribution type 2 is also referred to as the Fréchet distribution. The Fréchet distribution is also known as the inverse Weibull distribution (IWD).

5.10.1 Parameters of the distribution

Since the three-parameter Weibull distribution has three parameters and it's one of the three types of extreme value distribution, then by default all the other types of extreme value distribution will have three parameters.

- k is the shape parameter.
- σ is the scale parameter.
- μ is the location parameter.

Now the last two parameters may seem like the same parameters as the normal distribution. However they are absolutely not. The location parameter μ is not the mean but the "center" of the distribution, and the scale parameter σ is not the standard deviation but governs the size of the deviations about μ.

Now how do these three types differ from each other?

The extreme value distribution is equivalent to the:

- Type I/Gumbel when the shape parameter $k = 0$.
- Type II/Fréchet when the shape parameter $k > 0$.
- Type III/three-parameter Weibull when the shape parameter $k < 0$.

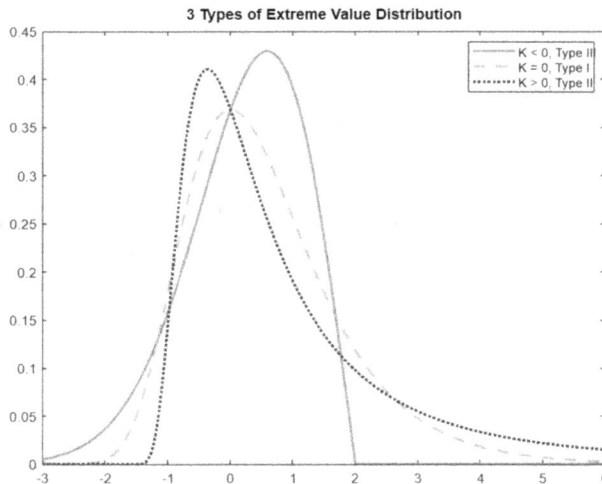

Figure 5.54

3 Types of Extreme Value Distribution CDF

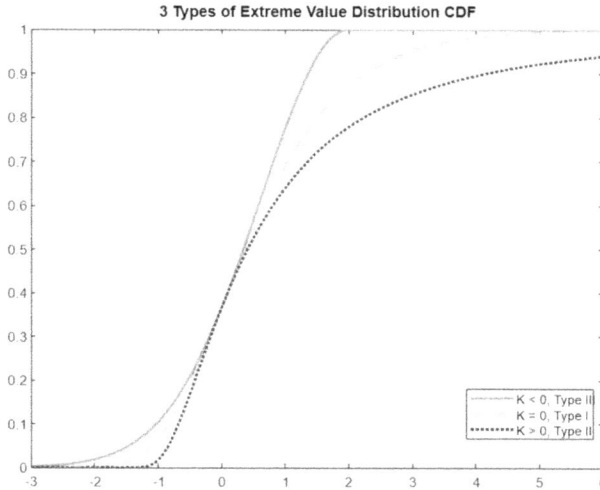

Figure 5.55

5.10.2 Shape of the distribution

5.10.2.1 Type I/Gumbel distribution

In Figure 5.56, we have two separate graphs. On the right, the Gumbel distribution's PDF is shown, and it is skewed to the right. This represents the maximum extreme value distribution. On the left, we see the minimum extreme value distribution, with the PDF skewed to the left. Examples of both the maximum extreme value distribution (Gumbel) and the minimum extreme value distribution are provided below

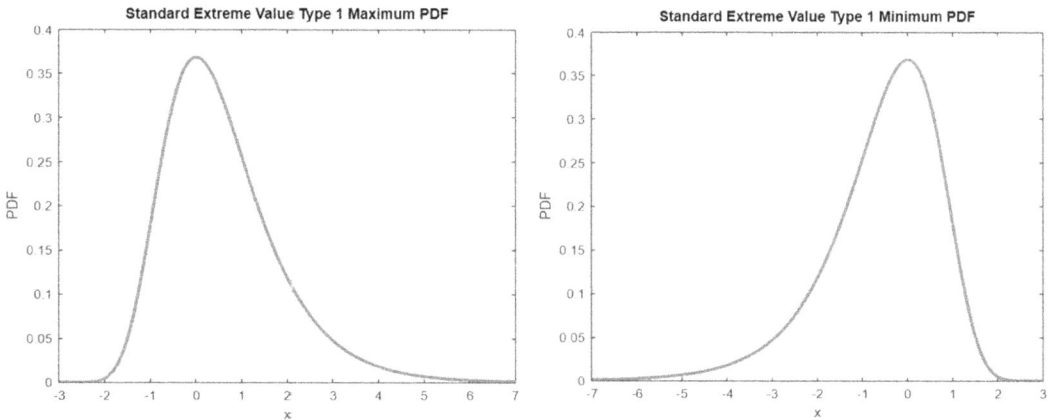

Figure 5.56

Since the shape parameter $k = 0$ this means the Gumbel PDF has no shape parameter which in turn means it has only one shape, which does not change. So if we hold the shape parameter constant at 0 and alter the location and scale parameters, we will notice that changing the location parameter shifts the distribution whereas changing the scale parameter either stretches or shrinks the distribution.

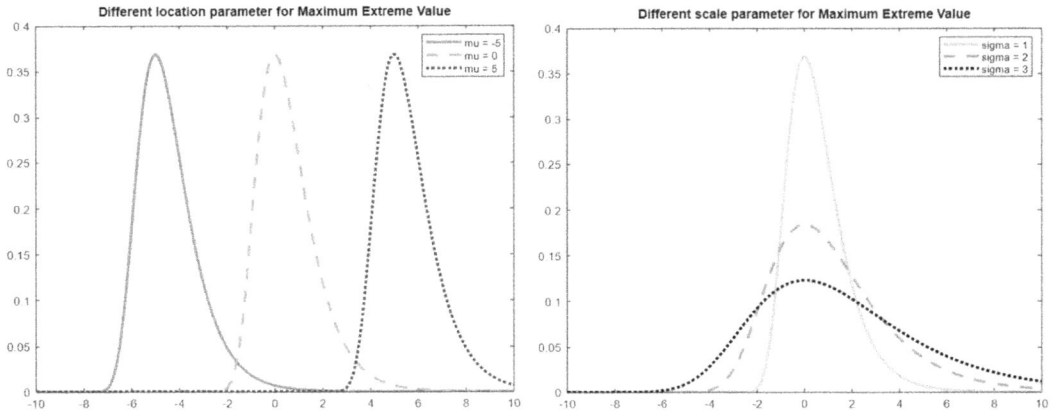

Figure 5.57

In Figure 5.57, we plot the maximum extreme value distribution with varying location and scale parameters. The figure contains two separate graphs: on the left, we examine the effect of changing the location parameter, and on the right, we observe the effect of changing the scale parameter.

Let's focus on the left-hand graph. As the location parameter μ decreases, the PDF shifts to the left. Conversely, as μ increases, the PDF shifts to the right.

Now, looking at the right-hand graph, as the scale parameter σ increases, the PDF becomes wider and shallower. On the other hand, when σ decreases, the PDF becomes taller and narrower.

Likewise if we plot the minimum extreme value different location and scale parameters.

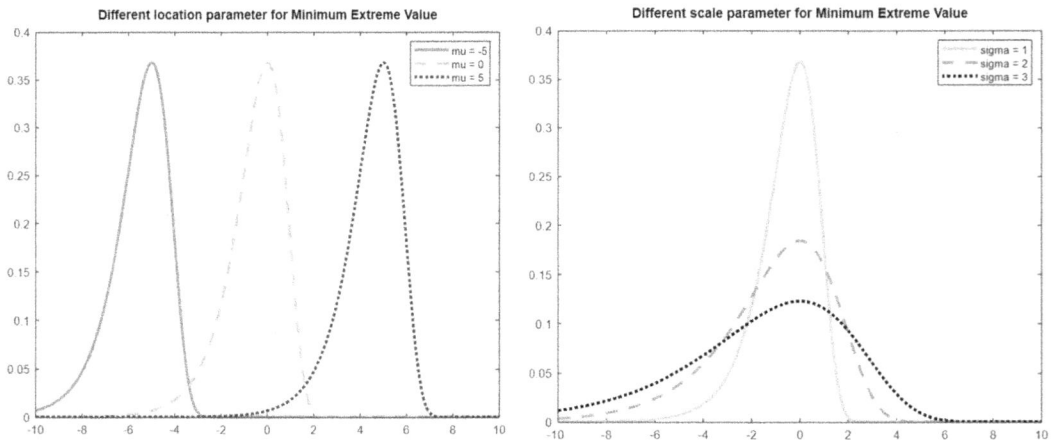

Figure 5.58

5.10.2.2 Type 2/Fréchet distribution

If we look back in our graph we will notice that when $k > 0$, the density has zero probability below a lower bound.

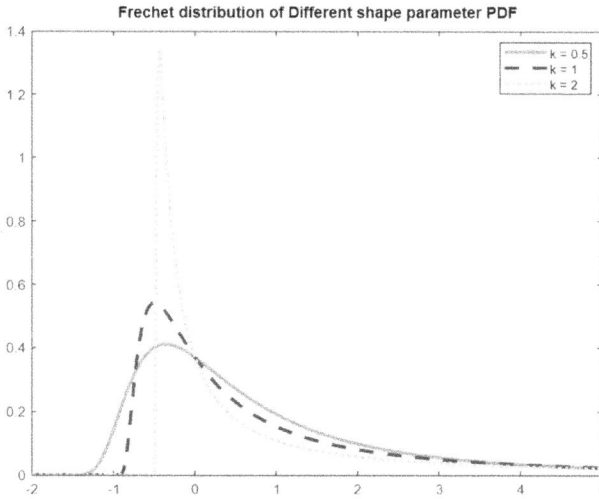

Figure 5.59

5.10.2.3 Type 3/three-parameter Weibull distribution

Similarly when $k < 0$, the density has zero probability above an upper bound.

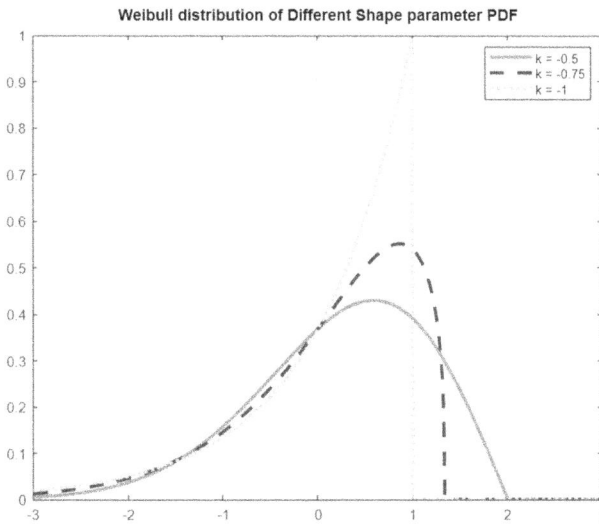

Figure 5.60

5.10.3 Calculations in theory

5.10.3.1 Type I/Gumbel distribution

5.10.3.1.1 Minimum

The general formula for the probability density function of the Gumbel minimum distribution is

$$f(x) = \left(\frac{1}{\sigma}\right) e^{\left(\frac{x-\mu}{\sigma}\right)} e^{-e^{\left(\frac{x-\mu}{\sigma}\right)}}$$

Subsequently, the formula for the cumulative distribution function of the Gumbel distribution minimum is

$$F(x) = 1 - e^{-e^{\left(\frac{x-\mu}{\sigma}\right)}}$$

5.10.3.1.2 Maximum

The formula for the probability density function of the Gumbel maximum distribution is

$$f(x) = \left(\frac{1}{\sigma}\right) e^{-\left(\frac{x-\mu}{\sigma}\right)} e^{-e^{-\left(\frac{x-\mu}{\sigma}\right)}}$$

Subsequently, the formula for the cumulative distribution function of the Gumbel distribution maximum is

$$F(x) = e^{-e^{-\left(\frac{x-\mu}{\sigma}\right)}}$$

The mean of the distribution $E[x] = \mu \pm 0.577215\sigma$.

The variance of the distribution $Var[x] = \dfrac{\sigma^2 \pi^2}{6}$.

5.10.3.2 Type 2/Fréchet distribution

5.10.3.2.1 Two parameters

Recall the probability density function for the two-parameter Weibull distribution for the scale and shape parameters α and β is

$$f(x) = \frac{\beta}{\alpha} \left(\frac{x}{\alpha}\right)^{\beta-1} e^{-\left(\frac{x}{\alpha}\right)^{\beta}}$$

Now since the Fréchet distribution is the inverse of the Weibull distribution, the probability density function of the Fréchet distribution is

$$f(x) = \frac{\beta}{\alpha} \left(\frac{\alpha}{x}\right)^{\alpha+1} e^{-\left(\frac{\alpha}{x}\right)^{\alpha}}$$

Similarly, the CDF is

$$F(x) = e^{-\left(\frac{\alpha}{x}\right)^{\alpha}}$$

5.10.3.2.2 Three parameters

The formula for the probability density function for the three-parameter inverse Weibull distribution for the location, scale, and shape parameters is

$$f(x) = \left(\frac{k}{\sigma}\right)\left(\frac{\sigma}{x-\mu}\right)^{k+1} e^{-\left(\frac{\sigma}{x-\mu}\right)^k}$$

Similarly, the formula for the cumulative distribution function is

$$F(x) = e^{-\left(\frac{\sigma}{x-\mu}\right)^k}$$

5.10.3.3 Type 3/three-parameter Weibull distribution

5.10.3.3.1 Two parameters

The probability density function for the two-parameter Weibull distribution for the scale and shape parameters α and β is

$$f(x) = \frac{\beta}{\alpha}\left(\frac{x}{\alpha}\right)^{\beta-1} e^{-\left(\frac{x}{\alpha}\right)^\beta}$$

The CDF is

$$F(x) = \int_0^k \frac{\beta}{\alpha}\left(\frac{x}{\alpha}\right)^{\beta-1} e^{-\left(\frac{x}{\alpha}\right)^\beta} dx = 1 - e^{-\left(\frac{k}{\alpha}\right)^\beta}$$

5.10.3.3.2 Three parameters

The formula for the probability density function for the three-parameter Weibull distribution for the location, scale, and shape parameters is

$$f(x) = \left(\frac{k}{\sigma}\right)\left(\frac{x-\mu}{\sigma}\right)^{k-1} e^{-\left(\frac{x-\mu}{\sigma}\right)^k}$$

Similarly, the formula for the cumulative distribution function is

$$F(x) = 1 - e^{-\left(\frac{x-\mu}{\sigma}\right)^k}$$

The probability distribution function of the three types of extreme value distribution with the parameters k, *sigma*, and *mu* is combined into a single distribution, the generalized extreme value distribution, whose distribution function is written as

$$f(x) = \left\{\left(\frac{1}{\sigma}\right)\left(1+k\left(\frac{x-\mu}{\sigma}\right)\right)^{-1-\frac{1}{k}} expexp\left(-\left(1+k\left(\frac{x-\mu}{\sigma}\right)\right)^{-\frac{1}{k}}\right), \quad if\ k \neq 0\right.$$

$$\left(\frac{1}{\sigma}\right)e^{\left(\frac{x-\mu}{\sigma}\right)}e^{-e^{\left(\frac{x-\mu}{\sigma}\right)}} or \left(\frac{1}{\sigma}\right)e^{-\left(\frac{x-\mu}{\sigma}\right)}e^{-e^{-\left(\frac{x-\mu}{\sigma}\right)}}, \qquad if\ k = 0$$

The cumulative distribution function of the three types of extreme value distribution with the parameters k, sigma, and mu is

$$F(x) = \{ \exp\left(-\left(1 + k\left(\frac{x-\mu}{\sigma} \right) \right)^{-\frac{1}{k}} \right), \quad if \ k \neq 0$$

$$1 - e^{-e^{\left(\frac{x-\mu}{\sigma} \right)}} \ or \ e^{-e^{-\left(\frac{x-\mu}{\sigma} \right)}}, \quad if \ k = 0$$

5.10.4 Calculations in MATLAB

5.10.4.1 Create an extreme value distribution object

5.10.4.1.1 Default parameters

```
pd = makedist('GeneralizedExtremeValue')
```

5.10.4.2 Use distribution-specific functions

gevpdf(x, k, sigma, mu)	returns the PDF of the generalized extreme value (GEV) distribution evaluated at the values in x with shape parameter k, location parameter mu, and scale parameter sigma, evaluated at the values in x.
gevcdf(x, k, sigma, mu)	returns the generalized extreme value CDF at each of the values in x using the corresponding shape parameter k, location parameter mu, and scale parameter sigma, evaluated at the values in x.
gevcdf(x, k, sigma, mu, 'upper')	• returns the complement of the CDF, evaluated at the values in x, using an algorithm that more accurately computes the extreme upper-tail probabilities.
gevinv(p, k, sigma, mu)	returns the inverse of the generalized extreme value CDF with shape parameter k, location parameter mu, and scale parameter sigma evaluated at the probability values in p.
gevstat(k, sigma, mu)	returns the mean and variance for the generalized extreme value CDF with shape parameter k, location parameter mu, and scale parameter sigma.
gevrnd(k, sigma, mu, [m, n, …])	generates an m-by-n-by-… array containing random numbers from the GEV distribution with parameters k, sigma, and mu. These parameters can each be scalars or arrays of the same size as R.

Using these functions we can now solve problems in MATLAB.

While the GEV distribution refers to a more general form that encompasses all three types of extreme value distributions, MATLAB has a separate function for the Type 1 extreme value distribution. Now remember the shape parameter k is zero in case of Gumbel; therefore this function takes out the shape parameter and only involves the location and scale parameters.

5.10.4.3 Create a Gumbel distribution object

5.10.4.3.1 Default parameters

```
pd = makedist('ExtremeValue')
```

5.10.4.3.2 Specified parameters

```
pd = makedist('ExtremeValue', 'mu', -0.5, 'sigma', 3)
```

5.10.4.4 Use distribution-specific functions

1) evpdf(x, mu, sigma) returns the PDF of the extreme value distribution with location parameter mu and scale parameter sigma, evaluated at the values in x.
2) evcdf(x, mu, sigma) returns the cumulative distribution function for the extreme value distribution, with location parameter mu and scale parameter sigma, evaluated at the values in x.
   • evcdf(x, mu, sigma, 'upper') returns the complement of the CDF, evaluated at the values in x, using an algorithm that more accurately computes the extreme upper-tail probabilities.
3) evinv(p, mu, sigma) returns the inverse of the Gumbel CDF with location parameter mu and scale parameter sigma evaluated at the probability values in p.
4) evstat(mu, sigma) returns the mean and variance for the extreme value CDF with location parameter mu and scale parameter sigma.
5) evrnd(mu, sigma, [m, n, …]) generates an m-by-n-by-… array containing random numbers from the extreme value distribution with parameters mu and sigma. These parameters can each be scalars or arrays of the same size as R.

Using these functions we can now solve problems in MATLAB.

In this chapter, we explored nine common continuous probability distributions and their implementation in MATLAB. However, numerous other continuous probability distributions are utilized across various fields and can be similarly implemented and calculated in MATLAB. In subsequent chapters, we will delve into two additional widely-used continuous probability distributions: the F distribution and the student's t distribution. We will also examine their significance in statistical analysis and hypothesis testing.

Chapter 6

Descriptive statistics

In Chapter 3 we introduced the topic of statistics and its two major branches: descriptive and inferential. In this chapter, we will look at descriptive statistics, including its definition and types, and finally wrap up with the differences between this and inferential statistics.

Descriptive statistics describes, shows, and summarizes the basic features of a given dataset which can be either a representation of an entire population or a sample of the population. It helps us to present and simplify large amounts of data into a simpler summary which makes it easier to visualize and interpret, especially if there is a lot of data.

A student's grade point average (GPA) is an excellent example for understanding the purpose of descriptive statistics. Students receive a grade for each assignment, project, and exam. Those grades are added together and divided by the number of units of work to calculate the average score for the semester. The idea of a GPA is that it takes data points from a wide range of exams, assignments, and projects and calculates the average score for the semester. So basically, this GPA represents the student's overall academic performance throughout the semester.

Think about a situation where you are given 1 million values to analyze and extract some useful information from. Now it is very unlikely that anyone would want to examine such a huge dataset. This is why descriptive statistics is such a valuable tool. It "summarizes" such immense quantities of data by extracting salient characteristics and conveying general trends.

6.1 MAIN BRANCHES OF DESCRIPTIVE STATISTICS

Descriptive statistics breaks down into several types, characteristics, or measures which we will summarize in a table shown in Figure 6.1.

In Chapter 3 we have already learned about frequency distribution tables and other graphs for visualizing data in MATLAB. In this chapter we shall discuss the next three measures of descriptive statistics in detail and learn to calculate them both in theory and in MATLAB.

6.2 CENTRAL TENDENCY

Once we have finished the step of data collection, the next step is to analyze the collected data. The first aspect of data analysis is to measure the central tendency which is referred to as the central location of a distribution. It is the point where the majority of the values fall, and therefore it is also seen as the representative of the collected data. Once we know the centrality, we have the basic idea of the data. There are three major types of estimates of central tendency (Figure 6.2).

 DOI: 10.1201/9781003399582-6

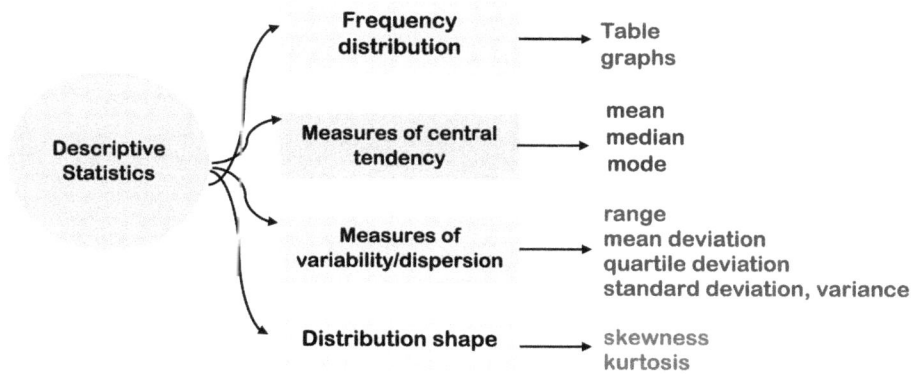

Figure 6.1

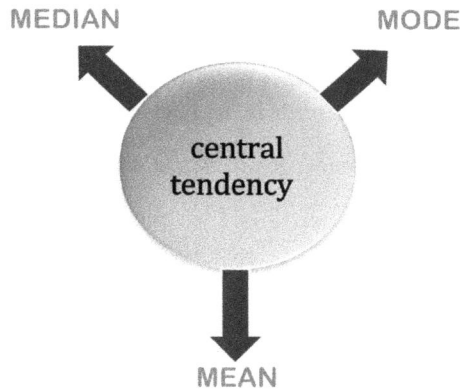

Figure 6.2

In Chapters 3, 4, and 5 we learned about different types of distributions and their corresponding graphs and probabilities. These distributions are closely related to central tendency.

Remember the symmetric distribution—the normal distribution! These three central tendency measures are equal for the normal distribution.

As distributions become more skewed the difference between these different measures of central tendency gets larger. The mean is pulled in the direction of the skewness (i.e., the direction of the tail).

We are going to talk more about this later in this chapter.

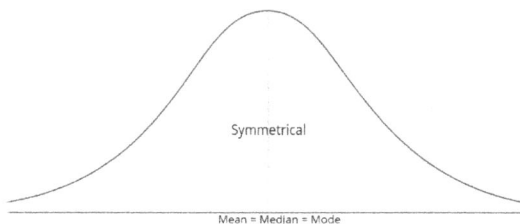

Figure 6.3

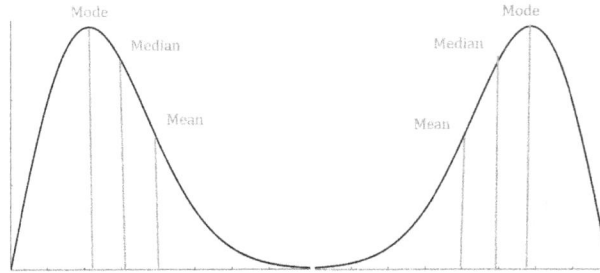

Figure 6.4

Each of these measures uses a different method to find the central location of a dataset. Depending on the type of data available, one of these three measures may be more suitable than the other two. For example:

- If you have a symmetrical distribution of continuous data, all three measures of central tendency hold good.
- If the distributions have outliers or are skewed, the median is often the most preferred measure of central tendency since it is more resistant to outliers than the mean.
- If you have categorical data, the mode is the best choice for finding the central tendency.

Now let's talk about each of these measures in detail.

6.2.1 Mean

The **mean** or average is probably the most commonly used method of describing the central tendency. In a symmetric data distribution, the mean value is located precisely at the center. But in a skewed continuous data distribution, the extreme values in the extended tail pull the mean value away from the center.

As mentioned previously, the preferred measure of central tendency often depends on the shape of the distribution. Of the three measures of tendency, the mean is most heavily influenced by any outliers or skewness.

Mean in statistics basically refers to the average of values in a given dataset. The simplest way to compute the mean is to add up all the values and divide by the number of values in the dataset. For example, Stephanie took five classes in the fall semester and had the following grades, along with their respective credit hours.

Grade	A	A+	B	B−	B+
Credit hours	3	4	3	4	3

If we want to calculate her GPA, we will follow the steps below:

1) We will convert the five letter grades into the standard GPA scale which gives us

$$A = 4.0, A+ = 4.0, B = 3.0, B- = 2.70, B+ = 3.3$$

2) Then we multiply each of the grade points by the number of credit hours for each class and add them up to find the total.

$$(4.0 \times 3) + (4.0 \times 4) + (3.0 \times 3) + (2.70 \times 4) + (3.3 \times 3) = 57.7$$

3) Finally we divide the total by the number of credit hours to calculate the average score for the semester.

$$\frac{57.7}{3+4+3+4+3} = 3.394$$

6.2.1.1 Arithmetic mean

The above example calculated the arithmetic mean. Arithmetic mean is nothing but the average. It is computed by adding all the values in the dataset divided by the number of observations in it.

6.2.1.1.1 Arithmetic mean for a sample dataset

The general formula for determining the arithmetic mean of a given dataset is

$$AM = \frac{\Sigma x}{n}$$

where
Σx = summation of the individual value
n = number of observations in the sample

6.2.1.1.2 Arithmetic mean for a frequency distribution table

We compute the mean by calculating it from the frequency distribution (if provided).
  In order to calculate the mean from a frequency table:

1. Multiply the number values by the frequencies to find the total.
2. Divide the total by the total frequencies which is the total number of observations.

$$Mean = \frac{\Sigma f X}{\Sigma f}$$

6.2.1.1.2.1 UNGROUPED FREQUENCY TABLE

Say we have recorded the number of errors in a piece of code during a software testing phase. We have the following data:

Number of errors	Tally marks	Frequency/count
1	||	2
2	||||	4
3	|||	3
4	⊥⊦⊦	5
5	|||	3
6	||	2

Number of errors	Frequency	x.f
1	2	$1 \times 2 = 2$
2	4	$2 \times 4 = 8$
3	3	$3 \times 3 = 9$
4	6	$4 \times 6 = 24$
5	3	$5 \times 3 = 15$
6	2	$6 \times 2 = 12$
	$\Sigma 20$	$\Sigma 70$

$$Mean = \frac{70}{20} = 3.5$$

6.2.1.1.2.2 GROUPED FREQUENCY TABLE

We cannot find the actual mean because we only have a range of possible values.

As we do not have the actual times taken we can only estimate the mean by using the midpoints of the class intervals. Instead we can find an estimate for the mean using the midpoints of each group. Hence the X in a grouped frequency table is the midpoint of each of the class intervals.

The calculation of the arithmetic mean of an inclusive series is the same as for an exclusive series.

6.2.1.1.3 Example 1

Let's go back to the engine power (in horsepower) of 30 vehicles.

Horsepower	Frequency(f)
149.5 – 159.5	8
159.5 – 169.5	12
169.5 – 179.5	5
179.5 – 189.5	7
189.5 – 199.5	10

Horsepower	Frequency (f)	Midpoint	Midpoint frequency
149.5 – 159.5	8	$\dfrac{149.5 + 159.5}{2} = 154.5$	$154.5 \times 8 = 1236$
159.5 – 169.5	12	$\dfrac{159.5 + 169.5}{2} = 164.5$	$164.5 \times 12 = 1974$
169.5 – 179.5	5	$\dfrac{169.5 + 179.5}{2} = 174.5$	$164.5 \times 12 = 1974$
179.5 – 189.5	7	$\dfrac{179.5 + 189.5}{2} = 184.5$	$184.5 \times 7 = 1291.5$
189.5 – 199.5	10	$\dfrac{189.5 + 199.5}{2} = 194.5$	$194.5 \times 10 = 1945$
	$\Sigma 42$		$\Sigma 7319$

$$Mean = \frac{7319}{42} = 174.262$$

However there are different types of mean, each differing by the formula used. The main types which we will discuss in this chapter are:

1) Geometric
2) Harmonic

6.2.1.2 Geometric mean

Like the arithmetic mean, the geometric mean is another type of average which shows the central tendency of a set of data. What differentiates the arithmetic mean from the geometric mean is that in the former we add n number of data values and then divide by n. In geometric mean, we multiply the given n number of data values and then take the nth root.

For example, if we have two data points, such as 4 and 9, and multiply the two numbers and take the square root of the product, then we get our geometric mean of 6. Similarly, if we have three data points, then take the cube root of the product of the three data, and so on.

6.2.1.2.1 Geometric mean for a sample dataset

The formula to calculate the geometric mean is given below:

$$G.M = \sqrt[n]{x_1 \times x_2 \times x_3 \times \ldots \ldots \times x_n} = \left(x_1 \times x_2 \times x_3 \times \ldots \ldots \times x_n \right)^{\frac{1}{n}}$$

This can also be rewritten as

$$\log(G.M) = \frac{1}{n} \log\left(x_1 \times x_2 \times \ldots \ldots x_n \right) = \frac{1}{n} \left(\log x_1 + \log x_2 + \ldots \log x_n \right)$$

$$= \frac{\Sigma \log x_i}{n}$$

Let's say we have a dataset of the average monthly temperature in Texas given below:

Month	Temperature (°F)	Month	Temperature (°F)
January	35	July	75
February	38	August	73
March	44	September	66
April	56	October	54
May	63	November	48
June	71	December	37

$$G.M = \sqrt[12]{35 \times 38 \times 44 \times 56 \times 63 \times 71 \times 75 \times 73 \times 66 \times 54 \times 48 \times 37}$$

$$= 53.1481$$

6.2.1.2.2 Geometric mean for a frequency distribution table

6.2.1.2.2.1 UNGROUPED FREQUENCY TABLE

Let's now find the geometric mean for the same example we did for the arithmetic mean.

Number of errors	Frequency	$log(x).f$
1	2	$log(1) \times 2 = 0$
2	4	$log(2) \times 4 = 1.20412$
3	3	$log(3) \times 3 = 1.431364$
4	6	$log(4) \times 6 = 3.61236$
5	3	$log(5) \times 3 = 2.097$
6	2	$log(6) \times 2 = 1.5563$
	$\Sigma 20$	$\Sigma 9.9011465$

$$\log G.M = \frac{\Sigma \log(x_i) f_i}{n}$$

$$= \frac{9.9011465}{20} = 0.495057325$$

$$GM = 10^{0.495057325} = 3.1265$$

6.2.1.2.2.2 GROUPED FREQUENCY TABLE

$$\log(G.M) = \frac{1}{n}(\log x_1 + \log x_2 + \ldots \log x_n)$$

$$= \frac{1}{f_1 + f_2 \ldots \ldots + f_n}(f_1 \log x_1 + f_2 \log x_2 + \ldots f_n \log x_n)$$

$$= \frac{1}{\Sigma f_i} \Sigma f_i \log(x_i)$$

Compute the geometric mean for the following frequency distribution.

Horsepower	Freq.	Mid	Log (mid)	log(mid)$\times f$
149.5 – 159.5	8	154.5	$log(154.5) = 2.189$	$2.189 \times 8 = 17.51$
159.5 – 169.5	12	164.5	$log(164.5) = 2.216166$	$2.216166 \times 12 = 26.594$
169.5 – 179.5	5	174.5	$log(174.5) = 2.2418$	$2.2418 \times 5 = 11.21$
179.5 – 189.5	7	184.5	$log(184.5) = 2.266$	$2.266 \times 7 = 15.862$
189.5 – 199.5	10	194.5	$log(194.5) = 2.289$	$2.289 \times 10 = 22.8892$
	$\Sigma 42$			$\Sigma 94.0652$

$$\log G.M = \frac{\Sigma f_i \log(x_i)}{n} = \frac{94.0652}{42} = 2.23964$$

$$GM = 10^{2.23964} = 173.6361$$

- If each object in the dataset is substituted by the GM, then the product of the objects remains unchanged.
- The ratio of the corresponding observations of the GM in two series is equal to the ratio of their geometric means.
- The products of the corresponding items of the GM in two series are equal to the product of their geometric mean.

6.2.1.2.3 When is the geometric mean better than the arithmetic mean?

From the discussion above we can actually conclude that the arithmetic mean is an additive mean and the geometric mean is a multiplicative mean. Even though it's less commonly used, the geometric mean is more useful and accurate than the arithmetic mean when the data values are skewed or include outliers and are dependent on each other. The arithmetic mean is more useful and accurate when it is used to calculate the average of a dataset where the numbers are not skewed and are independent.

In a dataset of 10, 15, 20, and 1000, the geometric mean is 41.62, while the arithmetic means is 261.25. The effect highlights prominent.

This significant difference highlights the effect of the extreme value (1000) on the arithmetic mean. Hence, no range dominates the weights, and any percentage does not significantly affect the dataset. The geometric mean is not influenced by skewed distributions as the arithmetic average is.

6.2.1.2.4 Geometric mean vs. arithmetic mean

The results of the geometric mean and arithmetic mean clearly imply that these two mean values will not only be different but the former will always be less than the arithmetic mean for the same data due to the log function or the root. Taking our first values as an example, we have seen that the geometric mean of 4 and 9 is 6; however the arithmetic mean rounds up to 6.50. Similarly we calculated the geometric mean for the grouped frequency which came to 39.98545 lower than the arithmetic mean, which was 51. So one of the most important properties of the geometric mean is that it is always less than the arithmetic mean. Another property is that the geometric mean can only be found for positive values; it doesn't take on any negative values. If any value in the dataset is zero, the geometric mean is zero since we need the product of the data values and any number multiplied by zero gives a zero. In contrast, the arithmetic mean can be calculated for positive and negative numbers.

Arithmetic mean	Geometric mean
Also known as the additive mean	Known as the multiplicative mean
Always higher than the geometric mean	Lower than the arithmetic mean
$Mean = \dfrac{x_1 + x_2 + x_3 + \ldots\ldots + x_n}{n}$	$Mean = \sqrt[n]{x_1 \times x_2 \times x_3 \times \ldots\ldots \times x_n}$
Example:	Using the same dataset we now calculate the geometric mean
We have a dataset	$GM = \sqrt[5]{25 \times 50 \times 75 \times 100 \times 200}$
$25, 50, 75, 100, 200$	$= 71.55$
$AM = \dfrac{25 + 50 + 75 + 100 + 200}{5} = 90$	

The arithmetic mean can be calculated for positive and negative numbers	The geometric mean can only be calculated for positive numbers
	It is useful in cases where the dataset is logarithmic or varies by multiples of 10
	Cannot be calculated when the value of any observations is zero or negative
The arithmetic mean is used by statisticians but for datasets with no significant outliers	The geometric mean is not influenced by skewed distributions as the arithmetic average is

6.2.1.3 Harmonic mean

The harmonic mean is the last of the three Pythagorean means (the other two are arithmetic mean and geometric mean) which is defined as the reciprocal of the arithmetic mean of the reciprocals. Basically it is calculated by dividing the number of values n by the sum of the reciprocals of each value in the dataset. Just like the geometric mean, the harmonic mean also gives a value lower than the arithmetic mean; in fact it gives the lowest value among the three Pythagorean means.

The harmonic mean gives less weightage to the large values and large weightage to the small values to balance the values correctly. In general, the harmonic mean is used when there is a need to give greater weight to the smaller items. It is applied in the case of times and average rates.

It is the most appropriate measure for ratios and rates because it equalizes the weights of each data point.

6.2.1.3.1 Harmonic mean for a sample dataset

$$HM = \frac{n}{\dfrac{1}{x_1} + \dfrac{1}{x_2} + \ldots\ldots + \dfrac{1}{x_n}} = \frac{n}{\Sigma \dfrac{1}{x_i}}$$

Suppose a trip from London to New York has five 100 km segments. We travel 80 km/h for the first 100 km, 120 km/h for the second 100 km, 150 km/h for the third 100 km, 100 km/h for the fourth 100 km, and then 65 km/h for the last 100 km. What is our average speed?

$$HM = \frac{5}{\dfrac{1}{80} + \dfrac{1}{120} + \dfrac{1}{150} + \dfrac{1}{100} + \dfrac{1}{65}} = 94.5454545455$$

6.2.1.3.2 Harmonic mean for a frequency distribution table

$$HM = \frac{\Sigma f_i}{\left(f_1 \times \dfrac{1}{x_1}\right) + \left(f_2 \times \dfrac{1}{x_2}\right) + \ldots\ldots + \left(f_n \times \dfrac{1}{x_n}\right)} = \frac{\Sigma f_i}{\Sigma f_i \dfrac{1}{x_i}}$$

6.2.1.3.2.1 UNGROUPED FREQUENCY TABLE

Let's calculate the harmonic mean for the number of errors.

Number of errors	Frequency	$\dfrac{f_i}{x_i}$
1	2	$\dfrac{2}{1} = 2$
2	4	$\dfrac{4}{2} = 2$
3	3	$\dfrac{3}{3} = 1$
4	6	$\dfrac{6}{4} = 1.5$
5	3	$\dfrac{3}{5} = 0.60$
6	2	$\dfrac{2}{6} = \dfrac{1}{3}$
	$\Sigma 20$	$\Sigma 7.433$

$$Mean = \frac{20}{7.433} = 2.69058$$

6.2.1.3.2.2 GROUPED FREQUENCY TABLE

Let's go back to the engine power (in horsepower) of 30 vehicles.

Horsepower	Frequency (f)
149.5 – 159.5	8
159.5 – 169.5	12
169.5 – 179.5	5
179.5 – 189.5	7
189.5 – 199.5	10

Horsepower	f	Mid	$\dfrac{f_i}{mid_i}$
149.5 – 159.5	8	154.5	$\dfrac{8}{154.5}$

159.5 – 169.5	12	164.5	$\dfrac{12}{164.5}$
169.5 – 179.5	5	174.5	$\dfrac{5}{174.5}$
179.5 – 189.5	7	184.5	$\dfrac{7}{184.5}$
189.5 – 199.5	10	194.5	$\dfrac{10}{194.5}$
	$\Sigma 42$		$\Sigma 0.24273$

$$Mean = \frac{42}{0.24273} = 173.03176$$

Arithmetic mean	Geometric mean	Harmonic mean
It is the sum of all observations divided by the no. of observations.	We multiply all the observations in the sample and then take the nth root of the product.	We divide n by the reciprocals of the arithmetic mean of the reciprocals.
$AM = \dfrac{\Sigma x_i}{n}$	$GM = \sqrt[n]{x_1 \times x_2 \times x_3 \times \ldots\ldots \times x_n}$	$HM = \dfrac{n}{\Sigma \dfrac{1}{x_i}}$
The arithmetic mean requires addition.	The geometric mean employs multiplication.	The harmonic mean utilizes reciprocals,
Works well to produce an "average" number of a dataset when there is a linear additive relationship between the numbers in the dataset.	Works well when there is a multiplicative or exponential relationship between the data values.	
		Always gives the smallest mean.
If the numbers in the group are all the same, the three means will be the same.		

6.2.1.3.3 Relationship between arithmetic, geometric, and harmonic mean

Let's take two data values x and y.

The arithmetic mean of the two numbers is $AM = \dfrac{x+y}{2}$ .... (6.1)

The geometric mean becomes $GM = \sqrt{x \times y}$ .... (6.2)

Similarly, the harmonic mean is then $HM = \dfrac{2}{\dfrac{1}{x}+\dfrac{1}{y}} = \dfrac{2}{\dfrac{y+x}{xy}} = \dfrac{2xy}{x+y}$ .... (6.3)

Now, substituting Equation 6.1 and Equation 6.2 in Equation 6.3, we get

$$HM = \frac{2x\,y}{x+y} = \left(\frac{2}{x+y}\right)(x\,y) = \left(\frac{1}{AM}\right)GM^2$$

These three discussed means are also known as the three classical Pythagorean means.

6.2.2 Median

Next we discuss the second type of central tendency—median. As mentioned, the arithmetic mean is really useful when there are no extreme values in the dataset. But when need to analyze a dataset which is extremely skewed and has outliers, the median is considered a better measure.

The median is the value that sits at the exact middle of the set of values in a dataset.

One way to compute the median is to list all scores in numerical order, and then locate the score in the center of the sample. Now to determine the location we first need to check whether the number of terms is odd or even.

6.2.2.1 Median for a sample dataset

If n is odd then the median can be calculated from $\left(\dfrac{n+1}{2}\right)^{th} obs$.

If n is even then the center point of the sample is $\dfrac{\left(\dfrac{n}{2}\right)^{th} obs. + \left[\left(\dfrac{n}{2}\right)+1\right]^{th} obs}{2}$.

6.2.2.2 Example

$$19, 21, 23, 20, 23, 27, 25, 24, 31$$

Ascending order: $19, 20, 21, 23, 23, 24, 25, 27, 31$

Since the number of samples $n = 9$ which is odd,

$$Median = \left(\frac{n+1}{2}\right)^{th} obs = \left(\frac{9+1}{2}\right)^{th} obs = \left(\frac{10}{2}\right)^{th} obs \text{ so } 5^{th} \text{ term hence } 23$$

$$50, 67, 24, 34, 78, 43$$

Ascending order: $24, 34, 43, 50, 67, 78$

Since $n = 6$ which is even,

$$Median = \frac{\left(\frac{n}{2}\right)^{th} obs. + \left[\left(\frac{n}{2}\right)+1\right]^{th} obs}{2} = \frac{\left(\frac{6}{2}\right)^{th} obs. + \left[\left(\frac{6}{2}\right)+1\right]^{th} obs}{2}$$

$$= \frac{3^{rd} obs. + 4^{th} obs}{2} = \frac{43 + 50}{2} = \frac{93}{2} = 46.50$$

6.2.2.3 Median for a frequency distribution table

$$Median / second\ quartile\ Q_2\ = L + \left[\left(\frac{2 \times \frac{n}{4} - cf}{f}\right) * h\right] = L + \left[\left(\frac{\frac{n}{2} - cf}{f}\right) * h\right]$$

where

- L = lower limit of the median class interval
- f = frequency of the median class
- cf = cumulative frequency of the class preceding the median class
- $n = \sum f$ = total frequency
- h = median class's interval size = $\dfrac{\text{range}}{\text{number of intervals in the table}}$

6.2.2.3.1 Ungrouped frequency table

Let's consider seven observations collected from prototype engine connectors.

Observations	Frequency/count
12.6	8
11.9	4
13.4	3
14.2	7
15.3	3
16.5	12
17.8	4

Observations	Frequency/count	Cumulative freq.	Cumulative freq. interval
12.6	8	8	≤ 8
11.9	4	12	≤ 12
13.4	3	$12 + 3 = 15$	≤ 15
14.2	7	$15 + 7 = 22$	≤ 22
15.3	3	$22 + 3 = 25$	≤ 25
16.5	12	$25 + 12 = 37$	≤ 37
17.8	4	$37 + 4 = 41$	≤ 41

Since the number of samples $n = 41$ which is odd,

$$Median = \left(\frac{n+1}{2}\right)^{th} obs = \left(\frac{41+1}{2}\right)^{th} obs = \left(\frac{42}{2}\right)^{th} obs \text{ so } 21^{st} \text{ term}$$

If the 41 observations were arranged in ascending order, the 21st observation would be the middle most one, since there are 20 observations on either side of this observation. Fifteen observations (see cumulative frequency) have values less than or equal to 13.4. Twenty-two observations have values less than or equal to 14.2. This means that the 21st observation is 14.2.

Therefore, median = 14.2.

6.2.2.3.2 Grouped frequency table

6.2.2.3.2.1 EXAMPLE

Let's say we have a dataset of power consumption.

Let's draw the table.

Consumption	Frequency	Cumulative freq.	Cumulative freq. interval
420 – 430	336	336	≤ 336
430 – 440	2112	$336 + 2112 = 2448$	≤ 2448
440 – 450	2107	$2448 + 2107 = 4555$	≤ 4555
450 – 460	1304	$4555 + 1304 = 5859$	≤ 5859
460 – 470	1553	$5859 + 1553 = 7412$	≤ 7412
470 – 480	1336	$7412 + 1336 = 8748$	≤ 8748
480 – 490	736	$8748 + 736 = 9484$	≤ 9484
490 – 500	84	$9484 + 84 = 9568$	≤ 9568

Median class $= \dfrac{9568}{2} = 4784$ which is in the interval $450 - 460$.

$L = 450$

$f = 1304$

$cf = 4555$

$n = 9568$

$h = 10$

$$\text{Median } = 450 + \left[\left(\frac{4784 - 4555}{1304} \right) * 10 \right] = 451.7561.$$

6.2.3 Mode

The mode is the most frequently occurring value in the set of scores. On a histogram it represents the highest bar in a bar chart or histogram. You can, therefore, sometimes consider the mode as being the most popular option. It is defined as the most frequently occurring value. Datasets can have a single mode or multiple modes.

- No mode: all numbers in a dataset occur the same number of times.
- Unimodal: only one mode, meaning one number appears more than all the others.
- Bimodal: two modes, which means two numbers appear more than the rest and with the same frequency as each other.
- Trimodal: three modes, which means three numbers appear the same number of times and more than the rest of the numbers.
- Multimodal: more than two modes.

6.2.3.1 Mode for frequency distribution table

Let's go back to the same example.

6.2.3.1.1 Ungrouped frequency table

Let's consider seven observations collected from the prototype engine connectors.

Observations	Frequency/count
12.6	8
11.9	4
13.4	3
14.2	7
15.3	3
16.5	12
17.8	4

Here the modal class is the observation with the highest frequency which is 12 and it occurs in the observation 16.5.

6.2.3.1.2 Grouped frequency table

$$Mode = L + \left[\frac{f_1 - f_0}{2f_1 - f_0 - f_2} \times h \right]$$

where

- L = lower limit of the modal class interval.
- f_1 is the frequency of the modal class.
- f_0 is the frequency of the class preceding the modal class.
- f_2 is the frequency of the class succeeding the modal class.
- h is the size of the class intervals.

Consumption	Frequency	Cumulative freq.	Cumulative freq. interval
420 – 430	336	336	– 336
430 – 440	2112	336 + 2112 = 2448	336 – 2448
440 – 450	2107	2448 + 2107 = 4555	2449 – 4555
450 – 460	1304	4555 + 1304 = 5859	4556 – 5859
460 – 470	1553	5859 + 1553 = 7412	5860 – 7412
470 – 480	1336	7412 + 1336 = 8748	7413 – 8748
480 – 490	736	8748 + 736 = 9484	8749 – 9484
490 – 500	84	9484 + 84 = 9568	9485 – 9568

Here the modal class in the interval with the highest frequency is 2112, and it occurs in the interval 430 – 440.

$$L = 430$$
$$f_1 = 2112$$
$$f_0 = 336$$
$$f_2 = 2107$$
$$h = 10$$

$$Mode = 430 + \left[\frac{2112 - 336}{2(2112) - 336 - 2107} \times 10 \right] = 439.9719$$

The mean, median, and mode are all valid measures of central tendency, but, under different conditions, some measures of central tendency become more appropriate to use than others. In the following sections, we will look at the mean, mode, and median and learn how to calculate them and under what conditions they are most appropriate.

6.2.4 Central tendency

The following table lists the functions that calculate the measures of central tendency.

Function name	Description
mean	Arithmetic average
geomean	Geometric mean
harmmean	Harmonic mean
median	50th percentile
mode	Most frequent value

6.2.4.1 Example

```
Errors = 1:6;
Frequency = [2, 4, 3, 6, 3, 2];
Data = repelem(Errors, Frequency);
AM = mean(Data)
GM = geomean(Data)
HM = harmmean(Data)
```

                              AM =

                                  3.5000

                              GM =

                                  3.1265

                              HM =

                                  2.6906

6.2.4.2 *Example*

```
Observation = [12.6, 11.9, 13.4, 14.2, 15.3, 16.5, 17.8];
Frequency = [8, 4, 3, 7, 3, 12, 4];
Data = repelem(Observation, Frequency);
Median = median(Data);
Mode = mode(Data);
```

Median =

14.2000

Mode =

16.5000

6.3 MEASURES OF DISPERSION/VARIABILITY

Measures of central tendency focus on the average or middle values of datasets, whereas measures of variability focus on the dispersion of data.

Dispersion refers to the spread of the values around the central tendency. It explains the extent to which data points are dispersed from each other.

It comprises four methods of calculation.

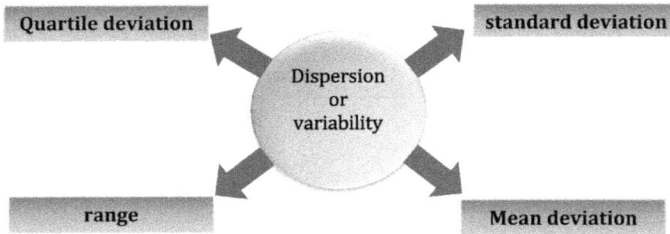

Figure 6.5

6.3.1 Range

The range is simply the largest value in the dataset minus the smallest value.

6.3.2 Mean deviation

1) Mean deviation about the mean
2) Mean deviation about the median

Mean deviation is used to compute how far the values in a dataset are from the center point. In other words, the mean deviation is used to calculate the average deviation from the data's center point, the mean deviation.

6.3.2.1 Sample dataset

The mean deviation is calculated using the formula

$$\frac{\Sigma|\text{Midpoint} - \text{mean}|}{\Sigma n}$$

6.3.2.2 Frequency distribution table

This is calculated using the formula

$$\frac{\Sigma|\text{Midpoint} - \text{mean}| \times \text{freq}}{\Sigma\text{freq}}$$

In calculation algebraic signs are ignored while taking the deviations of the items. If the signs of the deviations are not ignored, the net sum of the deviations will be zero.

Class limits	freq	Mid	Mid× freq.	|Mid mean|	|Mid mean| × freq.
1000 – 1200	27	1100	29,700	537.821	14,521.2
1200 – 1400	74	1300	96,200	337.821	24,998.7
1400 – 1600	42	1500	63,000	137.821	5788.46
1600 – 1800	65	1700	110,500	62.17949	4041.667
1800 – 2000	58	1900	110,200	262.1795	15,206.41
2000 – 2200	32	2100	67,200	462.1795	14,789.74
2200 – 2400	4	2300	9200	662.1795	2648.718
2400 – 2600	10	2500	25,000	862.1795	8621.795
	Σ312		511,000		Σ90,616.67

6.3.2.3 Example

$$\text{Mean} = \frac{511000}{312} = 1637.82051$$

$$\text{Mean deviation about the mean} = \frac{90616.67}{312} = 290.438$$

Class limits	freq	cf	
1000 – 1200	27	27	29,700
1200 – 1400	74	27 + 74 =	96,200
1400 – 1600	42	1500	63,000
1600 – 1800	65	1700	110,500
1800 – 2000	58	1900	110,200
2000 – 2200	32	2100	67,200
2200 – 2400	4	2300	9200
2400 – 2600	10	2500	25,000
	Σ312		511,000

6.3.3 Standard deviation

The standard deviation tells us how far the mean is from each observation in a given dataset.

6.3.3.1 Sample mean vs. population mean

When data from a sample (of size n) is used to estimate parameters of the population, the sample standard deviation is calculated. First the deviations of data values from the sample mean are calculated. Since the sample mean is used in place of the population mean (which is unknown), taking the quadratic mean is not appropriate. In order to compensate for the use of the sample mean, the sum of squares of deviations is divided by $(n-1)$ instead of n.

6.3.3.2 Sample dataset

The sample standard deviation is calculated using the formula

$$\sqrt{\frac{\Sigma\left(x_i - mean\right)^2}{n-1}}$$

6.3.3.3 Frequency distribution table

Similarly,

$$\sqrt{\frac{\Sigma\left(x_i - mean\right)^2 \times f_i}{\Sigma f_i - 1}}$$

Class limits	freq	Mid	$\left(\mathbf{Mid} - \mathbf{mean}\right)^2$	$\left(\mathbf{Mid} - \mathbf{mean}\right)^2 \times$ freq.
1000 – 1200	27	1100	289,250.9	7,809,774
1200 – 1400	74	1300	114,122.7	8,445,080
1400 – 1600	42	1500	18,994.49	797,768.7
1600 – 1800	65	1700	3866.289	251,308.8
1800 – 2000	58	1900	68,738.08	3,986,809
2000 – 2200	32	2100	213,609.9	6,835,516
2200 – 2400	4	2300	438,481.7	1,753,927
2400 – 2600	10	2500	743,353.5	7,433,535
	$\Sigma 312$			$\Sigma 37,313,717.9$

Sample standard deviation $= \sqrt{\dfrac{37313717.9}{312-1}} = 346.381$

Consequently, the population standard deviation $= \sqrt{\dfrac{37313717.9}{312}} = 345.82546$

6.3.4 Quartile deviation

6.3.4.1 Sample dataset

Quartiles can be obtained using the following formulas:

$$Q_1 = \left[\frac{(n+1)}{4}\right]^{th} \text{observation}$$

$$Q_2 = \left[\frac{(n+1)}{2}\right]^{th} \text{observation}$$

$$Q_3 = \left[\frac{3(n+1)}{4}\right]^{th} \text{observation } 3(n+1)/4]\text{th item}$$

Where n represents the total number of observations in the given dataset.

6.3.4.2 Frequency distribution table

Similarly,

$$Q_i = L + \left[\frac{\left(i \times \dfrac{n}{4}\right) - cf}{f} \times h\right]$$

where

L = lower limit of the quartile class interval

f = frequency of quartile class

a) $cf = cumulative\ frequency\ of\ the\ class\ preceding\ the\ quartile\ class$
b) $n = \Sigma f = total\ frequency$
c) $h = quartile\ class's\ interval\ size$

$$\text{Quartile deviation} = \frac{Q_3 - Q_1}{2} = \frac{\text{interquartile range}}{2}$$

Class limits	freq	Cumulative freq.	Cumulative freq. interval
1000–1200	27	27	–27
1200–1400	74	27+74=101	28–101
1400 – 1600	42	$101 + 42 = 143$	$102 - 143$
1600 – 1800	65	$143 + 65 = 208$	$144 - 208$
1800 – 2000	58	$208 + 58 = 266$	$209 - 266$
2000 – 2200	32	$266 + 32 = 298$	$267 - 298$
2200 – 2400	4	$298 + 4 = 302$	$299 - 302$
2400 – 2600	10	$302 + 10 = 312$	$303 - 312$
	£ 312		

Q_1 or first quartile

- First quartile class $\dfrac{312}{4} = 78$ and this is in the interval $1200 - 1400$.

- $Q_1 = L + \left[\dfrac{\left(\dfrac{n}{4}\right) - cf}{f} \times h \right] = 1200 + \left[\dfrac{78 - 27}{74} \times 200 \right] = 1337.8378$.

6.3.4.3 Q_3 or third quartile

−Third quartile class $3 * \dfrac{312}{4} = 234$ and this is in the interval $1800 - 2000$.

$$-Q_3 = L + \left[\dfrac{3\left(\dfrac{n}{4}\right) - cf}{f} \times h \right] = 1800 + \left[\dfrac{234 - 208}{58} \times 200 \right] = 1889.6552.$$

Quartile deviation

$$\dfrac{1889.6552 - 1337.8378}{2} = 275.9087$$

6.3.4.4 Empirical relation between measures of variation

Quartile deviation (QD) $= \dfrac{2}{3}$ standard deviation (σ)

Mean deviation (MD) $= \dfrac{4}{5}$ standard deviation (σ)

Quartile deviation (QD) $= \dfrac{5}{6}$ mean deviation (MD)

6.3.4.5 In MATLAB

The following table lists the functions that calculate the measures of dispersion.

Function name	Description
range	Range of a dataset
mad	mad(x, flag) specifies whether to compute the mean absolute deviation (flag = 0, the default) or the median absolute deviation (flag = 1)
std	Standard deviation
var	Variance
iqr	Interquartile range

6.4 DISTRIBUTION SHAPE

The study of central tendency provides us with valuable information relating to the central value, and measures of variation provides us the variability of the distribution. Unfortunately, these measures fail to describe the shape of the data. Understanding the shape not only helps us to understand where the most information is lying but also helps us to analyze the outliers present in the data. The two statistics that give proper insights into the shape of the distribution are skewness and kurtosis. So what exactly are they?

6.4.1 Skewness

In the previous chapters we introduced normal distribution. One of the most important properties of the distribution includes its symmetry about the mean—the same proportion of data on each side of the distribution where half the values fall below the mean and half above the mean.

However not all probability distributions are symmetrical or have a bell-shaped curve. The type of distributions where the right side and the left side of the curve are not mirror images of each other are called skewed distributions. The term skewness refers to lack of symmetry or departure from symmetry. In a skewed distribution values are more spread out on one side of the center than on the other.

Distributions that have values more widely spread on the left side are called left skewed or negative skewed. Left-skewed distributions have longer tails extended to the left of the distribution which means the lower values (toward the left on a number line) are more spread out than the higher values.

Similarly, distributions with values more widely spread to the right are called right skewed or positive skewed. The higher values (toward the right on a number line) are more spread out than the lower values; therefore the distribution has longer tails extended to the right of the distribution.

In both cases, the plot of the distribution is stretched to one side rather than to the other. This means that, in the case of skewness, we can say that the mean, median, and mode of our dataset are not equal and do not follow the assumptions of a normally distributed curve. A symmetrical dataset will always have a skewness equal to 0. So, a normal distribution will have a skewness of 0.

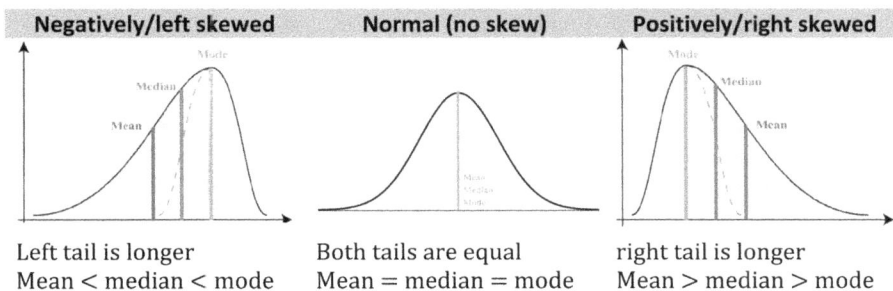

Negatively/left skewed	Normal (no skew)	Positively/right skewed
Left tail is longer	Both tails are equal	right tail is longer
Mean < median < mode	Mean = median = mode	Mean > median > mode

Figure 6.6

In Figure 6.6, the mean always gets pulled the furthest to the tail of skewed distributions. The reason for this is that the mean is much more affected by extreme outliers than the median. The median is simply the boundary which divides the top 50% of the data from the bottom 50% of the data. The mean has to include all values in its calculation, and so can be more affected by extreme values than the median is.

The value for skewness can range from negative infinity to positive infinity.

- If skewness is positive, the data is positively skewed or skewed right, meaning that the right tail of the distribution is longer than the left.
- If skewness is negative, the data is negatively skewed or skewed left, meaning that the left tail is longer.
- If skewness = 0, the data is perfectly symmetrical.

It is important to realize that skewness of exactly zero is quite unlikely for real-world data, so how can we interpret the skewness number? We use the following rule of thumb:

- If skewness < -1 or skewness $> +1$, the distribution is **highly skewed.**
- If $-1 <$ skewness < -0.5 or $+0.5 <$ skewness $< +1$, the distribution is **moderately skewed.**
- If $-0.5 <$ skewness < 0.5 the distribution is **approximately symmetric.**

6.4.2 Kurtosis

Another measure of the shape of a data distribution is called kurtosis. Kurtosis relates to how heavy or thin the two tails of the distribution are. Another way of describing it is the degree to which a given distribution is more or less "peaked" relative to the normal distribution which is taken as the standard reference. The kurtosis of a normal distribution equals 3. Therefore, the excess kurtosis is found using the formula below:

Excess kurtosis = kurtosis − 3

6.4.2.1 Types of kurtosis

The types of kurtosis are determined by the excess kurtosis of a particular distribution. The excess kurtosis can take positive or negative values, as well as values close to zero.

6.4.2.1.1 Mesokurtic

Data that follows a mesokurtic distribution shows an excess kurtosis of zero or close to zero. This means that if the data follows a normal distribution, it follows a mesokurtic distribution.

6.4.2.1.2 Leptokurtic

If the kurtosis is greater than 3, then leptokurtic distribution indicates a positive excess kurtosis. This means the dataset has heavier tails than a normal distribution indicating large outliers. This also means it is more peaked than the normal distribution.

6.4.2.1.3 Platykurtic

If the kurtosis is less than 3, then platykurtic distribution shows a negative excess kurtosis. This means the dataset has lighter tails than a normal distribution (less in the tails). The kurtosis reveals a distribution with lower peak when compared to the normal distribution. The flat tails indicate the small outliers in a distribution.

6.4.3 Moments

Before we can get into the calculation of skewness and kurtosis we need to first understand the concept of moments. For starters, moments are an important tool used to measure the characteristics of a distribution. They can provide valuable information about the center, spread, and shape of a distribution. The concept of moments is crucial in descriptive statistics because it offers a mathematical framework to precisely quantify the three essential characteristics of a dataset in descriptive statistics.

It's important to know that there are three different kinds of "moment":

- Moments about an arbitrary point
- Moments about the origin which are called raw moments
- Moments about the mean which are called central moments

Suppose that we have a dataset $x_1, x_2, x_3, \ldots, x_n$ of sample size n.

One important calculation, which is actually several numbers, is called the sth moment. The sth moment of a distribution (or set of data) about a number is the expected value of the sth power of the deviations about that number. Mathematically, the sth moment of the dataset with values $x_1, x_2, x_3, \ldots, x_n$ is given by the formula:

$$E\left((X-a)^s\right) = \frac{(x_1-a)^s + (x_2-a)^s + (x_3-a)^s + ..(x_n-a)^s}{n} = \frac{\sum_{i=1}^n (x_i-a)^s}{n}$$

Where

x_i = each value

a could be any value depending on the type of moment, for example:

- Moment about an arbitrary point so any value of a
- Moment about the origin so $a = 0$
- Moment about the mean so $a = \bar{x}$

(We shall only discuss the moment about the mean in this textbook.)

As stated earlier, there are several values of s, each providing different information about the dataset. Let's take a look at some of these moments and how they can be used in statistical analysis.

When we take the deviation from the actual mean and calculate the moments, these are known as moments about the mean or central moments.

6.4.3.1 First moment

For the first moment, we set $s = 1$. The formula for the first moment is thus:

$$E\left((X-\mu)^1\right) = \frac{(x_1-\bar{x})^1 + (x_2-\bar{x})^1 + (x_3-\bar{x})^1 + ..(x_n-\bar{x})^1}{n}$$

$$= \frac{x_1 - \bar{x} + x_2 - \bar{x} + x_3 - \bar{x} + ..x_n - \bar{x}}{n} = \frac{(x_1 + x_2 + x_3 + ..x_n) - n\bar{x}}{n}$$

$$= \frac{(x_1 + x_2 + x_3 + ..x_n)}{n} - \frac{n\bar{x}}{n} = \bar{x} - \bar{x} = 0$$

Therefore, the first central moment is zero. That is

$$E(X - \mu) = 0$$

6.4.3.2 Second moment

For the second moment we set $s = 2$. The formula for the second moment is:

$$E\left((X - \mu)^2\right) = \frac{(x_1 - \bar{x})^2 + (x_2 - \bar{x})^2 + (x_3 - \bar{x})^2 + ..(x_n - \bar{x})^2}{n} = \frac{\sum_{i=1}^{n}(x_i - \bar{x})^2}{n}$$

We saw that the formula of standard deviation for the population is

$$SD = \sqrt{\frac{\sum_{i=1}^{n}(x_i - \bar{x})^2}{n}}$$

This means

$$Variance = \frac{\sum_{i=1}^{n}(x_i - \bar{x})^2}{n}$$

This formula is equivalent to that for the variance. Therefore, the second central moment is the variance. That is

$$E(X - \mu)^2 = \sigma^2$$

6.4.3.3 Third moment

For the third moment we set $s = 3$. The formula for the third moment is:

$$E(X - \mu)^3 = \frac{(x_1 - \bar{x})^3 + (x_2 - \bar{x})^3 + (x_3 - \bar{x})^3 + ..(x_n - \bar{x})^3}{n} = \frac{\sum_{i=1}^{n}(x_i - \bar{x})^3}{n}$$

To get the formula of skewness we need the third and the second moments. Skewness is the ratio of the third moment and the second moment—variance raised to the power of 3/2. So mathematically we can write,

$$\frac{E(X - \mu)^3}{\left(E(X - \mu)^2\right)^{3/2.}} = \frac{E\left((X - \mu)^3\right)}{\left(\sigma^2\right)^{3/2.}} = \frac{E\left((X - \mu)^3\right)}{\sigma^3} = \frac{\frac{1}{n}\sum_{i=1}^{n}(x_i - \bar{x})^3}{\left(\sqrt{\frac{1}{n}\sum_{i=1}^{n}(x_i - \bar{x})^2}\right)^3} = \frac{\frac{1}{n}\sum_{i=1}^{n}(x_i - \bar{x})^3}{\left(\frac{1}{n}\sum_{i=1}^{n}(x_i - \bar{x})^2\right)^{\frac{3}{2}}}$$

6.4.3.4 Fourth moment

For the fourth moment we set $s = 4$. The formula for the fourth moment is:

$$E(X-\mu)^4 = \frac{(x_1-\bar{x})^4+(x_2-\bar{x})^4+(x_3-\bar{x})^4+..(x_n-\bar{x})^4}{n} = \frac{\sum_{i=1}^{n}(x_i-\bar{x})^4}{n}$$

The kurtosis of a dataset is computed almost the same way as the skewness with a slight difference. While skewness involves the third moment of the distribution (numerator), kurtosis involves the fourth moment in the numerator and

$$\frac{E(X-\mu)^4}{\left(E(X-\mu)^2\right)^{2.}} = \frac{E\left((X-\mu)^4\right)}{\left(\sigma^2\right)^{2.}} = \frac{E\left((X-\mu)^4\right)}{\sigma^4} = \frac{\frac{1}{n}\sum_{i=1}^{n}(x_i-\bar{x})^4}{\left(\sqrt{\frac{1}{n}\sum_{i=1}^{n}(x_i-\bar{x})^2}\right)^4} = \frac{\frac{1}{n}\sum_{i=1}^{n}(x_i-\bar{x})^4}{\left(\frac{1}{n}\sum_{i=1}^{n}(x_i-\bar{x})^2\right)^2}$$

6.4.3.5 For a sample dataset

6.4.3.5.1 First moment about the mean

$$E(X-\mu) = 0$$

6.4.3.5.2 Second moment about the mean

$$E(X-\mu)^2 = \sigma^2 = \frac{\sum_{i=1}^{n}(x_i-\bar{x})^2}{n}$$

6.4.3.5.3 Third moment about the mean

$$E(X-\mu)^3 = \frac{\sum_{i=1}^{n}(x_i-\bar{x})^3}{n}$$

6.4.3.5.4 Fourth moment about the mean

$$E(X-\mu)^4 = \frac{\sum_{i=1}^{n}(x_i-\bar{x})^4}{n}$$

6.4.3.6 Example

Let's assume we have a dataset of velocity measurements made along a cross section of a fluid flow field.

2.9, 3.5, 4.4, 1.5, 5.2

1. First, let's calculate the mean of the values.

$$\bar{x} = \frac{2.9+3.5+4.4+1.5+5.2}{5} = 3.5$$

Therefore:

$$\text{Skewness} = \frac{\frac{1}{5}\sum_{i=1}^{n}\left(x_i - 3.5\right)^3}{\left(\sqrt{\frac{1}{5}\sum_{i=1}^{n}\left(x_i - 3.5\right)^2}\right)^3}$$

$$\text{Kurtosis} = \frac{\frac{1}{5}\sum_{i=1}^{n}\left(x_i - 3.5\right)^4}{\left(\frac{1}{5}\sum_{i=1}^{n}\left(x_i - 3.5\right)^2\right)^2}$$

Let's first calculate the skewness.

- So first we calculate the third moment.

We subtract this mean from each value of the dataset. Then raise each of these differences to the power 3 and add the numbers. Then we divide this sum by the number of values we started with.

$$\frac{\left(2.9-3.5\right)^3+\left(3.5-3.5\right)^3+\left(4.4-3.5\right)^3+\left(1.5-3.5\right)^3+\left(5.2-3.5\right)^3}{5}=-0.5148$$

$$E\left(X-\mu\right)^3 = -0.5148$$

- Now we calculate the second central moment.

$$\frac{\left(2.9-3.5\right)^2+\left(3.5-3.5\right)^2+\left(4.4-3.5\right)^2+\left(1.5-3.5\right)^2+\left(5.2-3.5\right)^2}{5}=\frac{8.06}{5}=1.612$$

$$E\left(X-\mu\right)^2 = -0.5148$$

- Then we divide the third central moment by the second central moment raised to the power of 3/2 so:

$$\frac{-0.5148}{\left(1.612\right)^{\frac{3}{2}}}=-0.2515$$

Let's now calculate the kurtosis.
- Let's calculate the fourth central moment.

$$\frac{\left(2.9-3.5\right)^4+\left(3.5-3.5\right)^4+\left(4.4-3.5\right)^4+\left(1.5-3.5\right)^4+\left(5.2-3.5\right)^4}{5}=\frac{25.1378}{5}=5.02756$$

$$E\left(X-\mu\right)^4 = 5.02756$$

- We already have our second central moment:

$$\frac{\left(2.9-3.5\right)^2+\left(3.5-3.5\right)^2+\left(4.4-3.5\right)^2+\left(1.5-3.5\right)^2+\left(5.2-3.5\right)^2}{5}=\frac{8.06}{5}=1.612$$

$$E\left(X-\mu\right)^{2}=-0.5148$$

- Then we divide the fourth central moment by the second central moment squared so:

$$\frac{5.02756}{\left(1.612\right)^{2}}=1.9348$$

6.4.3.7 Population vs. sample

The formulas used for skewness and kurtosis above are valid when our dataset includes the entire population. Very often, we don't have data for the whole population, we only have access to a sample from a population, and we use that sample to estimate the two measures of the entire population. When calculating **sample skewness and kurtosis**, you need to make a small adjustment to the existing formula (the function of the adjustment is to correct a bias inherent in small samples):

$$\text{Sample skewness} = \frac{\sqrt{n\left(n-1\right)}}{n-2}\left(\text{population skewness}\right)$$

$$\text{Sample kurtosis} = \frac{\left(n-1\right)}{\left(n-2\right)\left(n-3\right)}\left(\left(n+1\right)\left(\text{population kurtosis}\right)-3\left(n-1\right)\right)+3$$

Therefore:

$$\text{Sample skewness} = \frac{\sqrt{5\left(5-1\right)}}{5-2}\left(-0.2515\right)=-0.3749$$

$$\text{Sample skewness} = \frac{\left(5-1\right)}{\left(5-2\right)\left(5-3\right)}\left(6*1.9348-3*4\right)+3=2.739$$

6.4.3.8 Interpretation

- Since the skewness $=-0.2515$, which is between -0.5 and 0.5, the sample data for velocity is approximately symmetric.
- Our kurtosis value is less than 3; therefore the distribution is platykurtic.

6.4.3.9 Example

Say we have a dataset of coefficient of friction (COF) of a particular material:

0.21, 0.28, 0.26, 0.53, 0.16, 0.52, 0.48, 0.19, 0.22, 0.35

1. Calculate the mean.

$$\bar{x}=\frac{0.21+0.28+0.25+0.33+0.16+0.52+0.48+0.19+0.22+0.35}{10}=0.3$$

2. Calculate the skewness.
    - Third central moment.

$$\frac{\left(0.21-0.3\right)^{3}+\left(0.28-0.3\right)^{3}+\ldots+\left(0.22-0.3\right)^{3}+\left(0.35-0.3\right)^{3}}{10}=0.0011244$$

$$E(X-\mu)^3 = 0.0011244$$

- Second central moment.

$$\frac{(0.21-0.3)^2 + (0.28-0.3)^2 + \ldots + (0.22-0.3)^2 + (0.35-0.3)^2}{10} = 0.01324$$

$$E(X-\mu)^2 = 0.01324$$

- Divide the third central moment by the second central moment raised to the power of 3/2 so:

$$\frac{0.0011244}{(0.01324)^{\frac{3}{2}}} = 0.73805$$

Sample skewness $= \dfrac{\sqrt{10(10-1)}}{10-2}(0.73805) = 0.87522$

3. Let's now calculate the kurtosis.
    - Fourth central moment.

$$\frac{(0.21-0.3)^4 + (0.28-0.3)^4 + \ldots + (0.22-0.3)^4 + (0.35-0.3)^4}{10} = 0.000403924$$

$$E(X-\mu)^4 = 0.000403924$$

- Our second central moment.

$$E(X-\mu)^2 = 0.01324$$

- Divide the fourth central moment by the second central moment squared so:

$$\frac{0.000403924}{(0.01324)^2} = 2.30421$$

Sample kurtosis $= \dfrac{(10-1)}{(10-2)(10-3)}(11*2.30421-3*9)+3 = 2.73423$

6.4.3.10 Interpretation

- With a skewness of 0.73805, the sample data for COF is **moderately skewed**.
- The distribution of the dataset is also platykurtic.

6.4.3.11 Example

Dataset of pressure:

34.2, 33.8, 31.7, 32.9, 34.5, 43.2, 33.5, 32.1, 38.8, 32.4, 34.9, 36.2

1. Calculating the mean of the values.

$$\bar{x} = \frac{34.2 + 33.8 + 31.7 + \ldots + 32.4 + 34.9 + 36.2}{12} = 34.85$$

2. Calculating the skewness.
   - Third central moment.

$$\frac{(34.2 - 34.85)^3 + (33.8 - 34.85)^3 + \ldots + (34.9 - 34.85)^3 + (36.2 - 34.85)^3}{12} = 47.347$$

$$E(X - \mu)^3 = 47.347$$

   - Second central moment.

$$\frac{(34.2 - 34.85)^2 + (33.8 - 34.85)^2 + \ldots + (34.9 - 34.85)^2 + (36.2 - 34.85)^2}{12} = 9.82583$$

$$E(X - \mu)^2 = 9.82583$$

   - Dividing the third central moment by the second central moment raised to the power of 3/2.

$$\frac{47.347}{(9.82583)^{\frac{3}{2}}} = 1.53722832$$

Sample skewness = $\dfrac{\sqrt{12(12-1)}}{12-2}(1.53722832) = 1.766141$

3. Calculating the kurtosis.
   - Fourth central moment.

$$\frac{(34.2 - 34.85)^4 + (33.8 - 34.85)^4 + \ldots + (34.9 - 34.85)^4 + (36.2 - 34.85)^4}{12} = 443.2378$$

$$E(X - \mu)^4 = 443.2378$$

   - Second central moment.

$$\frac{(34.2 - 34.85)^2 + (33.8 - 34.85)^2 + \ldots + (34.9 - 34.85)^2 + (36.2 - 34.85)^2}{12} = 9.82583$$

$$E(X - \mu)^2 = 9.82583$$

   - Dividing the fourth central moment by the second central moment squared.

$$\frac{443.2378}{(9.825831)^2} = 4.5909$$

$$\text{Sample kurtosis} = \frac{(12-1)}{(12-2)(12-3)}(13*4.5909 - 3*11) + 3 = 6.2611$$

6.4.3.12 Interpretation

- With a skewness of 1.537 the sample data is **highly skewed**.
- Since our kurtosis value is greater than 3, the distribution is leptokurtic.

6.4.3.13 Example

Temperature dataset:

29.7, 27.4, 30.1, 37.8, 30.8, 28.6, 45.9

1. Calculating the mean of the values.

$$\bar{x} = \frac{29.7 + 27.4 + 30.1 + 37.8 + 30.8 + 28.6 + 45.9}{7} = 32.9$$

2. Calculating the skewness.
    - Third central moment.

$$\frac{(29.7-32.9)^3 + (27.4-32.9)^3 + .. + (28.6-32.9)^3 + (45.9-32.9)^3}{7} = 286.398$$

$$E(X-\mu)^3 = 286.398$$

    - Second central moment.

$$\frac{(29.7-32.9)^2 + (27.4-32.9)^2 + .. + (28.6-32.9)^2 + (45.9-32.9)^2}{7} = 37.74857$$

$$E(X-\mu)^2 = 37.74857$$

    - Dividing the third central moment by the second central moment raised to the power of 3/2.

$$\frac{286.398}{(37.74857)^{\frac{3}{2}}} = 1.23486$$

$$\text{Sample skewness} = \frac{\sqrt{7(7-1)}}{7-2}(1.23486) = 1.60057$$

3. Calculate the kurtosis.
    - The fourth central moment.

$$\frac{(29.7-32.9)^4 + (27.4-32.9)^4 + .. + (28.6-32.9)^4 + (45.9-32.9)^4}{7} = 4368.59914$$

$$E(X-\mu)^4 = 4368.59914$$

- The second central moment .

$$E(X - \mu)^2 = 37.74857$$

- Dividing the fourth central moment by the second central moment squared.

$$\frac{4368.59914}{(37.74857)^2} = 3.06578$$

Sample skewness = $\dfrac{(7-1)}{(7-2)(7-3)}(8*3.06578 - 3*6) + 3 = 4.9579$

6.4.3.14 Interpretation

- With a skewness of 1.60057 the sample data is **highly skewed.**
- Since our kurtosis value is greater than 3, the distribution is leptokurtic.

6.4.3.15 For frequency distribution

$$\mu_s = \frac{\Sigma(x_i - \bar{x})^s . f_i}{\Sigma f_i}$$

6.4.3.15.1 First moment about the mean

$$\mu_1 = \frac{\Sigma(x_i - \bar{x})^1 . f_i}{\Sigma f_i} = 0$$

6.4.3.15.2 Second moment about the mean

$$\mu_2 = \frac{\Sigma(x_i - \bar{x})^2 . f_i}{\Sigma f_i}$$

6.4.3.15.3 Third moment about the mean

$$\mu_3 = \frac{\Sigma(x_i - \bar{x})^3 . f_i}{\Sigma f_i}$$

6.4.3.15.4 Fourth moment about the mean

$$\mu_4 = \frac{\Sigma(x_i - \bar{x})^4 . f_i}{\Sigma f_i}$$

6.4.3.16 Example

Class intervals	Frequency
17.5 – 19.5	3
19.5 – 21.5	4
21.5 – 23.5	2
23.5 – 25.5	6
25.5 – 27.5	4
27.5 – 29.5	5
29.5 – 31.5	2
31.5 – 33.5	3
33.5 – 35.5	1

First, we calculate the mean.

Class intervals	Midpoint	Frequency	Midpoint* Freq
17.5 – 19.5	18.5	3	55.5
19.5 – 21.5	20.5	4	82
21.5 – 23.5	22.5	2	45
23.5 – 25.5	24.5	6	147
25.5 – 27.5	26.5	4	106
27.5 – 29.5	28.5	5	142.5
29.5 – 31.5	30.5	2	61
31.5 – 33.5	32.5	3	97.5
33.5 – 35.5	34.5	1	34.5
		$\Sigma 30$	$\Sigma 771$

$$\text{Mean} = \frac{771}{30} = 25.7$$

Second, we calculate the moments.

Intervals	Mid	f	$(x_i - 25.7)^1 \cdot f_i$	$(x_i - 25.7)^2 \cdot f_i$	$(x_i - 25.7)^3 \cdot f_i$	$(x_i - 25.7)^4 \cdot f_i$
17.5 – 19.5	18.5	3	−21.6	155.52	−1119.744	8062.1568
19.5 – 21.5	20.5	4	−20.8	108.16	−562.432	2924.6464
21.5 – 23.5	22.5	2	−6.4	20.48	−65.536	209.7152
23.5 – 25.5	24.5	6	−7.2	8.64	−10.368	12.4416
25.5 – 27.5	26.5	4	3.2	2.56	2.048	1.6384
27.5 – 29.5	28.5	5	14	39.2	109.76	307.328
29.5 – 31.5	30.5	2	9.6	46.08	221.184	1061.6832
31.5 – 33.5	32.5	3	20.4	138.72	943.296	6414.4128
33.5 – 35.5	34.5	1	8.8	77.44	681.472	5996.9536
			0	596.8	199.68	24,990.976

a) The first moment about the mean $= 0$

b) The second moment about the mean $= \dfrac{596.8}{30} = 19.89333$

c) The third moment about the mean $= \dfrac{199.68}{30} = 6.656$

d) The fourth moment about the mean $= \dfrac{24990.976}{30} = 833.03253$

Hence skewness $= \dfrac{E(X-\mu)^3}{\left(E(X-\mu)^2\right)^{3/2.}} = \dfrac{6.656}{(19.89333)^{3/2.}} = 0.075$

Kurtosis $= \dfrac{E(X-\mu)^4}{\left(E(X-\mu)^2\right)^2} = \dfrac{833.03253}{(19.89333)^2} = 2.10497$

6.4.3.17 Example

Class limits	Frequency
9 – 12	7
12 – 15	46
15 – 18	54
18 – 21	64
21 – 24	40
24 – 27	50
27 – 30	41
30 – 33	38
33 – 36	21
36 – 39	20
39 – 42	5
42 – 45	5
45 – 48	1

Class intervals	Midpoint	Frequency	Midpoint* Freq
9 – 12	10.5	7	73.5
12 – 15	13.5	46	621
15 – 18	16.5	54	891
18 – 21	19.5	64	1248
21 – 24	22.5	40	900
24 – 27	25.5	50	1275

27 – 30	28.5	41	1168.5
30–33	31.5	38	1197
33–36	34.5	21	724.5
36–39	37.5	20	750
39–42	40.5	5	202.5
42–45	43.5	5	217.5
45–48	46.5	I	46.5
		£ 392	£ 9315

Mean $\bar{x} = \dfrac{9315}{392}$

Second, we calculate the moments.

Intervals	Mid	f	$(x_i - \bar{x})^1 . f_i$	$(x_i - \bar{x})^2 . f_i$	$(x_i - \bar{x})^3 . f_i$	$(x_i - \bar{x})^4 . f_i$
9 – 12	10.5	7	-92.8393	1231.305	−16,330.5	216,587.3
12 – 15	13.5	46	-472.087	4844.911	−49,722.1	510,286
15 – 18	16.5	54	-392.189	2848.371	−20,687	150,244.8
18 – 21	19.5	64	-272.816	1162.949	−4957.37	21,132.04
21 – 24	22.5	40	-50.5102	63.78202	−80.5411	101.7036
24 – 27	25.5	50	86.86224	150.901	262.152	455.4222
27 – 30	28.5	41	194.227	920.1011	4358.744	20648.44
30 – 33	31.5	38	294.0153	2274.868	17601.21	136184.9
33 – 36	34.5	21	225.4821	2421.057	25995.48	279119.9
36 – 39	37.5	20	274.7449	3774.238	51847.63	712243.6
39 – 42	40.5	5	83.68622	1400.677	23443.47	392379.1
42 – 45	43.5	5	98.68622	1947.794	38444.09	758780.4
45 – 48	46.5	I	22.73724	516.9823	11754.75	267270.7
				23557.94	81929.985	3465434.371

The first moment about the mean $= 0$

The second moment about the mean $= \dfrac{23557.94}{392} = 60.09678$

The third moment about the mean $= \dfrac{81929.985}{392} = 209$

The fourth moment about the mean $= \dfrac{3465434.371}{392} = 8840.3563$

Hence skewness $= \dfrac{E(X-\mu)^3}{\left(E(X-\mu)^2\right)^{3/2.}} = \dfrac{209}{(60.09678)^{3/2}} = 0.44861$

Kurtosis $= \dfrac{E(X-\mu)^4}{\left(E(X-\mu)^2\right)^2} = \dfrac{8840.3563}{(60.09678)^2} = 2.44775$

6.4.3.18 *Calculations in MATLAB*

1) To calculate the moment, we use the function:

```
m = moment(Data, order)
```

This returns the central moment of Data for the order specified by order.
   Therefore to calculate the four moments:

```
m1 = moment(Data, 1)
m2 = moment(Data, 2)
m3 = moment(Data, 3)
m4 = moment(Data, 4)
```

Let's calculate the four moments for all four sample datasets.

```
Velocity = [2.9, 3.5, 4.4, 1.5, 5.2];
MV1 = moment(Velocity, 1);
MV2 = moment(Velocity, 2);
MV3 = moment(Velocity, 3);
MV4 = moment(Velocity, 4);

COF = [0.21, 0.28, 0.26, 0.33, 0.16, 0.52, 0.48, 0.19, 0.22, 0.35];
MCOF1 = moment(COF, 1);
MCOF2 = moment(COF, 2);
MCOF3 = moment(COF, 3);
MCOF4 = moment(COF, 4);

Pressure = [34.2, 33.8, 31.7, 32.9, 34.5, 43.2, 33.5, 32.1, 38.8, 32.4, 34.9, 36.2];
MP1 = moment(Pressure, 1);
MP2 = moment(Pressure, 2);
MP3 = moment(Pressure, 3);
MP4 = moment(Pressure, 4);

Temperature = [29.7, 27.4, 30.1, 37.8, 30.8, 28.6, 45.9];
MT1 = moment(Temperature, 1);
MT2 = moment(Temperature, 2);
MT3 = moment(Temperature, 3);
MT4 = moment(Temperature, 4);
```

```
MV1 =                MCOF1 =              MP1 =               MT1 =

    0                    0                    0                   0


MV2 =                MCOF2 =              MP2 =               MT2 =

    1.6120               0.0132               9.8258              37.7486


MV3 =                MCOF3 =              MP3 =               MT3 =

   -0.5148               0.0011              47.3470             286.3980


MV4 =                MCOF4 =              MP4 =               MT4 =

    5.0276               4.0392e-04          443.2378            4.3686e+03
```

2) To calculate the skewness, we use the following syntax:

```
y = skewness(Data, flag)
```

Recall that flag specifies whether to correct for bias (flag = 0) or not (flag = 1, the default). When our data represents a sample from a population, the skewness is biased, meaning it tends to differ from the population skewness by a systematic amount based on the sample size. We set flag to 0 to correct for this systematic bias.

```
Velocity = [2.9, 3.5, 4.4, 1.5, 5.2];
Sample_V = skewness(Velocity, 0);
Population_V = skewness(Velocity, 1);

COF = [0.21, 0.28, 0.26, 0.33, 0.16, 0.52, 0.48, 0.19, 0.22, 0.35];
Sample_COF = skewness(COF, 0);
Population_COF = skewness(COF, 1);

Pressure = [34.2, 33.8, 31.7, 32.9, 34.5, 43.2, 33.5, 32.1, 38.8, 32.4, 34.9, 36.2];
Sample_P = skewness(Pressure, 0);
Population_P = skewness(Pressure, 1);

Temperature = [29.7, 27.4, 30.1, 37.8, 30.8, 28.6, 45.9];
Sample_T = skewness(Temperature, 0);
Population_T = skewness(Temperature, 1);
```

Sample_V =	Sample_COF =	Sample_P =	Sample_T =
-0.3750	0.8752	1.7661	1.6006

Population_V =	Population_COF =	Population_P =	Population_T =
-0.2515	0.7381	1.5372	1.2349

3) To calculate the kurtosis, we use

```
y = kurtosis(Data, flag)
```

Just like skewness, kurtosis too uses flag to specify whether to correct for bias (flag = 0) or not (flag = 1, the default).

```
Velocity = [2.9, 3.5, 4.4, 1.5, 5.2];
Sample_V = kurtosis(Velocity, 0);
Population_V = kurtosis(Velocity, 1);

COF = [0.21, 0.28, 0.26, 0.33, 0.16, 0.52, 0.48, 0.19, 0.22, 0.35];
Sample_COF = kurtosis(COF, 0);
Population_COF = kurtosis(COF, 1);

Pressure = [34.2, 33.8, 31.7, 32.9, 34.5, 43.2, 33.5, 32.1, 38.8, 32.4, 34.9, 36.2];
Sample_P = kurtosis(Pressure, 0);
Population_P = kurtosis(Pressure, 1);

Temperature = [29.7, 27.4, 30.1, 37.8, 30.8, 28.6, 45.9];
Sample_T = kurtosis(Temperature, 0);
Population_T = kurtosis(Temperature, 1);
```

Sample_V =	Sample_COF =	Sample_P =	Sample_T =
2.7390	2.7342	6.2611	4.9579

Population_V =	Population_COF =	Population_P =	Population_T =
1.9348	2.3042	4.5909	3.0658

6.4.3.19 MATLAB

6.4.3.19.1 For ungrouped data

6.4.3.19.1.1 Example 1

```
%Number of errors
Num_Errors = 1:6;
%Frequency
Value = repelem(Num_Errors, [2 4 3 6 3 2]);
Data = categorical(Value);
t = cell2table(tabulate(Data),'VariableNames', {'Number_Errors','Count','Percentage'});
t.Number_Errors = categorical(t.Number_Errors);
Arithmetic_mean = mean(Value);
Geometric_mean = geomean(Value);
Harmonic_mean = harmmean(Value);

t =

  6×3 table
```

Number_Errors	Count	Percentage
1	2	10
2	4	20
3	3	15
4	6	30
5	3	15
6	2	10

```
Arithmetic_mean =

    3.5000


Geometric_mean =

    3.1265


Harmoic_mean =

    2.6906
```

6.4.3.20 MATLAB

```
tblstats = grpstats(tbl, groupvars)
```

This function returns a table with group summary statistics for the variables in the table tbl, where the function determines the groups according to the grouping variables in tbl specified by groupvars.

If all variables in tbl (other than the grouping variables) are numeric or logical, then the summary statistic is the mean of each group for each variable in tbl. Otherwise, the summary statistic is the number of elements in each group. tblstats contains a row for each observed unique value or combination of values in the grouping variables.

```
1    x = readtable("autompg.csv",'VariableNamingRule','preserve');
2    Model_Year = x.ModelYear;
3    Displacement = x.displacement;
4    Horsepower = x.horsepower;
5    Weight = x.weight;
6    tbl = table(Model_Year, Displacement, Horsepower, Weight);
7    tblstats = grpstats(tbl,"Model_Year");
8
```

Command Window

```
tblstats =

  13×5 table
```

Model_Year	GroupCount	mean_Displacement	mean_Horsepower	mean_Weight
70	70	29	281.41	147.83	3372.8
71	71	27	213.89	107.04	3030.6
72	72	28	218.38	120.18	3237.7
73	73	40	256.88	130.47	3419
74	74	26	170.65	94.231	2878
75	75	30	205.53	101.07	3176.8
76	76	34	197.79	101.12	3078.7
77	77	28	191.39	105.07	2997.4
78	78	36	177.81	99.694	2861.8
79	79	29	206.69	101.21	3055.3
80	80	27	116.07	77.481	2441.6
81	81	28	136.57	81.036	2530.2
82	82	30	128.13	81.467	2434.2

```
1    x = readtable("autompg.csv",'VariableNamingRule','preserve');
2    Model_Year = x.ModelYear;
3    Displacement = x.displacement;
4    Horsepower = x.horsepower;
5    Weight = x.weight;
6    Num_Cylinders = x.cylinders;
7    tbl = table(Model_Year, Num_Cylinders, Displacement, Horsepower, Weight);
8    tblstats2 = grpstats(tbl,["Model_Year","Num_Cylinders"]);
```

	Model_Year	Num_Cylinders	GroupCount	mean_Displacement	mean_Horsepower	mean_Weight
70_4	70	4	7	107	87.714	2292.6
70_6	70	6	4	199	91.75	2710.5
70_8	70	8	18	367.56	183.67	3940.1
71_4	71	4	12	102.17	77.583	2057.2
71_6	71	6	8	243.38	98.875	3171.9
71_8	71	8	7	371.71	166.86	4537.7
72_3	72	3	1	70	97	2330
72_4	72	4	14	111.54	85.143	2382.6
72_8	72	8	13	344.85	159.69	4228.4
73_3	73	3	1	70	90	2124
73_4	73	4	11	109.27	82.909	2338.1
73_6	73	6	8	212.25	102.12	2917.1
73_8	73	8	20	365.25	170	4279.1
74_4	74	4	15	96.533	74	2151.5
74_6	74	6	6	235.5	101.67	3394.2
74_8	74	8	5	315.2	146	4438.4
75_4	75	4	12	114.83	84.917	2489.2
75_6	75	6	12	233.75	96.75	3398.3
75_8	75	8	6	330.5	142	4108.8
76_4	76	4	15	106.33	75.6	2306.6
76_6	76	6	10	221.4	98.7	3349.6
76_8	76	8	9	324	146.33	4064.7
77_3	77	3	1	80	110	2720
77_4	77	4	14	106.5	78.786	2205.1
77_6	77	6	5	220.4	102	3383
77_8	77	8	8	335.75	152.38	4177.5
78_4	78	4	17	112.12	79.706	2296.8
78_5	78	5	1	131	103	2830
78_6	78	6	12	213.25	109.83	3314.2
78_8	78	8	6	300.83	135.5	3563.3
79_4	79	4	12	113.58	75.75	2357.6
79_5	79	5	1	183	77	3530
79_6	79	6	6	205.67	105	3025.8
79_8	79	8	10	321.4	131.9	3862.9
80_3	80	3	1	70	100	2420
80_4	80	4	23	110.87	74.043	2359.2
80_5	80	5	1	121	67	2950
80_6	80	6	2	196.5	111	3145.5
81_4	81	4	20	109.3	72.95	2273.2
81_6	81	6	7	184	100.71	3093.6
81_8	81	8	1	350	105	3725
82_4	82	4	27	117.37	79.148	2378.9
82_6	82	6	3	225	102.33	2931.7

```
tblstats = grpstats(tbl, groupvars, whichstats)
```

This function specifies the summary statistic types whichstats.

Statistic	Description
"mean"	Mean
"sem"	Standard error of the mean
"std"	Standard deviation
"var"	Variance
"min"	Minimum
"max"	Maximum
"range"	Range
"meanci"	95% confidence interval for the mean. You can specify different significance levels using the Alpha name-value argument
"predci"	95% prediction interval for a new observation. You can specify different significance levels using the Alpha name-value argument

```
7       tblstats1 = grpstats(tbl,"Model_Year", ["min","max"], "DataVars","Weight");
8
```

Command Window

```
tblstats1 =

  13×4 table
```

	Model_Year	GroupCount	min_Weight	max_Weight
70	70	29	1835	4732
71	71	27	1613	5140
72	72	28	2100	4633
73	73	40	1867	4997
74	74	26	1649	4699
75	75	30	1795	4668
76	76	34	1795	4380
77	77	28	1825	4335
78	78	36	1800	4080
79	79	29	1915	4360
80	80	27	1845	3381
81	81	28	1755	3725
82	82	30	1965	3015

```
1    x = readtable("autompg.csv",'VariableNamingRule','preserve');
2    Model_Year = x.ModelYear;
3    Displacement = x.displacement;
4    Horsepower = x.horsepower;
5    Weight = x.weight;
6    Num_Cylinders = x.cylinders;
7    tbl = table(Model_Year, Num_Cylinders, Displacement, Horsepower, Weight);
8    tblstats2 = grpstats(tbl, "Num_Cylinders", ["mean", "min", "max", "std"], "DataVars","Horsepower");
9
```

Command Window

```
tblstats2 =

  5×6 table
```

	Num_Cylinders	GroupCount	mean_Horsepower	min_Horsepower	max_Horsepower	std_Horsepower
3	3	4	99.25	90	110	8.3016
4	4	199	78.281	46	115	14.523
5	5	3	82.333	67	103	18.583
6	6	83	101.51	72	165	14.31
8	8	103	158.3	90	230	28.454

Descriptive statistics and inferential statistics, or what you're doing with the data, vary considerably. Let's take a look to learn more about the two terms.

Descriptive analysis	Inferential analysis
It is focused on providing a description of the population being studied.	It is concentrated on inferring information about the community from sample analysis and findings.
To summarize the sample, it describes the data previously known.	It makes an effort to draw a sample from the population that goes beyond the evidence at hand.
Descriptive statistics is limited to the sample set only. It helps in visualizing and summarizing the current sample set.	Inferential statistics helps in analyzing the current sample set and helps in predicting or making conclusions about the whole population.
	The sample should be a representative of the whole population, otherwise the conclusions/inferences will not be effective enough.

Chapter 7

Inferential statistics

Before we can jump into this chapter, we need to get familiar with some concepts and terms.

A **parameter** is any quantity from a population whereas a **statistic** is any quantity from the sample of a population. Now what are the quantities we are talking about here? When we say quantities we basically mean numerical descriptive measures like mean, median, variance, standard deviation, etc. (Chapter 6). So any time we are describing the population we are talking about its parameters, and the features that define the sample data are known as **statistics**.

Now understanding these two terms is extremely important as inferential statistics attempts to estimate population parameters using its sample statistics. Understand that it is almost impossible to estimate the parameters of an entire population. Even though the most accurate results can be obtained if the entire population is considered, it is neither feasible nor practical. It takes an infinite amount of time and effort so why go to the trouble of measuring every individual in the population when just a small sample is sufficient to accurately estimate the parameters? But here is the catch. Pulling one sample from a population can never produce a statistic that is a good estimator of the corresponding population parameter.

For example, let's say there are 50 athletes on a college football team. Now we calculate the mean height of the players on the team and find it to be 70 inches. Now if we take a sample of three it is very likely that their mean height will be a lot different than the mean height of the entire class. Why? Samples are random and you can pick anyone, so in this sample if you happen to pick the three tallest team members, the chances are that the mean height from your sample will be significantly higher than the mean height of the entire team. Similarly, if you instead just happened to choose the three shortest players for your sample, your sample mean would be much lower than the actual population mean.

If that is the case, then your next question could be how do we go from our sample to the population we are interested in? Put another way, how do we ensure that the sample chosen represents the entire population?

Let's say we have a population with parameters mean μ and standard deviation σ. Now from this population we are going to take a sample of a specific size and calculate its statistics—mean and standard deviation.

Just like an individual score will differ from its mean, an individual sample mean will differ from the true population mean. This deviation in the sampled values or, in other words, the difference between the real values of the population and the values derived by using samples from the population is referred to as **sampling error**. So we can say that due to sampling error the results found in the sample do not represent the results that would be obtained from the entire population.

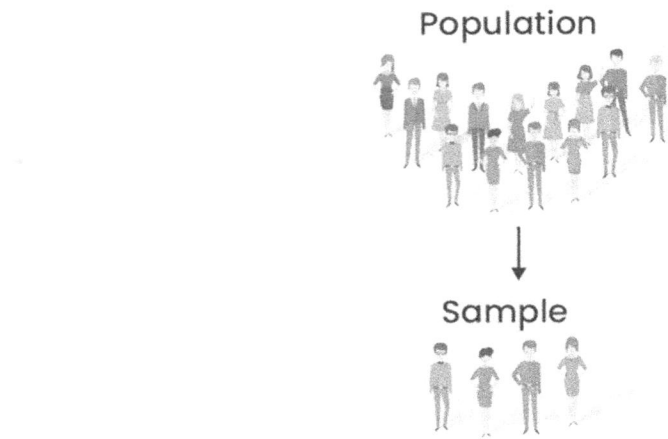

Figure 7.1

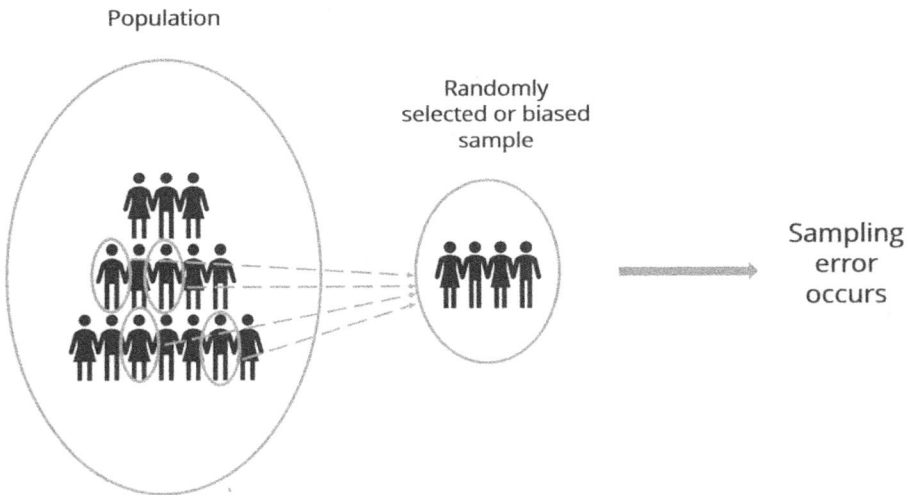

Figure 7.2

Sampling errors occur because the sample is not representative of the population.

Sampling errors occur when the numerical parameters of an entire population are derived from a sample of the entire population. Since the whole population is not included in the sample, the parameters derived from the sample differ from those of the actual population. This is because a sample is only an approximation of the population from which it is drawn. Remember our goal is to have the sample estimates as close as possible to the true value of the population parameter which means reducing sampling error is an important step toward better measurement. But how are we going to do that?

Realize we have a situation here—our first task is to ensure that our sample establishes representative results of a comparatively larger population, and the second task is to minimize the sampling error. What do we do to ensure that both of these criteria have been fulfilled?

The answer to both the **sampling distribution**.

7.1 SAMPLING DISTRIBUTION

This is a statistical method of obtaining samples from a group. A sampling distribution is a probability distribution of a statistic that results from obtaining an infinite number of random samples of the same size from a population.

So going back to the same population with parameters μ and σ, we take one sample and calculate its mean and standard deviation, denoted by $\overline{x_1}$ and s_1. Now we take another sample of the same size and do the same calculation as the first one; we will notice that this sample has a different mean and standard deviation. Let's denote the second statistics as $\overline{x_2}$ and s_2. The reason is that the samples are always chosen at random, and hence random samples selected would produce different mean values. If we keep taking a substantial number of random samples of a fixed size, let's say n, and compute the sample mean for each sample, we'd observe a broad spectrum of sample means. This list or distribution of sample means is known as the sampling distribution of the mean.

To illustrate the statement let's provide a diagram (Figure 7.3).

So basically we can conclude that one can obtain a sampling distribution by drawing many random samples of the same sample size from the same population. The idea of sampling distribution is based on sample statistics (here sample means), not on individual scores. The distribution of sample means is formed by statistics obtained by selecting all possible samples of a specific size from a population.

So how does this sampling distribution help us in estimating the parameters of the population distribution?

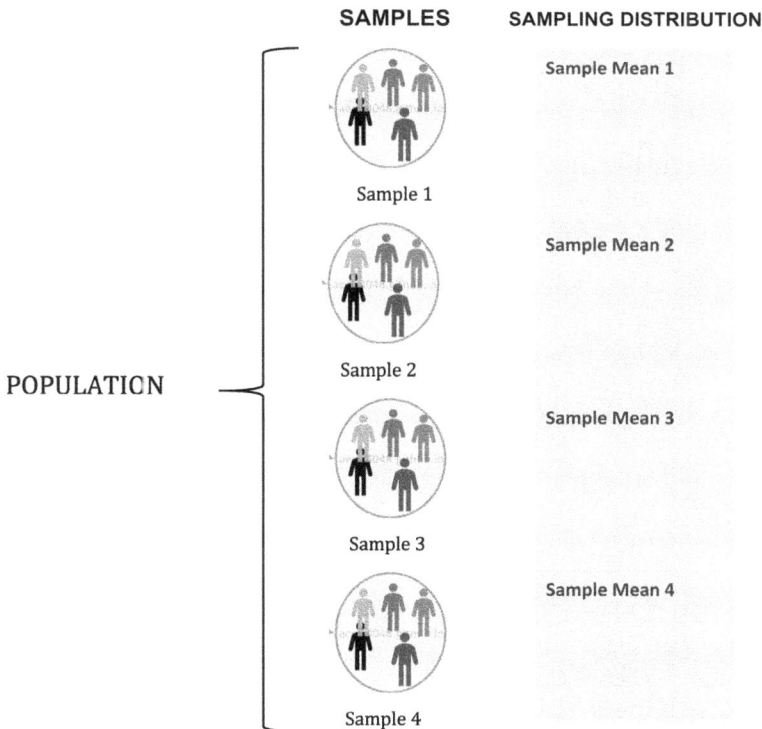

Figure 7.3

Even though this list of sample means are all different from each other if we plot them on a graph, we will notice a pattern that makes it possible to predict an estimated mean of the entire population with some degree of accuracy. How?

Let's use an example to see how that is even possible. Take a unique population of test scores. This population is very small and consists of only four scores: 82, 86, 90, and 94. Now in this example, our population is extremely small so we can easily calculate the population mean by adding all the values in the given population divided by a total number of values in the population which here is just four; therefore our population mean is computed by

$$\text{Population mean } \mu = \frac{82 + 86 + 90 + 94}{4} = 88$$

Understand that most populations are much larger than our simple example. In real life, to calculate the population mean we might even have to add up 10 trillion numbers which is close to impossible. So, the idea of the sampling distribution is to take multiple samples of a specific size and calculate the mean of each sample taken from the population.

The method of mean computation is the same in both cases. But, when the population is that large, we are better off with calculating the mean of samples. Computing the sample mean is easy. It not only saves time but it is also feasible to get an approximate value that represents the entire population.

So going back to the four-test score population example, we take a bunch of samples from this population. Each of our samples will consist of two scores. That is, the sample size is two ($n = 2$). Because this population is so small we can take every sample possible from the population. Below is a table showing all 16 possible samples of $n = 2$.

Sample #	First test score	Second test score	Mean
1	82	82	82
2	82	86	84
3	82	90	86
4	82	94	88
5	86	82	84
6	86	86	86
7	86	90	88
8	86	94	90
9	90	82	86
10	90	86	88
11	90	90	90
12	90	94	92
13	94	82	88
14	94	86	90
15	94	90	92
16	94	94	94

The far-right column shows the mean of each sample taken. From this column we can observe that there are seven possible values of the sample mean. The value 82 happens only one way (when the test score 82 is selected both times), as does the value 94, but the other values occur multiple times and hence are more likely to be observed than 82 and 94 are. Simply by counting we can easily obtain the probability distribution of 16 sample means and create their own frequency distribution which we call a **distribution of sample means or sampling distribution of sample means**.

Sample mean $\bar{x}$	82	84	86	88	90	92	94
Frequency	1	2	3	4	3	2	1
Probability $P(\bar{x})$	$\dfrac{1}{16}$	$\dfrac{2}{16}$	$\dfrac{3}{16}$	$\dfrac{4}{16}$	$\dfrac{3}{16}$	$\dfrac{2}{16}$	$\dfrac{1}{16}$

Now that we have the data let's draw a graph or a histogram representing the distribution.

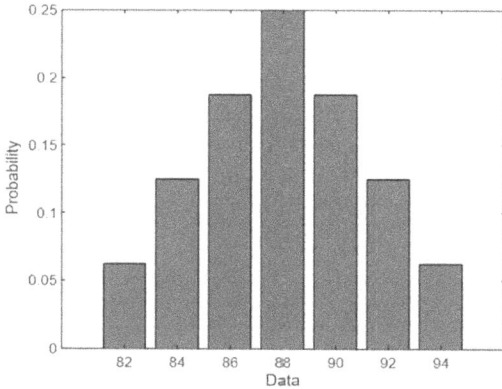

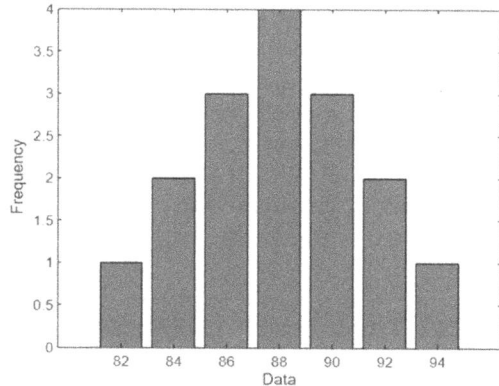

What do we notice? A bell curve. What does this bell curve remind us of? Welcome back, normal distribution! As we can clearly see, the shape of our sampling distribution is normal: a bell-shaped curve with a single peak and two tails extending symmetrically in either direction with the mean of the sampling distribution of the mean located at the center just like we saw in previous chapters.

This distribution is always normal (as long as we have enough samples; more on this later), and this normal distribution is called the **sampling distribution of the sample mean**.

Because the sampling distribution of the sample mean is normal, we can of course find a mean and standard deviation for the distribution.

The most intriguing part is that this center also happens to be the true center of the population.

In other words, the mean of a sampling distribution of the mean (located at the center) denoted by $\mu_{\bar{x}}$ is actually the mean of the population μ:

$$\mu = \mu_x = 88$$

Similarly, we can also use the sample **standard deviation** to approximate the population standard deviation:

$$\sigma = \frac{\sigma_x}{\sqrt{n}}$$

7.1.1 Example

The mean and standard deviation of the tax value of all vehicles registered in a certain state are $\mu = \$15,000$ and $\sigma = \$6200$. Suppose random samples of size 100 are drawn from the population of vehicles. What are the mean and standard deviation of the sample mean?

$$\mu_x = \mu = \$15,000$$

$$\sigma_x = \frac{\sigma}{\sqrt{n}} = \frac{6200}{\sqrt{100}}$$

Hence we have proved that the sampling distribution can estimate any parameters (in this example mean and standard deviation) without having to make thousands, millions, and trillions of calculations.

Now let's take a step back for a second and ask ourselves the question:

Is the sampling distribution of the sample mean always normal?
Guess what? It actually always is.

If the population is normal to begin with then the sample mean also has a normal distribution, regardless of the sample size.

But what if the population is not normal? In fact, in reality, most population distributions do not approximate the bell-shaped curve of a normal distribution. So if the population is not normal will the sampling distribution of the sample mean still be normal? Absolutely! This is profound! Why or how? This is the part where we introduce one of the most powerful concepts in statistics—the central limit theorem. The central limit theorem is our justification for why this is true.

7.2 CENTRAL LIMIT THEOREM

The central limit theorem states that the sampling distribution will be normal as long as the sample size of the samples is large enough. The theorem is the idea that the shape of the sampling distribution will be normalized as the sample size increases. What this means is that any sampling distribution will approach normality as n increases.

As bigger samples will create a more normal distribution, we are better able to use the techniques we developed for normal distributions and probabilities. This means if sample size n taken from a population with a given mean (μ) and standard deviation (σ) is large enough then the sampling distribution of sample means will have a mean (μ_M) equal to the population mean and a standard deviation (σ_M) equal to the standard error.

So how large is large enough? In general, a sampling distribution will be normal if the sample size is equal to or greater than 30. In other words, plotting the data that you get will result in a shape closer to a bell curve if the sample size n is ≥ 30.

This theorem is extremely powerful because it enables us to use methods developed for normal distributions even if the true population distribution is skewed.

What our discussion makes clear is the fact that the shape of the sampling distribution does not depend on the population. The central limit theorem says that no matter what the distribution of the population is, as long as the sample is "large", meaning of size 30 or more, the sampling distribution of the sample mean becomes increasingly bell-shaped, practically the same as a normal distribution centered on the population mean.

We saw an example of a sampling distribution of the mean but remember sampling distributions can describe the assortment of values for other sample statistics as well. For example, instead of the mean, medians can also be computed for each sample. The infinite number of medians would be called the sampling distribution of the median. While the sampling distribution of the mean is the most common type, the median, standard deviation, range, correlation, and test statistics can be characterized in hypothesis tests. Understanding the distribution of sample means will be enough to see how all other sampling distributions work.

The more sample groups we take, the less variability in the means for the sample groups. When the sample size increases, the standard error decreases. Therefore, the center of the sampling distribution is fairly close to the actual mean of the population.

Sampling errors can be prevented if the analysts select subsets or samples of data to represent the whole population effectively. Sampling errors are affected by factors such as the size and design of the sample, population variability, and sampling fraction. To reduce the

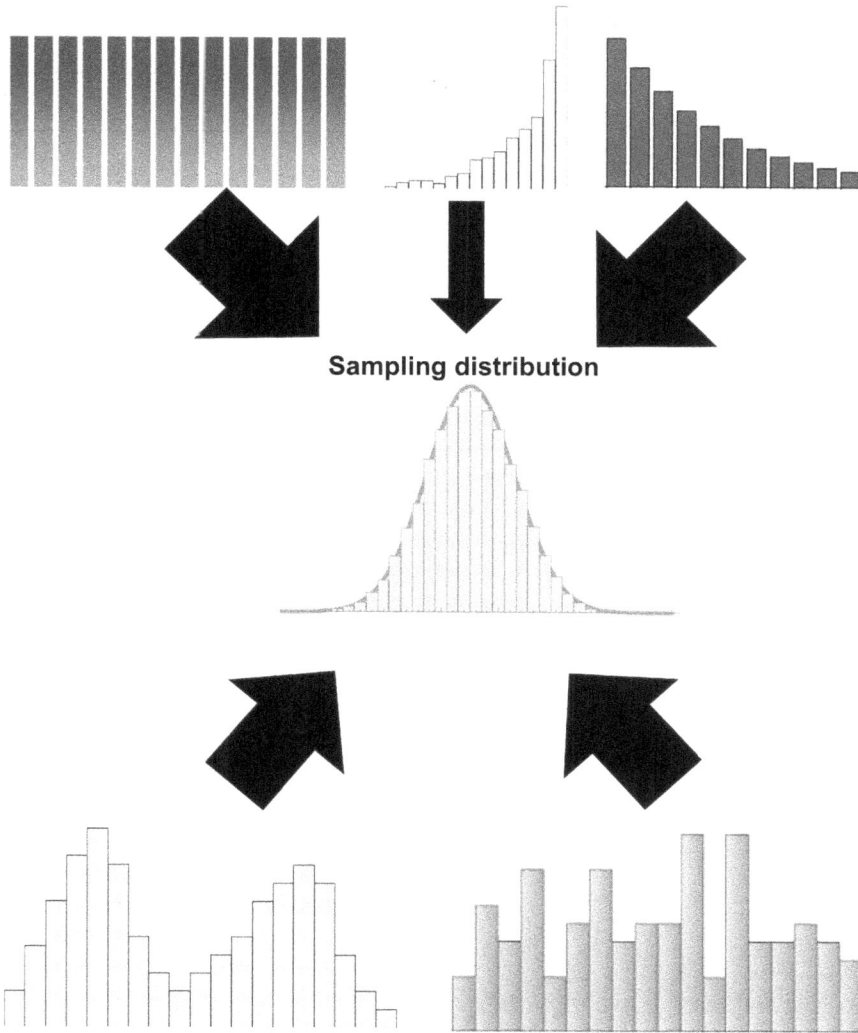

Sampling distribution

Figure 7.4

errors by half, the sample size needs to be increased by four times. The population variability causes variations in the estimates derived from different samples, leading to larger errors. The effect of population variability can be reduced by increasing the size of the samples so that these can more effectively represent the population. Now the greater the difference between the sample statistics and the population parameters, the greater the sampling error.

Understanding the relationship between sampling distributions, probability distributions, and hypothesis testing is crucial for inferential statistics.

7.3 HYPOTHESIS TESTING

The first area of statistical inference—hypothesis testing—tries to compare two populations, population 1 and population 2. We use population 1, also known as the research population, to draw conclusions or assess claims about population 2.

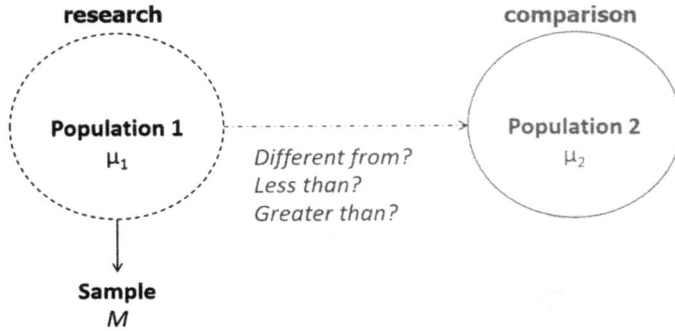

Figure 7.5

Note that we do not have access to the actual research population so instead we use its sample data, which is our best proxy for population 1, and then compare it with population 2. This is the part where we need to understand the sampling distribution because it allows us to use the sample statistics to approximate a specific population parameter. In other words, due to the sampling distribution we can use a sample to make an inference about a population.

7.3.1 Hypothesis test terminology

There are many different hypothesis tests that we shall discuss in the upcoming chapters. Even though each of them is different and makes different assumptions about the distribution, they share the same overall structure. The structure involves setting up two **competing hypotheses**, the null and alternative, generated about a population parameter, followed by conducting a statistical test of significance. So, the test is performed to determine which of the two is true. These statistical tests examine and analyze the sample taken from the population to assess the likelihood that the hypothesis is true. Let's discuss each of the steps in detail.

7.3.1.1 First step: Stating the hypothesis

Hypotheses are assumptions about one or more of the population parameters on the basis of the information obtained from the sample. The most important step is to set up the hypotheses of which only one is accepted. Now as mentioned earlier, there are two types of hypothesis used in the testing—the null hypothesis and the alternative hypothesis. These two hypotheses will always be mutually exclusive. If the null hypothesis is rejected, the statistical conclusion is that the alternative hypothesis is true and vice versa. Let's discuss each of the hypothesis individually:

a) **Null hypothesis**

Sometimes referred to simply as the "null", it is represented as H_0. It proposes that there is no statistical significance between the parameters of two or more population data points. These parameters could be the mean, median variance, proportion, etc., depending on the data.

So if we are comparing the means of two groups our null hypothesis would be:
$\mu = \mu_0$ which means the population mean equals the hypothesized mean.

b) **Alternative hypothesis**

An alternative hypothesis is a contradictory theory to that taken by a null hypothesis about a specified research parameter. In contrast to the null, the alternative hypothesis proposes that there is a difference between the groups that are being compared. It is notated as H_A. There are two main types of alternative hypotheses—one-tailed and two-tailed tests.

In a one-tailed test, the alternative hypothesis only tests one direction, so it could be either greater than or less than the value specified in the null hypothesis.

If the alternative hypothesis results are less than the value of the null, then it is called a left-tailed hypothesis.

$$\mu < \mu_0$$

Contrarily, when the value of the results is greater than the null statement, it's called right-tailed.

$$\mu > \mu_0$$

To make it more clear let's look into two representations of two different scenarios.

7.3.1.1.1 Situation 1

A manufacturer claims that a particular additive has improved the tensile strength of a particular material, which had a mean tensile strength of 50 megapascals (MPa). As an engineer, you want to test this claim and determine if the new additive has significantly increased the tensile strength.
The hypotheses would be:
  • Null hypothesis: the mean tensile strength is 50 MPa so we write

$$H_0 : \mu = 50$$

  • Alternative hypothesis: the mean tensile strength is greater than 50 MPa so

$$H_A : \mu > 50$$

7.3.1.1.2 Situation 2

A manufacturer claims that a new cooling system has reduced the average operating temperature of a machine to below 50°C. As an engineer, you want to test this claim and determine if the new cooling system has significantly decreased the operating temperature.
The hypotheses would be:

  • Null hypothesis: the mean operating temperature is 80°C or higher.

$$H_0 : \mu \geq 80$$

- Alternative hypothesis: the mean operating temperature is less than 80°C.

$H_A : \mu < 80$

Unlike one-tailed tests, two-tailed tests do not specify a direction. It is basically the exact opposite of the null hypothesis.

7.3.1.1.3 Situation 3

A manufacturer claims that a new production process has completely changed the mean diameter of circular rods of 20 millimeters (mm). As an engineer, you want to test this claim and determine if the diameter is significantly different from 20 mm.
   The hypotheses would be:

- Null hypothesis: the mean diameter is 20 mm.

$H_0 : \mu = 20$

- Alternative hypothesis: the mean diameter is different from 20 mm.

$H_A : \mu \neq 20$

The most crucial difference between one- and two-tailed tests is that one-tailed tests have only one critical or rejection region and two-tailed tests have two. We will talk more about this in the upcoming steps.

7.3.1.2 Second step: Choose the correct alpha/significance level α

In **hypothesis testing**, the **level of significance** is a measure of how likely you are to reject the null hypothesis when it is actually true. The significance level is the probability of making a type I error or rejecting the null hypothesis when it in fact is true. A common significance level is 0.05, meaning there is a 5% chance of rejecting the null hypothesis when it is true.
   The area of the critical/rejection region depends on the level of significance α.
   The critical/region is an area of a graph where you would reject the null hypothesis if your test statistics fall into that area. So any value which falls in that region is sufficient evidence to reject the null hypothesis. For example, if we use our normal criterion of $\alpha = .05$, then 5% of the area under the curve becomes the rejection region.
   Other popular choices are 0.01 (1%) and 0.1 (10%). However, the most common value is 0.05 or 5%. In fact it is also used when no other level of significance is given or mentioned. For the two-tailed test, the rejection region is present on both ends.
   If the critical/rejection region is specified by alpha/significance level then the acceptance region is determined by the *confidence interval*. The confidence interval is a set of values for the test statistic for which the null hypothesis is accepted, i.e. if the observed test statistic is in the confidence interval then we accept the null hypothesis and reject the alternative hypothesis.
   Confidence level = 1 – alpha.

7.3.1.3 Third step: Calculate the test statistic

A test statistic is a number that describes how much the research results differ from the null hypothesis. Therefore, the test statistic is a hypothesis test that helps you determine whether to support or reject a null hypothesis in your study.

The test statistic summarizes your observed data into a single number using the central tendency, variation, and sample size in your statistical model.

When we compute the statistic for a given sample, we obtain an outcome of the test statistic. In order to perform a statistical test we should know the distribution of the test statistic under the null hypothesis. This distribution depends largely on the assumptions made in the model. If the specification of the model includes the assumption of normality, then the appropriate statistical distribution is the normal distribution or any of the distributions associated with it, such as the chi-squared, Student's t, or Snedecor's F.

7.3.1.4 Fourth step: Choosing the approach

There are basically two approaches to hypothesis testing—the probability or p-value approach and the critical value approach.

They both do the same thing: enable you to support or reject the null hypothesis in a test. But they differ in how you get to make that decision. In other words, they are two different approaches that give the same conclusion or results when testing hypotheses.

Regardless of the approach you use, the test statistics will always be needed. We either use it to compare to a critical value or to calculate the p-value.

7.3.1.4.1 Critical value approach

7.3.1.4.1.1 WHAT IS A CRITICAL VALUE?

A critical value(s) is a barrier that divides the domain of the distribution into a region where the null hypothesis will be accepted or not accepted. In hypothesis testing, the critical region/rejection region is represented by a set of values where the null hypothesis is rejected. It takes different boundary values for different levels of significance. So a critical value separates the critical/rejection region from the noncritical/acceptance region. It acts as a sort of cutoff value.

Remember that in step 1, we discussed the differences between one-tailed and two-tailed tests and their critical/rejection region. The graphical representation is shown in Figure 7.6.

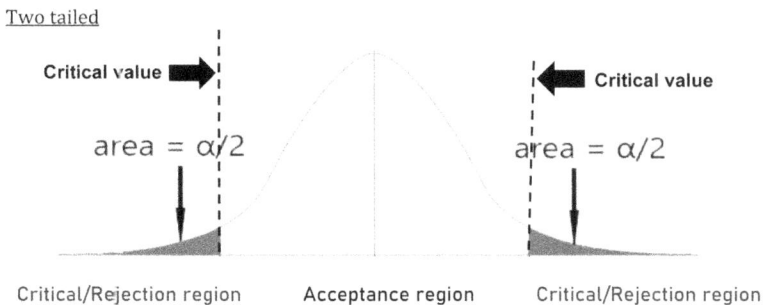

Figure 7.6

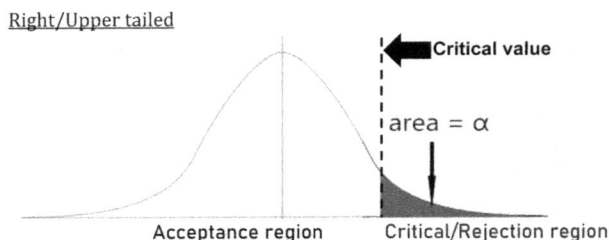

Figure 7.7

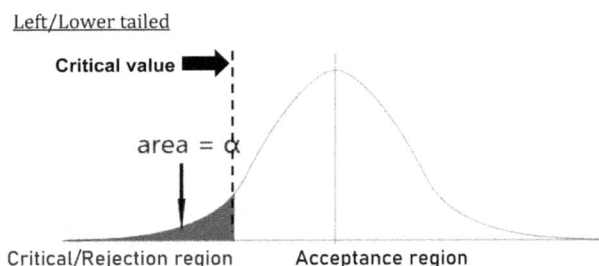

Figure 7.8

TWO-TAILED

Here the shaded region is called the critical region or the rejection region and as we can see for the one-tailed test (left or right) there is only one critical region but in the two-tailed test there are two. For a two-tailed test, there are two critical values. The absolute values for both critical values are same, they only differ by sign.

The shaded area under the curve is equal to the level of significance/alpha or α (from step 2). Moreover, the critical value splits the graph into the acceptance region and the rejection region for hypothesis testing.

7.3.1.4.1.2 HOW DO WE CALCULATE THE CRITICAL VALUE?

In this step we formally lay out the criteria we will use to calculate our critical value. From step 3 onwards basically we will see the differences in calculating critical values for different tests used for different data. Regardless of the test applied, a critical value is derived based on

- The level of significance
- The directionality of the test—whether it is two-tailed or one-tailed

The critical values are always tabulated and thus obtained from the corresponding distribution table. This means the values can be attained straight from the standard normal table in the case of a large sample or from the t-distribution table in the case of a small sample.

7.3.1.4.1.3 NEXT HOW DO WE USE OUR CRITICAL VALUE IN THE HYPOTHESIS TESTING?

For the critical value approach, you need to compute the test statistic and find the critical value to test the hypothesis.

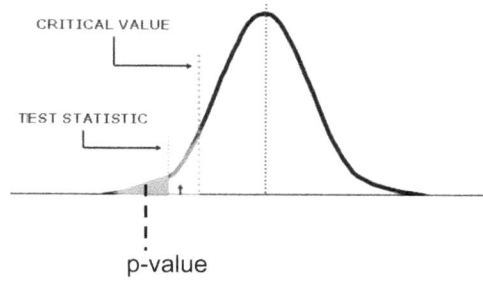

Figure 7.9

7.3.1.4.2 Probability or p-value approach

7.3.1.4.2.1 WHAT IS A P-VALUE?

The second approach to hypothesis testing is the probability value or p-value approach. Just like alpha level, the p-value is also a probability, but while the alpha level defines the probability of incorrectly rejecting a true null hypothesis (Type 1 error), a p-value measures the probability of obtaining the observed results, assuming that the null hypothesis is true. The p-value measures the probability of getting a more extreme value than the one you got from the experiment. It measures the probability of observing a test statistic at least as large as the one observed, by random chance, assuming that the null hypothesis is true.

Now recall that the alpha level is the area under the curve to the right of the positive critical value for the right-tailed test and to the left of the negative critical value for left-tailed test or both right and left in the case of the two-tailed test (remember, for the two-tailed test, the total alpha level is divided by the two sides). The p-value is the area under the curve to the right or to the left of the test statistic value. If shown graphically for left-tailed (Figure 7.9)

7.3.1.4.2.2 HOW DO WE CALCULATE THE P- VALUE?

To calculate the p-value we follow the steps stated below:

- Determine the test statistic.
- Find the probability or p-value from its corresponding test statistic.

7.3.1.4.2.3 FINALLY, HOW DO WE USE OUR P-VALUE IN HYPOTHESIS TESTING?

In this method, instead of calculating a critical value you calculate a probability or p- value and reject or fail to reject the null hypothesis based on the calculation and a predetermined significance/alpha level. The p-value approach has the advantage in that you just need to compute one value, the p-value, to do the test. Because the p-value approach requires just one computation, most statistical software and calculators use the p-value approach for hypothesis testing.

7.3.1.5 Fifth step: Making the decision

We have learned two methods for making a statistical decision—the critical value and p-value methods.

Regardless of the method applied, the conclusions from the two approaches are the same.

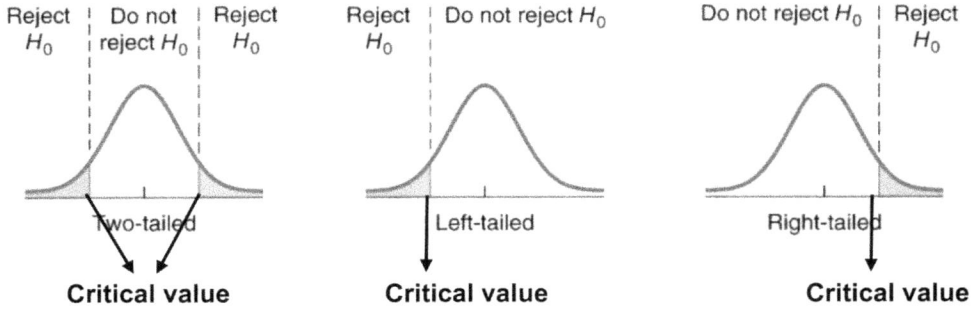

Figure 7.10

So if we use the critical approach then our decision will be based on whether the test statistic lies in the rejection region or not.

If the test statistic is more extreme than the critical value, this means that the value falls in the rejection region indicating that there is a significant difference and that the null hypothesis should be rejected. In other words, if the test statistic falls in the rejection/critical region then we reject the null hypothesis; otherwise we do not.

For the p-value approach, if the p-value is less than (or equal to) α, then the null hypothesis is rejected in favor of the alternative hypothesis.

And, if the p-value is greater than α, then the null hypothesis is not rejected.

If we want to present it in the form of a diagram

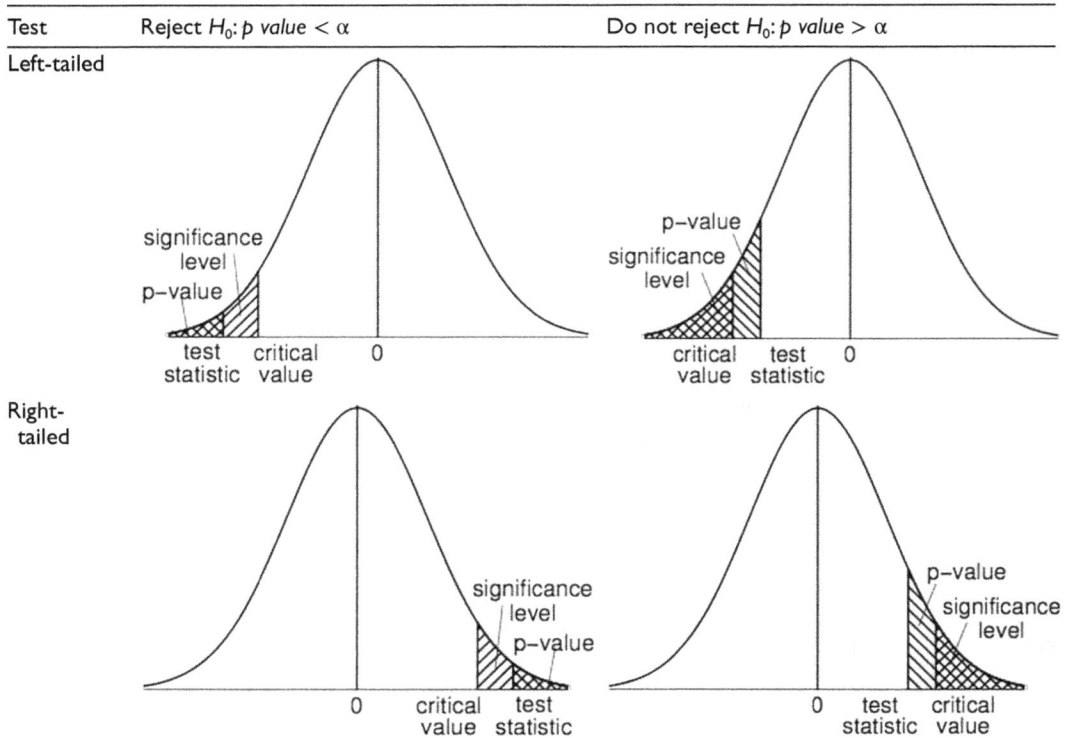

Two-tailed

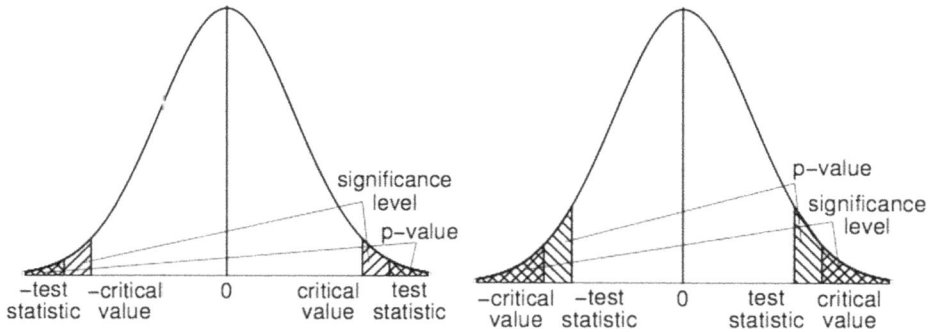

Finally our last task is to make the decision in relation to our research question, stating what we concluded, what we based our conclusion on, and the specific statistics we obtained.

7.3.1.6 Comparing critical value and p-value approaches

Critical value	p-value
Step 1: state the null and alternative hypotheses	Step 1: state the null and alternative hypotheses
Step 2: decide on the significance level α	Step 2: decide on the significance level α
Step 3: compute the test statistics	Step 3: compute the test statistics
Step 4: determine the critical value	Step 4: determine the p-value
• Significance level α	
} z-test } t-test	
• Directionality of test	
• Degrees of freedom	
Step 5: decision	Step 5: decision
Comparing the critical value with test statistics	Comparing the p-value with significance level α
Reject H_0: if the test statistic falls in the rejection region	Reject H_0: $p\ value < \alpha$
Do not reject H_0: if the test statistic falls within the acceptance region	Do not reject H_0: $p\ value > \alpha$

7.3.2 Types of tests

So far, we have discussed how statistical methods help us reject or not reject our null hypothesis, based on probability distributions, which can be one-tailed or two-tailed, depending on the hypotheses that we've chosen. Moreover, based on the assumptions they make about the underlying population distribution and the types of data they can handle, the hypothesis test falls into two broad categories, namely:

1) Parametric tests
2) Non-parametric tests

To understand this, let's take a look at these two tests in detail.

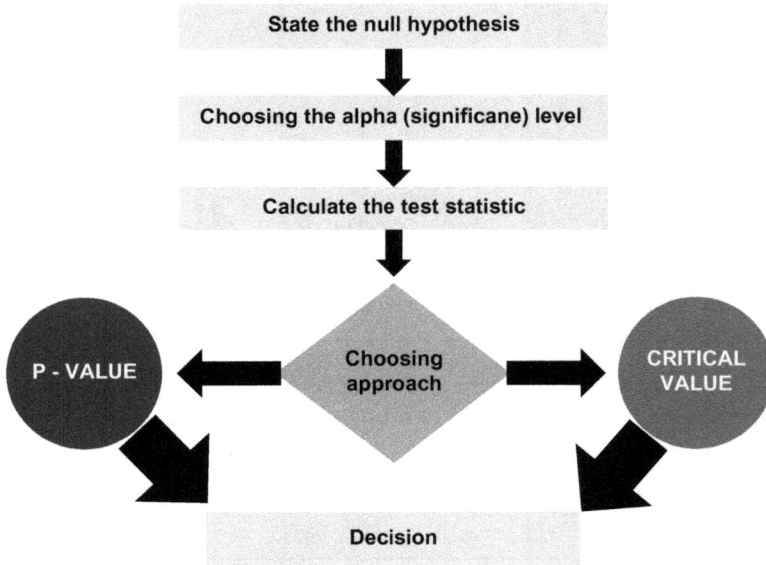

Figure 7.11

7.3.2.1 Parametric tests

Parametric hypothesis testing is the most common type of testing done to understand the characteristics of the population from a sample. While there are many parametric test types, and they have certain differences, **some properties are shared across all the tests** that make them a part of "parametric tests".

- **Level of data to be examined**

    For parametric tests, the study variable must be numerical. In these tests, the dependent variable is always quantitative and measured on a scale that is at least interval in nature. It is even better if the variable follows a ratio scale

- **Assume that your data follows a certain distribution,** such as normal, exponential, or binomial. Recall that hypothesis tests use and analyze a sample (and not an entire population) to draw conclusions about the same population the sample was taken from. Parametric tests are those that assume that this sample data comes from a population that follows normal distribution with a fixed set of parameters.

    Note: in reality no data is ever normal, and perfect parametric normal populations only exist in mathematical and simulation models. Consequently, the term **approximately normal** is often applied, even though there is no quantitative criterion as to what "approximately" might mean.

Now even though this criterion is one of the main characteristics of parametric tests, parametric analyses can still produce reliable results even when our continuous data is nonnormally distributed. This is only possible if the sample size is large enough.

- **The sample "n"**

The "n" is the size of the sample taken from the population. Theoretically, the sample size should be more than 30 so that the central limit theorem can come into effect, making the sample normally distributed even if the population is not.

So we conclude that parametric tests are those tests for which we have prior knowledge of the population distribution, in this case normal, or if not then we can easily approximate it to a normal distribution which is possible with the help of the central limit theorem.

7.3.2.2 Non-parametric tests

Like parametric tests, there are many non-parametric tests available in MATLAB. Let's look at each property of non-parametric tests that differentiates them from parametric tests.

- Unlike parametric tests, non-parametric tests are used when the levels of measures being used have no meaning, such as the color of eyes. In other words, nominal and ordinal measures require a non-parametric test in most cases.
- As one would guess, in non-parametric tests, we don't make any assumption about the parameters for the given population or the population we are studying. In fact, these tests don't depend on the population. Hence, there is no fixed set of parameters available, and also there is no distribution (normal distribution, etc.) of any kind available for use. This is also the reason that non-parametric tests are also referred to as **distribution-free tests**.
- Sample size
    Usually when we have a small sample size ($n<30$), and the researcher has no idea regarding the population parameter, there is a higher possibility that we may be forced to use a non-parametric test. This is because with a small sample it can be difficult to tell whether the data is normal or not. For instance, the histogram may not be smooth even if the data is normal. Also, the data displayed in a histogram with a small sample is most often asymmetrical but there are occasions when there is no significant evidence of symmetry or asymmetry, and thus it is impossible to tell.

7.3.2.2.1 Note

We usually can't know a parameter with certainty, because our data represents only a sample of the population. We can, however, produce an estimate of a parameter by computing the corresponding statistical value based on the sample. This is why knowing the population distribution is so important.

7.4 ASSESSING THE POPULATION DISTRIBUTION

We have already learned that statistical techniques rely heavily on observations from a population that follows a normal distribution. But how do we test that? There are several methods to assess whether data is normally distributed, and they fall under two broad categories: graphical and analytical. We have already discussed the different types of graphical representation in Chapter 3 but now we shall discuss how we can assess normality from these graphs.

7.4.1 Visual methods

The graphical test for normality is a visual method of assessing normality. This method does not always provide accurate results since it is solely based on one's judgment. Thus, this method is unreliable and does not guarantee the existence of normal distribution for a variable. The normality of data can be graphically tested in a number of ways. Let's discuss each of them.

7.4.1.1 Histogram

The first method that almost everyone thinks of to assess normality is the histogram. If sample data is normally distributed, its distribution curve approximately matches the normal distribution curve. In other words, the frequency distribution should be bell-shaped. Let's look at some examples and determine if they follow a normal distribution.

7.4.1.1.1 Example 1

We have a dataset of MPG values so let's create a histogram to visualize the distribution of those values. The histogram will show the frequency or relative frequency of MPG values in different bins.

```
data = readtable("autompg.csv", VariableNamingRule = "preserve");
MPG = data.MPG;
histogram(MPG)
xlabel("MPG")
ylabel("Frequency")
title("Histogram of MPG values")
```

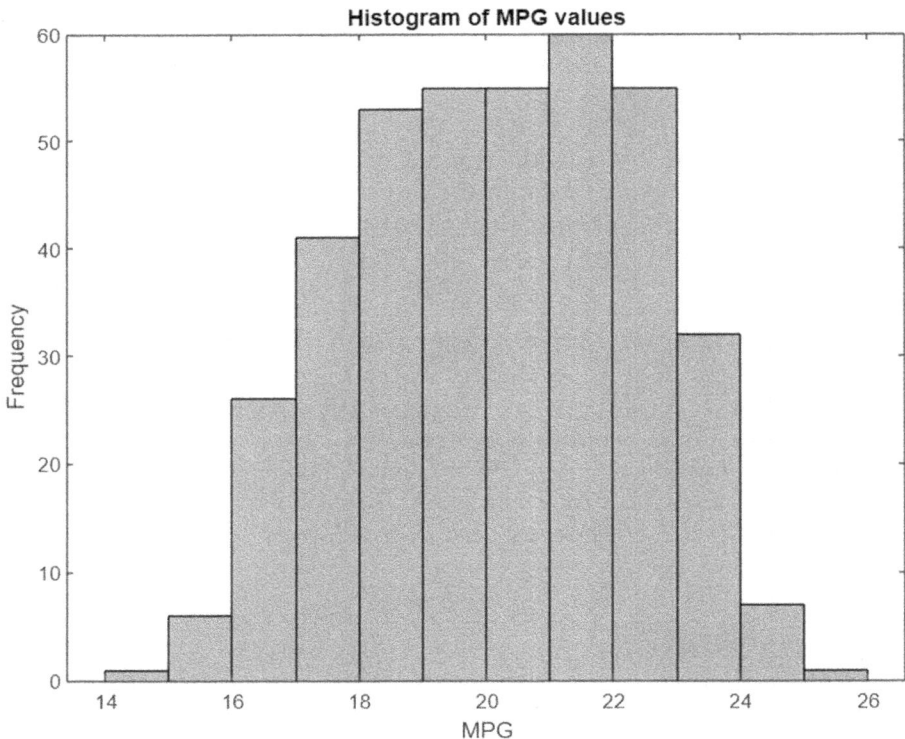

Figure 7.12

The histogram in Figure 7.12 almost looks perfectly symmetrical and bell-shaped and has no violation of normality.

7.4.1.1.2 Example 2

Let's now create a histogram to visualize the distribution of horsepower values.

```
data = readtable("autompg.csv", VariableNamingRule = "preserve");
Horsepower = data.horsepower;
histogram(Horsepower)
xlabel("Horsepower")
ylabel("Frequency")
title("Histogram of Horsepower values")
```

There is a clear indication that the data in Figure 7.13 is right-skewed with some strong outliers. The assumption of normality is clearly violated.

7.4.1.1.3 Example 3

```
Data = readtable("Capacitance.csv", VariableNamingRule = "preserve");
Capacitance = Data.Capacitance;
histogram(Capacitance)
xlabel("Capacitance (×10^-14)")
ylabel("Frequency")
title("Histogram of Capacitance values")
```

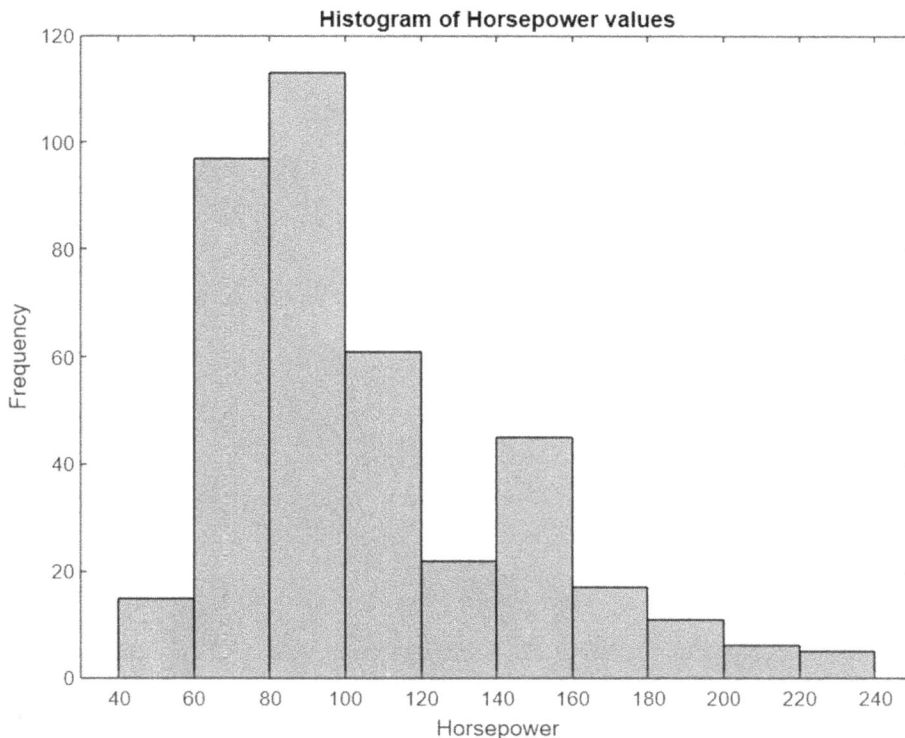

Figure 7.13

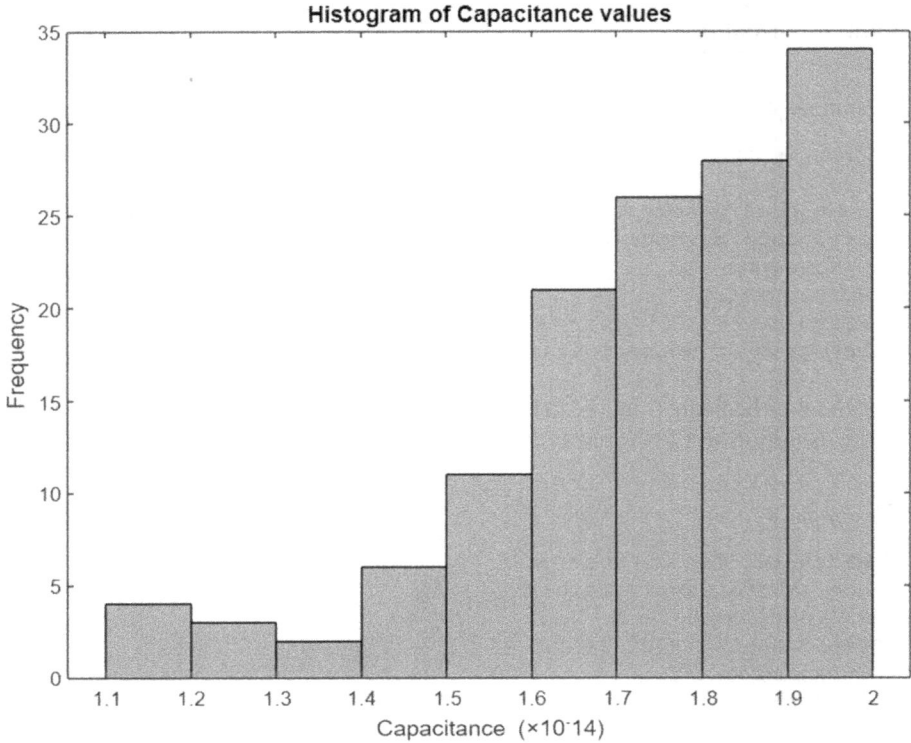

Figure 7.14

As we can see in Figure 7.14 the data is left skewed so the assumption of normality is also violated.

7.4.1.1.4 Example 4

Let's look at another example of strength values.
```
Data = readtable("concrete_data.csv", VariableNamingRule = "preserve");
Strength = Data.Strength;
histogram(Strength)
xlabel("Tensile Strength")
ylabel("Frequency")
title("Histogram of Tensile Strength values")
```

Although the histogram in Figure 7.15 is not perfectly symmetric and bell-shaped as in Example 1, there is no clear violation of normality.

7.4.1.1.5 Disadvantages

Even though *histograms* might seem to be the best graph for assessing normality. they do have some limitations and potential disadvantages:

1. **Bin selection**: the shape of the histogram can be influenced by the choice of bin width. Selecting an appropriate bin width can be subjective. Small changes can alter the visual impression and impact the interpretation of the histogram quite drastically.

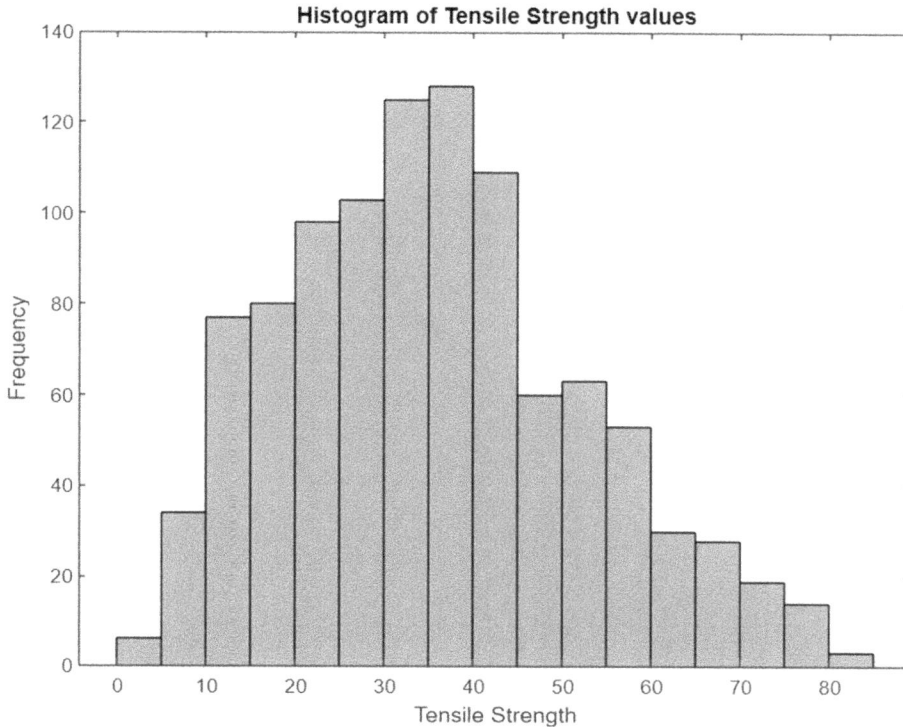

Figure 7.15

2. **Sample size:** the appearance of a histogram can be largely affected by the sample size. Histograms are not useful for small sample sizes, as they may not accurately reflect the true shape of the distribution. As the sample size increases, the histogram tends to provide a more reliable representation.
3. **Visual interpretation:** the interpretation of a histogram can be subjective. Different individuals may interpret the same histogram differently. If we look at the histogram for tensile strength, its non-symmetricity can led to different interpretations of the histogram. In such cases is important to consider multiple graphical methods and statistical tests to make a more informed assessment of normality.

7.4.1.2 Q-Q plot

Because of the limitations, many prefers the normal Q-Q plot over histograms to assess normality. A Q-Q plot compares the theoretical quantiles that the data should have if it was perfectly normal with the quantiles of the sample data values. In other words, if the sample data is normally distributed, its values should fall on the expected normal distribution line. This means that the further the sample data deviates from the line, the less normally distributed it is. Let's look at some examples and determine if they follow a normal distribution.

7.4.1.2.1 Example 1

```
Data = readtable("Resistance.csv", VariableNamingRule = "preserve")
Resistance = Data. Resistance;
qqplot(Resistance)
```

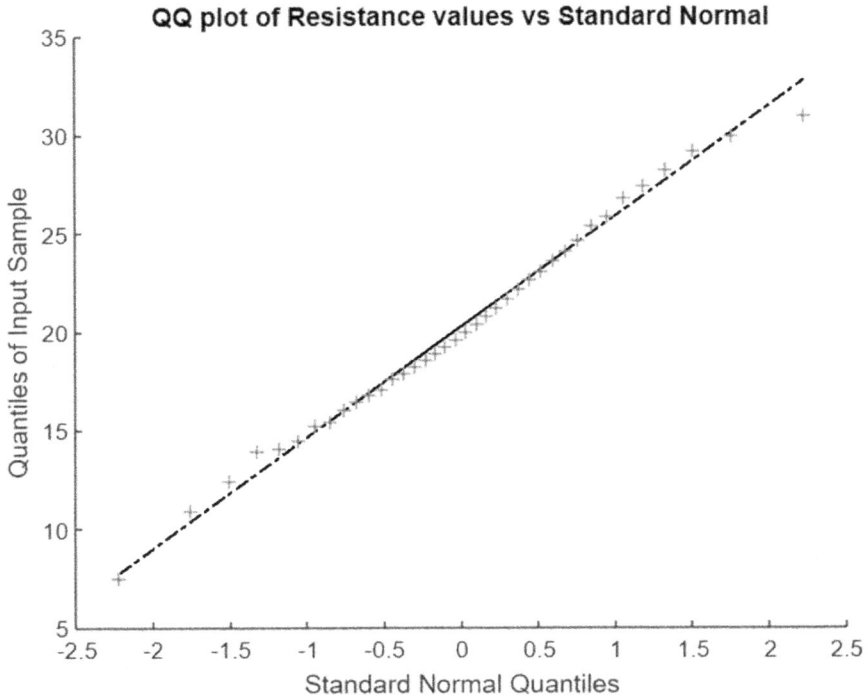

Figure 7.16

The observations fall very closely, almost perfectly, on the line. There is almost no indication of non-normality.

7.4.1.2.2 Example 2

```
Data = readtable("Temperature.csv", VariableNamingRule = "preserve")
Temperature = Data. Temperature;
qqplot(Temperature)
title("QQ plot of Temperature vs Standard Normal")
```

There is a clear indication that the data does not follow the line. The Q-Q plot curves away from the straight line at both ends, suggesting a departure from normality.

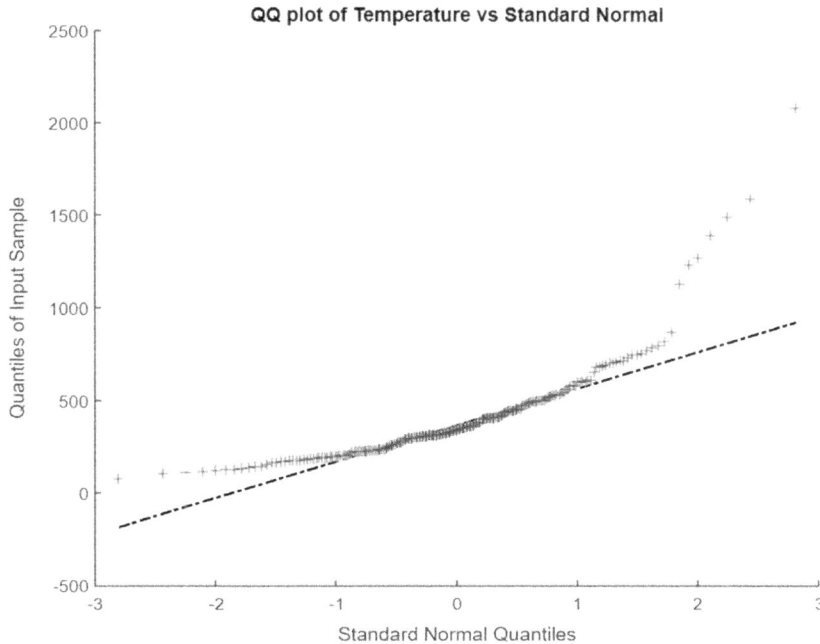

QQ plot of Temperature vs Standard Normal

Figure 7.17

7.4.1.3 Boxplot

Boxplots can act as another graphical display test to see if our data is normally distributed. We have already discussed the structure of a boxplot where the center line of the box represents the median of the sample data. **Data which follows a normal distribution** shows the median roughly in the middle of the box. Therefore if a distribution is normal, we would expect to see the line (representing the median) roughly in the middle of the box. In contrast, a skewed distribution produces an uneven boxplot, with one side of the box longer or wider than the other. For a distribution that is positively skewed, the boxplot will show the median closer to the lower or bottom quartile. For a distribution that is negatively skewed, the boxplot will show the median closer to the upper or top quartile.

7.4.1.3.1 Example 1

```
Data = readtable("Voltage.csv", VariableNamingRule = "preserve")
Voltage = Data. Voltage;
boxplot(Voltage)
ylabel("Voltage")
title("Boxplot of Voltage")
```

7.4.1.3.2 Example 2

```
Data = readtable("Diameters.csv", VariableNamingRule = "preserve")
Diameter = Data.A;
boxplot(Diameter)
ylabel("Diameter(mm)")
title("Dimaters of cast-iron pipe")
```

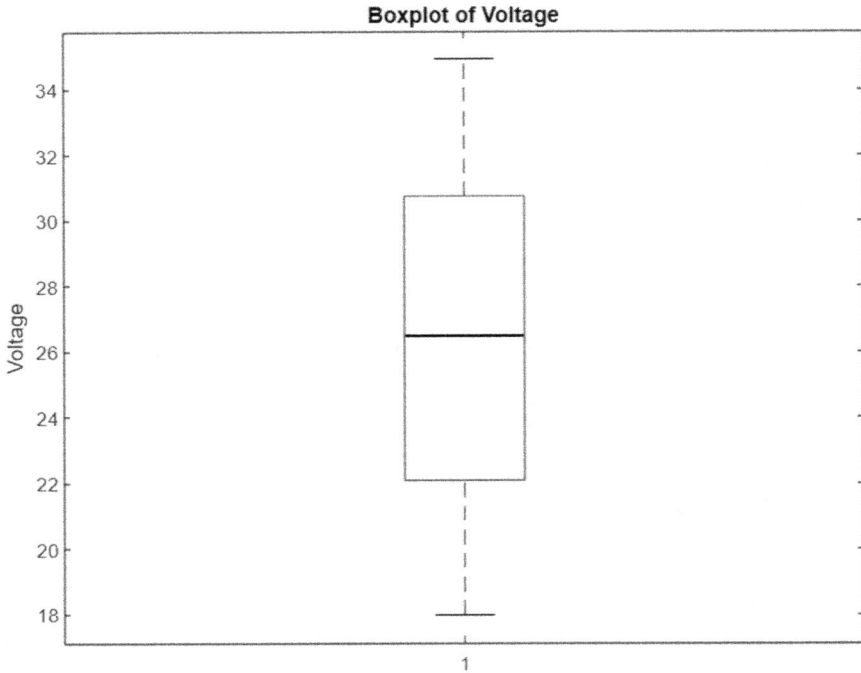

Figure 7.18

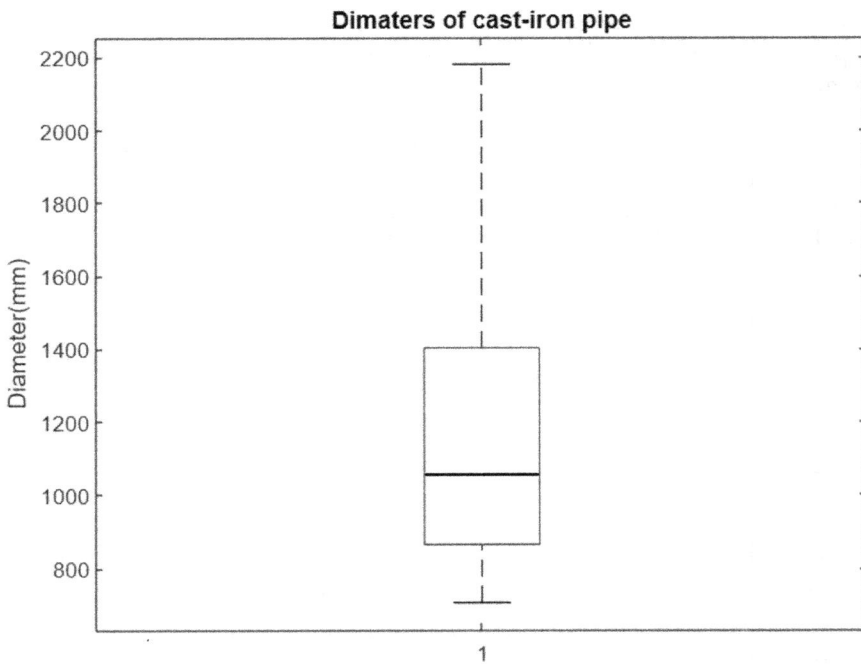

Figure 7.19

The top whisker far exceeds the bottom whisker and the line is gravitating toward the bottom of the box indicating that the distribution is skewed right (Figure 7.19).

7.4.2 Statistical tests

The normality tests are supplementary to the graphical assessment of normality. In this section we shall discuss some statistical tests that can be used to check whether our sample data is likely to be from a normal distribution. We have already mentioned earlier that all statistical tests have a similar structure which means they involve hypotheses, test statistics, critical or p-values, and then finally their comparison to get the conclusion. If that is the case, we can simply state that our hypotheses for checking the normality of the population the sample is taken from are:

- Null hypothesis: sample data come from the normal distribution.
- Alternative hypothesis: sample data **do not** come from the normal distribution.

7.4.2.1 Specifying the alpha level

As mentioned earlier, one is free to choose any alpha level, although 5% is the most commonly used.

Now even though the underlying hypotheses are the same, different statistical tests will have slightly different ways of calculating these test statistics. So it is important to decide which statistical tests to perform to determine if the sample comes from a normal distribution.

7.5 STATISTICAL TESTS FOR NORMALITY

MATLAB provides several normality tests that you can use to assess whether a dataset follows a normal distribution. These tests help you determine whether your data deviates significantly from normality. Here's a list of some common normality tests available in MATLAB:

1) **Chi-squared goodness-of-fit test**: tests if a sample comes from a specified distribution, against the alternative that it does not come from that distribution.
2) **Anderson–Darling**: just like Chi-square, this statistical test is also used to check if a sample of data came from a population with a specific distribution.
3) **One-sample Kolmogorov–Smirnov test**: tests if a sample comes from a continuous distribution with specified parameters, against the alternative that it does not come from that distribution.
4) **Lilliefors test**: tests if a sample comes from a distribution in the normal family, against the alternative that it does not come from a normal distribution.
5) **Jarque–Bera test**: tests if a sample comes from a normal distribution with unknown mean and variance, against the alternative that it does not come from a normal distribution.

7.5.1 The chi-squared goodness of fit

Before we get into the hypothesis testing, let's first talk about the distribution.

When $\beta = \dfrac{df}{2}$, where df = degrees of freedom where $n-1$ and $\dfrac{1}{\alpha} = \dfrac{1}{2}$, then a gamma distribution becomes a chi-squared distribution.

The χ^2 distribution is an asymmetric curve that reaches a peak to the right then gradually declines in height. For each degree of freedom there is a different χ^2 distribution curve. This implies that the χ^2 distribution is more spread out, with a peak farther to the right, for larger than for smaller degrees of freedom. As a result, for any given level of significance, the critical region begins at a larger chi-square value, the larger the degree of freedom. As the degrees of freedom increase, the chi-squared curve approaches a normal distribution. Let's now look at the plots of the densities of some chi-squared random variables. These plots help us to understand how the shape of the chi-squared distribution changes by changing the degrees of freedom parameter.

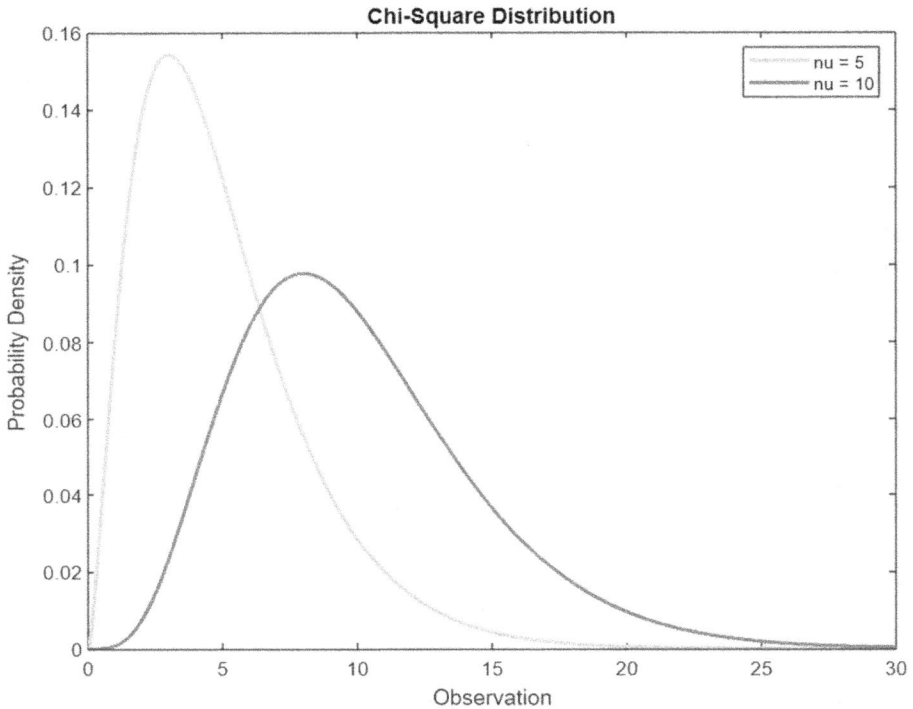

Figure 7.20

```
x = 0:0.01:30;
y1 = chi2pdf(x, 5);
y2 = chi2pdf(x, 10)
figure;
plot(x, y1, 'Color', [0.9330 0.7450 0.3010], 'LineWidth', 2)
hold on;
plot(x, y2, 'Color', [0.3010 0.6330 0.5450], 'LineWidth', 2)
hold off;
xlabel('Observation')
ylabel('Probability Density')
title("Chi-Square Distribution")
legend("nu = 5", "nu = 10")
```

```
x = 0:0.01:30;
y1 = chi2cd=(x, 5);
y2 = chi2cd=(x, 10)
figure;
plot(x, y1, 'Color', [0.9330 0.7450 0.3010], 'LineWidth', 2)
hold on;
plot(x, y2, 'Color', [0.3010 0.6330 0.5450], 'LineWidth', 2)
hold off;
xlabel('Observation')
ylabel('Cumulative Density Function')
title("Chi-Square Distribution")
legend("nu = 5", "nu = 10")
```

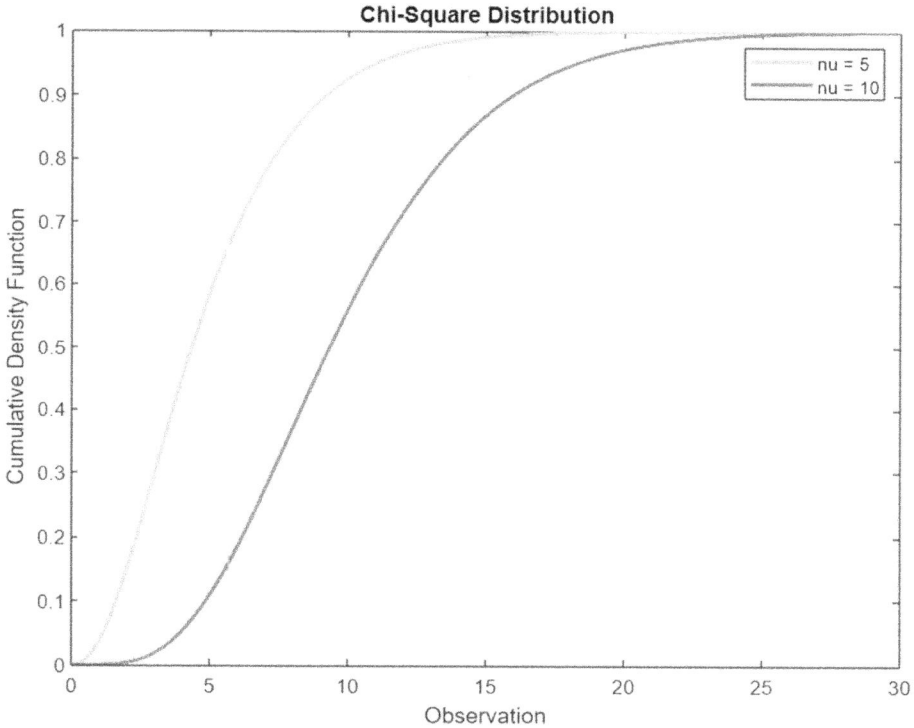

Figure 7.21

Going back to gamma the PDF is

$$f(x) = \begin{cases} \dfrac{1}{\alpha^\beta} \dfrac{1}{\Gamma(\beta)} x^{\beta-1} \epsilon^{-\frac{1}{\alpha}.x} & x > 0 \\ 0 & \text{otherwise} \end{cases}$$

Now substituting β and $\dfrac{1}{\alpha}$ with $\dfrac{df}{2}$ and $\dfrac{1}{2}$:

The probability density function of chi-squared distribution $f(x) = \dfrac{1}{2^{\frac{df}{2}}} \dfrac{1}{\Gamma\left(\dfrac{df}{2}\right)} x^{\frac{df}{2}-1} e^{-\frac{1}{2}.x}$

The cumulative distribution function of chi-squared distribution $F(x) = \int_0^k f(x)\,dx$

$$= \int_0^k \frac{1}{2^{\frac{df}{2}} \Gamma\left(\frac{df}{2}\right)} x^{\frac{df}{2}-1} e^{-\frac{1}{2}\cdot x}\,dx$$

The mean of the distribution is equal to the number of degrees of freedom: $\mu = df$.
The variance is equal to two times the number of degrees of freedom: $\sigma^2 = 2*df$.

7.5.1.1 Example

Assume that the lengths that Arizona State students drive from campus back home (in miles) can be approximated by a chi-squared distribution with 30 degrees of freedom. What is the probability that a random Arizona State student drives more than 50 miles to get home?

$$P(X > 50) = 1 - P(X \le 50) = 1 - \int_0^{50} \frac{1}{2^{\frac{30}{2}} \Gamma\left(\frac{30}{2}\right)} x^{\frac{30}{2}-1} e^{-\frac{1}{2}\cdot x}\,dx$$

$$= 1 - \int_0^{50} \frac{1}{2^{15} \Gamma(15)} x^{14} e^{-\frac{1}{2}\cdot x}\,dx = 0.0124$$

7.5.1.2 Calculation in MATLAB

Below are the functions:

1) chi2pdf(x, nu) returns the PDF of the chi-squared distribution with nu degrees of freedom, evaluated at the values in x.
2) chi2cdf(x, nu) returns the CDF of the chi-squared distribution with degrees of freedom nu, evaluated at the values in x.
   - chi2cdf(x, nu, 'upper') returns the complement of the CDF, evaluated at the values in x with degrees of freedom nu, using an algorithm that more accurately computes the extreme upper-tail probabilities than subtracting the lower tail value from 1.
3) chi2inv(p, nu) returns the inverse CDF of the chi-squared distribution with degrees of freedom nu, evaluated at the probability values in p.
4) chi2stat(nu) returns the mean and variance for the chi-squared distribution with degrees of freedom parameters specified by nu.
5) chi2rnd(nu, size) generates a random number from the chi-squared distribution with nu degrees of freedom.

Now we have discussed a little about the distribution.
  The chi-squared goodness-of-fit test checks whether your sample data is likely to be from any theoretical distribution specified by its parameters.
  It determines whether the observed sample distribution matches or fits the expected values; hence we use the term goodness of fit.

7.5.1.3 Conditions

1) A chi-square **goodness-of-fit test can** be conducted when there is one categorical **variable** with more than two levels.

2) Ensure that observations are independent.
3) Ensure that the expected frequency of each group of the categorical variable ≥ 5.

7.5.1.4 Test statistic

For the chi-squared goodness-of-fit computation, the data is grouped into N bins and the test statistic is defined as

$$\chi^2 = \sum_{i=1}^{N} \frac{(O_i - E_i)^2}{E_i}$$

Where
O_i are the observed counts
E_i are the expected counts

Note: they are based on the hypothesized distribution.

7.5.1.5 Critical value approach

This approach uses the alpha value and the df to find the critical chi-squared value from a chi-squared table. The degrees of freedom in the chi-squared test is equal to

df = class intervals $-1-$ number of parameters.

When we are testing for normality then the number of parameters will clearly be two; hence our df = class intervals -3.

7.5.1.6 Compare test statistics and critical value

If our chi-squared statistic > critical value, we may conclude our data is *not* normal.

7.5.1.7 Example

The table below shows a frequency distribution of grouped data from observing the car battery life in years of a particular type manufactured by a company. Test if the car battery life follows a normal distribution. Assume 95% as your confidence level.

Life of battery (in hours)	Number of batteries
1.0–1.5	26
1.5–2.0	45
2.0–2.5	34
2.5–3.0	18
3.0–3.5	28
3.5–4.0	12
4.0–4.5	16
4.5–5.0	21

First let's calculate the parameters of the data—mean and standard deviation.

Interval	Mid	Freq	Mid × Freq	(Mid−2.705)² × Freq
1.0–1.5	1.25	26	32.5	55.04265
1.5–2.0	1.75	45	78.75	41.04113
2.0–2.5	2.25	34	76.5	7.03885
2.5–3.0	2.75	18	49.5	0.03645
3.0–3.5	3.25	28	91	8.3167
3.5–4.0	3.75	12	45	13.1043
4.0–4.5	4.25	16	68	38.1924
4.5–5.0	4.75	21	99.75	87.82253
		$\Sigma 200$	$\Sigma 541$	$\Sigma 250.595$

$$\text{Mean} = \frac{541}{200} = 2.705$$

$$\text{Standard deviation } s = \sqrt{\frac{250.595}{200-1}} = 1.1222$$

Now we are ready to perform the goodness-of-fit hypotheses test.

Remember the frequency given is the observed frequency. Now we shall calculate the expected frequency. So how do we compute the expected frequency?

Compute the expected frequencies by considering areas under the normal distribution that has the same mean and standard deviation as our data. We'll get those areas from the standard normal distribution table after transforming our data. So let's transform our data using $= \dfrac{x - \bar{x}}{s}$.

Interval	Z score
1.0–1.5	$\dfrac{1.0 - 2.705}{1.1222}$ to $\dfrac{1.5 - 2.705}{1.1222} = -\infty$ to -1.074
1.5–2.0	$\dfrac{1.5 - 2.705}{1.1222}$ to $\dfrac{2.0 - 2.705}{1.1222} = -1.074$ to -0.62823
2.0–2.5	$\dfrac{2.0 - 2.705}{1.1222}$ to $\dfrac{2.5 - 2.705}{1.1222} = -0.62823$ to -0.1827
2.5–3.0	$\dfrac{2.5 - 2.705}{1.1222}$ to $\dfrac{3.0 - 2.705}{1.1222} = -0.1827$ to 0.2629
3.0–3.5	$\dfrac{3.0 - 2.705}{1.1222}$ to $\dfrac{3.5 - 2.705}{1.1222} = 0.2629$ to 0.70843
3.5–4.0	$\dfrac{3.5 - 2.705}{1.1222}$ to $\dfrac{4.0 - 2.705}{1.1222} = 0.70843$ to 1.154
4.0–4.5	$\dfrac{4.0 - 2.705}{1.1222}$ to $\dfrac{4.5 - 2.705}{1.1222} = 1.154$ to 1.60
4.5–5.0	$\dfrac{4.5 - 2.705}{1.1222}$ to $\dfrac{5.0 - 2.705}{1.1222} = 1.60$ to ∞

Now let's find the areas from the standard normal distribution table.

For the first class interval:

STANDARD NORMAL DISTRIBUTION: Table Values Represent AREA to the LEFT of the Z score.

Z	.00	.01	.02	.03	.04	.05	.06	.07	.08	.09
-3.9	.00005	.00005	.00004	.00004	.00004	.00004	.00004	.00004	.00003	.00003
-3.8	.00007	.00007	.00007	.00006	.00006	.00006	.00006	.00005	.00005	.00005
-3.7	.00011	.00010	.00010	.00010	.00009	.00009	.00008	.00008	.00008	.00008
-3.6	.00016	.00015	.00015	.00014	.00014	.00013	.00013	.00012	.00012	.00011
-3.5	.00023	.00022	.00022	.00021	.00020	.00019	.00019	.00018	.00017	.00017
-3.4	.00034	.00032	.00031	.00030	.00029	.00028	.00027	.00026	.00025	.00024
-3.3	.00048	.00047	.00045	.00043	.00042	.00040	.00039	.00038	.00036	.00035
-3.2	.00069	.00066	.00064	.00062	.00060	.00058	.00056	.00054	.00052	.00050
-3.1	.00097	.00094	.00090	.00087	.00084	.00082	.00079	.00076	.00074	.00071
-3.0	.00135	.00131	.00126	.00122	.00118	.00114	.00111	.00107	.00104	.00100
-2.9	.00187	.00181	.00175	.00169	.00164	.00159	.00154	.00149	.00144	.00139
-2.8	.00256	.00248	.00240	.00233	.00226	.00219	.00212	.00205	.00199	.00193
-2.7	.00347	.00336	.00326	.00317	.00307	.00298	.00289	.00280	.00272	.00264
-2.6	.00466	.00453	.00440	.00427	.00415	.00402	.00391	.00379	.00368	.00357
-2.5	.00621	.00604	.00587	.00570	.00554	.00539	.00523	.00508	.00494	.00480
-2.4	.00820	.00798	.00776	.00755	.00734	.00714	.00695	.00676	.00657	.00639
-2.3	.01072	.01044	.01017	.00990	.00964	.00939	.00914	.00889	.00866	.00842
-2.2	.01390	.01355	.01321	.01287	.01255	.01222	.01191	.01160	.01130	.01101
-2.1	.01786	.01743	.01700	.01659	.01618	.01578	.01539	.01500	.01463	.01426
-2.0	.02275	.02222	.02169	.02118	.02068	.02018	.01970	.01923	.01876	.01831
-1.9	.02872	.02807	.02743	.02680	.02619	.02559	.02500	.02442	.02385	.02330
-1.8	.03593	.03515	.03438	.03362	.03288	.03216	.03144	.03074	.03005	.02938
-1.7	.04457	.04363	.04272	.04182	.04093	.04006	.03920	.03836	.03754	.03673
-1.6	.05480	.05370	.05262	.05155	.05050	.04947	.04846	.04746	.04648	.04551
-1.5	.06681	.06552	.06426	.06301	.06178	.06057	.05938	.05821	.05705	.05592
-1.4	.08076	.07927	.07780	.07636	.07493	.07353	.07215	.07078	.06944	.06811
-1.3	.09680	.09510	.09342	.09176	.09012	.08851	.08691	.08534	.08379	.08226
-1.2	.11507	.11314	.11123	.10935	.10749	.10565	.10383	.10204	.10027	.09853
-1.1	.13567	.13350	.13136	.12924	.12714	.12507	.12302	.12100	.11900	.11702
-1.0	.15866	.15625	.15386	.15151	.14917	.14686	.14457	.14231	.14007	.13786
-0.9	.18406	.18141	.17879	.17619	.17361	.17106	.16853	.16602	.16354	.16109
-0.8	.21186	.20897	.20611	.20327	.20045	.19766	.19489	.19215	.18943	.18673
-0.7	.24196	.23885	.23576	.23270	.22965	.22663	.22363	.22065	.21770	.21476
-0.6	.27425	.27093	.26763	.26435	.26109	.25785	.25463	.25143	.24825	.24510
-0.5	.30854	.30503	.30153	.29806	.29460	.29116	.28774	.28434	.28096	.27760

Since the value −1.074 is in the middle of −1.07 and −1.08, we take the average of 0.14231 and 0.14007 which is around 0.141.

Similarly let's now calculate the area for the second interval and then the rest of the intervals.

$$P(Z < -0.62823) - P(Z < -1.074) = 0.26493 - 0.14141 = 0.12352$$

$$P(Z < -0.1827) - P(Z < -0.62823) = 0.42752 - 0.26493 = 0.16259$$

$$P(Z < 0.2629) - P(Z < -0.1827) = 0.60369 - 0.42752 = 0.17617$$

$$P(Z < 0.70843) - P(Z < 0.2629) = 0.76066 - 0.60369 = 0.15697$$

$$P(Z < 1.154) - P(Z < 0.70843) = 0.87575 - 0.76066 = 0.11509$$

$$P(Z < 1.60) - P(Z < 1.154) = 0.9452 - 0.87575 = 0.06945$$

$$1 - P(Z < 1.154) = 1 - 0.9452 = 0.0548$$

Now that we have the area for each interval, we convert it to the expected frequency by multiplying the area by the total sample size, in this case 200. Therefore our expected values for each interval would be:

Interval	Z score	Areas	Expected
1.0–1.5	$-\infty$ to -1.074	0.14141	28.282
1.5–2.0	-1.074 to -0.62823	0.12352	24.704
2.0–2.5	-0.62823 to -0.1827	0.16259	32.518
2.5–3.0	-0.1827 to 0.2629	0.17617	35.234
3.0–3.5	0.2629 to 0.70843	0.15697	31.394
3.5–4.0	0.70843 to 1.154	0.11509	23.018
4.0–4.5	1.154 to 1.60	0.06945	13.89
4.5–5.0	1.60 to ∞	0.0548	10.96

So, finally, the observed and expected frequencies that we'll use for the hypothesis test are:

Interval	Observed	Expected
1.0–1.5	26	28.282
1.5–2.0	45	24.704
2.0–2.5	34	32.518
2.5–3.0	18	35.234
3.0–3.5	28	31.394
3.5–4.0	12	23.018
4.0–4.5	16	13.89
4.5–5.0	21	10.96
	$\Sigma 200.$	

$$\chi^2 = \frac{(26 - 28.282)^2}{28.282} + \frac{(45 - 24.704)^2}{24.704} + \frac{(34 - 32.518)^2}{32.518} + \frac{(18 - 35.234)^2}{35.234}$$

$$+ \frac{(28 - 31.394)^2}{31.394} + \frac{(12 - 23.018)^2}{23.018} + \frac{(16 - 13.89)^2}{13.89} + \frac{(21 - 10.96)^2}{10.96}$$

$$\chi^2 = 40.5145$$

Let's now calculate the critical value.

Let's take $\alpha = 0.05$ and $df = 8 - 3 = 5$ from the chi-square distribution table.

Significance level (α)

Degrees of freedom (*df*)	.99	.975	.95	.9	.1	.05	.025	.01
1	-------	0.001	0.004	0.016	2.706	3.841	5.024	6.635
2	0.020	0.051	0.103	0.211	4.605	5.991	7.378	9.210
3	0.115	0.216	0.352	0.584	6.251	7.815	9.348	11.345
4	0.297	0.484	0.711	1.064	7.779	9.488	11.143	13.277
5	0.554	0.831	1.145	1.610	9.236	11.070	12.833	15.086
6	0.872	1.237	1.635	2.204	10.645	12.592	14.449	16.812
7	1.239	1.690	2.167	2.833	12.017	14.067	16.013	18.475
8	1.646	2.180	2.733	3.490	13.362	15.507	17.535	20.090
9	2.088	2.700	3.325	4.168	14.684	16.919	19.023	21.666
10	2.558	3.247	3.940	4.865	15.987	18.307	20.483	23.209
11	3.053	3.816	4.575	5.578	17.275	19.675	21.920	24.725
12	3.571	4.404	5.226	6.304	18.549	21.026	23.337	26.217
13	4.107	5.009	5.892	7.042	19.812	22.362	24.736	27.688
14	4.660	5.629	6.571	7.790	21.064	23.685	26.119	29.141
15	5.229	6.262	7.261	8.547	22.307	24.996	27.488	30.578
16	5.812	6.908	7.962	9.312	23.542	26.296	28.845	32.000
17	6.408	7.564	8.672	10.085	24.769	27.587	30.191	33.409
18	7.015	8.231	9.390	10.865	25.989	28.869	31.526	34.805
19	7.633	8.907	10.117	11.651	27.204	30.144	32.852	36.191
20	8.260	9.591	10.851	12.443	28.412	31.410	34.170	37.566

Here we get our $\chi_{critical}^{2} = 11.070$.

7.5.1.8 Decision

Since $\chi^{2} = 40.5145 > \chi_{critical}^{2} = 11.070$ we may conclude our data is *not* normal.

Let's work on some examples in MATLAB

```
[h, p] = chi2gof(Data, Name, Value)
```

This function returns a test decision, *p*-value, and statistics for the chi-squared goodness-of-fit test with additional options specified by one or more name-value pair arguments.

- NBins: **number of bins**
- Ctrs: **bin centers**
- Edges: **bin edges**

In MATLAB, we cannot specify more than one of 'nbins', 'ctrs', and 'edges'.

- Expected: **expected counts**
- Frequency: **frequency**
- Alpha: **significance level**

```
Edges = 1:0.5:5;
[h, p, stats] = chi2gof(Data, 'Edges', Edges)


        h =

            1


        p =

          1.1749e-07


        stats =

          struct with fields:

            chi2stat: 40.5164
                  df: 5
               edges: [1 1.5000 2 2.5000 3 3.5000 4 4.5000 5]
                   O: [26 45 34 18 28 12 16 21]
                   E: [28.2908 24.6935 32.5205 35.2311 31.3973 23.0172 13.8803 10.9693]
```

If we check for normality visually, i.e. with a histogram, such as Figure 7.22.

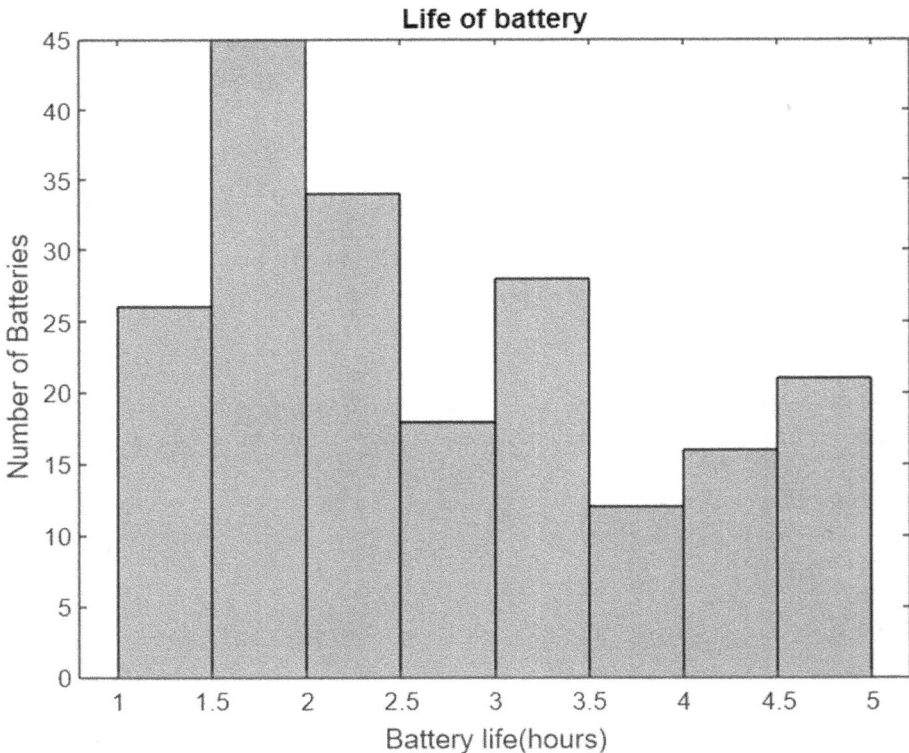

Figure 7.22

Evidently this dataset is not normal.

7.5.1.9 Example 2

Suppose a civil engineer is studying the compressive strength of concrete samples produced by a construction company.

```
x = readtable("hardness and Tensile Strength.csv",'VariableNamingRule','preserve');
Strength = x.T_Strength;
Lower = min(Strength);
Upper = max(Strength);
Width = 250;
histogram(Strength)
xlabel("Class intervals")
ylabel("Frequency")
[Frequency, Edges] = histcounts(Strength);
```

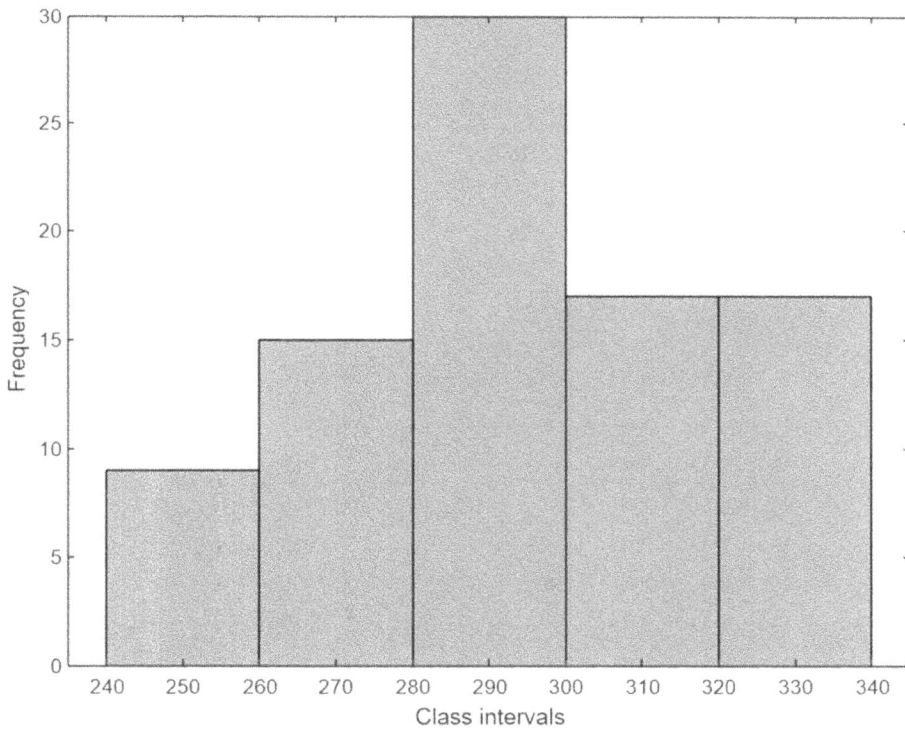

Figure 7.23

```
Lower =

    240.6540


Upper =

    333.8456


Frequency =

    9    15    30    17    17


Edges =

    240    260    280    300    320    340
```

The maximum and minimum values confirm that our histogram has intervals which do not lose any data points (all in the bins). Therefore let's draw our table and perform the testing.

Strength	Observed frequency
240 – 260	9
260 – 280	15
280 – 300	30
300 – 320	17
320 – 340	17

First let's calculate the parameters of the data—mean and standard deviation.

Interval	Mid	Freq	Mid×Freq	$(\text{Mid} - 294.1)^2 \times \text{Freq}$
240 – 260	250	9	2250	17496.07
260 – 280	270	15	4050	8705.579
280 – 300	290	30	8700	502.0661
300 – 320	310	17	5270	4302.686
320 – 340	330	17	5610	21920.87
		$\Sigma 88$	$\Sigma\ 25880$	$\Sigma 52927.27$

$$Mean = \frac{25880}{88} = 294.0909091$$

$$\text{Standard deviation } s = \sqrt{\frac{52927.27}{88-1}} = 24.66494$$

Transforming our data using $=\dfrac{x-\bar{x}}{s}$:

Interval	Z score
240 – 260	$\dfrac{240-294.091}{24.665}$ to $\dfrac{260-294.091}{24.665}$ $=-\infty$ to -1.38216
260 – 280	$\dfrac{260-294.091}{24.665}$ to $\dfrac{280-294.091}{24.665}$ $=-1.38216$ to -0.57129
280 – 300	$\dfrac{280-294.091}{24.665}$ to $\dfrac{300-294.091}{24.665}$ $=-0.57129$ to 0.23957
300 – 320	$\dfrac{300-294.091}{24.665}$ to $\dfrac{320-294.091}{24.665}$ $=0.23957$ to 1.050442
320 – 340	$\dfrac{320-294.091}{24.665}$ to $\dfrac{340-294.091}{24.665}$ $=1.050442$ to ∞

Now let's find the areas from the standard normal distribution table.

$$P(Z<-1.38216)=0.083461$$

$$P(Z<-0.57129)-P(Z<-1.38216)=0.2839-0.083461=0.200439$$

$$P(Z<0.23957)-P(Z<-0.57129)=0.59467-0.2839=0.31077$$

$$P(Z<1.050442)-P(Z<0.23957)=0.85324-0.59467=0.25857$$

$$1-P(Z<1.050442)=1-0.85324=0.14676$$

Calculating our expected values for each interval would give 151*areas:

Interval	Z score	Areas	Expected
240 – 260	$-\infty$ to -1.38216	0.083461	7.344568
260 – 280	-1.38216 to -0.57129	0.200439	17.63863
280 – 300	-0.57129 to 0.23957	0.31077	27.34776
300 – 320	0.23957 to 1.05043	0.25857	22.75416
320 – 340	1.05043 to ∞	0.14676	12.91488

So, finally, the observed and expected frequencies that we'll use for the hypothesis test are:

Strength	Observed frequency	Expected frequency
240 – 260	9	7.344568
260 – 280	15	17.638632
280 – 300	30	27.34776
300 – 320	17	22.75416
320 – 340	17	12.91488

$$\chi^2 = \frac{(9-7.344568)^2}{7.344568} + \frac{(15-17.638632)^2}{17.638632} + \frac{(30-27.34776)^2}{27.34776}$$

$$+ \frac{(17-22.75416)^2}{22.75416} + \frac{(17-12.91488)^2}{12.91488}$$

$$\chi^2 = 3.77236$$

Let's now calculate the critical value.

Let's take $\alpha = 0.01$ and $df = 5 - 3 = 2$ From the chi-square distribution table:

d.f.	.995	.99	.975	.95	.9	.1	.05	.025	.01
1	0.00	0.00	0.00	0.00	0.02	2.71	3.84	5.02	6.63
2	0.01	0.02	0.05	0.10	0.21	4.61	5.99	7.38	9.21
3	0.07	0.11	0.22	0.35	0.58	6.25	7.81	9.35	11.34
4	0.21	0.30	0.48	0.71	1.06	7.78	9.49	11.14	13.28
5	0.41	0.55	0.83	1.15	1.61	9.24	11.07	12.83	15.09
6	0.68	0.87	1.24	1.64	2.20	10.64	12.59	14.45	16.81
7	0.99	1.24	1.69	2.17	2.83	12.02	14.07	16.01	18.48
8	1.34	1.65	2.18	2.73	3.49	13.36	15.51	17.53	20.09
9	1.73	2.09	2.70	3.33	4.17	14.68	16.92	19.02	21.67
10	2.16	2.56	3.25	3.94	4.87	15.99	18.31	20.48	23.21
11	2.60	3.05	3.82	4.57	5.58	17.28	19.68	21.92	24.72
12	3.07	3.57	4.40	5.23	6.30	18.55	21.03	23.34	26.22
13	3.57	4.11	5.01	5.89	7.04	19.81	22.36	24.74	27.69

Here we get our $\chi_{critical}^2 = 9.21$.

7.5.1.10 Decision

Since $\chi^2 < \chi_{critical}^2$ we may conclude our data is normal.

```
[h, p, stats] = chi2gof(Strength,'Ctrs',[250 270 290 310 330], 'Alpha',
0.01)

        h =

            0


        p =

            0.1344


        stats =

        struct with fields:

            chi2stat: 4.0135
                  df: 2
               edges: [240.0000 260.0000 280.0000 300.0000 320.0000 340.0000]
                   O: [9 15 30 17 17]
                   E: [7.0307 17.5007 27.5668 23.0191 12.8826]
```

7.5.2 Anderson–Darling

7.5.2.1 Test statistic

The Anderson–Darling test statistic is defined as

$$AD = -n - \frac{1}{n}\sum_{i=1}^{n}(2i-1)\Big[\ln\big(F(X_i)\big) + \ln\big(1 - F(X_{n+1-i})\big)\Big]$$

Where

F is the cumulative distribution function of the specified distribution.

i = the ith sample, calculated when the data is sorted in ascending order.

Now if the sample size is small then the value of AD needs to be adjusted. The adjusted AD value is given by

$$AD^* = AD\left(1 + \frac{0.75}{n} + \frac{2.25}{n^2}\right)$$

As we can see, the Anderson–Darling test statistic is extremely cumbersome to calculate by hand; therefore these tests are generally conducted using computer software.

7.5.2.2 Approach

7.5.2.2.1 Critical value

The critical values for the Anderson–Darling test are dependent on the specific distribution that is being tested. Note that for the normal distribution, the critical value is given by

$$crit = a\left(1 - \frac{b}{n} - \frac{d}{n^2}\right)$$

where a, b, and d are shown in the table of critical values in the Anderson–Darling statistical table.

	α	0.005	0.01	0.025	0.05	0.10	0.20
Normal	a	1.1578	1.0348	0.8728	0.7514	0.6305	0.5091
Lognormal	b	1.063	1.013	0.881	0.795	0.750	0.756
	d	1.34	0.93	0.94	0.89	0.80	0.39

As we can see, the values of a, b, and d *solely depend on the alpha.*

7.5.2.2.2 p-value

For the normal distribution, p-values can be obtained from Figure 7.3.

AD	p-value
$AD \leq 0.2$	$1 - exp\left(-13.436 + 101.14AD - 223.73\,AD^2\right)$
$0.2 < AD \leq 0.34$	$1 - exp\left(-8.318 + 42.796AD - 59.938\,AD^2\right)$
$0.34 < AD < 0.6$	$exp\left(0.9177 - 4.279\,AD - 1.38\,AD^2\right)$
$AD \geq 0.6$	$exp\left(1.2937 - 5.709\,AD + 0.0186\,AD^2\right)$

7.5.2.3 Example

We have a dataset of weights of vehicles to determine if our data is normally distributed.

4732, 4615, 4376, 4382, 4633, 3664, 4997, 4951, 4952, 4955

First step: calculate the mean and standard deviation of the values.

$$\bar{x} = \frac{4732 + 4615 + 4376 + 4382 + 4633 + 3664 + 4997 + 4951 + 4952 + 4955}{10} = 4625.7$$

$$s = \sqrt{\frac{\left(4732 - 4625.7\right)^2 + \left(4615 - 4625.7\right)^2 + \ldots + \left(4952 - 4625.7\right)^2 + \left(4955 - 4625.7\right)^2}{10 - 1}}$$

$$= 410.5103$$

Second step: next we shall sort the weight values in ascending order.

3664, 4376, 4382, 4615, 4633, 4732, 4951, 4952, 4955, 4997

Third step: Next we shall transform the sorted weights like we did in chi-square using the mean and standard deviation calculated in step 1.

Next we find the cumulative function of this transformed value $F\left(X_i\right)$.

Then we calculate $1 - F\left(X_i\right)$.

Finally we sort the values again in ascending order which is $1 - F\left(X_{n+1-i}\right)$.

Weight	Z score	$F\left(X_i\right)$	$1 - F\left(X_i\right)$	$1 - F\left(X_{n+1\ i}\right)$
3664	$\dfrac{3664 - 4625.7}{410.5103} = -2.34269$	0.0095726	0.990427	0.18287
4376	$\dfrac{4376 - 4625.7}{410.5103} = -0.60826$	0.27151	0.72849	0.21123

4382	$\dfrac{4382-4625.7}{410.5103}=-0.59365$	0.27637	0.72363	0.21335
4615	$\dfrac{4615-4625.7}{410.5103}=-0.02606$	0.4896	0.5104	0.21406
4633	$\dfrac{4633-4625.7}{410.5103}=0.01778$	0.50709	0.49291	0.39784
4732	$\dfrac{4732-4625.7}{410.5103}=0.25894$	0.60216	0.39784	0.49291
4951	$\dfrac{4951-4625.7}{410.5103}=0.79242$	0.78594	0.21406	0.5104
4952	$\dfrac{4952-4625.7}{410.5103}=0.79486$	0.78665	0.21335	0.72363
4955	$\dfrac{4955-4625.7}{410.5103}=0.80217$	0.78877	0.21123	0.72849
4997	$\dfrac{4997-4625.7}{410.5103}=0.90448$	0.81713	0.18287	0.990427

We are now ready to calculate the summation portion of the equation.

$F(X_i)$	$\ln F(X_i)$	$1-F(X_{n+1-i})$	$ln(1-F(X_{n+1-i}))$	Col 2 + Col 4
0.0095726	−4.64885	0.18287	−1.69898	−6.34783
0.27151	−1.30376	0.21123	−1.55481	−2.85856
0.27637	−1.28601	0.21335	−1.54482	−2.83084
0.4896	−0.71417	0.21406	−1.5415	−2.25567
0.50709	−0.67907	0.39784	−0.92171	−1.60077
0.60216	−0.50723	0.49291	−0.70743	−1.21466
0.78594	−0.24087	0.5104	−0.67256	−0.91344
0.78665	−0.23997	0.72363	−0.32348	−0.56345
0.78877	−0.23728	0.72849	−0.31678	−0.55406
0.81713	−0.20196	0.990427	−0.00962	−0.21158

$$\sum_{i=1}^{n}(2i-1)\Big[\ln\big(F(X_i)\big)+\ln\big(1-F(X_{n+1-i})\big)\Big]=(2(1)-1)(-6.34783)+(2(2)-1)(-2.85856)$$

$$+(2(3)-1)(-2.83084)+(2(4)-1)(-2.25567)+(2(5)-1)(-1.60077)+(2(6)-1)(-1.21466)$$

$$+(2(7)-1)(-0.91344)+(2(8)-1)(-0.56345)+(2(9)-1)(-0.55406)+(2(10)-1)(-0.21158)$$

$$=-106.4011$$

$$AD = -10 - \frac{1}{10}(-106.4011) = 10.64011 - 10 = 0.64011$$

7.5.2.3.1 Critical value

Let's take $\alpha = 0.05$. Therefore our values $a, b,$ and d are $0.7514, 0.795,$ and 0.89.

$$Critical = 0.7514\left(1 - \frac{0.795}{10} - \frac{0.89}{10^2}\right) = 0.685$$

7.5.2.3.2 p-value

Since our estimated AD = 0.64011 ≥ 0.6 therefore

$$p - value = exp\left(1.2937 - 5.709(0.64011) + 0.0186(0.64011)^2\right) =$$

Let's work on some examples in MATLAB

```
[h, p, adstat, cv] = adtest(Data, 'Distribution', '', 'Alpha', )
```

'norm'	Normal distribution
'exp'	Exponential distribution
'ev'	Extreme value distribution
'logn'	Lognormal distribution
'weibull'	Weibull distribution

This means that the Anderson–Darling test can be used to test if the sample data comes from a normal distribution, exponential distribution, extreme value distribution, lognormal distribution, or Weibull distribution.

```
Weight = [4732, 4615, 4376, 4382, 4633, 3664, 4997, 4951, 4952, 4955];
[h,p,adstat,cv] = adtest(Weight);

              h =

                logical

                   0


              p =

                 0.0672


           adstat =

                 0.6401


              cv =

                 0.6857
```

7.5.2.4 Example

Let's say we have a dataset of MPG and we want to see if the variable MPG is normally distributed or not. To visualize the distribution of values, we first generate a boxplot, and we see that it is skewed with a longer whisker at the top and some outliers.

```
Data = readtable("Autompg1.csv", VariableNamingRule = "preserve");
MPG = Data.MPG;
[h,p,adstat,cv] = adtest(MPG);
```

```
h =

  logical

   1


p =

   5.0000e-04


adstat =

   2.7029


cv =

   0.7490
```

7.5.3 Kolmogorov–Smirnov test

Another widely used goodness-of-fit test is the Kolmogorov–Smirnov test or K–S test. It is very versatile which means that any continuous distribution can be fit with it and it works extremely well with small samples. **This test (the K–S test) is used to decide if a sample comes from a population with a completely specified continuous distribution.** The test relies on empirical distribution functions (ECDF) and the CDF of the reference distribution. For instance, if we wish to check whether a sample is normally distributed, we compare the ECDF of the sample with the theoretical CDF of the normal distribution.

7.5.3.1 Conditions

To conduct this test, it is assumed that the population distribution is fully specified. In practical situations where the mean and SD of the population distribution is not specified in advance, one can use a modification of the K–S test for checking the normality assumption, which is called the Lilliefors test.

7.5.3.2 Test statistic

The Kolmogorov–Smirnov test statistic is defined as

$$D = Max\left(D_i^+, D_i^-\right) = Max\left(F_n - F_0, F_0 - F_{n-1}\right) = Max\left(\frac{i}{n} - F_0, F_0 - \frac{i-1}{n}\right)$$

Where

F_0 is the theoretical CDF of the model and, in case of a normality test, the normal CDF.

7.5.3.3 Approach

7.5.3.3.1 Critical value

The critical value depends on the sample size n and α.

$n \backslash \alpha$	0.001	0.01	0.02	0.05	0.1	0.15	0.2
1		0.99500	0.99000	0.97500	0.95000	0.92500	0.90000
2	0.97764	0.92930	0.90000	0.84189	0.77639	0.72614	0.68377
3	0.92063	0.82900	0.78456	0.70760	0.63604	0.59582	0.56481
4	0.85046	0.73421	0.68887	0.62394	0.56522	0.52476	0.49265
5	0.78137	0.66855	0.62718	0.56327	0.50945	0.47439	0.44697
6	0.72479	0.61660	0.57741	0.51926	0.46799	0.43526	0.41035
7	0.67930	0.57580	0.53844	0.48343	0.43607	0.40497	0.38145
8	0.64098	0.54180	0.50654	0.45427	0.40962	0.38062	0.35828
9	0.60846	0.51330	0.47960	0.43001	0.38746	0.36006	0.33907
10	0.58042	0.48895	0.45662	0.40925	0.36866	0.34250	0.32257
11	0.55588	0.46770	0.43670	0.39122	0.35242	0.32734	0.30826
12	0.53422	0.44905	0.41918	0.37543	0.33815	0.31408	0.29573
13	0.51490	0.43246	0.40362	0.36143	0.32548	0.30233	0.28466
14	0.49753	0.41760	0.38970	0.34890	0.31417	0.29181	0.27477
15	0.48182	0.40420	0.37713	0.33760	0.30397	0.28233	0.26585
16	0.46750	0.39200	0.36571	0.32733	0.29471	0.27372	0.25774
17	0.45440	0.38085	0.35528	0.31796	0.28627	0.26587	0.25035
18	0.44234	0.37063	0.34569	0.30936	0.27851	0.25867	0.24356
19	0.43119	0.36116	0.33685	0.30142	0.27135	0.25202	0.23731
20	0.42085	0.35240	0.32866	0.29407	0.26473	0.24587	0.23152
25	0.37843	0.31656	0.30349	0.26404	0.23767	0.22074	0.20786
30	0.34672	0.28988	0.27704	0.24170	0.21756	0.20207	0.19029
35	0.32187	0.26898	0.25649	0.22424	0.20184	0.18748	0.17655
40	0.30169	0.25188	0.23993	0.21017	0.18939	0.17610	0.16601
45	0.28482	0.23780	0.22621	0.19842	0.17881	0.16626	0.15673
50	0.27051	0.22585	0.21460	0.18845	0.16982	0.15790	0.14886
OVER 50	$\dfrac{1.94947}{\sqrt{n}}$	$\dfrac{1.62762}{\sqrt{n}}$	$\dfrac{1.51743}{\sqrt{n}}$	$\dfrac{1.35810}{\sqrt{n}}$	$\dfrac{1.22385}{\sqrt{n}}$	$\dfrac{1.13795}{\sqrt{n}}$	$\dfrac{1.07275}{\sqrt{n}}$

If $D_{test\ statistic} > D_{critical\ value}$ we conclude that the data is not a good fit for the normal distribution.

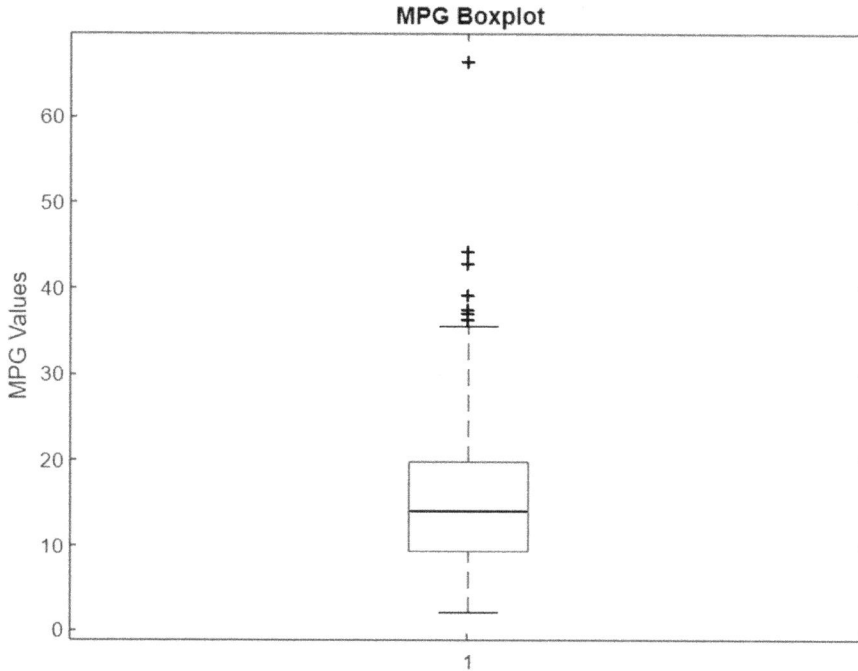

Figure 7.24

7.5.3.4 Example

Suppose a certain mechanical component produced by a company has a width that is normally distributed with a mean $\mu = 300$ and a standard deviation $\sigma = 50$.

255.5984, 305.0046, 272.7736, 315.1760, 269.9837, 324.4983,

336.9682, 385.5944, 290.2938, 193.0822

Test the null hypothesis that the data comes from a normal distribution with given parameters.

- First we sort the data in ascending order.
- According to the formula of the test statistic we need to first estimate the theoretical CDF. To obtain this we need the parameters of the distribution we are testing. For instance, we need to know the mean and standard deviation to get the theoretical CDF of the normal distribution. Since parameters are specified (they always have to be specified in the Kolmogorov–Smirnov test), let's calculate the CDF.

Width	Z score	CDF
193.0822	193.0822-300$\backslash$50=-2.138356	0.01624393
255.5984	255.5984-300$\backslash$50=-0.888032	0.187261769
269.9837	269.9837-300$\backslash$50=-0.600326	0.274144497
272.7736	272.7736-300$\backslash$50=-0.544528	0.293039093

290.2938	$\dfrac{290.2938 - 300}{50} = -0.194124$	0.423039395
305.0046	$\dfrac{305.0046 - 300}{50} = 0.100092$	0.539864357
315.1760	$\dfrac{315.1760 - 300}{50} = 0.30352$	0.619253196
324.4983	$\dfrac{324.4983 - 300}{50} = 0.489966$	0.687921021
336.9682	$\dfrac{336.9682 - 300}{50} = 0.739364$	0.770157002
385.5944	$\dfrac{385.5944 - 300}{50} = 1.711888$	0.956541345

Second step:

Width	$F_0(x)$	$F_n(x)$	$F_{n-1}(x)$	$D^+ = \lvert F_n - F_0 \rvert$	$D^- = F_0 - F_{n-1}$
193.0822	0.01624393	$\dfrac{1}{10} = 0.10$	$\dfrac{1-1}{10} = 0$	0.083756	0.0162439
255.5984	0.187261769	$\dfrac{2}{10} = 0.20$	$\dfrac{2-1}{10} = 0.10$	0.012738	0.0872618
269.9837	0.274144497	$\dfrac{3}{10} = 0.30$	$\dfrac{3-1}{10} = 0.20$	0.025856	0.0741445
272.7736	0.293039093	$\dfrac{4}{10} = 0.40$	$\dfrac{4-1}{10} = 0.30$	0.106961	0.0069609
290.2938	0.423039395	$\dfrac{5}{10} = 0.50$	$\dfrac{5-1}{10} = 0.40$	0.076961	0.0230394
305.0046	0.539864357	$\dfrac{6}{10} = 0.60$	$\dfrac{6-1}{10} = 0.50$	0.060136	0.0398644
315.1760	0.619253196	$\dfrac{7}{10} = 0.70$	$\dfrac{7-1}{10} = 0.60$	0.080747	0.0192532
324.4983	0.687921021	$\dfrac{8}{10} = 0.80$	$\dfrac{8-1}{10} = 0.70$	0.112079	0.012079
336.9682	0.770157002	$\dfrac{9}{10} = 0.90$	$\dfrac{9-1}{10} = 0.80$	0.129843	0.029843
385.5944	0.956541345	$\dfrac{10}{10} = 1$	$\dfrac{10-1}{10} = 0.90$	0.043459	0.0565413
				$Max = 0.129843$	$Max = 0.087262$

$$D_{test\ statistic} = Max(0.129843, 0.087262) = 0.129843$$

7.5.3.4.1 Critical value

Our sample size $n = 10$ and take $\alpha = 0.05$. Therefore $D_{critical\ value} = 0.40925$.

7.5.3.4.2 Decision

Since $D_{test\ statistic} = 0.129843 < 0.40925 = D_{critical\ value}$ we conclude that the data is normally distributed.

7.5.3.5 Example 2

Suppose an engineer is analyzing vibration data collected from a machine in an industrial setting. The engineer wants to determine if the vibration measurements follow a normal distribution. The engineer has a sample of nine vibration measurements taken at different time intervals.

3.78, 2.51, 4.24, 2.13, 4.98, 5.57, 5.19, 3.72, 4.22

Test the null hypothesis that the data comes from a normal distribution with mean = 2.5 and standard deviation = 0.8.

Vibration	$Z = \dfrac{x - 2.5}{0.8}$	$F_0(x)$	$F_n(x)$	$F_{n-1}(x)$	$D^+ = F_n - F_0$	$D^- = F_0 - F_{n-1}$
2.13	-0.4625	0.321861401	$\dfrac{1}{9}$	$\dfrac{1-1}{9} = 0$	0.21075	0.321861
2.51	0.0125	0.504986649	$\dfrac{2}{9}$	$\dfrac{2-1}{9} = \dfrac{1}{9}$	0.282764	0.393876
3.72	1.525	0.936370451	$\dfrac{3}{9}$	$\dfrac{3-1}{9} = \dfrac{2}{9}$	0.603037	0.714148
3.78	1.6	0.945200708	$\dfrac{4}{9}$	$\dfrac{4-1}{9} = \dfrac{3}{9}$	0.500756	0.611867
4.22	2.15	0.984222393	$\dfrac{5}{9}$	$\dfrac{5-1}{9} = \dfrac{4}{9}$	0.428667	0.539778
4.24	2.175	0.985184942	$\dfrac{6}{9}$	$\dfrac{6-1}{9} = \dfrac{5}{9}$	0.318518	0.429629
4.98	3.1	0.999032397	$\dfrac{7}{9}$	$\dfrac{7-1}{9} = \dfrac{6}{9}$	0.221255	0.332366
5.19	3.3625	0.999613799	$\dfrac{8}{9}$	$\dfrac{8-1}{9} = \dfrac{7}{9}$	0.110725	0.221836
5.57	3.8375	0.999937853	$\dfrac{9}{9} = 1$	$\dfrac{9-1}{9} = \dfrac{8}{9}$	$6.21E-05$	0.111049
					Max =0.603	Max =0.71414

$$D_{test\ statistic} = Max(0.603037, 0.71414) = 0.71414$$

7.5.3.5.1 Critical value

Our sample size $n = 9$ and take $\alpha = 0.01$. Therefore $D_{critical\ value} = 0.5133$.

7.5.3.5.2 Decision

Since $D_{test\ statistic} = 0.71414 > 0.5133 = D_{critical\ value}$ we conclude that the data is not normally distributed.

7.5.3.6 MATLAB

```
x = (Value - mu)/sigma;
[h, p, kstat, critval] = kstest(Data, 'Alpha', )
```

This function returns a test decision for the null hypothesis that the data comes from a normal distribution with parameters mu and sigma, against the alternative that it does not come from such a distribution. Along with the p-value, this function also represents the test statistic and the critical value of the test.

7.5.3.7 Example I

```
Strength = [255.5984 305.0046 272.7736 315.1760 269.9837
      324.4983 336.9682 385.5944 290.2938 193.0822];
mu = 300;
sigma = 50;
x = (Strength - mu)/sigma;
[h, p, kstat, critval] = kstest(x)
                          h =

                        logical

                           0


                          p =

                        0.9876
                       kstat =

                        0.1298


                      critval =

                        0.4093
```

The result h is 0; hence the test does not reject the null hypothesis at the 5% significance level.

7.5.3.8 Example 2

```
Vibration = [2.13 2.51 3.72 3.78 4.22 4.24 4.98 5.19 5.57];
mu = 2.5;
sigma = 0.8;
x = (Vibration − mu)/sigma;
[h,p,kstat,critval] = kstest(x, 'Alpha', 0.01)
```

h =

logical

1

p =

3.6863e-05

kstat =

0.7141

critval =

0.5133

The result h is 1; hence the test rejects the null hypothesis at the 1% significance level.

7.5.4 Lilliefors test

The Lilliefors test is a special case of the Kolmogorov–Smirnov goodness-of-fit test. The K–S test is only appropriate where the parameters of the hypothesized distribution—in this case, the mean and standard deviation—are already known. However, when they are not, the parameters are estimated based on the sample data. In the Lilliefors test, the Kolmogorov–Smirnov test is implemented using the sample mean and standard deviation as the mean and standard deviation of the theoretical (benchmark) population against which the observed sample is compared.

In summary, when the population mean and standard deviation are known, we can use the one-sample Kolmogorov–Smirnov test to test for normality, but when they are not known and instead are estimated from the sample data then we use the **Lilliefors** test to test for normality.

7.5.4.1 Test statistic

Not surprisingly the Lilliefors test has the same test statistics as the Kolmogorov–Smirnov test which is defined as

$$D = Max\left(D_i^+, D_i^-\right) = Max\left(F_n - F_0, F_0 - F_{n-1}\right) = Max\left(\frac{i}{n} - F_0, F_0 - \frac{i-1}{n}\right)$$

Where F_0 is the CDF of the model and, in the case of a normality test, the normal CDF.
F_n *is the cumulative frequencies of the observations.*

7.5.4.1.1 Steps to calculate the test statistics

- Since both Lilliefors and Kolmogorov–Smirnov have the same test statistics, here too we need both the empirical cumulative distribution function and the theoretical CDF. As we already know, **to obtain the theoretical distribution, we need the parameters. But now that we are not given those, we get them instead from the sample. Therefore the first step is to get compute the mean and standard deviation from the sample.**
- We follow the exact same step for **Lilliefors as we did for Kolmogorov–Smirnov.**
- Sort the data in ascending order.
- Next, we find the z score for each value.
- Then using these standardized values we calculate the cumulative distribution function values $F_0(x)$ assuming that the original data is normally distributed.

7.5.4.2 Approach

7.5.4.2.1 Critical value

Even though both Lilliefors and Kolmogorov–Smirnov have the same statistic, the table for the critical values is different which leads to a different conclusion about the normality of a dataset. Like Kolmogorov–Smirnov, the Lilliefors test has critical values which depend on the sample size n and α.

$n\backslash\alpha$	0.01	0.05	0.10	0.15	0.20
4	0.4129	0.3754	0.3456	0.3216	0.3027
5	0.3959	0.3427	0.3188	0.3027	0.2893
6	0.3728	0.3245	0.2982	0.2816	0.2694
7	0.3504	0.3041	0.2802	0.2641	0.2521
8	0.3331	0.2875	0.2649	0.2502	0.2387
9	0.3162	0.2744	0.2522	0.2382	0.2273
10	0.3037	0.2616	0.2410	0.2273	0.2171
11	0.2905	0.2506	0.2306	0.2179	0.2080
12	0.2812	0.2426	0.2228	0.2101	0.2004
13	0.2714	0.2337	0.2147	0.2025	0.1932
14	0.2627	0.2257	0.2077	0.1959	0.1869
15	0.2545	0.2196	0.2016	0.1899	0.1811
16	0.2477	0.2128	0.1956	0.1843	0.1758
17	0.2408	0.2071	0.1902	0.1794	0.1711
18	0.2345	0.2018	0.1852	0.1747	0.1666
19	0.2285	0.1965	0.1803	0.1700	0.1624
20	0.2226	0.1920	0.1764	0.1666	0.1589
21	0.2190	0.1881	0.1726	0.1629	0.1553
22	0.2141	0.1840	0.1690	0.1592	0.1517
23	0.2090	0.1798	0.1650	0.1555	0.1484
24	0.2053	0.1766	0.1619	0.1527	0.1458
25	0.2010	0.1726	0.1589	0.1498	0.1429

26	0.1985	0.1699	0.1562	0.1472	0.1406
27	0.1941	0.1665	0.1533	0.1448	0.1381
28	0.1911	0.1641	0.1509	0.1423	0.1358
29	0.1886	0.1614	0.1483	0.1398	0.1334
30	0.1848	0.1590	0.1460	0.1378	0.1315
31	0.1820	0.1559	0.1432	0.1353	0.1291
32	0.1798	0.1542	0.1415	0.1336	0.1274
33	0.1770	0.1518	0.1392	0.1314	0.1254
34	0.1747	0.1497	0.1373	0.1295	0.1236
35	0.1720	0.1478	0.1356	0.1278	0.1220
36	0.1695	0.1454	0.1336	0.1260	0.1203
37	0.1677	0.1436	0.1320	0.1245	0.1188
38	0.1653	0.1421	0.1303	0.1230	0.1174
39	0.1634	0.1402	0.1288	0.1214	0.1159
40	0.1616	0.1386	0.1275	0.1204	0.1147
41	0.1599	0.1373	0.1258	0.1186	0.1131
42	0.1573	0.1353	0.1244	0.1172	0.1119
43	0.1556	0.1339	0.1228	0.1159	0.1106
44	0.1542	0.1322	0.1216	0.1148	0.1095
45	0.1525	0.1309	0.1204	0.1134	0.1083
46	0.1512	0.1293	0.1189	0.1123	0.1071
47	0.1499	0.1282	0.1180	0.1113	0.1062
48	0.1476	0.1269	0.1165	0.1098	0.1047
49	0.1463	0.1256	0.1153	0.1089	0.1040
50	0.1457	0.1246	0.1142	0.1079	0.1030
OVER 50	$\dfrac{1.035}{f(n)}$	$\dfrac{0.895}{f(n)}$	$\dfrac{0.819}{f(n)}$	$\dfrac{0.775}{f(n)}$	$\dfrac{0.741}{f(n)}$

Where

$$f(n) = \frac{0.83 + n}{\sqrt{n}} - 0.01$$

7.5.4.3 Example 1

We have a dataset representing tensile strengths:

240.65, 243.14, 245.48. 248.43, 250.52, 252.51, 254.41, 256.82, 258.55, 260.21

Test the null hypothesis that the data comes from a normal distribution.
  First step: calculate the mean and standard deviation.

$$Mean = \frac{240.65 + 243.14 + 245.48 + \ldots + 256.82 + 258.55 + 260.21}{10}$$

$$= 251.072$$

$$s = \sqrt{\frac{(240.65 - 251.072)^2 + (243.14 - 251.072)^2 + \ldots + (258.55 - 251.072)^2 + (260.21 - 251.072)^2}{10 - 1}}$$

$$= 6.63126534$$

Second step: calculate the standardized values and, using these, compute the cumulative distribution function values $F_0(x)$. So these two would be the second and third columns of this table.

Strength	$Z = \dfrac{x - 251.072}{6.63126534}$	$F_0(x)$	$F_n(x)$	$F_{n-1}(x)$	$D^+ = F_n - F_0$	$D^- = F_0 - F_{n-1}$
240.65	−1.57164575	0.058016	$\dfrac{1}{10} = 0.10$	$\dfrac{1-1}{10} = 0$	0.041984	0.058016
243.14	−1.1961518	0.11582	$\dfrac{2}{10} = 0.20$	$\dfrac{2-1}{10} = 0.10$	0.08418	0.01582
245.48	−0.84327797	0.19954	$\dfrac{3}{10} = 0.30$	$\dfrac{3-1}{10} = 0.20$	0.10046	−0.00046
248.43	−0.39841567	0.34516	$\dfrac{4}{10} = 0.40$	$\dfrac{4-1}{10} = 0.30$	0.05484	0.04516
250.52	−0.08324203	0.46683	$\dfrac{5}{10} = 0.50$	$\dfrac{5-1}{10} = 0.40$	0.03317	0.06683
252.51	0.216851525	0.58584	$\dfrac{6}{10} = 0.60$	$\dfrac{6-1}{10} = 0.50$	0.01416	0.08584
254.41	0.503373011	0.69265	$\dfrac{7}{10} = 0.70$	$\dfrac{7-1}{10} = 0.60$	0.00735	0.09265
256.82	0.866802896	0.80697	$\dfrac{8}{10} = 0.80$	$\dfrac{8-1}{10} = 0.70$	−0.00697	0.10697
258.55	1.127688249	0.87027	$\dfrac{9}{10} = 0.90$	$\dfrac{9-1}{10} = 0.80$	0.02973	0.07027
260.21	1.378017547	0.9159	$\dfrac{10}{10} = 1$	$\dfrac{10-1}{10} = 0.90$	0.0841	0.0159
£ 2510.72					Max = 0.10046	Max = 0.107

$$D_{test\ statistic} = Max\left(0.10046, 0.107071\right) = 0.107071$$

7.5.4.3.1 Critical value

Our sample size $n = 10$ and take $\alpha = 0.01$. Therefore $D_{critical\ value} = 0.3037$.

7.5.4.3.2 Decision

Since $D_{test\ statistic} = 0.107071 < 0.3037 = D_{critical\ value}$ we conclude that the data is normally distributed.

7.5.4.4 *Example 2*

Suppose an engineer is studying temperature readings from a thermal process in a manufacturing plant. The engineer wants to determine if the temperature readings follow a normal distribution. The engineer has a sample of 12 temperature measurements taken at different time intervals.

$$34.15, 19.85, 20.23, 24.01, 44.42, 23.89, 43.50, 23.52,$$

$$47.54, 27.29, 21.88, 23.77$$

First step: calculate the mean and standard deviation.

$$Mean = \frac{34.15 + 19.85 + 20.23 + \ldots + 27.29 + 21.88 + 23.77}{12} = 29.5041$$

$$s = \sqrt{\frac{(34.15 - 29.5041)^2 + (19.85 - 29.5041)^2 + \ldots + (21.88 - 29.5041)^2 + (23.77 - 29.5041)^2}{12 - 1}}$$

$=10.17071876$

Second step:

Temp.	$Z = \frac{x - 29.5}{10.17}$	$F_0(x)$	$F_n(x)$	$F_{n-1}(x)$	$D^+ = F_n - F_0$	$D^- = F_0 - F_{n-1}$
19.85	−0.949211435	0.171256543	$\frac{1}{12}$	$\frac{1-1}{12} = 0$	0.08792321	0.171256543
20.23	−0.911849293	0.180924029	$\frac{2}{12}$	$\frac{2-1}{12} = \frac{1}{12}$	0.014257362	0.097590695
21.88	−0.74961894	0.22674212	$\frac{3}{12}$	$\frac{3-1}{12} = \frac{2}{12}$	0.02325788	0.060075454
23.52	−0.588371801	0.27814138	$\frac{4}{12}$	$\frac{4-1}{12} = \frac{3}{12}$	0.055191953	0.02814138
23.77	−0.563791445	0.286448037	$\frac{5}{12}$	$\frac{5-1}{12} = \frac{4}{12}$	0.130218629	0.046885296
23.89	−0.551992874	0.290476618	$\frac{6}{12}$	$\frac{6-1}{12} = \frac{5}{12}$	0.209523382	0.126190049
24.01	−0.540194302	0.294531521	$\frac{7}{12}$	$\frac{7-1}{12} = \frac{6}{12}$	0.288801813	0.205468479
27.29	−0.217700025	0.413831422	$\frac{8}{12}$	$\frac{8-1}{12} = \frac{7}{12}$	0.252835244	0.169501911
34.15	0.456784959	0.676087191	$\frac{9}{12}$	$\frac{9-1}{12} = \frac{8}{12}$	0.073912809	0.009420524

43.5	1.376090293	0.91560316	$\dfrac{10}{12}$	$\dfrac{10-1}{12}=\dfrac{9}{12}$	0.082269826	0.16560316
44.42	1.466546005	0.928750201	$\dfrac{11}{12}$	$\dfrac{11-1}{12}=\dfrac{10}{12}$	0.012083535	0.095416868
47.54	1.773308855	0.961911229	$\dfrac{12}{12}=1$	$\dfrac{12-1}{12}=\dfrac{11}{12}$	0.038088771	0.045244562
					$Max=0.2888$	$Max=0.2055$

$$D_{test\ statistic} = \text{Max}(0.288801813,\ 0.205468479) = 0.288801813$$

7.5.4.4.1 Critical value

Our sample size $n=12$ and take $\alpha=0.05$. Therefore $D_{critical\ value}=0.242$.

7.5.4.4.2 Decision

Since $D_{test\ statistic}=0.288801813 > 0.242 = D_{critical\ value}$ we conclude that the data is *not* normally distributed.

7.5.4.5 MATLAB

```
[h, p, kstat, critval] = lillietest(Value, 'Alpha', )
```

This function returns a test decision for the null hypothesis that the data comes from a normal distribution against the alternative that it does not come from such a distribution.

7.5.4.6 Example 1

```
Strength = [240.65, 243.14, 245.48, 248.43, 250.52, 252.51,
            254.41, 256.82, 258.55, 260.21];
[h, p, kstat, critval] = lillietest(Strength, 'Alpha', 0.01);
                        h =

                            0


                        p =

                          0.5000


                        kstat =

                          0.1070


                        critval =

                          0.3034
```

The result h is 0; hence the test does not reject the null hypothesis at the 1% significance level.

7.5.4.7 Example 2

```
Temperature = [34.15,   19.85,   20.23,   24.01,   44.42,   23.89,
               43.50,   23.52,   47.54,   27.29,   21.88,   23.77] ;
[h, p, kstat, critval]  = lillietest (Temperature)
```

$$h =$$

$$1$$

$$p =$$

$$0.0069$$

$$kstat =$$

$$0.2888$$

$$critval =$$

$$0.2418$$

The result h is 1; hence the test rejects the null hypothesis at the 5% significance level.

7.5.5 Jarque–Bera test

Unlike the chi-square and Anderson-Darling tests which can be used to test if the sample data comes from any specified distribution, the Jarque–Bera test on the other hand is specific to normality. It is a goodness-of-fit test that measures if sample data has skewness and kurtosis like that of the normal distribution (skewness of 0 and kurtosis of 3).

The test statistics therefore is

$$JB = \frac{n}{6}\left(skew^2 + \frac{(kurt - 3)^2}{4} \right)$$

The data follows a normal distribution if the test statistic is close to zero and the p-value is larger than our standard 0.05. The p-value relates to a null hypothesis that the data has a normal distribution. If the test statistic is large and the p-value is less than 0.05, the data does not follow a normal distribution.

7.5.5.1 Example

Suppose an engineer is studying the turbulence intensity levels in a wind tunnel to assess the aerodynamic characteristics of an aircraft design. The engineer wants to determine if the turbulence intensity data follows a normal distribution. The engineer has a sample of 15 turbulence intensity measurements obtained at various locations within the wind tunnel.

$$8.30, 5.80, 11.63, 10.11, 7.67, 4.53, 24.46, 17.37, 13.96,$$

$$4.1, 5.6, 2.00, 5.83, 6.92, 4.86$$

Recall that to calculate the skewness and kurtosis we need the third and fourth central moments along with the second central moment.

$$E\left((X-\mu)^2\right) = \frac{(x_1-\bar{x})^2 + (x_2-\bar{x})^2 + (x_3-\bar{x})^2 + ..(x_n-\bar{x})^2}{n} = \frac{\sum_{i=1}^{n}(x_i-\bar{x})^2}{n}$$

$$E(X-\mu)^3 = \frac{(x_1-\bar{x})^3 + (x_2-\bar{x})^3 + (x_3-\bar{x})^3 + ..(x_n-\bar{x})^3}{n} = \frac{\sum_{i=1}^{n}(x_i-\bar{x})^3}{n}$$

$$E(X-\mu)^4 = \frac{(x_1-\bar{x})^4 + (x_2-\bar{x})^4 + (x_3-\bar{x})^4 + ..(x_n-\bar{x})^4}{n} = \frac{\sum_{i=1}^{n}(x_i-\bar{x})^4}{n}$$

Let's compute each of these moments.

First step: now all three moments need the mean value, so our first step would be to calculate the mean of the dataset.

$$\text{Mean } \bar{x} = \frac{8.3 + 5.8 + 11.63 + ... + 5.83 + 6.92 + 4.86}{15} = 8.876$$

Second step:

Intensity	$x_1 - \bar{x}$	$(x_1 - \bar{x})^2$	$(x_1 - \bar{x})^3$	$(x_1 - \bar{x})^4$
8.30	−0.576	0.331776	−0.191102976	0.110075314
5.80	−3.076	9.461776	−29.10442298	89.52520507
11.63	2.754	7.584516	20.88775706	57.52488295
10.11	1.234	1.522756	1.879080904	2.318785836
7.67	−1.206	1.454436	−1.754049816	2.115384078
4.53	−4.346	18.887716	−82.08601374	356.7458157
24.46	15.584	242.861056	3784.746697	58981.49252
17.37	8.494	72.148036	612.8254178	5205.339099
13.96	5.084	25.847056	131.4064327	668.0703039
4.10	−4.776	22.810176	−108.9414006	520.3041292
5.60	−3.276	10.732176	−35.15860858	115.1796017
2.00	−6.876	47.279376	−325.0929894	2235.339395
5.83	−3.046	9.278116	−28.26114134	86.08343651
6.92	−1.956	3.825936	−7.483530816	14.63778628
4.86	−4.016	16.128256	−64.7710761	260.1206416
		Σ490.15316	Σ3868.901049	Σ68,594.90706

$$Skewness = \frac{\frac{1}{n}\sum_{i=1}^{n}(x_i - \bar{x})^3}{\left(\frac{1}{n}\sum_{i=1}^{n}(x_i - \bar{x})^2\right)^{\frac{3}{2}}} = \frac{\frac{1}{15}\times 3868.901049}{\left(\frac{1}{15}\times 490.15316\right)^{\frac{3}{2}}} = \frac{257.9267366}{186.793} = 1.3808$$

$$Kurtosis = \frac{\frac{1}{n}\sum_{i=1}^{n}(x_i - \bar{x})^4}{\left(\frac{1}{n}\sum_{i=1}^{n}(x - \bar{x})^2\right)^{2}} = \frac{\frac{1}{15}\times 68594.90706}{\left(\frac{1}{15}\times 490.15316\right)^{2}} = \frac{4572.99380}{1067.77783} = 4.28272$$

The test statistic therefore is

$$JB = \frac{15}{6}\left((1.3808)^2 + \frac{(4.28272 - 3)^2}{4}\right) = 5.7949$$

7.5.5.2 Example

Suppose an engineer is conducting a study on the noise levels generated by different machines in a factory. The engineer wants to determine if the noise level data follows a normal distribution. The engineer has a sample of 20 noise level measurements obtained at various locations within the factory.

75.3, 78.2, 80.5, 76.8, 77.9, 79.1, 74.6, 76.5, 77.2, 75.9,

79.6, 78.3, 76.7, 77.8, 80.2, 76.4, 78.7, 75.8,

79.9, 77.5

First step:

$$\bar{x} = \frac{75.3 + 78.2 + 80.50 + \ldots + 75.8 + 79.9 + 77.5}{20} = 77.645$$

Second step:

Noise	$x_1 - \bar{x}$	$(x_1 - \bar{x})^2$	$(x_1 - \bar{x})^3$	$(x_1 - \bar{x})^4$
75.3	−2.345	5.499025	−12.89521363	30.23927595
78.2	0.555	0.308025	0.170953875	0.094879401
80.5	2.855	8.151025	23.27117638	66.43920855
76.8	−0.845	0.714025	−0.603351125	0.509831701
77.9	0.255	0.065025	0.016581375	0.004228251
79.1	1.455	2.117025	3.080271375	4.481794851
74.6	-3.045	9.272025	−28.23331613	85.9704476
76.5	−1.145	1.311025	−1.501123625	1.718786551

77.2	−0.445	0.198025	−0.088121125	0.039213901
75.9	−1.745	3.045025	−5.313568625	9.272177251
79.6	1.955	3.822025	7.472058875	14.6078751
78.3	0.655	0.429025	0.281011375	0.184062451
76.7	−0.945	0.893025	−0.843908625	0.797493651
77.8	0.155	0.024025	0.003723875	0.000577201
80.2	2.555	6.528025	16.67910388	42.6151104
76.4	−1.245	1.550025	−1.929781125	2.402577501
78.7	1.055	1.113025	1.174241375	1.238824651
75.8	−1.845	3.404025	−6.280426125	11.5873862
79.9	2.255	5.085025	11.46673138	25.85747925
77.5	−0.145	0.021025	−0.003048625	0.000442051
		Σ53.5495	Σ5.923995	Σ298.0616725

$$Skewness = \frac{\frac{1}{n}\sum_{i=1}^{n}(x_i - \bar{x})^3}{\left(\frac{1}{n}\sum_{i=1}^{n}(x_i - \bar{x})^2\right)^{\frac{3}{2}}} = \frac{\frac{1}{20} \times 5.923995}{\left(\frac{1}{20} \times 53.5495\right)^{\frac{3}{2}}} = \frac{0.29619}{4.38113} = 0.0676$$

$$Kurtosis = \frac{\frac{1}{n}\sum_{i=1}^{n}(x_i - \bar{x})^4}{\left(\frac{1}{n}\sum_{i=1}^{n}(x_i - \bar{x})^2\right)^{2}} = \frac{\frac{1}{20} \times 298.0616725}{\left(\frac{1}{20} \times 53.5495\right)^{2}} = \frac{14.90308363}{7.168872376} = 2.078860224$$

The test statistic therefore is

$$JB = \frac{20}{6}\left((0.0676)^2 + \frac{(2.078860224 - 3)^2}{4}\right) = 0.7223$$

7.5.5.3 MATLAB

[h, p, jbstat, critval] = jbtest(___)

7.5.5.4 Example I

```
Intensity = [8.3 5.8 11.63 10.11 7.67 4.53 24.46 17.37 13.96 4.1
    5.6 2 5.83 6.92 4.86];
[h, p, kstat, critval] = jbtest(Intensity)
```

```
                              h =

                                 1


                              p =

                                   0.0199


                              kstat =

                                   5.7950


                              critval =

                                   3.2985
```

7.5.5.5 Example

```
Noise = [75.3, 78.2, 80.5, 76.8, 77.9, 79.1, 74.6, 76.5, 77.2, 75.9,
    79.6, 78.3, 76.7, 77.8, 80.2, 76.4, 78.7, 75.8, 79.9, 77.5];
[h, p, kstat, critval] = jbtest(Noise)

                              h =

                                 0


                              p =

                                   0.5000


                              kstat =

                                   0.7223


                              critval =

                                   3.8011
```

7.6 SUMMARY

The normal distribution holds great importance in statistical analysis since the resulting information will lead the statistical analysis on the pathway of parametric or non-parametric tests. Therefore, assessing the normality of data is essential for verifying the assumptions of parametric tests. If the data significantly deviates from normality, alternative non-parametric tests or data transformations may be required.

However, we also learned that, with large enough sample sizes (≥ 30), the violation of the normality assumption does not cause significant issues, as stated by the central limit theorem (CLT). According to this theorem, for a sufficiently large sample size, the sampling distribution of the mean tends to approximate a normal distribution, even if the underlying population distribution is non-normal.

Chapter 8

Parametric tests

In the previous chapter we presented the topic of statistical inference which is a **method of drawing conclusions and making decisions about the parameters of a population, based on the statistics of a sample.** Then we discussed one of the key applications of statistical inference—hypothesis testing—along with their types based on whether or not the data collected from a sample follows a specific distribution.

In this chapter we will learn about different types of parametric tests. As previously stated, parametric statistical tests are basically concerned with making assumptions regarding the population parameters and the distributions the data comes from. In this textbook we will discuss the three most commonly used parametric tests and their conditions followed by examples both in theory and in MATLAB.

8.1 Z TEST

This is the most basic type of hypothesis test and is widely used.

Now remember that a hypothesis test tests if a population parameter is different to a hypothesized value. Here, the parameter associated with the test (z and t test) is the mean.

To be able to use this test certain conditions need to be met first. These include

- Population distribution is normal. Put simply, a z test is carried out if and only if the test statistic follows a normally distributed dataset.
- Samples are random and independent.
- A large sample size: the sample size should be above 30. Thus, z tests will never work if the experiments feature less than 30 subjects.
- Known population standard deviation. However, if it is not known and the sample size is large enough, we can use the central limit theorem which states that when the population standard deviation σ is unknown, the sample standard deviation s can be used in the formula as long as the sample size is 30 or more.

Ideally, there are two types of z tests: the one-sample z test and the two-sample z test.

8.1.1 One-sample z test

In this section we will discuss the easiest and most well-known z test, the one-sample mean test. This is used to determine if the difference between the mean of a sample and the mean of a population is statistically significant. Hence the test focuses on testing a value of a single mean against what we would expect from the population.

DOI: 10.1201/9781003399582-8

Previously, we said that from step 3 onwards basically we will see the differences between the different tests used for different data.

1. To calculate the test statistic:

Recall from Chapter 7 we introduced the concept of the z-score and learned the equation of a data point calculated by subtracting the population mean from the data point (referred to as x) and then dividing by the population standard deviation. Mathematically,

$$z = \frac{x - \mu}{\sigma}$$

In hypothesis testing we use the term z statistic which is basically an interchangeable term with z-score.

Now the formula of the z-score we learned previously was for a sample dataset but remember in hypothesis testing we consider the sampling distribution so clearly the z-score formula (z statistic) will differ a bit.

This makes sense because in the case of sample data, the objective is to find the number of standard deviations an observation is away from the population mean, while in the case of sampling distribution, the goal is find the number of standard deviations between the sample mean and the population mean which is used to determine whether to reject the null hypothesis or not. Therefore the formula would be

$$z = \frac{x - \mu_0}{\left(\dfrac{\sigma}{\sqrt{n}}\right)}$$

where
x = sample mean
μ_0 = hypothesized population mean
σ = population standard deviation

Now as mentioned earlier, in the absence of population, **the sample standard deviation can be used as an estimate of the population standard deviation,**

$$z = \frac{x - \mu_0}{\left(\dfrac{s}{\sqrt{n}}\right)}$$

where
x = sample mean
μ_0 = hypothesized population mean
s = sample standard deviation

2. The two approaches:

- Comparing the test statistic with the critical value

As z tests are based on a normal probability distribution, to determine the critical region we use the table for the standard normal distribution. As mentioned earlier, the critical value is computed based on the given significance level α and the directionality of the test, determined by your alternative hypothesis. Here α or alpha determines how much of the area under the curve composes our rejection region, and the directionality of the test determines where the region will be.

Now how do we find critical values from the normal distribution table?

Recall from Chapter 5 and Chapter 7 that we learned how to calculate the z-score from a normal distribution.

Calculating the critical value is same as calculating the z-score for any area under the curve. In fact, here critical value and z-score refer to the same thing because the sampling distribution of a z test is normal, or close to normal (the population standard deviation is known or, when you have larger sample sizes, you can refer to the central limit theorem).

Even though they are interchangeable, the reason we have a different term is because the concept of critical value is also applicable to other types of distributions such as Student's t-score distribution, the chi-squared distribution, etc.

Let's say we want to find the critical value for the 95% confidence level assuming a two-tailed test.

Recall that confidence level = 1 – alpha so alpha $= 1 - 0.95 = 0.05$.

This means a total of 5% of the area under the curve is considered the critical/rejection region.

Since this is a two-tailed test, we need to put the critical region in both tails, but we don't want to increase the overall size of the rejection region. To do this, we simply split the alpha level in half so that an equal proportion of the area under the curve falls in each tail's critical/rejection region.

If we represent the above statement graphically:

So looking up the z-score associated with 0.025 on the normal distribution we find −1.96 which is the critical value of the left tail (Figure 8.1).

z	0.00	0.01	0.02	0.03	0.04	0.05	0.06	0.07	0.08	0.09
-3.4	0.0003	0.0003	0.0003	0.0003	0.0003	0.0003	0.0003	0.0003	0.0003	0.0002
-3.3	0.0005	0.0005	0.0005	0.0004	0.0004	0.0004	0.0004	0.0004	0.0004	0.0003
-3.2	0.0007	0.0007	0.0006	0.0006	0.0006	0.0006	0.0006	0.0005	0.0005	0.0005
-3.1	0.0010	0.0009	0.0009	0.0009	0.0008	0.0008	0.0008	0.0008	0.0007	0.0007
-3.0	0.0013	0.0013	0.0013	0.0012	0.0012	0.0011	0.0011	0.0011	0.0010	0.0010
-2.9	0.0019	0.0018	0.0018	0.0017	0.0016	0.0016	0.0015	0.0015	0.0014	0.0014
-2.8	0.0026	0.0025	0.0024	0.0023	0.0023	0.0022	0.0021	0.0021	0.0020	0.0019
-2.7	0.0035	0.0034	0.0033	0.0032	0.0031	0.0030	0.0029	0.0028	0.0027	0.0026
-2.6	0.0047	0.0045	0.0044	0.0043	0.0041	0.0040	0.0039	0.0038	0.0037	0.0036
-2.5	0.0062	0.0060	0.0059	0.0057	0.0055	0.0054	0.0052	0.0051	0.0049	0.0048
-2.4	0.0082	0.0080	0.0078	0.0075	0.0073	0.0071	0.0069	0.0068	0.0066	0.0064
-2.3	0.0107	0.0104	0.0102	0.0099	0.0096	0.0094	0.0091	0.0089	0.0087	0.0084
-2.2	0.0139	0.0136	0.0132	0.0129	0.0125	0.0122	0.0119	0.0116	0.0113	0.0110
-2.1	0.0179	0.0174	0.0170	0.0166	0.0162	0.0158	0.0154	0.0150	0.0146	0.0143
-2.0	0.0228	0.0222	0.0217	0.0212	0.0207	0.0202	0.0197	0.0192	0.0188	0.0183
-1.9	0.0287	0.0281	0.0274	0.0268	0.0262	0.0256	0.0250	0.0244	0.0239	0.0233
-1.8	0.0359	0.0351	0.0344	0.0336	0.0329	0.0322	0.0314	0.0307	0.0301	0.0294
-1.7	0.0446	0.0436	0.0427	0.0418	0.0409	0.0401	0.0392	0.0384	0.0375	0.0367
-1.6	0.0548	0.0537	0.0526	0.0516	0.0505	0.0495	0.0485	0.0475	0.0465	0.0455
-1.5	0.0668	0.0655	0.0643	0.0630	0.0618	0.0606	0.0594	0.0582	0.0571	0.0559
-1.4	0.0808	0.0793	0.0778	0.0764	0.0749	0.0735	0.0721	0.0708	0.0694	0.0681
-1.3	0.0968	0.0951	0.0934	0.0918	0.0901	0.0885	0.0869	0.0853	0.0838	0.0823
-1.2	0.1151	0.1131	0.1112	0.1093	0.1075	0.1056	0.1038	0.1020	0.1003	0.0985
-1.1	0.1357	0.1335	0.1314	0.1292	0.1271	0.1251	0.1230	0.1210	0.1190	0.1170
-1.0	0.1587	0.1562	0.1539	0.1515	0.1492	0.1469	0.1446	0.1423	0.1401	0.1379

Now we need to find the z-score for the other half of the area in our right tail. Recall that our normal distribution table provides the area of the region located under the bell curve and to the left of a given z-score. So the area of the right tail is also 0.025; hence the total

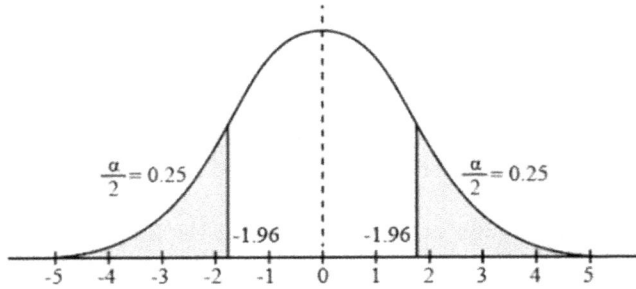

Figure 8.1

area to the left is $1 - 0.025 = 0.975$ corresponding to the critical value of $+1.96$ of the right tail.

z	0.00	0.01	0.02	0.03	0.04	0.05	0.06	0.07	0.08	0.09
0.0	0.5000	0.5040	0.5080	0.5120	0.5160	0.5199	0.5239	0.5279	0.5319	0.5359
0.1	0.5398	0.5438	0.5478	0.5517	0.5557	0.5596	0.5636	0.5675	0.5714	0.5753
0.2	0.5793	0.5832	0.5871	0.5910	0.5948	0.5987	0.6026	0.6064	0.6103	0.6141
0.3	0.6179	0.6217	0.6255	0.6293	0.6331	0.6368	0.6406	0.6443	0.6480	0.6517
0.4	0.6554	0.6591	0.6628	0.6664	0.6700	0.6736	0.6772	0.6808	0.6844	0.6879
0.5	0.6915	0.6950	0.6985	0.7019	0.7054	0.7088	0.7123	0.7157	0.7190	0.7224
0.6	0.7257	0.7291	0.7324	0.7357	0.7389	0.7422	0.7454	0.7486	0.7517	0.7549
0.7	0.7580	0.7611	0.7642	0.7673	0.7704	0.7734	0.7764	0.7794	0.7823	0.7852
0.8	0.7881	0.7910	0.7939	0.7967	0.7995	0.8023	0.8051	0.8078	0.8106	0.8133
0.9	0.8159	0.8186	0.8212	0.8238	0.8264	0.8289	0.8315	0.8340	0.8365	0.8389
1.0	0.8413	0.8438	0.8461	0.8485	0.8508	0.8531	0.8554	0.8577	0.8599	0.8621
1.1	0.8643	0.8665	0.8686	0.8708	0.8729	0.8749	0.8770	0.8790	0.8810	0.8830
1.2	0.8849	0.8869	0.8888	0.8907	0.8925	0.8944	0.8962	0.8980	0.8997	0.9015
1.3	0.9032	0.9049	0.9066	0.9082	0.9099	0.9115	0.9131	0.9147	0.9162	0.9177
1.4	0.9192	0.9207	0.9222	0.9236	0.9251	0.9265	0.9279	0.9292	0.9306	0.9319
1.5	0.9332	0.9345	0.9357	0.9370	0.9382	0.9394	0.9406	0.9418	0.9429	0.9441
1.6	0.9452	0.9463	0.9474	0.9484	0.9495	0.9505	0.9515	0.9525	0.9535	0.9545
1.7	0.9554	0.9564	0.9573	0.9582	0.9591	0.9599	0.9608	0.9616	0.9625	0.9633
1.8	0.9641	0.9649	0.9656	0.9664	0.9671	0.96	0.9686	0.9693	0.9699	0.9706
1.9	0.9713	0.9719	0.9726	0.9732	0.9738	0.9744	0.9750	0.9756	0.9761	0.9767
2.0	0.9772	0.9778	0.9783	0.9788	0.9793	0.9798	0.9803	0.9808	0.9812	0.9817
2.1	0.9821	0.9826	0.9830	0.9834	0.9838	0.9842	0.9846	0.9850	0.9854	0.9857
2.2	0.9861	0.9864	0.9868	0.9871	0.9875	0.9878	0.9881	0.9884	0.9887	0.9890
2.3	0.9893	0.9896	0.9898	0.9901	0.9904	0.9906	0.9909	0.9911	0.9913	0.9916
2.4	0.9918	0.9920	0.9922	0.9925	0.9927	0.9929	0.9931	0.9932	0.9934	0.9936

Let's do another one.

What would be the critical value for a right-tailed test with $\alpha = 0.01$?

Again, interpreting the alpha, this means a total of 1% of the area under the curve is considered the critical/rejection region. Since this is one-tailed, the entire area is congregated on the right. So if we represent the information with a graph

Again our normal distribution table provides areas to the left of the z-score; therefore we want to find the critical value corresponding to the acceptance region which is $1 - 0.01 = 0.99$. So from the normal table, the z-score corresponding to 99% of the area under the curve is approximately 2.33 (Figure 8.2).

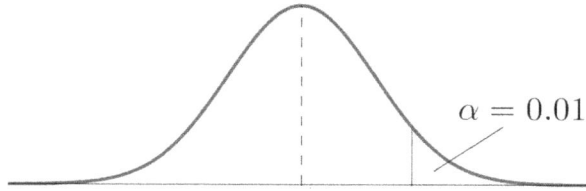

Figure 8.2

z	0.00	0.01	0.02	0.03	0.04	0.05	0.06	0.07	0.08	0.09
0.0	0.5000	0.5040	0.5080	0.5120	0.5160	0.5199	0.5239	0.5279	0.5319	0.5359
0.1	0.5398	0.5438	0.5478	0.5517	0.5557	0.5596	0.5636	0.5675	0.5714	0.5753
0.2	0.5793	0.5832	0.5871	0.5910	0.5948	0.5987	0.6026	0.6064	0.6103	0.6141
0.3	0.6179	0.6217	0.6255	0.6293	0.6331	0.6368	0.6406	0.6443	0.6480	0.6517
0.4	0.6554	0.6591	0.6628	0.6664	0.6700	0.6736	0.6772	0.6808	0.6844	0.6879
0.5	0.6915	0.6950	0.6985	0.7019	0.7054	0.7088	0.7123	0.7157	0.7190	0.7224
0.6	0.7257	0.7291	0.7324	0.7357	0.7389	0.7422	0.7454	0.7486	0.7517	0.7549
0.7	0.7580	0.7611	0.7642	0.7673	0.7704	0.7734	0.7764	0.7794	0.7823	0.7852
0.8	0.7881	0.7910	0.7939	0.7967	0.7995	0.8023	0.8051	0.8078	0.8106	0.8133
0.9	0.8159	0.8186	0.8212	0.8238	0.8264	0.8289	0.8315	0.8340	0.8365	0.8389
1.0	0.8413	0.8438	0.8461	0.8485	0.8508	0.8531	0.8554	0.8577	0.8599	0.8621
1.1	0.8643	0.8665	0.8686	0.8708	0.8729	0.8749	0.8770	0.8790	0.8810	0.8830
1.2	0.8849	0.8869	0.8888	0.8907	0.8925	0.8944	0.8962	0.8980	0.8997	0.9015
1.3	0.9032	0.9049	0.9066	0.9082	0.9099	0.9115	0.9131	0.9147	0.9162	0.9177
1.4	0.9192	0.9207	0.9222	0.9236	0.9251	0.9265	0.9279	0.9292	0.9306	0.9319
1.5	0.9332	0.9345	0.9357	0.9370	0.9382	0.9394	0.9406	0.9418	0.9429	0.9441
1.6	0.9452	0.9463	0.9474	0.9484	0.9495	0.9505	0.9515	0.9525	0.9535	0.9545
1.7	0.9554	0.9564	0.9573	0.9582	0.9591	0.9599	0.9608	0.9616	0.9625	0.9633
1.8	0.9641	0.9649	0.9656	0.9664	0.9671	0.9678	0.9686	0.9693	0.9699	0.9706
1.9	0.9713	0.9719	0.9726	0.9732	0.9738	0.9744	0.9750	0.9756	0.9761	0.9767
2.0	0.9772	0.9778	0.9783	0.9788	0.9793	0.9798	0.9803	0.9808	0.9812	0.9817
2.1	0.9821	0.9826	0.9830	0.9834	0.9838	0.9842	0.9846	0.9850	0.9854	0.9857
2.2	0.9861	0.9864	0.9868	0.9871	0.9875	0.9878	0.9881	0.9884	0.9887	0.9890
2.3	0.9893	0.9896	0.9898	0.9901	0.9904	0.9906	0.9909	0.9911	0.9913	0.9916
2.4	0.9918	0.9920	0.9922	0.9925	0.9927	0.9929	0.9931	0.9932	0.9934	0.9936

Below is a table containing the critical values of some of the most commonly used alpha levels used in hypothesis testing.

		Critical value	
Alpha level	Two-tailed test	Right-tailed test	Left-tailed test
0.10	±1.645	1.282	−1.282
0.05	±1.960	1.645	−1.645
0.01	±2.576	2.326	−2.326
0.005	±2.81	2.576	− 2.576
0.001	±3.291	3.090	−3.090
0.0001	±3.819	3.719	−3.719

- Using the test statistic to find the p-value and compare it with the alpha level

The p-value is the area under the distribution curve beyond the value of the test statistic.

To calculate the p-value we need to have the test statistic and know the directionality of the test from the alternative hypothesis.

Directionality	Alternative hypothesis	p-value
Two-tailed test	$\mu \neq \mu_o$	$2P(z > |test\ statistics|)$
Right-tailed test	$\mu > \mu_o$	$P(z > test\ statistics)$
Left-tailed test	$\mu < \mu_o$	$P(z < test\ statistics)$

- Find the p-value for a two-tailed test:

What is the p-value that corresponds to the test statistic of –0.54 of a two-tailed test? Since it is a two-tailed test, our p-value $= 2P(z > |-0.54|) = 2P(z > 0.54) = 2(1 - P(z < 0.54))$.

To find the p-value, we can first locate the value 0.54 in the z table.

z	0.00	0.01	0.02	0.03	0.04	0.05	0.06	0.07	0.08	0.09
0.0	0.5000	0.5040	0.5080	0.5120	0.5160	0.5199	0.5239	0.5279	0.5319	0.5359
0.1	0.5398	0.5438	0.5478	0.5517	0.5557	0.5596	0.5636	0.5675	0.5714	0.5753
0.2	0.5793	0.5832	0.5871	0.5910	0.5948	0.5987	0.6026	0.6064	0.6103	0.6141
0.3	0.6179	0.6217	0.6255	0.6293	0.6331	0.6368	0.6406	0.6443	0.6480	0.6517
0.4	0.6554	0.6591	0.6628	0.6664	0.6700	0.6736	0.6772	0.6808	0.6844	0.6879
0.5	0.6915	0.6950	0.6985	0.7019	0.7054	0.7088	0.7123	0.7157	0.7190	0.7224
0.6	0.7257	0.7291	0.7324	0.7357	0.7389	0.7422	0.7454	0.7486	0.7517	0.7549
0.7	0.7580	0.7611	0.7642	0.7673	0.7704	0.7734	0.7764	0.7794	0.7823	0.7852
0.8	0.7881	0.7910	0.7939	0.7967	0.7995	0.8023	0.8051	0.8078	0.8106	0.8133
0.9	0.8159	0.8186	0.8212	0.8238	0.8264	0.8289	0.8315	0.8340	0.8365	0.8389
1.0	0.8413	0.8438	0.8461	0.8485	0.8508	0.8531	0.8554	0.8577	0.8599	0.8621
1.1	0.8643	0.8665	0.8686	0.8708	0.8729	0.8749	0.8770	0.8790	0.8810	0.8830
1.2	0.8849	0.8869	0.8888	0.8907	0.8925	0.8944	0.8962	0.8980	0.8997	0.9015
1.3	0.9032	0.9049	0.9066	0.9082	0.9099	0.9115	0.9131	0.9147	0.9162	0.9177
1.4	0.9192	0.9207	0.9222	0.9236	0.9251	0.9265	0.9279	0.9292	0.9306	0.9319
1.5	0.9332	0.9345	0.9357	0.9370	0.9382	0.9394	0.9406	0.9418	0.9429	0.9441
1.6	0.9452	0.9463	0.9474	0.9484	0.9495	0.9505	0.9515	0.9525	0.9535	0.9545
1.7	0.9554	0.9564	0.9573	0.9582	0.9591	0.9599	0.9608	0.9616	0.9625	0.9633
1.8	0.9641	0.9649	0.9656	0.9664	0.9671	0.9678	0.9686	0.9693	0.9699	0.9706
1.9	0.9713	0.9719	0.9726	0.9732	0.9738	0.9744	0.9750	0.9756	0.9761	0.9767
2.0	0.9772	0.9778	0.9783	0.9788	0.9793	0.9798	0.9803	0.9808	0.9812	0.9817
2.1	0.9821	0.9826	0.9830	0.9834	0.9838	0.9842	0.9846	0.9850	0.9854	0.9857
2.2	0.9861	0.9864	0.9868	0.9871	0.9875	0.9878	0.9881	0.9884	0.9887	0.9890
2.3	0.9893	0.9896	0.9898	0.9901	0.9904	0.9906	0.9909	0.9911	0.9913	0.9916
2.4	0.9918	0.9920	0.9922	0.9925	0.9927	0.9929	0.9931	0.9932	0.9934	0.9936
2.5	0.9938	0.9940	0.9941	0.9943	0.9945	0.9946	0.9948	0.9949	0.9951	0.9952
2.6	0.9953	0.9955	0.9956	0.9957	0.9959	0.9960	0.9961	0.9962	0.9963	0.9964
2.7	0.9965	0.9966	0.9967	0.9968	0.9969	0.9970	0.9971	0.9972	0.9973	0.9974
2.8	0.9974	0.9975	0.9976	0.9977	0.9977	0.9978	0.9979	0.9979	0.9980	0.9981
2.9	0.9981	0.9982	0.9982	0.9983	0.9984	0.9984	0.9985	0.9985	0.9986	0.9986
3.0	0.9987	0.9987	0.9987	0.9988	0.9988	0.9989	0.9989	0.9989	0.9990	0.9990
3.1	0.9990	0.9991	0.9991	0.9991	0.9992	0.9992	0.9992	0.9992	0.9993	0.9993
3.2	0.9993	0.9993	0.9994	0.9994	0.9994	0.9994	0.9994	0.9995	0.9995	0.9995

The value that corresponds to a test statistic of 0.54 is 0.7054.
Hence p-value $= 2(1 - 0.7054) = 0.5892$.

- Find the p-value for a right-tailed test:

Suppose we conduct a left-tailed hypothesis test and get a test statistic of 1.58. What is the p-value that corresponds to this value?
   Since it is a right-tailed test, our p-value $= P(z > 1.58)$

$$= 1 - P(z < 1.58)$$

To find the p-value, we can first locate the value 1.58 in the z table:

z	0.00	0.01	0.02	0.03	0.04	0.05	0.06	0.07	0.08	0.09
0.0	0.5000	0.5040	0.5080	0.5120	0.5160	0.5199	0.5239	0.5279	0.5319	0.5359
0.1	0.5398	0.5438	0.5478	0.5517	0.5557	0.5596	0.5636	0.5675	0.5714	0.5753
0.2	0.5793	0.5832	0.5871	0.5910	0.5948	0.5987	0.6026	0.6064	0.6103	0.6141
0.3	0.6179	0.6217	0.6255	0.6293	0.6331	0.6368	0.6406	0.6443	0.6480	0.6517
0.4	0.6554	0.6591	0.6628	0.6664	0.6700	0.6736	0.6772	0.6808	0.6844	0.6879
0.5	0.6915	0.6950	0.6985	0.7019	0.7054	0.7088	0.7123	0.7157	0.7190	0.7224
0.6	0.7257	0.7291	0.7324	0.7357	0.7389	0.7422	0.7454	0.7486	0.7517	0.7549
0.7	0.7580	0.7611	0.7642	0.7673	0.7704	0.7734	0.7764	0.7794	0.7823	0.7852
0.8	0.7881	0.7910	0.7939	0.7967	0.7995	0.8023	0.8051	0.8078	0.8106	0.8133
0.9	0.8159	0.8186	0.8212	0.8238	0.8264	0.8289	0.8315	0.8340	0.8365	0.8389
1.0	0.8413	0.8438	0.8461	0.8485	0.8508	0.8531	0.8554	0.8577	0.8599	0.8621
1.1	0.8643	0.8665	0.8686	0.8708	0.8729	0.8749	0.8770	0.8790	0.8810	0.8830
1.2	0.8849	0.8869	0.8888	0.8907	0.8925	0.8944	0.8962	0.8980	0.8997	0.9015
1.3	0.9032	0.9049	0.9066	0.9082	0.9099	0.9115	0.9131	0.9147	0.9162	0.9177
1.4	0.9192	0.9207	0.9222	0.9236	0.9251	0.9265	0.9279	0.9292	0.9306	0.9319
1.5	0.9332	0.9345	0.9357	0.9370	0.9382	0.9394	0.9406	0.9418	0.9429	0.9441
1.6	0.9452	0.9463	0.9474	0.9484	0.9495	0.9505	0.9515	0.9525	0.9535	0.9545
1.7	0.9554	0.9564	0.9573	0.9582	0.9591	0.9599	0.9608	0.9616	0.9625	0.9633
1.8	0.9641	0.9649	0.9656	0.9664	0.9671	0.9678	0.9686	0.9693	0.9699	0.9706
1.9	0.9713	0.9719	0.9726	0.9732	0.9738	0.9744	0.9750	0.9756	0.9761	0.9767
2.0	0.9772	0.9778	0.9783	0.9788	0.9793	0.9798	0.9803	0.9808	0.9812	0.9817
2.1	0.9821	0.9826	0.9830	0.9834	0.9838	0.9842	0.9846	0.9850	0.9854	0.9857
2.2	0.9861	0.9864	0.9868	0.9871	0.9875	0.9878	0.9881	0.9884	0.9887	0.9890
2.3	0.9893	0.9896	0.9898	0.9901	0.9904	0.9906	0.9909	0.9911	0.9913	0.9916
2.4	0.9918	0.9920	0.9922	0.9925	0.9927	0.9929	0.9931	0.9932	0.9934	0.9936
2.5	0.9938	0.9940	0.9941	0.9943	0.9945	0.9946	0.9948	0.9949	0.9951	0.9952
2.6	0.9953	0.9955	0.9956	0.9957	0.9959	0.9960	0.9961	0.9962	0.9963	0.9964

The value that corresponds to a test statistic of 1.58 is 0.9429.
Hence our p-value is $1 - 0.9429 = 0.0571$.

- Find the p-value for a left-tailed test:

Suppose we conduct a left-tailed hypothesis test and get a test statistic of −2.52. What is the p-value that corresponds to this value?
   Since it is a left-tailed test, our p-value $= P(z < -2.52)$.
   To find the p-value, we can first locate the value −2.52 in the z table:

z	0.00	0.01	0.02	0.03	0.04	0.05	0.06	0.07	0.08	0.09
-3.4	0.0003	0.0003	0.0003	0.0003	0.0003	0.0003	0.0003	0.0003	0.0003	0.0002
-3.3	0.0005	0.0005	0.0005	0.0004	0.0004	0.0004	0.0004	0.0004	0.0004	0.0003
-3.2	0.0007	0.0007	0.0006	0.0006	0.0006	0.0006	0.0006	0.0005	0.0005	0.0005
-3.1	0.0010	0.0009	0.0009	0.0009	0.0008	0.0008	0.0008	0.0008	0.0007	0.0007
-3.0	0.0013	0.0013	0.0013	0.0012	0.0012	0.0011	0.0011	0.0011	0.0010	0.0010
-2.9	0.0019	0.0018	0.0018	0.0017	0.0016	0.0016	0.0015	0.0015	0.0014	0.0014
-2.8	0.0026	0.0025	0.0024	0.0023	0.0023	0.0022	0.0021	0.0021	0.0020	0.0019
-2.7	0.0035	0.0034	0.0033	0.0032	0.0031	0.0030	0.0029	0.0028	0.0027	0.0026
-2.6	0.0047	0.0045	0.0044	0.0043	0.0041	0.0040	0.0039	0.0038	0.0037	0.0036
-2.5	0.0062	0.0060	0.0059	0.0057	0.0055	0.0054	0.0052	0.0051	0.0049	0.0048
-2.4	0.0082	0.0080	0.0078	0.0075	0.0073	0.0071	0.0069	0.0068	0.0066	0.0064
-2.3	0.0107	0.0104	0.0102	0.0099	0.0096	0.0094	0.0091	0.0089	0.0087	0.0084
-2.2	0.0139	0.0136	0.0132	0.0129	0.0125	0.0122	0.0119	0.0116	0.0113	0.0110
-2.1	0.0179	0.0174	0.0170	0.0166	0.0162	0.0158	0.0154	0.0150	0.0146	0.0143
-2.0	0.0228	0.0222	0.0217	0.0212	0.0207	0.0202	0.0197	0.0192	0.0188	0.0183
-1.9	0.0287	0.0281	0.0274	0.0268	0.0262	0.0256	0.0250	0.0244	0.0239	0.0233
-1.8	0.0359	0.0351	0.0344	0.0336	0.0329	0.0322	0.0314	0.0307	0.0301	0.0294
-1.7	0.0446	0.0436	0.0427	0.0418	0.0409	0.0401	0.0392	0.0384	0.0375	0.0367
-1.6	0.0548	0.0537	0.0526	0.0516	0.0505	0.0495	0.0485	0.0475	0.0465	0.0455
-1.5	0.0668	0.0655	0.0643	0.0630	0.0618	0.0606	0.0594	0.0582	0.0571	0.0559
-1.4	0.0808	0.0793	0.0778	0.0764	0.0749	0.0735	0.0721	0.0708	0.0694	0.0681
-1.3	0.0968	0.0951	0.0934	0.0918	0.0901	0.0885	0.0869	0.0853	0.0838	0.0823
-1.2	0.1151	0.1131	0.1112	0.1093	0.1075	0.1056	0.1038	0.1020	0.1003	0.0985
-1.1	0.1357	0.1335	0.1314	0.1292	0.1271	0.1251	0.1230	0.1210	0.1190	0.1170
-1.0	0.1587	0.1562	0.1539	0.1515	0.1492	0.1469	0.1446	0.1423	0.1401	0.1379

The p-value that corresponds to a z-score of -2.52 is 0.0059.

3. Decision:

Compare the test statistic with the critical value or compare the p-value with α to arrive at a conclusion. In other words, decide if the null hypothesis is to be rejected or not.

Let's do an example using two approaches.

8.1.1.1 Example

Suppose we randomly sampled 50 students from an honors program and found their IQ score to be of mean 106. We want to determine whether their mean IQ score differs from the general population. The general population's IQ scores are defined as having a mean of 100 and a standard deviation of 20.

8.1.1.1.1 Stating the hypotheses

Null hypothesis: H_0: $\mu = 100$.
Alternative hypothesis: H_A: $\mu \neq 100$.

8.1.1.1.2 Choose significance level

Since no specific α is given, we take the conventional one: 5%.

8.1.1.1.3 Calculate the z test statistic

$$z = \frac{x - \mu_0}{\left(\dfrac{\sigma}{\sqrt{n}}\right)} = \frac{106 - 100}{\left(\dfrac{20}{\sqrt{50}}\right)} = 2.12132$$

8.1.1.1.4 Using either of the two approaches

8.1.1.1.4.1 CRITICAL APPROACH

It is a two-tailed test with confidence level 0.05 so our critical value is ±1.960.

8.1.1.1.4.2 P-VALUE APPROACH

Now we need to find the p-value that corresponds to the test statistic of 2.12132.
Since it is a two-tailed test, our p-value $= 2P(z > |2.12132|) = 2P(z > 2.12132)$

$$= 2(1 - P(z < 2.12132)$$

$$= 0.033897 \text{ for two tails}$$

Since the test statistic $= 2.12132 > 1.960 =$ critical value which clearly implies that it lies in the rejection region which is also shown in the graph; hence we reject the null hypothesis.

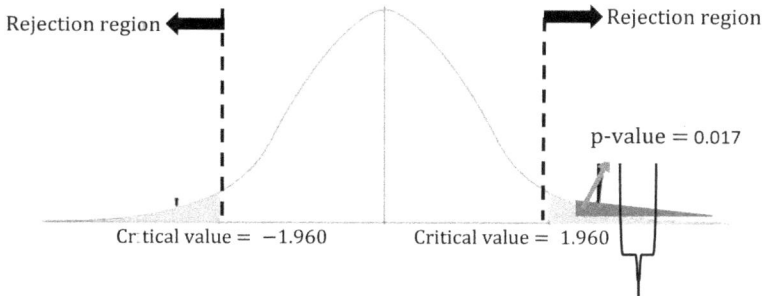

Similarly, p-value $0.033897 < 0.05 = \alpha$; hence we reject the null hypothesis.
Hence, we proved that regardless of what approach we use the conclusion does not change. Let's do another example.

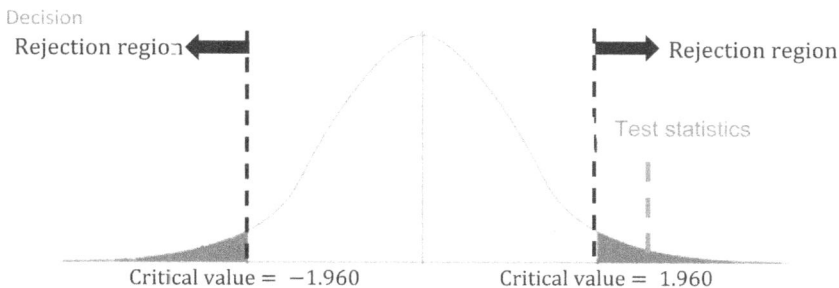

Figure 8.3

8.1.1.2 Example

Incandescent bulbs, generally, have the shortest lifespans. The average incandescent bulb lifespan is approximately 1000 hours. A company wants to determine if the average life of their bulbs is less than that of another incandescent bulb. So to test this claim, 250 new bulbs were allowed to burn out. They were found to have lifespan of 987 hours with a standard deviation of 71 hours. Use a 1% level of significance.

8.1.1.2.1 Stating the hypotheses

Null hypothesis: H_0: $\mu = 1000$.
Alternative hypothesis: H_A: $\mu < 1000$.

8.1.1.2.2 Calculate the z test statistic

Notice the population standard deviation of the light bulbs is not given and since the sample size is sufficiently large, the sample can be used instead.

$$z = \frac{987 - 1000}{\left(\dfrac{71}{\sqrt{250}}\right)} = -2.895$$

8.1.1.2.3 Using either of the two approaches

8.1.1.2.3.1 CRITICAL APPROACH

It is a one-tailed test with confidence level 0.01 so our critical value is −2.33.

8.1.1.2.3.2 P-VALUE APPROACH

Now we need to find the p-value that corresponds to the test statistic of −2.895.
Since it is a left-tailed test, our p-value = $P(z < -2.895)$

$= .001896.$

Critical value $= -2.33$
Since the test statistic falls in the rejection region, we reject the null hypothesis.
Likewise, p-value .001896 $< 0.01 = \alpha$; hence we reject the null hypothesis.

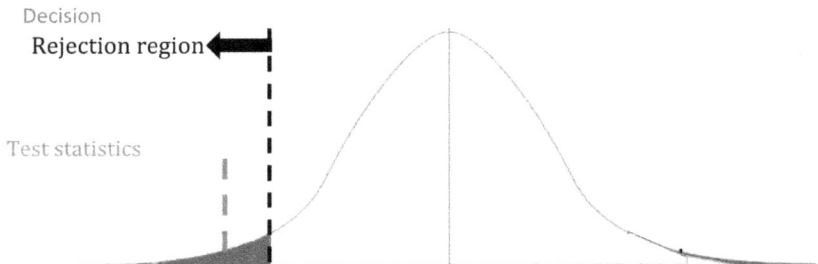

Figure 8.4

8.1.1.3 MATLAB

Now that we have learned to perform the test by hand, let's get on to MATLAB.

Notice that we used two different approaches—critical value and p-value—to come to a conclusion.

8.1.1.3.1 Critical value in MATLAB

Since the z test is based on normal distribution, we will use the norminv function.

The function x = norminv(p) returns the inverse of the standard normal CDF, evaluated at the probability values in p.

Let's say we want to find the critical value for a two-tailed test with alpha level 0.05.

Since it is a two-tailed test, each tail will get $0.05 / 2 = 0.025$.

From the left side 0.025 and from the right $1 - 0.025 = 0.975$.

```
critical_value = norminv([0.025 0.975])
```

Now say we want to find the critical value for a left-tailed test with alpha level 0.05.

Since it is a one-tailed test, the left tail will get the entire area, 0.05.

```
critical_value = norminv(0.05)
```

Since normal distribution provides the area of the region located to the left of a given z-score to represent probabilities of occurrence in a given population, our alpha level would be $1 - 0.05 = 0.95$.

```
critical_value = norminv(0.95)
```

8.1.1.3.2 P-value in MATLAB

Two-tailed: p_value = 2*(1-normcdf(abs(test_statistics)))
Right-tailed: p_value = 1-normcdf(abs(test_statistics))
Left-tailed: p_value = normcdf(abs(test_statistics))

These approaches might seem a bit of a handful; however, fortunately, MATLAB has in-built functions in the Statistics Toolbox that handle the details of the z test for us.

The function:

```
h = ztest(x, m, sigma, 'Alpha', value , 'Tail','type')
```

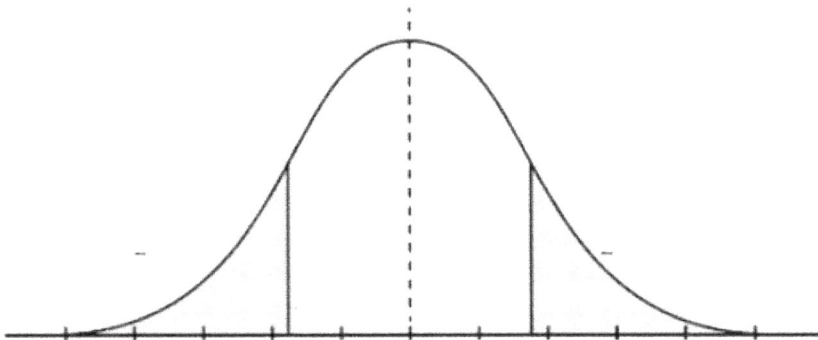

Figure 8.5

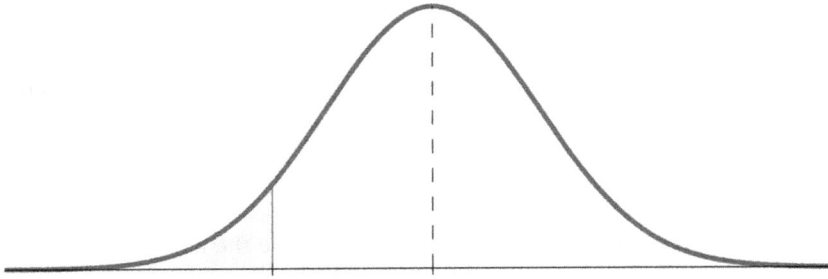

Figure 8.6

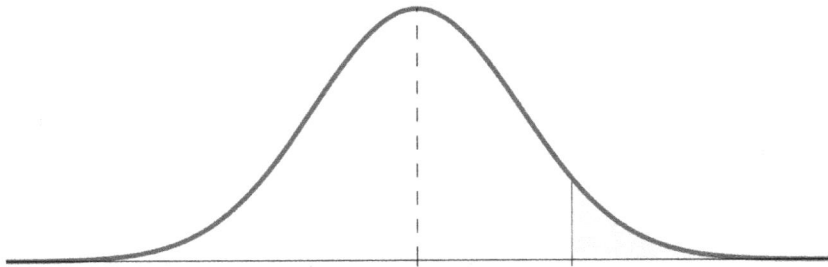

Figure 8.7

8.1.1.3.2.1 THE INPUT ARGUMENTS

x = sample mean

m = hypothesized population mean

sigma = population standard deviation. As mentioned, if the population is not given then the sample can be used instead as long as the sample size is > 30 (central limit theorem).

The next two arguments are optional, meaning if they are left out, then there is a default value to conduct the test.

'Alpha' = significance level followed by its value in the range $(0,1)$. We have already seen some of the most commonly used alpha levels earlier. Here the default alpha value is 0.05.

'Tail' = followed by the type of test (two-tailed/right/left). Here the default is both, which means a two-tailed test.

8.1.1.4 Example

```
h = ztest(x, m, sigma)
```
This specifies a two-tailed hypothesis test at the 5% alpha level.

```
h = ztest(x, m, sigma, 'Alpha', 0.01 , 'Tail','right')
```
This specifies a right-tailed hypothesis test at the 1% alpha level.

```
h = ztest(x, m, sigma, 'Alpha', 0.001 , 'Tail','left')
```
This specifies a left-tailed hypothesis test at the 0.001 alpha level.

Now what do these two functions return? What is h?

In MATLAB the test result is returned as 1 or 0.

- If h = 1, this indicates the rejection of the null hypothesis at a given significance level.
- If h = 0, this indicates a failure to reject the null hypothesis at a given significance level.

In addition to h it also returns:

- The p-value of the test. For simplicity we can denote it by p.
- The confidence interval of the population mean. We can denote it by ci.
- The value of test statistics. Let's say we denote it by zstat.

To get all of the outputs together we use the function:

```
[h,p,ci,zstat] = ztest()
```

However, if we don't want one or the other we will simply replace that function with ~.

For example if we want to leave out the ci we can write [h,p,~,zstat] = ztest.

Same thing could be done if we decide to keep the ci and skip the zstat.

Yet if we only want to include the h and p in the output we can simply write it as [h,p].

Notice that the inbuilt functions for the z test in MATLAB only use the p-value to make the comparison. That is, it uses the zstat to calculate the p-value and then compares this with the significance level (0.05 by default) to draw conclusions.

Let's now move on to the example we have previously done by hand.

But before we can jump into the calculation we first need to check if the z test is actually applicable in this example.

Since it fulfils the criteria of larger sample size $(n = 50 > 30)$ and known population standard deviation 20, we now need to check for normality—if the sample data has been drawn from a normal distribution.

```
%importing the Excel data into MATLAB
data = readtable('IQ.csv');
IQ = table2array(data(:, 'IQ'));

%Displaying histogram and Q-Q plot
tiledlayout(2,1)

% Top axes
ax1 = nexttile;
histogram(IQ,8)
xlabel('IQ')

% Bottom axes
ax2 = nexttile;
qqplot(ax2,IQ)
ylabel(ax2,'IQ scores')
title(ax2,'QQ Plot of IQ scores in 50 people vs. Standard Normal')
```

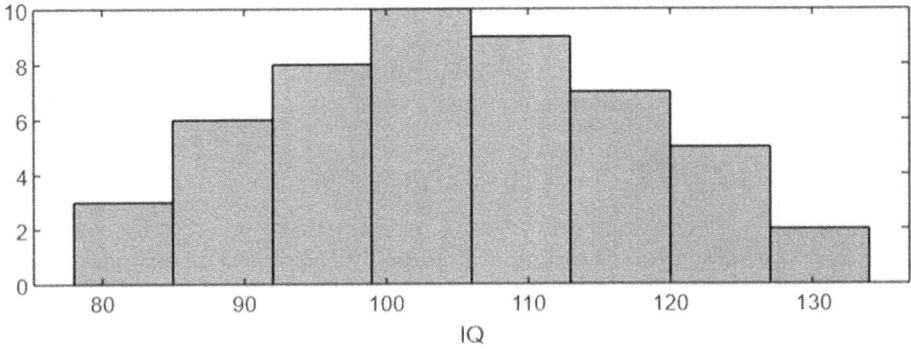

Figure 8.8

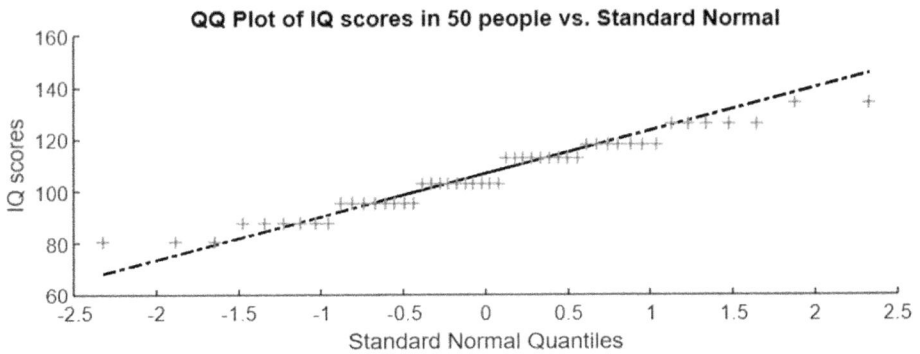

Figure 8.9

From the graphs it is evident that the data collected is roughly normally distributed; therefore we can perform the z test.

```
data = readtable('IQ.csv');
IQ = table2array(data(:, 'IQ'));
[h,p,ci,zstat] = ztest(IQ,100,20)
```

```
h =

     1


p =

    0.0339


ci =

   100.4564
   111.5436


zstat =

    2.1213
```

The output aligns with our calculated result which implies that we reject the null hypothesis and that

8.1.1.5 Example 2

There has been a long discussion about the life expectancy of Texans and Californians. Suppose the life expectancy of Californians is 77.10 with a standard deviation of 8.20. We wish to know if, on the basis of this data, we may conclude that the life expectancy of Texans is just about the same. To test this out we randomly selected 100 Texans and found their mean life expectancy to be 76. Use $\alpha = 0.01$.

First to ensure if our data satisfies normality:

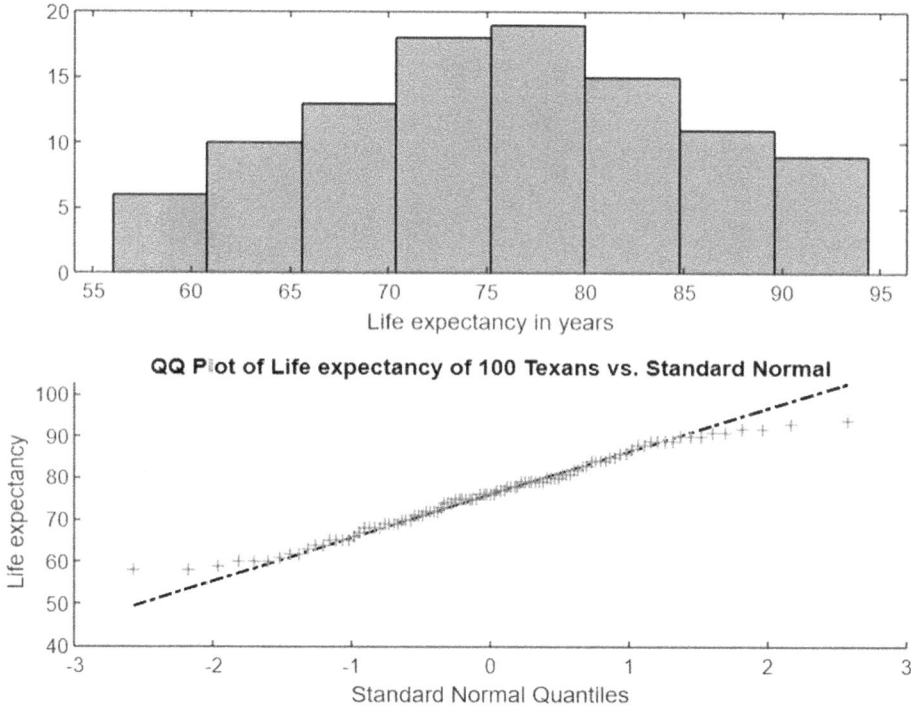

Figure 8.10

We can also use another visualization technique—the boxplot in Figure 8.11:

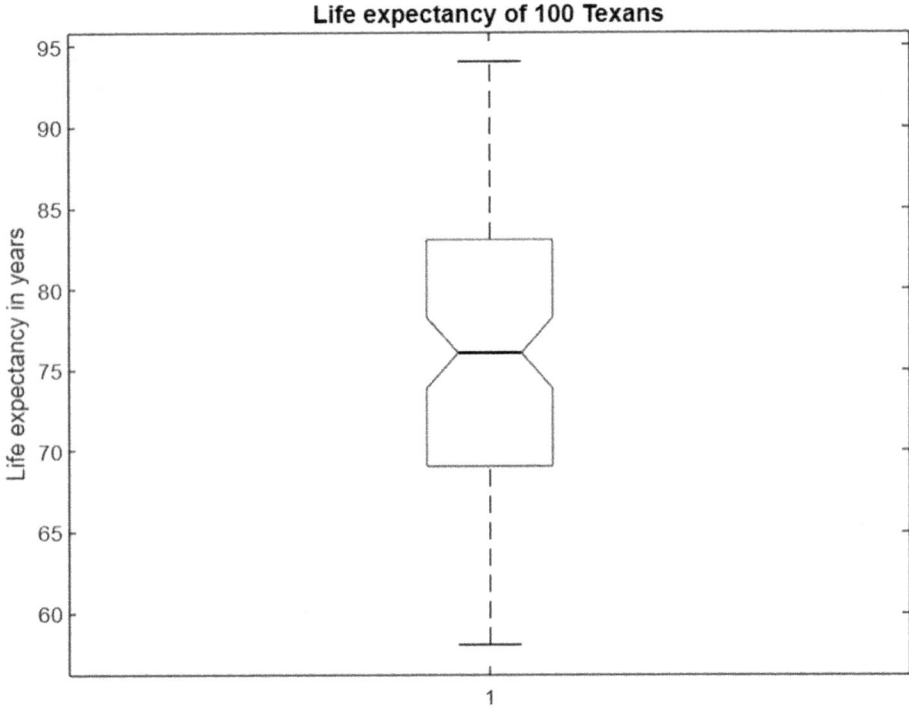

Figure 8.11

Graphically we can confirm that the data is normally distributed. Now let's prove that analytically.

```
data = readtable('Life Expectancy.csv');
Life_expectancy = table2array(data(:, 'Expectancy'));

[h,p,~,zstat] = ztest(Life_expectancy,77.10,8.20,'Alpha',0.01 )
```

```
h =

     0


p =

   0.2485


zstat =

  -1.1540
```

8.1.1.6 Example 3

A professor at Stanford currently teaches students to use a studying method that results in an average exam score of 66. However, his unsatisfied students developed a new studying method among themselves which they believed would produce exam scores with an average value greater than 66. To test this, the professor noted down the test scores of all his students and found their mean and standard deviation.

8.1.1.6.1 Stating the hypotheses

Null hypothesis: H_0: the average score was 66 so $\mu = 66$.

Alternative hypothesis: H_A: the average score was more than 66 so $\mu > 66$.

8.1.1.6.2 Significance level

Here too = 5%.

```
data = readtable('exam scores.csv');
scores = table2array(data(:, 'scores'));

%using sample standard deviation
[h,p] = ztest(scores,66,'Tail','right' )
```

```
                        mean =

                          64.8739


                        median =

                          63.5000


                        mode =

                          61.9900


                        skewness =

                          -0.0432


                        kurtosis =

                          2.4591
```

From the table we can see that the three measures of central tendency are relatively similar, the skewness is close to 0, and the kurtosis rounds to 3.0. The values confirm that the data is roughly normally distributed making the z test applicable.

8.1.1.7 Practice questions

1) According to the Centers for Disease Control and Prevention (CDC) Trusted Source, the average height of 21-year-old Hispanic men is 68.10 inches. We wish to know if, on the basis of these data, we may conclude that African American men are taller than Hispanic men. To test this out we selected 500 African American men and found their mean height and standard deviation to be 69.50 inches and 13.50. Use $\alpha = 0.01$.

2) The average score of an eight-grader in a Dallas school district on a general science aptitude exam is 80 with $\sigma = 9.75$. A random sample of 75 students in one school was taken. The mean score of these students was 77. Does this indicate that the students of this school are significantly lower in their scientific ability?

3) A normal resting heart rate for adults is known to be 70 beats per minute on average, with a standard deviation of 10.4 beats per minute. A set of researchers believe that a high level of caffeine has multiple effects on the central nervous system especially on the heart rate. To test this hypothesis, the researchers measured the pulse rates of a group of 150 regular caffeine drinkers and found a mean pulse rate of 72.16 beats per minute. Researchers want to know if the mean pulse rate for coffee drinkers is any different than the current standard of normal resting heart rate. Conduct the test at the 5% level and interpret your result.

4) CS major Monica believes that students majoring in computer science are the most sleep-deprived, getting 5.85 hours a night on average with a standard deviation of 1.05 hours. However her best friend, Rachel, who is studying architecture, argues that students in her degree are the most sleep-deprived. To prove her claim she collected the data of the number of hours sleep per night for 80 students and found the mean hours of sleep to be 6.05, which was averaged over a one-month monitoring period. Is there any evidence to suggest that Rachel's belief is incorrect? Conduct the test at the 10% level and interpret your result.

5) According to the CDC the cholesterol levels in our bodies vary by race and ethnicity as well as gender. They claim that Black men have the lowest level of cholesterol with an average of 4.10 mmol / L with standard deviation of 2.20. However in one study **non-Hispanic white men were also claimed to have cholesterol levels as low as Black men**. To verify the study a researcher took a sample of 220 individuals who were non-Hispanic white men and noted their average cholesterol level and found it to be around 4.50 mmol / L. Test the hypothesis at the 1% level.

8.1.2 Two-sample z test

As previously stated, the other type of z test which will now discuss is the two-sample z test. While the one-sample test checks the difference between the population mean and sample mean, the two-sample test is used to find if a difference exists between the averages of two different populations. In this test we get two samples from each population and determine if they are equal. For example we can use the test to compare:

- The performance of students in two different classes
- The average salaries of men and women in a company
- The KPIs of two different teams
- The performance of employees in two different departments
- The average IQ scores of two groups of people
- The body mass index of men and women

8.1.2.1 Calculating the test statistic

Two-sample $z = \dfrac{(x_1 - x_2) - (\mu_1 - \mu_2)}{\sqrt{\dfrac{\sigma_1^2}{n_1} + \dfrac{\sigma_2^2}{n_2}}}$

Where

x_1 = mean of the first sample

x_2 = mean of the second sample

μ_1 = hypothesized mean of the first population

μ_2 = hypothesized mean of the second population

$\mu_1 - \mu_2$ = hypothesized difference between the population means

σ_1 = standard deviation of the first population

σ_2 = standard deviation of the second population

n_1 = sample size of the first sample

n_2 = sample size of the second sample

8.1.2.2 Example

A researcher wants to investigate whether or not a given drug has any effect on
the performance on a task of ESP sensitivity. The experiment was assessed via the scores
of the human subjects performing the task.

Prior to testing, 500 subjects in group 1 (the experimental group) receive an oral adminis-
tration of the drug. In contrast, 750 subjects in group 2 (the control group) receive a placebo.

The results of the study found that the mean score of the subjects in group 1 on the ESP
test was 9.48 with standard deviation 4.01, and for the no-drug group (placebo), the mean
score was 12.75, and the standard deviation was 4.21.

8.1.2.2.1 State the null and alternative hypotheses

Null hypothesis H_0 : there is no difference between the population means of the drug group
and the no-drug group on the test of ESP sensitivity.

$$H_0 : \mu_1 = \mu_2$$

Alternative hypothesis H_A: there is a difference between the population means of the drug
group and the no-drug group on the test of ESP sensitivity.

$$H_A : \mu_1 \neq \mu_2$$

Level of significance: $\alpha = 0.05$

8.1.2.2.2 Calculate the test statistic

$$z = \frac{(9.48 - 12.75) - (0)}{\sqrt{\dfrac{4.01^2}{500} + \dfrac{4.21^2}{750}}} = \frac{-3.27}{\sqrt{0.032 + 0.023632}} = -13.844$$

8.1.2.2.3 Using either of the two approaches

8.1.2.2.3.1 CRITICAL APPROACH

It is a two-tailed test with confidence level 0.05 so our critical value is ± 1.960.

8.1.2.2.3.2 P-VALUE APPROACH

Now we need to find the p-value that corresponds to the test statistic of -13.844.

Since it is a two-tailed test, our p-value $= 2P(z > |-13.844|) = 2P(z > 13.844)$

$$= 2(1 - P(z < 13.844)$$

almost negligible

8.1.2.2.4 Decision

8.1.2.2.4.1 CRITICAL APPROACH

Since the test statistic is far to the left of the critical value, i.e. $-13.844 < -1.960$, this clearly implies that it will lie in the rejection region; hence we reject the null hypothesis.

8.1.2.2.4.2 P-VALUE APPROACH

Again since p-value associated with a difference this strong or stronger is extremely rare, we *reject* the plausibility of our null hypothesis.

We therefore conclude that indeed there is a difference between the two groups. In fact the group that got the drug did worse than the control group. We know this from the negative sign of the difference between the two means. It appears the drug hampers ESP ability.

Now unfortunately MATLAB does not have any inbuilt functions to perform the two-sample hypothesis test. So if we still want to do the testing we will need to use the traditional approach. For example:

```
sample_size1 = 500
sample_mean1 = 9.48
sample_std1 = 4.01

sample_size2 = 750
sample_mean2 = 12.75
sample_std2 = 4.21

significance_level = 0.05

test_statistics = (sample_mean1- sample_mean2 ) /
 sqrt(((sample_std1)^2 / sample_size1) + ((sample_std2)^2/sample_size2))
```

```
%Using critical value approach
critical_value = norminv([0.025 0.975])

if(test_statistics < critical_value)
  disp('Do not reject the null hypothesis')
else
  disp('reject the null hypothesis')
end
```

```
%Using p-value approach
p_value = 2*(1-normcdf(abs(test_statistics)))
if(p_value > significance_level)
  disp('Do not reject the null hypothesis')
else
  disp('reject the null hypothesis')
end
```

8.1.2.3 Practice questions

1) Thomas Jefferson High School physics teacher Robert claims that students in his section will score higher marks than those in his colleague Sue's section. The mean score for 62 students in Robert's section is 21.50, and the standard deviation is 4.20. The mean score for 45 of the colleagues' sections is 19.20, and the standard deviation is 6.10. At $\alpha = 0.05$, can the teacher's claim be supported?

2) A dietician made a new weight loss program and claims that this new program is way more effective than the old one he gave to his clients. To test his claim, he calculated the weight reduction each program yielded. The mean weight reduction for 120 participants in the old weight loss program was 15.8 lb, with standard deviation 4.2. Similarly, for the new program, the mean weight reduction for 150 participants is 17.10 lb, and the standard deviation is 6.40. At the 1% level of significance, can you support the dietician's claim?

3) Gregory sports center is planning to compare the ages from a random sample of male and female players in their center. The coordinator collected data on 75 female runners and calculated their mean age to be 23.1 with a standard deviation of 3.5. Similarly, collected data on 46 male swimmers and the mean average is 19.2 with a standard deviation of 4.8. Assume the population follows a standard normal distribution. At a 5% significance level, test whether there is a significant difference in age between the sexes.

The z test can also be used to test the population proportion. But since there are no inbuilt function for proportion in MATLAB we shall not discuss it in this textbook.

8.2 T TEST

When it comes to statistical tests, another widely used test is the t test. This is similar to the z test and is also used to determine whether the mean calculated from sample data collected from a single group is different from the mean of the larger population. But while the z test is based on a standard normal distribution, the t test makes use of the t distribution. It also differs from a z test in the following aspects:

- Unlike the z test which is used for large datasets, t tests work with relatively smaller sample size data which ideally contains only 5 to 30 sample units. So does that mean we cannot use the t test for larger sample sizes? Absolutely we can; however it is not preferred. The reason behind this is that if the size of the sample is more than 30, then the distribution of the t test and the normal distribution will not be distinguishable.
- In the Z test, if the population standard deviation is **unknown,** we can use a sample standard deviation given that the sample size is large enough. But what if it's not? This is when we use a t distribution instead of a normal distribution. Simply stated, in a t test, the population variance is unknown.
- While the z test uses the standard normal distribution to compute the critical values, the t test makes use of degree of freedoms for the calculation.

Just like the z test, t-tests are too classified into two: one-sample and two-sample tests.

8.2.1 One-sample t test

Just like the one-sample z test, the t test is also carried out to determine whether the mean of a population is statistically different from a known or hypothesized value.

As we know already every hypothesis test requires the test statistic to make a decision. In the t test, the statistic is calculated by using the formula:

$$\text{test statistic} = \frac{\bar{x} - \mu_0}{\left(\dfrac{s}{\sqrt{n}}\right)}$$

where

$\bar{x}$ = sample mean
μ_0 = hypothesized population mean
s = sample standard deviation

The two approaches:

- Comparing the test statistic with the critical value

To find the critical value we need:

- A significance level (common choices are 0.01, 0.05, and 0.10).
- The degrees of freedom: can be found by subtracting 1 from the sample size $df = n - 1$.
- The type of test (one-tailed or two-tailed): if the hypothesis test is one-tailed then use the one-tailed t distribution table. Otherwise, use the two-tailed t distribution table for a two-tailed test.

After we get all three, we can now use our t distribution table and find the value which corresponds to all of the three above (with the intersection of the row representing the df and the alpha column value matching with the tail).

8.2.1.1 Example

Find the critical value for a two-tailed test with $\alpha = 0.05$ and 30 degrees of freedom.

t Table

cum. prob	$t_{.50}$	$t_{.75}$	$t_{.80}$	$t_{.85}$	$t_{.90}$	$t_{.95}$	$t_{.975}$	$t_{.99}$	$t_{.995}$	$t_{.999}$	$t_{.9995}$
one-tail	0.50	0.25	0.20	0.15	0.10	0.05	0.025	0.01	0.005	0.001	0.0005
two-tails	1.00	0.50	0.40	0.30	0.20	0.10	0.05	0.02	0.01	0.002	0.001
df											
1	0.000	1.000	1.376	1.963	3.078	6.314	12.71	31.82	63.66	318.31	636.62
2	0.000	0.816	1.061	1.386	1.886	2.920	4.303	6.965	9.925	22.327	31.599
3	0.000	0.765	0.978	1.250	1.638	2.353	3.182	4.541	5.841	10.215	12.924
4	0.000	0.741	0.941	1.190	1.533	2.132	2.776	3.747	4.604	7.173	8.610
5	0.000	0.727	0.920	1.156	1.476	2.015	2.571	3.365	4.032	5.893	6.869
6	0.000	0.718	0.906	1.134	1.440	1.943	2.447	3.143	3.707	5.208	5.959
7	0.000	0.711	0.896	1.119	1.415	1.895	2.365	2.998	3.499	4.785	5.408
8	0.000	0.706	0.889	1.108	1.397	1.860	2.306	2.896	3.355	4.501	5.041
9	0.000	0.703	0.883	1.100	1.383	1.833	2.262	2.821	3.250	4.297	4.781
10	0.000	0.700	0.879	1.093	1.372	1.812	2.228	2.764	3.169	4.144	4.587
11	0.000	0.697	0.876	1.088	1.363	1.796	2.201	2.718	3.106	4.025	4.437
12	0.000	0.695	0.873	1.083	1.356	1.782	2.179	2.681	3.055	3.930	4.318
13	0.000	0.694	0.870	1.079	1.350	1.771	2.160	2.650	3.012	3.852	4.221
14	0.000	0.692	0.868	1.076	1.345	1.761	2.145	2.624	2.977	3.787	4.140
15	0.000	0.691	0.866	1.074	1.341	1.753	2.131	2.602	2.947	3.733	4.073
16	0.000	0.690	0.865	1.071	1.337	1.746	2.120	2.583	2.921	3.686	4.015
17	0.000	0.689	0.863	1.069	1.333	1.740	2.110	2.567	2.898	3.646	3.965
18	0.000	0.688	0.862	1.067	1.330	1.734	2.101	2.552	2.878	3.610	3.922
19	0.000	0.688	0.861	1.066	1.328	1.729	2.093	2.539	2.861	3.579	3.883
20	0.000	0.687	0.860	1.064	1.325	1.725	2.086	2.528	2.845	3.552	3.850
21	0.000	0.686	0.859	1.063	1.323	1.721	2.080	2.518	2.831	3.527	3.819
22	0.000	0.686	0.858	1.061	1.321	1.717	2.074	2.508	2.819	3.505	3.792
23	0.000	0.685	0.858	1.060	1.319	1.714	2.069	2.500	2.807	3.485	3.768
24	0.000	0.685	0.857	1.059	1.318	1.711	2.064	2.492	2.797	3.467	3.745
25	0.000	0.684	0.856	1.058	1.316	1.708	2.060	2.485	2.787	3.450	3.725
26	0.000	0.684	0.856	1.058	1.315	1.706	2.056	2.479	2.779	3.435	3.707
27	0.000	0.684	0.855	1.057	1.314	1.703	2.052	2.473	2.771	3.421	3.690
28	0.000	0.683	0.855	1.056	1.313	1.701	2.048	2.467	2.763	3.408	3.674
29	0.000	0.683	0.854	1.055	1.311	1.699	2.045	2.462	2.756	3.396	3.659
30	0.000	0.683	0.854	1.055	1.310	1.697	2.042	2.457	2.750	3.385	3.646
40	0.000	0.681	0.851	1.050	1.303	1.684	2.021	2.423	2.704	3.307	3.551

Since it is two-tailed, our critical value is ±2.042.
Using the test statistic to find the p-value and compare it with the alpha level

- After calculating the test statistic now we calculate the *df* from its sample size.
- In the t-distribution table, we need to look at the row that corresponds to *df* from the preceding step and attempt to look for our test statistic **from step 2**.
- It is unlikely that we will find the exact test statistic value from the table so we will try to look for the closest one.

For example let's say we get a test statistic of 2.13 and *df* = 14. If we get the table and look for the row representing *df* = 14, we will notice that the value closest to our test statistic is 2.145. If our test is one sided then our p-value will be close to 0.025 and if its two-tailed then it will be close to 0.05.

8.2.1.2 Example

In a certain population, the average IQ is 100. CDER wants to test a new medication to see if it has either a positive or negative effect on intelligence, or no effect at all. A sample of 20 participants who have taken the medication has a mean of 124.55 with standard deviation of 8.89. Did the medication affect intelligence? Use alpha = 0.01. .

8.2.1.2.1 Stating the hypotheses

H_0 (null hypothesis): $\mu = 100$.
 (alternative hypothesis): $\mu \neq 100$.

8.2.1.2.2 Choose significance level

Given $1\% = 0.01$.

8.2.1.2.3 Calculate the t test statistic

T-statistic value $= \dfrac{124.55 - 100}{\dfrac{8.39}{\sqrt{20}}} = 12.3527$.

8.2.1.2.4 Using critical approach

It is a two-tailed t test, $df = 20 - 1 = 19$, alpha level 0.01, so our critical value is ±2.861.

8.2.1.2.5 Conclusion/decision

Since the t test statistic value $= 12.4226 > 2.861$, it falls in the rejection region and thus we reject the null hypothesis. Therefore we conclude that the medication significantly affected intelligence.

8.2.1.3 MATLAB

8.2.1.3.1 How to find the t critical value in MATLAB

Unlike the z test, the t test is based on Student's t distribution; hence we will use the tinv function that returns the inverse of Student's t cumulative distribution.

Just like norminv, this function also evaluates at the probability values in p but using the corresponding degrees of freedom:

```
x = tinv(p,df)
```

The syntax to find the critical value of the t test at a particular alpha level is almost identical to that of the z test with the addition of the degrees of freedom.

To find the critical value for a two-tailed test with alpha level 0.05 and df = 25:

```
critical _ value = tinv([0.025 0.975],25).
```

Similarly for left-tailed: critical_value = tinv(0.05,25).

Similarly for right-tailed: critical_value = tinv(0.95,25).

8.2.1.3.2 P-value in MATLAB

Like the z test, the t test uses *cdf* to calculate the p-value, making the syntax almost the same.

MATLAB also provides functions to perform the t test which has the exact same input arguments and output.

The function:

h = ttest(x,m,Name,Value) returns a test decision for the one-sample *t* test with additional options specified by one or more name-value pair arguments. For example, you can change the significance level or conduct a one-sided test.

```
[h,p,ci,stats] = ttest(x, m, sigma, 'Alpha', value , 'Tail','type')
```

Doing the same problem:

First make sure the data is normally distributed using the same graphical technique:

```
data = readtable('IQ vs Medication.csv');
IQ = table2array(data(:, 'IQT'));

tiledlayout(2,2)
% Top axes
ax1 = nexttile;
histogram(IQ, 5)
xlabel("IQ")
title("IQ of 20 participants after medication")

% Bottom axes
ax2 = nexttile;
qqplot(IQ)
ylabel('IQ')
title('QQ Plot of IQ of 20 participants after medication vs. Standard
Normal')
```

```
ax3 = nexttile;
boxplot(IQ)
ylabel('IQ')
title('Boxplot Plot of IQ of 20 participants after medication')

[h,p,ci,stats] = ttest(IQ, 100)
```

```
                        h =

                          1


              p =

                1.5861e-10


              ci =

                120.3903
                128.7097


              stats =

                struct with fields:

                  tstat: 12.3527
                     df: 19
                     sd: 8.8880
```

8.2.1.4 Example

PCOS is a very common hormone problem that affects around 116 million women globally. Many studies have suggested that women with PCOS have higher cholesterol levels than the average with normal cholesterol levels of 170 mg/dL. To test the claim, the cholesterol levels of 18 women with PCOS are measured. Using the data below determine if the average cholesterol level of these women is higher than 170 mg / dL.

Women	Cholesterol level
1	167.11
2	171.82
3	174.92
4	165.49
5	179.95
6	172.79
7	161.51
8	172.63
Women	Cholesterol level
9	175.3
10	169.7
11	188.72

12	176.47
13	168.93
14	183.85
15	177.59
16	182.5

8.2.1.4.1 Defining the hypotheses

H_0 (null hypothesis): $\mu \leq 170$.
H_A (alternative hypothesis): $\mu > 170$.

```
[h,~,~,stats] = ttest(x,20,'Alpha',0.01,'Tail','right')

                            h =

                               0

                         stats =

                           struct with fields:

                              tstat: 2.4185
                                 df: 15
                                 sd: 7.1616
```

From the given data we get the mean $= 174.33$ and standard deviation $= 7.1616$ so our test statistic becomes $\dfrac{174.33 - 170}{\dfrac{7.16155}{\sqrt{16}}} = 2.4185$ which is what we got in the output.

8.2.1.4.2 Calculating the critical value

- A significance level of 0.01
- The degrees of freedom $df = 16 - 1 = 15$
- Right-tailed test

For a t-distribution with 15 degrees of freedom and a significance level of 0.01 the critical value would indeed be approximately 2.602

8.2.1.4.3 Decision

Since *test statistics* $= 2.4185 < 2.602 = critical\ value$, we fail to reject the null hypothesis which is why we got $h = 0$, and we can conclude that women with PCOS do not have high cholesterol.

8.2.1.5 Example

In 2021, it was found that the average annual electricity consumption for a US residential utility customer was around 875 kWh per month. A manufacturer has invented a type of LED light and claims that it uses at least 75% less energy which would in turn reduce the current consumption. To test the claim we fit these lights in 13 households and noted their consumption in the following month using those lights given below. Should we accept or reject the manufacturer's claim?

Household	1	2	3	4	5	6	7	8	9
Consumption kWh / month	977	829	798	894	1051	1200	1089	950	1019

8.2.1.5.1 Defining the hypotheses

H_0 (null hypothesis): $\mu \geq 875$.
  H_A (alternative hypothesis): $\mu < 875$.

```
h =

      0


p =

   0.9793


stats =

  struct with fields:

     tstat: 2.4272
       df: 8
       sd: 127.9913
```

From the given data we get the *mean* $= 978.56$ and standard deviation $= 127.9913$ so our test statistic becomes $\dfrac{978.56 - 875}{\dfrac{127.9913}{\sqrt{9}}} = 2.4272$ which is what we got in the output.

8.2.1.5.2 Calculating the p-value

We got our test statistic of 2.4272 with $df = 8$. So now we look at the row that corresponds to $df = 8$. The value closest to our statistics is between 2.306 and 2.896 corresponding to 0.025 and 0.01. The middle value is around 2.601 corresponding to 0.0175. Then to get more precise we find the middle value of 2.306 and 2.601 in parallel to 0.025 and 0.0175. The middle value is 2.4535 corresponding to the value 0.02125, which now makes more sense. Since it is a left-tailed test, our p-value $= 1 - 0.02125 = 0.97875 \sim 0.979$ which again is roughly the same as in the output.

8.2.1.5.3 Decision

Since our p-value $= 0.979 > 0.05 =$ alpha value, we accept the null hypothesis $(h = 0)$ and reject the manufacturer's claim.

8.2.1.6 Exercise problems

1) To be considered a good golfer, the player has to drive a golf ball of a minimum of 200 yards. The one constant in golf is that every player wants to optimize their distance. One of the factors that can get the players the most yardage is their golf ball. Hence choosing the right golf ball is a must for ultimate distance. A company claims that their golf balls can drive over 200 yards, and to test their claim 12 of their balls are

tested and the overall distance is measured. Below is the distance each ball could drive. Use the data to determine if the manufacturer's claim should be accepted.

Golf ball	1	2	3	4	5	6	7	8	9	10	11	12
Distance	189	196	211	218	207	192	224	228	212	236	198	205

2) It was decided that protein bars which contained less than 20 grams of protein would be removed from the Amazon.co.uk 1 Marketplace. A company which makes different flavors of protein bar claimed that they contain more than 20 grams. To test this, a random sample of 15 different-flavored protein bars were collected. The protein content of each of the bars is given below. Does this average from our sample of 15 bars imply that the company should not go bankrupt for selling protein bars that had less protein content?

Protein bar	Protein content
1	18.50
2	20.40
3	21.20
4	19.50
5	20.50
6	24.80
7	22.40
8	21.80

Protein bar	Protein content
1	22.30
2	19.80
3	20.80
4	21.50
5	23.60
6	22.30
7	22.70
8	21.50

3) Karen, a car dealer, claims that BMW cars give a highway mileage of 20 kmpl on average. However his friend, Josh, who owns a Porsche car dealership, argues that his cars return a higher mileage. To test the claim, Karen samples ten of Josh's Porsches, measures their mileage, and notes the data below.

Porsche	Mileage
1	19.5
2	22.8
3	21.5
4	21.8
5	22.4

Porsche	Mileage
1	23
2	22
3	24.2
4	24.8
5	26.3

4) It takes the average person about **three hours** to solve a Rubik's cube for the first time. However, a school claims that their students can solve it faster than the average person. To test their claim 23 of their students were asked to solve the cube and the time taken was recorded.

Student	Time taken (hours)
1	3.20
2	2.82
3	2.59
4	2.75
5	3.23
6	2.68
7	3.09
8	3.33
9	2.89
10	2.96
11	3.45
12	3.68
Student	Time taken (hours)
13	2.99
14	3.21
15	2.92
16	3.76
17	3.48
18	3.62
19	3.42
20	3.29
21	3.84
22	3.56
23	3.12

8.2.2 Two-sample t test

When it comes to testing hypotheses regarding two population means, the most commonly used test is the two-sample t test.

8.2.2.1 Independent test

The two-sample *t* test (also known as the independent-samples *t* test) is a method used to test whether the unknown population means of two groups are equal or not. You can use the test when your data values are independent, are randomly sampled from two normal populations, and the two independent groups have equal variances (pooled test). But the mandatory requirement is that you need to have two **independent samples**. This means that there is no relationship between the two groups.

The values in one sample have no association with the values in the other sample and are not related in any way.

For example:

- Comparing the BMI between men and women
- Comparing the average test scores of two classes from two different schools
- Measuring the difference in height between two different races
- Assessing if there is a difference in blood pressure levels between patients taking medication and those not taking medication
- Comparing the difference in income between undergraduates and postgraduates

There are two versions of this test; one is used when the variances of the two populations are equal or similar (the pooled test) and the other one is used when the variances of the two populations are unequal (the unpooled test). The calculation of the t-statistic differs slightly for the two scenarios.

8.2.2.1.1 T-statistics when homogeneity of variances does not hold (variances not equal)

When the variances of populations are not equal, the following formula is used to calculate the T-statistic and degrees of freedom:

$$test\ statistic = \frac{(x_1 - x_2) - (\mu_1 - \mu_2)}{\sqrt{\dfrac{s_1^2}{n_1} + \dfrac{s_2^2}{n_2}}}$$

where

x_1 = mean of the first sample

x_2 = mean of the second sample

μ_1 = hypothesized mean of the first population

μ_2 = hypothesized mean of the second population

$\mu_1 - \mu_2$ = hypothesized difference between the population means

s_1 = standard deviation of the first population

s_2 = standard deviation of the second population

n_1 = sample size of the first sample

n_2 = sample size of the second sample

The degrees of freedom can be calculated using the formula

$$df = \frac{\left(\dfrac{s_1^2}{n_1} + \dfrac{s_2^2}{n_2}\right)^2}{\left(\dfrac{1}{n_1-1}\right)\left(\dfrac{s_1^2}{n_1}\right)^2 + \left(\dfrac{1}{n_2-1}\right)\left(\dfrac{s_2^2}{n_2}\right)^2}.$$

8.2.2.1.2 T-statistics when population variances or standard deviations are equal

$$\text{test statistic} = \frac{(x_1 - x_2) - (\mu_1 - \mu_2)}{s_p\sqrt{\dfrac{1}{n_1} + \dfrac{1}{n_2}}} \quad \text{and } s_p = \sqrt{\frac{(n_1-1)s_1^2 + (n_2-1)s_2^2}{n_1 + n_2 - 2}}$$

where

s_p = pooled standard deviation

x_1 = mean of the first sample

x_2 = mean of the second sample

μ_1 = hypothesized mean of the first population

μ_2 = hypothesized mean of the second population

$\mu_1 - \mu_2$ = hypothesized difference between the population means

n_1 = sample size of the first sample

n_2 = sample size of the second sample

The degrees of freedom can be calculated using the formula $df = n_1 + n_2 - 2$.

8.2.2.2 Summary

Two sample t test

Statistics with pooled variance	Statistics with unpooled variance
where	

8.2.2.3 Example

Sam, a freshman, wants to test the differences in BMI between males and females. To perform the test he takes a sample of 25 males and females and records their BMI. The data is provided below:

Group	BMI								
Men	15.58	22.46	25.82	19.85	14.80	21.75	16.96	23.98	23.34
	13.45	20.47	24.62	20.71	17.91	27.28	17.36	18.69	21.03
	19.68	18.10	17.63	19.23	20.33	21.55	23.19		
Women	17.56	14.07	21.85	15.85	16.85	18.89	18.64	20.14	21.48
	23.39	21.51	19.58	19.81	19.64	20.84	20.40	21.27	27.12
	22.22	23.09	23.89	22.96	23.68	24.96	25.94		

Now that we have our data and we want to check if the mean BMI is the same for both genders.

However, before jumping into the calculations, our first task is to check if the two-sample t test is indeed an appropriate method to evaluate the difference in body mass index between men and women.

- The data values are independent. The BMI for any one person does not depend on the BMI of another person.
- We assume the people measured represent a simple random sample.
- Let's check if the data is normally distributed.

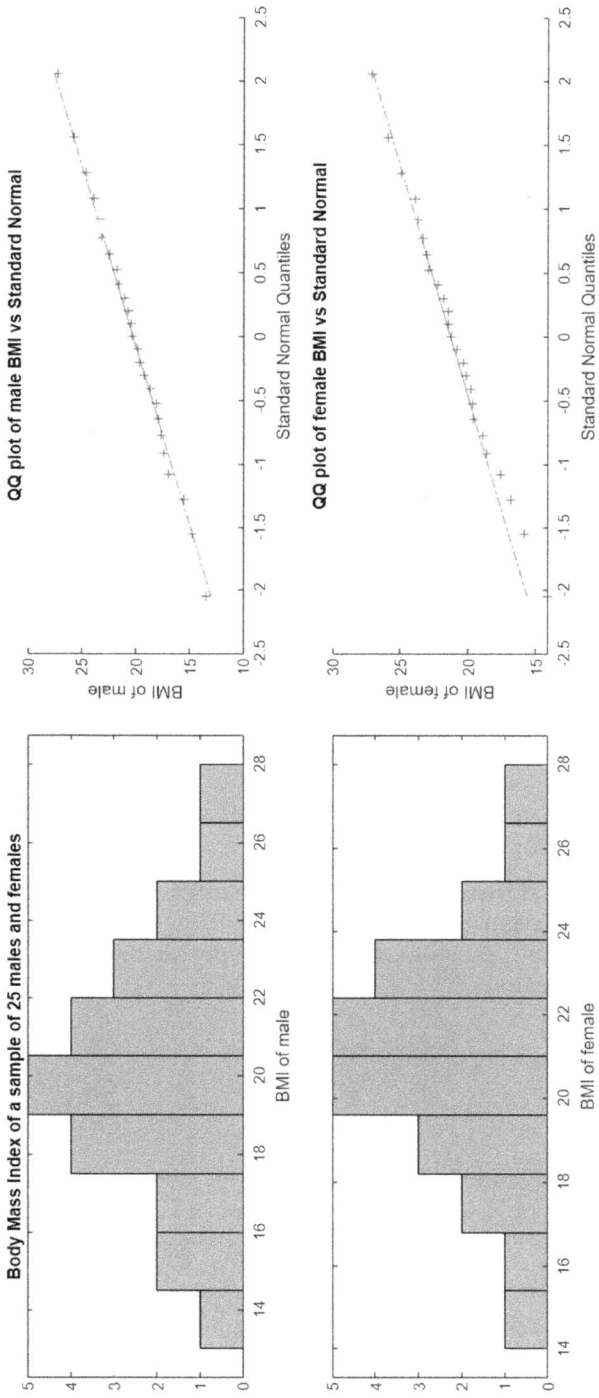

Figure 8.12

The two histograms are on the same scale (Figure 8.12). From a quick look, we can see that there are no very unusual points, or *outliers*. The data looks roughly bell-shaped, so our initial idea of a normal distribution seems reasonable. Similarly the Q-Q plot appears to be relatively straight; therefore the data can be concluded to be normally distributed.

- The data values are body fat measurements. The measurements are continuous.
- Finally, we need to check the homogeneity of the variances—pooled or unpooled.

Minimum	min =	13.45
Maximum	max =	27.28
Range	R =	13.83
Size	n =	25
Sum	sum =	505.77
Mean	$\bar{x}$ =	20.2308
Median	$\tilde{x}$ =	20.33
Mode	mode =	13.45, 14.80, 15.58, 16.96, 17.36, 17.63, 17.91, 18.10, 18.69, 19.23, 19.68, 19.85, 20.33, 20.47, 20.71, 21.03, 21.55, 21.75, 22.46, 23.19, 23.34, 23.98, 24.62, 25.82, 27.28
Standard Deviation	s =	3.41068165
Variance	s^2 =	11.6327493

Minimum	min =	14.07
Maximum	max =	27.12
Range	R =	13.05
Size	n =	25
Sum	sum =	525.63
Mean	$\bar{x}$ =	21.0252
Median	$\tilde{x}$ =	21.27
Mode	mode =	14.07, 15.85, 16.85, 17.56, 18.89, 18.64, 19.58, 19.81, 19.64, 20.14, 20.40, 20.84, 21.27, 21.85, 21.48, 21.51, 22.22, 22.96, 23.39, 23.09, 23.68, 24.96, 23.89, 25.94, 27.12
Standard Deviation	s =	3.09426098
Variance	s^2 =	9.574451

Examining the summary statistics, we see that the standard deviations are similar. This supports the idea of equal variances.

Based on these observations, the two-sample t test appears to be an appropriate method to test for a difference in means.

8.2.2.3.1 Stating the hypotheses

$$H_0 : \mu_1 = \mu_2$$

This means that the underlying population means are the same—the BMI of both males and females are the same.

$$H_A : \mu_1 \neq \mu_2$$

The alternative hypothesis is that the means are not equal.

8.2.2.3.2 Calculating the test statistic

Since the data has equal variance we are going to be using the formula:

$$\text{test statistics} = \frac{(x_1 - x_2) - (\mu_1 - \mu_2)}{s_p \sqrt{\frac{1}{n_1} + \frac{1}{n_2}}}$$

$$\text{where } s_p = \sqrt{\frac{(n_1 - 1)s_1^2 + (n_2 - 1)s_2^2}{n_1 + n_2 - 2}}$$

$$df = n_1 + n_2 - 2$$

If we calculate it by hand then we get the following values:

$$df = 25 + 25 - 2 = 48$$

$$s_p = \sqrt{\frac{(25-1)9.574451 + (25-1)11.632749}{48}} = 3.25631$$

$$\text{Test statistic} = \frac{(20.2308 - 21.0252) - (0)}{3.25631\sqrt{\frac{1}{25} + \frac{1}{25}}} = -0.862527$$

8.2.2.3.3 Using either of the two approaches

8.2.2.3.3.1 CRITICAL APPROACH

We set $\alpha = 0.05$.

The degrees of freedom (df) for this case (pooled variance):$25 + 25 - 2 = 48$.
Therefore our critical value is 1.677.

8.2.2.3.3.2 P-VALUE APPROACH

$df = 48$ unfortunately is not given in the table; therefore we can only approximate the value of p. Since 48 falls between 40 and 60 let's look at the two rows. The values closest to the test statistic between $df = 40$ and $df = 60$ are 0.851 and 0.848 which are both under the 0.40 column. Hence we can conclude that our p-value is somewhere around 0.40.

8.2.2.3.4 Decision

8.2.2.3.4.1 CRITICAL APPROACH

Since our *test statistics* = 0.862527 < 1.677 = *critical value*, we fail to reject the null hypothesis.

8.2.2.3.4.2 P-VALUE APPROACH

Since our p-value= 0.40 > 0.05 = *alpha value*, we fail to reject the null hypothesis.

So therefore we assume that the null hypothesis is true; hence the average BMI of males and females in this sample is the same.

8.2.2.4 In MATLAB

Now let's work on the same problem in MATLAB.

To perform the independent-samples t test we use the function:

```
[h,p,ci,stats] = ttest2(x, y, 'Alpha', , 'Tail' , '' , 'Vartype', '')
```

We can see that, in addition to the arguments we discussed earlier, there is the argument Vartype which specifies if the variances of the two datasets are equal or not. It can have one of two values.

Vartype	What it means	Formula
'equal' (default)	x_1 and x_2 have equal variances	$$test\ stat = \frac{(x_1 - x_2) - (\mu_1 - \mu_2)}{s_p\sqrt{\dfrac{1}{n_1} + \dfrac{1}{n_2}}}$$ $$s_p = \sqrt{\frac{(n_1 - 1)s_1^2 + (n_2 - 1)s_2^2}{n_1 + n_2 - 2}}$$ $$df = n_1 + n_2 - 2$$
'unequal'	x_1 and x_2 have unequal variances	$$test\ stat = \frac{(x_1 - x_2) - (\mu_1 - \mu_2)}{\sqrt{\dfrac{s_1^2}{n_1} + \dfrac{s_2^2}{n_2}}}$$ $$df = \frac{\left(\dfrac{s_1^2}{n_1} + \dfrac{s_2^2}{n_2}\right)^2}{\left(\dfrac{1}{n_1 - 1}\right)\left(\dfrac{s_1^2}{n_1}\right)^2 + \left(\dfrac{1}{n_2 - 1}\right)\left(\dfrac{s_2^2}{n_2}\right)^2}$$

Now using the same problem:

Alpha = 0.05
Tail= *Two tailed*
Vartype= *Equal*
All of the values of the arguments are default; hence:
[h,p,ci,stats] = ttest2(BMI_M, BMI_F)

```
                    h =

                         0


                    p =

                         0.3927


                    ci =

                         -2.6462     1.0574


                    tstats =

                         struct with fields:

                         tstat: -0.8625
                            df: 48
                            sd: 3.2563
```

As we can see, the output aligns with our results.

8.2.2.5 Example

Thomas Jefferson High School physics teacher Sue claims that students in her section will score higher marks than those in her colleague Robert's section. At $\alpha = 0.01$, can the teacher's claim be supported?

Class	Scores							
Robert	85	58	71	62	75	55	85	52
	68	40	80	79	58	67	78	71
	49	98	75	68	65			
Sue	95	100	75	79	80	75	70	92
	84	78	82	90	89	85	81	86
	88	88	86					

Doing the normality check:

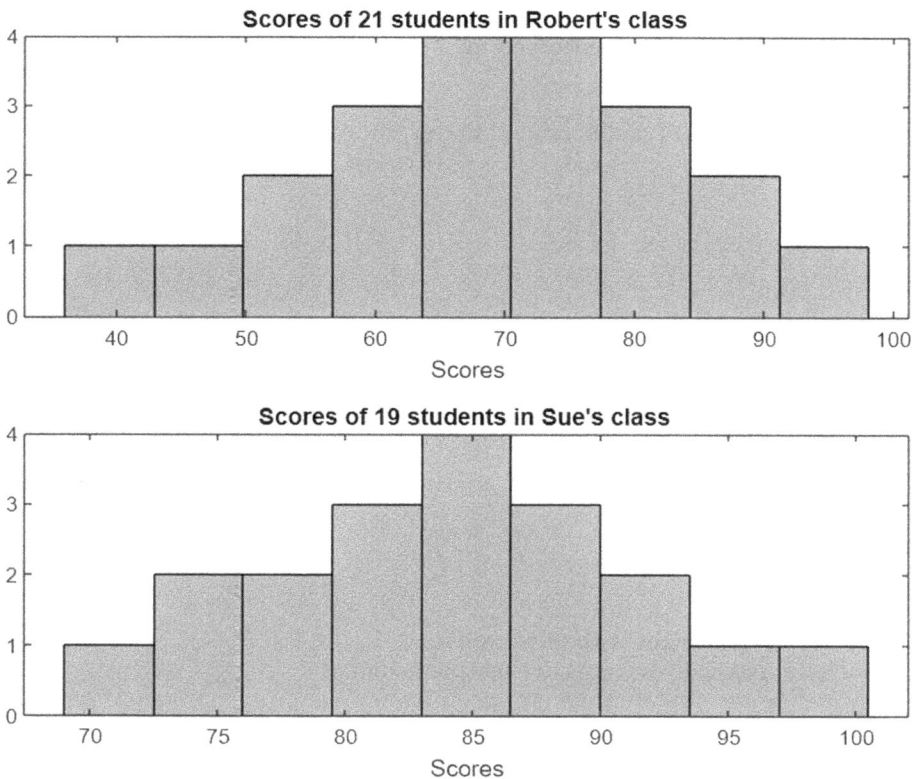

Figure 8.13

Now let's check the homogeneity of the variances—pooled or unpooled.

Answer:		
Variance	s^2 =	54.912281
Standard Deviation	s =	7.4102821
Count	n =	19
Mean	$\bar{x}$ =	84.368421

Answer:		
Variance	s^2 =	188.4619
Standard Deviation	s =	13.728143
Count	n =	21
Mean	$\bar{x}$ =	68.52381

As we can see there is a huge gap between the two variances; therefore we have strong evidence that they have unequal variances.

Performing the test with unequal variances:

8.2.2.5.1 Stating the hypotheses

$$H_0 : \mu_{Sue} = \mu_{Robert}$$

This means that the average scores of both sections are the same.

$$H_A : \mu_{Sue} > \mu_{Robert}$$

This means that Sue's class performed better.

Using the function:

```
[h,p,~,tstat] = ttest2(Sue,Robert,'Alpha',0.01, 'Tail','right',
                       'Vartype','unequal')
```

```
h =

        1


p =

    3.2960e-05


tstat =

    struct with fields:

        tstat: 4.6000
           df: 31.3440
           sd: [7.4103 13.7281]
```

Let's see if the output aligns with our results.

Since they have unequal variances, we use the formula:

$$test\ stat = c\ \frac{\left(x_{Sue} - x_{Robert}\right) - \left(0\right)}{\sqrt{\dfrac{s_{Robert}^2}{n_{Robert}} + \dfrac{s_{Sue}^2}{n_{Sue}}}} = \frac{\left(84.368421 - 68.52381\right) - \left(0\right)}{\sqrt{\dfrac{188.4619}{21} + \dfrac{54.912281}{19}}} = 4.5999 \sim 4.60$$

$$df = \frac{\left(\dfrac{s_{Robert}^2}{n_{Robert}} + \dfrac{s_{Sue}^2}{n_{Sue}}\right)^2}{\left(\dfrac{1}{n_{Robert}-1}\right)\left(\dfrac{s_{Robert}^2}{n_{Robert}}\right)^2 + \left(\dfrac{1}{n_{sue}-1}\right)\left(\dfrac{s_{Sue}^2}{n_{Sue}}\right)^2}$$

$$= \frac{\left(\dfrac{188.4619}{21} + \dfrac{54.912281}{19}\right)^2}{\left(\dfrac{1}{21-1}\right)\left(\dfrac{188.4619}{21}\right)^2 + \left(\dfrac{1}{19-1}\right)\left(\dfrac{54.912281}{19}\right)^2} = 31.344$$

Since $df = 31.344 \sim 31$ is close to 30 look at the row. The value closest to the test statistic is 3.646 which is under the 0.0005 column. Hence we can conclude that our p-value is < 0.0005.

8.2.2.5.2 Decision

Since our p-value$< 0.0005 < alpha\ value = 0.01$, we reject the null hypothesis. Therefore our calculated result goes with the output and we conclude that

8.2.2.6 Example

1) According to the National Association of Colleges and Employers, those who had a master's degree earned 22% more than those with a bachelor's degree. Joe, who is having second thoughts about pursuing a master's degree, decided to test the study himself and collected data on the annual income of his friends and families with different degrees. Should he go to graduate school?

Level of education	Annual income ($)						
Undergraduate	75,949.36	69,957.16	83,074.56	59,200.00	66,278.40	68,590.36	79,915.68
	82,100	52,681.20	76,818.04	65,573.04	45,183.84	72,150.00	
	64,947.48	93,593.76	56,491.76				
Postgraduate	102,007.60	80,575.00	78,452.12	78,992.00	86,961.84	57,250.56	72,585.00
	75,478.00	69,185.68	84,178.80	89,443.80	92,173.9	62,173.95	

8.2.2.6.1 Stating the hypotheses

$H_0 : \mu_{undergrad} = \mu_{postgrad}$

The average earnings of both groups are the same.

$H_A : \mu_{undergrad} < \mu_{postgrad}$

The average earnings of someone with an advanced degree are higher than someone with only a bachelor's degree.

Answer:		
Variance	s^2 = **155412830**	
Standard Deviation	s = 12466.468	
Count	n = 16	
Mean	$\bar{x}$ = 69531.54	

Answer:		
Variance	s^2 = **150877740**	
Standard Deviation	s = 12283.23	
Count	n = 13	
Mean	$\bar{x}$ = 79189.096	

Again, from the data, it is clear that they have similar variances. Hence we use the function:

```
[h,p,~,tstat] = ttest2(Undergraduate, Postgraduate ,'Tail','left')

h =

     1

p =

     0.0232

tstat =

  struct with fields:

     tstat: -2.0883
        df: 27
        sd: 1.2385e+04
```

Checking for ourselves:

$$df = 16 + 13 - 2 = 27$$

$$s_p = \sqrt{\frac{(16-1)155412830 + (13-1)150877740}{27}} = 12385.36371$$

$$\text{Test statistic} = \frac{(69531.54 - 9189.096) - (0)}{12385.36371\sqrt{\frac{1}{16} + \frac{1}{13}}} = -2.08827$$

Given the df = 27 we look at row 27. The value closest to the test statistic is 2.052 which is under the 0.025 column. Hence we can conclude that our p-value is somewhere around 0.025.

Since our p-value is *around* 0.025 < 0.05 = *alpha value*, we reject the null hypothesis.

8.2.2.7 Exercise problems

A student in an anthropology class claims that ethnicity plays a very small role in adult height and that moving to a country with better access to nutritious food, healthcare, and employment opportunities can have a substantial influence on the height of the next generation. To test the claim, the height of two male populations from the United States—Black and Asian—a sample of 15 and 12 males from each race is randomly selected and the measured heights are provided.

Ethnicity	Height							
Black	70.25152	69.43894	68.39874	68.69784	71.0916	67.4998	70.63033	69.17896
	70.78881	70.80592	72.22442	72.4337	69.71947	70.1766	71.48752	
Asian	66.57831	66.1833	66.6893	68.0312	67.29265	66.78236	67.56238	67.0125
	67.1656	66.461	67.13118	67.58379				

2) For years, there has been an ongoing study about the difference in brain size between right-handed and left-handed people. Drake, a neuroscience student, decided to test if there is some relationship between handedness and the brain. In his study he decided to include 21 right- and left-handers of the same age and use $\alpha = 0.01$ to come to a conclusion.

Handedness	Brainsize					
Right	1260	1215	1149	1137	1234	1226
	1282	1175	1128	1248	1161	1220
	1184	1210	1179	1218	1190	1159
	1195	1203	1198			
Left	1287	1125	1146	1263	1155	1215
	1169	1235	1181	1228	1176	1245
	1198	1209	1220	1251	1192	1210
	1230	1237	1206			

3) A company hires an engineer to build a connector for their automotive engine. The engineer builds two prototypes for the company. However, the company thinks prototype A is better than prototype B. To test the claim the engineer records the pull-off forces of the two prototypes and the data is given below. Is there sufficient evidence to support the company's claim?

Prototype	Pull-offforces					
A	12.62	12.98	12.49	12.57	12.43	11.86
	12.19	13.15	12.24	12.79		
B	12.85	13.18	11.76	12.64	12.45	11.99
	12.37	12.15	11.90	12.29		

4) A manufacturer builds resistors from two materials X and Y. He checks the resistance of both resistors at different temperatures and concludes that resistor X is weaker than resistor Y. Test the claim using $\alpha = 0.01$.

Resistor	Resistance (ohm)							
	20 °C	25 °C	30 °C	35 °C	40 °C	45 °C	50 °C	60 °C
X	20.30	22.56	23.27	23.42	23.72	24.97	24.98	26.02
Y	21.10	22.78	22.80	23.86	24.14	24.55	25.04	26.53

8.2.2.8 Dependent/paired t test

While the independent t test determines whether there is a difference between two unrelated groups, the paired t test compares the means of two measurements taken from the same individuals, objects, or related units. For instance, students who took a math test passed an additional course for math and it would be interesting to find whether their results became better if they completed the course. It is possible to take a sample from the same group and use the paired t test.

The paired t test is a method used to test whether the mean difference between pairs of measurements is zero or not.

With the paired t test, the null hypothesis is that the pairwise difference between the two tests is equal (H_0: $\mu_d = 0$).

Assumptions:

1. Since we are measuring before and after treatment, the sample size should be the same.
2. The differences between the pairs should be approximately normally distributed.

8.2.2.9 Example

Fuel economy is the measurement of the distance a vehicle can travel on a given amount of fuel—usually per gallon. Even though auto manufacturers these days are phasing out diesel engines there are cars that still offer both petrol and diesel engines. Joe, a chemical engineer, wishes to compare the fuel economy of petrol and diesel fuel. To differentiate between the two, he selects pairs of cars of the same model where the first car in each pair is operated on petrol and the second car is operated on diesel.

Car model	Car 1	Car 2
1	17.6	18.2
2	20.4	20.6
3	25.6	25.9
4	16.2	17.5
5	15.9	15.4
6	25.8	25.5
7	19.7	19.1
8	23.5	22.3
9	18.1	17.7
10	17.5	16.8
11	21.4	19.2

H_0: there is no difference in fuel economy between petrol and diesel.
H_A: there is a significant difference in fuel economy between petrol and diesel.

8.2.2.9.1 Calculating the test statistic

$$\text{Test statistic} = \frac{d_i}{\frac{s_d}{\sqrt{n}}}$$

Where d_i = mean of the difference between the two groups.

Car 1	17.6	20.4	25.6	16.2	15.9	25.8	19.7	23.5	18.1	17.5	21.4
Car 2	18.2	20.6	25.9	17.5	15.4	25.5	19.1	22.3	17.7	16.8	19.3
d_i	−0.60	−0.20	−0.30	−1.30	0.50	0.30	0.60	1.20	0.40	0.70	2.10

Note here we calculated the difference d_i using the formula = before − after.

However we can calculate it the other way too, i.e. after − before.

So it does not matter in what order the subtraction is done, as long as it is done in the same order for all pairs.

Therefore, $d_i = \dfrac{\Sigma d_i}{n} = \dfrac{(-0.60)+(-0.20)+(-0.30)....0.70+2.20}{11} \sim 0.31.$

$$s_d = \frac{\sqrt{(-0.6-)^2 +(-0.2-)^2 +(-0.3-)^2 +...(0.7-)^2 +(2.2-)^2}}{11-1} \sim 0.913$$

Test statistic ~ 1.1233.

8.2.2.9.2 Using either of the two approaches

8.2.2.9.2.1 CRITICAL APPROACH

We are given the significance level ($\alpha = 0.05$) and the degrees of freedom $df = 11-1 = 10$. Therefore our t value with $\alpha = 0.05$ and 10 degrees of freedom is 2.228.

8.2.2.9.2.2 P-VALUE APPROACH

Since our $df = 24$ we look at row 10 and look for the value closest to our test statistic which falls between 1.093 and 1.372; hence our p-value is between 0.20 and 0.30.

8.2.2.9.3 Decision

8.2.2.9.3.1 CRITICAL APPROACH

Because $1.1233 < 2.228$, we fail to reject our null hypothesis.

8.2.2.9.3.2 P-VALUE APPROACH

Since $0.20 < p\text{-value} < 0.30$, it is clearly > 0.05 alpha. Hence we fail to reject the null hypothesis.

8.2.2.10 MATLAB

In MATLAB, to perform the paired t test, we use the function:

```
[h,p,ci,stats] = ttest(x,y, 'Alpha', , 'Tail','')
```

(Note: here the difference d_i depends on the orientation of x and y; d_i will be calculated by $x_i - y_i$ and if the places were reversed then $y_i - x_i$.)

If we do the same problem in MATLAB:

```
[h,p,ci,stats] = ttest(Car1,Car2)


                                    h =

                                        0


                        p =

                            0.2876


                        ci =

                            -0.3040      0.9222


                        stats =

                            struct with fields:

                                tstat: 1.1233
                                   df: 10
                                   sd: 0.9126
```

As we can see, the output perfectly aligns with our calculation; hence we can conclude that there is no difference in fuel economy between petrol and diesel.

8.2.2.11 Example

A pharmaceutical scientist has produced a drug which he claims can decrease cholesterol levels. The scientist asked 24 volunteers to take the drug for 12 weeks. Their cholesterol (in mmol/L) was measured before and after using the drug.

Cholesterol	
Before drug	After drug
251	247
232	235
248	236
229	254
233	249
229	228
260	233
242	240
230	222
228	217
219	205
224	196

	Cholesterol
Before drug	After drug
215	188
210	189
198	203
186	199
175	187
178	189
169	175
167	181
158	178
150	169
147	150
213	202

H_0: there is no difference in mean cholesterol before and after 12 weeks.

H_A: the cholesterol level decreases after 12 weeks using the drug.

This could be written in two ways:

$\mu_{before} > \mu_{after}$ =in this case it will be a right-tailed test.

$\mu_{after} < \mu_{before}$ =in this case it will be a left-tailed test.

8.2.2.11.1 Calculating the test statistic

Before drug	After drug	Difference d_i
251	247	−4
232	235	3
248	239	−9
229	254	25
233	249	16
229	228	1
260	235	−25
242	240	−2
230	222	−8
228	217	−11
219	205	−14
224	196	−28
239	227	−12
Before drug	After drug	Difference d_i
215	206	−9
210	189	−21

198	203	5
186	199	13
175	187	12
178	189	11
169	175	6
167	181	14
158	178	20
150	169	19
147	150	3
213	202	−11

Note here we calculated the difference d_i using the formula: after − before.

$$\text{Test statistic} = \frac{-0.79167}{\frac{15.723091}{\sqrt{24}}} \sim -0.2467.$$

8.2.2.11.2 Calculating p-value approach

Since our $df = 23$ we look at row 23 and look for the value closest to the absolute value of our test statistic which is between 0.000 and 0.685 which falls under the column between 0.05 and 0.25. Now since we took the absolute value of the test statistic, the p-value we get will be for the left-tailed test.

Hence for the right-tailed test, the p-value $= 1$ p-value of the left-tailed test.

8.2.2.11.3 P-value approach

Because p-value > 0.01 alpha, we fail to reject the null hypothesis.

Doing the same problem in MATLAB:

Since there are two versions (right-tailed test):

```
[h,p,ci,stats] = ttest(After_D,Before_D,'Alpha',0.01,'Tail','right')
                            h =

                               0


                            p =

                              0.5963


                         stats =

                          struct with fields:

                            tstat: -0.2467
                               df: 23
                               sd: 15.7231

[h,p,ci,stats] = ttest(After_D,Before_D,'Alpha',0.01,'Tail','left')
```

```
h =

    0


p =

    0.4037


stats =

    struct with fields:

        tstat: -0.2467
           df: 23
           sd: 15.7231
```

8.2.2.12 Example

After getting their standardized test results, 12 students asked the course coordinator for a specialized program that offers an intensive preparation for the next test. After taking the second test the coordinator wanted to see if the program had an effect on the performance of the students. Using $\alpha=0.01$ test if the program is indeed effective.

Student	1	2	3	4	5	6	7	8	9	10	11	12
Before the program	65	58	71	85	69	55	92	67	82	91	79	80
After the program	72	63	75	88	70	67	90	72	86	93	88	89

8.2.2.12.1 Stating the hypotheses

H0:d=0
  This means that the mean difference between the scores is zero.
HA:d>0
  Using the function:

```
[h,p,ci,stats] = ttest(after_prog,before_prog, 'Alpha', 0.01 ,
'Tail','right')
```

```
h =

    1


p =

    0.0011


ci =

    2.4565    7.3768


stats =

    struct with fields:

        tstat: 4.3987
           df: 11
           sd: 3.8720
```

Student	Before the program	After the program	Difference di
1	65	72	72-65=7
2	58	63	63-58=5
3	71	75	75-71=4
4	85	88	88-85=3
5	69	70	70-69=1
6	55	67	67-55=12
7	92	90	90-92=-2
8	67	72	72-67=5
9	82	86	86-82=4
10	91	93	93-91=2
11	79	88	88-79=9
12	80	89	89-80=9
			di=Σdin=5912=4.9167

$$s_{\underline{d}} = \frac{\sqrt{(7-4.9167)^2 + (5-4.9167)^2 + (4-4.9167)^2 + ...(9-4.9167)^2 + (9-4.9167)^2}}{12-1}$$

$$= 3.8720$$

Test statistic $= \dfrac{4.9167}{\dfrac{3.872}{\sqrt{12}}} \sim 4.3987$.

8.2.2.12.2 Calculating the p-value

Since our df=11 we look at row 11 and look for the value closest to our test statistic which is 4.437 which falls under 0.001; hence our p-value is close to 0.001.

8.2.2.12.3 Decision

8.2.2.12.3.1 CRITICAL APPROACH

Because 4.398747 > 2.201 we reject our null hypothesis and make the conclusion that the program is indeed effective.

8.2.2.12.3.2 P-VALUE APPROACH

Since p-value 0.001<0.05 alpha, we reject the null hypothesis.

As we can see the output perfectly aligns with our calculation, and we can conclude that the program definitely was effective for the students.

8.2.2.13 Exercise problems

A dietician made a new weight loss program and claims that this new program is way more effective than the old one he gave to his clients. To test his claim, he calculated the weight reduction each program yielded. At the 1% level of significance, can you support the dietician's claim?

Program	Weightloss									
Old program	12.6	19.0	8.8	13.7	12.9	15.2	11.8	10.5	7.8	14.6
	5.8	11.1	17.3	10.8	9.2	13.5	12.2	20.1	9.2	15.3
	14.4	12.9	14.5	16.3	15.4	11.2	16.1			
New program	13.2	18.5	10.3	13.5	14.8	16.1	12.9	11.2	9.6	15.0
	7.2	12.3	17.5	11.7	12.4	14.2	12.8	20.4	10.1	16.2
	15.3	13.9	14.6	17.2	16.3	12.7	16.4			

In chemistry, the amount of product that *may be* produced by a reaction under specified conditions is called the theoretical yield of the reaction. Now a chemical engineer is trying to determine if a special catalyst affects the yield of the reaction. He runs a test in the pilot plant which results in the data shown below. Is there any difference between the yields? Use a 5% significance level and assume equal variances.

Reaction	Catalyst yield data								
Before catalyst	91.60	90.25	89.78	90.78	92.51	91.34	91.78	93.12	92.54
After catalyst	92.58	90.12	89.82	91.20	92.48	91.09	90.89	93.43	91.15

There are unlimited applications of solar irradiance from climate modeling and weather forecasting to heating and cooling systems for homes and buildings. An engineer wants to determine if a booster reflector can increase solar irradiance density. Does the solar irradiance increase when the booster reflector is used?

Booster reflector	Solar irradiance (W/m^2)					
Without	743	735	762	698	820	726
	672	754	653	741	780	691
With	794	789	810	867	888	750
	670	862	756	790	841	809

8.2.3 The chi-square test for one variance

This is a statistical test used to compare the variance of a sample to a known population variance. It is similar to the z test where we compare the mean of a sample to a known population mean. Since it is a parametric test, it is based on the assumption that the sample is drawn from a normally distributed population.

8.2.3.1 Steps in the chi-square test for one variance

Stating the hypotheses:

The null hypothesis is usually that the population variance is equal to a specific value. The alternative hypothesis is that the population variance is not equal to that value.

Calculate the test statistic:

Test statistic = sample variance/known population variance.

Determine the critical value of the test statistic based on the significance level (alpha) of the test and the degrees of freedom. The degrees of freedom are calculated as the sample size minus 1.

Compare the calculated test statistic to the critical value to determine whether to reject or fail to reject the null hypothesis.

If the calculated test statistic exceeds the critical value, the null hypothesis is rejected, and the alternative hypothesis is accepted.

8.3 f-TEST

The last type of parametric test we are going to walk through in this book is the f test. In previous chapters we saw how to test hypotheses concerning population means and population proportions. The idea of testing hypotheses can be extended to many other situations that involve different parameters and use different test statistic. While the z test and t test were used in testing the mean of a population or comparing the means from two continuous populations, the f test is used for comparing more than two means and equality of variance. Recall that in the two-sample t test we had two formulas for the test statistic—one for pooled variance (equal variance) and the other for unpooled variance (unequal variance). The f test determines if the samples are pooled or unpooled.

Another key point is that the test statistic that appeared in earlier chapters followed either a normal or Student t distribution, but in this chapter the tests will involve F-distributions. So it makes sense that in an f test, the data follows an F-distribution.

Now to learn how to conduct an f test it is vital to understand the F-distribution first.

The F-distribution is a distribution differentiated by the number of degrees of freedom, similar to the t distribution and chi-square distribution (in Chapter 5). However, unlike the t distribution which is defined by only one degree of freedom, the F-distribution is specified by two, called the numerator and denominator degrees of freedom denoted by df_1 and df_2. Moreover, the number of degrees of freedom for this distribution is determined in a different manner than for a t distribution. To understand this, we need to first know how this distribution was derived.

Years ago, statisticians discovered that when pairs of samples are taken from a normal population, the ratios of the variances of the samples in each pair will always follow the same distribution. This means there is a sample from each of these populations and thus there are degrees of freedom for both of these samples. In fact, we subtract one from both of the sample sizes to determine our two numbers of degrees of freedom. Now comes the interesting part; now that we have two degrees of freedom, Rather than combining these two numbers into another single number, we retain both of them. Therefore, any use of an F-distribution table requires us to look up two different degrees of freedom.

Let's talk about the shape of the distribution.

As previously stated, the F-distribution is the ratio of two sample variances. Now suppose we take our two samples with n_1 and n_2 observations; then the ratio will be

$$F = \frac{s_1^2}{s_2^2}$$ where s_1^2 and s_2^2 are our sample variances.

Recall that sample variance is measured by summing the squared differences of the individual values and the mean value and dividing by $n - 1$ for sampled data, where n equals the number of measured samples.

$$\text{sample variance} = s^2 = \frac{\Sigma\left(x - \underline{x}\right)^2}{n-1}$$

When deriving an individual F ratio:

$$F = \frac{s_1^{\ 2}}{s_2^{\ 2}} = \frac{\dfrac{\Sigma\left(x_i - \underline{x_1}\right)^2}{n_1 - 1}}{\dfrac{\Sigma\left(x_j - \underline{x_2}\right)^2}{n_2 - 1}}$$

We can see that the term $n-1$ will appear in both the numerator and denominator. This value represents the **degrees of freedom** in each of the samples. This evidently proves that the shape of the distribution depends on the degrees of freedom of the numerator and denominator. Now since samples could be of any size, there is a different curve for each set of dfs.

However just like in the t distribution, as the degrees of freedom for the numerator and for the denominator get larger, the curve approximates the normal.

Now for each sample variance the numerator in the formula is the sum of squares, so it will be positive, and the denominator is the sample size minus one, which will also be positive. Therefore, variance is always positive, and since f is the ratio of variances, the value of the F-distribution is always positive. Since it spans only non-negative numbers, it is not symmetrical (unlike the normal and t distributions which include both positive and negative values).

If s_1^2 and s_2^2 come from samples from the same population, meaning they have the same variance, the ratio would be close to one. But if they are different and the larger variance is at the top, the ratio will be greater than 1. In fact if s_1^2 is a lot larger than s_2^2, F can be quite large. It is equally possible for s_2^2 to be a lot larger than s_1^2, and then F would be very close to zero. Since F goes from zero to very large, with most of the values around one, it is obviously not symmetric; there is a long tail to the right and a steep descent to zero on the left.

To summarize everything, we can say that:

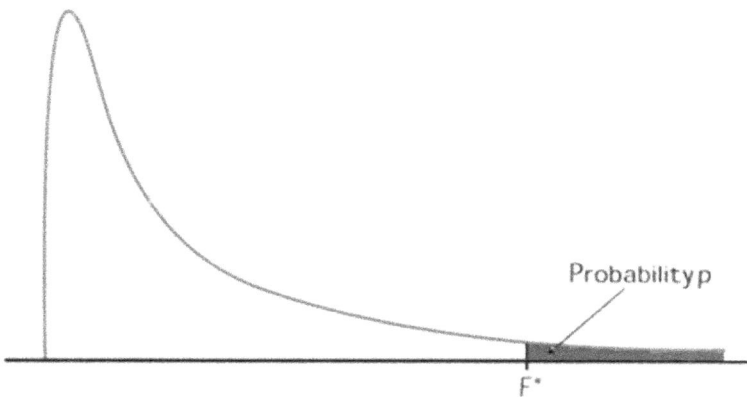

1. The F-distribution curve is not symmetrical but is always skewed to the right.
2. Even though there is a different curve for each set of *dfs*, it will always follow the same skewed trend.
3. The *F* statistic is greater than or equal to zero but with most of the values around one.
4. As the degrees of freedom for the numerator and for the denominator get larger, the curve approximates the normal.

As previously mentioned, in the f test the data follows an F-distribution, so it was necessary to get to know about the distribution first. Now that we have learned about the F-distribution, we can now move on to the f test. Just like the other tests we have described previously, to conduct a test, certain conditions must be met. Below are the conditions for the f test:

1. The samples must be drawn randomly from normally distributed populations.
2. Each observation in each sample must be independent of the others.
3. The larger sample variance always goes in the numerator for a right-tailed test, and the right-tailed tests are always easy to calculate. (More discussion below.)

8.3.1 Steps in the F test for equality of two variances

1) Specify the null and alternative hypotheses.
2) Calculate the test statistic.
   - The value is derived as the ratio of the variances between two samples.

   $$f \text{ statistic} = \frac{s_1^2}{s_2^2}$$

   - For a right-tailed and a two-tailed f test, the variance with the greater value will be in the numerator and the smaller value variance will be the denominator.
   - For a left-tailed test, the smaller variance becomes the numerator and the higher variance goes in the denominator.
4) Using either of the two approaches:

8.3.2 Determine the p-value of the test statistic

So far we have learned to calculate the p-value for the z test and t test which depended on their test statistic. The z test used the z test statistic, and the t test used the t test statistic, so clearly the f test will use the f test statistic. Moreover, in both tests we needed tables of the corresponding distribution in order to obtain the parallel p-values. So even if the p-value always denotes the same thing, it always depends on which hypothesis test and test statistic you are using.

Having obtained the f test statistic, we need tables of the F-distribution in order to obtain the corresponding P-values.

The 10% table gives us the value f such that $P(F \geq f) = 0.10$.

For fixed df_1, df_2, the tabulated value is the number $f = F_{df_1, df_2}(0.10)$ such that for

$$F \sim F(df_1, df_2), P(F \geq f) = 0.10$$

For example, suppose that we look up the entry in the column of the 10% table defined by $df_1 = 7$ and the row defined by $df_2 = 10$. We find that it is $2.41397 \sim 2.414$. Thus, for $df_1 = 7$ and $df_2 = 10$ $F \sim F(7,10)$, $P(F \geq 2.414) = 0.10$.

The 5% table, 2.5% table, and 1% table work in the same way.
From the $df_1 = 7$ column and $df_2 = 10$ row of each of the latter tables we find that

$$P\left(F \geq 3.1355\right) = 0.05$$

$$P\left(F \geq 3.9498\right) = 0.025$$

$$P\left(F \geq 5.200\right) = 0.01$$

Let us now use this information to bracket a p-value.

Suppose our test statistic is 3.54. Clearly the f lies between 3.1355 and 3.9498 and we have that the p-value lies between 0.05 and 0.025, so our p-value is

$$0.025 < p-value < 0.05$$

To be more precise it is closer to 0.05 than it is to 0.025.

8.3.3 Example

Suppose now that $df_1 = 25$, $df_2 = 40$, and the f test statistic is 2.50.

Let's look at the 10% table first. We find that we have $df_1 = 24$ with $f = 1.57411$ and $df_1 = 30$ with $f = 1.54108$, but there is no entry for $df_1 = 25$. Without access to more detailed tables we have to choose one of these two entries. Which one? Since we always act conservatively, we choose the tabulated df_1 with the bigger f value (which makes it harder to get significance), namely $df_1 = 24$. We therefore use the tabulated value of df_1 immediately below the value required.

What happens if df_2 is not tabulated? Inspecting the tables we see that we do the same thing with df_2 as with df_1. For example, if $df_2 = 45$ we enter the table with $df_2 = 40$.

- Left-tailed F-test:
p-value = cdf$_{F,d2,d2}$(F$_{score}$)

- Right-tailed F-test:
p-value = 1 − cdf$_{F,d2,d2}$(F$_{score}$)

- Two-tailed F-test:
p-value = 2 × min{cdf$_{F,d1,d2}$(F$_{score}$), 1 − cdf$_{F,d1,d2}$(F$_{score}$)}

8.3.4 Determine the critical value of the test statistic

The critical F-value is determined using an F-table. There are several different F-tables. Each one has a different level of significance. So find the correct level of significance first, and then look up the numerator degrees of freedom and the denominator degrees of freedom to find the critical value.

In conclusion, to use the F-distribution table, you only need two values:

- Degrees of freedom

As mentioned earlier we are going to have two degrees of freedom df_1 and df_2. The parameter df_1 is often referred to as the *numerator* degrees of freedom and the parameter df_2 as the *denominator* degrees of freedom. It is important to keep in mind that they are not interchangeable. For example, the F-distribution with degrees of freedom $df_1 = 4$ and $df_2 = 10$ is a different distribution from the F-distribution with degrees of freedom $df_1 = 10$ and $df_2 = 4$.

For a right-/two-tailed test, the degrees of freedom with larger variance will be in the numerator while for the left-tailed test, they will be in the denominator.

- The alpha level (common choices are 0.01, 0.05, and 0.10)

If it is a right-tailed test then α is the significance level denoted by $F\alpha$.

If it is a two-tailed test then the significance level is given by $\alpha/2$ denoted by $F\alpha/2$.

For a left-tailed test $1 - \alpha$ is the alpha level denoted by $F1-\alpha$.

8.3.4.1 Finding the critical values for a right-tailed test

The graph of the F-distribution is always positive and skewed right. These F-tables provide the critical values for right-tailed F-tests.

8.3.4.2 Finding the critical values for a two-tailed test

When we are conducting a two-tailed test the alpha α is split and the right-tailed table of that split alpha is used.

For example, let's say we want to find the critical value of a two-tailed test with alpha 0.05 and degrees of freedom 12 in the numerator and 15 in the denominator. So what we do is we split 0.05; now we find the critical value of the right-tailed test for alpha 0.025 with the degrees of freedom assigned.

/	df₁=1	2	3	4	5	6	7	8	9	10	12	15	20	24	30	40	60	120	∞
df₂=1	647.7890	799.5000	864.1630	899.5833	921.8479	937.1111	948.2169	956.6562	963.2846	968.6274	976.7079	984.8668	993.1028	997.2492	1001.414	1005.598	1009.800	1014.020	1018.258
2	38.5063	39.0000	39.1655	39.2484	39.2982	39.3315	39.3552	39.3730	39.3869	39.3980	39.4146	39.4313	39.4479	39.4562	39.465	39.473	39.481	39.490	39.498
3	17.4434	16.0441	15.4392	15.1010	14.8848	14.7347	14.6244	14.5399	14.4731	14.4189	14.3366	14.2527	14.1674	14.1241	14.081	14.037	13.992	13.947	13.902
4	12.2179	10.6491	9.9792	9.6045	9.3645	9.1973	9.0741	8.9796	8.9047	8.8439	8.7512	8.6565	8.5599	8.5109	8.461	8.411	8.360	8.309	8.257
5	10.0070	8.4336	7.7636	7.3879	7.1464	6.9777	6.8531	6.7572	6.6811	6.6192	6.5245	6.4277	6.3286	6.2780	6.227	6.175	6.123	6.069	6.015
6	8.8131	7.2599	6.5988	6.2272	5.9876	5.8198	5.6955	5.5996	5.5234	5.4613	5.3662	5.2687	5.1684	5.1172	5.065	5.012	4.959	4.904	4.849
7	8.0727	6.5415	5.8898	5.5226	5.2852	5.1186	4.9949	4.8993	4.8232	4.7611	4.6658	4.5678	4.4667	4.4150	4.362	4.309	4.254	4.199	4.142
8	7.5709	6.0595	5.4160	5.0526	4.8173	4.6517	4.5286	4.4333	4.3572	4.2951	4.1997	4.1012	3.9995	3.9472	3.894	3.840	3.784	3.728	3.670
9	7.2093	5.7147	5.0781	4.7181	4.4844	4.3197	4.1970	4.1020	4.0260	3.9639	3.8682	3.7694	3.6669	3.6142	3.560	3.505	3.449	3.392	3.333
10	6.9367	5.4564	4.8256	4.4683	4.2361	4.0721	3.9498	3.8549	3.7790	3.7168	3.6209	3.5217	3.4185	3.3654	3.311	3.255	3.198	3.140	3.080
11	6.7241	5.2559	4.6300	4.2751	4.0440	3.8807	3.7586	3.6638	3.5879	3.5257	3.4296	3.3299	3.2261	3.1725	3.118	3.061	3.004	2.944	2.883
12	6.5538	5.0959	4.4742	4.1212	3.8911	3.7283	3.6065	3.5118	3.435	3.3736	3.2773	3.1772	3.0728	3.0187	2.963	2.906	2.848	2.787	2.725
13	6.4143	4.9653	4.3472	3.9959	3.7667	3.6043	3.4827	3.3880	3.		3.1532	3.0527	2.9477	2.8932	2.837	2.780	2.720	2.659	2.595
14	6.2979	4.8567	4.2417	3.8919	3.6634	3.5014	3.3799	3.2853	3.2093		3.0502	2.9493	2.8437	2.7888	2.732	2.674	2.614	2.552	2.487
15	6.1995	4.7650	4.1528	3.8043	3.5764	3.4147	3.2934	3.1987	3.1227		2.9633	2.8621	2.7559	2.7006	2.644	2.585	2.524	2.461	2.395
16	6.1151	4.6867	4.0768	3.7294	3.5021	3.3406	3.2194	3.1248	3.0488	2.9862	2.8890	2.7875	2.6808	2.6252	2.568	2.509	2.447	2.383	2.316
17	6.0420	4.6189	4.0112	3.6648	3.4379	3.2767	3.1556	3.0610	2.9849	2.9222	2.8249	2.7230	2.6158	2.5598	2.502	2.442	2.380	2.315	2.247
18	5.9781	4.5597	3.9539	3.6083	3.3820	3.2209	3.0999	3.0053	2.9291	2.8664	2.7689	2.6667	2.5590	2.5027	2.445	2.384	2.321	2.256	2.187
19	5.9216	4.5075	3.9034	3.5587	3.3327	3.1718	3.0509	2.9563	2.8801	2.8172	2.7196	2.6171	2.5089	2.4523	2.394	2.333	2.270	2.203	2.133
20	5.8715	4.4613	3.8587	3.5147	3.2891	3.1283	3.0074	2.9128	2.8365	2.7737	2.6758	2.5731	2.4645	2.4076	2.349	2.287	2.223	2.156	2.085

So our critical value is 2.9633.

8.3.4.3 Finding the critical values for left-tailed test

You will notice that all of the tables only give the level of significance for right-tailed tests. Because the F-distribution is not symmetric, and there are no negative values, you may not simply take the opposite of the right critical value to find the left critical value like we did for the z test in the normal distribution table. The way to find a left critical value is to reverse the degrees of freedom, look up the right critical value, and then take the reciprocal of this value.

So let's use the same degrees of freedom and alpha level as we did for the two-tailed test but now we want to find the critical value of a left-tailed test. So what we do is we find the critical value of the right-tailed test for the same alpha level but instead interchange the degrees of freedom, so basically we have 15 in the numerator and 12 in the denominator, and then take the reciprocal of that critical value. So basically our $df1$ will be for the smaller variance and $df2$ will be for the larger variance.

	df₁=1	2	3	4	5	6	7	8	9	10	12	15	20	24	30	40	60	120	∞
df₂=1	161.4476	199.5000	215.7073	224.5832	230.1319	233.9860	236.7684	238.8827	240.5433	241.8817	243.9060	245.9499	248.0131	249.0518	250.0951	251.1432	252.1957	253.2529	254.3144
2	18.5128	19.0000	19.1643	19.2468	19.2964	19.3295	19.3532	19.3710	19.3848	19.3959	19.4125	19.4291	19.4458	19.4541	19.4624	19.4707	19.4791	19.4874	19.4957
3	10.1280	9.5521	9.2766	9.1172	9.0135	8.9406	8.8867	8.8452	8.8123	8.7855	8.7446	8.7029	8.6602	8.6385	8.6166	8.5944	8.5720	8.5494	8.5264
4	7.7086	6.9443	6.5914	6.3882	6.2561	6.1631	6.0942	6.0410	5.9988	5.9644	5.9117	5.8578	5.8025	5.7744	5.7459	5.7170	5.6877	5.6581	5.6281
5	6.6079	5.7861	5.4095	5.1922	5.0503	4.9503	4.8759	4.8183	4.7725	4.7351	4.6777	4.6188	4.5581	4.5272	4.4957	4.4638	4.4314	4.3985	4.3650
6	5.9874	5.1433	4.7571	4.5337	4.3874	4.2839	4.2067	4.1468	4.0990	4.0600	3.9999	3.9381	3.8742	3.8415	3.8082	3.7743	3.7398	3.7047	3.6689
7	5.5914	4.7374	4.3468	4.1203	3.9715	3.8660	3.7870	3.7257	3.6767	3.6365	3.5747	3.5107	3.4445	3.4105	3.3758	3.3404	3.3043	3.2674	3.2298
8	5.3177	4.4590	4.0662	3.8379	3.6875	3.5806	3.5005	3.4381	3.3881	3.3472	3.2839	3.2184	3.1503	3.1152	3.0794	3.0428	3.0053	2.9669	2.9276
9	5.1174	4.2565	3.8625	3.6331	3.4817	3.3738	3.2927	3.2296	3.1789	3.1373	3.0729	3.0061	2.9365	2.9005	2.8637	2.8259	2.7872	2.7475	2.7067
10	4.9646	4.1028	3.7083	3.4780	3.3258	3.2172	3.1355	3.0717	3.0204	2.9782	2.9130	2.8450	2.7740	2.7372	2.6996	2.6609	2.6211	2.5801	2.5379
11	4.8443	3.9823	3.5874	3.3567	3.2039	3.0946	3.0123	2.9480	2.8962	2.8536		2.7186	2.6464	2.6090	2.5705	2.5309	2.4901	2.4480	2.4045
12	4.7472	3.8853	3.4903	3.2592	3.1059	2.9961	2.9134	2.8486	2.7964	2.7534		2.6169	2.5436	2.5055	2.4663	2.4259	2.3842	2.3410	2.2962
13	4.6672	3.8056	3.4105	3.1791	3.0254	2.9153	2.8321	2.7669	2.7144	2.6710	2.6037	2.5331	2.4589	2.4202	2.3803	2.3392	2.2966	2.2524	2.2064
14	4.6001	3.7389	3.3439	3.1122	2.9582	2.8477	2.7642	2.6987	2.6458	2.6022	2.5342	2.4630	2.3879	2.3487	2.3082	2.2664	2.2229	2.1778	2.1307
15	4.5431	3.6823	3.2874	3.0556	2.9013	2.7905	2.7066	2.6408	2.5876	2.5437	2.4753	2.4034	2.3275	2.2878	2.2468	2.2043	2.1601	2.1141	2.0658
16	4.4940	3.6337	3.2389	3.0069	2.8524	2.7413	2.6572	2.5911	2.5377	2.4935	2.4247	2.3522	2.2756	2.2354	2.1938	2.1507	2.1058	2.0589	2.0096
17	4.4513	3.5915	3.1968	2.9647	2.8100	2.6987	2.6143	2.5480	2.4943	2.4499	2.3807	2.3077	2.2304	2.1898	2.1477	2.1040	2.0584	2.0107	1.9604
18	4.4139	3.5546	3.1599	2.9277	2.7729	2.6613	2.5767	2.5102	2.4563	2.4117	2.3421	2.2686	2.1906	2.1497	2.1071	2.0629	2.0166	1.9681	1.9168
19	4.3807	3.5219	3.1274	2.8951	2.7401	2.6283	2.5435	2.4768	2.4227	2.3779	2.3080	2.2341	2.1555	2.1141	2.0712	2.0264	1.9795	1.9302	1.8780
20	4.3512	3.4928	3.0984	2.8661	2.7109	2.5990	2.5140	2.4471	2.3928	2.3479	2.2776	2.2033	2.1242	2.0825	2.0391	1.9938	1.9464	1.8963	1.8432
21	4.3248	3.4668	3.0725	2.8401	2.6848	2.5727	2.4876	2.4205	2.3660	2.3210	2.2504	2.1757	2.0960	2.0540	2.0102	1.9645	1.9165	1.8657	1.8117
22	4.3009	3.4434	3.0491	2.8167	2.6613	2.5491	2.4638	2.3965	2.3419	2.2967	2.2258	2.1508	2.0707	2.0283	1.9842	1.9380	1.8894	1.8380	1.7831
23	4.2793	3.4221	3.0280	2.7955	2.6400	2.5277	2.4422	2.3748	2.3201	2.2747	2.2036	2.1282	2.0476	2.0050	1.9605	1.9139	1.8648	1.8128	1.7570
24	4.2597	3.4028	3.0088	2.7763	2.6207	2.5082	2.4226	2.3551	2.3002	2.2547	2.1834	2.1077	2.0267	1.9838	1.9390	1.8920	1.8424	1.7896	1.7330
25	4.2417	3.3852	2.9912	2.7587	2.6030	2.4904	2.4047	2.3371	2.2821	2.2365	2.1649	2.0889	2.0075	1.9643	1.9192	1.8718	1.8217	1.7684	1.7110

So the value for $df_1 = 15$ and $df_2 = 12$ is 2.6169.

Therefore the critical value for the left-tailed test will be $\dfrac{1}{2.6169} = 0.38213$.

Let's do a quick exercise problem combining all three tests.

Suppose F is a random variable with degrees of freedom $df1 = 20$ and $df2 = 24$. Let $\alpha = 0.05$. Use the tables to find:

$F\alpha$

$F\alpha / 2$

$F1 - \alpha$

$F1 - \alpha / 2$

1) We look for the table for which $\alpha = 0.05$ and find the value which is in the intersection of the column with heading $df1 = 20$ and the row with the heading $df2 = 24$.

df2 \ df1=1	1	2	3	4	5	6	7	8	9	10	12	15	20	24	30	40	60	120	∞
1	161.4476	199.5000	215.7073	224.5832	230.1619	233.9860	236.7684	238.8827	240.5433	241.8817	243.9060	245.9499	248.0131	249.0518	250.0951	251.1432	252.1957	253.2529	254.3144
2	18.5128	19.0000	19.1643	19.2468	19.2964	19.3295	19.3532	19.3710	19.3848	19.3959	19.4125	19.4291	19.4458	19.4541	19.4624	19.4707	19.4791	19.4874	19.4957
3	10.1280	9.5521	9.2766	9.1172	9.0135	8.9406	8.8867	8.8452	8.8123	8.7855	8.7446	8.7029	8.6602	8.6385	8.6166	8.5944	8.5720	8.5494	8.5264
4	7.7086	6.9443	6.5914	6.3882	6.2561	6.1631	6.0942	6.0410	5.9988	5.9644	5.9117	5.8578	5.8025	5.7744	5.7459	5.7170	5.6877	5.6581	5.6281
5	6.6079	5.7861	5.4095	5.1922	5.0503	4.9503	4.8759	4.8183	4.7725	4.7351	4.6777	4.6188	4.5581	4.5272	4.4957	4.4638	4.4314	4.3985	4.3650
6	5.9874	5.1433	4.7571	4.5337	4.3874	4.2839	4.2067	4.1468	4.0990	4.0600	3.9999	3.9381	3.8742	3.8415	3.8082	3.7743	3.7398	3.7047	3.6689
7	5.5914	4.7374	4.3468	4.1203	3.9715	3.8660	3.7870	3.7257	3.6767	3.6365	3.5747	3.5107	3.4445	3.4105	3.3758	3.3404	3.3043	3.2674	3.2298
8	5.3177	4.4590	4.0662	3.8379	3.6875	3.5806	3.5005	3.4381	3.3881	3.3472	3.2839	3.2184	3.1503	3.1152	3.0794	3.0428	3.0053	2.9669	2.9276
9	5.1174	4.2565	3.8625	3.6331	3.4817	3.3738	3.2927	3.2296	3.1789	3.1373	3.0729	3.0061	2.9365	2.9005	2.8637	2.8259	2.7872	2.7475	2.7067
10	4.9646	4.1028	3.7083	3.4780	3.3258	3.2172	3.1355	3.0717	3.0204	2.9782	2.9130	2.8450	2.7740	2.7372	2.6996	2.6609	2.6211	2.5801	2.5379
11	4.8443	3.9823	3.5874	3.3567	3.2039	3.0946	3.0123	2.9480	2.8962	2.8536	2.7876	2.7186	2.6464	2.6090	2.5705	2.5309	2.4901	2.4480	2.4045
12	4.7472	3.8853	3.4903	3.2592	3.1059	2.9961	2.9134	2.8486	2.7964	2.7534	2.6866	2.6169	2.5436	2.5055	2.4663	2.4259	2.3842	2.3410	2.2962
13	4.6672	3.8056	3.4105	3.1791	3.0254	2.9153	2.8321	2.7669	2.7144	2.6710	2.6037	2.5331	2.4589	2.4202	2.3803	2.3392	2.2966	2.2524	2.2064
14	4.6001	3.7389	3.3439	3.1122	2.9582	2.8477	2.7642	2.6987	2.6458	2.6022	2.5342	2.4630	2.3879	2.3487	2.3082	2.2664	2.2229	2.1778	2.1307
15	4.5431	3.6823	3.2874	3.0556	2.9013	2.7905	2.7066	2.6408	2.5876	2.5437	2.4753	2.4034	2.3275	2.2878	2.2468	2.2043	2.1601	2.1141	2.0658
16	4.4940	3.6337	3.2389	3.0069	2.8524	2.7413	2.6572	2.5911	2.5377	2.4935	2.4247	2.3522	2.2756	2.2354	2.1938	2.1507	2.1058	2.0589	2.0096
17	4.4513	3.5915	3.1968	2.9647	2.8100	2.6987	2.6143	2.5480	2.4943	2.4499	2.3807	2.3077	2.2304	2.1898	2.1477	2.1040	2.0584	2.0107	1.9604
18	4.4139	3.5546	3.1599	2.9277	2.7729	2.6613	2.5767	2.5102	2.4563	2.4117	2.3421	2.2686	2.1906	2.1497	2.1071	2.0629	2.0166	1.9681	1.9168
19	4.3807	3.5219	3.1274	2.8951	2.7401	2.6283	2.5435	2.4768	2.4227	2.3779	2.3080	2.2341	2.1555	2.1141	2.0712	2.0264	1.9795	1.9302	1.8780
20	4.3512	3.4928	3.0984	2.8661	2.7109	2.5990	2.5140	2.4471	2.3928	2.3479	2.2776	2.2033	2.1242	2.0825	2.0391	1.9938	1.9464	1.8963	1.8432
21	4.3248	3.4668	3.0725	2.8401	2.6848	2.5727	2.4876	2.4205	2.3660	2.3210	2.2504	2.1757	2.0960	2.0540	2.0102	1.9645	1.9165	1.8657	1.8117
22	4.3009	3.4434	3.0491	2.8167	2.6613	2.5491	2.4638	2.3965	2.3419	2.2967	2.2258	2.1508	2.0707	2.0283	1.9842	1.9380	1.8894	1.8380	1.7831
23	4.2793	3.4221	3.0280	2.7955	2.6400	2.5277	2.4422	2.3748	2.3201	2.2747	2.2036	2.1282	2.0476	2.0050	1.9605	1.9139	1.8648	1.8128	1.7570
24	4.2597	3.4028	3.0088	2.7763	2.6207	2.5082	2.4226	2.3551	2.3002	2.2547	2.1834	2.1077	2.0267	1.9838	1.9390	1.8920	1.8424	1.7896	1.7330
25	4.2417	3.3852	2.9912	2.7587	2.6030	2.4904	2.4047	2.3371	2.2821	2.2365	2.1649	2.0889	2.0075	1.9643	1.9192	1.8718	1.8217	1.7684	1.7110
26	4.2252	3.3690	2.9752	2.7426	2.5868	2.4741	2.3883	2.3205	2.2655	2.2197	2.1479	2.0716	1.9898	1.9464	1.9010	1.8533	1.8027	1.7488	1.6906
27	4.2100	3.3541	2.9604	2.7278	2.5719	2.4591	2.3732	2.3053	2.2501	2.2043	2.1323	2.0558	1.9736	1.9299	1.8842	1.8361	1.7851	1.7306	1.6717

Our critical value for $\alpha = 0.05$ with $df1 = 20$ and $df2 = 24$ is 2.0267.

2) We look for the table for which $\alpha/2 = 0.025$ and find the value which is in the intersection of the column with heading $df1 = 20$ and the row with heading $df2 = 24$.

$df_2 \backslash df_1$	1	2	3	4	5	6	7	8	9	10	12	15	20	24	30	40	60	120	∞
1	647.7890	799.5000	864.1630	899.5833	921.8479	937.1111	948.2169	956.6562	963.2846	968.6274	976.7079	984.8668	993.1028	997.2492	1001.414	1005.598	1009.800	1014.020	1018.258
2	38.5063	39.0000	39.1655	39.2484	39.2982	39.3315	39.3552	39.3730	39.3869	39.3980	39.4146	39.4313	39.4479	39.4562	39.465	39.473	39.481	39.490	39.498
3	17.4434	16.0441	15.4392	15.1010	14.8848	14.7347	14.6244	14.5399	14.4731	14.4189	14.3366	14.2527	14.1674	14.1241	14.081	14.037	13.992	13.947	13.902
4	12.2179	10.6491	9.9792	9.6045	9.3645	9.1973	9.0741	8.9796	8.9047	8.8439	8.7512	8.6565	8.5599	8.5109	8.461	8.411	8.360	8.309	8.257
5	10.0070	8.4336	7.7636	7.3879	7.1464	6.9777	6.8531	6.7572	6.6811	6.6192	6.5245	6.4277	6.3286	6.2780	6.227	6.175	6.123	6.069	6.015
6	8.8131	7.2599	6.5988	6.2272	5.9876	5.8198	5.6955	5.5996	5.5234	5.4613	5.3662	5.2687	5.1684	5.1172	5.065	5.012	4.959	4.904	4.849
7	8.0727	6.5415	5.8898	5.5226	5.2852	5.1186	4.9949	4.8993	4.8232	4.7611	4.6658	4.5678	4.4667	4.4150	4.362	4.309	4.254	4.199	4.142
8	7.5709	6.0595	5.4160	5.0526	4.8173	4.6517	4.5286	4.4333	4.3572	4.2951	4.1997	4.1012	3.9995	3.9472	3.894	3.840	3.784	3.728	3.670
9	7.2093	5.7147	5.0781	4.7181	4.4844	4.3197	4.1970	4.1020	4.0260	3.9639	3.8682	3.7694	3.6669	3.6142	3.560	3.505	3.449	3.392	3.333
10	6.9367	5.4564	4.8256	4.4683	4.2361	4.0721	3.9498	3.8549	3.7790	3.7168	3.6209	3.5217	3.4185	3.3654	3.311	3.255	3.198	3.140	3.080
11	6.7241	5.2559	4.6300	4.2751	4.0440	3.8807	3.7586	3.6638	3.5879	3.5257	3.4296	3.3299	3.2261	3.1725	3.118	3.061	3.004	2.944	2.883
12	6.5538	5.0959	4.4742	4.1212	3.8911	3.7283	3.6065	3.5118	3.4358	3.3736	3.2773	3.1772	3.0728	3.0187	2.963	2.906	2.848	2.787	2.725
13	6.4143	4.9653	4.3472	3.9959	3.7667	3.6043	3.4827	3.3880	3.3120	3.2497	3.1532	3.0527	2.9477	2.8932	2.837	2.780	2.720	2.659	2.595
14	6.2979	4.8567	4.2417	3.8919	3.6634	3.5014	3.3799	3.2853	3.2093	3.1469	3.0502	2.9493	2.8437	2.7888	2.732	2.674	2.614	2.552	2.487
15	6.1995	4.7650	4.1528	3.8043	3.5764	3.4147	3.2934	3.1987	3.1227	3.0602	2.9633	2.8621	2.7559	2.7006	2.644	2.585	2.524	2.461	2.395
16	6.1151	4.6867	4.0768	3.7294	3.5021	3.3406	3.2194	3.1248	3.0488	2.9862	2.8890	2.7875	2.6808	2.6252	2.568	2.509	2.447	2.383	2.316
17	6.0420	4.6189	4.0112	3.6648	3.4379	3.2767	3.1556	3.0610	2.9849	2.9222	2.8249	2.7230	2.6158	2.5598	2.502	2.442	2.380	2.315	2.247
18	5.9781	4.5597	3.9539	3.6083	3.3820	3.2209	3.0999	3.0053	2.9291	2.8664	2.7689	2.6667	2.5590	2.5027	2.445	2.384	2.321	2.256	2.187
19	5.9216	4.5075	3.9034	3.5587	3.3327	3.1718	3.0509	2.9563	2.8801	2.8172	2.7196	2.6171	2.5089	2.4523	2.394	2.333	2.270	2.203	2.133
20	5.8715	4.4613	3.8587	3.5147	3.2891	3.1283	3.0074	2.9128	2.8365	2.7737	2.6758	2.5731	2.4645	2.4076	2.349	2.287	2.223	2.156	2.085
21	5.8266	4.4199	3.8183	3.4754	3.2501	3.0895	2.9686	2.8740	2.7977	2.7348	2.6368	2.5338	2.4247	2.3675	2.308	2.246	2.182	2.114	2.042
22	5.7863	4.3828	3.7829	3.4401	3.2151	3.0546	2.9338	2.8392	2.7628	2.6998	2.6017	2.4984	2.3890	2.3315	2.272	2.210	2.145	2.076	2.003
23	5.7498	4.3492	3.7505	3.4083	3.1835	3.0232	2.9023	2.8077	2.7313	2.6682	2.5699	2.4665	2.3567	2.2989	2.239	2.176	2.111	2.041	1.968
24	5.7166	4.3187	3.7211	3.3794	3.1548	2.9946	2.8738	2.7791	2.7027	2.6396	2.5411	2.4110	2.3273	2.2693	2.209	2.146	2.080	2.010	1.935
25	5.6864	4.2909	3.6943	3.3550	3.1287	2.9685	2.8478	2.7531	2.6766	2.6135	2.5149	2.4110	2.3005	2.2422	2.182	2.118	2.052	1.981	1.906
26	5.6586	4.2655	3.6697	3.3289	3.1048	2.9447	2.8240	2.7293	2.6528	2.5896	2.4908	2.3867	2.2759	2.2174	2.157	2.093	2.026	1.954	1.878
27	5.6331	4.2421	3.6472	3.3067	3.0828	2.9228	2.8021	2.7074	2.6309	2.5676	2.4688	2.3644	2.2533	2.1946	2.133	2.069	2.002	1.930	1.853

Our critical value for $\alpha/2 = 0.025$ with $df1 = 20$ and $df2 = 24$ is 2.3273 .

3) We look for the table for $\alpha = 0.05$ and find the value which is in the intersection of the column with heading $df1 = 24$ and the row with heading $df2 = 20$.

df_2	$df_1=1$	2	3	4	5	6	7	8	9	10	12	15	20	24	30	40	60	120	∞
1	161.4476	199.5000	215.7073	224.5832	230.1619	233.9860	236.7684	238.8827	240.5433	241.8817	243.9060	245.9499	248.0131	249.0518	250.0951	251.1432	252.1957	253.2529	254.3144
2	18.5128	19.0000	19.1643	19.2468	19.2964	19.3295	19.3532	19.3710	19.3848	19.3959	19.4125	19.4291	19.4458	19.4541	19.4624	19.4707	19.4791	19.4874	19.4957
3	10.1280	9.5521	9.2766	9.1172	9.0135	8.9406	8.8867	8.8452	8.8123	8.7855	8.7446	8.7029	8.6602	8.6385	8.6166	8.5944	8.5720	8.5494	8.5264
4	7.7086	6.9443	6.5914	6.3882	6.2561	6.1631	6.0942	6.0410	5.9988	5.9644	5.9117	5.8578	5.8025	5.7744	5.7459	5.7170	5.6877	5.6581	5.6281
5	6.6079	5.7861	5.4095	5.1922	5.0503	4.9503	4.8759	4.8183	4.7725	4.7351	4.6777	4.6188	4.5581	4.5272	4.4957	4.4638	4.4314	4.3985	4.3650
6	5.9874	5.1433	4.7571	4.5337	4.3874	4.2839	4.2067	4.1468	4.0990	4.0600	3.9999	3.9381	3.8742	3.8415	3.8082	3.7743	3.7398	3.7047	3.6689
7	5.5914	4.7374	4.3468	4.1203	3.9715	3.8660	3.7870	3.7257	3.6767	3.6365	3.5747	3.5107	3.4445	3.4105	3.3758	3.3404	3.3043	3.2674	3.2298
8	5.3177	4.4590	4.0662	3.8379	3.6875	3.5806	3.5005	3.4381	3.3881	3.3472	3.2839	3.2184	3.1503	3.1152	3.0794	3.0428	3.0053	2.9669	2.9276
9	5.1174	4.2565	3.8625	3.6331	3.4817	3.3738	3.2927	3.2296	3.1789	3.1373	3.0729	3.0061	2.9365	2.9005	2.8637	2.8259	2.7872	2.7475	2.7067
10	4.9646	4.1028	3.7083	3.4780	3.3258	3.2172	3.1355	3.0717	3.0204	2.9782	2.9130	2.8450	2.7740	2.7372	2.6996	2.6609	2.6211	2.5801	2.5379
11	4.8443	3.9823	3.5874	3.3567	3.2039	3.0946	3.0123	2.9480	2.8962	2.8536	2.7876	2.7186	2.6464	2.6090	2.5705	2.5309	2.4901	2.4480	2.4045
12	4.7472	3.8853	3.4903	3.2592	3.1059	2.9961	2.9134	2.8486	2.7964	2.7534	2.6866	2.6169	2.5436	2.5055	2.4663	2.4259	2.3842	2.3410	2.2962
13	4.6672	3.8056	3.4105	3.1791	3.0254	2.9153	2.8321	2.7669	2.7144	2.6710	2.6037	2.5331	2.4589	2.4202	2.3803	2.3392	2.2966	2.2524	2.2064
14	4.6001	3.7389	3.3439	3.1122	2.9582	2.8477	2.7642	2.6987	2.6458	2.6022	2.5342	2.4630	2.3879	2.3487	2.3082	2.2664	2.2229	2.1778	2.1307
15	4.5431	3.6823	3.2874	3.0556	2.9013	2.7905	2.7066	2.6408	2.5876	2.5437	2.4753	2.4034	2.3275	2.2878	2.2468	2.2043	2.1601	2.1141	2.0658
16	4.4940	3.6337	3.2389	3.0069	2.8524	2.7413	2.6572	2.5911	2.5377	2.4935	2.4247	2.3522	2.2756	2.2354	2.1938	2.1507	2.1058	2.0589	2.0096
17	4.4513	3.5915	3.1968	2.9647	2.8100	2.6987	2.6143	2.5480	2.4943	2.4499	2.3807	2.3077	2.2304	2.1898	2.1477	2.1040	2.0584	2.0107	1.9604
18	4.4139	3.5546	3.1599	2.9277	2.7729	2.6613	2.5767	2.5102	2.4563	2.4117	2.3421	2.2686	2.1906	2.1497	2.1071	2.0629	2.0166	1.9681	1.9168
19	4.3807	3.5219	3.1274	2.8951	2.7401	2.6283	2.5435	2.4768	2.4227	2.3779	2.3080	2.2341	2.1555	2.1141	2.0712	2.0264	1.9795	1.9302	1.8780
20	4.3512	3.4928	3.0984	2.8661	2.7109	2.5990	2.5140	2.4471	2.3928	2.3479	2.2776	2.2033	2.1242	2.0825	2.0391	1.9938	1.9464	1.8963	1.8432
21	4.3248	3.4668	3.0725	2.8401	2.6848	2.5727	2.4876	2.4205	2.3660	2.3210	2.2504	2.1757	2.0960	2.0540	2.0102	1.9645	1.9165	1.8657	1.8117
22	4.3009	3.4434	3.0491	2.8167	2.6613	2.5491	2.4638	2.3965	2.3419	2.2967	2.2258	2.1508	2.0707	2.0283	1.9842	1.9380	1.8894	1.8380	1.7831
23	4.2793	3.4221	3.0280	2.7955	2.6400	2.5277	2.4422	2.3748	2.3201	2.2747	2.2036	2.1282	2.0476	2.0050	1.9605	1.9139	1.8648	1.8128	1.7570
24	4.2597	3.4028	3.0088	2.7763	2.6207	2.5082	2.4226	2.3551	2.3002	2.2547	2.1834	2.1077	2.0267	1.9838	1.9390	1.8920	1.8424	1.7896	1.7330
25	4.2417	3.3852	2.9912	2.7587	2.6030	2.4904	2.4047	2.3371	2.2821	2.2365	2.1649	2.0889	2.0075	1.9643	1.9192	1.8718	1.8217	1.7684	1.7110

The value we get is 2.0825.

So our critical value is $\dfrac{1}{2.0825} = 0.48$.

5) **Compare the calculated test statistic to the critical value.**

The F-statistic will be compared to the critical F-value when deciding to accept or reject the null hypothesis. If the calculated test statistic exceeds the critical value, the null hypothesis is rejected.

8.3.4.4 Decision

Terminology	**Alternative hypothesis**	**Rejection region**
Right-tailed	$H_A{:}s_1^2 > s_2^2 \text{ or } \sigma_1^2 > \sigma_2^2$	

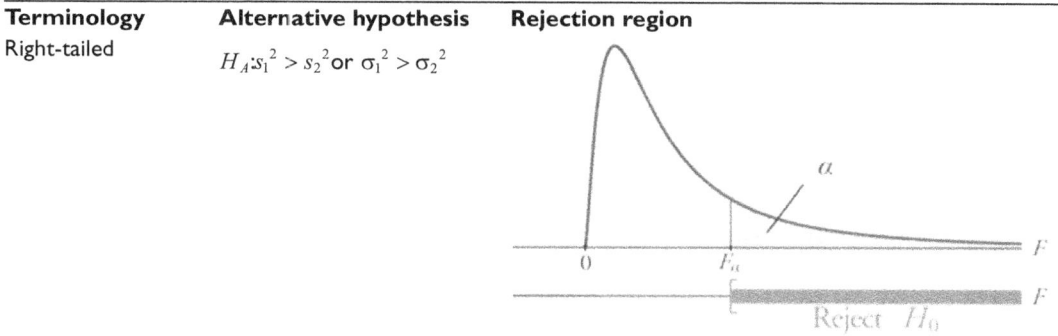

Figure 8.16

$$F_{test\ statistics} > F_{critical\ value}$$

Left-tailed	$H_A{:}s_1^2 < s_2^2 \text{ or } \sigma_1^2 < \sigma_2^2$	

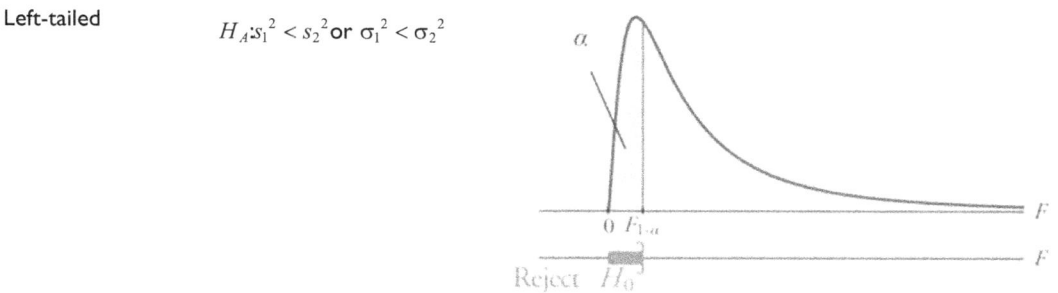

Figure 8.17

Two-tailed	$H_A{:}s_1^2 \neq s_2^2 \text{ or } \sigma_1^2 \neq \sigma_2^2$	

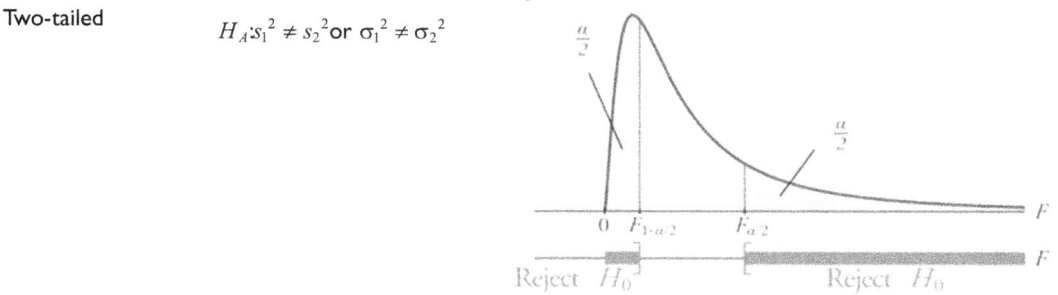

Terminology	Alternative Hypothesis	Rejection Region
Right-tailed	$H_a : \sigma_1^2 > \sigma_2^2$	$F \geq F_\alpha$
Left-tailed	$H_a : \sigma_1^2 < \sigma_2^2$	$F \leq F_{1-\alpha}$
Two-tailed	$H_a : \sigma_1^2 \neq \sigma_2^2$	$F \leq F_{1-\alpha/2}$ or $F \geq F_{\alpha/2}$

The critical values within the table are often compared to the F statistic of an F test. If the F statistic is greater than the critical value found in the table, then you can reject the null hypothesis of the F test and conclude that the results of the test are statistically significant.

There are two uses of the F-distribution that will be discussed in this chapter. The first is a very simple test to see if two samples come from populations with the same variance. The second is one-way analysis of variance (ANOVA), which uses the F-distribution to test whether three or more samples come from populations with the same mean.

8.3.5 F test to match two variances

This test is to check whether the variances of two populations or two samples are equal or not. Because the F-distribution is generated by drawing two samples from the same normal population, it can be used to test the hypothesis that two samples come from populations with the same variance. You would have two samples (one of size n_1 and one of size n_2) and the sample variance from each.

8.3.5.1 Example

Two random samples were drawn from two normal populations and their values are:

A	65	74	76	78	80	82	85	88	92			
B	55	62	65	70	71	72	73	76	78	80	82	84	94

Test whether the two populations have the same variance at the 5% level of significance.

8.3.5.1.1 Hypothesis statements

H_0 : no difference in variances $s_1^2 = s_2^2$.
H_A: difference in variances $s_1^2 \neq s_2^2$.

8.3.5.1.2 Calculate your F statistic

Let's calculate the variances of each group.
    For group A:

Values	$x_i - \underline{x_1}$	$\left(x_i - \underline{x_1}\right)^2$
65	$65 - 80 = -15$	225
74	$74 - 80 = -6$	36
76	$76 - 80 = -4$	16

Values	$x_i - \underline{x_1}$	$\left(x_i - \underline{x_1}\right)^2$
78	$78 - 80 = -2$	4
80	$80 - 80 = 0$	0
82	$82 - 80 = 2$	4
85	$85 - 80 = 5$	25
88	$88 - 80 = 8$	64
92	$92 - 80 = 12$	144
$\Sigma720$		$\Sigma518$

$$s_1^2 = \frac{\Sigma\left(x_i - \underline{x_1}\right)^2}{n_1 - 1} = \frac{518}{9-1} = 64.75$$

For group B:

Values	$x_j - \underline{x_2}$	$\left(x_j - \underline{x_2}\right)^2$
55	$55 - 74 = -19$	361
62	$62 - 74 = -12$	144
65	$65 - 74 = -9$	81
70	$70 - 74 = 4$	16
71	$71 - 74 = 3$	9
72	$72 - 74 = 2$	4
73	$73 - 74 = 1$	1
76	$76 - 74 = 2$	4
78	$78 - 74 = 4$	16
80	$80 - 74 = 6$	36
82	$82 - 74 = 8$	64
84	$84 - 74 = 10$	100
94	$94 - 74 = 20$	400
$\Sigma962$		$\Sigma1236$

$$s_2^2 = \frac{\Sigma\left(x_j - \underline{x_2}\right)^2}{n_2 - 1} = \frac{1236}{13-1} = 103$$

Since it is a two-tailed test, we divide the bigger variance by the smaller one. F statistic $= \frac{103}{64.75} = 1.59$.

8.3.5.1.3 Calculate the critical value

Our alpha level is 0.05 for a two-tailed test; therefore we look at the table for 0.025. Also our df_1 will be for the larger variance; therefore $df_1 = 13-1 = 12$ and $df_2 = 9-1 = 8$.

The value approximates to 4.20.

8.3.5.1.4 P-value

The $df_1 = 12$, $df_2 = 8$, and the F-test statistic is 1.59.

The value which comes the closest to the test statistic is 2.50 so $P(F \geq 2.50) = 0.100$. Therefore we can conclude that the p-value is > 0.100.

8.3.5.1.5 Decision

8.3.5.1.5.1 CRITICAL VALUE APPROACH

Since the test statistic $= 1.7384 < 4.20$ we do not reject the null hypothesis and we can conclude that there is no difference in variances.

8.3.5.1.5.2 P-VALUE APPROACH

Since the p-value >0.10 hence we do not reject null hypothesis.

8.3.5.2 In MATLAB

Just like the other two tests we have previously discussed, MATLAB has an inbuilt function for the f test of equal variances. The syntax is:

```
[h,p,ci,stats] = vartest2(x, y, 'Alpha', ,'Tail', '')
```

So using the same example we get the output:

```
x = [65 74 76 78 80      82 85 88 92];
y = [55 62 65 70 71 72 73 76 78 80 82 84 94];
[h,p,ci,stats] = vartest2(x,y);

                    h =

                        0


                    p =

                        0.5194


                    ci =

                        0.1790    2.6401


                    stats =

                    struct with fields:

                        fstat: 0.6286
                          df1: 8
                          df2: 12
```

Now notice something interesting in the output. Instead of using 12 which gives the larger variance, our df_1 is 8. Similarly we divided the smaller variance by the larger one reciprocating the value we found by hand: $\dfrac{64.75}{103} = 0.6286$.

However let's try change the positions

```
x = [55 62 65 70 71 72 73 76 78 80 82 84 94];
y = [65 74 76 78 80       82 85 88 92];
[h,p,ci,stats] = vartest2(x,y);


                    h =

                        0


                    p =

                        0.5194


                    ci =

                        0.3788      5.5863


                    stats =

                        struct with fields:

                            fstat: 1.5907
                              df1: 12
                              df2: 8
```

Do you see any difference now?

Clearly this time the output aligns with our results but how?

Notice that, instead of setting values that give smaller variance in x, we are setting it to values that return larger variance.

This means MATLAB cannot by default keep track of which variance is larger and this is something we need to do beforehand.

$$s_x^2 = 103$$

$$s_y^2 = 64.75$$

However all this would not have been necessary if our test was left-tailed since it requires smaller variance to be in the numerator.

Let's look at the examples we used in the t test but instead of using descriptive statistics we are going to use the f test to determine if the two samples have the same variance.

8.3.5.3 Example: difference in BMI of males and females

Group					BMI				
Men	15.58	22.46	25.82	19.85	14.80	21.75	16.96	23.98	23.34
	13.45	20.47	24.62	20.71	17.91	27.28	17.36	18.69	21.03
	19.68	18.10	17.63	19.23	20.33	21.55	23.19		
Women	17.56	14.07	21.85	15.85	16.85	18.89	18.64	20.14	21.48
	23.39	21.51	19.58	19.81	19.64	20.84	20.40	21.27	27.12
	22.22	23.09	23.89	22.96	23.68	24.96	25.94		

The first thing we need to do is check and compare the variances of the two groups. Since it is a two-tailed problem, we first need to check which variance is larger which will be the first argument (in the numerator).

```
if var(men) > var(women)
  % Use men as the first input argument in the variance test
  [h,p,ci,stats] = vartest2(men, women)
else
  % Use women as the first input argument in the variance test
  [h,p,ci,stats] = vartest(women, men)
end
```

```
h =

     0


p =

     0.6372


ci =

     0.5354     2.7571


stats =

  struct with fields:

     fstat: 1.2150
       df1: 24
       df2: 24
```

Let's test the output.
For group men:

BMI	$x_i - \underline{x_1}$	$\left(x_i - \underline{x_1}\right)^2$
15.58	−4.93333	24.33778
22.46	1.946667	3.789511
25.82	5.306667	28.16071

19.85	−0.66333	0.440011
14.80	−5.71333	32.64218
21.75	1.236667	1.529344
16.96	−3.55333	12.62618
23.98	3.466667	12.01778
23.34	2.826667	7.990044
13.45	−6.7808	45.97925
20.47	−0.04333	0.001878
24.62	4.106667	16.86471
20.71	0.196667	0.038678

BMI	$x_i - \underline{x_1}$	$\left(x_i - \underline{x_1}\right)^2$
17.91	−2.60333	6.777344
27.28	6.766667	45.78778
17.36	−3.15333	9.943511
18.69	−1.82333	3.324544
21.03	0.516667	0.266944
19.68	−0.83333	0.694444
18.10	−2.41333	5.824178
17.63	−2.88333	8.313611
19.23	−1.28333	1.646944
20.33	−0.18333	0.033611
21.55	1.036667	1.074678
23.19	2.676667	7.164544

$$s_{men}^2 = \frac{\Sigma\left(x_i - \underline{x_1}\right)^2}{n_1 - 1} = \frac{279.186}{25 - 1} = 11.63275$$

For group women:

BMI	$x_i - \underline{x_1}$	$\left(x_i - \underline{x_1}\right)^2$
17.56	−3.4652	12.00761
14.07	−6.9552	48.37481
21.85	0.8248	0.680295
15.85	−5.1752	26.7827
16.85	−4.1752	17.4323
18.89	−2.1352	4.559079
18.64	−2.3852	5.689179

20.14	−0.8852	0.783579
21.48	0.4548	0.206843
23.39	2.3648	5.592279
21.51	0.4848	0.235031
19.58	−1.4452	2.088603
19.81	−1.2152	1.476711

BMI	$x_i - \underline{x_1}$	$\left(x_i - \underline{x_1}\right)^2$
19.64	−1.3852	1.918779
20.84	−0.1852	0.034299
20.40	−0.6252	0.390875
21.27	0.2448	0.059927
27.12	6.0948	37.14659
22.22	1.1948	1.427547
23.09	2.0648	4.263399
23.89	2.8648	8.207079
22.96	1.9348	3.743451
23.68	2.6548	7.047963
24.96	3.9348	15.48265
25.94	4.9148	24.15526

$$s_{women}^2 = \frac{\Sigma\left(x_i - \underline{x_1}\right)^2}{n_1 - 1} = \frac{229.7868}{25 - 1} = 9.574451$$

Since it is a two-tailed test, we divide the bigger variance by the smaller one: F statistic

$$= \frac{11.63275}{9.574451} = 1.21497 \sim 1.2150$$

8.3.5.3.1 P-value

The $df_1, df_2 = 25$ and the F-test statistic is 1.59.

We look at the row that corresponds to $df_1, df_2 = 25$. The value which comes the closest to the test statistic is 1.68 so $P(F \geq 1.68) = 0.100$. Therefore we can conclude that the p-value is > 0.100.

8.3.5.3.2 Decision

Since the p-value > 0.100 > 0.05 = $alpha$, we do not reject the null hypothesis.

8.3.5.4 Example

Class			Scores					
Robert	85	58	71	62	75	55	85	52
	68	40	80	79	58	67	78	71
	49	98	75	68	65			
Sue	95	100	75	79	80	75	70	92
	84	78	82	90	89	85	81	86
	88	88	86					

Since this is a right-tailed problem, here too we need the larger variance to be the first argument (in the numerator).

```
if var(Robert) > var(Sue)
  % Use Robert as the first input argument in the variance test
  [h,p,~,stats] = vartest2(Robert, Sue,'Alpha',0.01,'Tail','right')
else
  % Use Sue as the first input argument in the variance test
  [h,p,~,stats] = vartest(Sue, Robert,'Alpha',0.01,'Tail','right')
end
```

```
h =

    1

p =

    0.0056

stats =

    struct with fields:

      fstat: 3.4321
        df1: 20
        df2: 18
```

Again let's test the output.
The scores in Robert's class:

Scores	$x_i - x_1$	$\left(x_i - x_1\right)^2$
85	16.47619	271.4649
58	−10.5238	110.7506
71	2.47619	6.131519
62	−6.52381	42.56009
75	6.47619	41.94104
55	−13.5238	182.8934
85	16.47619	271.4649
52	−16.5238	273.0363

68	−0.52381	0.274376
40	−28.5238	813.6077
80	11.47619	$x_i - \underline{x_1}$

Scores	$\left(x_i - \underline{x_1}\right)^2$	79
10.47619	109.7506	58
−10.5238	110.7506	67
−1.52381	2.321995	78
9.47619	89.79819	71
2.47619	6.131519	49
−19.5238	381.1791	98
29.47619	868.8458	75
6.47619	41.94104	68
−0.52381	0.274376	65
−3.52381	12.41723	$s_{Robert}^{2} =$

$$\frac{\Sigma\left(x_i - \underline{x_1}\right)^2}{n_1 - 1} = \frac{3769.238}{21-1} = 188.4619 \quad Scores$$

The scores in Sue's class:

$x_i - \underline{x_1}$	$\left(x_i - \underline{x_1}\right)^2$	95
10.63158	113.0305	100
15.63158	244.3463	75
−9.36842	87.76729	79
−5.36842	28.81993	80`
−4.36842	19.08309	75
−9.36842	87.76729	70
−14.3684	206.4515	92
7.631579	58.24101	84
−0.36842	0.135733	78
−6.36842	40.55677	82
−2.36842	5.609413	*Scores*

$x_i - \underline{x_1}$	$\left(x_i - \underline{x_1}\right)^2$	90
5.631579	31.71469	89
4.631579	21.45153	85

0.631579	0.398893	81
−3.36842	11.34625	86
1.631579	2.662053	88
3.631579	13.18837	88
3.631579	13.18837	86
1.631579	2.662053	$s_{women}^2 =$

$$\frac{\Sigma\left(x_i - \underline{x_1}\right)^2}{n_1 - 1} = \frac{988.4211}{19 - 1} = 54.91228 \quad = \frac{188.4619}{54.91228} = 3.43205 \sim 3.432$$

Since this is a right-tailed test, we divide the bigger variance by the smaller one: F statistic $df_1 = 20, df_2 = 18$

8.3.5.4.1 P-value

The 3.4321, and the F-test statistic is $df_1 = 20, df_2 = 18..$

We look at the row that corresponds to 3.08 The value which comes the closest to the test statistic is $P\left(F \geq 3.08\right) = 0.01$ sop. Therefore we can conclude that the < 0.01.-value is p

8.3.5.4.2 Decision

Since the $< 0.0100 = alpha$,-value 30we reject the null hypothesis.

8.3.5.5 Exercise problems

1) A manufacturer wants to place an order for batteries for laptops. But he is not sure which supplier he should choose. So, he decides to collect 28 samples from supplier A and 4.5, 6.2, 6.5, 5.6, 6.9, 7.0, 5.9, 7.3, 9.4, 5.0, from supplier B.

Supplier A:

6.8, 7.5, 8.1, 11.1, 9.5, 8.8, 7.2, 10.2, 9.5,

7.4,   8.2,   9.3,   10.1,   7.5,   10.5,   8.2,   9.8,   7.8,

8.5,   7.8

7.2, 15.2, 7.8, 8.9, 5.2, 6.4,   8.8,   9.9,   16.5

Supplier B:

$$8.1, 10.9, 11.2, 9.5, 13.8, \qquad 11.9, \qquad 9.9, \qquad 10.5, \qquad 14,$$

$$11.4, \qquad 12.5, \qquad 14.5, \qquad 11.4, \quad 12.7, \quad 13.3, \qquad 12.7,$$

$$0.05$$

At the $= 1\%$ level of significance, determine if there is a difference in battery life between the two suppliers.

2) In one study, the emissions of carbon monoxide (CO) from two different heated tobacco products (HTPs) were compared. Use alpha$Y$ to test if the CO emission of Brand X is higher than that of Brand X given the following data:

Cigarettes				CO emissions				
Brand 12.56	17.21	13.18	14.52	15.33	16.58	13.67	14.03	14.21
	15.27	11.24	14.78	15.13	13.25	12.72	16.41	Brand Y
12.70	14.78	15.13	11.16	13.67	17.88	12.55	14.52	13.25
	14.21	16.61	15.22	13.17	15.29	16.44	15.33	50 Hz

3) An engineer calculates the synchronous speed and typical full load speed of an AC motor with frequency 10% to determine whether the RPM of the full load is higher than that of the synchronous. Test, at the 489 level of significance, whether the data provides sufficient evidence to conclude that his assumption is true.

RPM = revolutions per minute	
Synchronous	AC
478	425
417	395
384	356
335	332
321	310
298	281
269	253
236	198
177	92.6

Parametric tests 507

4) A technician has experimented with a different type of steel which he claims to have stronger tensile strength in comparison to the ones he generally uses in the elevator in the sense that they will accept greater tension loads. In order to test the performance of the new cables in comparison with the old cables, samples are tested for failure under tension. The following results were obtained:

Failure tension	
90.8	91.9
92.3	93.7
93.7	92.2
92.5	94.8
94.9	93.1
94.6	92.6
92.5	91.4
91.9	95.3
95.7	96.1
96.2	1%

The technician decides to perform the test at the 1% level of significance to distinguish between the cables. On the basis of the results given, should the manufacturer replace the old cable with the new one? You may assume that the populations from which the samples are drawn have equal variances.

8.3.6 Analysis of variance (ANOVA)

ANOVA can determine whether the means of three or more groups are different. ANOVA uses f tests to statistically test the equality of means.

The tests we have done so far are only applicable for estimating the test of significance between two sample means, while ANOVA allows three or more independent sample means at a time. This test uses the F-statistic (F_0) and critical values of F (F_e) from the F-distribution table to test the hypothesis of variances. The f test is used in the analysis of one-way or two-way ANOVA to check if the null hypothesis is accepted or rejected at a stated level of significance. The ratio of variance between the sample means and the variance within the sample is called the F-statistic.

ANOVA stands for analysis of variance and tests for differences in the effects of independent variables on a dependent variable.

There are two main types of analysis of variance: one-way (or unidirectional) and two-way (bidirectional). One-way or two-way refers to the number of independent variables in your analysis of variance test.

8.3.6.1 One-way ANOVA

A one-way ANOVA is a type of statistical test that compares the variance in the group means within a sample whilst considering only one independent variable or factor.

This test is used to see if there is a variation in the mean values of three or more groups. Such a test is used where the dataset has only one categorical independent variable (also known as a factor) and a normally distributed continuous (i.e., interval or ratio level) dependent variable. The independent variable divides cases into two or more mutually exclusive levels, categories, or groups. The one-way ANOVA test for differences in the means of the dependent variable is broken down by the levels of the independent variable.

An example to understand this is prescribing medicines. Suppose there is a group of patients who have high blood pressure. They are given three different β-adrenergic blockers—**bisoprolol, carvedilol, and metoprolol succinate**—that have the same functionality, i.e. to decrease the rate and force of heart contractions, reducing blood pressure. To understand the effectiveness of each drug and choose the best among them, the ANOVA test is used.

8.3.6.1.1 Assumptions

The one-way ANOVA test is carried out based on these assumptions:

- The observations from which samples are drawn should be normally distributed.
- Independence of cases: the sample cases should be independent of each other.
- Homogeneity of variance: homogeneity means that the variance among the groups should be approximately equal. By rule of thumb: the largest sample standard deviation should not be more than twice the smallest sample standard deviation.

Steps in hypothesis testing:

1) Stating the null and alternative hypotheses.

The null hypothesis H_0: $\mu_1 = \mu_2 = \ldots = \mu_k$.
   Alternative hypothesis H_1: $\mu_1 \neq \mu_2 \neq \ldots \neq \mu_k$.

2) Calculating the f statistic.

In order to calculate the F test statistic, we need to define some notation and determine several quantities beforehand. Let's represent our data, the group means, and the grand mean as follows:

Groups	Number of elements	Values in each group			Sample mean
1	n_1	x_{11} x_{12}	$\cdots$	x_{n_1}	$\underline{x_1} = \dfrac{x_{11} + x_{12} + \ldots x_{n_1}}{n_1}$
2	n_2	x_{21} x_{22}	$\cdots$	x_{n_2}	$\underline{x_2} = \dfrac{x_{21} + x_{22} + \ldots x_{n_2}}{n_2}$
.		.		.	
.		.		.	
.		.		.	
.					

Groups	Number of elements	Values in each group				Sample mean
m	n_m	x_{m1}	x_{m2}	....	x_{n_m}	$\underline{x_m} = \dfrac{x_{m1} + x_{m2} + \ldots x_{n_m}}{n_m}$
	Grand mean $\overline{\overline{x}}$					$\dfrac{(x_{11} + \ldots x_{n_1}) + (x_{21} + \ldots x_{n_2}) + \left(x_{m1} + \ldots x_{n_m}\right)}{n_1 + n_2 + n_m}$

In the ANOVA test, there are two types of mean that are calculated: the grand and sample mean.

The sample mean x_1 represents the average value for a group while the grand mean $\overline{\overline{x}}$ represents the average value of sample means of different groups or the mean of all the observations combined.

Source	df	Sum of squares (SS)	Mean squares (MS)	F
Treatment or between groups	$m-1$	$SST \,/\, SSB = \Sigma n_{ij}\left(x_{ij} - \overline{\overline{x}}\right)^2$	$\mathbf{MSB} = \dfrac{SSB}{m-1}$	$\dfrac{MSB}{MSE}$
Error	$n-m$	$SSE \,/\, SSW = \Sigma(n_{ij} - 1)s_{ij}^{\,2}$ $= \Sigma\Sigma\left(x_i - \underline{x}\right)^2$	$\mathbf{MSE} = \dfrac{SSE}{n \quad m}$	
Total	$n-1$	$TSS = SST + SSE$		

Where
   $m =$ the number of groups being compared/number of treatments
   $n =$ total data points collected
   $SST \,/\, SSB =$ sum of squares for treatment or the between-group sum of squares
   $SSE \,/\, SSW =$ sum of squares for error or the within-group sum of squares
   $TSS =$ total sum of squares
   $MSB =$ mean sum of squares between groups
   $MSE =$ error mean sum of squares

As we can see, the ANOVA partitions the TSS into sum of squares due to treatment or the between-groups effect and the sum of squared errors or within groups.

Let's look at each of the columns separately.

8.3.6.1.2 Sum of squares (SS) column

This column represents the sum of squared deviations.

8.3.6.1.3 Mean squares (MS) column

Each SS has a corresponding df which is a measure of the number of independent pieces of information present in the deviations that are used to compute the corresponding SS.

And each MS is the SS divided by the df for that line. Each MS is a variance estimate or a variance-like quantity, and as such its units are the squares of the outcome units.

8.3.6.2 Example

```
A = [8 10 1 7  5 3 6 4 5];
B = [8 1 9 6  9 3 5 7 4 5 8 10 13 9 7 11];
C = [2 4 5 11 3 10 6];
```

$$x_1 = 8,10,1,7,5,3,6,4,5$$

$$x_2 = 8,1,9,6,9,3,5,7,4,5,8,10,13,9,7,11 \quad x_3 = 2,4,5,11,3,10,6$$

	n	Summation Σx_i	Mean $\dfrac{\Sigma x_i}{n}$
x_1	9	$8+10+1+7+5+3+6+4+5 = 49$	$\underline{x_1} = \dfrac{49}{9}$
x_2	16	$8+1+9+6+9...+13+9+7+11 = 115$	$\underline{x_2} = \dfrac{115}{16}$
x_3	7	$2+4+5+11+3+10+6 = 41$	$\underline{x_3} = \dfrac{41}{7}$
	$\Sigma 32$	$\Sigma 205$	$\overline{\overline{x}} = \dfrac{205}{32}$

	n	Mean $\dfrac{\Sigma x_i}{n}$	$n_{ij}\left(x_{ij} - \overline{\overline{x}}\right)^2$
x_1	9	$\underline{x_1} = \dfrac{49}{9}$	$9*\left(\dfrac{49}{9} - \dfrac{205}{32}\right)^2$
x_2	16	$\underline{x_2} = \dfrac{115}{16}$	$16*\left(\dfrac{115}{16} - \dfrac{205}{32}\right)^2$
x_3	7	$\underline{x_3} = \dfrac{41}{7}$	$7*\left(\dfrac{41}{7} - \dfrac{205}{32}\right)^2$
	$\Sigma 32$	$\overline{\overline{x}} = \dfrac{205}{32}$	

$$SST = n_1 \left(\underline{x_1} - \overline{\overline{x}} \right)^2 + n_2 \left(\underline{x_2} - \overline{\overline{x}} \right)^2 + n_3 \left(\underline{x_3} - \overline{\overline{x}} \right)^2$$

$$= 9 * \left(\frac{49}{9} - \frac{205}{32} \right)^2 + 16 * \left(\frac{115}{16} - \frac{205}{32} \right)^2 + 7 * \left(\frac{41}{7} - \frac{205}{32} \right)^2 = 20.20188$$

	Mean $\dfrac{\Sigma x_i}{n}$	$\left(x_i - \underline{x} \right)^2$
x_1	$\underline{x_1} = \dfrac{49}{9}$	$\left(8 - \dfrac{49}{9} \right)^2 + \left(10 - \dfrac{49}{9} \right)^2 + \ldots \left(4 - \dfrac{49}{9} \right)^2 + \left(5 - \dfrac{49}{9} \right)^2$
x_2	$\underline{x_2} = \dfrac{115}{16}$	$\left(8 - \dfrac{115}{16} \right)^2 + \left(1 - \dfrac{115}{16} \right)^2 + \ldots \left(7 - \dfrac{115}{16} \right)^2 + \left(11 - \dfrac{115}{16} \right)^2$
x_3	$\underline{x_3} = \dfrac{41}{7}$	$\left(2 - \dfrac{41}{7} \right)^2 + \left(4 - \dfrac{41}{7} \right)^2 + \ldots \left(10 - \dfrac{41}{7} \right)^2 + \left(6 - \dfrac{41}{7} \right)^2$
	$\overline{\overline{x}} = \dfrac{205}{32}$	

$$SSE = \left(x_1 - \underline{x_1} \right)^2 + \left(x_2 - \underline{x_2} \right)^2 + \left(x_3 - \underline{x_3} \right)^2$$

$$= \left[\left(8 - \frac{49}{9} \right)^2 + \left(10 - \frac{49}{9} \right)^2 + \ldots \left(4 - \frac{49}{9} \right)^2 + \left(5 - \frac{49}{9} \right)^2 \right]$$

$$+ \left[\left(8 - \frac{115}{16} \right)^2 + \left(1 - \frac{115}{16} \right)^2 + \ldots \left(7 - \frac{115}{16} \right)^2 + \left(11 - \frac{115}{16} \right)^2 \right]$$

$$+ \left[\left(2 - \frac{41}{7} \right)^2 + \left(4 - \frac{41}{7} \right)^2 + \ldots \left(10 - \frac{41}{7} \right)^2 + \left(6 - \frac{41}{7} \right)^2 \right]$$

$$= 273.5169$$

$$TSS = SST + SSE = 20.20188 + 273.5169 = 293.71878$$

Source	df	Sum of squares (SS)	Mean squares (MS)		F
Treatment	$3-1=2$	20.20188	$\dfrac{20.20188}{3-1}=10.101$		$\dfrac{10.101}{9.43161}=1.07097$
Error	$32-3=29$	273.5169	$\dfrac{273.5169}{32-3}=9.43161$		
Total	$32-1=31$	293.71878			

Hence our test statistic $=1.07097$.

8.3.6.2.1 Calculating p-value approach

The p-value that corresponds to $F \sim F\left(df_1 = 2, df_2 = 29\right)$. Now if we check the table at $df_1 = 2, df_2 = 29$ and the value closest to our statistics is 2.50 which means the p-value is > 0.100.

8.3.6.2.2 Decision

Since our p-value $> 0.100 > 0.05 = alpha$, we fail to reject the null hypothesis.

8.3.6.3 MATLAB

The syntax for using the anova1 function is as follows:

```
[p,tbl,stats] = anova1(data, group, Indicator)
```

There are three input arguments:

1) Data: a vector or matrix of data values for all groups combined.
2) Group: a vector or cell array of grouping variables that indicate which group each observation belongs to.
1) Indicator: indicator to display the ANOVA table and boxplot. It can be either of the two values:

'on' (default)	It displays the standard ANOVA table and boxplot.
'off'	anova1 returns the output arguments, only. It does not display the standard ANOVA table and boxplot.

There are three output arguments:

1) p-value
2) tbl ANOVA table, returned as a cell array
3) Stats

gnames	Names of the groups
n	Number of observations in each group
source	Source of the stats output
means	Estimated values of the means
df	Error (within-groups) degrees of freedom ($N - k$, where N is the total number of observations and k is the number of groups)
s	Square root of the mean squared error

```
% Unequal sample sizes
A = [8 10 1 7  5 3 6 4 5];
B = [8 1 9 6  9 3 5 7 4 5 8 10 13 9 7 11];
C = [2 4 5 11 3 10 6];
data = [A, B, C];


% Create a grouping variable
groups = [repmat({'Group 1'}, 1, length(A)),
          repmat({'Group 2'}, 1, length(B)),
          repmat({'Group 3'}, 1, length(C))];


% Perform ANOVA test
[p, tbl, stats] = anova1(data, groups)
```

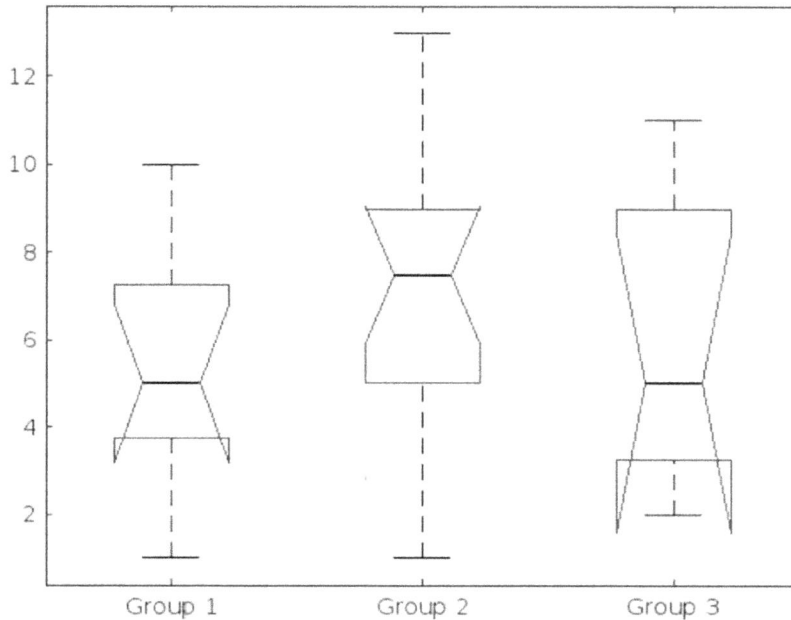

Figure 1: One-way ANOVA   — ✕

ANOVA Table

Source	SS	df	MS	F	Prob>F
Groups	20.202	2	10.1009	1.07	0.3558
Error	273.517	29	9.4316		
Total	293.719	31			

```
p =

    0.3558


tbl =

   4×6 cell array

    {'Source'}    {'SS'        }    {'df'}    {'MS'        }    {'F'        }    {'Prob>F'   }
    {'Groups'}    {[  20.2019]}    {[  2]}    {[ 10.1009]}    {[  1.0710]}    {[  0.3558]}
    {'Error' }    {[273.5169]}    {[29]}    {[  9.4316]}    {0×0 double}    {0×0 double}
    {'Total' }    {[293.7188]}    {[31]}    {0×0 double}    {0×0 double}    {0×0 double}


stats =

   struct with fields:

    gnames: {3×1 cell}
         n: [9 16 7]
    source: 'anova1'
     means: [5.4444 7.1875 5.8571]
        df: 29
         s: 3.0711
```

The output perfectly aligns with our calculation; hence we fail to reject the null hypothesis and the mean values of the three datasets are equal.

8.3.6.4 Example

A mechanical engineer measures the tensile strength of four different metal bars to determine if their mean strength is the same when given the same heat treatment. The mean tensile strengths of each type of metal bar are given below. At $\alpha = 0.01$ does there appear to be a difference in mean tensile strength among the four types of metal bars?

Metal bar 1	Metal bar 2	Metal bar 3	Metal bar 4
375	385	402	358
388	397	398	364
400	402	407	362
394	399	401	373
389	401	408	360
392	409	411	365

Metal bar 1	Metal bar 2	Metal bar 3	Metal bar 4
380	398	408	370
382	395	404	
386	389	405	
	392	404	
	394		
	392		

8.3.6.4.1 Hypotheses

The null hypothesis H_0: $\mu_1 = \mu_2 = \mu_3 = \mu_4$. Alternative hypothesis H_1: $\mu_1 \neq \mu_2 \neq \mu_3 \neq \mu_4$.

8.3.6.4.2 Compute the value of the test statistic

$m = 4$ since we are comparing the tensile strength of four different metal bars.
$n = 9 + 12 + 10 + 7 = 38$ since there 38 data points.

Metal bar	n	Summation Σx_i	Mean $\dfrac{\Sigma x_i}{n}$
1	9	$375 + 388 + \ldots + 382 + 386 = 3486$	$\underline{x_1} = \dfrac{3486}{9}$
2	12	$385 + 397 + \ldots + 394 + 392 = 4753$	$\underline{x_2} = \dfrac{4753}{12}$
3	10	$402 + 398 + \ldots + 405 + 404 = 4048$	$\underline{x_3} = \dfrac{4048}{10}$
4	7	$358 + 364 + \ldots 373 + 360 = 2552$	$\underline{x_4} = \dfrac{2552}{7}$
	$\Sigma 38$	$\Sigma 14839$	$\overline{\overline{x}} = \dfrac{14839}{38}$

Metal bar	n	Mean $\dfrac{\Sigma x_i}{n}$	$n_{ij}\left(x_{ij}-\overline{\overline{x}}\right)^2$
I	9	$\underline{x_1}=\dfrac{3486}{9}$	$9*\left(\dfrac{3486}{9}-\dfrac{14839}{38}\right)^2$
2	12	$\underline{x_2}=\dfrac{4753}{12}$	$12*\left(\dfrac{4753}{12}-\dfrac{14839}{38}\right)^2$
3	10	$\underline{x_3}=\dfrac{4048}{10}$	$10*\left(\dfrac{4048}{10}-\dfrac{14839}{38}\right)^2$
4	7	$\underline{x_4}=\dfrac{2552}{7}$	$7*\left(\dfrac{2552}{7}-\dfrac{14839}{38}\right)^2$
	$\Sigma 38$	$\overline{\overline{x}}=\dfrac{14839}{38}$	

$$SST = n_1\left(\underline{x_1}-\overline{\overline{x}}\right)^2 + n_2\left(\underline{x_2}-\overline{\overline{x}}\right)^2 + n_3\left(\underline{x_3}-\overline{\overline{x}}\right)^2 + n_4\left(\underline{x_4}-\overline{\overline{x}}\right)^2$$

$$= 9*\left(\dfrac{3486}{9}-\dfrac{14839}{38}\right)^2 + 12*\left(\dfrac{4753}{12}-\dfrac{14839}{38}\right)^2 + 10*\left(\dfrac{4048}{10}-\dfrac{14839}{38}\right)^2$$

$$+ 7*\left(\dfrac{2552}{7}-\dfrac{14839}{38}\right)^2$$

$$= 7215.269$$

bar	Mean $\dfrac{\Sigma x_i}{n}$	$\left(x_i-\underline{x}\right)^2$
I	$\underline{x_1}=\dfrac{3486}{9}$	$\left(375-\dfrac{3486}{9}\right)^2+\left(388-\dfrac{3486}{9}\right)^2+\ldots+\left(382-\dfrac{3486}{9}\right)^2+\left(386-\dfrac{3486}{9}\right)^2$
2	$\underline{x_2}=\dfrac{4753}{12}$	$\left(385-\dfrac{4753}{12}\right)^2+\left(397-\dfrac{4753}{12}\right)^2+\ldots+\left(394-\dfrac{4753}{12}\right)^2+\left(392-\dfrac{4753}{12}\right)^2$

bar	Mean $\dfrac{\Sigma x_i}{n}$	$\left(x_i - \underline{x}\right)^2$
3	$\underline{x_3} = \dfrac{4048}{10}$	$\left(402 - \dfrac{4048}{10}\right)^2 + \left(398 - \dfrac{4048}{10}\right)^2 + \ldots + \left(405 - \dfrac{4048}{10}\right)^2 + \left(404 - \dfrac{4048}{10}\right)^2$
4	$\underline{x_4} = \dfrac{2552}{7}$	$\left(358 - \dfrac{2552}{7}\right)^2 + \left(364 - \dfrac{2552}{7}\right)^2 + \ldots + \left(365 - \dfrac{2552}{7}\right)^2 + \left(370 - \dfrac{2552}{7}\right)^2$

$$SSE = \left(x_1 - \underline{x_1}\right)^2 + \left(x_2 - \underline{x_2}\right)^2 + \left(x_3 - \underline{x_3}\right)^2$$

$$= \left[\left(375 - \dfrac{3486}{9}\right)^2 + \left(388 - \dfrac{3486}{9}\right)^2 + \ldots + \left(382 - \dfrac{3486}{9}\right)^2 + \left(386 - \dfrac{3486}{9}\right)^2\right]$$

$$+ \left[\left(385 - \dfrac{4753}{12}\right)^2 + \left(397 - \dfrac{4753}{12}\right)^2 + \ldots + \left(394 - \dfrac{4753}{12}\right)^2 + \left(392 - \dfrac{4753}{12}\right)^2\right]$$

$$+$$

$$\left[\left(402 - \dfrac{4048}{10}\right)^2 + \left(398 - \dfrac{4048}{10}\right)^2 + \ldots + \left(405 - \dfrac{4048}{10}\right)^2 + \left(404 - \dfrac{4048}{10}\right)^2\right]$$

$$+ \left[\left(358 - \dfrac{2552}{7}\right)^2 + \left(364 - \dfrac{2552}{7}\right)^2 + \ldots + \left(365 - \dfrac{2552}{7}\right)^2 + \left(370 - \dfrac{2552}{7}\right)^2\right]$$

$$= 466 + 450.9167 + 133.6 + 171.7143 = 1222.231$$

$$TSS = SST + SSE = 7215.269 + 1222.231 = 8437.5$$

Source	df	Sum of squares (SS)	Mean squares (MS)	F
Treatment	$4 - 1 = 3$	7215.269	$\dfrac{7215.269}{4-1} = 2405.08966$	$\dfrac{2405.08966}{35.94797}$
Error	$38 - 4 = 34$	1222.231	$\dfrac{1222.231}{38-4} = 35.94797$	$= 66.905$
Total	$38 - 1 = 37$	8437.5		

8.3.6.4.3 Using the function

```
data = [Bar1, Bar2, Bar3, Bar4];

% Create a grouping variable
groups = [repmat({'Group 1'}, 1, length(Bar1)),
          repmat({'Group 2'}, 1, length(Bar2)),
          repmat({'Group 3'}, 1, length(Bar3)),
          repmat({'Group 4'}, 1, length(Bar4))];

% Perform ANOVA test
[p, tbl, stats] = anova1(data, groups,'off')
```

8.3.6.4.4 The output

```
p =

   2.4049e-14


tbl =

  4×6 cell array

    {'Source'}    {'SS'        }    {'df'}    {'MS'        }    {'F'       }    {'Prob>F'   }
    {'Groups'}    {[7.2153e+03]}    {[ 3]}    {[2.4051e+03]}    {[ 66.9047]}    {[2.4049e-14]}
    {'Error' }    {[1.2222e+03]}    {[34]}    {[   35.9480]}    {0×0 double }    {0×0 double }
    {'Total' }    {[8.4375e+03]}    {[37]}    {0×0 double  }    {0×0 double }    {0×0 double }


stats =

  struct with fields:

    gnames: {4×1 cell}
         n: [9 12 10 7]
    source: 'anova1'
     means: [387.3333 396.0833 404.8000 364.5714]
        df: 34
         s: 5.9957
```

As we can see, it aligns with our value of the test statistic.

Since our p-value < 0.01, we reject the null hypothesis and conclude that at least one of the means is different.

8.3.6.5 Exercise problems

1) A medical researcher wants to know if four different medications lead to different mean blood pressure reductions in patients. He randomly assigns ten patients to use each medication for one month, then measures the blood pressure reduction in each patient. At $\alpha = 0.01$, is there evidence of any difference in the effect of the four medications?

	Blood pressure reduction in mmHg		
Medication 1	Medication 2	Medication 3	Medication 4
5.2	3.6	5.2	4.2
4.8	3.2	4.8	4.8
4.0	2.7	4.7	4.0
3.7	3.1	5.0	3.9
4.5	4.0	4.9	4.5
4.2	3.3	5.2	4.1
4.4	2.9	5.5	4.4
4.7	3.5	4.5	3.8
4.5	3.2	5.0	4.2

2) An electric engineer uses semiconductors of different material to make five diodes and wants to determine which of the diodes will have the maximum power output. Tests are carried out on the five diodes and the results are as follows. Is there any difference between the measured power outputs at the 5% level of significance?

Diode	Maximum power output (mW)											
1	280	274	268	282	270	288	275	285	299	281	290	262
2	304	289	302	277	285	312	294	288	295	292		
3	289	295	299	305	308	310	292	317	301			
4	258	262	250	265	267	270	272	260	267	284	275	
5	251	246	254	240	257	268						

3) A manufacturer wants to develop a circuit to feed current to a particular component in a laptop display screen. He is unsure which metal wire to use to get the maximum current. Therefore he decides to test the current flow of three different metal wires. In tests involving the three circuits, the following results are obtained.

Metal wire	Current (mA)									
1	80.15	82.32	84.24	82.68	85.07	81.43	83.16	82.49	82.1	
2	84.21	82.34	84.22	82.65	85.13	82.03	83.24	82.40	82.12	80.17
3	80.36	81.99	82.19	83.52	84.77	82.60	82.98	83.72	81.01	

On the assumption that the populations from which the samples are drawn haveequal variances, is the manufacturer free to choose any of the metal wires or does he need to perform further tests to determine which metal wire would be the best? Use the 5% level of significance.

8.3.6.6 Two-way ANOVA

We have seen how the one-way ANOVA can be used to compare three or more sample means in studies involving a single independent variable. Two independent variables are used in the two-way ANOVA. As a result, it can be viewed as an extension of a one-way ANOVA. With a one-way ANOVA, you have one independent variable affecting a dependent variable which is a measurement variable (i.e. a quantitative variable). With a two-way ANOVA, there are two categorical/nominal independent variables (also known as factors) and a normally distributed continuous (i.e., interval or ratio level) dependent variable. The independent variables divide cases into two or more mutually exclusive levels, categories, or groups.

The two-way ANOVA compares the mean differences between groups that have been split into two independent variables (called factors). The primary purpose of a two-way ANOVA is to understand if there is an interaction between the two independent variables on the dependent variable. So graphically:

Factors/independent variable

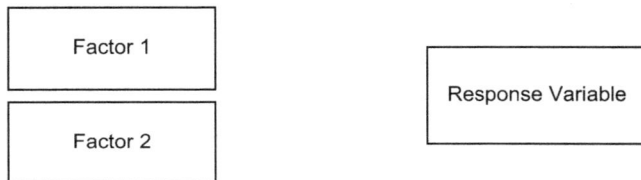

For example, you could use a two-way ANOVA to determine the effect of physical activity and BMI on blood cholesterol concentration. Here the dependent variable would be "blood cholesterol concentration", measured on a continuous scale in mmol/L, and the two independent variables would be "physical activity level", which has three groups—"low", "moderate", and "high"—and "BMI", which has four groups: "underweight", "healthy", "overweight", and "obesity".

Alternately, you may want to understand whether there is an interaction between the effects of degree type and educational level on the annual income of a person, where degree type ("business studies", "architecture", "biological sciences", "engineering", and "law") and education level (high school/undergraduate/postgraduate) are your independent variables, and annual income is your dependent variable.

8.3.6.6.1 Stating the hypotheses

Because the two-way ANOVA considers the effect of two categorical factors, and the effect of the categorical factors on each other, there are three pairs of null or alternative hypotheses for the two-way ANOVA.

A two-way ANOVA with interaction tests three null hypotheses at the same time:

- There is no difference in group means at any level of the first independent variable.
- There is no difference in group means at any level of the second independent variable.
- The effect of one independent variable does not depend on the effect of the other independent variable (a.k.a. no interaction effect).

Or

There is no significant interaction effect between the two factors.
The alternative hypotheses for each factor and their interaction are:

- There is a significant difference in between at least one pair of the.
- There is a significant interaction effect between the two factors.

8.3.6.6.2 Calculating the test statistic

The overall structure of the two-way ANOVA table is:

Factor A	Factor B			
	1	*2*	....	*b*
1				
2				
.				
.				
.				
.				
a				

Factor A has 1, 2, ..., a levels.
Factor B has 1, 2, ..., b levels.
Now before we look at the two-way ANOVA table, let's remember the ANOVA summary table for one independent variable.

Source	df	Sum of squares (SS)	Mean squares (MS)	F
Treatment or between groups	$m-1$	$SST / SSB = \Sigma n_{ij} \left(\overline{\overline{x_{ij}}} - \overline{\overline{x}} \right)^2$	$MSB = \dfrac{SSB}{m-1}$	$\dfrac{MSB}{MSE}$
Error or within groups	$n-m$	$SSE / SSW = \Sigma(n_{ij} - 1)s_{ij}^{2}$ $= \Sigma\Sigma\left(x_i - \underline{x}\right)^2$	$MSE = \dfrac{SSE}{n \quad m}$	
Total	$n-1$	$TSS = SST + SSE$		

There we are dealing with only one independent variable but in two-way ANOVA we will be dealing with two, so clearly we will have to add more to the table. Since we have two factors in play, we have two rows for each factor plus the interaction between the two.

Just like in one-way ANOVA, here too we will have df, sum of squares, mean squares, and f statistic for the factors and their interaction.

Source	df	Sum of squares (SS)	Mean squares (MS)	F
Between group for Factor A		SS_A	MS_A	
Between group for Factor B		SS_B	MS_B	
Interaction AB		SS_{AB}	MS_{AB}	
Error or within group		SSE	MSE	
Total				

However this time we have different formulas for each of them. Let's discuss each of them separately.

8.3.6.6.3 Degrees of freedom

As mentioned earlier, each factor has certain number of categories and levels so if we assume factor A has k_1 number of levels and factor B has k_2 number of levels then

df for Factor A = $k_1 - 1$.

df for Factor B = $k_2 - 1$.

Now df for the interaction would be Factor B = $(k_1 - 1)(k_2 - 1)$.

df for the total is the same as in one-way ANOVA = $n - 1$ where $n =$ total number of data points.

df for error follows the same formula as in one-way ANOVA:

$$= df_{total} - df_A - df_B - df_{AB}$$

$$= (n-1) - (k_1 - 1) - (k_2 - 1) - (k_1 - 1)(k_2 - 1)$$

8.3.6.6.4 Mean squares (MS)

As seen in the table, in two-way ANOVAs, we have four mean squares (for Factor A, B, AB, and interaction).

Regardless of which mean aquare we are calculating, it will be the same as all of the mean squares that we've covered in one-way ANOVA:

$$MS = \frac{SS}{df}$$

Therefore

$$MS_A = \frac{SS_A}{df_A}$$

$$MS_B = \frac{SS_B}{df_B}$$

$$MS_{AB} = \frac{SS_{AB}}{df_{AB}}$$

$$MSE = \frac{SSE}{df_E}$$

8.3.6.6.5 F statistics

As per the table in two-way ANOVAs, we have three f statistics, one for each factor and one for the interaction. Each F-score is calculated the same way as the F in a between groups in one-way ANOVA where we divide the mean square of the variable by the mean square of the error.

$$F = \frac{MS}{MSE}$$

Therefore the f statistic for Factor A $= \dfrac{MS_A}{MSE}$

f statistic for Factor B $= \dfrac{MS_B}{MSE}$

f statistic for interaction AB $= \dfrac{MS_{AB}}{MSE}$

Source	df	Sum of squares (SS)	Mean squares (MS)	F
Factor A	$a-1$	SS_A	$MS_A = \dfrac{SS_A}{a-1}$	$\dfrac{MS_A}{MSE}$
Factor B	$b-1$	SS_B	$MS_B = \dfrac{SS_B}{b-1}$	$\dfrac{MS_B}{MSE}$
Interaction AB	$(a-1)(b-1)$	SS_{AB}	$MS_{AB} = \dfrac{SS_{AB}}{(a-1)(b-1)}$	$\dfrac{MS_{AB}}{MSE}$
Error	$df_{total} - df_A - df_B - df_{AB}$	SSE	$MSE = \dfrac{SSE}{df_{total} - df_A - df_B - df_{AB}}$	
Total	$n-1$			

Now you must be wondering why we have not discussed the sum of squares. Since the formula of the sum of squares is difficult to understand and visualize, we will use an example to make it seem simple.

The data shown in the table below consists of Factor A which has two levels (level 1 and level 2), and factor B which has three levels (level 1, level 2, and level 3).

A	B		
	1	2	3
1	4.10	3.10	3.50
	3.90	2.80	3.20
	4.30	3.30	3.60
2	2.70	1.90	2.70
	3.10	2.20	2.30
	2.60	2.30	2.50

8.3.6.6.5.1 FIRST STEP

Calculate the means for each combination of factor levels:

A		B	
1	1	2	3
	4.10	3.10	3.50
	3.90	2.80	3.20
	4.30	3.30	3.60
Means	$\frac{4.10+3.90+4.30}{3}=4.10$	$\frac{3.10+2.80+3.30}{3}=\frac{46}{15}$	$\frac{3.50+3.20+3.60}{3}=\frac{103}{30}$
2	2.70	1.90	2.70
	3.10	2.20	2.30
	2.60	2.30	2.50
Means	$\frac{2.70+3.10+2.60}{3}=2.80$	$\frac{1.90+2.20+2.30}{3}=\frac{32}{15}$	$\frac{2.70+2.30+2.50}{3}=2.50$

Mean for Level 1 of A and Level 1 of B: 4.10.
Mean for Level 1 of A and Level 2 of B: 3.067.
Mean for Level 1 of A and Level 3 of B: 3.433.
Mean for Level 2 of A and Level 1 of B: 2.80.
Mean for Level 2 of A and Level 2 of B: 2.133.
Mean for Level 2 of A and Level 3 of B: 2.50.

8.3.6.6.5.2 SECOND STEP

A				B
	1	**2**	**3**	Means of means
1	4.10	$\dfrac{46}{15}$	$\dfrac{103}{30}$	$\dfrac{4.10+\dfrac{46}{15}+\dfrac{103}{30}}{3}=\dfrac{53}{15}$
2	2.80	$\dfrac{32}{15}$	2.50	$\dfrac{2.80+\dfrac{32}{15}+2.50}{3}=\dfrac{223}{90}$
Means of means	$\dfrac{4.10+2.80}{2}=3.45$	$\dfrac{\dfrac{46}{15}+\dfrac{32}{15}}{2}=2.60$	$\dfrac{\dfrac{103}{30}+2.50}{2}=\dfrac{89}{30}$	Grand mean $=\dfrac{541}{180}$

8.3.6.6.5.3 THIRD STEP

Calculate the SS for each factor and their interaction:
  (SS)A
  For Group 1 For Group 2

B		
1 2.70	$\left(\dfrac{223}{90}-\dfrac{541}{180}\right)^2=$	$\dfrac{361}{1296}$
3.10	$\left(\dfrac{223}{90}-\dfrac{541}{180}\right)^2=$	$\dfrac{361}{1296}$
2.60	$\left(\dfrac{223}{90}-\dfrac{541}{180}\right)^2=$	$\dfrac{361}{1296}$
2 1.90	$\left(\dfrac{223}{90}-\dfrac{541}{180}\right)^2=$	$\dfrac{361}{1296}$
2.20	$\left(\dfrac{223}{90}-\dfrac{541}{180}\right)^2=$	$\dfrac{361}{1296}$
2.30	$\left(\dfrac{223}{90}-\dfrac{541}{180}\right)^2=$	$\dfrac{361}{1296}$

3   2.70 $\left(\dfrac{223}{90}-\dfrac{541}{180}\right)^2=\dfrac{361}{1296}$

2.30 $\left(\dfrac{223}{90}-\dfrac{541}{180}\right)^2=\dfrac{361}{1296}$

2.50 $\left(\dfrac{223}{90}-\dfrac{541}{180}\right)^2=\dfrac{361}{1296}$

$\Sigma\ \dfrac{361}{144}$

B

1   4.10 $\left(\dfrac{53}{15}-\dfrac{541}{180}\right)^2=\dfrac{361}{1296}$

3.90 $\left(\dfrac{53}{15}-\dfrac{541}{180}\right)^2=\dfrac{361}{1296}$

4.30 $\left(\dfrac{53}{15}-\dfrac{541}{180}\right)^2=\dfrac{361}{1296}$

2   3.10 $\left(\dfrac{53}{15}-\dfrac{541}{180}\right)^2=\dfrac{361}{1296}$

2.80 $\left(\dfrac{53}{15}-\dfrac{541}{180}\right)^2=\dfrac{361}{1296}$

3.30 $\left(\dfrac{53}{15}-\dfrac{541}{180}\right)^2=\dfrac{361}{1296}$

3   3.50 $\left(\dfrac{53}{15}-\dfrac{541}{180}\right)^2=\dfrac{361}{1296}$

3.20 $\left(\dfrac{53}{15}-\dfrac{541}{180}\right)^2=\dfrac{361}{1296}$

3.60 $\left(\dfrac{53}{15}-\dfrac{541}{180}\right)^2=\dfrac{361}{1296}$

$\Sigma\dfrac{361}{144}$

$$(SS)A = \left(Mean\ for\ Level\ 1\ of\ A - Grand\ Mean \right)^2$$

$$+ \left(Mean\ for\ Level\ 2\ of\ A - Grand\ Mean \right)^2$$

$$= \frac{361}{144} + \frac{361}{144} = \frac{361}{72} = 5.0139$$

$(SS)B$

For Group 1   For Group 2

A

I   4.10 $\left(3.45 - \dfrac{541}{180} \right)^2 = \dfrac{16}{81}$

3.90 $\left(3.45 - \dfrac{541}{180} \right)^2 = \dfrac{16}{81}$

4.30 $\left(3.45 - \dfrac{541}{180} \right)^2 = \dfrac{16}{81}$

2   2.70 $\left(3.45 - \dfrac{541}{180} \right)^2 = \dfrac{16}{81}$

3.10 $\left(3.45 - \dfrac{541}{180} \right)^2 = \dfrac{16}{81}$

2.60 $\left(3.45 - \dfrac{541}{180} \right)^2 = \dfrac{16}{81}$

A

I   3.10 $\left(2.60 - \dfrac{541}{180} \right)^2 = \dfrac{5329}{32400}$

2.80 $\left(2.60 - \dfrac{541}{180} \right)^2 = \dfrac{5329}{32400}$

3.30 $\left(2.60 - \dfrac{541}{180} \right)^2 = \dfrac{5329}{32400}$

2   1.90 $\left(2.60 - \dfrac{541}{180}\right)^2 = \dfrac{5329}{32400}$

2.20 $\left(2.60 - \dfrac{541}{180}\right)^2 = \dfrac{5329}{32400}$

2.30· $\left(2.60 - \dfrac{541}{180}\right)^2 = \dfrac{5329}{32400}$

For Group 3

A		
I 3.50	$\left(\dfrac{89}{30} - \dfrac{541}{180}\right)^2 = \dfrac{49}{32400}$	
3.20	$\left(\dfrac{89}{30} - \dfrac{541}{180}\right)^2 = \dfrac{49}{32400}$	
3.60	$\left(\dfrac{89}{30} - \dfrac{541}{180}\right)^2 = \dfrac{49}{32400}$	
2 2.70	$\left(\dfrac{89}{30} - \dfrac{541}{180}\right)^2 = \dfrac{49}{32400}$	
2.30	$\left(\dfrac{89}{30} - \dfrac{541}{180}\right)^2 = \dfrac{49}{32400}$	
2.50	$\left(\dfrac{89}{30} - \dfrac{541}{180}\right)^2 = \dfrac{49}{32400}$	

$$(SS)B = \left(Mean\ for\ Level\ 1\ of\ B - Grand\ Mean\right)^2$$
$$+ \left(Mean\ for\ Level\ 2\ of\ B - Grand\ Mean\right)^2$$
$$+ \left(Mean\ for\ Level\ 3\ of\ B - Grand\ Mean\right)^2$$

$$= \left(\frac{16}{81} \times 6\right) + \left(\frac{5329}{32400} \times 6\right) + \left(\frac{49}{32400} \times 6\right) = \frac{1963}{900}$$

$(SS)E$

A
1   1

4.10 $(4.10 - 4.10)^2 = 0$

3.90 $(3.90 - 4.10)^2 = 0.04$

4.30 $(4.30 - 4.10)^2 = 0.04$

A
2   1

2.70 $(2.70 - 2.80)^2 = 0.01$

3.10 $(3.10 - 2.80)^2 = 0.09$

2.60 $(2.60 - 2.80)^2 = 0.04$

A
1   2

3.10 $\left(3.10 - \dfrac{46}{15}\right)^2 = \dfrac{1}{900}$

2.80 $\left(2.80 - \dfrac{46}{15}\right)^2 = \dfrac{16}{225}$

3.30 $\left(3.30 - \dfrac{46}{15}\right)^2 = \dfrac{49}{900}$

A
1   2

1.90 $\left(1.90 - \dfrac{32}{15}\right)^2 = \dfrac{49}{900}$

2.20 $\left(2.20 - \dfrac{32}{15}\right)^2 = \dfrac{1}{225}$

2.30 $\left(2.30 - \dfrac{32}{15}\right)^2 = \dfrac{1}{36}$

A
I   3

3.50 $\left(3.50 - \dfrac{103}{30}\right)^2 = \dfrac{1}{225}$

3.20 $\left(3.20 - \dfrac{103}{30}\right)^2 = \dfrac{49}{900}$

3.60 $\left(3.60 - \dfrac{103}{30}\right)^2 = \dfrac{1}{36}$

A
I   3

2.70 $\left(2.70 - 2.50\right)^2 = 0.04$

2.30 $\left(2.30 - 2.50\right)^2 = 0.04$

2.50 $\left(2.50 - 2.50\right)^2 = 0$

$$(SS)E = \Sigma\left(Data\ value - mean\ of\ intersection\right)^2 = 0.6($$
(SS)T

4.10 $\left(4.10 - \dfrac{541}{180}\right)^2 = \left(\dfrac{197}{180}\right)^2$

3.90 $\left(3.90 - \dfrac{541}{180}\right)^2 = \left(\dfrac{161}{180}\right)^2$

4.30 $\left(4.30 - \dfrac{541}{180}\right)^2 = \left(\dfrac{233}{180}\right)^2$

3.10 $\left(3.10 - \dfrac{541}{180}\right)^2 = \left(\dfrac{17}{180}\right)^2$

2.80 $\left(2.80 - \dfrac{541}{180}\right)^2 = \left(\dfrac{37}{180}\right)^2$

3.30 $\left(3.30 - \dfrac{541}{180}\right)^2 = \left(\dfrac{53}{180}\right)^2$

3.50 $\left(3.50-\dfrac{541}{180}\right)^2=\left(\dfrac{89}{180}\right)^2$

3.20 $\left(3.20-\dfrac{541}{180}\right)^2=\left(\dfrac{7}{36}\right)^2$

3.60 $\left(3.60-\dfrac{541}{180}\right)^2=\left(\dfrac{107}{180}\right)^2$

2.70 $\left(2.70-\dfrac{541}{180}\right)^2=\left(\dfrac{11}{36}\right)^2$

3.10 $\left(3.10-\dfrac{541}{180}\right)^2=\left(\dfrac{17}{180}\right)^2$

2.60 $\left(2.60-\dfrac{541}{180}\right)^2=\left(\dfrac{73}{180}\right)^2$

1.90 $\left(1.90-\dfrac{541}{180}\right)^2=\left(\dfrac{199}{180}\right)^2$

2.20 $\left(2.20-\dfrac{541}{180}\right)^2=\left(\dfrac{29}{36}\right)^2$

2.30 $\left(2.30-\dfrac{541}{180}\right)^2=\left(\dfrac{127}{180}\right)^2$

2.70 $\left(2.70-\dfrac{541}{180}\right)^2=\left(\dfrac{11}{36}\right)^2$

2.30 $\left(2.30-\dfrac{541}{180}\right)^2=\left(\dfrac{127}{180}\right)^2$

2.50 $\left(2.50-\dfrac{541}{180}\right)^2=\left(\dfrac{91}{180}\right)^2$

$\Sigma\,7.9294$

Hence

$$SS\ First\ Factor + SS\ Second\ Factor + SS\ Both\ Factors + SS\ Error = Total\ SS$$

$$\frac{361}{72} + \frac{1963}{900} + SS\ Both\ Factors + 0.60 = 7.9294$$

$$SS\ Both\ Factors = 0.1344$$

Source	df	Sum of squares (SS)	Mean squares (MS)	F
Factor A (row)	$2 - 1 = 1$	5.0139	$\dfrac{5.0139}{1} = 5.0139$	$\dfrac{5.0139}{0.05}$
Factor B (column)	$3 - 1 = 2$	2.1811	$\dfrac{2.1813}{2} = 1.0906$	$\dfrac{1.0906}{0.05}$
Interaction AB	$1 \times 2 = 2$	0.1344	$\dfrac{0.1344}{2} = 0.0672$	$\dfrac{0.0672}{0.05}$
Error	$17 - 1 - 2 - 2 = 12$	0.60	$\dfrac{0.60}{12} = 0.05$	
Total	$18 - 1 = 17$	7.9294		

Hence our test statistics are:

- The F-statistic for the effect of $= \dfrac{5.0139}{0.05} = 100.28.$

- The F-statistic for the effect of $= \dfrac{1.0906}{0.05} = 21.81.$

- The F-statistic for the interaction effect $= \dfrac{0.0672}{0.05} = 1.34.$

8.3.6.6.6 CALCULATING THE P-VALUE

The table is complete and we have found the test statistics, so next we will find the p-value
    To make a decision about the hypothesis test, we really need a p-value. The numerator df is the df for the source and the denominator df is the df for the error.

- A: F has $df = 1$ numerator and denominator $df = 12$.
- B: F has $df = 2$ numerator and denominator $df = 12$.
- Interaction: F has $df = 2$ numerator and denominator $df = 12$.

The p-value is the area to the right of the test statistic since this is always a right-tailed test.

- The value which is closest to our calculated test statistic is 12.97 which clearly does not come close to 100.3; therefore we can conclude that the p-value $<<< 0.001$.
- The value closest to our calculated test statistic is 18.64 which is roughly close to 21.82; therefore we can conclude that the p-value < 0.001.
- Finally the value close to the test statistic 1.332 is 2.81 therefore the p-value > 0.100.

8.3.6.6.7 Decision

- Since the p-value $<<< 0.001 < 0.05$, hence
- Since the p-value $< 0.001 < 0.05$, hence
- The p-value $> 0.100 > 0.05$.

8.3.6.7 In MATLAB

```
[p,tbl,stats] = anova2(Data,reps,displayopt)
```

The input arguments

- Data

This represents the sample data, specified as a matrix.

- reps

This signifies replications which are basically the number of elements or observations for each combination of groups (levels) of the row and column factors. In our example the row factor A has two levels, the column factor B has three levels, and there are three elements in each combination of factors A and B.

8.3.6.8 Example

Suppose you have data on the mileage of three different tire tread designs (symmetric or multi-directional, asymmetric, directional) and two types of transmission (automatic and manual). The data is stored in a matrix Y with dimensions 2×3, where each row represents a different transmission and each column represents a different tire design. Here's an example dataset:

| Transmission | Tire tread | | |
	Symmetric	Asymmetrical	Directional
Automatic	17.23	15.96	17.85
Manual	15.18	14.31	16.12

As we can see there is only one observation for each combination of factor, e.g. the gas mileage for the vehicle with symmetrical tires and automatic transmission is 17.23.

Therefore:

```
Data = [17.23 15.96 17.85;
        15.18 14.31 16.12];
[p,tbl,stats] = anova2(Data)
```

But in our worked example we have three observations for each combination of factors:

```
      B = 1  B = 2  B = 3
```

```
Data = [4.10 3.10 3.50;

       3.90 2.80 3.20;  A = 1
       4.30 3.30 3.60;

       2.70 1.90 2.70;
       3.10 2.20 2.30;  A = 2
       2.60 2.30 2.50]
```

```
% Perform the two-way ANOVA
[p, tbl, stats] = anova2(Data, 3)
```

Figure 1: Two-way ANOVA — ×

ANOVA Table

Source	SS	df	MS	F	Prob>F
Columns	2.18111	2	1.09056	21.81	0.0001
Rows	5.01389	1	5.01389	100.28	0
Interaction	0.13444	2	0.06722	1.34	0.2973
Error	0.6	12	0.05		
Total	7.92944	17			

```
p =

   0.0001    0.0000    0.2973


tbl =

  6×6 cell array

    {'Source'      }    {'SS'    }    {'df'}    {'MS'      }    {'F'       }    {'Prob>F'   }
    {'Columns'     }    {[2.1811]}    {[ 2]}    {[  1.0906]}    {[ 21.8111]}    {[1.0083e-04]}
    {'Rows'        }    {[5.0139]}    {[ 1]}    {[  5.0139]}    {[100.2778]}    {[3.5280e-07]}
    {'Interaction'}     {[0.1344]}    {[ 2]}    {[  0.0672]}    {[  1.3444]}    {[  0.2973]}
    {'Error'       }    {[0.6000]}    {[12]}    {[  0.0500]}    {0×0 double}    {0×0 double }
    {'Total'       }    {[7.9294]}    {[17]}    {0×0 double}    {0×0 double}    {0×0 double }


stats =

  struct with fields:

     source: 'anova2'
    sigmasq: 0.0500
   colmeans: [3.4500 2.6000 2.9667]
       coln: 6
   rowmeans: [3.5333 2.4778]
       rown: 9
      inter: 1
       pval: 0.2973
         df: 12
```

As we can see the output perfectly aligns with our calculation.

8.3.6.8.1 Decision

- Since the p-value $<<< 0.001 < 0.05$, **we reject the null hypothesis in favor of the alternative hypothesis and conclude that** there is a significant difference in group means at any level of the first independent variable A.
- Since the p-value $< 0.001 < 0.05$, here too there is a significant difference in group means at any level of the first independent variable B.
- Since the p-value $> 0.100 > 0.05$, we fail to reject the null hypothesis and there is no significant interaction effect between the two factors.

8.3.6.9 Example

There has been a significant upward trend in the number of electric vehicles being both manufactured and purchased on a daily basis. Since these vehicles run on batteries, they are the core of their functionality. Three different weights of electric vehicles were tested at three different temperatures to measure the effects on battery life.Determine at $\alpha = 0.01$ whether or not there are any statistical differences between the weight and the temperature with respect to the battery life (in months) of the vehicles.

Weight	Temperature		
	Low	Medium	High
Light	173,149,146,	117,131,129,	97,85,93,
	152,159,165,	137,140,159,	80,88,86,
	161,182,175	150,138,145	76,65,109
Average	113,132,138,	104,84,91,	95,73,59,
	142,129,155,	80,81,86,	79,87,65,
	127,136,144	76,60,75	77,55,69
Heavy	93,74,79,	84,71,69,	75,45,63,
	70,82,65,	65,62,57,	59,54,47,
	62,74,51	54,58,41	42,31,52

8.3.6.9.1 Null hypotheses (H_0)

There is no difference in battery life for any weight of the vehicle.

There is no difference in battery life at either temperature.

The effect of one independent variable on battery life does not depend on the effect of the other independent variable (a.k.a. no interaction effect).

8.3.6.9.2 Alternate hypotheses (H_A)

There is a difference in battery life by weight.

There is a difference in battery life by temperature.

There is an interaction effect between weight and temperature on battery life.

```
          Low    Medium      High
Data = 173      117         97;
          149    131         85;


          146    129         93;
          152    137         80;
          159    145         88; A = Light
          165    159         76;
          161    150         65;
          182    138         109;

          175    140         86;


          113    104         95;
```

$$
\begin{array}{ccc}
132 & 184 & 73; \\
139 & 91 & 59; \\
142 & 80 & 87; \\
129 & 76 & 65; \quad \text{A = Average} \\
155 & 81 & 79; \\
127 & 60 & 69;
\end{array}
$$

$$
\begin{array}{ccc}
 & 136 \quad 75 \quad 55;
\end{array}
$$

$$
\begin{array}{ccc}
144 & 86 & 77; \\
\end{array}
$$

$$
\begin{array}{ccc}
93 & 71 & 47; \\
74 & 84 & 75; \\
79 & 65 & 63; \\
82 & 57 & 31; \\
51 & 69 & 45; \quad \text{A = Heavy} \\
65 & 58 & 54; \\
70 & 41 & 42;
\end{array}
$$

$$
\begin{array}{ccc}
 & 62 \quad 62 \quad 52;
\end{array}
$$

$$
\begin{array}{ccc}
74 & 54 & 59;
\end{array}
$$

```
% Perform the two-way ANOVA
[p, tbl, stats] = anova2(Data, 9)
```

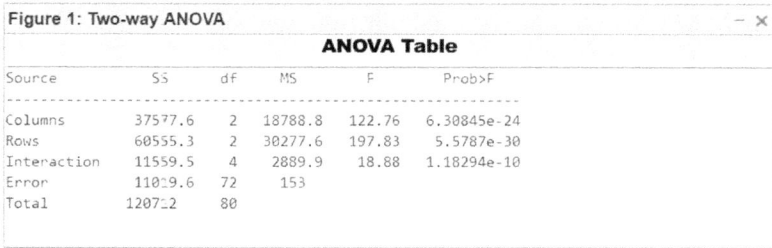

Figure 1: Two-way ANOVA   – ×

ANOVA Table

Source	SS	df	MS	F	Prob>F
Columns	37577.6	2	18788.8	122.76	6.30845e-24
Rows	60555.3	2	30277.6	197.83	5.5787e-30
Interaction	11559.5	4	2889.9	18.88	1.18294e-10
Error	11019.6	72	153		
Total	120712	80			

```
p =

   1.0e-09 *

   0.0000   0.0000   0.1183


tbl =

   6×6 cell array

   {'Source'     }  {'SS'        }  {'df'}  {'MS'        }  {'F'       }  {'Prob>F'    }
   {'Columns'    }  {[3.7578e+04]}  {[ 2]}  {[1.8789e+04]}  {[122.7629]}  {[6.3084e-24]}
   {'Rows'       }  {[6.0555e+04]}  {[ 2]}  {[3.0278e+04]}  {[197.8292]}  {[5.5787e-30]}
   {'Interaction'}  {[1.1560e+04]}  {[ 4]}  {[2.8899e+03]}  {[ 18.8820]}  {[1.1829e-10]}
   {'Error'      }  {[1.1020e+04]}  {[72]}  {[ 153.0494]}  {0×0 double}  {0×0 double  }
   {'Total'      }  {[1.2071e+05]}  {[80]}  {0×0 double  }  {0×0 double}  {0×0 double  }


stats =

   struct with fields:

       source: 'anova2'
      sigmasq: 153.0494
     colmeans: [123.2593 94.2222 70.5926]
         coln: 27
     rowmeans: [129.1481 96.7407 62.1852]
         rown: 27
        inter: 1
         pval: 1.1829e-10
           df: 72
```

From the output we can draw our own table to make clear conclusions.

Factor	F test statistics	p-value	Reject/fail to reject	Significant difference
Weight	197.8292	~ 0.0	**Reject**	**Yes**
Temperature	122.7629	~ 0.0	**Reject**	**Yes**
Interaction	18.882	~ 0.0	**Reject**	**Yes**

Hence:

There is a difference in battery life by weight.

There is a difference in battery life by temperature.

There is an interaction effect between weight and temperature on battery life.

8.3.6.10 N-way ANOVA

In first two types of ANOVA we discussed, one-way and two-way, one and two indicate the number of independent variables affecting the dependent variable. Now if, instead of one or two, we are analyzing multiple (three or more) independent variables, we are performing n-way ANOVA.

Using the same example of determining the effect of degree type and level of education, here we also add gender to understand whether there is an interaction between the three independent variables on the annual income of a person.

While it was possible to perform one- and two-way ANOVA by hand, it is difficult to do so with more than two observations. Therefore we will not get into the topic in this textbook. However MATLAB does offer an inbuilt function for n-way ANOVA using almost the same syntax as in the previous ones.

Chapter 9

Non-parametric testing

In terms of selecting a statistical test, the question that always comes up is if the test is parametric or not. Now from previous chapters we have learned that parametric tests are those that make assumptions that the data from which the sample is drawn has a known distribution. We also saw different distributions (binomial, geometric, normal, etc.) described with mathematical formulas. These formulas include parameters that dictate the shape and/or location of the distribution. For example, the binomial probability distribution is characterized by two parameters, **the number of independent trials n and the probability of success p**, while the variance and mean are the two parameters of the normal distribution that dictate its shape and location, respectively. Since the Wilcoxon Rank Sum Test does not assume known distributions, it does not deal with parameters, and therefore we call it a non-parametric test.

One of the most common assumptions of the parametric test is that the data used must be normally distributed. For example, if a t test or an ANOVA is to be calculated, it must first be tested whether the data or variables are normally distributed. However, in contrast to parametric tests, non-parametric tests, as their name tells us, are statistical tests without assumptions about the parameters, especially the assumption about normally distributed data. Simply stated, if that assumption does not hold, the non-parametric test is a better safeguard against drawing erroneous conclusions about the population.

Simply stated, we use a parametric t test for normally distributed data and a non-parametric test for skewed data.

Now does that mean that we only use non-parametric tests when the data is not normal? Well surprise! Non-parametric tests are actually valid for both non-normally distributed data and normally distributed data.

If that is the case then you must be wondering why we have the parametric tests in the first place. Why read an entire chapter when we could already work with this one?

Now since non-parametric tests are considered a parallel universe to parametric tests, below there is a list of non-parametric tests equivalent to a number of parametric ones.

Parametric tests (means)	Non-parametric tests (medians)
One-sample t test One-sample t test	One-sample sign test One-sample Wilcoxon signed rank test
Two-sample independent t test	Wilcoxon rank-sum test
Two-sample paired t test	Two-sample Wilcoxon signed rank test
One-way ANOVA	Kruskal–Wallis test

DOI: 10.1201/9781003399582-9

All the non-parametric tests described in this chapter may be thought of as direct competitors of the t test and F test when normality can be assumed. Essentially, you should remember that in cases where it is difficult or impossible to justify normality but it is known that the underlying distribution is continuous, non-parametric methods remain valid while parametric methods may not be. You should also bear in mind that in terms of practical application it may be difficult to decide whether to use parametric or non-parametric tests since both the t test and the F test are relatively insensitive to small departures from normality.

The table above illustrates that, for example, if we aim to compare variances across the means of five groups and our data isn't normally distributed, we can't use ANOVA. Instead, we would apply a non-parametric alternative—the Kruskal–Wallis test. As in the previous chapter, we'll review several key non-parametric tests, examining their purposes, null hypotheses, test statistics, and decision rules. The first non-parametric test we'll cover is the Wilcoxon test, which comes in two forms: the rank-sum test and the signed-rank test. We'll discuss each in detail.

9.1 WILCOXON RANK-SUM TEST

Recall that in previous chapter we learned to perform the two-sample t test to compare the average or the mean of two groups, where the data was sampled from a normally distributed population. Now since the normality assumption made by the latter is in doubt, the Wilcoxon rank-sum test, sometimes called the Mann–Whitney U test or the *Wilcoxon–Mann–Whitney test*, is to be used as an alternative to the t test. Now recall the procedure of the t-test which uses real magnitudes of the observations. Now understand that this was accurate since there were no outliers present. Because we are now allowing the population distributions to be non-normal, the test needs to deal with this non-normality in the form of extreme outliers. In statistics this is vital because outliers have a **large influence on** the results of our conclusions; in fact the effect is so strong that it can cause the actual value to deviate from its stated value. Since the test considers the median instead of the mean, these outliers have no effect, making the test an excellent option when dealing with skewed data.

Just like its parametric counterpart, the **Wilcoxon rank-sum test** is used to compare differences between **one independent variable** that consists of **two categorical, independent groups and one dependent variable** that is measured at the **continuous** or **ordinal** level.

For example, you could use it to determine how often an activity is performed, where the frequency of the activity is measured on an ordinal scale (the scale includes "never", "rarely", "sometimes", "often", and "always"] or differs based on gender (i.e., the dependent variable would be frequency and the independent variable would be "gender", which has two groups: "male" and "female").

Alternately, you could use the test to understand whether salaries, measured on a continuous scale, differed based on educational level (i.e., your dependent variable would be "salary" and your independent variable would be "educational level", which has two groups: "high school" and "university").

Although the test does not require normally distributed data, that does not mean it is assumption free. For the **Wilcoxon rank-sum test**, data from each population must be an independent random sample.

Let's now discuss how the test is performed.

To conduct a **Wilcoxon rank-sum test**, we follow the standard five-step hypothesis testing procedure.

9.1.1 Stating the null hypothesis

It turns out the test is valid under a range of different sets of possible null and alternative hypotheses. Here, we will use it to interpret whether there are differences either in the "**distributions**" of two groups or in the "**medians**" of two groups.

So how do we know which one or what exactly we should be testing? Clearly, we are dealing with two situations here. To make it simpler, the best way to tell them apart is by the shape of the distributions of the two datasets.

- If the two distributions are the **same shape**, the **Wilcoxon rank-sum test** is used to determine whether there are differences in the **medians** of our two groups.
- If they are different, then we use the test to determine whether there are differences in the **distributions** of our two groups.

Put simply, we use **Wilcoxon rank-sum test** to determine whether the distribution of our dependent variable for both groups of our independent variable have the same shape with regards to their data. So, we write:

H_0: the two populations are equal.
H_A: the two populations are not equal—one group has larger (or smaller) values than the other.

9.1.2 Calculating the test statistic

Unlike the t tests we have seen earlier, the Mann–Whitney U test is agnostic to outliers and concentrates on the center of the distribution.

While the t-test uses real values to compare the means of the two groups, the U test uses the ranks of the real values to compare the medians of the two groups or, in case of different distributions, to compare the distributions of the two groups. To calculate the test statistic, the U test will replace each data value with its rank (from lowest to highest) in the combined sample—that is, the sample consisting of the data from both populations. The steps:

1) We will first rank each value of the dependent variable irrespective of the group it is in, according to its size, with the smallest rank assigned to the smallest value. So the smallest value in the combined sample is assigned the rank of 1, the next smallest value gets the rank of 2, and so on, with the largest value assigned the rank of n which is the total number of combined samples $(n_1 + n_2)$. The ranks are not affected by how far the smallest (largest) data value is from next smallest (largest) data value making the test much more robust against outliers and heavy tail distributions.

For our illustration, suppose we want to test whether the scores of girls and boys in an exam differ from each other.

Boys	12	20	45	58	69	75	82	91	100
Girls	41	55	59	60	78	95	84		

We need to combine the observations from both groups and rank them in order from least to greatest:

Ranks	Boys	Girls
1	12	
2	20	
3		41
4	45	
5		55
6	58	
7		59
8		60
9	69	
10	75	
11		78
12	82	
13		84
14	91	
15		95
16	100	

Now let's look at another dataset of the class test marks obtained in Lisa's class:

Boys	08	25	41	54	63	67	86	88	92
Girls	38	51	58	69	71	91	63		

Notice the score 63 has been repeated twice; they cannot have different ranks, so what would it be?

Let's first rank them in order from least to greatest:

Ranks	Boys	Girls
1	08	
2	25	
3		38
4	41	
5		51
6	54	
7		58
8	63	
9		63
10	67	
11		69
12		71
13	86	
14	88	
15		91
16	92	

Now realize that since 63 is a repeated value, it clearly they can't have two different ranks. So now the question is: will it be 8 or will it be 9 ?

If two or more scores are identical, this is a "tie". They get the average of the ranks that they would have obtained, had they been different from each other. Since their ranks initially are 8 and 9 , the average rank is $\dfrac{8+9}{2} = 8.50$. Therefore if we were to redraw the table:

Ranks	Boys	Girls
1	08	
2	25	
3		38
4	41	
5		51
6	54	
7		58
8.5	63	
8.5		63
10	67	
11		69
12		71
13	86	
14	88	
15		91
16	92	

So from the table we can see that the ranks of the boys' group are:

$$1, 2, 4, 6, 8.50, 10, 13, 14, 16$$

And the ranks of the girls group are:

$$3, 5, 7, 8.50, 11, 12, 15$$

2) Now that we have learned to sort the two groups based on their ranks, next we will add the ranks obtained from the groups separately. So now we have two rank sums for the two groups.

9.1.2.1 For the untied ranks example

The rank sum of the scores for boys:

$$1 + 2 + 4 + 6 + 8.5 + 10 + 13 + 14 + 16 = 74.50$$

And the rank sum of the scores for girls:

$$3 + 5 + 7 + 8.5 + 11 + 12 + 15 = 61.5$$

So with the rank sum denoted by R, we get $R_{boys} = 74.5$ and $R_{girls} = 61.5$.

4) In the final step we calculate the test statistic. Now the test statistic in the Wilcoxon rank-sum test is the **sum of the ranks for one of the two samples**. Therefore, for the tied ranks example, the test statistic could be 74.5 or 61.5. Theoretically,
   - For equal sample size: the choice of which sum to calculate is arbitrary since both samples are the same size. In other words we are free to choose either of the rank sums.
   - For unequal sample size: the test statistic is the sum of the ranks of the smaller sample size.

Considering this, since the sample size of the girls is smaller, the Wilcoxon test statistic should be $R_{girls} = 61.5$.

This is the classic way to calculate the Wilcoxon rank-sum test statistic. However, in MATLAB, things are a little different. While in theory the sample size of the groups plays an important role in determining the test statistic, it does not hold that significance in MATLAB. Put simply, in MATLAB the test statistic is the rank sum of the sample put as the first input argument, regardless of the sample size.

So if we had put the scores of the girls as the first argument we would have gotten the test statistic we anticipated. This will get clearer in the next section when we start discussing MATLAB.

9.1.3 Finding the p-value

There are two ways of obtaining p-values in non-parametric tests. One way calculates the exact probability of obtaining observed or more extreme results under the null hypothesis, which is suitable for small sample size studies, and is referred to as the exact non-parametric test. The other way calculates the p-value based on the asymptotic property, which is suitable for large sample size studies, and is referred to as the asymptotic non-parametric test.

9.1.3.1 The exact method

We use this method when the sample size of both groups is small, $n \leq 20$. This method is computationally intensive and may not be feasible for large sample sizes. However, it is more accurate and reliable than the approximate method.

To calculate the p-value, we need to randomly permute the dataset many times, recalculating the test statistic each time, and record the values of the test statistic for each permutation. We count the number of permutations that result in a test statistic as extreme as or more extreme than the observed value of the test statistic. This count is the number of times that the null hypothesis would have been rejected if we had used that permuted dataset instead of the original dataset. Then finally, we divide the count of extreme test statistics by the total number of permutations to obtain the p-value. Therefore for groups n_1 and n_2 the total permutation would be $\binom{n_1 + n_2}{n_1}$.

Let's look at the cases for each type of test.

9.1.3.1.1 If our test is right-tailed

Let's say we have two groups of sample size n_1 and n_2.

After evaluating the test statistics by summing the ranks of the first group we count the number of permutations of sample size n_1 that result in a rank sum $\geq$ test statistics and divide it by $\begin{pmatrix} n_1 + n_2 \\ n_1 \end{pmatrix}$.

9.1.3.1.2 If our test is left-tailed

For this case, we count the number of permutations of sample size n_1 that result in a rank sum $\leq$ test statistics and divide it by $\begin{pmatrix} n_1 + n_2 \\ n_1 \end{pmatrix}$.

9.1.3.1.3 If our test is two-tailed

The number of permutations differ for the rank sum of the two groups of data.

- If the rank sum of the first group > the rank sum of the second group, then we count

the number of permutations of sample size n_1 that result in a rank sum $\geq$ test statistics and divide it by $\begin{pmatrix} n_1 + n_2 \\ n_1 \end{pmatrix}$

- If the rank sum of the first group < rank sum of the second group, then we count

the number of permutations of sample size n_1 that result in a rank sum $\leq$ test statistics and divide it by $\begin{pmatrix} n_1 + n_2 \\ n_1 \end{pmatrix}$.

Let's work on some examples to illustrate the explanations.

9.1.3.1.4 Example

Suppose a mechanical engineer is studying the effect of a new lubricant and wants to determine if it increases the friction coefficient of two different types of metal surfaces commonly used in a particular manufacturing process. The engineer takes random samples of five and six metal pieces from each surface type, applies the new lubricant to one group and a standard lubricant to the other, and measures the friction coefficient for each sample. Test at the 0.10 significance level.

| New lubricant | 0.85 | 0.76 | 0.82 | 0.79 | |
| Standard lubricant | 0.75 | 0.68 | 0.80 | 0.78 | 0.65 |

H_0: there is no difference in the effect of the two lubricants on the friction coefficient.
H_A: the new lubricant increases the friction coefficient.

According to the procedure for finding the p-value, the first step is to find the test statistics. First step: ranking each group we get:
New lubricant: 9, 4, 8, 6
Standard lubricant: 3, 2, 7, 5, 1

Second step: sum of ranks of each group:
New lubricant: $9 + 4 + 8 - 6 = 27$
Standard lubricant: $3 + 2 + 7 + 5 + 1 = 18$

So our Wilcoxon test statistic is 27.

Since it is a right-tailed test, we count the number of permutations that result in a rank sum ≥ 27 and divide it by $\begin{pmatrix} 4+5 \\ 4 \end{pmatrix} = \dfrac{9!}{4!(9-4)!} = 126$.

So there are 126 possible ways to assign ranks to the 9 observations in the 2 groups. To find the p-value using the exact method, we need to consider all possible ways of assigning the total ranks to each group while keeping the number of ranks assigned to each sample fixed.

New lubricant				Rank-sum statistics
9.0	8.0	7.0	6.0	30
9.0	8.0	7.0	5.0	29
9.0	8.0	7.0	4.0	28
9.0	8.0	7.0	3.0	29
9.0	8.0	6.0	5.0	28
9.0	8.0	6.0	4.0	27
9.0	7.0	6.0	5.0	27

As we can see, we get a total of 7 permutations with a rank sum ≥ 27.

Therefore, our p-value = $\dfrac{7}{126} = 0.0556$.

Since p-value $= 0.0556 < 0.10 = \alpha$, we reject the null hypothesis and conclude that the new lubricant does increase the friction coefficient.

9.1.3.1.5 Example

Let's say you're working on a project to compare the concentrations of a particular pollutant (e.g., arsenic) in two different rivers (River A and River B) to determine if there is a significant difference in pollutant levels between the two rivers. You collect water samples from both rivers and measure the concentration of arsenic in each sample.

The data you collect is as follows:

River A : 0.085 0.080 0.062 0.091 0.072 (in mg / L)

River B : 0.076 0.083 0.050 0.067 0.058 (in mg / L)

H_0: there is no significant difference in pollutant levels between the two rivers.
H_A: there is a difference in pollutant levels between the rivers.

Sum of ranks of each group:
River A : $9+7+3+10+5 = 34$
River B :: $6+8+1+4+2 = 21$

The Wilcoxon rank-sum statistic is therefore 34.

Since the rank sum of River A > rank sum of River B, we count the number of permutations that result in a rank sum ≥ 34.

Formula A					Rank-sum statistics
10.0	9.0	8.0	7.0	6.0	40
10.0	9.0	8.0	7.0	5.0	39
10.0	9.0	8.0	7.0	4.0	38
10.0	9.0	8.0	7.0	3.0	37
10.0	9.0	8.0	7.0	2.0	36
10.0	9.0	8.0	7.0	1.0	35
10.0	9.0	8.0	6.0	5.0	38
10.0	9.0	8.0	6.0	4.0	37
10.0	9.0	8.0	6.0	3.0	36
10.0	9.0	8.0	6.0	2.0	35
10.0	9.0	8.0	6.0	1.0	34
10.0	9.0	8.0	5.0	4.0	36
10.0	9.0	8.0	5.0	3.0	35
10.0	9.0	8.0	5.0	2.0	34
10.0	9.0	8.0	4.0	3.0	34
10.0	9.0	7.0	6.0	5.0	37
10.0	9.0	7.0	6.0	4.0	36
10.0	9.0	7.0	6.0	3.0	35
10.0	9.0	7.0	6.0	2.0	34
10.0	9.0	7.0	5.0	4.0	35
10.0	9.0	7.0	5.0	3.0	34
10.0	9.0	6.0	5.0	4.0	34
10.0	8.0	7.0	6.0	5.0	36
10.0	8.0	7.0	6.0	4.0	35
10.0	8.0	7.0	6.0	3.0	34
10.0	8.0	7.0	5.0	4.0	34
10.0	8.0	7.0	6.0	5.0	35
10.0	8.0	7.0	6.0	4.0	34

Total combinations $\binom{5+5}{5} = 252$.

The number of permutations that result in a rank sum $\geq 34 = 28$.

Hence our p-value for two-tailed $= 2\left(\dfrac{28}{252}\right) = \dfrac{2}{9}$.

Since p-value $= 0.222 > 0.05 = \alpha$, we fail to reject the null hypothesis and conclude that there is no significant difference in pollutant levels between the two rivers.

9.1.3.1.6 Example

A petroleum company is evaluating two different drilling techniques, Technique A and Technique B, to determine which one is more effective in terms of drilling speed. The company has two teams of petroleum engineers, Team A and Team B, and they want to compare the drilling speeds of the two teams when using the two different techniques. The drilling speeds (in meters per hour) for each team using each technique are as follows:

Team A (Technique A): 21, 19, 21, 20, 17
Team B (Technique B): 22, 18, 22, 21, 23

$H_0 = 0.222 > 0.05 = \alpha$: there is no significant difference in the drilling speeds of the two techniques.
H_A: there is a difference in the drilling speeds.

Sum of ranks of each group:

Technique A : $6 + 3 + 6 + 4 + 1 = 20$
Technique B : $8.5 + 2 + 8.5 + 6 + 10 = 35$

The Wilcoxon rank-sum statistic is therefore 20.

Since the rank sum of technique A < the rank sum of technique B, we count the number of permutations that result in a rank sum ≤ 20.

Formula A	Rank-sum statistics
1.0 2.0 3.0 4.0 6.0a	16
1.0 2.0 3.0 4.0 6.0b	16
1.0 2.0 3.0 4.0 6.0c	16
1.0 2.0 3.0 4.0 8.5a	18.5
1.0 2.0 3.0 4.0 8.5b	18.5
1.0 2.0 3.0 4.0 10.0	20
1.0 2.0 3.0 6.0a 6.0b	18
1.0 2.0 3.0 6.0a 6.0c	18
1.0 2.0 3.0 6.0b 6.0c	18
1.0 2.0 4.0 6.0a 6.0b	19
1.0 2.0 4.0 6.0a 6.0c	19
1.0 2.0 4.0 6.0b 6.0c	19
1.0 3.0 4.0 6.0a 6.0b	20
1.0 3.0 4.0 6.0a 6.0b	20
1.0 3.0 4.0 6.0b 6.0c	20

The number of permutations that result in a rank sum $\leq 20 = 15$.

Hence our p-value for two-tailed $= 2 \left(\dfrac{15}{252} \right) = 0.1190$.

Since p-value $= 0.222 > 0.01 = \alpha$, we fail to reject the null hypothesis and conclude that there is no significant difference in the drilling speeds of the two techniques.

9.1.3.1.7 Example

A mechanical engineer is investigating a new type of alloy for potential use in one of his automotive components. However his company claims that the current alloy has better mechanical properties. To find out he conducts a series of tests to measure the ultimate tensile strength (UTS) of each formulation. The results are as follows:

New alloy	28	27	21	22
Old alloy	22	20	18	16

H_0: there is no significant difference in the tensile strength of the two alloys.
H_A: the new alloy has better mechanical properties—higher tensile strength.

Sum of ranks of each group:
Old alloy: $8 + 7 + 4 + 5.5 = 24.50$
New alloy: $5.5 + 3 + 2 + 1 = 11.50$

So, our Wilcoxon test statistic is 24.50.

Since it is a left-tailed test, we count the number of permutations that result in a rank sum ≤ 24.5 and divide it by $\binom{4+4}{4} = \dfrac{8!}{4!(8-4)!} = 70$.

So there are 70 possible ways to assign ranks to the 8 observations in the 2 groups. To find the p-value using the exact method, we need to consider all possible ways of assigning the total ranks to each group while keeping the number of ranks assigned to each sample fixed.

	New alloy				Old alloy			
Strength	28	27	21	22	22	20	18	16
	↓	↓	↓	↓	↓	↓	↓	↓
Ranks	8	7	4	5.5	5.5	3	2	1

New alloy				Rank-sum statistics
8.0	7.0	5.5a	4.0	24.50
8.0	7.0	5.5a	3.0	23.50
8.0	7.0	5.5a	2.0	22.50
8.0	7.0	5.5a	1.0	21.50
8.0	7.0	5.5b	4.0	24.50
8.0	7.0	5.5b	3.0	23.50
8.0	7.0	5.5b	2.0	22.50
8.0	7.0	5.5b	1.0	21.50
8.0	7.0	4.0	3.0	22
8.0	7.0	4.0	2.0	21
8.0	7.0	4.0	1.0	20
8.0	7.0	3.0	2.0	20
8.0	7.0	3.0	1.0	19
8.0	7.0	2.0	1.0	18
8.0	5.5a	5.5b	4.0	22
8.0	5.5a	5.5b	3.0	21
8.0	5.5a	5.5b	2.0	20
8.0	5.5a	5.5b	1.0	19

New alloy				Rank-sum statistics
8.0	5.5a	4.0	3.0	20
8.0	5.5a	4.0	2.0	19
8.0	5.5a	4.0	1.0	18
8.0	5.5a	3.0	2.0	18
8.0	5.5a	3.0	1.0	17
8.0	5.5a	2.0	1.0	16
8.0	5.5b	4.0	3.0	20
8.0	5.5b	4.0	2.0	19
8.0	5.5b	4.0	1.0	18
8.0	5.5b	3.0	2.0	18
8.0	5.5b	3.0	1.0	17
8.0	5.5b	2.0	1.0	16
8.0	4.0	3.0	2.0	17
8.0	4.0	3.0	1.0	16
8.0	4.0	2.0	1.0	15
8.0	3.0	2.0	1.0	14
7.0	5.5a	5.5b	4.0	21
7.0	5.5a	5.5b	3.0	20
7.0	5.5a	5.5b	2.0	19
7.0	5.5a	5.5b	1.0	18
7.0	5.5a	4.0	3.0	19.5
7.0	5.5a	4.0	2.0	18.5
7.0	5.5a	4.0	1.0	17.5
7.0	5.5a	3.0	2.0	17.5
7.0	5.5a	3.0	1.0	16.5
7.0	5.5a	2.0	1.0	15.5
7.0	5.5b	4.0	3.0	19.5
7.0	5.5b	4.0	2.0	18.5
7.0	5.5b	4.0	1.0	17.5
7.0	5.5b	3.0	2.0	17.5
7.0	5.5b	3.0	1.0	16.5
7.0	5.5b	2.0	1.0	15.5
7.0	4.0	3.0	2.0	16
7.0	4.0	3.0	1.0	15
7.0	4.0	2.0	1.0	14
7.0	3.0	2.0	1.0	13
5.5a	5.5b	4.0	3.0	18
5.5a	5.5b	4.0	2.0	17
5.5a	5.5b	4.0	1.0	16
5.5a	5.5b	3.0	2.0	16
5.5a	5.5b	3.0	1.0	15
5.5a	5.5b	2.0	1.0	14
5.5a	4.0	3.0	2.0	14.5
5.5a	4.0	3.0	1.0	13.5

New alloy				Rank-sum statistics
5.5a	4.0	2.0	1.0	12.5
5.5a	3.0	2.0	1.0	11.5
5.5b	4.0	3.0	2.0	14.5
5.5b	4.0	3.0	1.0	13.5
5.5b	4.0	2.0	1.0	12.5
5.5b	3.0	2.0	1.0	11.5
4.0	3.0	2.0	1.0	10

If we make a list, we get a total of 69 permutations with a rank sum ≤ 24.5.

Therefore, our p-value $= \dfrac{69}{70} = 0.9857$.

If the test was right-tailed then we would have counted the number of permutations that result in a rank sum ≥ 24.5 and divided it by 70.

To find the number of the permutations we simply had to subtract the permutations with a rank sum < 24.5 from the total permutations and clearly there are 67 permutation < 24.5 so we get $\dfrac{70 - 67}{70} = \dfrac{3}{70}$.

As we can see from this example, counting the number of permutations with extreme statistics can be an extremely tedious process. Unlike the critical value approach, when trying to take the p-value route it is best to leave the job to our computer and even better to MATLAB!

9.1.3.2 The normal approximation

If at least 1 of the samples has more than 20 elements, it approximates the p-value from a Gaussian approximation by default. Here, the term Gaussian has to do with the distribution of the sum of ranks and does not imply that your data needs to follow a Gaussian distribution. It approximates the distribution of the test statistic to a normal distribution using the central limit theorem. This method is faster and more practical for large sample sizes, but it may not be accurate when the sample size is small or the distribution is highly skewed.

Now the technique to calculate the p-value by normal approximation is slightly different when there are ties and when there are not. Let's walk through the two methods individually.

9.1.3.2.1 When there are no ties

- Calculate the Wilcoxon test statistics.
- Calculate the mean/expected value using $\dfrac{1}{2}n_1(n_1 + n_2 + 1)$.
- Calculate the variance using $\dfrac{1}{12}n_1 n_2(n_1 + n_2 + 1)$.

9.1.3.2.1.1 IF OUR TEST IS LEFT-TAILED

Our statistics T_L = Wilcoxon test statistic .

$$Z \text{ statistics} = \frac{(T_L + 0.50) - \text{mean}}{\sqrt{\text{variance}}}$$

In this situation the p-value is equal to the p-value for the lower-tailed p-value.

$$P(Z < z)$$

9.1.3.2.1.2 IF OUR TEST IS RIGHT- OR TWO-TAILED

Our statistics $T_{RT} = n_1(n_1 + n_2 + 1) - \text{Wilcoxon test statistics}$

Adjusted statistics $T^* = $ If $T_{RT} < \text{mean}$ use $(T_{RT} + 0.50)$ else use $(T_{RT} - 0.50)$

$$Z \text{ statistics} = \frac{\text{mean} - T^*}{\sqrt{\text{variance}}}$$

9.1.3.2.1.2.1 Right-tailedIn this situation the p-value is equal to the p-value for the upper-tailed p-value.

$$P(Z > z)$$

9.1.3.2.1.2.2 Two-tailedRecall that if the calculated value of the test statistic from the sample is positive, calculate an upper-tailed p-value. The p-value is equal to two times the upper-tailed p-value.

In contrast, when the calculated value of the test statistic from the sample is negative, the p-value is equal to two times the lower-tailed p-value.

9.1.3.2.2 When there are ties

- Calculate the Wilcoxon test statistics.
- Calculate the mean/expected value using $\frac{1}{2}n_1(n_1 + n_2 + 1)$.
- Calculate the variance using $\frac{1}{12}n_1 n_2 (n_1 + n_2 + 1)$.
- However due to the presence of ties an adjustment is needed for variance:

$$\text{adjusted variance} = \frac{1}{12}n_1 n_2 (n_1 + n_2 + 1) - \frac{n_1 n_2 \sum_{i=1}^{k} (t_i^3 - t_i)}{12(n_1 + n_2)(n_1 + n_2 - 1)}$$

The rest of the calculation stays the same.

9.1.3.2.3 Example

An aeronautical engineer is interested in comparing the performance of two different materials for building aircraft wings. He decides to conduct an experiment where the manufacturer builds a set of wings out of Material A and another set out of Material B for a Boeing 747. He then tests the wings to see how much weight they can support before breaking.

After conducting the experiment, the engineer has the following data: (note the weight is in $\times 10^3$)

Material A : 180 lbs, 182 lbs, 188 lbs, 185 lbs, 178 lbs

Material B : 175 lbs, 181 lbs, 176 lbs, 179 lbs, 183 lbs

The engineer wants to know if there is a significant difference in the weight that the two materials can support at $\alpha = 0.05$.

H_0: there is no significant difference in the performance of the two materials.

H_A: there is a significant difference in the weight that the two materials can support.

The ranked maximum supporting weight of wings out of Material A:

$$\{5, 7, 10, 9, 3\}$$

And the ranked maximum supporting weight of wings out of Material B:

$$\{1, 6, 2, 4, 8\}$$

We then sum up the ranks for each group:

A = 34
B = 21

NOTE:

Remember the sum of the ranks will always be equal:

$$\frac{n(n+1)}{2}$$

Since $A = 34$ and $B = 21$, the sum is $34 + 21 = 55$.

So as a check on our assignment of ranks, we need to check if the value

$$\frac{n(n+1)}{2} = 55$$

The total number of data points is 5 + 5 = 10; hence our total sample size is 10. So now if we

check $\dfrac{10(10+1)}{2} = 55$

Therefore our assigned ranks are considered accurate.

Calculating the Wilcoxon test statistics = 34.

Calculate the mean/expected value using $\dfrac{1}{2} \times 5 \times (5 + 5 + 1) = 27.5$.

Calculate the variance using $\dfrac{1}{12} \times 5 \times 5 \times (5 + 5 + 1) = \dfrac{275}{12}$.

The test statistic for a two-tailed test is

$$T^* = 5(5 + 5 + 1) - \text{Wilcoxon test statistics} = 55 - 34 = 21$$

Adjusted statistics = since $21 < 27.5$, we use $21 + 0.50 = 21.50$.

$$Z \text{ statistics} = \frac{27.5 - 21.5}{\sqrt{\dfrac{275}{12}}} = 1.2534$$

Since the test statistic is positive, the p-value is equal to two times the upper-tailed p-value.

$$2P(Z > 1.2534)$$

So if we pull up our z-score table, the p-value for the lower tail is around 0.8948 and since it is two-tailed, $2(1 - 0.8948) \sim 0.2104$.

Since the p-value $= 0.2104 > 0.05 = \alpha$, we fail to reject the null hypothesis and conclude that there is no significant difference in the weight that the two materials can support.

9.1.3.2.4 Example

Let's say an engineering firm is testing the effectiveness of two different materials in reducing sound transmission in building walls. They are claiming that material A absorbs more sound than material B. To test the claim, the firm takes measurements of sound transmission loss (in decibels) for each material and performs the Wilcoxon rank-sum test at 0.10.

Material A	43	46	47	48	52	48	50	49	
Material B	45	49	45	42	39	41	44	50	51

H_0: there is no significant difference in the sound transmission loss (in decibels) for the two materials.
H_A: material A absorbs more sound than material B.
Rank material A $= 4 + 8 + 9 + 10.5 + 17 + 10.5 + 14.5 + 12.5 = 86$.
Rank material B $= 6.5 + 12.5 + 6.5 + 3 + 1 + 2 + 5 + 14.5 + 16 = 67$.

Hence the Wilcoxon test statistic $= 86$.

The mean/expected value $= \dfrac{1}{2} \times 8 \times (8 + 9 + 1) = 72$.

The variance using $\dfrac{1}{12} \times 8 \times 9 \times (8 + 9 + 1) = 108$.

Since there are ties, an adjustment is needed for variance.

Notice there are ties of ranks $6.5, 10.5, 12.5, 14.5$ repeated twice each.

$$\text{Adjusted variance} = \frac{1}{12} \times 8 \times 9 \times (8 + 9 + 1) - \frac{8 \times 9 \times [(2^3 - 2) + (2^3 - 2) + (2^3 - 2) + (2^3 - 2)]}{12 \times (9 + 8)(9 + 8 - 1)}$$

$$= 108 - \frac{9}{17} = \frac{1827}{17}$$

The test statistic for a right-tailed test is

$$T^* = 8(9 + 8 + 1) - \text{Wilcoxon test statistic} = 144 - 86 = 58$$

Adjusted statistics $=$ since $58 < 72$, we use $58 + 0.50 = 58.50$.

$$\frac{72 - 58.5}{\sqrt{\frac{1827}{17}}} = 1.3022$$

In this situation the p-value is equal to the upper-tailed p-value.

$$P(Z > 1.3022) = 1 - P(Z < 1.3022) = 1 - 0.9036 \sim 0.0964$$

Since p-value $= 0.0964 < 0.10 = \alpha$, we reject the null hypothesis and conclude that material A has higher sound transmission loss.

9.1.3.2.5 Example

Suppose that a mechanical engineer is working on designing a new type of piston for an engine. The engineer has two different design options in mind, but his client thinks Design B piston will perform better in terms of durability. To test this, the engineer decides to conduct a test on a sample of 10 pistons for Design A and 12 for Design B, where the pistons are put through a stress test to measure their durability. After conducting the test, the engineer obtains the following data:

Design A: 210, 220, 205, 215, 225, 220, 200, 210, 215, 205, 200, 220
Design B: 225, 230, 220, 235, 240, 225, 230, 220, 225, 230

H_0: there is no significant difference in the durability of the two designs.
H_A: Design B has better durability.

Design A	Rank	Design B	Rank
210	5.5	225	15.5
220	11	230	19
205	3.5	220	11
215	7.5	235	21
225	15.5	240	22
220	11	225	15.5
200	1.5	230	19
210	5.5	220	11
215	7.5	225	15.5
205	3.5	230	19
200	1.5		
220	11		

Rank Design A $= 5.5 + 11 + 3.5 + 7.5 + 15.5 + 11 + 1.5 + 5.5 + 7.5 + 3.5 + 1.5 + 11 = 84.5$

Rank Design B $= 15.5 - 19 + 11 + 21 + 22 + 15.5 + 19 + 11 + 15.5 + 19 = 168.5$

Hence the Wilcoxon test statistic $= 84.5$.

The mean/expected value $= \frac{1}{2} \times 12 \times (12 + 10 + 1) = 138.$

the variance using $\frac{1}{12} \times 12 \times 10 \times (12 + 10 + 1) = 230$.

Since there are ties an adjustment is needed for variance.
Notice there are ties of ranks 1.5, 3.5, 5.5, 7.5, 11, 15.5, and 19 repeated.

$$\text{Adjusted variance} = 230 - \frac{12 \times 10 \times \left[4(2^3 - 2) + (3^3 - 3) + (4^3 - 4) + (5^3 - 5) \right]}{12 \times (12 + 10)(12 + 10 - 1)}$$

$$= 230 - \frac{380}{77} = 225.065$$

The test statistic for a left-tailed test is $T_L = 84.5$.

$$\text{Z statistics} = \frac{(84.5 + 0.50) - 138}{\sqrt{225.065}} = -3.5328$$

Here the p-value is equal to the lower-tailed p-value.

$$P(Z < -3.5328) \sim 0.0002$$

Since the p-value $= 0.0002 < 0.01 = \alpha$, we reject the null hypothesis and conclude that piston with Design B has higher durability.

9.1.3.2.6 Example

Suppose a marine engineer is testing a new fuel additive for a ship's engines and wants to determine if the new additive increases fuel efficiency compared to the old additive. The engineer collects data on the fuel consumption (in gallons) for ten different voyages using the old additive, and ten using the new. The data is summarized in the table below:

Old additive	New additive
800	789
788	782
821	796
792	802
815	779
795	800
780	791
786	801
794	795
819	787

H_0: there is no significant difference in the fuel efficiency between the two additives.
H_A: the new additive increases fuel efficiency compared to the old additive.
Old additive $= 14.5 + 6 + 20 + 9 + 18 + 11.5 + 2 + 4 + 10 + 19 = 114$.
New additive $= 7 + 3 + 13 + 17 + 1 + 14.5 + 8 + 16 + 11.5 + 5 = 96$.
Hence the Wilcoxon test statistics $= 114$.

The mean/expected value $=\dfrac{1}{2}\times10\times(10+10+1)=105.$

The variance using $\dfrac{1}{12}\times10\times10\times(10+10+1)=175.$

Since there are ties an adjustment is needed for variance.

Notice there are ties of ranks 11.5 and 14.5 repeated.

$$\text{Adjusted variance}=175-\dfrac{10\times10\times\left[(2^3-2)+(2^3-2)\right]}{10\times(10+10)(10+10-1)}$$

$$=175-\dfrac{6}{19}=174.6842$$

The test statistic for a left-tailed test is $T_L=114.$

$$\dfrac{(114+0.50)-105}{\sqrt{174.6842}}=0.7187$$

In this situation the p -value is equal to the lower-tailed p -value.

$$P(Z<0.7187)=\sim0.7640$$

Since the p- value $=0.7640>0.05=\alpha,$ we fail to reject the null hypothesis and conclude that both additives provide the same fuel efficiency.
   Now that we have the theory discussion out of the way we are all ready to get to MATLAB.

9.1.3.2.7 MATLAB

Like all the other tests we have discussed, MATLAB has inbuilt function to calculate the p-value for the Wilcoxon rank-sum test.

```
[p,h,stats] = ranksum(x, y, 'alpha',  ,'method', ' ', 'tail', ' ')
```

Let's discuss each of the input arguments:

   x - sample data, specified as a vector.
   y - sample data, specified as a vector.

As we have seen in the previous examples, the two vectors do not
   have to have the same length. Also the test statistic depends on the first argument added in the function. Here since x is the first argument, the test statistic would be the rank sum of the sample x.
   The input arguments alpha and tail have the same meaning as the rest of the tests.

'exact'	Exact computation of the p-value, p.
'approximate'	Normal approximation while computing the p-value, p.

So, this argument 'method' specifies which method is going to be used to calculate the p-value. Like all functions this too has a default value which is exact.

Let's run an example in MATLAB and see if the results align with our calculations.

9.1.3.2.8 Example 1

```
[p,h,stats] = ranksum(New_Lubricant, Old_Lubricant, 'tail','right',
'alpha', 0.10, 'method','exact')

p =

    0.0556


h =

  logical

   1


stats =

  struct with fields:

    ranksum: 27
```

9.1.3.2.9 Example 2

```
[p,h,stats] = ranksum(RiverA, RiverB, 'method','exact')
p =

    0.2222


h =

  logical

   0

stats =

  struct with fields:

    ranksum: 34
```

9.1.3.2.10 Example 3

```
[p,h,stats] = ranksum(Team_A, Team_B, 'tail','both', 'alpha', 0.01,
                'method','exact')

p =

    0.1190


h =

  logical

    0


stats =

  struct with fields:

    ranksum: 20
```

9.1.3.2.11 Example 4

```
[p,h,stats] = ranksum(Old_Alloy, New_Alloy, 'tail','left','method','exact')
p =

    0.9857


h =

  logical

    0


stats =

  struct with fields:

    ranksum: 24.5000
```

Now let's test for approximate p-values.

9.1.3.2.12 Example 1

```
p,h,stats] = ranksum(MaterialA, MaterialB, 'tail','both',
                'method','approximate')
```

```
p =

    0.2101


h =

  logical

   0


stats =

  struct with fields:

        zval: 1.2534
     ranksum: 34
```

9.1.3.2.13 Example 2

```
[p,h,stats] = ranksum(Material_A, Material_B, 'tail','right',
'alpha', 0.10, 'method','approximate')

p =

    0.0964


h =

  logical

   1


stats =

  struct with fields:

        zval: 1.3022
     ranksum: 86
```

9.1.3.2.14 Example 3

```
[p,h,stats] = ranksum(DesignA, DesignB, 'tail','left', 'alpha', 0.01,
'method','approximate')
```

```
p =

   2.0557e-04

h =

  logical

   1

stats =

  struct with fields:

        zval: -3.5328
     ranksum: 84.5000
```

9.1.3.2.15 Example 4

```
[p,h,stats] = ranksum(OldAdd, NewAdd, 'tail', 'left', 'method',
'approximate')
p =

   0.7638

h =

  logical

   0

stats =

  struct with fields:

        zval: 0.7187
     ranksum: 114
```

As we can see the output perfectly aligns with our calculations.
Let's perform a couple of more examples.

9.1.3.2.16 Example

Suppose a mechanical engineer is testing a new alloy and wants to determine if it has better mechanical properties. To come to a conclusion he selects a random sample of 12 specimens from each metal and measures their tensile strength in pounds per square inch (psi). The results are as follows:

New alloy: 520 500 600 620 650 670 700 720 750 800 780 690
Old alloy: 580 420 450 480 500 580 550 580 600 620 650 550

Test at $\alpha = 0.01$ to see if he can replace the new alloy with the old one.

9.1.3.2.16.1 SOLUTION

H_0: there is no significant difference in the tensile strength of the two alloys.
H_A: the new alloy has higher tensile strength.

Since the sample size is small, we should use the exact method.

```
[p,h,stats] = ranksum(New,Old,'tail','right','alpha', 0.01,
'method','exact')

p =

    0.0011


h =

  logical

   1


stats =

  struct with fields:

    ranksum: 201
```

Since the p- value = $0.0011 < 0.01 = \alpha$, we reject the null hypothesis and conclude that the new alloy does indeed have higher tensile strength.

9.1.3.2.17 Example

An electrical engineer is working on a project that involves two different methods of signal processing for a piece of technological equipment. You want to determine which method produces better results based on a particular metric, such as signal-to-noise ratio (SNR).

Method A	19.6	17.8	20.1	17.2	18.2	17.8	19.3	18.7	19.5
Method B	20.1	18.2	20.4	18.5	19.3	19.9	20.8		

Test at $\alpha = 0.05$ to see if method A is less efficient than B.

H_0: there is no significant difference in the signal-to-noise ratio of the two methods.
H_A: method B has a higher ratio.

```
[p,h,stats] = ranksum(MaterialA, MaterialB,'tail','left', 'method', 'exact')

p =

    0.0484


h =

  logical

   1


stats =

  struct with fields:

    ranksum: 60.5000
```

Since the p-value = $0.0484 < 0.05 = \alpha$, we reject the null hypothesis and conclude that method B has a higher signal-to-noise ratio.

9.1.3.2.18 Example

An electrical engineer is working on designing a new type of battery. He has two different materials for the battery's electrodes and wants to determine if there is any difference between the batteries' performances. He assigns 10 batteries to the first material group, and 12 to the second, and then charges each battery fully and measures their discharge times (in minutes) until they reach a predetermined cut-off voltage. Here are the results:

Material A: 23, 20, 21, 20, 21, 23, 24, 23, 22, 23
Material B: 21, 20, 20, 22, 20, 21, 20, 21, 22, 22, 20, 22

H_0: there is no significant difference in the batteries' performances.
H_A: there is a significant difference in the batteries' performances.

```
[p,h,stats] = ranksum(Material_A, Material_B, 'method','exact')
p =

   0.0643


h =

  logical

   0


stats =

  struct with fields:

   ranksum: 143
```

Since the p-value = $0.0643 > 0.05 = \alpha$, we fail to reject the null hypothesis and conclude that there is no difference between the batteries' performances.

9.1.3.2.19 Example

Suppose a computer engineer invents two programming languages and wants to compare their performances in terms of their speed of execution. He selects an algorithm and runs it in both languages on a set of benchmark data and measures the time it takes for each program to run. At $\alpha = 0.01$ test whether there is a significant difference in the speed of execution between the two languages.

Programming language X						Programming language Y					
8	12	16	19	20	20	9	12	13	15	18	21
9	13	17	15	12	21	9	10	14	15	17	20
10	14	18	10	17	15	10	13	12	12	17	12
11	14	18	18	22	20	12	11	14	13	15	19
12	15	19	16	23	15	12	12	14	16	18	17

H_0: there is no significant difference in the speed of execution between the two languages.
H_A: there is a significant difference in the speed.

```
p =

    0.1055


h =

  logical

    0


stats =

  struct with fields:

        zval: 1.6188
     ranksum: 1.0245e+03
```

Since the p-value $= 0.1055 > 0.01 = \alpha$, we fail to reject the null hypothesis and conclude that there is no difference in the speed of execution.

9.2 WILCOXON SIGNED-RANK TEST

In the previous section we discussed the Wilcoxon rank-sum test which is analogous to the unpaired t-test. In this section we will talk about the Wilcoxon signed-rank test, which is a non-parametric alternative to both the one-sample and paired t test when the data cannot be assumed to be normally distributed.

9.2.1 One sample

The **one-sample Wilcoxon test is used to compare our observations to specified value** as a non-parametric equivalent to a z test or t test. Like the other tests this test too is used to determine if a group is significantly different from a known or hypothesized population value on the variable of interest.

9.2.1.1 Stating the hypothesis

While the one-sample t test compares the mean of the population to a known value, its non-parametric counterpart, the one-sample Wilcoxon rank-sum test, compares the **median** of the population to the known value. In other words, it tests whether there is no significant difference between the sample and the hypothesized distribution.

Remember that in all the tests we have discussed so far (z-test, t-test, f-test) μ is used to represent the mean of a distribution. So even though there is no widely accepted standard notation for the median, we will use the $\mu\sim$ notation for the population median and $x\sim$ for the hypo median throughout this chapter.

Therefore for three different situations our hypothesis are as follows.

9.2.1.1.1 Two-tailed

H_0: the population median is equal to the hypothesized median: $\mu\sim = x\sim$.
H_A: the population median is equal to the hypothesized median: $\mu\sim \neq x\sim$.

9.2.1.1.2 Right-tailed

H_0 : the population median is less than or equal to the hypothesized median: $\mu \sim \le x \sim$.
H_A : the population median is greater than the hypothesized median: $\mu \sim > x \sim$.

9.2.1.1.3 Left-tailed

H_0 : the population median is greater than or equal to the hypothesized median: $\mu \sim \ge x \sim$.
H_A : the population median is less than the hypothesized median: $\mu \sim < x \sim$.

9.2.1.2 Calculating the test statistics

As with the Wilcoxon rank-sum test, the Wilcoxon signed-rank test is based on ranks assigned to the data points rather than the actual observed data. However in this test, we rank the data values on the basis **of the absolute difference** between the observations and the hypothesized median value.

Now each difference between the two is equally likely to be positive or negative. So we have the sum of the ranks of the positive differences and the sum of the ranks of the negative differences. However, in order to test a particular pair of hypotheses, we need only test the number of plus signs.

Let's consider an example.

A manufacturer of car batteries claims that his batteries last at least an average of four years before needing to be replaced. A team of engineers wants to test this claim to ensure that the batteries meet their quality standards. They randomly select a sample of five batteries giving the following results:

Observation	Battery life
I	3.7
2	4.3
3	4.1
4	4.5
5	3.8

Test (at the 5% level of significance) the null hypothesis that the median battery life is four years against the alternative that it is greater.

The hypotheses are
H_0 : $\mu \sim \le 4$.
H_A : $\mu \sim > 4$.

To calculate the test statistic, we compute the difference of the battery life for each battery. Then we rank the absolute values difference scores where 1 is assigned to the smallest/most negative and n to the largest/most positive values of the difference scores. In presence of ties we deal with them the same way we did for Wilcoxon rank-sum and for the difference of 0 we just ignore it.

Difference	Sign	Absolute diff.	Ranks	Signed ranks
3.7 − 4 = −0.3	-	0.3	3.5	−3.5
4.3 - 4 = 0.3	+	0.4	3.5	+3.5
4.1 - 4 = 0.1	+	0.1	1	+1
4.5- 4 = 0.5	+	0.5	5	+5
3.8 - 4 = −0.2	-	0.2	2	−2

Rank sum for positive signs $= 3.5 + 1 + 5 = 9.5$.
Rank sum for negative signs $= 3.5 + 2 = 5.5$.

9.2.1.3 Finding the p-value

9.2.1.3.1 Exact value method

An exact p-value for the signed-rank test is the proportion of all permutations of outcomes within each pair that lead to a test statistic as extreme as or more extreme than the one observed. The p-value is therefore the number of possible permutations that provide a sum of positive ranks as extreme as or more extreme than the observed one, divided by 2^n.

9.2.1.3.1.1 IF OUR TEST IS RIGHT-TAILED

For two groups of sample size n_1 and n_2, after evaluating the test statistics by summing the ranks of the first group we count
the number of permutations of sample size n_1 that result in a sum of positive ranks $\geq$ test statistics and divide it by 2^n.

9.2.1.3.1.2 IF OUR TEST IS LEFT-TAILED

For this case, we count the number of permutations of sample size n_1 that result in a sum of positive ranks $\leq$ test statistics and divide it by 2^n.

9.2.1.3.1.3 IF OUR TEST IS TWO-TAILED

Here too, the number of permutations differ for the rank sum of the two group of data.

- If the rank sum for positive signs $<$ rank sum for negative signs then we count the number of permutations of sample size n_1 that result in a sum of positive ranks $\geq$ rank sum for negative signs.
- If the rank sum for positive signs $>$ rank sum for negative signs then we count the number of permutations of sample size n_1 that result in a sum of positive ranks $\geq$ rank sum for positive signs .

So if we look at our previous example the ranks of our five data points will always be the numbers $1, 2, 3.5, \text{and} 5$. When the null hypothesis is true, each data point is equally likely to be positive or negative. Since our test is right-tailed, if we list out the number of permutations whose sum rank is greater than or equal to 9.5 we get:

r_1	r_2	r_3	r_4	r_5	sum of+ranks	r_1	r_2	r_3	r_4	r_5	sum of+ranks
−3.5	−3.5	−1	−5	−2	0	3.5	−3.5	−1	−5	−2	3.5
−3.5	−3.5	−1	−5	2	2	3.5	−3.5	−1	−5	2	5.5
−3.5	−3.5	−1	5	−2	5	3.5	−3.5	−1	5	−2	8.5
−3.5	−3.5	−1	5	2	7	3.5	−3.5	−1	5	2	10.5
−3.5	−3.5	1	5	2	8	3.5	−3.5	1	−5	−2	4.5
−3.5	−3.5	1	−5	2	3	3.5	−3.5	1	−5	2	6.5

r_1	r_2	r_3	r_4	r_5	sum of+ranks	r_1	r_2	r_3	r_4	r_5	sum of+ranks
-3.5	-3.5	1	5	-2	6	3.5	-3.5	1	5	-2	9.5
-3.5	-3.5	1	-5	-2	1	3.5	-3.5	1	5	2	11.5
-3.5	3.5	-1	-5	-2	3.5	3.5	3.5	-1	-5	-2	7
-3.5	3.5	-1	-5	2	5.5	3.5	3.5	-1	-5	2	9
-3.5	3.5	-1	5	-2	8.5	3.5	3.5	-1	5	-2	12
-3.5	3.5	-1	5	2	10.5	3.5	3.5	-1	5	2	14
-3.5	3.5	1	-5	-2	4.5	3.5	3.5	1	-5	-2	8
-3.5	3.5	1	-5	2	6.5	3.5	3.5	1	-5	2	10
-3.5	3.5	1	5	-2	9.5	3.5	3.5	1	5	-2	13
-3.5	3.5	1	5	2	11.5	3.5	3.5	1	5	2	15

The total number of possibilities is $2^5 = 32$.

As you can see, we have 11 of the 32 possibilities that are greater than or equal to 9.5. Therefore our p-value $= \dfrac{11}{32} = 0.34375$.

9.2.1.3.1.4 EXAMPLE

A petroleum engineer wants to determine if the average porosity of a rock formation in a particular oil field is larger than the reference value of 0.25. The engineer collects a sample of porosity measurements from the rock formation and gets the following data. Test at the 5% level of significance.

Observation	Porosity
1	0.28
2	0.24
3	0.21
4	0.27
5	0.19

The hypotheses are:

H_0 : the null hypothesis in this case would be that the average porosity of the rock formation is less than or equal to the reference value of 0.25.

the he alternative hypothesis would be that the average porosity of the rock formation is greater than the reference value of 0.25.

Difference	Sign	Absolute diff.	Ranks	Signed ranks
0.28 - 0.25 = 0.03	+	0.03	3	+3
0.24 - 0.25 = -0.01	-	0.01	1	-1
0.21 - 0.25 = -0.04	-	0.04	4	-4
0.27 - 0.25 = 0.02	+	0.02	2	+2
0.19 - 0.24 = -0.05	-	0.05	5	-5

Rank sum for positive signs $= 3 + 2 = 5$.
Rank sum for negative signs $= 1 + 4 + 5 = 10$.

Here, the ranks of our data points are the numbers $1, 2, 3, 4,$ and 5. Since our test is right-tailed, we list out the number of permutations whose sum rank is greater than or equal to 5:

r_1	r_2	r_3	r_4	r_5	sum of+ranks	r_1	r_2	r_3	r_4	r_5	sum of+ranks
−1	−2	−3	−4	−5	0	1	−2	−3	−4	−5	1
−1	−2	−3	−4	5	5	1	−2	−3	−4	5	6
−1	−2	−3	4	−5	4	1	−2	−3	4	−5	5
−1	−2	−3	4	5	9	1	−2	−3	4	5	10
−1	−2	3	−4	−5	3	1	−2	3	−4	−5	4
−1	−2	3	−4	5	8	1	−2	3	−4	5	9
−1	−2	3	4	−5	7	1	−2	3	4	−5	8
−1	−2	3	4	5	12	1	−2	3	4	5	13
−1	2	−3	−4	−5	2	1	2	−3	−4	−5	3
−1	2	−3	−4	5	7	1	2	−3	−4	5	8
−1	2	−3	4	−5	6	1	2	−3	4	−5	7
−1	2	−3	4	5	11	1	2	−3	4	5	12
−1	2	3	−4	−5	5	1	2	3	−4	−5	6
−1	2	3	−4	5	10	1	2	3	−4	5	11
−1	2	3	4	−5	9	1	2	3	4	−5	10
−1	2	3	4	5	14	1	2	3	4	5	15

The total number of possibilities is $2^5 = 32$.

As you can see, we have 25 of the 32 possibilities that are greater than or equal to 5. Therefore our p-value $= \dfrac{25}{32} = 0.78125$.

9.2.1.3.1.5 EXAMPLE

A manufacturer of LED light bulbs claims that their bulbs have an average lifespan of 10,000 hours. A quality control engineer wants to test this claim by taking a sample of 10 light bulbs and measuring their lifespans. The engineer obtains the following data:

9,800 9,900 10,200 9,700 10,100 10,300 10,100 10,000 9,900 10,100

The engineer wants to test whether the average lifespan of the bulbs in the sample is less than the manufacturer's claim of 10,000 hours, using a significance level of 0.05.

The hypotheses are:

H_0: $\mu \sim\geq 10,000$.
H_A: $\mu \sim< 10,000$.

Bulb	Lifespan	Difference	Sign	Absolute diff.	Ranks	Signed ranks
1	9,800	9,800 − 10,000 = −200	−	200	6.5	−6.5
2	10,000	10,000 − 10,000 = 0	N/A	0		
3	10,200	10,200 − 10,000 = 200	+	200	6.5	+6.5
4	9,700	9,700 − 10,000 = −300	−	300	8	−8

Bulb	Lifespan	Difference	Sign	Absolute diff.	Ranks	Signed ranks
5	10,100	10,100 − 10,000 = 100	+	100	3	+3
6	9,900	9,900 − 10,000 = −100	-	100	3	−3
7	10,100	10,100 − 10,000 = 100	+	100	3	+3
8	10,000	10,000 − 10,000 = 0	N/A	0		
9	9,900	9,900 − 10,000 = −100	-	100	3	−3
10	10,100	10,100 − 10,000 = 100	+	100	3	+3

Rank sum for positive signs $= 6.5 + 3 + 3 + 3 = 15.5$.
Rank sum for negative signs $= 6.5 + 8 + 3 + 3 = 20.5$.

The ranks of our 9 data points will always be the numbers 3, 6.5, and 8.5. Since our test is left-tailed, we list out the number of permutations whose sum rank is less than or equal to 15.5.

r_1	r_2	r_3	r_4	r_5	r_6	r_7	r_8	sum of +ranks
−8	−6.5	−6.5	−3	−3	−3	−3	−3	0
−8	−6.5	−6.5	−3	−3	−3	−3	3	3
−8	−6.5	−6.5	−3	−3	3	−3	−3	3
−8	−6.5	−6.5	−3	−3	−3	3	−3	3
−8	−6.5	−6.5	−3	−3	−3	3	3	6
−8	−6.5	−6.5	−3	−3	3	−3	3	6
−8	−6.5	−6.5	−3	−3	3	3	−3	6
−8	−6.5	−6.5	−3	−3	3	3	3	9
−8	−6.5	−6.5	−3	3	−3	−3	−3	3
−8	−6.5	−6.5	−3	3	−3	−3	3	6
−8	−6.5	−6.5	−3	3	3	−3	−3	6
−8	−6.5	−6.5	−3	3	−3	3	−3	6
−8	−6.5	−6.5	−3	3	−3	3	3	9
−8	−6.5	−6.5	−3	3	3	−3	3	9
−8	−6.5	−6.5	−3	3	3	3	−3	9
−8	−6.5	−6.5	−3	3	3	3	3	12
−8	−6.5	−6.5	3	−3	−3	−3	−3	3
−8	−6.5	−6.5	3	−3	−3	−3	3	6
−8	−6.5	−6.5	3	−3	3	−3	−3	6
−8	−6.5	−6.5	3	−3	−3	3	−3	6
−8	−6.5	−6.5	3	−3	−3	3	3	9
−8	−6.5	−6.5	3	−3	3	−3	3	9
−8	−6.5	−6.5	3	−3	3	3	−3	9
−8	−6.5	−6.5	3	−3	3	3	3	12
−8	−6.5	−6.5	3	3	−3	−3	−3	6
−8	−6.5	−6.5	3	3	−3	−3	3	9
−8	−6.5	−6.5	3	3	3	−3	−3	9
−8	−6.5	−6.5	3	3	−3	3	−3	9
−8	−6.5	−6.5	3	3	−3	3	3	12
−8	−6.5	−6.5	3	3	3	−3	3	12
−8	−6.5	−6.5	3	3	3	3	−3	12

r_1	r_2	r_3	r_4	r_5	r_6	r_7	r_8	sum of+ranks
−8	−6.5	−6.5	3	3	3	3	3	15
−8	−6.5	6.5	−3	−3	−3	−3	−3	6.5
−8	−6.5	6.5	−3	−3	−3	−3	3	9.5
−8	−6.5	6.5	−3	−3	3	−3	−3	9.5
−8	−6.5	6.5	−3	−3	−3	3	−3	9.5
−8	−6.5	6.5	−3	−3	−3	3	3	12.5
−8	−6.5	6.5	−3	−3	3	−3	3	12.5
−8	−6.5	6.5	−3	−3	3	3	−3	12.5
−8	−6.5	6.5	−3	−3	3	3	3	15.5
−8	−6.5	6.5	−3	3	−3	−3	−3	9.5
−8	−6.5	6.5	−3	3	−3	−3	3	12.5
−8	−6.5	6.5	−3	3	3	−3	−3	12.5
−8	−6.5	6.5	−3	3	−3	3	−3	12.5
−8	−6.5	6.5	−3	3	−3	3	3	15.5
−8	−6.5	6.5	−3	3	3	−3	3	15.5
−8	−6.5	6.5	−3	3	3	3	−3	15.5
−8	−6.5	6.5	−3	3	3	3	3	18.5
−8	−6.5	6.5	3	−3	−3	−3	−3	9.5
−8	−6.5	6.5	3	−3	−3	−3	3	12.5
−8	−6.5	6.5	3	−3	3	−3	−3	12.5
−8	−6.5	6.5	3	−3	−3	3	−3	12.5
−8	−6.5	6.5	3	−3	−3	3	3	15.5
−8	−6.5	6.5	3	−3	3	−3	3	15.5
−8	−6.5	6.5	3	−3	3	3	−3	15.5
−8	−6.5	6.5	3	−3	3	3	3	18.5
−8	−6.5	6.5	3	3	−3	−3	−3	12.5
−8	−6.5	6.5	3	3	−3	−3	3	15.5
−8	−6.5	6.5	3	3	3	−3	−3	15.5
−8	−6.5	6.5	3	3	−3	3	−3	15.5
−8	−6.5	6.5	3	3	−3	3	3	18.5
−8	−6.5	6.5	3	3	3	−3	3	18.5
−8	−6.5	6.5	3	3	3	3	−3	18.5
−8(64)	−6.5	6.5	3	3	3	3	3	21.5
−8	6.5	−6.5	−3	−3	−3	−3	−3	6.5
−8	6.5	−6.5	−3	−3	−3	−3	3	9.5
−8	6.5	−6.5	−3	−3	3	−3	−3	9.5
−8	6.5	−6.5	−3	−3	−3	3	−3	9.5
−8	6.5	−6.5	−3	−3	−3	3	3	12.5
−8	6.5	−6.5	−3	−3	3	−3	3	12.5
−8	6.5	−6.5	−3	−3	3	3	−3	12.5
−8	6.5	−6.5	−3	−3	3	3	3	15.5
−8	6.5	−6.5	−3	3	−3	−3	−3	9.5
−8	6.5	−6.5	−3	3	−3	−3	3	12.5
−8	6.5	−6.5	−3	3	3	−3	−3	12.5
−8	6.5	−6.5	−3	3	−3	3	−3	12.5
−8	6.5	−6.5	−3	3	−3	3	3	15.5

r_1	r_2	r_3	r_4	r_5	r_6	r_7	r_8	sum of+ranks
−8	6.5	−6.5	−3	3	3	−3	3	15.5
−8	6.5	−6.5	−3	3	3	3	−3	15.5
−8	6.5	−6.5	−3	3	3	3	3	18.5
−8	6.5	−6.5	3	−3	−3	−3	−3	9.5
−8	6.5	−6.5	3	−3	−3	−3	3	12.5
−8	6.5	−6.5	3	−3	3	−3	−3	12.5
−8	6.5	−6.5	3	−3	−3	3	−3	12.5
−8	6.5	−6.5	3	−3	−3	3	3	15.5
−8	6.5	−6.5	3	−3	3	−3	3	15.5
−8	6.5	−6.5	3	−3	3	3	−3	15.5
−8	6.5	−6.5	3	−3	3	3	3	18.5
−8	6.5	−6.5	3	3	−3	−3	−3	12.5
−8	6.5	−6.5	3	3	−3	−3	3	15.5
−8	6.5	−6.5	3	3	3	−3	−3	15.5
−8	6.5	−6.5	3	3	−3	3	−3	15.5
−8	6.5	−6.5	3	3	−3	3	3	18.5
−8	6.5	−6.5	3	3	3	−3	3	18.5
−8	6.5	−6.5	3	3	3	3	−3	18.6
−8	6.5	−6.5	3	3	3	3	3	21.5
−8	6.5	6.5	−3	−3	−3	−3	−3	13
−8	6.5	6.5	−3	−3	−3	−3	3	16
−8	6.5	6.5	−3	−3	3	−3	−3	16
−8	6.5	6.5	−3	−3	−3	3	−3	16
−8	6.5	6.5	−3	−3	−3	3	3	19
−8	6.5	6.5	−3	−3	3	−3	3	19
−8	6.5	6.5	−3	−3	3	3	−3	19
−8	6.5	6.5	−3	−3	3	3	3	22
−8	6.5	6.5	−3	3	−3	−3	−3	16
−8	6.5	6.5	−3	3	−3	−3	3	19
−8	6.5	6.5	−3	3	3	−3	−3	19
−8	6.5	6.5	−3	3	−3	3	−3	19
−8	6.5	6.5	−3	3	−3	3	3	22
−8	6.5	6.5	−3	3	3	−3	3	22
−8	6.5	6.5	−3	3	3	3	−3	22
−8	6.5	6.5	−3	3	3	3	3	25
−8	6.5	6.5	3	−3	−3	−3	−3	16
−8	6.5	6.5	3	−3	−3	−3	3	19
−8	6.5	6.5	3	−3	3	−3	−3	19
−8	6.5	6.5	3	−3	−3	3	−3	19
−8	6.5	6.5	3	−3	−3	3	3	22
−8	6.5	6.5	3	−3	3	−3	3	22
−8	6.5	6.5	3	−3	3	3	−3	22
−8	6.5	6.5	3	−3	3	3	3	25
−8	6.5	6.5	3	3	−3	−3	−3	19
−8	6.5	6.5	3	3	−3	−3	3	22
−8	6.5	6.5	3	3	3	−3	−3	22
−8	6.5	6.5	3	3	−3	3	−3	22

r_1	r_2	r_3	r_4	r_5	r_6	r_7	r_8	sum of +ranks
−8	6.5	6.5	3	3	−3	3	3	25
−8	6.5	6.5	3	3	3	−3	3	25
−8	6.5	6.5	3	3	3	3	−3	25
−8(64)	6.5	6.5	3	3	3	3	3	28
8	−6.5	−6.5	−3	−3	−3	−3	−3	8
8	−6.5	−6.5	−3	−3	−3	−3	3	11
8	−6.5	−6.5	−3	−3	3	−3	−3	11
8	−6.5	−6.5	−3	−3	−3	3	−3	11
8	−6.5	−6.5	−3	−3	−3	3	3	14
8	−6.5	−6.5	−3	−3	3	−3	3	14
8	−6.5	−6.5	−3	−3	3	3	−3	14
8	−6.5	−6.5	−3	−3	3	3	3	17
8	−6.5	−6.5	−3	3	−3	−3	−3	11
8	−6.5	−6.5	−3	3	−3	−3	3	14
8	−6.5	−6.5	−3	3	3	−3	−3	14
8	−6.5	−6.5	−3	3	−3	3	−3	14
8	−6.5	−6.5	−3	3	−3	3	3	17
8	−6.5	−6.5	−3	3	3	−3	3	17
8	−6.5	−6.5	−3	3	3	3	−3	17
8	−6.5	−6.5	−3	3	3	3	3	20
8	−6.5	−6.5	3	−3	−3	−3	−3	11
8	−6.5	−6.5	3	−3	−3	−3	3	14
8	−6.5	−6.5	3	−3	3	−3	−3	14
8	−6.5	−6.5	3	−3	−3	3	−3	14
8	−6.5	−6.5	3	−3	−3	3	3	17
8	−6.5	−6.5	3	−3	3	−3	3	17
8	−6.5	−6.5	3	−3	3	3	−3	17
8	−6.5	−6.5	3	−3	3	3	3	20
8	−6.5	−6.5	3	3	−3	−3	−3	14
8	−6.5	−6.5	3	3	−3	−3	3	17
8	−6.5	−6.5	3	3	3	−3	−3	17
8	−6.5	−6.5	3	3	−3	3	−3	17
8	−6.5	−6.5	3	3	−3	3	3	20
8	−6.5	−6.5	3	3	3	−3	3	20
8	−6.5	−6.5	3	3	3	3	−3	20
8	−6.5	−6.5	3	3	3	3	3	23
8	−6.5	6.5	−3	−3	−3	−3	−3	14.5
8	−6.5	6.5	−3	−3	−3	−3	3	17.5
8	−6.5	6.5	−3	−3	3	−3	−3	17.5
8	−6.5	6.5	−3	−3	−3	3	−3	17.5
8	−6.5	6.5	−3	−3	−3	3	3	20.5
8	−6.5	6.5	−3	−3	3	−3	3	20.5
8	−6.5	6.5	−3	−3	3	3	−3	20.5
8	−6.5	6.5	−3	−3	3	3	3	23.5
8	−6.5	6.5	−3	3	−3	−3	−3	17.5
8	−6.5	6.5	−3	3	−3	−3	3	20.5

r_1	r_2	r_3	r_4	r_5	r_6	r_7	r_8	sum of+ranks
8	−6.5	6.5	−3	3	3	−3	−3	20.5
8	−6.5	6.5	−3	3	−3	3	−3	20.5
8	−6.5	6.5	−3	3	−3	3	3	23.5
8	−6.5	6.5	−3	3	3	−3	3	23.5
8	−6.5	6.5	−3	3	3	3	−3	23.5
8	−6.5	6.5	−3	3	3	3	3	26.5
8	−6.5	6.5	3	−3	−3	−3	−3	17.5
8	−6.5	6.5	3	−3	−3	−3	3	20.5
8	−6.5	6.5	3	−3	3	−3	−3	20.5
8	−6.5	6.5	3	−3	−3	3	−3	20.5
8	−6.5	6.5	3	−3	−3	3	3	23.5
8	−6.5	6.5	3	−3	3	−3	3	23.5
8	−6.5	6.5	3	−3	3	3	−3	23.5
8	−6.5	6.5	3	−3	3	3	3	26.5
8	−6.5	6.5	3	3	−3	−3	−3	20.5
8	−6.5	6.5	3	3	−3	−3	3	23.5
8	−6.5	6.5	3	3	3	−3	−3	23.5
8	−6.5	6.5	3	3	−3	3	−3	23.5
8	−6.5	6.5	3	3	−3	3	3	26.5
8	−6.5	6.5	3	3	3	−3	3	26.5
8	−6.5	6.5	3	3	3	3	−3	26.5
8(64)	−6.5	6.5	3	3	3	3	3	29.5
8	6.5	−6.5	−3	−3	−3	−3	−3	14.5
8	6.5	−6.5	−3	−3	−3	−3	3	17.5
8	6.5	−6.5	−3	−3	3	−3	−3	17.5
8	6.5	−6.5	−3	−3	−3	3	−3	17.5
8	6.5	−6.5	−3	−3	−3	3	3	20.5
8	6.5	−6.5	−3	−3	3	−3	3	20.5
8	6.5	−6.5	−3	−3	3	3	−3	20.5
8	6.5	−6.5	−3	−3	3	3	3	23.5
8	6.5	−6.5	−3	3	−3	−3	−3	17.5
8	6.5	−6.5	−3	3	−3	−3	3	20.5
8	6.5	−6.5	−3	3	3	−3	−3	20.5
8	6.5	−6.5	−3	3	−3	3	−3	20.5
8	6.5	−6.5	−3	3	−3	3	3	23.5
8	6.5	−6.5	−3	3	3	−3	3	23.5
8	6.5	−6.5	−3	3	3	3	−3	23.5
8	6.5	−6.5	−3	3	3	3	3	26.5
8	6.5	−6.5	3	−3	−3	−3	−3	17.5
8	6.5	−6.5	3	−3	−3	−3	3	20.5
8	6.5	−6.5	3	−3	3	−3	−3	20.5
8	6.5	−6.5	3	−3	−3	3	−3	20.5
8	6.5	−6.5	3	−3	−3	3	3	23.5
8	6.5	−6.5	3	−3	3	−3	3	23.5
8	6.5	−6.5	3	−3	3	3	−3	23.5
8	6.5	−6.5	3	−3	3	3	3	26.5

r_1	r_2	r_3	r_4	r_5	r_6	r_7	r_8	sum of+ranks
8	6.5	−6.5	3	3	−3	−3	−3	20.5
8	6.5	−6.5	3	3	−3	−3	3	23.5
8	6.5	−6.5	3	3	3	−3	−3	23.5
8	6.5	−6.5	3	3	−3	3	−3	23.5
8	6.5	−6.5	3	3	−3	3	3	26.5
8	6.5	−6.5	3	3	3	−3	3	26.5
8	6.5	−6.5	3	3	3	3	−3	26.5
8	6.5	−6.5	3	3	3	3	3	29.5
8	6.5	6.5	−3	−3	−3	−3	−3	21
8	6.5	6.5	−3	−3	−3	−3	3	24
8	6.5	6.5	−3	−3	3	−3	−3	24
8	6.5	6.5	−3	−3	−3	3	−3	24
8	6.5	6.5	−3	−3	−3	3	3	27
8	6.5	6.5	−3	−3	3	−3	3	27
8	6.5	6.5	−3	−3	3	3	−3	27
8	6.5	6.5	−3	−3	3	3	3	30
8	6.5	6.5	−3	3	−3	−3	−3	24
8	6.5	6.5	−3	3	−3	−3	3	27
8	6.5	6.5	−3	3	3	−3	−3	27
8	6.5	6.5	−3	3	−3	3	−3	27
8	6.5	6.5	−3	3	−3	3	3	30
8	6.5	6.5	−3	3	3	−3	3	30
8	6.5	6.5	−3	3	3	3	−3	30
8	6.5	6.5	−3	3	3	3	3	33
8	6.5	6.5	3	−3	−3	−3	−3	24
8	6.5	6.5	3	−3	−3	−3	3	27
8	6.5	6.5	3	−3	3	−3	−3	27
8	6.5	6.5	3	−3	−3	3	−3	27
8	6.5	6.5	3	−3	−3	3	3	30
8	6.5	6.5	3	−3	3	−3	3	30
8	6.5	6.5	3	−3	3	3	−3	30
8	6.5	6.5	3	−3	3	3	3	33
8	6.5	6.5	3	3	−3	−3	−3	27
8	6.5	6.5	3	3	−3	−3	3	30
8	6.5	6.5	3	3	3	−3	−3	30
8	6.5	6.5	3	3	−3	3	−3	30
8	6.5	6.5	3	3	−3	3	3	33
8	6.5	6.5	3	3	3	−3	3	33
8	6.5	6.5	3	3	3	3	−3	33
8	6.5	6.5	3	3	3	3	3	36

Out of 256 combinations only 103 arrangements have statistics ≤ 15.5. Therefore our p-value $\frac{103}{256} = 0.4023$.

Since the p- value = $0.1055 > 0.01 = \alpha$, we fail to reject the null hypothesis and conclude that there is no difference in the speed of execution.

9.2.1.3.1.6 EXAMPLE

A mechanical engineer is conducting a study to determine the effectiveness of a new material for use in the manufacturing of a specific type of turbine blade. The engineer collects a sample of eight turbine blades and measures their performance in terms of turbine efficiency (in percentage) using a standardized testing procedure. The engineer wants to determine if the mean efficiency of the turbine blades using the new material is significantly different from a theoretical value of 80%, which is the benchmark efficiency for the existing turbine blades currently in production.

Turbine blade	Efficiency
1	76
2	82
3	85
4	78
5	80
6	72
7	76
8	83

The null hypothesis: the mean efficiency of the turbine blades using the new material is equal to 80%.

The alternative hypothesis: the mean efficiency of the turbine blades using the new material is not equal to 80%.

Turbine blade	Efficiency	Difference	Sign	Absolute diff.	Ranks	Signed ranks
1	76	76 − 80 = −4	-	4	4.5	−4.5
2	82	82 − 80 = 2	+	2	1.5	+1.5
3	85	85 − 80 = 5	+	5	6	+6
4	78	78 − 80 = −2	-	2	1.5	−1.5
5	80	80 − 80 = 0	N/A	0		
6	72	72 − 80 = −8	-	8	7	−7
7	76	76 − 80 = −4	-	4	4.5	−4.5
8	83	83 − 80 = 3	+	3	3	+3

Rank sum for positive signs $= 1.5 + 6 + 3 = 10.5$.
Rank sum for negative signs $= 4.5 + 1.5 + 7 + 4.5 = 17.5$.

The ranks of our 7 data points will always be the numbers 1.5, 3, 4.5, 6, and 7. When H_0 is true, each data point is equally likely to be positive or negative, and so, the expected value.

r_1	r_2	r_3	r_4	r_5	r_6	r_7	sum of +ranks
−1.5	−1.5	−3	−4.5	−4.5	−6	−7	0
−1.5	−1.5	−3	−4.5	−4.5	−6	7	7
−1.5	−1.5	−3	−4.5	−4.5	6	−7	6
−1.5	−1.5	−3	−4.5	−4.5	6	7	13
−1.5	−1.5	−3	−4.5	4.5	−6	−7	4.5

r_1	r_2	r_3	r_4	r_5	r_6	r_7	sum of+ranks
−1.5	−1.5	−3	−4.5	4.5	−6	7	11.5
−1.5	−1.5	−3	−4.5	4.5	6	−7	10.5
−1.5	−1.5	−3	−4.5	4.5	6	7	17.5
−1.5	−1.5	−3	4.5	−4.5	−6	−7	4.5
−1.5	−1.5	−3	4.5	−4.5	−6	7	11.5
−1.5	−1.5	−3	4.5	−4.5	6	−7	10.5
−1.5	−1.5	−3	4.5	−4.5	6	7	17.5
−1.5	−1.5	−3	4.5	4.5	−6	−7	9
−1.5	−1.5	−3	4.5	4.5	−6	7	16
−1.5	−1.5	−3	4.5	4.5	6	−7	15
−1.5	−1.5	−3	4.5	4.5	6	7	22
				8			
−1.5	−1.5	3	−4.5	−4.5	−6	−7	3
−1.5	−1.5	3	−4.5	−4.5	−6	7	10
−1.5	−1.5	3	−4.5	−4.5	6	−7	9
−1.5	−1.5	3	−4.5	−4.5	6	7	16
−1.5	−1.5	3	−4.5	4.5	−6	−7	7.5
−1.5	−1.5	3	−4.5	4.5	−6	7	14.5
−1.5	−1.5	3	−4.5	4.5	6	−7	13.5
−1.5	−1.5	3	−4.5	4.5	6	7	20.5
−1.5	−1.5	3	4.5	−4.5	−6	−7	7.5
−1.5	−1.5	3	4.5	−4.5	−6	7	14.5
−1.5	−1.5	3	4.5	−4.5	6	−7	13.5
−1.5	−1.5	3	4.5	−4.5	6	7	20.5
−1.5	−1.5	3	4.5	4.5	−6	−7	12
−1.5	−1.5	3	4.5	4.5	−6	7	19
−1.5	−1.5	3	4.5	4.5	6	−7	18
−1.5	−1.5	3	4.5	4.5	6	7	25
				5			
−1.5	1.5	−3	−4.5	−4.5	−6	−7	1.5
−1.5	1.5	−3	−4.5	−4.5	−6	7	8.5
−1.5	1.5	−3	−4.5	−4.5	6	−7	7.5
−1.5	1.5	−3	−4.5	−4.5	6	7	14.5
−1.5	1.5	−3	−4.5	4.5	−6	−7	5.5
−1.5	1.5	−3	−4.5	4.5	−6	7	13
−1.5	1.5	−3	−4.5	4.5	6	−7	12
−1.5	1.5	−3	−4.5	4.5	6	7	19
−1.5	1.5	−3	4.5	−4.5	−6	−7	6
−1.5	1.5	−3	4.5	−4.5	−6	7	13
−1.5	1.5	−3	4.5	−4.5	6	−7	12
−1.5	1.5	−3	4.5	−4.5	6	7	19
−1.5	1.5	−3	4.5	4.5	−6	−7	10.5
−1.5	1.5	−3	4.5	4.5	−6	7	17.5
−1.5	1.5	−3	4.5	4.5	6	−7	16.5
−1.5	1.5	−3	4.5	4.5	6	7	23.5
				6			
−1.5	1.5	3	−4.5	−4.5	−6	−7	4.5

r_1	r_2	r_3	r_4	r_5	r_6	r_7	sum of +ranks
−1.5	1.5	3	−4.5	−4.5	−6	7	11.5
−1.5	1.5	3	−4.5	−4.5	6	−7	10.5
−1.5	1.5	3	−4.5	−4.5	6	7	17.5
−1.5	1.5	3	−4.5	4.5	−6	−7	9
−1.5	1.5	3	−4.5	4.5	−6	7	16
−1.5	1.5	3	−4.5	4.5	6	−7	15
−1.5	1.5	3	−4.5	4.5	6	7	22
−1.5	1.5	3	4.5	−4.5	−6	−7	9
−1.5	1.5	3	4.5	−4.5	−6	7	16
−1.5	1.5	3	4.5	−4.5	6	−7	15
−1.5	1.5	3	4.5	−4.5	6	7	22
−1.5	1.5	3	4.5	4.5	−6	−7	13.5
−1.5	1.5	3	4.5	4.5	−6	7	20.5
−1.5	1.5	3	4.5	4.5	6	−7	19.5
−1.5	1.5	3	4.5	4.5	6	7	26.5
				4			
1.5	−1.5	−3	−4.5	−4.5	−6	−7	1.5
1.5	−1.5	−3	−4.5	−4.5	−6	7	8.5
1.5	−1.5	−3	−4.5	−4.5	6	−7	7.5
1.5	−1.5	−3	−4.5	−4.5	6	7	14.5
1.5	−1.5	−3	−4.5	4.5	−6	−7	6
1.5	−1.5	−3	−4.5	4.5	−6	7	13
1.5	−1.5	−3	−4.5	4.5	6	−7	12
1.5	−1.5	−3	−4.5	4.5	6	7	19
1.5	−1.5	−3	4.5	−4.5	−6	−7	6
1.5	−1.5	−3	4.5	−4.5	−6	7	13
1.5	−1.5	−3	4.5	−4.5	6	−7	12
1.5	−1.5	−3	4.5	−4.5	6	7	19
1.5	−1.5	−3	4.5	4.5	−6	−7	10.5
1.5	−1.5	−3	4.5	4.5	−6	7	17.5
1.5	−1.5	−3	4.5	4.5	6	−7	16.5
1.5	−1.5	−3	4.5	4.5	6	7	23.5
				6			
1.5	−1.5	3	−4.5	−4.5	−6	−7	4.5
1.5	−1.5	3	−4.5	−4.5	−6	7	11.5
1.5	−1.5	3	−4.5	−4.5	6	−7	10.5
1.5	−1.5	3	−4.5	−4.5	6	7	17.5
1.5	−1.5	3	−4.5	4.5	−6	−7	15.5
1.5	−1.5	3	−4.5	4.5	−6	7	16
1.5	−1.5	3	−4.5	4.5	6	−7	15
1.5	−1.5	3	−4.5	4.5	6	7	22
1.5	−1.5	3	4.5	−4.5	−6	−7	9
1.5	−1.5	3	4.5	−4.5	−6	7	16
1.5	−1.5	3	4.5	−4.5	6	−7	15
1.5	−1.5	3	4.5	−4.5	6	7	22
1.5	−1.5	3	4.5	4.5	−6	−7	13.5
1.5	−1.5	3	4.5	4.5	−6	7	20.5

r_1	r_2	r_3	r_4	r_5	r_6	r_7	sum of+ranks
1.5	−1.5	3	4.5	4.5	6	−7	19.5
1.5	−1.5	3	4.5	4.5	6	7	26.5
				3			
1.5	1.5	−3	−4.5	−4.5	−6	−7	3
1.5	1.5	−3	−4.5	−4.5	−6	7	10
1.5	1.5	−3	−4.5	−4.5	6	−7	9
1.5	1.5	−3	−4.5	−4.5	6	7	16
1.5	1.5	−3	−4.5	4.5	−6	−7	7.5
1.5	1.5	−3	−4.5	4.5	−6	7	14.5
1.5	1.5	−3	−4.5	4.5	6	−7	13.5
1.5	1.5	−3	−4.5	4.5	6	7	20.5
1.5	1.5	−3	4.5	−4.5	−6	−7	7.5
1.5	1.5	−3	4.5	−4.5	−6	7	14.5
1.5	1.5	−3	4.5	−4.5	6	−7	13.5
1.5	1.5	−3	4.5	−4.5	6	7	20.5
1.5	1.5	−3	4.5	4.5	−6	−7	12
1.5	1.5	−3	4.5	4.5	−6	7	19
1.5	1.5	−3	4.5	4.5	6	−7	18
1.5	1.5	−3	4.5	4.5	6	7	25
				5			
1.5	1.5	3	−4.5	−4.5	−6	−7	6
1.5	1.5	3	−4.5	−4.5	−6	7	13
1.5	1.5	3	−4.5	−4.5	6	−7	12
1.5	1.5	3	−4.5	−4.5	6	7	19
1.5	1.5	3	−4.5	4.5	−6	−7	10.5
1.5	1.5	3	−4.5	4.5	−6	7	17.5
1.5	1.5	3	−4.5	4.5	6	−7	16.5
1.5	1.5	3	−4.5	4.5	6	7	23.5
1.5	1.5	3	4.5	−4.5	−6	−7	10.5
1.5	1.5	3	4.5	−4.5	−6	7	17.5
1.5	1.5	3	4.5	−4.5	6	−7	16.5
1.5	1.5	3	4.5	−4.5	6	7	23.5
1.5	1.5	3	4.5	4.5	−6	−7	15
1.5	1.5	3	4.5	4.5	−6	7	22
1.5	1.5	3	4.5	4.5	6	−7	21
1.5	1.5	3	4.5	4.5	6	7	28

Total number of permutations $= 2^7 = 128$.

Number of arrangements having statistics $\geq 17.5 = 41$.

Out of 128 combinations 41 arrangements have statistics ≥ 17.5.

Therefore our p-value $\dfrac{41}{128} = 0.3203$.

Since it is two-tailed, the p-value $2 \times \dfrac{41}{128} = 0.640625$.

Since the p- value $= 0.640625 > 0.05 = \alpha$, we fail to reject the null hypothesis and conclude that there is no difference.

9.2.1.3.1.7 EXAMPLE

A petroleum engineering company wants to determine if the average drilling depth of their wells is significantly different from the industry average of 8000 feet. They collect data on the drilling depths of a random sample of ten wells and want to conduct a one-sample t test to determine if there is a significant difference.

Here are the drilling depths (in feet) for the ten wells in the sample:

7984, 8000, 8010, 8000, 8021, 7995, 8013, 8000, 8018

Null hypothesis (H_0): the average drilling depth of the wells is equal to 8000 feet.

Alternative hypothesis (H_A): the average drilling depth of the wells is not equal to 8000 feet.

Well	Depth	Difference	Sign	Absolute diff.	Ranks	Signed ranks
1	7984	7984 − 8000 = −16	-	16	4	−4
2	8000	8000 − 8000 = 0	N/A			
3	8010	8010 − 8000 = 10	+	10	2	+2
4	8000	8000 − 8000 = 0	N/A			
5	8021	8021 − 8000 = 21	+	21	6	+6
6	7995	7995 − 8000 = −5	-	5	1	−1
7	8013	8013 − 8000 = 13	+	13	3	+3
8	8000	8000 − 8000 = 0	N/A			
9	8018	8018 − 8000 = 18	+	18	5	+5

Rank sum for positive signs = 2 + 6 + 3 + 5 = 16.
Rank sum for negative signs = 4 + 1 = 5.

r_1	r_2	r_3	r_4	r_5	r_6	sum of+ranks
−1	−2	−3	−4	−5	−6	0
−1	−2	−3	−4	5	−6	5
−1	−2	−3	−4	−5	6	6
−1	−2	−3	−4	5	6	11
−1	−2	−3	4	−5	−6	4
−1	−2	−3	4	5	−6	9
−1	−2	−3	4	−5	6	10
−1	−2	−3	4	5	6	15
−1	−2	3	−4	−5	−6	3
−1	−2	3	−4	5	−6	8
−1	−2	3	−4	−5	6	9
−1	−2	3	−4	5	6	14
−1	−2	3	4	−5	−6	7
−1	−2	3	4	5	−6	12
−1	−2	3	4	−5	6	13
−1	−2	3	4	5	6	18
−1	2	−3	−4	−5	−6	2
−1	2	−3	−4	5	−6	7
−1	2	−3	−4	−5	6	8
−1	2	−3	−4	5	6	13

r_1	r_2	r_3	r_4	r_5	r_6	sum of +ranks
−1	2	−3	4	−5	−6	6
−1	2	−3	4	5	−6	11
−1	2	−3	4	−5	6	12
−1	2	−3	4	5	6	17
−1	2	3	−4	−5	−6	5
−1	2	3	−4	5	−6	10
−1	2	3	−4	−5	6	11
−1	2	3	−4	5	6	16
−1	2	3	4	−5	−6	9
−1	2	3	4	5	−6	14
−1	2	3	4	−5	6	15
−1	2	3	4	5	6	20
1	−2	−3	−4	−5	−6	1
1	−2	−3	−4	5	−6	6
1	−2	−3	−4	−5	6	7
1	−2	−3	−4	5	6	12
1	−2	−3	4	−5	−6	5
1	−2	−3	4	5	−6	10
1	−2	−3	4	−5	6	11
1	−2	−3	4	5	6	16
1	−2	3	−4	−5	−6	4
1	−2	3	−4	5	−6	9
1	−2	3	−4	−5	6	10
1	−2	3	−4	5	6	15
1	−2	3	4	−5	−6	8
1	−2	3	4	5	−6	13
1	−2	3	4	−5	6	14
1	−2	3	4	5	6	19
1	2	−3	−4	−5	−6	3
1	2	−3	−4	5	−6	8
1	2	−3	−4	−5	6	9
1	2	−3	−4	5	6	14
1	2	−3	4	−5	−6	7
1	2	−3	4	5	−6	12
1	2	−3	4	−5	6	13
1	2	−3	4	5	6	18
1	2	3	−4	−5	−6	6
1	2	3	−4	5	−6	11
1	2	3	−4	−5	6	12
1	2	3	−4	5	6	17
1	2	3	4	−5	−6	10
1	2	3	4	5	−6	15
1	2	3	4	−5	6	16
1	2	3	4	5	6	21

Total number of permutations = $2^6 = 64$.
Number of arrangements having statistics $\geq 16 = 10$.

Out of 64 combinations 10 arrangements have statistics ≥ 16.

Therefore our p-value $\dfrac{10}{64} = 0.15625$.

Since it is two-tailed, the p-value $2 \times 0.15625 = 0.3125$.

Since the p- value = $0.3125 > 0.05 = \alpha$, we fail to reject the null hypothesis and conclude that there is no difference.

9.2.1.3.2 The normal approximation

Even though the conditions are the same as the rank-sum test, the method varies with and without ties. Let's discuss the two methods.

9.2.1.3.2.1 WHEN THERE ARE NO TIES

- 9.2.1.3.2.1.1 First method of calculating z statisticsCalculate the mean using $\dfrac{1}{4} \times n \times (n+1)$.
- Calculate the variance using $\dfrac{1}{24} \times n \times (n+1)(2n+1)$.

9.2.1.3.2.1.2 Another method of calculating z statistics Calculate the variance using $\dfrac{1}{6} \times n \times (n+1)(2n+1)$.

9.2.1.3.2.2 WHEN THERE ARE TIES

- Calculate the mean/expected using $\dfrac{1}{4} \times n \times (n+1)$.

- Calculate the variance using $\dfrac{1}{24} \times n \times (n+1)(2n+1)$.

- However due to the presence of ties an adjustment is needed for variance:

$$\text{Wilcoxon signed test statistics} - \dfrac{\sum_{i=1}^{k} t_i^{\,3} - t_i}{48}$$

where

t_i is the frequency of the ith unique rank

We use the same method as in the rank-sum test to calculate the p-value for different tests.

9.2.1.3.2.2.1 If our test is right-tailed Our statistics T_R = Wilcoxon test statistics.

$$Z \text{ statistic} = \dfrac{(T_R - 0.50) - \text{mean}}{\sqrt{\text{variance}}}$$

In this situation the p -value is equal to the upper-tailed p -value.

$$P(Z < z)$$

9.2.1.3.2.2.2 If our test is left-tailed Our statistics T_L = Wilcoxon test statistics.

$$Z \text{ statistic} = \frac{(T_L + 0.50) - \text{mean}}{\sqrt{\text{variance}}}$$

In this situation the p-value is equal to the lower-tailed p-value.

$$P(Z < z)$$

9.2.1.3.2.2.3 If our test is two-tailed Our statistics T_R = Wilcoxon test statistics.

$$Z \text{ statistic} = \frac{T_R - \text{mean}}{\sqrt{\text{variance}}}$$

- If the z statistic is negative then the p-value is equal to two times the lower-tailed p-value: $2P(Z < z)$.
- If the z statistic is positive then the p-value is equal to two times the upper-tailed p-value: $2P(Z > z)$.

9.2.1.3.2.2 First example Suppose a civil engineer is designing a bridge and needs to determine if the average weight limit of the bridge is greater 50 tons. The engineer takes a random sample of 20 trucks and records their weights. The weights are as follows (in tons):

52, 54, 48, 49, 51, 52, 48, 53, 51, 48, 52, 49, 53, 51, 52

The hypotheses are:
H_0: $\mu \sim\leq 50$, the population mean weight is less than or equal to 50 tons.
H_A: $\mu \sim> 50$, the population mean weight is greater than 50 tons.

Truck	Weight	Difference	Signs	Absolute diff.	Ranks
1	52	52 – 50 = 2	+	2	9
2	54	54 - 50 = 4	+	4	15
3	48	48 - 50 = –2	-	2	9
4	49	49 - 50 = –1	-	1	3
5	51	51 - 50 = 1	+	1	3
6	52	52 – 50 = 2	+	2	9
7	48	48 - 50 = –2	-	2	9
8	53	53 – 50 = 3	+	3	13.5
9	51	51 - 50 = 1	+	1	3
10	48	48 - 50 = –2	-	2	9
11	52	52 – 50 = 2	+	2	9
12	49	49 - 50 = –1	-	1	3
13	53	53 – 50 = 3	+	3	13.5
14	51	51 - 50 = 1	+	1	3
15	52	52 – 50 = 2	+	2	9

Rank sum for positive signs $= 9 + 15 + 3 + 9 + 13.5 + 3 + 9 + 13.5 + 3 + 9 = 87$.
Rank sum for negative signs $= 9 + 3 + 9 + 9 + 3 = 33$.

Using the above example:

- Mean/expected $= \dfrac{1}{4} \times 15 \times (15+1) = 60.$

- Variance $= \dfrac{1}{24} \times 15 \times (15+1)(30+1) = 310.$

- Adjusted variance $= 310 - \dfrac{(5^3 - 5) + (7^3 - 7) + (2^3 - 2)}{48} = 300.375.$

REMEMBER OUR WILCOXON SIGNED-RANK TEST STATISTIC $= 87.$

Since our test is right-tailed

$$Z \text{ statistic} = \frac{(87 - 0.50) - 60}{\sqrt{300.375}} = 1.5290$$

In this situation the p -value is equal to the upper-tailed p -value.

$$P(Z > 1.5290) = 1 - P(Z < 1.5290) = 1 - 0.9369 = 0.063$$

Since the p- value $=\ 0.0631 > 0.01 = \alpha$, we fail to reject the null hypothesis and conclude that the population mean weight is less than or equal to 50 tons.

9.2.1.3.2.3 SECOND EXAMPLE

Suppose a company wants to determine if the average time it takes for a group of aeronautical engineers to design a new airplane part is less than ten weeks. The company randomly selects a sample of 12 aeronautical engineers and records the time it took each engineer to design the part. The data is as follows:

9.7, 11.2, 8.5, 10.0, 9.4, 9.9, 10.5, 11.1, 10.4,

11.0, 9.6, 10.2

The hypotheses are:

H_0 : $\mu \sim \geq 10$. The average time is greater than or equal to ten weeks.

H_A : $\mu \sim < 10$. The average time to design a new airplane part is less than ten weeks.

Engineer	Time taken	Difference	Engineer	Time taken	Difference
1	9.7	9.7 – 10 = –0.3	7	10.5	10.5 - 10 = 0.5
2	11.2	11.2 – 10 = 1.2	8	11.1	11.1- 10 = 1.1
3	8.5	8.5 – 10 = –1.5	9	10.4	10.4 - 10 = 0.4
4	10.0	10.0 – 10 = 0	10	11.0	11.0- 10 = 1.0
5	9.4	9.4 – 10 = –0.6	11	9.6	9.6- 10 = –0.4
6	9.9	9.9 – 10 = –0.1	12	10.2	10.2- 10 = 0.2

Difference	Absolute diff.	Sign	Ranks	Difference	Absolute diff.	Sign	Ranks
9.7 – 10 = –0.3	0.3	-	3	10.5 - 10 = 0.5	0.5	+	6
11.2 – 10 = 1.2	1.2	+	10	11.1- 10 = 1.1	1.1	+	9
8.5 – 10 = –1.5	1.5	-	11	10.4 - 10 = 0.4	0.4	+	4.5
10.0 – 10 = 0	C	N/A		11.0- 10 = 1.0	1.0	+	8
9.4 – 10 = –0.6	0.6	-	7	9.6- 10 = –0.4	0.4	-	4.5
9.9 – 10 = –0.1	0.1	-	1	10.2- 10 = 0.2	0.2	+	2

Rank sum for positive signs $= 10+6+9+4.5+8+2 = 39.5$.
Rank sum for negative signs $= 3+11+7+1+4.5 = 26.5$.

- Mean/expected $= \dfrac{1}{4} \times 11 \times (11+1) = 33$.
- Variance $= \dfrac{1}{24} \times 11 \times (11+1)(22+1) = 126.5$.
- Adjusted variance $= 126.5 - \dfrac{(2^3-2)}{48} = 126.375$.

REMEMBER OUR WILCOXON SIGNED-RANK TEST STATISTIC $= 39.50$.
Since our test is left-tailed:

$$Z \text{ statistics} = \frac{(39.50+0.50)-33}{\sqrt{126.375}} = 0.6227$$

In this situation the p-value is equal to the lower-tailed p-value.

$$P(Z < 0.6227) \sim 0.734$$

Since the p-value $= 0.734 > 0.10 = \alpha$, we fail to reject the null hypothesis and conclude that the population mean time taken is less than or equal to ten days.

9.2.1.3.2.4 THIRD EXAMPLE

Let's say you work for a manufacturing company that produces metal brackets, and you're interested in whether the average thickness of a particular type of bracket meets the required specifications. You have a sample of 20 brackets, and you know that the required thickness is 5 millimeters.

Bracket	Thickness	Bracket	Thickness
1	4.7	11	4.6
2	5.2	12	4.5
3	5.1	13	4.4
4	5.0	14	5.2
5	5.1	15	5.4
6	4.8	16	4.6
7	5.3	17	5.1
8	4.9	18	4.7
9	4.6	19	5.0
10	5.1	20	4.8

The hypotheses are:

H_0: $\mu \sim= 5$. The average thickness of the metal bracket is 5 mm.
H_A: $\mu \sim\neq 5$. The average thickness of the metal bracket is not equal to 5 mm.

Bracket	Difference	Bracket	Difference
4.7	4.7 - 5 = −0.30	4.6	4.6 - 5 = −0.40
5.2	5.2 - 5 = 0.20	4.5	4.5 - 5 = −0.50
5.1	5.1 - 5 = 0.10	4.4	4.4 - 5 = −0.60
5.0	5.0 - 5 = 0	5.2	5.2 - 5 = 0.20
5.1	5.1 - 5 = 0.10	5.4	5.4 - 5 = 0.40
4.8	4.8 - 5 = −0.20	4.6	4.6 - 5 = −0.40
5.3	5.3 - 5 = 0.30	5.1	5.1 - 5 = 0.10
4.9	4.9 - 5 = −0.10	4.7	4.7 - 5 = −0.30
4.6	4.6 - 5 = −0.40	5.0	5.0 - 5 = 0.0
5.1	5.1 - 5 = 0.10	4.8	4.8 - 5 = −0.20

Difference	Sign	Absolute diff.	Ranks	Difference	Sign	Absolute diff.	Ranks
4.7 - 5 = −0.30	−	0.3	11	4.6 - 5 = −0.40	−	0.4	14.5
5.2 - 5 = 0.20	+	0.2	7.5	4.5 - 5 = −0.50	−	0.5	17
5.1 - 5 = 0.10	+	0.1	3	4.4 - 5 = −0.60	−	0.6	18
5.0 - 5 = 0	N/A	0		5.2 - 5 = 0.20	+	0.2	7.5
5.1 - 5 = 0.10	+	0.1	3	5.4 - 5 = 0.40	+	0.4	14.5
4.8 - 5 = −0.20	−	0.2	7.5	4.6 - 5 = −0.40	−	0.4	14.5
5.3 - 5 = 0.30	+	0.3	11	5.1 - 5 = 0.10	+	0.1	3
4.9 - 5 = −0.10	−	0.1	3	4.7 - 5 = −0.30	−	0.3	11
4.6 - 5 = −0.40	−	0.4	14.5	5.0 - 5 = 0.0	N/A		
5.1 - 5 = 0.10	+	0.1	3	4.8 - 5 = −0.20	−	0.2	7.5

Rank sum for positive signs = $7.5 + 3 + 3 + 11 + 3 + 7.5 + 14.5 + 3 = 52.5$.
Rank sum for negative signs = $11 + 7.5 + 3 + 14.5 + 14.5 + 17 + 18 + 14.5 + 11 + 7.5 = 118.5$.

Using the above example:

- Mean/expected $= \frac{1}{4} \times 18 \times (18 + 1) = 85.5$.

- Variance $= \frac{1}{24} \times 18 \times (18 + 1)(36 + 1) = 527.25$.

- Adjusted variance = $527.25 - \frac{(3^3 - 3) + (4^3 - 4) + (5^3 - 5) + (4^3 - 4)}{48} = 521.75$.

REMEMBER OUR WILCOXON SIGNED-RANK TEST STATISTIC = 52.50.
Since our test is two-tailed:

$$Z \text{ statistic} = \frac{52.50 - 85.50}{\sqrt{521.75}} = -1.4447$$

In this situation the p-value is equal to two times the lower-tailed p-value.

$$2P(Z < -1.4447) \sim 2(0.0742) \sim 0.1484$$

Since the p-value = $0.1484 > 0.05 = \alpha$, we fail to reject the null hypothesis and conclude that the average thickness of the metal bracket is 5 mm.

9.2.1.3.2.5 EXAMPLE

A mining company wants to assess whether the average concentration of gold in a particular ore deposit is significantly different from a target value of 10 grams per ton (g/t). The company has collected 30 ore samples from the deposit and measured the gold concentration in each sample.

H_0: the average gold concentration in the ore deposit = 10 g/t.
H_A: the average gold concentration in the ore deposit $\neq$ 10 g/t.

Ore sample	Gold concentration	Ore sample	Gold concentration	Ore sample	Gold concentration
1	9.5	11	10.7	21	8.8
2	11.8	12	10.8	22	10
3	9.7	13	9.4	23	11.4
4	10	14	12.1	24	10.6
5	11.5	15	10	25	12.1
6	12.1	16	10.5	26	10
7	10.3	17	9.5	27	9.2
8	12.4	18	10	28	11.9
9	9.9	19	12.2	29	8.5
10	8.6	20	8.5	30	10.7

Difference	Sign	Absolute diff.	Ranks	Difference	Sign	Absolute diff.	Ranks
9.5 – 10 = –0.5	-	0.5	4	10.5 – 10 = 0.5	+	0.5	4
11.8 – 10 = 1.8	+	1.8	17	9.5 – 10 = –0.5	-	0.5	4
9.7 – 10 = –0.3	-	0.3	2	10 – 10 = 0	N/A		
10 – 10 = 0	N/A			12.2 – 10 = 2.2	+	2.2	21
11.5 – 10 = 1.5	+	1.5	14	11.5 – 10 = 1.5	+	1.5	14
12.1 – 10 = 2.1	+	2.1	19.5	8.8 – 10 = –1.2	-	1.2	11
10.9 – 10 = 0.9	+	0.9	10	10 – 10 = 0	N/A		
12.4 – 10 = 2.4	+	2.4	22	11.4 – 10 = 1.4	+	1.4	12
9.9 – 10 = –0.1	-	0.1	1	10.6 – 10 = 0.6	+	0.6	6.5
11.6 – 10 = 1.6	+	1.6	16	12.1 – 10 = 2.1	+	2.1	19.5
12.7 – 10 = 2.7	+	2.7	23.5	10 – 10 = 0	N/A		
10.8 – 10 = 0.8	+	0.8	8.5	9.2 – 10 = –0.8	-	0.8	8.5
9.4 – 10 = –0.6	-	0.6	6.5	11.9 – 10 = 1.9	+	1.9	18
13.1 – 10 = 3.1	+	3.1	25	8.5 – 10 = –1.5	-	1.5	14
10 – 10 = 0	N/A	0		12.7 – 10 = 2.7	+	2.7	23.5

Rank sum for positive signs:

$$17+14+19.5+10+22+16+23.5+8.5+25+4+21+14+12+6.5+19.5+18+23.5=274$$

Rank sum for negative signs:

$$4+2+1+6.5+4+11+8.5+14=51$$

Using the above example:

$$\text{Mean/expected} = \frac{1}{4} \times 25 \times (25+1) = 162.5.$$

$$\text{Variance} = \frac{1}{24} \times 25 \times (25+1)(50+1) = 1381.25.$$

Adjusted variance:

$$1381.25 - \frac{(3^3-3)+(3^3-3)+(2^3-2)+(2^3-2)+(2^3-2)+(2^3-2)}{48} = 1379.75.$$

REMEMBER OUR WILCOXON SIGNED-RANK TEST STATISTIC $= 274$.
Since our test is two-tailed:

$$Z \text{ statistic} = \frac{274-162.5}{\sqrt{1379.75}} \sim 3.0018$$

In this situation the p-value is equal to two times the upper-tailed p-value.

$$2P(Z > 3.0018) \sim 2(1 - P(Z < 3.0018) \sim 2(1 - 0.9987) \sim 0.0026$$

Since the p-value $= 0.0026 \; < 0.05 = \alpha$, we reject the null hypothesis and conclude that the average gold concentration in the ore deposit is not equal to 10.

Let's now work in MATLAB.

Similar to Wilcoxon rank-sum, the Wilcoxon sign-rank sum test can be performed in MATLAB with the inbuilt function:

```
[p, h, stats] = signrank(x, m, 'alpha', ,'method', ' ', 'tail', ' ')
```

Here:

x = the single dataset we are comparing.
M = hypothesized value of the median.

The rest of the arguments are the same as the Wilcoxon rank-sum test.

Let's perform the example in MATLAB using the function and compare it with our results.

```
[p,h,stats] = signrank(BatteryLife,4, 'tail', 'right', 'method', 'exact')

p =

   0.3438


h =

  logical

   0


stats =

  struct with fields:

    signedrank: 9.5000
```

```
[p,h,stats] = signrank(porosity, 0.25,'tail','right', 'alpha', 0.01,
                        'method','exact')
p =

    0.7812


h =

  logical

    0


stats =

  struct with fields:

    signedrank: 5
```
```
[p,h,stats] = signrank(Lifespan, 10000, 'alpha', 0.10, 'tail','left',
                        'method','exact')
p =

    0.4023


h =

  logical

    0


stats =

  struct with fields:

    signedrank: 15.5000
```
```
[p,h,stats] = signrank(Efficiency, 80, 'tail', 'both', 'method', 'exact')
p =

    0.6406


h =

  logical

    0


stats =

  struct with fields:

    signedrank: 10.5000
```
```
[p,h,stats] = signrank(Depth, 8000,'tail','both', 'alpha', 0.01
                        'method','exact')
```

```
p =

    0.3125

h =

  logical

   0

stats =

  struct with fields:

    signedrank: 16

[p,h,stats] = signrank(Weight, 50,'tail','right','method', 'approximate')
p =

    0.0631

h =

  logical

   0

stats =

  struct with fields:

        zval: 1.5290
    signedrank: 87

[p,h,stats] = signrank(Thickness, 5,'alpha',0.01,'method','approximate')
p =

    0.1485

h =

  logical

   0

stats =

  struct with fields:

        zval: -1.4447
    signedrank: 52.5000

[p,h,stats] = signrank(Time, 10,'tail','left','alpha',0.10, 'method',
                       'approximate');
```

```
p =

    0.7333


h =

  logical

   0


stats =

  struct with fields:

        zval: 0.6227
  signedrank: 39.5000

[p,h,stats] = signrank(Conc, 10, 'alpha', 0.01, 'method','approximate')
p =

    0.0027


h =

  logical

   1


stats =

  struct with fields:

        zval: 3.0018
  signedrank: 274
```

9.2.2 Two sample

Like in the paired t-test, here too we find the differences between the pairs of data, but unlike the parametric test, this test is used when these differences are not normally distributed.

The **Wilcoxon matched-pairs signed-rank test** computes the difference between each set of matched pairs, then follows the same procedure as the signed-rank test to compare the sample against some median.

9.2.3 Example

Let's say a team of engineers is working on improving the fuel efficiency of a particular engine. They run a series of tests on the engine before and after making some changes to the fuel injection system. The results of these tests are given in the following table:

Fuel efficiency (before)	Fuel efficiency (after)
10.3	12.2
11.7	11.2
10.3	11.4
12.5	11.9
11.8	12.4
11.2	11.9

H_0: there is no significant difference in the fuel efficiency of the engine.
H_A: there is a significant difference in the fuel efficiency of the engines after some changes to the fuel injection system.

Fuel efficiency (before)	Fuel Efficiency (after)	Difference (after − before)	Absolute difference	Ranks	Signed ranks
10.3	12.2	+1.9	1.9	6	+6
11.7	11.2	−0.5	0.5	1	−1
10.3	11.4	+1.1	1.1	5	+5
12.5	11.9	−0.6	0.6	2.5	−2.5
11.8	12.4	+0.6	0.6	2.5	+2.5
11.2	11.9	+0.7	0.7	4	+4

Theoretically we can find the difference either way as long as it is specified in the input argument in MATLAB.

Rank sum for positive signs $= 6 + 5 + 2.5 + 4 = 17.5$.
Rank sum for negative signs $= 1 + 2.5 = 3.5$.

Since we always use the rank sum for positive signs as Wilcoxon signed-rank test statistics , the Wilcoxon signed-rank test statistic $= 17.50$.

9.2.4 Finding the p-value

9.2.4.1 If we want to use the exact method

Just like in the one-sample t test here too the p-value is the number of possible permutations that provide a sum of positive ranks as extreme as or more extreme than the observed one, divided by 2^n.
Using the previous example:

r_1	r_2	r_3	r_4	r_5	r_6	sum of +ranks
−1	−2.5	−2.5	−4	−5	−6	0
−1	−2.5	−2.5	−4	5	−6	5
−1	−2.5	−2.5	−4	−5	6	6
−1	−2.5	−2.5	−4	5	6	11
−1	−2.5	−2.5	4	−5	−6	4
−1	−2.5	−2.5	4	5	−6	9
−1	−2.5	−2.5	4	−5	6	10
−1	−2.5	−2.5	4	5	6	15
−1	−2.5	2.5	−4	−5	−6	2.5
−1	−2.5	2.5	−4	5	−6	7.5
−1	−2.5	2.5	−4	−5	6	8.5
−1	−2.5	2.5	−4	5	6	13.5
−1	−2.5	2.5	4	−5	−6	6.5
−1	−2.5	2.5	4	5	−6	11.5
−1	−2.5	2.5	4	−5	6	12.5
−1	−2.5	2.5	4	5	6	17.5

r_1	r_2	r_3	r_4	r_5	r_6	sum of+ranks
−1	2.5	−2.5	−4	−5	−6	2.5
−1	2.5	−2.5	−4	5	−6	7.5
−1	2.5	−2.5	−4	−5	6	8.5
−1	2.5	−2.5	−4	5	6	13.5
−1	2.5	−2.5	4	−5	−6	6.5
−1	2.5	−2.5	4	5	−6	11.5
−1	2.5	−2.5	4	−5	6	12.5
−1	2.5	−2.5	4	5	6	17.5
−1	2.5	2.5	−4	−5	−6	5
−1	2.5	2.5	−4	5	−6	10
−1	2.5	2.5	−4	−5	6	11
−1	2.5	2.5	−4	5	6	16
−1	2.5	2.5	4	−5	−6	9
−1	2.5	2.5	4	5	−6	14
−1	2.5	2.5	4	−5	6	15
−1	2.5	2.5	4	5	6	20
1	−2.5	−2.5	−4	−5	−6	1
1	−2.5	−2.5	−4	5	−6	6
1	−2.5	−2.5	−4	−5	6	7
1	−2.5	−2.5	−4	5	6	12
1	−2.5	−2.5	4	−5	−6	5
1	−2.5	−2.5	4	5	−6	10
1	−2.5	−2.5	4	−5	6	11
1	−2.5	−2.5	4	5	6	16
1	−2.5	2.5	−4	−5	−6	3.5
1	−2.5	2.5	−4	5	−6	8.5
1	−2.5	2.5	−4	−5	6	9.5
1	−2.5	2.5	−4	5	6	14.5
1	−2.5	2.5	4	−5	−6	7.5
1	−2.5	2.5	4	5	−6	12.5
1	−2.5	2.5	4	−5	6	13.5
1	−2.5	2.5	4	5	6	18.5
1	2.5	−2.5	−4	−5	−6	3.5
1	2.5	−2.5	−4	5	−6	8.5
1	2.5	−2.5	−4	−5	6	9.5
1	2.5	−2.5	−4	5	6	14.5
1	2.5	−2.5	4	−5	−6	7.5
1	2.5	−2.5	4	5	−6	12.5
1	2.5	−2.5	4	−5	6	13.5
1	2.5	−2.5	4	5	6	18.5
1	2.5	2.5	−4	−5	−6	6
1	2.5	2.5	−4	5	−6	11
1	2.5	2.5	−4	−5	6	12
1	2.5	2.5	−4	5	6	17
1	2.5	2.5	4	−5	−6	10

r_1	r_2	r_3	r_4	r_5	r_6	sum of +ranks
1	2.5	2.5	4	5	−6	15
1	2.5	2.5	4	−5	6	16
1	2.5	2.5	4	5	6	21

Total number of permutations = $2^6 = 64$.

Number of arrangements having statistics $\geq 17.5 = 6$.

Out of 64 combinations 6 arrangements have statistics ≥ 17.5.

Therefore our p-value $\dfrac{6}{64} = 0.09375$.

Since it is two-tailed, the p-value $2 \times 0.09375 = 0.1875$.

9.2.4.2 Example

A naval engineering team has developed a new propulsion system for naval vessels and wants to test its performance against the existing propulsion system. They collect data from a sample of six naval vessels, measuring the speed (in knots) of each vessel using both the new and existing propulsion systems. The data is collected over a period of one month, with each vessel using both propulsion systems on different days in random order. The team wants to determine if there is a statistically significant difference in the average speed between the two propulsion systems.

Speed (before)	Speed (after)
23	25
27	26
23	24
26	23
25	24
27	28

H_0: there is no significant difference in the speed of the vessels.
H_A: there is a significant difference in the speed of the vessels.

Fuel efficiency (before)	Fuel efficiency (after)	Difference (after − before)	Absolute difference	Ranks	Signed ranks
23	25	+2	2	4	+4
27	26	−1	1	2	−2
28	24	−4	4	6	−6
26	23	−3	3	5	−5
25	24	−1	1	2	−2
27	28	+1	1	2	+2

Theoretically we can find the difference either way as long as it is specified in the input argument in MATLAB.

Rank sum for positive signs $= 4 + 2 = 6$.
Rank sum for negative signs $= 2 + 6 + 5 + 2 = 15$.

Since we always use the rank sum for positive signs as the Wilcoxon signed-rank test statistic, the Wilcoxon signed-rank test statistic = 6.

r_1	r_2	r_3	r_4	r_5	r_6	sum of+ranks
-2	-2	-2	-4	-5	-6	0
-2	-2	-2	-4	5	-6	5
-2	-2	-2	-4	-5	6	6
-2	-2	-2	-4	5	6	11
-2	-2	-2	4	-5	-6	4
-2	-2	-2	4	5	-6	9
-2	-2	-2	4	-5	6	10
-2	-2	-2	4	5	6	15
-2	-2	2	-4	-5	-6	2
-2	-2	2	-4	5	-6	7
-2	-2	2	-4	-5	6	8
-2	-2	2	-4	5	6	13
-2	-2	2	4	-5	-6	6
-2	-2	2	4	5	-6	11
-2	-2	2	4	-5	6	12
-2	-2	2	4	5	6	17
-2	2	-2	-4	-5	-6	2
-2	2	-2	-4	5	-6	7
-2	2	-2	-4	-5	6	8
-2	2	-2	-4	5	6	13
-2	2	-2	4	-5	-6	6
-2	2	-2	4	5	-6	11
-2	2	-2	4	-5	6	12
-2	2	-2	4	5	6	17
-2	2	2	-4	-5	-6	4
-2	2	2	-4	5	-6	9
-2	2	2	-4	-5	6	10
-2	2	2	-4	5	6	15
-2	2	2	4	-5	-6	8
-2	2	2	4	5	-6	13
-2	2	2	4	-5	6	14
-2	2	2	4	5	6	19
2	-2	-2	-4	-5	-6	2
2	-2	-2	-4	5	-6	7
2	-2	-2	-4	-5	6	8
2	-2	-2	-4	5	6	13
2	-2	-2	4	-5	-6	6
2	-2	-2	4	5	-6	11
2	-2	-2	4	-5	6	12
2	-2	-2	4	5	6	17
2	-2	2	-4	-5	-6	4
2	-2	2	-4	5	-6	9

r_1	r_2	r_3	r_4	r_5	r_6	sum of +ranks
2	−2	2	−4	−5	6	10
2	−2	2	−4	5	6	15
2	−2	2	4	−5	−6	8
2	−2	2	4	5	−6	13
2	−2	2	4	−5	6	14
2	−2	2	4	5	6	19
2	2	−2	−4	−5	−6	4
2	2	−2	−4	5	−6	9
2	2	−2	−4	−5	6	10
2	2	−2	−4	5	6	15
2	2	−2	4	−5	−6	8
2	2	−2	4	5	−6	13
2	2	−2	4	−5	6	14
2	2	−2	4	5	6	19
2	2	2	−4	−5	−6	6
2	2	2	−4	5	−6	11
2	2	2	−4	−5	6	12
2	2	2	−4	5	6	17
2	2	2	4	−5	−6	10
2	2	2	4	5	−6	15
2	2	2	4	−5	6	16
2	2	2	4	5	6	21

Total number of permutations = $2^6 = 64$.

Number of arrangements with statistics $\geq 15 = 14$.

Out of 64 combinations 14 arrangements have statistics ≥ 15.

Therefore our p-value $\dfrac{14}{64} = 0.21875$.

Since it is two-tailed, the p-value $2 \times 0.21875 = 0.4375$.

If we want to use the normal approximation

9.2.4.3 Example

A civil engineering firm wants to assess the effectiveness of a new road construction technique that claims to reduce road cracking. They collect data on road crack measurements from a sample of 20 road segments before and after implementing the new technique. The crack measurements are recorded in millimeters (mm) and represent the crack width. The data collected from the 20 road segments is as follows:

Before	10	8	12	9	11	14	7	11	9	14	13	8	14	9	13	15	8
After	8	7	9	5	12	8	5	13	13	11	6	10	9	11	9	10	13

H_0: there is no significant difference in the width of the crack.

H_A: there is a significant difference in the width of the crack before and after implementing the new technique.

Before	After	Difference (before − after)	Absolute difference	Ranks	Signed ranks
10	8	2	2	5	+
8	7	1	1	1.5	+
12	9	3	3	8.5	+
9	5	4	4	11	+
11	12	−1	1	1.5	-
14	8	6	6	16	+
7	5	2	2	5	+
11	13	−2	2	5	-
9	13	−4	4	11	-
14	11	3	3	8.5	+
13	6	7	7	17	+
8	10	−2	2	5	-
14	9	5	5	14	+
9	11	−2	2	5	-
13	9	4	4	11	+
15	10	5	5	14	+
8	13	−5	5	14	-

Theoretically we can find the difference either way as long as it is specified in the input argument in MATLAB.

Rank sum for positive signs = $5 + 1.5 + 8.5 + 11 + 16 + 5 + 8.5 + 17 + 14 + 11 + 14 = 111.5$.

Rank sum

Since we always use the rank sum for positive signs as the Wilcoxon signed-rank test statistic , the Wilcoxon signed-rank test statistic $= 111.5$.

- The mean/expected value $= \frac{1}{4} \times 17 \times (17 + 1) = 76.5$.

- Calculate the variance using $\frac{1}{24} \times 17 \times (17 + 1)(34 + 1) = 446.25$.

- Adjusted variance $= 446.25 - \dfrac{(2^3 - 2) + (5^3 - 5) + (2^3 - 2) + (3^3 - 3) + (3^3 - 3)}{48} = 442.5$.

REMEMBER OUR WILCOXON SIGNED-RANK TEST STATISTIC $= 111.5$.

$$Z \text{ statistic} = \frac{111.5 - 76.5}{\sqrt{442.5}} = 1.6638$$

Since our test statistics is positive and the test is two-tailed, our p-value equals two times the upper-tailed p -value.

$$2P(Z > 1.6638) \sim 2(1 - P(Z < 1.6638) \sim 2(1 - 0.9519) \sim 0.0962$$

Since the p- value $= 0.0962 < 0.10 = \alpha$, we reject the null hypothesis and conclude that there is a significant difference in the width of the crack before and after implementing the new technique.

9.2.4.4 Example

Suppose you are working in a manufacturing plant that produces bolts. You came up with a new manufacturing technique which is supposed to increase the length of the bolt. To assess the capability of the process, you collect a sample of 30 bolts before and after implementing the new technique and measure their lengths. You record the data in a spreadsheet given as follows:

Before	12.1	10.8	11.4	10.9	11.1	11.9	12.0	12.3	10.9	10.2
	10.4	11.3	12.2	11.6	10.5	11.2	11.7	12.2	11.8	12.3
	11.2	10.4	12.3	11.9	10.8					
After	11.6	12.2	11.7	10.7	12.3	11.6	12.4	12.1	11.4	10.9
	10.8	11.4	12.0	11.5	10.7	11.5	12.3	13.1	11.7	12.4
	11.8	11.7	12.5	11.8	11.6					

H_0: there is no significant difference in the length of the bolts even after the technique is applied.
H_A: the length of the bolts increases after implementing the new technique.

Before	After	Difference (after – before)	Absolute difference	Ranks	Signed ranks
12.1	11.6	−0.5	0.5	16.5	−16.5
10.8	12.2	1.4	1.4	25	25
11.4	11.7	0.3	0.3	12	12
10.9	10.7	−0.2	0.2	8	−8
11.1	12.3	1.2	1.2	23	23
11.9	11.6	−0.3	0.3	12	−12
12.0	12.4	0.4	0.4	14.5	14.5
12.3	12.1	−0.2	0.2	8	−8
10.9	11.4	0.5	0.5	16.5	16.5
10.2	10.9	0.7	0.7	20	20
10.4	10.8	0.4	0.4	14.5	14.5
11.3	11.4	0.1	0.1	3	3
12.2	12.0	−0.2	0.2	8	−8
11.6	11.5	−0.1	0.1	3	−3
10.5	10.7	0.2	0.2	8	8
11.2	11.5	0.3	0.3	12	12
11.7	12.3	0.6	0.6	18.5	18.5
12.2	13.1	0.9	0.9	22	22
11.8	11.7	−0.1	0.1	3	−3
12.3	12.4	0.1	0.1	3	3
11.2	11.8	0.6	0.6	18.5	18.5
10.4	11.7	1.3	1.3	24	24
12.3	12.5	0.2	0.2	8	8
11.9	11.8	−0.1	0.1	3	−3
10.8	11.6	0.8	0.8	21	21

Rank sum for positive signs $= 5 + 1.5 + 8.5 + 11 + 16 + 5 + 8.5 + 17 + 14 + 11 + 14 = 263.5$.
Rank sum for negative signs $= 1.5 + 5 + 11 + 5 + 5 + 14 = 61.5$.

- The mean/expected value $= \frac{1}{4} \times 25 \times (25+1) = 162.5$.

- Calculate the variance using $\frac{1}{24} \times 25 \times (25+1)(50+1) = 1381.25$.

- Adjusted variance =
$$1381.25 - \frac{(2^3-2)+(3^3-3)+(5^3-5)+(2^3-2)+(5^3-5)+(2^3-2)}{48} = 1375.375.$$

REMEMBER OUR WILCOXON SIGNED-RANK TEST STATISTIC = 263.5.
Since our test was right-tailed:

$$Z \, \text{statistic} = \frac{(263.50 - 0.50) - 162.50}{\sqrt{1375.375}} = 2.7099$$

In this situation the p-value is equal to the upper-tailed p-value.

$$P(Z > 2.7099) = 1 - P(Z < 2.7099) = 1 - 0.9966 = 0.0034$$

Since the p-value $= 0.0034 < 0.01 = \alpha$, we reject the null hypothesis and conclude that the length of the bolts increases after applying the new technique.

9.2.4.5 Example

Suppose a company wants to assess the effectiveness of a new manufacturing process that is expected to improve the efficiency of their production line. The company collects data on the time taken to complete a specific task before and after implementing the new process for a sample of 30 production line workers. The data is as follows:

Before	After	Before	After	Before	After
25	24	26	23	32	30
26	26	24	25	29	27
28	27	29	26	25	22
26	28	22	25	28	28
30	31	25	22	30	30
29	31	28	30	33	30
28	25	26	24	28	28
27	24	30	27	21	20
25	25	24	26	26	25
28	24	30	29	30	31

H_0: there is no significant improvement in the efficiency of their production line.
H_A: the time taken to complete a specific task decreases after implementing the new process.

Before	After	Difference (before − after)	Ranks	Before	After	Difference (before − after)	Ranks
25	24	−1	3	40	30	10	26
26	26	0	N/A	26	24	2	8.5
28	27	1	3	30	27	3	15
26	28	−2	8.5	24	26	−2	8.5

Before	After	Difference (before − after)	Ranks	Before	After	Difference (before − after)	Ranks
30	31	1	3	30	24	6	21.5
29	31	−2	8.5	32	30	2	8.5
28	25	3	15	29	21	8	25
27	22	5	20	25	22	3	15
25	25	0	N/A	28	30	−2	8.5
28	24	4	19	30	30	0	N/A
26	23	3	15	33	30	3	15
24	25	−1	3	28	28	0	N/A
33	26	7	23.5	21	20	1	3
22	25	−3	15	26	20	6	21.5
25	22	3	15	30	23	7	23.5

Rank sum for positive signs = 296 .
Rank sum for negative signs = 55.

- The mean/expected value $= \dfrac{1}{4} \times 26 \times (26+1) = 175.5$.

- Calculate the variance using $\dfrac{1}{24} \times 26 \times (26+1)(52+1) = 1550.25$.

- Adjusted variance $= 1550.25 - \dfrac{(5^3 - 5) + (6^3 - 6) + (7^3 - 7) + (2^3 - 2) + (2^3 - 2)}{48} = 1536.125$.

9.2.4.6 If our test was left-tailed

$$Z \text{ statistic} = \frac{(296 + 0.50) - 175.5}{\sqrt{1536.125}} = 3.087252$$

In this situation the p -value is equal to the lower-tailed p -value.

$$P(Z < 3.087252) = 0.9990$$

Since the p- value = 0.9990 > 0.10 = α, we fail to reject the null hypothesis and conclude that there is no significant improvement in the efficiency of their production line.

Let's now work on MATLAB.

Similar to Wilcoxon rank sum, the Wilcoxon signed-rank sum test can be performed in MATLAB with the inbuilt function:

```
[p,h,stats] = signrank(x, y, 'alpha',  ,'method', ' ', 'tail', ' ')
```

The input arguments are the same as the Wilcoxon rank-sum test.

Let's perform the examples in MATLAB using the function and compare with our results.

9.2.4.7 Example 1

```
[p,h,stats] = signrank(After, Before, 'tail', 'both', 'alpha', 0.01,
                  'method','exact')
```

```
p =

    0.1875

h =

  logical

   0

stats =

  struct with fields:

    signedrank: 17.5000
```

Example 2

```
[p,h,stats] = signrank(After, Before, 'tail', 'both', 'method','exact')
p =

    0.4375

h =

  logical

   0

stats =

  struct with fields:

    signedrank: 6

[p,h,stats] = signrank(Before, After, 'tail', 'both', 'alpha', 0.10,
'method','approximate')
p =

    0.0961

h =

  logical

   1

stats =

  struct with fields:

        zval: 1.6638
    signedrank: 111.5000

[p,h,stats] = signrank(After, Before, 'tail', 'right', 'alpha', 0.01,
                       'method','approximate')
```

```
p =

    0.0034


h =

  logical

  1


stats =

  struct with fields:

        zval: 2.7099
  signedrank: 263.5000

[p,h,stats] = signrank(Before, After, 'tail', 'left', 'alpha', 0.10,
                 method','approximate')
p =

    0.9990


h =

  logical

  0


stats =

  struct with fields:

        zval: 3.0873
  signedrank: 296
```

9.3 SIGN TEST

The next non-parametric test we shall discuss is the sign test. Now the sign test and the Wilcoxon signed-ranked test look similar and both are used for one sample and two samples, but the Wilcoxon signed-rank test is more powerful than the sign test.

The **sign test** is a non-parametric test that is used to test whether or not two groups are equally sized. The sign test is used when dependent samples are ordered in pairs, where the bivariate random variables are mutually independent. It is based on the direction of the plus and minus signs of the observation, and not on their numerical magnitude. It is also called the binominal sign test, with $p = .5$. The sign test is considered a weaker test, because it tests the pair value below or above the median and it does not measure the pair difference. The sign test is available in SPSS: click "menu", select "analysis", then click on "nonparametric", and choose "two related sample" and "sign test".

9.3.1 One sample

We set up the hypothesis so that + and − signs are the values of random variables with equal size.

9.3.1.1 Stating the hypothesis

Similar to the one-sample t test, the sign test for a population median can be a one-tailed (right- or left-tailed) or two-tailed distribution based on the hypothesis.

9.3.1.2 Calculating the test statistics

9.3.2 Example 1

An electrical engineer is working on a project to develop a new type of battery for electric vehicles. One of the key performance metrics for the battery is its energy density, measured in watt-hours per kilogram (Wh/kg). He has developed a prototype of the battery and want to determine if its energy density is significantly different from the industry standard for electric vehicle batteries, which is 150 Wh/kg. He measures the energy density of 10 prototype batteries and gets the following values: 155,148,150,152,151,147,153,144,149,154.

H_0: there is no significant difference in the energy density.
H_A: the energy density of the prototype is significantly different from the industry standard for electric vehicle batteries.

Battery	Energy density	Hypothesized	Difference	Signs
1	155	150	155 – 150 = 5	+
2	148	150	148 – 150 = –2	-
3	150	150	150 – 150 = 0	N/A
4	152	150	152 –150 = 2	+
5	151	150	151 – 150 = 1	+
6	147	150	147 – 150 = –3	-
7	153	150	153 – 150 = 3	+
8	144	150	144 – 150 = –6	-
9	149	150	149 – 150 = –1	-
10	154	150	154 – 150 = 4	+

Number of positive signs = 5.
Number of negative signs = 4.

Since we use the rank sum for positive signs as sign test statistics, the sign test statistic = 5 .

9.3.3 Example 2

Let's say a company manufactures ball bearings for industrial use, and the manufacturing process is supposed to produce bearings with a diameter of 2 centimeters. However, the company wants to know if the mean diameter of ball bearings produced by the process is any different than the target diameter of 2 centimeters. To test this hypothesis, the company takes a random sample of 12 ball bearings and measures their diameters.

The data is given below:

Ball bearing	Diameter	Ball bearing	Diameter
1	1.88	7	1.83
2	1.59	8	2.04
3	2.00	9	1.86
4	1.75	10	2.02
5	2.02	11	2.00
6	2.05	12	1.91

H_0: there is no significant difference in the diameter of the ball bearings.
H_A: the diameter of ball bearings produced by the process is significantly different than the target diameter of 2 centimeters.

Ball bearing	Diameter	Hypothesized	Difference	Signs
1	1.88	2.00	1.88 − 2.00 = −0.12	-
2	1.59	2.00	1.59 − 2.00 = −0.41	-
3	2.00	2.00	2.00 - 2.00 = 0	N/A
4	1.75	2.00	1.75 - 2.00 = −0.25	-
5	2.02	2.00	2.02 - 2.00 = 0.02	+
6	2.05	2.00	2.05 - 2.00 = 0.05	+
7	1.83	2.00	1.83 - 2.00 = −0.17	-
8	2.04	2.00	2.04 - 2.00 = 0.04	+
9	1.86	2.00	1.86 - 2.00 = −0.14	-
10	2.02	2.00	2.02 - 2.00 = 0.02	+
11	2.00	2.00	2.00 - 2.00 = 0	N/A
12	1.91	2.00	1.91 − 2.00 = −0.09	-

Number of positive signs $= 4$.
Number of negative signs $= 6$.
Since we use the rank sum for positive signs as sign test statistics, the
sign test statistic $= 4$.

9.3.4 Example 3

An electrical engineer is testing a new type of electric motor and wants to determine if this motor produces a higher amount of power compared to the industry standard motor of power output of 1000 watts. To do this, he takes a sample of 15 motors of the new type and measures their power output.

Electric motor	Power output (watts)
1	997
2	980
3	1030
4	1000
5	1072
6	988
7	1046
8	991
9	1000
10	1100
11	1053
12	985
13	1062
14	1000
15	1058

H_0: there is no significant difference in the power output.

H_A: the motor produces a higher amount of power compared to the industry standard motor of power output of 1000 watts.

Electric motor	Power output	Hypothesized	Difference	Signs
1	997	1000	997 − 1000 = −3	-
2	980	1000	980 − 1000 = −20	-
3	1030	1000	1030 − 1000 = 30	+
4	1000	1000	1000 − 1000 = 0	N/A
5	1012	1000	1012 − 1000 = 12	+
6	988	1000	988 − 1000 = −12	-
7	1036	1000	1036 − 1000 = 36	+
8	991	1000	991 − 1000 = −9	-
9	1000	1000	1000 − 1000 = 0	N/A
10	1100	1000	1100 − 1000 = 100	+
11	1028	1000	1028 − 1000 = 28	+
12	985	1000	985 − 1000 = −15	-
13	1023	1000	1023 − 1000 = 23	+
14	1000	1000	1000 − 1000 = 0	N/A
15	1018	1000	1018 − 1000 = 18	+

Number of positive signs $= 7$.

Number of negative signs $= 5$.

Since we use the rank sum for positive signs as sign test statistics , the sign test statistic $= 7$.

9.3.5 Example 4

A nuclear engineering research team is investigating the efficiency of a new type of nuclear reactor coolant. The team collects temperature data from the coolant system over a 30-day period and wants to determine if the average temperature of the coolant is significantly different from the expected value of 80°C. The data is given below:

Electric motor	Power output
1	78
2	80
3	83
4	85
5	76
6	78
7	80
8	79
9	82
10	80

Null hypothesis: the average temperature of the coolant is equal to 80°C. Alternative hypothesis: the average temperature of the coolant is $>80°C$.

Electric motor	Power output	Hypothesized	Difference	Signs
1	78	80	78 – 80 = –2	-
2	80	80	80 – 80 = 0	N/A
3	83	80	83 – 80 = 3	+
4	85	80	85 – 80 = 5	+
5	76	80	76 – 80 = –4	-
6	78	80	78 – 80 = –2	-
7	80	80	80 – 80 = 0	N/A
8	79	80	79 – 80 = –1	-
9	82	80	82 – 80 = 2	+
10	80	80	80 – 80 = 0	N/A

Number of positive signs = 3.
Number of negative signs = 4.
Since we use the rank sum for positive signs as sign , the sign test statistic = 3.

9.3.6 Example 5

A mechanical engineer is working on designing a new cooling system for a particular type of engine. However, the company he is working for believes that the new cooling system is not sufficient enough to have the capacity of 1200 watts. To test this hypothesis, the engineer takes a sample of 20 engines that have been fitted with the new cooling system and measures their cooling capacity.

Engine	Capacity	Engine	Capacity
1	1187	11	1184
2	1229	12	1216
3	1210	13	1162
4	1175	14	1202
5	1212	15	1231
6	1205	16	1198
7	1178	17	1207
8	1231	18	1166
9	1164	19	1219
10	1223	20	1223

Engine	Capacity	Hypothesized	Difference	Signs
1	1187	1200	1187 – 1200 = –13	-
2	1229	1200	1229 – 1200 = 29	+
3	1210	1200	1210 – 1200 = 10	+
4	1175	1200	1175 – 1200 = –25	-
5	1212	1200	1212 – 1200 = 12	+
6	1205	1200	1205 – 1200 = 5	+
7	1178	1200	1178 – 1200 = –22	-

Engine	Capacity	Hypothesized	Difference	Signs
8	1231	1200	1231 − 1200 = 31	+
9	1164	1200	1164 − 1200 = −36	-
10	1223	1200	1223 − 1200 = 23	+
11	1184	1200	1184 − 1200 = −16	-
12	1216	1200	1216 − 1200 = 16	+
13	1162	1200	1162 − 1200 = −38	-
14	1202	1200	1202 − 1200 = 2	+
15	1231	1200	1231 − 1200 = 31	+
16	1198	1200	1198 − 1200 = −2	-
17	1207	1200	1207 − 1200 = 7	+
18	1166	1200	1166 − 1200 = −34	-
19	1219	1200	1219 − 1200 = 19	+
20	1223	1200	1223 − 1200 = 23	+

Number of positive signs $= 12$.

Number of negative signs $= 8$.

Since we use the rank sum for positive signs as sign test statistics, the sign test statistic $= 12$.

9.3.7 Example 6

A group of marine engineers is conducting research to determine the average underwater welding time for a particular type of marine structure. The engineers have a hypothesis that the average welding time is 75 minutes. To test this hypothesis, they randomly sample 20 underwater welding tasks and record the welding times in minutes:

72, 58, 61, 63, 59, 57, 60, 61, 63, 59, 61, 60, 62, 59, 58, 60, 61, 62, 59, 61

The engineers want to determine if the average welding time is significantly different from 60 minutes at a significance level of 0.05 (or 5%).

Null hypothesis: the average welding time is equal to 75 minutes.

Alternative hypothesis: the average welding time is less than 75 minutes.

Task	Time	Hypothesized	Difference	Signs
1	78	75	78 − 75 = 3	+
2	69	75	69 − 75 = −6	-
3	75	75	75 − 75 = 0	N/A
4	67	75	67 − 75 = −8	-
5	72	75	72 − 75 = −3	-
6	70	75	70 − 75 = −5	-
7	80	75	80 − 75 = 5	+
8	68	75	68 − 75 = −7	-
9	72	75	72 − 75 = −3	-
10	77	75	77 − 75 = 2	+
11	84	75	84 − 75 = 9	+
12	66	75	66 − 75 = −9	-
13	80	75	80 − 75 = 5	+

Task	Time	Hypothesized	Difference	Signs
14	81	75	81 − 75 = 6	+
15	75	75	75 − 75 = 0	N/A
16	68	75	68 − 75 = −7	-
17	60	75	60 − 75 = −15	-
18	64	75	64 − 75 = −11	-
19	72	75	72 − 75 = −3	-
20	71	75	71 − 75 = 0 − 4	-

Number of positive signs $= 6$.
Number of negative signs $= 12$.
Since we use the rank sum for positive signs as signed , the sign test statistic $= 6$.

9.3.8 Computing the p-value

9.3.8.1 Exact method

The exact distribution is the Binomial(n,p) with:

n = number of positive differences + number of negative differences

p = 0.5

We can calculate the p-value by calculating the probability of observing the number of positive differences found, given that the null hypothesis is true.

9.3.8.1.1 If our test is right-tailed

We calculate the binomial probability of getting the number of positive signs or more.

9.3.8.1.2 If our test is left-tailed

We calculate the binomial probability of getting the number of positive signs or less.

9.3.8.1.3 If our test is two-tailed

- If the number of positive signs > the number of negative sign then we calculate the binomial probability of getting the number of positive signs or more.
- If the number of positive signs < the number of negative sign then we calculate the binomial probability of getting the number of positive signs or less.

9.3.8.1.4 In the first example

We want to determine whether there is a significant difference in the energy density. Since we have an observation of difference 0 we omit that observation such that:
Our sample size $n = 10 - 1 = 9$

$p = 0.5$

Number of positive signs $= 5 > 4 =$ number of negative signs

$$2P(X \geq 5) = 2[P(X = 5) + P(X = 6) + P(X = 7) + P(X = 8) + P(X = 9)]$$

$$= 2[0.24609 + 0.16406 + 0.07031 + 0.01758 + 0.00195] = 1$$

9.3.8.1.5 In the second example

We want to determine whether there is a significant difference in the diameter of the ball bearings.

Since we have two observations of difference 0 we omit those observations such that our sample size $n = 12 - 2 = 10$.

$p = 0.5$

Number of positive signs $= 4 < 6 =$ number of negative signs

$$2P(X \leq 4) = 2[P(X = 0) + P(X = 1) + P(X = 2) + P(X = 3) + P(X = 4)]$$

$$= 2[0.00098 + 0.00977 + 0.04395 + 0.11719 + 0.20508] = 0.75394$$

9.3.8.1.6 In the third example

In this example we test to check if the electric motors have higher output than the standard which makes it a right-tailed test.

Since we have three observations of difference 0 we omit those observations such that our sample size $n = 15 - 3 = 12$.

$p = 0.5$

Number of positive signs $= 7$

$$P(X \geq 7) = P(X = 7) + P(X = 8) + P(X = 9) + P(X = 10) + P(X = 11) + P(X = 12)$$

$$= 0.19336 + 0.12085 + 0.05371 + 0.01611 + 0.00293 + 0.00024 = 0.3872$$

9.3.8.1.7 In the fourth example

In this example we test to check if the average temperature of the new coolant is greater than 80°C which makes it a right-tailed test.

Since we have three observations of difference 0 we omit those observations such that our sample size $n = 10 - 3 = 7$.

$p = 0.5$

Number of positive signs $= 3$

$$P(X \geq 3) = P(X = 3) + P(X = 4) + P(X = 5) + P(X = 6) + P(X = 7) = 0.77344$$

9.3.8.1.8 In the fifth example

In this example we test to check if the new cooling system has lower capacity. This makes it a left-tailed test.

$n = 20$

$p = 0.5$

Number of positive signs $= 12$

$$P(X \leq 12) = P(X = 0) + P(X = 1) + P(X = 2) + P(X = 3) + P(X = 4) + P(X = 5)$$
$$+ P(X = 6) + P(X = 7) + P(X = 8) + P(X = 9) + P(X = 10) + P(X = 11) + P(X = 12)$$
$$= 0 + 0.00002 + 0.00018 + 0.00109 + 0.00462 + 0.01479 + 0.03696 + 0.07393$$
$$+ 0.12013 + 0.16018 + 0.1762 + 0.16018 + 0.12013 = 0.86841$$

9.3.8.1.9 In the sixth example

In this example we test to check if the average welding time is lower than 75 minutes which makes it a left-tailed test.

$n = 18$

$p = 0.5$

Number of positive signs $= 6$

$$P(X \leq 6) = P(X = 0) - P(X = 1) + P(X = 2) + P(X = 3) + P(X = 4) + P(X = 5) + P(X = 6)$$
$$= 0.11894$$

9.3.8.2 Approximate method

If n is large, W is approximately normally distributed under the null hypothesis, with:

$$\text{mean} = (n_p + n_n) p = (n_p + n_n) \times 0.5$$

$$\text{standard deviation} = \sqrt{(n_p + n_n) \times p \times (1 - p)} = \sqrt{(n_p + n_n) \times 0.5 \times (1 - 0.5)}$$

9.3.8.2.1 If our test is right-tailed

$$\text{Z statistic} = \frac{\text{Signed statistics} - \text{mean} - 0.50}{\text{standard deviation}}$$

In this situation the p-value is equal to the upper-tailed p-value.

9.3.8.2.2 If our test is left-tailed

$$\text{Z statistic} = \frac{\text{Signed statistics} - \text{mean} + 0.50}{\text{standard deviation}}$$

In this situation the p-value is equal to the lower-tailed p-value.

9.3.8.2.3 If our test is two-tailed

- If the number of positive signs > the number of negative sign then

$$\text{Z statistic} = \frac{\text{Signed statistics} - \text{mean} - 0.50}{\text{standard deviation}}$$

- If the number of positive signs < the number of negative sign then

$$\text{Z statistic} = \frac{\text{Signed statistics} - \text{mean} + 0.50}{\text{standard deviation}}$$

Depending on the sign of the z statistic the p-value is equal to two times the lower-/upper-tailed p-value.

9.3.8.2.4 In the first example

$$\text{mean} = (5+4) \times 0.50 = 4.50$$

$$\text{standard deviation} = \sqrt{(5+4) \times 0.50(1-0.50)} = \sqrt{1.125}$$

Since the number of positive signs > number of positive,

$$= \frac{5 - 4.50 - 0.50}{\sqrt{1.125}} = 0$$

In this situation the p-value is equal to two times the upper-tailed p-value.

$$2P(Z > 0) = 2(1 - 0.50) \sim 1$$

9.3.8.2.5 In the second example

$$\text{mean} = (4+6) \times 0.50 = 5$$

$$\text{standard deviation} = \sqrt{(4+6) \times 0.50(1-0.50)} = \sqrt{2.5}$$

Since the number of positive signs < number of negative signs,

$$Z \text{ statistic} = \frac{4-5+0.50}{\sqrt{2.5}} = -0.3162$$

In this situation the p-value is equal to two times the lower-tailed p-value.

$$2P(Z < -0.3162) = 2(0.3762) \sim 0.7520$$

9.3.8.2.6 In the third example

$$\text{mean} = (7+5) \times 0.5 = 6$$

$$\text{standard deviation} = \sqrt{(7+5) \times 0.5 \times (1-0.5)} = \sqrt{3}$$

$$Z \text{ statistics} = \frac{7-6-0.50}{\sqrt{3}} = 0.2887$$

In this situation the p-value is equal to the upper-tailed p-value.

$$P(Z \geq 0.2887) \sim 1 - P(Z \leq 0.2887) \sim 1 - 0.6140 \sim 0.3860$$

9.3.8.2.7 In the fourth example

$$\text{mean} = (3+4) \times 0.5 = 3.5$$

$$\text{standard deviation} = \sqrt{(3+4) \times 0.5 \times (1-0.5)} = \sqrt{1.75}$$

$$Z \text{ statistic} = \frac{3-3.5-0.50}{\sqrt{1.75}} = -0.756$$

In this situation the p-value is equal to the upper-tailed p-value.

$$P(Z \geq -0.756) \sim 1 - P(Z \leq -0.756) \sim 1 - 0.2250 \sim 0.775$$

9.3.8.2.8 In the fifth example

$$\text{mean} = (12+8) \times 0.5 = 10$$

$$\text{standard deviation} = \sqrt{(12+8) \times 0.5 \times (1-0.5)} = \sqrt{5}$$

$$Z \text{ statistic} = \frac{12-10+0.50}{\sqrt{5}} = 1.118$$

In this situation the p-value is equal to the lower-tailed p-value.

$$P(Z \leq 1.118) \sim 0.8680$$

9.3.8.2.9 In the sixth example

$$\text{mean} = (6+12) \times 0.5 = 9$$

$$\text{standard deviation} = \sqrt{(6+12) \times 0.5 \times (1-0.5)} = \sqrt{4.5}$$

$$\text{Z statistic} = \frac{6-9+0.50}{\sqrt{4.5}} = -1.1785$$

In this situation the p-value is equal to the lower-tailed p-value.

$$P(Z \leq -1.1785) \sim 0.1189$$

9.3.8.3 MATLAB

The inbuilt function of the sign test has the following syntax:

```
[p, h, stats] = signtest(x, hypothesizedValue,
                'alpha', ,'method', ' ', 'tail', ' ')
```

All the input arguments are just about the same as all the previously discussed tests.

9.3.8.3.1 First example

```
[p,h,stats] = signtest(EnergyDensity, 150,'method','exact')
p =

    1

h =

   logical

   0

stats =

   struct with fields:

     zval: NaN
     sign: 5
```

```
[p,h,stats] = signtest(Diameter, 2, 'alpha', 0.10, 'method','exact')
p =

    0.7539


h =

  logical

   0


stats =

  struct with fields:

    zval: NaN
    sign: 4

[p,h,stats] = signtest(PowerOutput, 1000,'tail','right', 'alpha', 0.10,
                       'method','exact')
p =

    0.3872


h =

  logical

   0


stats =

  struct with fields:

    zval: NaN
    sign: 7

[p,h,stats] = signtest(Temperature, 80,'tail','right','method','exact')
p =

    0.7734


h =

  logical

   0


stats =

  struct with fields:

    zval: NaN
    sign: 3
```

```
[p,h,stats] = signtest(Capacity, 1200,'tail','left', 'alpha', 0.10,
                        'method','exact')
p =

    0.8684


h =

  logical

   0


stats =

  struct with fields:

    zval: NaN
    sign: 12

[p,h,stats] = signtest(Time, 75,'tail','left','alpha',
0.01,'method','exact')
p =

    0.1189


h =

  logical

   0


stats =

  struct with fields:

    zval: NaN
    sign: 6

[p,h,stats] = signtest(EnergyDensity, 150,'method','approximate')
p =

    1


h =

  logical

   0


stats =

  struct with fields:

    zval: 0
    sign: 5
```

```
[p,h,stats] = signtest(Diameter, 2,'alpha', 0.10,  'method','
approximate')
p =

   0.7518


h =

  logical

   0


stats =

  struct with fields:

    zval: -0.3162
    sign: 4

[p,h,stats] = signtest(PowerOutput, 1000,'tail','right', 'alpha', 0.10,
                    'method','approximate')
p =

   0.3864


h =

  logical

   0


stats =

  struct with fields:

    zval: 0.2887
    sign: 7

[p,h,stats] = signtest(Temperature, 80,'tail','right','method',
'approximate')
p =

   0.7752


h =

  logical

   0


stats =

  struct with fields:

    zval: -0.7559
    sign: 3
```

```
[p,h,stats] = signtest(Capacity, 1200,'tail','left', 'alpha', 0.10,
                        'method','approximate')
p =

    0.8682


h =

  logical

   0


stats =

  struct with fields:

    zval: 1.1180
    sign: 12
[p,h,stats] = signtest(Time, 75,'tail','left','alpha', 0.01, 'method',
                       'approximate')
p =

    0.1193


h =

  logical

   0


stats =

  struct with fields:

    zval: -1.1785
    sign: 6
```

9.3.9 Paired sample

Like the two-sample **Wilcoxon signed-rank test,** this test is also an alternative to the paired t test. This test uses the + and – signs in paired sample tests or in before-after study. In this test, the null hypothesis is set up so that the signs of + and – are of equal size, or the population means are equal to the sample mean.

9.3.9.1 Stating the hypothesis

9.3.9.1.1 The null hypothesis

The median difference is zero which means that approximately half of the differences are positive and half are negative.

9.3.9.1.2 The alternate hypothesis

9.3.9.2 Calculating the test statistic

In theory, the test statistic for the sign test is the number of positive signs or the number of negative signs, whichever is smaller. However in MATLAB, our statistic is the number of positive signs in the sample.

9.3.9.2.1 Example

Let's say we have a group of engineers and we want to determine if their average annual salary has increased after a recent job training program. We collect data on the annual salaries of ten engineers before and after the training, as shown below:

Before training (in USD): 60 K, 62 K, 61 K, 59 K, 63 K, 61.5 K, 60.5 K, 62.5 K, 59.5 K, 61 K
After training (in USD): 62 K, 63 K, 62 K, 60.5 K, 64 K, 63.5 K, 62.5 K, 64.5 K, 61.5 K, 63 K

From the data, can we conclude at the 0.01 significance level that there is a significant difference in the engineers' salaries before and after the training program?

Training	Engineers									
	1	2	3	4	5	6	7	8	9	10
Before	60	62	61	59	63	61.5	60.5	62.5	59.5	61
After	62	63	62	60.5	64	63.5	62.5	64.5	61.5	63

Notice that when we compute the difference in incomes using before training – after training then we get all negative signs, no positive signs; therefore our statistic would be zero. However understand that if our first sample was the earnings after training then we would have got all positive signs and our statistic would be 10.

9.3.9.2.2 Example 1

A marine engineer is working on designing a new propeller for a ship. He wants to test whether a new design of propeller produces a significant difference compared with the old design. To do this, he conducts a series of tests where he measures the thrust generated by each propeller design on the same ship under identical conditions.

Propeller	Thrust (in kN)											
	1	2	3	4	5	6	7	8	9	10	11	12
New design	2.2	1.8	1.7	2.4	1.9	2.1	2.3	2.4	1.9	1.6	2.4	2.6
Old design	2.1	2.0	1.5	2.2	1.6	2.3	2.3	2.5	1.7	1.6	1.9	2.3

Null hypothesis: there is no significant difference in the thrust generated by both propeller designs on the same ship.

Alternative hypothesis: there is a significant difference in the thrust.

Computing the difference:

New design	Old design	Difference	Signs	New design	Old design	Difference	Signs
2.2	2.1	0.10	+	2.3	2.3	0	N/A
1.8	2.0	−0.20	-	2.4	2.5	−0.10	-
1.7	1.5	0.20	+	1.9	1.7	0.20	+
2.4	2.2	0.20	+	1.6	1.6	0	N/A
1.9	1.6	0.30	+	2.4	1.9	0.50	+
2.1	2.3	−0.20	-	2.6	2.3	0.30	+

Notice there are two observations (even) whose difference in thrust is 0.
Number of positive signs $= 7$.
Number of negative signs $= 3$.
Since we use the rank sum for positive signs as sign test statistics, the sign test statistic $= 7$.

9.3.9.2.3 Example 2

An aeronautical engineer has developed a new design for an airplane wing, and he want to determine whether there is a significant difference in the lift-to-drag ratios between the old and new wing designs. To do this, he collects data on the lift-to-drag ratios of ten airplane models with the old wing design and the new wing design, as shown in the table below:

Airplane model	Old wing design	New wing design
1	24.2	25.1
2	24.7	24.2
3	23.8	24.4
4	24.6	24.6
5	23.5	24.3
6	25.1	24.8
7	23.2	23.6
8	24.9	24.7

Null hypothesis: there is no significant difference in the lift-to-drag ratios between the old and new wing designs.
Alternative hypothesis: there is a significant difference in the lift-to-drag ratios.
Computing the difference:

Old wing design	New wing design	Difference	Signs
24.2	25.1	−0.9	\-
24.7	24.2	0.5	+
23.8	24.4	−0.6	-
24.6	24.6	0	N/A
23.5	24.3	−0.8	-
25.1	24.8	0.3	+
23.2	23.6	−0.4	-
24.9	24.7	0.2	+

Notice there is a single observation whose difference in the lift-to-drag ratio is 0. Therefore we can drop that; hence:
    Number of positive signs = 3
    Number of negative signs = 4
    Since we use the rank sum for positive signs as sign test statistics, the sign test statistic = 3.

9.3.9.2.4 Example 3

Suppose a group of marine engineers is conducting an experiment to test the effectiveness of a new coating material for reducing marine biofouling on ship hulls. They collect data on the average amount of biofouling (in square centimeters) on a set of ship hulls before and after applying the new coating. The data for ten randomly selected hulls is as follows:

    Before coating: 11, 10, 9, 12, 13, 14, 11, 10, 12, 10
    After coating: 9, 10, 11, 9, 8, 11, 7, 10, 6, 11

Null hypothesis: there is no significant difference in the marine biofouling on ship hulls. Alternative hypothesis: the average amount of biofouling reduces after coating.
Computing the difference:

Before	After	Difference	Signs	Before	After	Difference	Signs
11	9	11 − 9 = 2	+	14	11	14 − 11 = 3	+
10	10	10 − 10 = 0	N/A	11	7	11 − 7 = 4	+
9	11	9 − 11 = −2	-	10	10	10 − 10 = 0	N/A
12	9	12 − 9 = 3	+	12	6	12 − 6 = 4	+
13	8	13 − 8 = 5	+	10	11	10 − 11 = −1	-

Here too, two observations (even) have difference 0.
    Number of positive signs = 6
    Number of negative signs = 2
    Since we use the rank sum for positive signs as sign test statistics, the sign test statistic = 6.

9.3.9.2.5 Example 4

A pharm tablet company wants to compare the performance of two different manufacturing processes for its products. They collected data on the number of defects produced by each process, and want to determine whether manufacturer process 1 produces a larger number of defects for each unit. The company randomly selects a sample of 15 units and produces them using both processes, recording the number of defects for each unit for both processes. They obtain the following data:

Process 1	5	4	6	5	8	7	3	7	5	3	5	8	5	7	6
Process 2	6	4	3	5	7	5	4	4	7	3	6	5	5	8	6

Null hypothesis: there is no significant difference in the number of defects for each unit. Alternative hypothesis: manufacturer process 1 produces a larger number of defects for each unit.

Process 1	Process 2	Difference	Signs
5	6	−1	-
4	4	0	N/A
6	3	3	+
5	5	0	N/A
8	7	1	+
7	5	2	+
3	4	−1	-
7	4	3	+
5	7	−2	-
3	3	0	N/A
5	6	−1	-
8	5	3	+
5	7	−2	-
7	8	−1	-
6	6	0	N/A

Number of positive signs = 5
   Number of negative signs = 6
   Since we use the rank sum for positive signs as sign test statistics , the
   sign test statistic = 5.

9.3.9.2.6 Example 5

Suppose a chemical engineer is testing a new type of catalyst for a chemical reaction. They conduct the experiment twice, once with the new catalyst and once with the old catalyst, and measure the yield of the desired product in each trial. The results are as follows:

New catalyst: 85%, 87%, 86%, 88%, 84%, 89%, 85%, 90%, 88%, 87%,
Old catalyst: 84%, 87%, 86%, 85%, 85%, 89%, 86%, 86%, 87%, 84%

Test whether the new catalyst is significantly better than the old one.
   The null hypothesis: there is no significant difference in the yield of the desired product.
   The alternative hypothesis: the new catalyst is significantly better than the old one.
   Computing the difference:

Old catalyst	New catalyst	Difference	Signs	Old catalyst	New catalyst	Difference	Signs
0.84	0.85	−0.10	-	0.89	0.89	0	N/A
0.87	0.87	0	N/A	0.86	0.85	0.10	+
0.86	0.86	0	N/A	0.86	0.90	−0.40	-
0.85	0.88	−0.30	-	0.87	0.88	−0.10	-
0.85	0.84	0.10	+	0.84	0.87	−0.30	-

Here there are three observations (odd) whose difference in yield is 0. Therefore we randomly drop one and reduce the sample size by 1.

Number of positive signs = 2
Number of negative signs = 5
Since we use the rank sum for positive signs as sign test statistics , the
sign test statistic = 2.

9.3.9.2.7 Example

A civil engineering firm is evaluating the effectiveness of a new road construction technique that is purported to reduce potholes. They decide to conduct a field study to compare the number of potholes before and after implementing the new technique on a stretch of road. The firm selects a random sample of ten road segments and measures the number of potholes on each segment before and after the implementation of the new technique. The data is shown in the table below:

Road segment	Number of potholes before	Number of potholes after
1	20	15
2	18	12
3	21	14
4	22	13
5	19	11
6	16	18
7	21	11
8	15	17
9	19	15
10	20	10

The null hypothesis: there is no significant difference in the mean number of potholes before and after implementing the new technique.

The alternative hypothesis: the new road construction technique reduces potholes on each road segment.

Computing the difference:

Before	After	Difference	Signs	Before	After	Difference	Signs
20	15	5	+	16	18	−2	-
18	12	6	+	21	11	10	+
21	14	7	+	15	17	-	-
22	13	9	+	19	15	4	+
15	16	−1	-	20	10	10	+

Number of positive signs = 7
Number of negative signs = 3
Since we use the rank sum for positive signs as sign , the sign test statistic = 7.

9.3.9.3 Computing the p-value

9.3.9.3.1 Exact method

The exact distribution is the Binomial(n,p) with:

n = number of positive differences + number of negative differences.

p = 0.5.

We can calculate the p-value by calculating the probability of observing the number of positive differences found, given that the null hypothesis is true.

9.3.9.3.1.1 IN THE FIRST EXAMPLE

We want to determine whether there is a significant difference in the thrust between the old and new designs of propeller.

Since we have two observations of difference 0, we omit those observations such that our sample size $n = 12 - 2 = 10$

$p = 0.5$

Number of positive signs = 7 > 3 = number of negative signs

$$2P(X \geq 7) = 2\left[P(X = 7) + P(X = 8) + P(X = 9) + P(X = 10)\right]$$
$$= 2\left[0.11719 + 0.04395 + 0.00977 + 0.00098\right] \sim 0.3438$$

9.3.9.3.1.2 IN THE SECOND EXAMPLE

We want to determine whether there is a significant difference in the lift-to-drag ratios between the old and new wing designs.

Since we have an observation of difference 0 we omit that observation such that our sample size $n = 8 - 1 = 7$

$p = 0.5$

Number of positive signs = 3 < 4 = number of negative signs

$$2P(X \leq 3) = 2\left[P(X = 0) + P(X = 1) + P(X = 2) + P(X = 3)\right]$$
$$= 2\left[0.00781 + 0.05469 + 0.16406 + 0.27344\right] = 2 \times 0.50 = 1.00$$

9.3.9.3.1.3 IN THE THIRD EXAMPLE

We want to determine whether the average amount of biofouling reduces after coating.

Since we have two observations of difference 0 we omit those observations such that our sample size $n = 10 - 2 = 8$

$p = 0.5$

Number of positive signs = 6

$$P(X \geq 6) = P(X = 6) + P(X = 7) + P(X = 8) = 0.14454$$

9.3.9.3.1.4 IN THE FOURTH EXAMPLE

In this example we test to check if the manufacturer process 1 produces a larger number of defects for each unit.

$$P(X \geq 5) = P(X = 5) + P(X = 6) + P(X = 7) + P(X = 8) + P(X = 9) + P(X = 10) + P(X = 11)$$

$$= 0.72559$$

9.3.9.3.1.5 IN THE FIFTH EXAMPLE

In this example we test to check if the new catalyst is significantly better than the old one. This makes it a left-tailed test; hence:

$$P(X \leq 2) = P(X = 0) + P(X = 1) + P(X = 2) = 0.22656$$

9.3.9.3.1.6 IN THE SIXTH EXAMPLE

In this example we test to check if the new road construction technique reduces potholes on each road segment.

$$P(X \leq 7) = P(X = 0) + P(X = 1) + P(X = 2) + P(X = 3) + P(X = 4) + P(X = 5) + P(X = 6) + P(X = 7)$$

$$= 0.94531$$

9.3.9.3.2 Approximate method

If n is large, W is approximately normally distributed under the null hypothesis, with:

$$\text{mean} = (n_p + n_n) p = (n_p + n_n) \times 0.5$$

$$\text{standard deviation} = \sqrt{(n_p + n_n) \times p \times (1 - p)} = \sqrt{(n_p + n_n) \times 0.5 \times (1 - 0.5)}$$

9.3.9.3.2.1 IN THE FIRST EXAMPLE

$$\text{mean} = (7 + 3) \times 0.50 = 5$$

$$\text{standard deviation} = \sqrt{(7 + 3) \times 0.50(1 - 0.50)} = \sqrt{2.5}$$

Since the number of positive signs > the number of negative signs,

$$Z \text{ statistic} = \frac{\text{Signed statistics} - \text{mean} - 0.50}{\text{standard deviation}}$$

$$Z \text{ statistic} = \frac{7 - 5 - 0.50}{\sqrt{2.5}} = 0.9486833 \sim 0.9487$$

In this situation the p-value is equal to two times the upper-tailed p-value.

$$2P(Z > 0.9487) = 2\left[1 - P(Z < 0.9487)\right] = 2(1 - 0.8286) \sim 0.3428$$

9.3.9.3.2.2 IN THE SECOND EXAMPLE

$$\text{mean} = (3 + 4) \times 0.50 = 3.5$$

$$\text{standard deviation} = \sqrt{(3 + 4) \times 0.50(1 - 0.50)} = \sqrt{1.75}$$

Since the number of positive signs < the number of negative signs,

$$Z\ \text{statistic} = \frac{3 - 3.5 + 0.50}{\sqrt{1.75}} = 0$$

In this situation the p-value is equal to two times the upper-tailed p-value.

$$2P(Z > 0) = 2\left[1 - P(Z < 0)\right] = 2(1 - 0.5) \sim 1$$

9.3.9.3.2.3 IN THE THIRD EXAMPLE

$$\text{mean} = (6 + 2) \times 0.5 = 4$$

$$\text{standard deviation} = \sqrt{(6 + 2) \times 0.5 \times (1 - 0.5)} = \sqrt{2}$$

$$Z\ \text{statistic} = \frac{6 - 4 - 0.50}{\sqrt{2}} = 1.06066$$

In this situation the p-value is equal to two times the upper-tailed p-value.

$$P(Z \geq 1.06066) \sim 1 - P(Z \leq 1.06066) \sim 1 - 0.8554 \sim 0.1446$$

9.3.9.3.2.4 IN THE FOURTH EXAMPLE

$$\text{mean} = (5 + 6) \times 0.5 = 5.50$$

$$\text{standard deviation} = \sqrt{(5 + 6) \times 0.5 \times (1 - 0.5)} = \sqrt{2.75}$$

$$Z\ \text{statistic} = \frac{\text{Signed statistics} - \text{mean} - 0.50}{\text{standard deviation}}$$

$$Z\ \text{statistic} = \frac{5 - 5.5 - 0.50}{\sqrt{2.75}} = -0.6030$$

In this situation the p-value is equal to the upper-tailed p-value.

$$(Z \geq -0.6030) = 1 - P(Z \leq -0.6030) = 1 - 0.2726 \sim 0.7270$$

9.3.9.3.2.5 IN THE FIFTH EXAMPLE

$$\text{mean} = (2+5) \times 0.5 = 3.50$$

$$\text{standard deviation} = \sqrt{(2+5) \times 0.5 \times (1-0.5)} = \sqrt{1.75}$$

$$\text{Z statistic} = \frac{2-3.5+0.50}{\sqrt{1.75}} = -0.7559$$

In this situation the p -value is equal to the lower-tailed p -value.

$$(Z \le -0.756) \sim 0.2250$$

9.3.9.3.2.6 IN THE SIXTH EXAMPLE

$$\text{mean} = (7+3) \times 0.5 = 5$$

$$\text{standard deviation} = \sqrt{(7+3) \times 0.5 \times (1-0.5)} = \sqrt{2.5}$$

$$\text{Z statistic} = \frac{7-5+0.50}{\sqrt{2.5}} = 1.5811$$

In this situation the p -value is equal to the lower-tailed p -value.

$$(Z \le 1.5811) \sim 0.9430$$

9.3.9.3.3 MATLAB

For the paired sample the hypothesized value is replaced by the second group y in the input arguments.

```
[p, h, stats] = signtest(x, y, 'alpha', ,'method', ' ', 'tail', ' ')
```

9.3.9.3.4 Examples

```
[p,h,stats] = signtest(NewDesign,OldDesign,'alpha',0.01,'method','exact')
p =

    0.3438

h =

  logical

   0                              .

stats =

  struct with fields:

    zval: NaN
    sign: 7
```

```
[p,h,stats] = signtest(OldWing, NewWing,'method','exact')
p =

    1


h =

  logical

   0


stats =

  struct with fields:

    zval: NaN
    sign: 3

[p,h,stats] = signtest(Before, After,'tail','right','method','exact')
p =

    0.1445


h =

  logical

   0


stats =

  struct with fields:

    zval: NaN
    sign: 6

[p,h,stats] = signtest(Process_1, Process_2,'tail','right',
                       'method','exact')
p =

    0.7256


h =

  logical

   0


stats =

  struct with fields:

    zval: NaN
    sign: 5
```

```
[p,h,stats] = signtest(OldCatalyst,NewCatalyst,'tail','left',
                        'method','exact')
p =

    0.2266


h =

  logical

   0


stats =

  struct with fields:

    zval: NaN
    sign: 2

[p,h,stats] = signtest(Before, After, 'tail', 'left',
                        'method', 'exact')
p =

    0.9453


h =

  logical

   0


stats =

  struct with fields:

    zval: NaN
    sign: 7

[p,h,stats] = signtest(NewDesign, OldDesign, 'alpha', 0.01, 'method',
                        'approximate')
p =

    0.3428


h =

  logical

   0


stats =

  struct with fields:

    zval: 0.9487
    sign: 7
```

```
[p,h,stats] = signtest(OldWing, NewWing,'method',' approximate')
p =

    1


h =

  logical

  0


stats =

  struct with fields:

    zval: 0
    sign: 3

[p,h,stats] = signtest(Before, After, 'tail', 'right', 'method',
                       'approximate')
p =

   0.1444


h =

  logical

  0


stats =

  struct with fields:

    zval: 1.0607
    sign: 6

[p,h,stats] = signtest(Process_1, Process_2,'tail','right',
                       'method','approximate')
p =

   0.7268


h =

  logical

  0


stats =

  struct with fields:

    zval: -0.6030
    sign: 5
```

```
[p,h,stats] = signtest(OldCatalyst,NewCatalyst,'tail','left',
                       'method','approximate')
p =

   0.2248


h =

  logical

   0


stats =

  struct with fields:

    zval: -0.7559
    sign: 2

[p,h,stats] = signtest(Before, After, 'tail', 'left',
                       'method', 'approximate')
p =

   0.9431


h =

  logical

   0


stats =

  struct with fields:

    zval: 1.5811
    sign: 7
```

9.4 KRUSKAL–WALLIS TEST

The next non-parametric test we shall discuss in this chapter is the Kruskal–Wallis test.

It is considered the non-parametric alternative to the one-way ANOVA, and an extension of the Mann–Whitney U test to allow the comparison of more than two independent groups. As we have seen earlier, while one-way ANOVA uses real values, Kruskal–Wallis uses the ranks. Therefore, the Kruskal–Wallis test is also often called the "Kruskal–Wallis one-way ANOVA by ranks test".

9.4.1 Stating the hypotheses

9.4.1.1 Null hypothesis

Population medians are equal.

There is no significant difference in the mean.

9.4.1.2 Alternative hypothesis

Population medians are not all equal.
There is a significant difference in the mean.

9.4.2 Using either of the two approaches

9.4.2.1 Critical value approach

9.4.2.1.1 Find the statistics

If the data contains no ties the statistic is denoted by

$$H = \left[\frac{12}{N(N+1)} \times \pounds \frac{R_i^2}{n_i} \right] - 3(N+1)$$

Here,

N = total sample size
R_i = sum of ranks for group i
n_i = sample size for group i

Now in the presence of ties the statistic formula above needs to be divided by

$$1 - \frac{\sum_{i=1}^{k} (t_i^3 - t_i)}{N(N^2 + 1)}$$

9.4.2.1.1.1 EXAMPLE

Suppose a chemical engineer is comparing the efficiency of four different types of catalysts (CX, CY, CZ, CW) in a chemical reaction. The engineer performs the reaction using each of the four catalysts multiple times and records the reaction rates (in units of moles per hour) for each trial. The data is given below:

Catalyst CX: 5.2, 4.4, 4.7, 5.7, 4.8, 4.2
Catalyst CY: 6.3, 5.4, 4.1, 4.6, 5.0, 5.5, 4.3
Catalyst CZ: 4.5, 6.2, 5.1, 5.3, 6.1, 5.6, 4.9, 5.8

The engineer wants to determine if there is a significant difference in the reaction rates among the three catalysts.

Catalyst CX	Rank	Catalyst CY	Rank	Catalyst CZ	Rank
5.2	12	6.3	21	4.5	5
4.4	4	5.4	14	6.2	20
4.7	7	4.1	1	5.1	11
5.7	17	4.6	6	5.3	13
4.8	8	5.0	10	6.1	19
4.2	2	5.5	15	5.6	16
		4.3	3	4.9	9
				5.8	18

Calculate the sum of the ranks for each group.
Rank sum for catalyst CX $= 12 + 4 + 7 + 17 + 8 + 2 = 50$
Rank sum for catalyst CY $= 20 + 14 + 1 + 6 + 10 + 15 + 3 = 70$
Rank sum for catalyst CZ $= 5 + 20 + 11 + 13 + 19 + 16 + 9 + 18 = 111$
Therefore in this example

$$N = 21 \text{ (Steel Type A = 6 + Steel Type B = 7 + Steel Type C = 8)}$$
$$R_{i=3}^{i=0} = \text{sum of ranks for 3 group so } 50, 70, 111 \text{ for each group}$$
$$n_{i=3}^{i=0} = \text{sample size for 3 group so } 6, 7, \text{ and } 8$$

$$H = \left[\frac{12}{21(21+1)} \times \left[\frac{50^2}{6} + \frac{70^2}{7} + \frac{111^2}{8} \right] \right] - 3(21+1) \sim 3.0076$$

9.4.2.1.1.2 EXAMPLE

An engineer is studying the impact resistance of three different types of steel. He performs impact tests on ten samples of each type of steel and records the results in terms of the amount of deformation observed. The data is given below:

Steel type A	Steel type B	Steel type C	Steel type D
8	5	9	11
6	6	7	10
7	4	8	12
9	4	10	11
5	5	6	9
6	6	7	13
8	7	8	12
7	4	9	10
7	5	9	9
6	6	7	10

Test at $\alpha = 0.01$ to see if there is any difference in the amount of deformation of each type of steel.

the first step is to rank all of the data points, regardless of which material they belong to:

Steel type A	Rank	Steel type B	Rank	Steel type C	Rank	Steel type D	Rank
8	23.5	5	5.5	9	28.5	11	36.5
6	11	6	11	7	18	10	33.5
7	18	4	2	8	23.5	12	38.5
9	28.5	4	2	10	33.5	11	36.5
5	5.5	5	5.5	6	11	9	28.5

Steel type A	Rank	Steel type B	Rank	Steel type C	Rank	Steel type D	Rank
6	11	6	11	7	18	13	40
8	23.5	7	18	8	23.5	12	38.5
7	18	4	2	9	28.5	10	33.5
7	18	5	5.5	9	28.5	9	28.5
6	11	6	11	7	18	10	33.5

Calculate the sum of the ranks for each group.

Rank sum for steel type A $= 23.5 + 11 + 18 + 28.5 + 5.5 + 11 + 23.5 + 18 + 18 + 11 = 168$

Rank sum for steel type B $= 5.5 + 11 + 2 + 2 + 5.5 + 11 + 18 + 2 + 5.5 + 11 = 73.50$

Rank sum for steel type C $= 28.5 + 18 + \ldots + 28.5 + 18 = 231$

Rank sum for steel type D $= 36.5 + 33.5 + \ldots + 28.5 + 33.5 = 347.5$

Therefore in this example:

$$N = 30 \text{ (steel type A = 10 + steel type B = 10 + steel type C = 10)}$$

$$R_{i=3}^{i=0} = \text{sum of ranks for 3 group so } 167, 73.5, 231, 347.5 \text{ for each group}$$

$$n_{i=3}^{i=0} = \text{sample size for 3 group so 10 for each group}$$

$$H = \left[\frac{12}{40(40+1)} \times \left[\frac{168^2}{10} + \frac{73.5^2}{10} + \frac{231^2}{10} + \frac{347.5^2}{10} \right] \right] - 3(40+1) = 27.8332$$

However since we have multiple ties for ranks 23.5, 11, 18, 28.5, 5.5, 2, 33.5, 36.5, 38.5 occurring 4, 7, 7, 6, 4, 3, 4, 2, and 2 times respectively, we divide the statistic 27.83323 by

$$1 - \frac{(4^3 - 4) + (7^3 - 7) + (7^3 - 7) + (6^3 - 6) + (4^3 - 4) + (3^3 - 3) + (2^3 - 2) + (2^3 - 2)}{40(40^2 + 1)}$$

$$= 1 - \frac{4349}{4495} = 0.9838$$

So our test statistic is now $\dfrac{27.83323}{0.9838} \sim 28.29155$.

9.4.2.1.2 Find the critical value

Now we must determine whether the observed test statistic H supports the null or research hypothesis. Once again, this is done by establishing a critical value of H. However there are two situations we are going to be dealing with when trying to find the critical value.

Now in the first two examples notice that each group has a somewhat large number of replicates. To put this into more specific terms, if we have more than five replicates per group, then the sampling distribution of H approximates that of chi squared, so to find the critical value of chi squared, we use $df = \#group - 1$.

Therefore in the first example: $df = 3 - 1 = 2$.

df	$\chi^2_{.995}$	$\chi^2_{.990}$	$\chi^2_{.975}$	$\chi^2_{.950}$	$\chi^2_{.900}$	$\chi^2_{.100}$	$\chi^2_{.050}$	$\chi^2_{.025}$	$\chi^2_{.010}$	$\chi^2_{.005}$
1	0.000	0.000	0.001	0.004	0.016	2.706	3.841	5.024	6.635	7.879
2	0.010	0.020	0.051	0.103	0.211	4.605	5.991	7.378	9.210	10.597
3	0.072	0.115	0.216	0.352	0.584	6.251	7.815	9.348	11.345	12.838
4	0.207	0.297	0.484	0.711	1.064	7.779	9.488	11.143	13.277	14.860
5	0.412	0.554	0.831	1.145	1.610	9.236	11.070	12.833	15.086	16.750
6	0.676	0.872	1.237	1.635	2.204	10.645	12.592	14.449	16.812	18.548
7	0.989	1.239	1.690	2.167	2.833	12.017	14.067	16.013	18.475	20.278
8	1.344	1.646	2.180	2.733	3.490	13.362	15.507	17.535	20.090	21.955
9	1.735	2.088	2.700	3.325	4.168	14.684	16.919	19.023	21.666	23.589
10	2.156	2.558	3.247	3.940	4.865	15.987	18.307	20.483	23.209	25.188
11	2.603	3.053	3.816	4.575	5.578	17.275	19.675	21.920	24.725	26.757
12	3.074	3.571	4.404	5.226	6.304	18.549	21.026	23.337	26.217	28.300
13	3.565	4.107	5.009	5.892	7.042	19.812	22.362	24.736	27.688	29.819
14	4.075	4.660	5.629	6.571	7.790	21.064	23.685	26.119	29.141	31.319
15	4.601	5.229	6.262	7.261	8.547	22.307	24.996	27.488	30.578	32.801
16	5.142	5.812	6.908	7.962	9.312	23.542	26.296	28.845	32.000	34.267
17	5.697	6.408	7.564	8.672	10.085	24.769	27.587	30.191	33.409	35.718
18	6.265	7.015	8.231	9.390	10.865	25.989	28.869	31.526	34.805	37.156
19	6.844	7.633	8.907	10.117	11.651	27.204	30.144	32.852	36.191	38.582
20	7.434	8.260	9.591	10.851	12.443	28.412	31.410	34.170	37.566	39.997
21	8.034	8.897	10.283	11.591	13.240	29.615	32.671	35.479	38.932	41.401
22	8.643	9.542	10.982	12.338	14.041	30.813	33.924	36.781	40.289	42.796
23	9.260	10.196	11.689	13.091	14.848	32.007	35.172	38.076	41.638	44.181
24	9.886	10.856	12.401	13.848	15.659	33.196	36.415	39.364	42.980	45.559
25	10.520	11.524	13.120	14.611	16.473	34.382	37.652	40.646	44.314	46.928
26	11.160	12.198	13.844	15.379	17.292	35.563	38.885	41.923	45.642	48.290
27	11.808	12.879	14.573	16.151	18.114	36.741	40.113	43.195	46.963	49.645
28	12.461	13.565	15.308	16.928	18.939	37.916	41.337	44.461	48.278	50.993
29	13.121	14.256	16.047	17.708	19.768	39.087	42.557	45.722	49.588	52.336
30	13.787	14.953	16.791	18.493	20.599	40.256	43.773	46.979	50.892	53.672
40	20.707	22.164	24.433	26.509	29.051	51.805	55.758	59.342	63.691	66.766
50	27.991	29.707	32.357	34.764	37.689	63.167	67.505	71.420	76.154	79.490
60	35.534	37.485	40.482	43.188	46.459	74.397	79.082	83.298	88.379	91.952

Look along the row that corresponds to your number of degrees of freedom.

So in this case, we look along the row for 2 df at $\alpha = 0.05$ which gives 5.99. Therefore $\chi^2 =$ (df = 2, p = .05) = 5.99.

In the second example: df = 4−1 = 3   so if we look along the row that corresponds to 3 at $\alpha = 0.01$ we get 11.345. Therefore $\chi^2 =$ (df = 3, p = .01) = 11.345.

However, in a situation where we have five or fewer participants in each group, then using the chi squared would not be valid. Instead, we need to use a special table called the Kruskal-Wallis table.

9.4.2.1.2.1 EXAMPLE

Suppose a group of electrical engineers is tasked with designing a new circuit board for a particular application. There are three different design approaches that the engineers could take, and they want to determine if there is a significant difference in performance between the three approaches. To do this, they collect data on the performance of each design approach by measuring the time it takes for the circuit board to complete a given task. Here are the time measurements for each design approach:

- Design approach 1: 2.5 seconds, 2.1 seconds, 2.4 seconds
- Design approach 2: 3.1 seconds, 3.5 seconds, 3.2 seconds, 3.4 seconds
- Design approach 3: 2.8 seconds, 2.9 seconds, 3.0 seconds, 2.7 seconds, 2.6 seconds

Design approach 1	Rank	Design approach 2	Rank	Design approach 3	Rank
2.5	3	3.0	8	2.8	6
2.7	5	2.9	7	2.3	1
2.4	2	3.2	10	3.1	9
		3.4	12	3.3	11
				2.6	4

Calculate the sum of the ranks for each group.
  Rank sum for design approach 1 $= 3 + 5 + 2 = 10$
  Rank sum for design approach 2 $= 8 + 7 + 10 + 12 = 37$
  Rank sum for design approach 3 $= 6 + 1 + 9 + 11 + 4 = 31$
  Therefore in this example:

$$N = 12 \ (3 + 4 + 5)$$
$$R_{i=3}^{i=0} = \text{sum of ranks for 3 group so } 10, 37, 31 \text{ for each group}$$
$$n_{i=3}^{i=0} = \text{sample size for 3 group so } 3, 4, 5 \text{ for each group}$$

$$H = \left[\frac{12}{12(12+1)} \times \left[\frac{10^2}{3} + \frac{37^2}{4} + \frac{31^2}{5} \right] \right] - 3(12+1) = 4.67564$$

So if we pull up the Kruskal–Wallis table and look for the row corresponding to sample size $n_1 = 5, n_2 = 4, n_3 = $ at standard $\alpha = 0.05$:

Sample Sizes			$\alpha = 0.05$	$\alpha = 0.01$
2	2	2	-	-
3	2	1	-	-
3	2	2	4.714	-
3	3	1	5.143	-
3	3	2	5.361	-
3	3	3	5.600	7.200
4	2	1	-	-
4	2	2	5.333	-
4	3	1	5.208	-
4	3	2	5.444	6.444
4	3	3	5.791	6.745
4	4	1	4.967	6.667
4	4	2	5.455	7.036
4	4	3	5.598	7.144
4	4	4	5.692	7.654
5	2	1	5.000	-
5	2	2	5.160	6.533
5	3	1	4.960	-
5	3	2	5.251	6.909
5	3	3	5.648	7.079
5	4	1	4.985	6.955
5	4	2	5.273	7.205
5	4	3	5.656	7.445

Therefore our critical value $= 5.656$.

9.4.2.1.3 Decision

Finally we compare our test statistic to our critical value.

If the observed value of H is greater than or equal to the critical value, we reject H_0 in favor of H_A; if the observed value of H is less than the critical value we do not reject H_0.

9.4.2.1.3.1 FOR EXAMPLE 1

Since our test statistic = 3.0076 < 5.99 = critical value, we do not reject the null hypothesis and we can conclude   that there is no significant difference in the reaction rates among the three catalysts.

9.4.2.1.3.2 FOR EXAMPLE 2

Since our test statistic = 28.29155 > 11.345 = critical value, we reject the null hypothesis and we can conclude   that there is a significant difference in the amount of deformation of at least one of the steel types.

9.4.2.2 p-value approach

Recall the table we discussed for one-way ANOVA:

Source	df	Sum of squares (SS)	Mean squares (MS)	F
Treatment or between groups	m- 1	$SST \,/\, SSB = \pounds n_{ij}\left(\overline{x_{ij}} - \overset{v}{x}\right)^2$	$MSB = \dfrac{SSB}{m-1}$	$\dfrac{MSB}{MSE}$
Error	n-m	$SSE \,/\, SSW = \pounds\,(n_{ij} - 1)s_{ij}^{\,2}$	$MSE = \dfrac{SSE}{n \quad m}$	
Total	n–1	$TSS = SST + SSE$		

Now we use a similar table for the Kruskal-Wallis test where df, sum of squares, and mean squares have the same definitions as in the ANOVA table but based on the ranks of the data. Recall that the F-statistic was used in classical one-way ANOVA. However, it is replaced by a chi-squared statistic, and the p-value measures the significance of the chi-squared statistic. The test statistic is then calculated as

$$\text{Chi squared statistic} = \frac{df_{total} \times SST}{TSS}$$

9.4.2.2.1 In the first example

$m = 3$ groups being compared – catalysts CX, CY, and CZ

$$n = 21\bigl(6 + 7 + 8\bigr)$$

$SST \,/\, SSB =$ sum of squares for treatment or the between-groups sum of squares

Catalyst	n	Summation $\pounds x_i$	Mean $\dfrac{\pounds x_i}{n}$
CX	6	$12+4+7+17+8+2 = 50$	$\bar{x}_1 = \dfrac{50}{6}$
CY	7	$21+14+1+6+10+15+3 = 70$	$\bar{x}_2 = \dfrac{70}{7}$
CZ	8	$5+20+11+13+19+16+9+18 = 111$	$\bar{x}_3 = \dfrac{111}{8}$
	$\bullet \ \Sigma\,21$	$\bullet \ \Sigma\,231$	$\overset{v}{x} = \dfrac{231}{21} = 1$

Catalyst	$\mathbf{n}\, n$	Mean $\dfrac{\pounds x_i}{n}$	$n_{ij}\left(\overline{x}_{ij} - \overset{\blacklozenge}{x}\right)^2$
CX	6	$\bar{x}_1 = \dfrac{50}{6}$	$6*\left(\dfrac{50}{6}-11\right)^2 = \dfrac{128}{3}$
CY	7	$\bar{x}_2 = \dfrac{70}{7}$	$7*\left(\dfrac{70}{7}-11\right)^2 = 7$
CZ	8	$\bar{x}_3 = \dfrac{111}{8}$	$8*\left(\dfrac{111}{8}-11\right)^2 = \dfrac{529}{8}$
	$\Sigma\,21$	$\overset{v}{x} = \dfrac{773}{72}$	

$$SST = n_1\left(\overline{x}_1 - \overline{\overline{x}}\right)^2 + n_2\left(\overline{x}_2 - \overline{\overline{x}}\right)^2 + n_3\left(\overline{x}_3 - \overline{\overline{x}}\right)^2$$

$$= 6*\left(\frac{50}{6}-11\right)^2 + 7*\left(\frac{70}{7}-11\right)^2 + 8*\left(\frac{111}{8}-11\right)^2$$

$$= \frac{128}{3} + 7 + \frac{529}{8} = \frac{2779}{24} = 115.792$$

$SSE\,/\,SSW$ = sum of squares for error or the within-group sum of squares

Catalyst	Mean $\dfrac{\pounds x_i}{n}$	$(x_i - \bar{x})^2$
CX	$\bar{x}_1 = \dfrac{50}{6}$	$\left(12 - \dfrac{50}{6}\right)^2 + \left(4 - \dfrac{50}{6}\right)^2 + \left(7 - \dfrac{50}{6}\right)^2 + \left(17 - \dfrac{50}{6}\right)^2 + \left(8 - \dfrac{50}{6}\right)^2 + \left(2 - \dfrac{50}{6}\right)^2 = \dfrac{448}{3}$
CY	$\bar{x}_2 = \dfrac{70}{7}$	$\left(21 - \dfrac{70}{7}\right)^2 + \left(14 - \dfrac{70}{7}\right)^2 \ldots\ldots + \left(10 - \dfrac{70}{7}\right)^2 + \left(15 - \dfrac{70}{7}\right)^2 + \left(3 - \dfrac{70}{7}\right)^2 = 308$
CZ	$\bar{x}_3 = \dfrac{111}{8}$	$\left(5 - \dfrac{111}{8}\right)^2 + \left(20 - \dfrac{111}{8}\right)^2 + \ldots \left(9 - \dfrac{111}{8}\right)^2 + \left(18 - \dfrac{111}{8}\right)^2 = \dfrac{1575}{8}$
	$\overset{v}{x} = \dfrac{773}{72}$	

SSE = x1-x12+x2-x22+x3-x32

$$= \left[\left(23.5 - \dfrac{167}{10}\right)^2 + \left(11 - \dfrac{167}{10}\right)^2 + \ldots \left(18 - \dfrac{167}{10}\right)^2 + \left(11 - \dfrac{167}{10}\right)^2\right]$$

$$+ \left[\left(5.5 - \dfrac{73.5}{10}\right)^2 + \left(11 - \dfrac{73.5}{10}\right)^2 + \ldots \left(5.5 - \dfrac{73.5}{10}\right)^2 + \left(11 - \dfrac{73.5}{10}\right)^2\right]$$

$$+ \left[\left(27.5 - \dfrac{224.5}{10}\right)^2 + \left(18 - \dfrac{224.5}{10}\right)^2 + \ldots \left(27.5 - \dfrac{224.5}{10}\right)^2 + \left(18 - \dfrac{224.5}{10}\right)^2\right]$$

$$= \dfrac{448}{3} + 308 + \dfrac{1575}{8} = 654.208333$$

$TSS =$ total sum of squares

$$= SST + SSE = 115.792 + 654.208333 \sim 770$$

$MSB =$ mean sum of squares between groups

$$= \dfrac{SST}{m-1} = \dfrac{115.792}{3-1} \sim 57.896$$

$MSE =$ error mean sum of squares

$$= \dfrac{SSE}{n-m} = \dfrac{654.208333}{21-3} \sim 36.3449$$

Therefore our statistic is $\dfrac{20 \times 115.792}{770} \sim 3.01.$

So now we have the table:

Source	df	Sum of squares (SS)	Mean squares (MS)	Chi sq.
Treatment or between groups	2	115.792	57.896	3.01
Error	18	654.208333	36.3449	
Total	20	770		

All we can do now is approximate the p-value since calculating the exact value can be quite tedious and error prone. It involves complex mathematical calculations, which are not easy to perform manually.

d.f.	.995	.99	.975	.95	.9	.1	.05	.025	.01
1	0.00	0.00	0.00	0.00	0.02	2.71	3.84	5.02	6.63
2	0.01	0.02	0.05	0.10	0.21	4.61	5.99	7.38	9.21
3	0.07	0.11	0.22	0.35	0.58	6.25	7.81	9.35	11.34
4	0.21	0.30	0.48	0.71	1.06	7.78	9.49	11.14	13.28
5	0.41	0.55	0.83	1.15	1.61	9.24	11.07	12.83	15.09
6	0.68	0.87	1.24	1.64	2.20	10.64	12.59	14.45	16.81
7	0.99	1.24	1.69	2.17	2.83	12.02	14.07	16.01	18.48
8	1.34	1.65	2.18	2.73	3.49	13.36	15.51	17.53	20.09
9	1.73	2.09	2.70	3.33	4.17	14.68	16.92	19.02	21.67
10	2.16	2.56	3.25	3.94	4.87	15.99	18.31	20.48	23.21
11	2.60	3.05	3.82	4.57	5.58	17.28	19.68	21.92	24.72
12	3.07	3.57	4.40	5.23	6.30	18.55	21.03	23.34	26.22
13	3.57	4.11	5.01	5.89	7.04	19.81	22.36	24.74	27.69
14	4.07	4.66	5.63	6.57	7.79	21.06	23.68	26.12	29.14
15	4.60	5.23	6.26	7.26	8.55	22.31	25.00	27.49	30.58
16	5.14	5.81	6.91	7.96	9.31	23.54	26.30	28.85	32.00
17	5.70	6.41	7.56	8.67	10.09	24.77	27.59	30.19	33.41
18	6.26	7.01	8.23	9.39	10.86	25.99	28.87	31.53	34.81
19	6.84	7.63	8.91	10.12	11.65	27.20	30.14	32.85	36.19
20	7.43	8.26	9.59	10.85	12.44	28.41	31.41	34.17	37.57
22	8.64	9.54	10.98	12.34	14.04	30.81	33.92	36.78	40.29
24	9.89	10.86	12.40	13.85	15.66	33.20	36.42	39.36	42.98
26	11.16	12.20	13.84	15.38	17.29	35.56	38.89	41.92	45.64
28	12.46	13.56	15.31	16.93	18.94	37.92	41.34	44.46	48.28
30	13.79	14.95	16.79	18.49	20.60	40.26	43.77	46.98	50.89
32	15.13	16.36	18.29	20.07	22.27	42.58	46.19	49.48	53.49
34	16.50	17.79	19.81	21.66	23.95	44.90	48.60	51.97	56.06

Since our $df = 3 - 1 = 2$, we look at the row with $df = 2$ and look for the value closest to our statistic $= 3.01$. We can see the value that seems the closest to 3.01 is 4.61 which falls under the significance level 0.01. Now since our test statistic is smaller than 4.61, our p-value > 0.10.

Since the p-value $> 0.10 > 0.05 = \alpha$, we do not reject the null hypothesis.

As we can see when we use the critical approach we get the same results.

To get the exact answer it is the best if we use software—MATLAB—which will not only save time but will also eliminate the risk of errors.

9.4.2.2.2 In the second example

$m = 3$ groups being compared – steel types A, B, and C

$$n = 30(10 + 10 + 10)$$

SST

Steel type	n	Summation £x_i	Mean
A	10	23.5+11+⋯+18+11 = 168	$\overline{x_1} = \dfrac{168}{10}$
B	10	5.5+11+⋯+5.5+11 = 73.50	$\overline{x_2} = \dfrac{73.5}{10}$
C	10	28.5+18+⋯+28.5+18 = 231	$\overline{x_3} = \dfrac{231}{10}$
D	10	36.5+33.5+⋯+28.5+33.5 = 347.5	$\overline{x_4} = \dfrac{347.5}{10}$
			$\overline{x} = 20.5$

Steel type	n	Mean $\dfrac{£x_i}{n}$	$n_{ij}\left(\overline{x_{ij}} - \overset{\blacklozenge}{\overline{x}}\right)^2$
A	10	$\overline{x_1} = \dfrac{168}{10}$	$10*\left(\dfrac{168}{10} - 20.5\right)^2 = 14.4$
B	10	$\overline{x_2} = \dfrac{73.5}{10}$	$10*\left(\dfrac{73.5}{10} - 20.5\right)^2 = 664.225$
C	10	$\overline{x_3} = \dfrac{231}{10}$	$10*\left(\dfrac{231}{10} - 20.5\right)^2 = 483.025$
D	10	$\overline{x_4} = \dfrac{347.5}{10}$	$10*\left(\dfrac{347.5}{10} - 20.5\right)^2 = 483.025$

$$SST = n_1\left(\overline{x_1} - \overline{\overline{x}}\right)^2 + n_2\left(\overline{x_2} - \overline{\overline{x}}\right)^2 + n_3\left(\overline{x_3} - \overline{\overline{x}}\right)^2 + n_4\left(\overline{x_4} - \overline{\overline{x}}\right)^2$$

$$= 10*\left(\dfrac{168}{10} - 20.5\right)^2 + 10*\left(\dfrac{73.5}{10} - 20.5\right)^2 + 10*\left(\dfrac{231}{10} - 20.5\right)^2 + 10*\left(\dfrac{347.5}{10} - 20.5\right)^2$$

$$= 3964.35$$

SSE

Steel type	Mean $\dfrac{\pounds x_i}{n}$	$(x_i - \bar{x})^2$
A	$\bar{x}_1 = \dfrac{168}{10}$	$\left(23.5 - \dfrac{168}{10}\right)^2 + \left(11 - \dfrac{168}{10}\right)^2 + \ldots\left(18 - \dfrac{167}{10}\right)^2 + \left(11 - \dfrac{167}{10}\right)^2 = 459.6$
B	$\bar{x}_2 = \dfrac{73.5}{10}$	$\left(5.5 - \dfrac{73.5}{10}\right)^2 + \left(11 - \dfrac{73.5}{10}\right)^2 + \ldots\left(5.5 - \dfrac{73.5}{10}\right)^2 + \left(11 - \dfrac{73.5}{10}\right)^2 = 249.525$
C	$\bar{x}_3 = \dfrac{231}{10}$	$\left(28.5 - \dfrac{231}{10}\right)^2 + \left(18 - \dfrac{231}{10}\right)^2 + \ldots\left(28.5 - \dfrac{231}{10}\right)^2 + \left(18 - \dfrac{231}{10}\right)^2 = 420.4$
D	$\bar{x}_4 = \dfrac{347.5}{10}$	$\left(36.5 - \dfrac{347.5}{10}\right)^2 + \left(33.5 - \dfrac{347.5}{10}\right)^2 + \ldots\left(28.5 - \dfrac{347.5}{10}\right)^2 + \left(33.5 - \dfrac{347.5}{10}\right)^2$
		$= 144.625$

$$SSE = \left(x_1 - \bar{x}_1\right)^2 + \left(x_2 - \bar{x}_2\right)^2 + \left(x_3 - \bar{x}_3\right)^2 + \left(x_4 - \bar{x}_4\right)^2$$

$$= 1274.15$$

$$TSS = 3964.35 + 1274.15 = 5238.5$$

$$MSB = \frac{3964.35}{4-1} = 1321.45$$

$$MSE = \frac{1274.15}{40-4} = 35.39$$

$$\text{Chi-squared statistic} = \frac{39 \times 3964.35}{5238.5} \sim 29.51$$

So now we have the table:

Source	df	Sum of squares (SS)	Mean squares (MS)	Chi
Treatment or between groups	3	3964.35	1321.45	29.51
Error	36	1274.15	35.39	
Total	39	5238.5		

d.f.	.995	.99	.975	.95	.9	.1	.05	.025	.01
1	0.00	0.00	0.00	0.00	0.02	2.71	3.84	5.02	6.63
2	0.01	0.02	0.05	0.10	0.21	4.61	5.99	7.3	9.21
3	0.07	0.11	0.22	0.35	0.58	6.25	7.81	9.35	11.34
4	0.21	0.30	0.48	0.71	1.06	7.78	9.49	11.14	13.28
5	0.41	0.55	0.83	1.15	1.61	9.24	11.07	12.83	15.09
6	0.68	0.87	1.24	1.64	2.20	10.64	12.59	14.45	16.81
7	0.99	1.24	1.69	2.17	2.83	12.02	14.07	16.01	18.48
8	1.34	1.65	2.18	2.73	3.49	13.36	15.51	17.53	20.09
9	1.73	2.09	2.70	3.33	4.17	14.68	16.92	19.02	21.67
10	2.16	2.56	3.25	3.94	4.87	15.99	18.31	20.48	23.21
11	2.60	3.05	3.82	4.57	5.58	17.28	19.68	21.92	24.72
12	3.07	3.57	4.40	5.23	6.30	18.55	21.03	23.34	26.22
13	3.57	4.11	5.01	5.89	7.04	19.81	22.36	24.74	27.69
14	4.07	4.66	5.63	6.57	7.79	21.06	23.68	26.12	29.14
15	4.60	5.23	6.26	7.26	8.55	22.31	25.00	27.49	30.58
16	5.14	5.81	6.91	7.96	9.31	23.54	26.30	28.85	32.00
17	5.70	6.41	7.56	8.67	10.09	24.77	27.59	30.19	33.41
18	6.26	7.01	8.23	9.39	10.86	25.99	28.87	31.53	34.81
19	6.84	7.63	8.91	10.12	11.65	27.20	30.14	32.85	36.19
20	7.43	8.26	9.59	10.85	12.44	28.41	31.41	34.17	37.57
22	8.64	9.54	10.98	12.34	14.04	30.81	33.92	36.78	40.29
24	9.89	10.86	12.40	13.85	15.66	33.20	36.42	39.36	42.98
26	11.16	12.20	13.84	15.38	17.29	35.56	38.89	41.92	45.64
28	12.46	13.56	15.31	16.93	18.94	37.92	41.34	44.46	48.28
30	13.79	14.95	16.79	18.49	20.60	40.26	43.77	46.98	50.89
32	15.13	16.36	18.29	20.07	22.27	42.58	46.19	49.48	53.49
34	16.50	17.79	19.81	21.66	23.95	44.90	48.60	51.97	56.06

Since our $df = 4 - 1 = 3$, we look at the row with $df = 3$ and look for the value closest to our statistic $= 29.51$. We can see the value that seems the closest to 29.51 is 11.34 which falls under the significance level 0.01. Now since our test statistic is greater than 11.34, our p- value < 0.01.

Since p- value $< 0.01 = \alpha$, we reject the null hypothesis which also aligns with our calculation when we used the critical approach.

9.4.2.2.3 MATLAB

To find the exact p-value in the Kruskal–Wallis test we use the following syntax in MATLAB:

```
[p,tbl,stats] = kruskalwallis(data, group, displayopt)
```

As we can see there is only one argument, data, which represents multiple groups of data like the other multiple comparison tests we have discussed before.

```
CX = [5.2, 4.4, 4.7, 5.7, 4.8, 4.2];
CY = [6.3, 5.4, 4.1, 4.6, 5.0, 5.5, 4.3];
CZ = [4.5, 6.2, 5.1, 5.3, 6.1, 5.6, 4.9, 5.8];
data = [CX, CY, CZ];
```

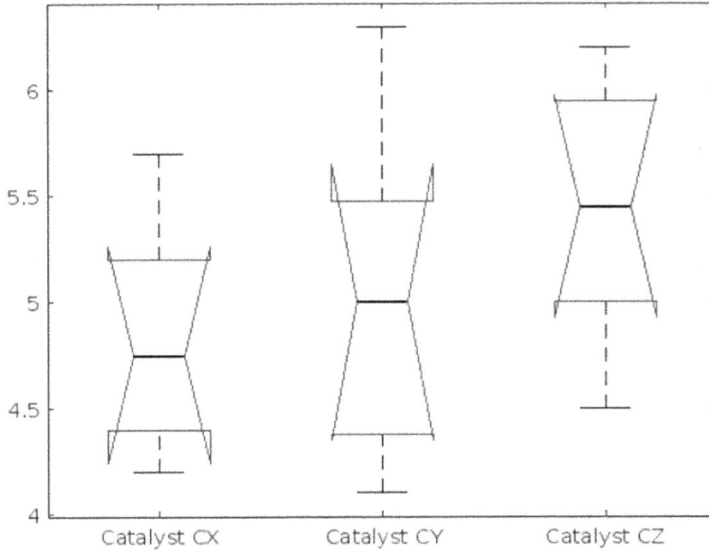

Figure 9.1

```
% Create a grouping variable
        groups = [repmat({'Catalyst CX'}, 1, length(CX)),
            repmat({'Catalyst CY'}, 1, length(CY)),
            repmat({'Catalyst CZ'}, 1, length(CZ))];
        % Perform ANOVA test
        [p, tbl, stats] = kruskalwallis(data, groups)
```

Kruskal-Wallis ANOVA Table					
Source	SS	df	MS	Chi-sq	Prob>Chi-sq
Groups	115.792	2	57.8958	3.01	0.2223
Error	654.208	18	36.3449		
Total	770	20			

```
p =

    0.2223


tbl =

    4×6 cell array

      {'Source'}    {'SS'       }    {'df'}    {'MS'       }    {'Chi-sq' }    {'Prob>Chi-sq'}
      {'Groups'}    {[115.7917]}    {[ 2]}    {[ 57.8958]}    {[ 3.0076]}    {[    0.2223]}
      {'Error' }    {[654.2083]}    {[18]}    {[ 36.3449]}    {0×0 double}    {0×0 double   }
      {'Total' }    {[    770]}    {[20]}    {0×0 double}    {0×0 double}    {0×0 double   }


stats =

    struct with fields:

        gnames: {3×1 cell}
             n: [6 7 8]
        source: 'kruskalwallis'
     meanranks: [8.3333 10 13.8750]
          sumt: 0
```

Let's discuss each of the three outputs individually.

9.4.2.2.4 The boxplot

Since we are comparing three catalysts—CX, CY, and CZ—we have three boxplots, one for each catalyst, displayed side by side. The x-axis represents the catalysts, and the y-axis represents the reaction rates of individuals.

As we discussed earlier in Chapter 3, a boxplot displays the summary statistics of a dataset, including the median, quartiles, and any outliers, providing a visual representation of the data's distribution.

The medians (line inside the box) of each catalyst are 4.75, 5.0, and 5.45, which are somewhat similar indicating that the datasets being compared come from the same population. However since the median is slightly different, this means they have slightly different central tendencies, i.e. means—8.33, 10, 13.875.

The whiskers extending from each of the boxplots indicate the range or variability of the data. Notice for Catalyst CX the upper whisker is longer than the lower indicating more spread or more outliers toward the upper extremes of the data. Similarly Catalyst CY is more spread out toward the upper extremes. For example if we look at the data again we see the extreme values 5.7 and 6.3 for Catalyst CX and CY respectively. However the lower whisker for Catalyst CZ is longer than the upper one indicating more spread toward the lower extremes of the data. Looking at the data again we will immediately find the value 4.5 for Catalyst CZ.

Now out of the three, Catalyst CY has the longest whisker specifying that the data in that boxplot has the largest spread or variability compared to the other boxplots.

If we calculate the range for each catalyst:

Range of Catalyst CX = 5.7 – 4.2 = 1.5
Range of Catalyst CY = 6.3 – 4.1 = 2.2
Range of Catalyst CZ = 6.2 – 4.5 = 1.7

The second output displays the ANOVA table perfectly aligning with our results along with the p-value.

Recall that in theory we approximated our p-value to be greater than > 0.01

and we get around 0.222, making it way easier to conclude that there is no significant difference in the reaction rates among the three catalysts.

The third and last output presents the data as a structure so it contains the p-value, the ANOVA table, and the statistics.

9.5.2.2.5 Example 2

```
A = [8 6 7 9 5 6 8 7 7 6];
B = [5 6 4 4 5 6 7 4 5 6];
C = [9 7 8 10 6 7 8 9 9 7];
D = [11 10 12 11 9 13 12 10 9 10];
data = [A, B, C, D];

% Create a grouping variable
        groups = [repmat({'Steel A'}, 1, length(A)),
            repmat({'Steel B'}, 1, length(B)),
            repmat({'Steel C'}, 1, length(C)),
            repmat({'Steel D'}, 1, length(D))];
% Perform ANOVA test
        [p, tbl, stats] = kruskalwallis(data, groups)
```

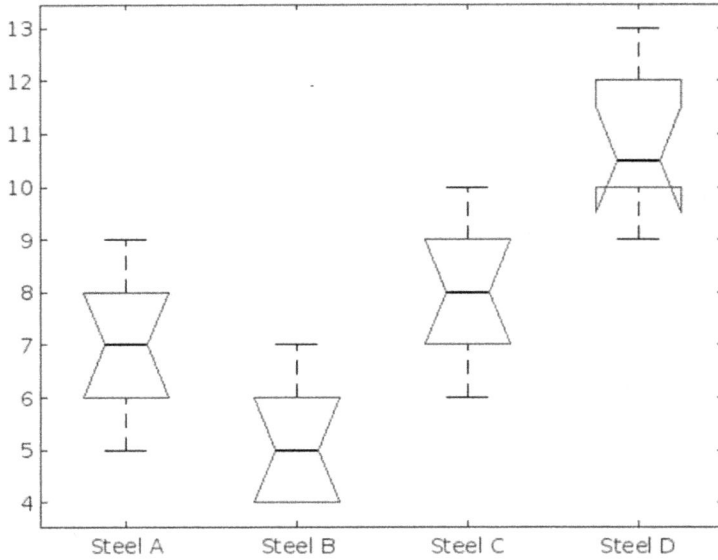

Figure 9.2

Kruskal-Wallis ANOVA Table					
Source	SS	df	MS	Chi-sq	Prob>Chi-sq
Groups	3964.35	3	1321.45	29.51	1.74614e-06
Error	1274.15	36	35.39		
Total	5238.5	39			

```
p =

   1.7461e-06


tbl =

  4×6 cell array

    {'Source'}    {'SS'       }    {'df'}    {'MS'        }    {'Chi-sq'  }    {'Prob>Chi-sq'}
    {'Groups'}    {[3.9643e+03]}    {[ 3]}    {[1.3215e+03]}    {[ 29.5141]}    {[ 1.7461e-06]}
    {'Error' }    {[1.2742e+03]}    {[36]}    {[   35.3931]}    {0×0 double}    {0×0 double   }
    {'Total' }    {[5.2385e+03]}    {[39]}    {0×0 double  }    {0×0 double}    {0×0 double   }


stats =

  struct with fields:

        gnames: {4×1 cell}
             n: [10 10 10 10]
        source: 'kruskalwallis'
     meanranks: [16.8000 7.3500 23.1000 34.7500]
          sumt: 1098
```

Using the exact p-value = 0.0000017461 < 0.01 = α;   hence we reject the null hypothesis which also aligns with our calculation when we used the critical approach.

9.4.2.2.6 Example

Suppose a group of engineers are testing the effect of five different lubricants on a particular type of machinery. They want to determine which one provides the best performance in terms of reducing wear on a particular machine part. The engineers take samples of the machinery and run each sample using one of the lubricants. The amount of wear on the machine part is measured in micrometers (µm) after a certain amount of use, and the results are recorded in the following table. Test, at $\alpha = 0.10$, if there is any significant difference in the mean wear rate between the three lubricants.

Lubricant X	Lubricant Y	Lubricant Z	Lubricant W	Lubricant K
40	35	38	36	39
35	34	33	32	37
38	37	34	29	42
41	35	39	38	29
36	38	35	37	34
42	40	31	35	38
45	37	37	39	36
	39	40	34	41
		36	28	38
		32		35
				33
				32

To conduct the Kruskal–Wallis test, the engineers will follow these steps.
   Null hypothesis:

   There is no significant difference in the mean wear rate between the three lubricants.

Alternative hypothesis:
   There is a significant difference in the mean wear rate between at least one pair of lubricants.

```
data = [X, Y, Z, W, K];
% Create a grouping variable
        groups = [repmat({'Lubricant X'}, 1, length(X)),
            repmat({'Lubricant Y'}, 1, length(Y)),
            repmat({'Lubricant Z'}, 1, length(Z)),
            repmat({'Lubricant W'}, 1, length(W)),
            repmat({'Lubricant K'}, 1, length(K))];
% Perform ANOVA test
        [p, tbl, stats] = kruskalwallis(data, groups)
```

```
Source      SS       df      MS      Chi-sq   Prob>Chi-sq
--------------------------------------------------------------
Groups    1393.07     4    348.269    7.8       0.0994
Error     6648.43    41    162.157
Total     8041.5     45
```

```
p =

   0.0994


tbl =

  4×6 cell array

    {'Source'}    {'SS'        }    {'df'}    {'MS'       }    {'Chi-sq' }    {'Prob>Chi-sq'}
    {'Groups'}    {[1.3931e+03]}    {[ 4]}    {[348.2685]}    {[ 7.7956]}    {[    0.0994]}
    {'Error' }    {[6.6484e+03]}    {[41]}    {[162.1567]}    {0×0 double}    {0×0 double   }
    {'Total' }    {[8.0415e+03]}    {[45]}    {0×0 double}    {0×0 double}    {0×0 double   }


stats =

  struct with fields:

       gnames: {5×1 cell}
            n: [7 8 10 9 12]
       source: 'kruskalwallis'
     meanranks: [34.6429 25.5625 20.2000 17 23.2500]
         sumt: 792
```

Since p-value = 0.0994 < 0.10 = α,   we reject the null hypothesis and conclude that there is a significant difference in the mean wear rate between at least one pair of lubricants.

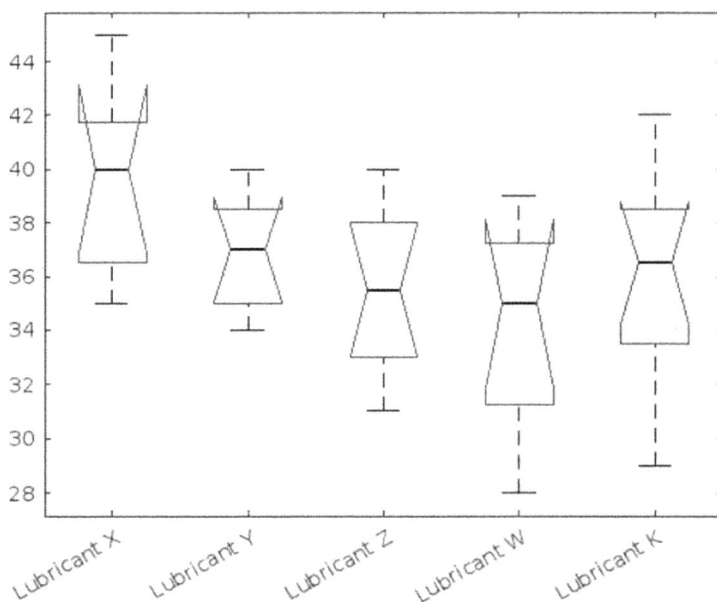

Figure 9.3

9.5 THE FRIEDMAN TEST

The last non-parametric test we shall discuss in this textbook is the Friedman test. It is an extension of the paired-data concept, comparing three or more **matched** groups.

Its relationship with the Kruskal–Wallis test is just like the relationship between the Mann–Whitney and Wilcoxon.

These are all tests for ordinal data.

Number of pairs	Two pairs	More than two pairs
Are they unmatched pairs?	**Mann–Whitney**	**Kruskal–Wallis**
Are they matched pairs?	**Wilcoxon**	**Friedman**

9.5.1 Using either of the two approaches

9.5.1.1 Critical value approach

9.5.1.1.1 Find the statistic

Just like any other non-parametric test, the first step in calculating the test statistic is to convert the original observation to ranks. However instead of ranking all the values from low to high, regardless of the group each value belongs to, in the Friedman test the observations within each subject are ranked separately from smallest to largest, so each subject contains a separate set of k ranks.

The second step is to obtain the sum of the ranks in each column. Therefore we have k sum of ranks. Now if none of the subjects contains ties, then the Friedman test statistic is defined as

$$F = \frac{12}{nK(K+1)} \sum_{i=1}^{k} R_i^2 - 3n(K+1)$$

Here,

n = number of subjects or like number of rows when drawn in a table

K = number of groups or treatment levels or like number of columns

R_i = sum of ranks for group

An equivalent computational form for the test statistic is

$$F = \frac{12}{nK(K+1)} \sum_{i=1}^{k} \left[\left(R_i - \frac{n(K+1)}{2} \right) \right]^2$$

Now in the presence of ties the statistic formula above needs to be divided by

$$1 - \frac{\sum_{i=1}^{k}(t_i^3 - t_i)}{nK(K^2-1)}$$

Which is equivalent to

$$F = \frac{n(K-1)\left[\sum_{i=1}^{k} \frac{R_i^2}{n} - \frac{1}{4}nK(K+1)^2\right]}{\sum r_{ij}^2 - \frac{1}{4}nK(K+1)^2}$$

Where r_{ij} = rank corresponding to subject j in column i.

Let's take the catalyst example of Kruskal-Wallis. If we perform the same chemical reaction thrice using each catalyst, we have the paired data concept. The data then looks like:

Chemical reaction	Catalyst CX	Catalyst CY	Catalyst CZ
1	4.7	5.2	5.0
2	4.5	4.6	4.7
3	4.4	4.8	4.5
4	4.5	4.3	4.9
5	4.6	5.1	5.3
6	4.8	4.7	5.0
7	5.0	4.9	5.4
8	4.8	5.1	4.9
9	4.7	4.3	4.8
10	5.1	5.0	4.6

- Null hypothesis: there is no significant difference in the reaction rate among the three catalysts.
- Alternative hypothesis: there is a significant difference in the reaction rate.

Catalyst CX	Rank	Catalyst CY	Rank	Catalyst CZ	Rank
4.7	1	5.2	3	5.0	2
4.5	1	4.6	2	4.7	3
4.4	1	4.8	3	4.5	2
4.5	2	4.3	1	4.9	3
4.6	1	5.1	2	5.3	3
4.8	2	4.7	1	5.0	3
5.0	2	4.9	1	5.4	3
4.8	1	5.1	3	4.9	2
4.7	2	4.3	1	4.8	3
5.1	3	5.0	2	4.6	1

Calculate the sum of the ranks for each group.

Rank sum for catalyst CX $= 1+1+1+2+1+2+2+1+2+3 = 16$
Rank sum for catalyst CY $= 3+2+3+1+2+1+1+3+1+2 = 19$
Rank sum for catalyst CZ $= 2+3+2+3+3+3+3+2+3+1 = 25$

Therefore in this example:

$$n = 30$$

$R_{i=3}^{i=0}$ = sum of ranks for 3 groups so $123.5, 159.5, 182$ for each group

$n_{i=3}^{i=0}$ = sample size for 3 groups

$$F = \frac{12}{10 \times 3 \times (3+1)}\left(16^2 + 19^2 + 25^2\right) - 3 \times 10 \times (3+1) \sim 4.2$$

Alternate way:

$$F = \frac{12}{10 \times 3 \times (3+1)} \sum_{i=1}^{k}\left(16 - \frac{10(3+1)}{2}\right)^2 + \left(19 - \frac{10(3+1)}{2}\right)^2 + \left(25 - \frac{10(3+1)}{2}\right)^2 \sim 4.2$$

9.5.1.1.1.1 EXAMPLE

In industrial applications, cooling can be critical to ensure that processes do not cause equipment or products to overheat. An industrial company is trying to determine which cooling system is more effective in engines. They have six cooling systems, System A, System B, System C, System D, System E, and System F . The engineers set up an experiment where they run the engine for a certain duration using each of the systems individually and record the temperature fluctuations for each system during that time.

Engine	System A	System B	System C	System D	System E
1	3.5	4.1	4.2	3.5	4.4
2	3.8	3.6	4.0	3.6	4.6
3	4.1	3.8	3.9	3.6	4.1
4	3.2	3.6	4.1	4.0	4.5
5	3.5	4.2	3.8	3.4	4.2
6	3.7	4.4	4.2	4.2	4.0
7	4.2	3.9	4.3	3.9	3.9
8	3.2	3.7	3.9	3.7	4.5
9	4.0	4.2	4.5	4.0	4.5
10	4.2	3.9	4.4	4.1	4.4
11	3.9	4.1	3.7	3.5	4.1
12	3.6	3.7	4.2	3.3	3.9
13	3.2	4.5	3.8	3.7	4.6
14	3.7	4.3	4.0	3.9	4.7
15	4.0	4.4	4.1	4.1	4.4
16	3.5	3.8	4.3	3.3	4.7
17	3.7	4.2	4.6	3.7	4.6
18	3.6	4.3	4.4	4.2	4.5
19	4.1	3.9	4.5	3.5	4.5
20	4.2	4.1	4.3	4.0	4.2

System A	Rank	System B	Rank	System C	Rank	System D	Rank	System E	Rank
3.5	1.5	4.1	3	4.2	4	3.5	1.5	4.4	5
3.8	3	3.6	1.5	4.0	4	3.6	1.5	4.6	5
4.1	4.5	3.8	2	3.9	3	3.6	1	4.1	4.5
3.2	1	3.6	2	4.1	4	4.0	3	4.5	5
3.5	2	4.2	4.5	3.8	3	3.4	1	4.2	4.5

3.7	1	4.4	5	4.2	3.5	4.2	3.5	4.0	2
4.2	4	3.9	2	4.3	5	3.9	2	3.9	2
3.2	1	3.7	2.5	3.9	4	3.7	2.5	4.5	5
4.0	1.5	4.2	3	4.5	4.5	4.0	1.5	4.5	4.5
4.2	3	3.9	1	4.4	4.5	4.1	2	4.4	4.5
3.9	3	4.1	4.5	3.7	2	3.5	1	4.1	4.5
3.6	2	3.7	3	4.2	5	3.3	1	3.9	4
3.2	1	4.5	4	3.8	3	3.7	2	4.6	5
3.7	1	4.3	4	4.0	3	3.9	2	4.7	5
4.0	1	4.4	4.5	4.1	2.5	4.1	2.5	4.4	4.5
3.5	2	3.8	3	4.3	4	3.3	1	4.7	5
3.7	1.5	4.2	3	4.6	4.5	3.7	1.5	4.6	4.5
3.6	1	4.3	3	4.4	4	4.2	2	4.5	5
4.1	3	3.9	2	4.5	4.5	3.5	1	4.5	4.5
4.2	3.5	4.1	2	4.3	5	4.0	1	4.2	3.5
$R_1 = 41.5$		$R_2 = 59.5$		$R_3 = 77$		$R_4 = 34.5$		$R_5 = 87.5$	

To find $\sum r_{ij}^2$ we square all the ranks and then sum them up.

System A	Rank	System B	Rank	System C	Rank	System D	Rank	System E	Rank
3.5	1.5^2	4.1	3^2	4.2	4^2	3.5	1.5^2	4.4	5^2
3.8	3^2	3.6	1.5^2	4.0	4^2	3.6	1.5^2	4.6	5^2
4.1	4.5^2	3.8	2^2	3.9	3^2	3.6	1^2	4.1	4.5^2
3.2	1^2	3.6	2^2	4.1	4^2	4.0	3^2	4.5	5^2
3.5	2^2	4.2	4.5^2	3.8	3^2	3.4	1^2	4.2	4.5^2
3.7	1^2	4.4	5^2	4.2	3.5^2	4.2	3.5^2	4.0	2^2
4.2	4^2	3.9	2^2	4.3	5^2	3.9	2^2	3.9	2^2
3.2	1^2	3.7	2.5^2	3.9	4^2	3.7	2.5^2	4.5	5^2
4.0	1.5^2	4.2	3^2	4.5	4.5^2	4.0	1.5^2	4.5	4.5^2
4.2	3^2	3.9	1^2	4.4	4.5^2	4.1	2^2	4.4	4.5^2
3.9	3^2	4.1	4.5^2	3.7	2^2	3.5	1^2	4.1	4.5^2
3.6	2^2	3.7	3^2	4.2	5^2	3.3	1^2	3.9	4^2
3.2	1^2	4.5	4^2	3.8	3^2	3.7	2^2	4.6	5^2
3.7	1^2	4.3	4^2	4.0	3^2	3.9	2^2	4.7	5^2
4.0	1^2	4.4	4.5^2	4.1	2.5^2	4.1	2.5^2	4.4	4.5^2
3.5	2^2	3.8	3^2	4.3	4^2	3.3	1^2	4.7	5^2
3.7	1.5^2	4.2	3^2	4.6	4.5^2	3.7	1.5^2	4.6	4.5^2
3.6	1^2	4.3	3^2	4.4	4^2	4.2	2^2	4.5	5^2
4.1	3^2	3.9	2^2	4.5	4.5^2	3.5	1^2	4.5	4.5^2
4.2	3.5^2	4.1	2^2	4.3	5^2	4.0	1^2	4.2	3.5^2
$r_1^2 = 110.25$		$r_2^2 = 201.25$		$r_3^2 = 310.5$		$r_4^2 = 69.75$		$r_5^2 = 398.25$	

$$\sum r_{ij}^2 = 110.25 + 201.25 + 310.5 + 69.75 + 398.25 = 1090$$

$$F = \frac{20 \times (5-1) \times \left[\frac{41.5^2}{20} + \frac{59.5^2}{20} + \frac{77^2}{20} + \frac{34.5^2}{20} + \frac{87.5^2}{20} - \frac{1}{4} \times 20 \times 5 \times (5+1)^2 \right]}{1090 - \frac{1}{4} \times 20 \times 5 \times (5+1)^2}$$

$$F = \frac{20(5-1)[1001.9 - 900]}{1090 - 900} = \frac{8152}{190} \sim 42.91$$

The alternate way:

$$F = \frac{12}{20 \times 5 \times (5+1)} \left[41.5^2 + 59.5^2 + 77^2 + 34.5^2 + 87.5^2 \right] - 3 \times 20 \times (5+1) = 40.76$$

Now if we count the number of observations tied for a given rank we will get:

16 times where an observation was tied twice .

1 time an observation was tied thrice.

Therefore $16(2^3 - 2) + 1(3^3 - 3) = 120$

$$1 - \frac{120}{20 \times 5 \times (5^2 - 1)} = 0.95$$

Therefore our F is now $\dfrac{40.76}{0.95} = 42.91.$

9.5.1.1.2 Find the critical value

After calculating the test statistic we need to then determine the critical value using the appropriate table.

9.5.1.1.2.1 FOR SMALL SAMPLES ($k < 6$ AND $n < 14$)

If the number of treatments $k < 6$ and the number of subjects $n < 14$ then we use the Friedman critical value table to get the critical value.
   Our catalyst example has $k = 3$ and $n = 10$; therefore our F($k = 3$, $N = 10$, $\alpha = .05$) $= 6.20$.

9.5.1.1.2.2 FOR LARGE SAMPLES ($k > 5$ OR $n > 13$)

If our k is over 5, or our n is over 13, we use the chi-squared critical value table to get the critical value. To determine our critical value we need to use $df = k - 1$.
   So in our second example $df = 5 - 1 = 4$; hence:
   χ^2 (df = 4, p = .05) = 9.488

9.5.1.1.2.3 IN SUMMARY

For the values of independent blocks (n) greater than 15 and/or values of groups (k) greater than 5, use the χ^2 table with $k - 1$ degrees of freedom; otherwise use the Friedman table.

9.5.1.1.2.4 DECISION

Since our test statistic $= 4.20 < 6.20 =$ critical value, we do not reject the null hypothesis and we can conclude that there is no significant difference in the reaction rates among the three catalysts.

9.5.1.1.2.5 IN THE SECOND EXAMPLE

Since our test statistic $= 42.91 > 9.488 =$ critical value we reject the null hypothesis and we can conclude that there is a significant difference between the cooling systems.

9.5.1.2 p-value approach

Using the same table structure as for ANOVA and Kruskal–Wallis, we'll perform similar calculations to obtain the test statistic and then determine the p-value. However, there are a few adjustments in the calculations, which we'll outline below

Source	df	Sum of squares (SS)	Mean squares (MS)	F
Treatment	• K- 1	SSB	$MSB = \dfrac{SSB}{K-1}$	
Error	$(n–1)(K–1)$	SSE/SSW	$MSB = \dfrac{SSE}{(n-1)(K-1)}$	
Total	$(n–1)$	TSS = SST+SSE		

The test statistic in the Friedman test is calculated as

$$\text{Chi squared statistic} = \frac{n \times (K-1) \times SST}{TSS}$$

9.5.1.2.1 First example

$k = 3$ treatments – Catalyst CX, Catalyst CY, and Catalyst CZ

$n = 10$ subjects

Catalyst	Summation $£x_i$	Mean
CX	1+1+1+2+1+2+2+1+2+3 = 16	$\overline{x_1} = \dfrac{16}{10}$
CY	3+2+3+1+2+1+1+3+1+2 = 19	$\overline{x_2} = \dfrac{19}{10}$
CZ	2+3+2+3+3+3+3+2+3+1 = 25	$\overline{x_3} = \dfrac{25}{10}$
		$\overline{\overline{x}} = 2$

Catalyst	Mean $\dfrac{\pounds x_i}{n}$	$n\left(\overline{x_{ij}}-\overline{\overline{x}}\right)^2$
CX	$\overline{x_1}=\dfrac{16}{10}$	$10*\left(\dfrac{16}{10}-2\right)^2=1.60$
CY	$\overline{x_2}=\dfrac{19}{10}$	$10*\left(\dfrac{19}{10}-2\right)^2=0.10$
CZ	$\overline{x_3}=\dfrac{25}{10}$	$10*\left(\dfrac{25}{10}-2\right)^2=2.50$

$$SST = n_1\left(\overline{x_1}-\overline{\overline{x}}\right)^2 + n_2\left(\overline{x_2}-\overline{\overline{x}}\right)^2 + n_3\left(\overline{x_3}-\overline{\overline{x}}\right)^2 = 1.60+0.10+2.50 = 4.2C$$

Alternate way:

$$\frac{16^2+19^2+25^2}{10}-\frac{\left(16+19+25\right)^2}{10*3}=4.20$$

Catalyst	Mean $\dfrac{\pounds x_i}{n}$	$\left(x_i-\overline{x}\right)^2$
CX	$\overline{x_1}=\dfrac{16}{10}$	$\left(1-\dfrac{16}{10}\right)^2+\left(1-\dfrac{16}{10}\right)^2+\dots\left(2-\dfrac{16}{10}\right)^2+\left(3-\dfrac{16}{10}\right)^2=4.40$
CY	$\overline{x_2}=\dfrac{19}{10}$	$\left(3-\dfrac{19}{10}\right)^2+\left(2-\dfrac{19}{10}\right)^2+\dots\left(1-\dfrac{19}{10}\right)^2+\left(2-\dfrac{19}{10}\right)^2=6.90$
CZ	$\overline{x_3}=\dfrac{25}{10}$	$\left(2-\dfrac{25}{10}\right)^2+\left(3-\dfrac{25}{10}\right)^2+\dots\left(3-\dfrac{25}{10}\right)^2+\left(1-\dfrac{25}{10}\right)^2=4.50$

$$SSE = \left(x_1-\overline{x_1}\right)^2+\left(x_2-\overline{x_2}\right)^2+\left(x_3-\overline{x_3}\right)^2 = 4.40+6.90+4.50=15.8$$

$TSS = 4.20+15.80 = 20$

Alternate way:

$$\left[1^2+1^2+\dots+2^2+3^2\right]+\left[3^2+2^2+\dots+1^2+2^2\right]+\left[2^2+3^2+\dots+3^2+1^2\right]$$

$$=140-\frac{60^2}{10*3}=20$$

$$MSB=\frac{4.20}{3-1}=2.10$$

$$MSE=\frac{15.80}{\left(10-1\right)\left(3-1\right)}=0.8778$$

Chi-squared statistic $= \dfrac{10 \times (3-1) \times 4.20}{20} \sim 4.20$

So now we have the table:

Source	df	Sum of squares	Mean squares	F
Treatment	3 – 1 = 2	4.20	2.10	4.20
Error	(10 – 1)(3 – 1) = 18	15.80	0.8778	
Total	30 – 1 = 29	20		

d.f.	.995	.99	.975	.95	.9	.1	.05	.025	.01
1	0.00	0.00	0.00	0.00	0.02	2.71	3.84	5.02	6.63
2	0.01	0.02	0.05	0.10	0.21	4.61	5.99	7.38	9.21
3	0.07	0.11	0.22	0.35	0.58	6.25	7.81	9.35	11.34
4	0.21	0.30	0.48	0.71	1.06	7.78	9.49	11.14	13.28
5	0.41	0.55	0.83	1.15	1.61	9.24	11.07	12.83	15.09
6	0.68	0.87	1.24	1.64	2.20	10.64	12.59	14.45	16.81
7	0.99	1.24	1.69	2.17	2.83	12.02	14.07	16.01	18.48
8	1.34	1.65	2.18	2.73	3.49	13.36	15.51	17.53	20.09
9	1.73	2.09	2.70	3.33	4.17	14.68	16.92	19.02	21.67
10	2.16	2.56	3.25	3.94	4.87	15.99	18.31	20.48	23.21
11	2.60	3.05	3.82	4.57	5.58	17.28	19.68	21.92	24.72
12	3.07	3.57	4.40	5.23	6.30	18.55	21.03	23.34	26.22
13	3.57	4.11	5.01	5.89	7.04	19.81	22.36	24.74	27.69
14	4.07	4.66	5.63	6.57	7.79	21.06	23.68	26.12	29.14
15	4.60	5.23	6.26	7.26	8.55	22.31	25.00	27.49	30.58
16	5.14	5.81	6.91	7.96	9.31	23.54	26.30	28.85	32.00
17	5.70	6.41	7.56	8.67	10.09	24.77	27.59	30.19	33.41
18	6.26	7.01	8.23	9.39	10.86	25.99	28.87	31.53	34.81
19	6.84	7.63	8.91	10.12	11.65	27.20	30.14	32.85	36.19
20	7.43	8.26	9.59	10.85	12.44	28.41	31.41	34.17	37.57
22	8.64	9.54	10.98	12.34	14.04	30.81	33.92	36.78	40.29
24	9.89	10.86	12.40	13.85	15.66	33.20	36.42	39.36	42.98
26	11.16	12.20	13.84	15.38	17.29	35.56	38.89	41.92	45.64
28	12.46	13.56	15.31	16.93	18.94	37.92	41.34	44.46	48.28
30	13.79	14.95	16.79	18.49	20.60	40.26	43.77	46.98	50.89
32	15.13	16.36	18.29	20.07	22.27	42.58	46.19	49.48	53.49
34	16.50	17.79	19.81	21.66	23.95	44.90	48.60	51.97	56.06

Since our $df = 3-1 = 2$, we look at the row with $df = 2$ and look for the value closest to our statistic $= 4.20$. We can see the value that seems the closest to 4.20 is 4.61 which falls under the significance level 0.10. Now since our test statistic is slightly lower than 4.61, our p-value is slightly greater than 0.10, around 0.12.

Since the p-value $> 0.05 = \alpha$, we do not reject the null hypothesis which aligns with our calculation when used the critical approach.

9.5.1.2.2 Example 2

$k = 5$ treatments

$n = 20$ subjects

System	Summation $\pounds x_i$	Mean
A	$1.5+3+4.5+\cdots+1+3+3.5 = 41.5$	$\overline{x}_1 = \dfrac{41.5}{20}$
B	$3+1.5+2+\cdots+3+2+2 = 59.5$	$\overline{x}_2 = \dfrac{59.5}{20}$
C	$4+4+3+\cdots+4+4.5+5 = 77$	$\overline{x}_3 = \dfrac{77}{20}$
D	$1.5+1.5+1+\cdots+2+1+1 = 34.5$	$\overline{x}_4 = \dfrac{34.5}{20}$
E	$5+5+4.5+\cdots+5+4.5+3.5 = 87.5$	$\overline{x}_5 = \dfrac{87.5}{20}$
		$\overline{\overline{x}} = 3$

System	Mean $\dfrac{\pounds x_i}{n}$	$n_{ij}\left(\overline{x}_{ij} - \overset{\blacklozenge}{x}\right)^2$
A	$\overline{x}_1 = \dfrac{41.5}{20}$	$20*\left(\dfrac{41.5}{20}-3\right)^2 = 17.1125$
B	$\overline{x}_2 = \dfrac{59.5}{20}$	$20*\left(\dfrac{59.5}{20}-3\right)^2 = 0.0125$
C	$\overline{x}_3 = \dfrac{77}{20}$	$20*\left(\dfrac{77}{20}-3\right)^2 = 14.45$
D	$\overline{x}_4 = \dfrac{34.5}{20}$	$20*\left(\dfrac{34.5}{20}-3\right)^2 = 32.5125$
E	$\overline{x}_5 = \dfrac{87.5}{20}$	$20*\left(\dfrac{87.5}{20}-3\right)^2 = 37.8125$

$$SST = 17.1125 + 0.0125 + 14.45 + 32.5125 + 37.8125 = 101.9$$

Alternate way:

$$\frac{41.5^2 + 59.5^2 + 77^2 + 34.5^2 + 87.5^2}{20} - \frac{(41.5 + 59.5 + 77 + 34.5 + 87.5)^2}{20*5} = 101.9$$

TSS

$$r_1^2 + r_2^2 + r_3^2 + r_4^2 + r_5^2 = 110.25 + 201.25 + 310.5 + 69.75 + 398.25 = 1090$$

$$1090 - \frac{\left(41.5 + 59.5 + 77 + 34.5 + 87.5\right)^2}{20 * 5} = 190$$

Therefore our SSE = $190 - 101.9 = 88.1$.

$$MSB = \frac{101.9}{5-1} = 25.475$$

$$MSE = \frac{88.1}{\left(20-1\right)\left(5-1\right)} = 1.1592$$

Chi-squared statistic $= \frac{20 \times \left(5-1\right) \times 101.9}{190} \sim 42.91$

So now we have the table:

Source	df	Sum of squares	Mean squares	F
Treatment	5 − 1 = 4	101.9	25.475	42.91
Error	(20 − 1)(5 − 1) = 76	88.1	1.1592	
Total	100 − 1 = 99	190		

d.f.	.995	.99	.975	.95	.9	.1	.05	.025	.01
1	0.00	0.00	0.00	0.00	0.02	2.71	3.84	5.02	6.63
2	0.01	0.02	0.05	0.10	0.21	4.61	5.99	7.38	9.21
3	0.07	0.11	0.22	0.35	0.58	6.25	7.81	9.35	11.34
4	0.21	0.30	0.48	0.71	1.06	7.78	9.49	11.14	13.28
5	0.41	0.55	0.83	1.15	1.61	9.24	11.07	12.83	15.09
6	0.68	0.87	1.24	1.64	2.20	10.64	12.59	14.45	16.81
7	0.99	1.24	1.69	2.17	2.83	12.02	14.07	16.01	18.48
8	1.34	1.65	2.18	2.73	3.49	13.36	15.51	17.53	20.09
9	1.73	2.09	2.70	3.33	4.17	14.68	16.92	19.02	21.67
10	2.16	2.56	3.25	3.94	4.87	15.99	18.31	20.48	23.21
11	2.60	3.05	3.82	4.57	5.58	17.28	19.68	21.92	24.72
12	3.07	3.57	4.40	5.23	6.30	18.55	21.03	23.34	26.22
13	3.57	4.11	5.01	5.89	7.04	19.81	22.36	24.74	27.69
14	4.07	4.66	5.63	6.57	7.79	21.06	23.68	26.12	29.14
15	4.60	5.23	6.26	7.26	8.55	22.31	25.00	27.49	30.58
16	5.14	5.81	6.91	7.96	9.31	23.54	26.30	28.85	32.00
17	5.70	6.41	7.56	8.67	10.09	24.77	27.59	30.19	33.41
18	6.26	7.01	8.23	9.39	10.86	25.99	28.87	31.53	34.81
19	6.84	7.63	8.91	10.12	11.65	27.20	30.14	32.85	36.19
20	7.43	8.26	9.59	10.85	12.44	28.41	31.41	34.17	37.57
22	8.64	9.54	10.98	12.34	14.04	30.81	33.92	36.78	40.29
24	9.89	10.86	12.40	13.85	15.66	33.20	36.42	39.36	42.98
26	11.16	12.20	13.84	15.38	17.29	35.56	38.89	41.92	45.64
28	12.46	13.56	15.31	16.93	18.94	37.92	41.34	44.46	48.28
30	13.79	14.95	16.79	18.49	20.60	40.26	43.77	46.98	50.89
32	15.13	16.36	18.29	20.07	22.27	42.58	46.19	49.48	53.49
34	16.50	17.79	19.81	21.66	23.95	44.90	48.60	51.97	56.06

Since our $df = 5 - 1 = 4$,   we look at the row with $df = 4$ and look for the value closest to our statistic $= 42.91$. We can see the value that seems the closest to 42.91 is 13.28 which falls under the significance level 0.01. Now since our test statistic is much bigger than 13.28, our p-value is way smaller than 0.01.

Since the p- value $< 0.05 = \alpha$,   we reject the null hypothesis.

9.5.1.2.3 MATLAB

Using the syntax:

```
[p,tbl,stats] = friedman(data, reps, displayopt)
```

In the syntax above, we observe three arguments: data, reps, and displayopt. The data argument retains the same meaning as in the other tests we have previously discussed.

The third argument, displayopt, enables the display of a figure that illustrates an ANOVA table, presenting the six columns as we have covered earlier.

The second argument, reps, represents the number of replicates for each combination of groups and must be specified as a positive integer. The variability of the ranks is influenced by the number of replicates. If reps = 1 (the default value), the ANOVA table divides the variability of the ranks into two components: one due to the differences among the column effects and the other due to unexplained sources (error). When reps > 1, an additional component of variability arises from the interaction between rows and columns.

9.5.1.2.4 Example 1

```
% Combine data into a matrix
        data = [Catalyst_CX' Catalyst_CY' Catalyst_CZ'];

% Perform the Friedman test
        [p, tbl, stats] = friedman(data)
```

Friedman's ANOVA Table

Source	SS	df	MS	Chi-sq	Prob>Chi-sq
Columns	4.2	2	2.1	4.2	0.1225
Error	15.8	18	0.87778		
Total	20	29			

```
p =

    0.1225


tbl =

    4×6 cell array

      {'Source' }    {'SS'     }    {'df'}    {'MS'     }    {'Chi-sq' }    {'Prob>Chi-sq'}
      {'Columns'}    {[ 4.2000]}    {[ 2]}    {[ 2.1000]}    {[ 4.2000]}    {[    0.1225]}
      {'Error'  }    {[15.8000]}    {[18]}    {[ 0.8778]}    {0×0 double}    {0×0 double   }
      {'Total'  }    {[    20]}     {[29]}    {0×0 double}    {0×0 double}    {0×0 double   }


stats =

    struct with fields:

          source: 'friedman'
               n: 10
       meanranks: [1.6000 1.9000 2.5000]
           sigma: 1
```

Since the p-value = $0.1225 > 0.05 = \alpha$,   we do not reject the null hypothesis and conclude that there is no significant difference in the reaction rates among the three catalysts.

9.5.1.2.5 Example 2

```
% Combine data into a matrix
        data = [System_A' System_B' System_C' System_D' System_E'];

% Perform the Friedman test
        [p, tbl, stats] = friedman(data,1)
```

Friedman's ANOVA Table					
Source	SS	df	MS	Chi-sq	Prob>Chi-sq
Columns	101.9	4	25.475	42.91	1.0827e-08
Error	88.1	76	1.1592		
Total	190	99			

```
p =

   1.0827e-08


tbl =

  4×6 cell array

    {'Source' }    {'SS'       }    {'df'}    {'MS'      }    {'Chi-sq' }    {'Prob>Chi-sq'}
    {'Columns'}    {[101.9000]}    {[ 4]}    {[ 25.4750]}    {[ 42.9053]}    {[ 1.0827e-08]}
    {'Error'  }    {[ 88.1000]}    {[76]}    {[  1.1592]}    {0×0 double}    {0×0 double    }
    {'Total'  }    {[     190]}    {[99]}    {0×0 double}    {0×0 double}    {0×0 double    }


stats =

  struct with fields:

        source: 'friedman'
             n: 20
     meanranks: [2.0750 2.9750 3.8500 1.7250 4.3750]
         sigma: 1.5411
```

Since the p-value $= 0.000000010827 < 0.10 = \alpha$,   we reject the null hypothesis and conclude that there is a significant difference between the cooling systems.

9.5.1.2.6 Note

The two examples we did had reps = 1; you can check the output you get when you increase the reps and see how your p-value changes.

Chapter 10

Correlation

Until now, we have conducted studies that involved univariate data. For example, in the previous chapter we studied a group of 50 students from an honors program to find out their average **IQ score**. In another case we studied a group of 40 African American men to find their average heights. Then we examined the BMI of males and females to check for the same variable—the mean. All of these examples involved the study of a single variable.

However what if we decided to study two variables instead of just one? Let's say instead of just comparing the IQ scores of the students we also want to study their college GPA, or we want to study both the height and weight of the African American men. At this point we are dealing with bivariate data. While univariate data involves the analysis of a single variable, bivariate analysis is where you are comparing two variables to study the nature of their relationship. These types of studies are quite common, and we use the concept of "correlation" to describe the relationship between the two variables.

For instance, we want to determine if the students with higher IQ also tend to have a higher GPA or vice versa. This relationship is defined by correlation which is what we are going to learn in this chapter.

10.1 WHAT IS CORRELATION?

In strict English term correlation means a connection, or some sort of relationship.

In the field of statistics, correlation is defined as a bivariate measure of the association (strength) of the relationship between two variables. It is very important to understand that correlation is not equivalent to causation, but an approximation of how likely it is that the two variables are related.

Unfortunately, the procedure of determining the relationship between variables is not as simple as it sounds. To describe the relationship between two variables both graphically and numerically we usually follow certain steps.

First step: construction of a scatter diagram.

The starting point of any such analysis should thus be the construction and subsequent examination of a scatterplot.

The first step to investigate the relationship between the two variables is to show the data values graphically on a scatter diagram. **A scatterplot** is just a visual representation of the correlation or relationship between the variables.

Let's discuss some of the properties we need to detect in a scatter diagram.

- The strength of the relationship between the two variables: on a scatter diagram, the closer the points lie on the line of best fit, the stronger the linear relationship between two variables (discussed in Chapter 3).

DOI: 10.1201/9781003399582-10

- The direction of the graph which divides the scatter diagrams into the following categories:
  - Positive correlation
    When the points on a scatterplot graph produce a lower-left-to-upper-right pattern, in other words, when both variables move in the same direction, we say that there is a **positive correlation** between the two variables. This pattern means that an increase in one variable causes another variable to increase and vice versa.
    – Negative correlation
    When the points on a scatterplot graph produce an upper-left-to-lower-right pattern (see below), simply stated, when the variables move in opposite directions, we say that there is a **negative correlation** between the two variables. This pattern means that an increase in one variable causes a decrease in the other **and vice versa**.
- The form of the relationship between the two variables:
  - A linear relationship. This means that the points on the scatterplot closely resemble a straight line. A relationship is linear if one variable increases at approximately the same rate as the other variables changes by one unit.
  - Nonlinear, rather than a straight line.

The following scatterplot examples illustrate the properties we discussed.

Second step: calculating the correlation coefficient which itself needs a separate section.

10.2 CORRELATION COEFFICIENT

Even though just by looking at a scatterplot we can immediately approximate the overall pattern of the data points, just having a rough idea of the pattern is not enough; we need a value to estimate the strength of the relationship.

The correlation coefficient gives a numerical value for the degree of correlation, or indeed, the degree of non-correlation.

However correlation coefficients can only indicate the strength of the linear relationship between two different variables. Remember we have seen that there can be three types of relationship between the two variables—linear, nonlinear, and no correlation. So if two variables have a nonlinear relationship, this correlation coefficient may not be a suitable measure of dependence. Therefore, it can provide a precise measurement of the strength of a relationship given that the form of the relationship is linear.

The correlation coefficient ranges between -1 and $+1$ and is denoted by r.

- A correlation coefficient greater than zero indicates a positive relationship while a value less than zero signifies a negative relationship.
- The closer the correlation coefficient is to ± 1 the more strongly it is linearly correlated. A correlation of exactly ± 1 indicates a perfect correlation. In other words, a correlation of -1.0 indicates a perfect negative correlation, and a correlation of $+1.0$ specifies a perfect positive correlation.
- A correlation of 0 indicates that there is no relationship between the two variables.

Absolute value of r	Strength of relationship
$r = 0$	No relationship between the variables
$r < 0.3$	Very weak
$0.3 < r < 0.5$	Weak
$0.5 < r < 0.7$	Moderate
$r > 0.7$	Strong

This demonstrates that the correlation coefficient describes both the magnitude, or the strength, and the direction of a linear relationship between two variables. The value measures the strength of the relationship and the sign indicates the direction of the relationship.

For example, a correlation coefficient of 0.20 indicates that there is a weak **linear relationship** between the variables, while a coefficient of –0.90 indicates that there is a strong linear relationship.

When there is no linear relationship between two variables, the correlation coefficient is 0. It is important to remember that a correlation coefficient of 0 indicates that there is no *linear* relationship, but there may still be a strong relationship between the two variables. For example, there could be a quadratic relationship between them. Correlation only describes linear relationships between two variables. Correlation does not describe curve relationships between variables, no matter how strong the relationship is.

There are several types of correlation coefficient. These include the Pearson product moment correlation (another term for the Pearson correlation coefficient) (r) which is commonly used in research when there are two continuous variables, the Spearman rank order correlation (rs) which is usually used when there are two ordinal variables, the Kendall rank correlation, popularly known as Kendal tau (Kt), and the rank biserial correlation (rrb) which are usually used to correlate an ordinal variable with a dichotomous variable, the point biserial correlation (rpb) which is usually used to correlate a continuous variable with a true dichotomy or to correlate a nominal variable and an internal variable, and phi coefficient of correlation (r_Φ) which is used to correlate two variables that are dichotomous.

10.2.1 Pearson correlation coefficient

This is the most common way of measuring a linear correlation.

1. **Level of measurement**: the two variables should be measured at the **interval** or **ratio** level.
2. **Linear relationship:**
   Pearson's correlation is only appropriate when there is a linear relationship between your two variables. To check this assumption we need to draw a scatterplot. When examining a scatterplot, we should study the overall pattern of the plotted points. If the data points approximately follow a straight line, you have a linear relationship. However, if the points are randomly scattered about the plot or if they exhibit some other shape other than a straight line, for example, a curved line, then a linear relationship does not exist between the variables.
3. **Normality**: both variables should be roughly normally distributed. We can check this assumption visually by creating a histogram Q-Q plot or even a boxplot for each variable.
4. **The data has no** outliers.

$$\text{Correlation coefficient} = \frac{n\Sigma(x_i y_i) - (\Sigma x_i)(\Sigma y_i)}{\sqrt{\left[n\Sigma x_i^2 - (\Sigma x_i)^2\right]\left[n\Sigma y_i^2 - (\Sigma y_i)^2\right]}}$$

Where

Σx_i =sum of all x scores of the sample: $x_1 + x_2 + x_3 + \ldots x_n$

Σx_i^2 = sum of all squared x scores of the sample: $(x_1)^2 + (x_2)^2 + \ldots (x_n)^2$

Σy_i =sum of all y scores of the sample: $y_1 + y_2 + y_3 + \ldots y_n$

Σy_i^2 = sum of all squared y scores of the sample: $(y_1)^2 + (y_2)^2 + \ldots (y_n)^2$

$\Sigma x_i y_i$= sum of the products of paired x and y scores of the sample: $(x_1 y_1) + (x_2 y_2) + \ldots (x_n y_n)$

As an illustration and application, let's look at the data used for the study of the relationship between the engine speed and fuel consumption.

Let's say an engineer is working on a project to improve the fuel efficiency of a car engine. He has collected data on various parameters such as engine speed, fuel consumption rate, and air intake temperature for a sample of cars. However he specifically wants to calculate the correlation coefficient between engine speed and fuel consumption rate. He collects the data of 15 cars and has the following values for engine speed and fuel consumption rate:

Engine speed	Consumption rate	Engine speed	Consumption rate	Engine speed	Consumption rate	Engine speed	Consumption rate
8000	10.5	6800	8.0	5800	8.2	4400	6.5
7800	9.8	6500	8.8	5500	7.1	4200	5.2
7500	8.8	6400	8.3	5200	7.5	3800	5.3
7300	10.1	6200	7.4	5000	6.3	3500	4.6
7000	9.1	5900	7.7	4600	5.9	3000	4.2

Now before we do the calculations, first we need to test if the data follows the Pearson correlation assumptions.

Level of measurement: both the speed and consumption rate are measured at **ratio** level.

Linear relationship: since we have already discussed the scatterplot in MATLAB, let's use it to check for linearity.

```
scatter(Speed, ConsumptionRate, 80, "magenta", "square","filled")

% Adding labels (optional)
xlabel('Speed')
ylabel('ConsumptionRate')
title('Scatterplot-Engine Speed vs Fuel Consumption Rate')
```

In this example, we see that the value of the consumption rate tends to increase as the speed increases. We can see an upward slope and a straight-line pattern in the plotted data points.

Normality: let's now carry out the normality test using a Q-Q plot for each variable.

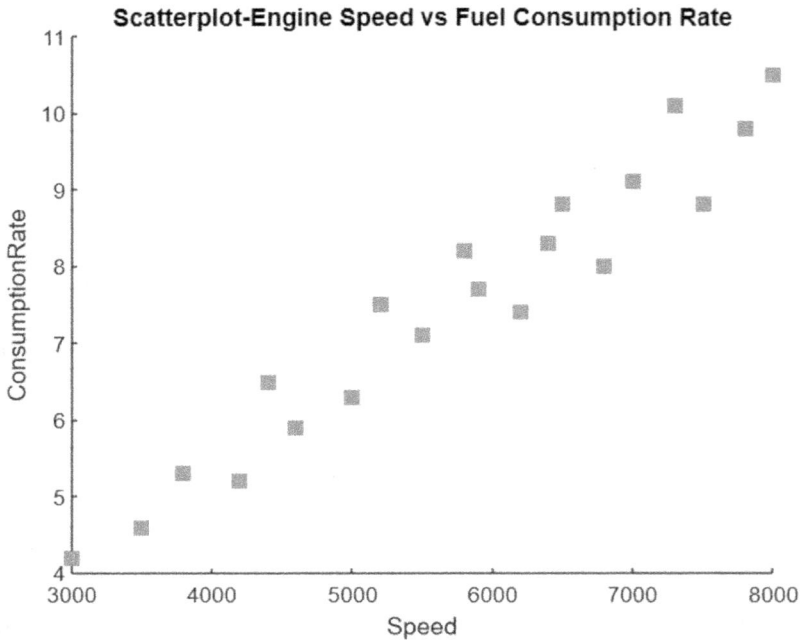

Figure 10.1

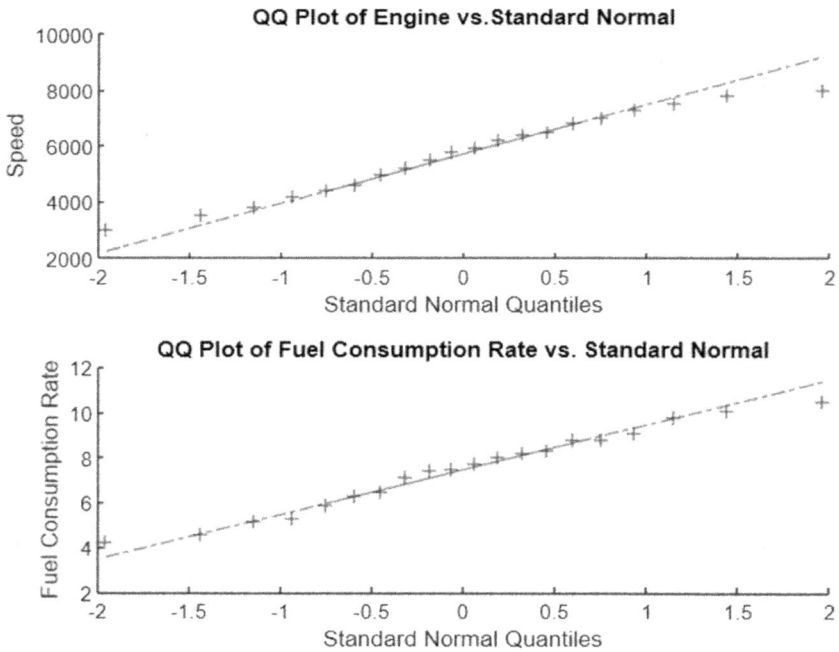

Figure 10.2

```
tiledlayout(2,1)

% Top axes
ax1 = nexttile;
qqplot(ax1,Speed)
ylabel(ax1,'Speed')
title(ax1,'QQ Plot of Engine vs.Standard Normal')

% Bottom axes
ax2 = nexttile;
qqplot(ax2,ConsumptionRate)
ylabel(ax2,'Fuel Consumption Rate')
title(ax2,'QQ Plot of Fuel Consumption Rate vs. Standard Normal')
```

The Q-Q plot in Figure 10.4 displays a straight line, suggesting that the observed quantiles align with the expected quantiles of the normal distribution. Therefore, the data meets the normality criterion.

x = speed	y = rate	xy	x^2	y^2
8000	10.5	84,000	64,000,000	110.25
7800	9.8	76,440	60,840,000	96.04
7500	8.8	66,000	56,250,000	77.44
7300	10.1	73,730	53,290,000	102.01
7000	9.1	63,700	49,000,000	82.81
6800	8.0	54,400	46,240,000	64
6500	8.8	57,200	42,250,000	77.44
6400	8.3	53,120	40,960,000	68.89
6200	7.4	45,880	38,440,000	54.76
5900	7.7	45,430	34,810,000	59.29
5800	8.2	47,560	33,640,000	67.24
5500	7.1	39,050	30,250,000	50.41
5200	7.5	39,000	27,040,000	56.25
5000	6.3	31,500	25,000,000	39.69
4600	5.9	27,140	21,160,000	34.81
4400	6.5	28,600	19,360,000	42.25
4200	5.2	21,840	17,640,000	27.04
3800	5.3	20,140	14,440,000	28.09
3500	4.6	16,100	12,250,000	21.16
3000	4.2	12,600	9,000,000	17.64
$\Sigma x_i = 114,400$	$\Sigma y_i = 149.3$	$\Sigma x_i y_i = 903,430$	$\Sigma x^2 = 695,860,000$	$\Sigma y^2 = 1177.51$

$$r = \frac{(20*903430)-(114400)(149.3)}{\sqrt{\left[20*695860000-(114400)^2\right]\left[20*1177.51-(149.3)^2\right]}} = 0.967$$

This is a strong positive correlation, which means **speed increases fuel consumption**.

Let's do another example.

Suppose an electrical engineer wants to investigate the relationship between the temperature of a certain material and its conductivity. The engineer collects data on the temperature and conductivity of the material at different points in time, and wants to determine if there is a correlation between these two variables.

Temperature (°C)	Thermal conductivity (W/m K)
20	2.10
25	2.15
30	2.40
35	2.64
40	2.30
45	2.88
50	2.64
55	2.37
60	2.90
65	3.20
70	2.73
75	3.00
80	2.40
85	2.75
90	2.56
95	2.62
100	2.60

Temperature is an interval variable while conductivity is a ratio.
Let's check for linearity.

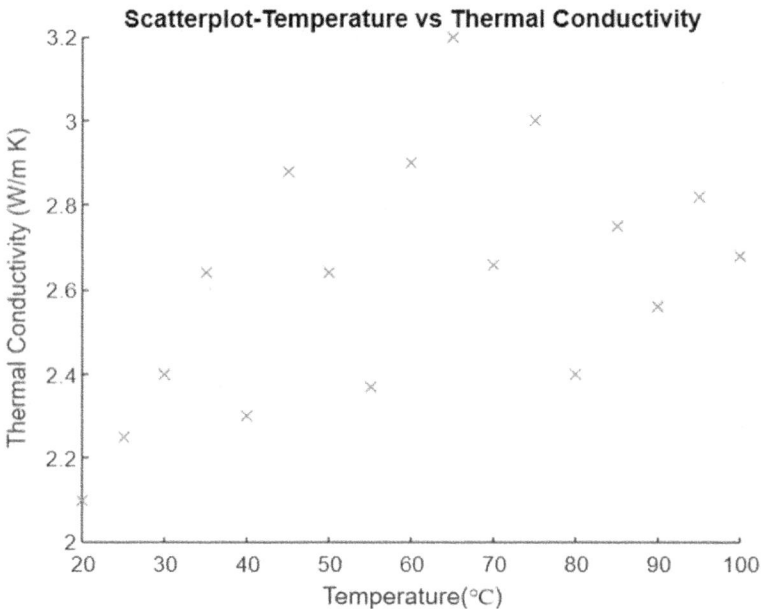

Figure 10.3

In Figure 10.3, we observe that, although the graph is not as perfectly straight as in the previous example, it is still considered linear. This suggests that the correlation coefficient may be relatively weak

Now let's test for normality.

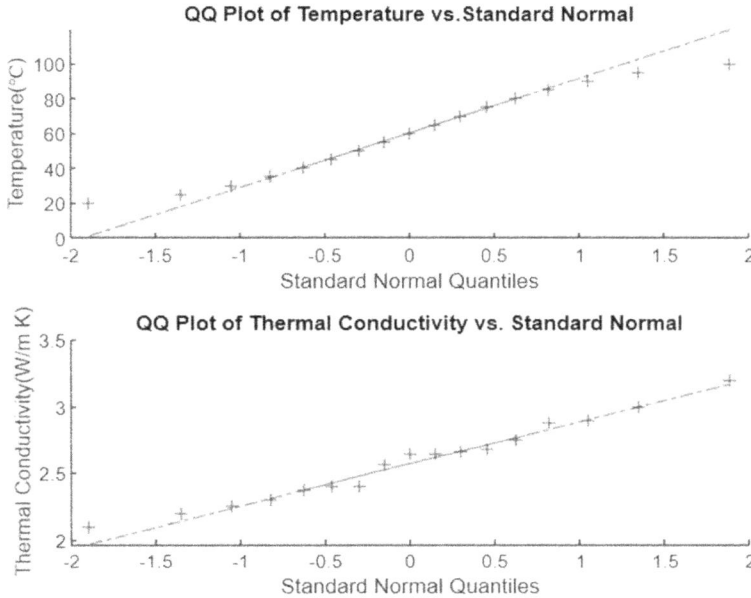

Figure 10.4

Again, even though not perfect, they are normally distributed with no extreme outliers.

x = temp	y = conductivity	xy	x^2	y^2
20	2.10	42	400	4.41
25	2.25	56.25	625	5.0625
30	2.40	72	900	5.76
35	2.64	92.4	1225	6.9696
40	2.30	92	1600	5.29
45	2.88	129.6	2025	8.2944
50	2.64	132	2500	6.9696
55	2.37	130.35	3025	5.6169
60	2.90	174	3600	8.41
65	3.20	208	4225	10.24
70	2.66	186.2	4900	7.0756
75	3.00	225	5625	9
80	2.40	192	6400	5.76
85	2.75	233.75	7225	7.5625
90	2.56	230.4	8100	6.5536
95	2.82	267.9	9025	7.9524
100	2.68	268	10000	7.1824
$\Sigma 1020$	$\Sigma 44.55$	$\Sigma 2731.85$	$\Sigma 71,400$	$\Sigma 118.1095$

Use the formula and the numbers we calculated in the previous steps:

$$r = \frac{(17*2731.85)-(1020)(44.55)}{\sqrt{\left[17*71400-(1020)^2\right]\left[17*118.1095-(44.55)^2\right]}} = 0.4992$$

As predicted the value of r is not that high, suggesting that the thermal conductivity could also be governed by the thermal properties of the material, such as its specific heat, its density, or its molecular structure.

Let's do an example of negative correlation.

In electrochemistry, the cell voltage is related to the overpotential, which describes the equilibrium potential of a cell as a function of the concentrations of the species involved and the temperature, while the current density, on the other hand, is a measure of the rate of electron transfer across the electrode–electrolyte interface. A student wants to analyze the relationship between the two variables and for that he sets up a circuit and records the measurements of the two quantities given below:

Voltage (V)	Current density	Voltage (V)	Current density
0.02	0.60	0.31	0.20
0.07	0.51	0.49	0.19
0.12	0.35	0.37	0.45
0.15	0.30	0.46	0.25
0.18	0.13	0.54	0.42
0.23	0.49	0.51	0.38
0.20	0.36	0.40	0.40
0.26	0.27	0.32	0.39
0.28	0.46	0.61	0.36
0.32	0.23	0.65	0.40

Let's check the linearity first.

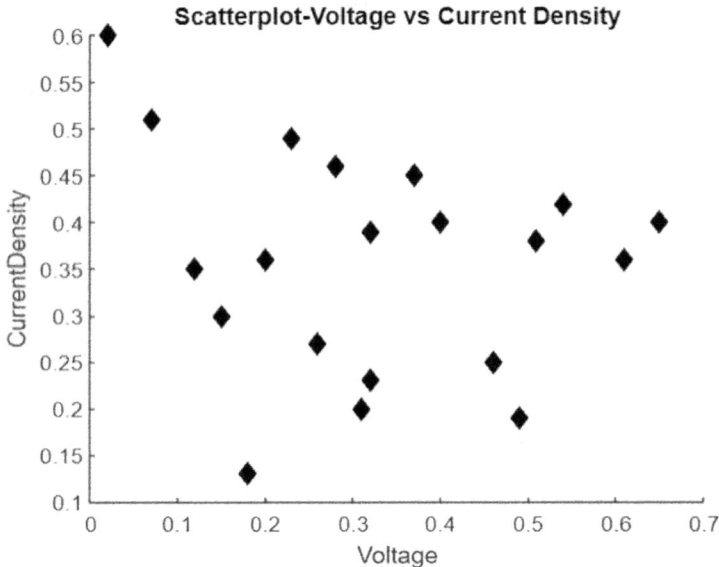

Figure 10.5

In Figure 10.5, we observe data points scattered with a general downward trend from left to right. The points do not form a clear line but show a roughly linear pattern, indicating a weak correlation.

So if we evaluate the correlation:

x = voltage	y =current density	xy	x^2	y^2
0.02	0.60	0.012	0.0004	0.36
0.07	0.51	0.0357	0.0049	0.2601
0.12	0.35	0.042	0.0144	0.1225
0.15	0.30	0.045	0.0225	0.09
0.18	0.13	0.0234	0.0324	0.0169
0.23	0.49	0.1127	0.0529	0.2401
0.20	0.36	0.072	0.04	0.1296
0.26	0.27	0.0702	0.0676	0.0729
0.28	0.46	0.1288	0.0784	0.2116
0.32	0.23	0.0736	0.1024	0.0529
0.31	0.20	0.062	0.0961	0.04
0.49	0.19	0.0931	0.2401	0.0361
0.37	0.45	0.1665	0.1369	0.2025
0.46	0.25	0.115	0.2116	0.0625
0.54	0.42	0.2268	0.2916	0.1764
0.51	0.38	0.1938	0.2601	0.1444
0.4	0.40	0.16	0.16	0.16
0.32	0.39	0.1248	0.1024	0.1521
0.61	0.36	0.2196	0.3721	0.1296
0.65	0.40	0.26	0.4225	0.16
Σ6.49	Σ7.14	Σ2.237	Σ2.7093	Σ2.8202

$$ r = \frac{(20*2.237)-(6.49)(7.14)}{\sqrt{\left[20*2.7093-(6.49)^2\right]\left[20*2.8202-(7.14)^2\right]}} = -0.19759. $$

The negative sign implies that as the current density increases, the voltage of the cell decreases. However such a small value implies that the relationship between cell voltage and current density is complex and depends on a variety of factors, such as the nature of the electrode and electrolyte, the temperature, or even the specific electrochemical system being studied.

10.2.2 Spearman correlation coefficient

Spearman's rank correlation coefficient is another widely used correlation coefficient. It is also considered the non-parametric version of the Pearson correlation coefficient. The paired variables are presented as ranks instead of interval or ratio data.

The best part of this test is that even though it relies on nearly all the same assumptions as the Pearson correlation, **it doesn't rely on normality** or linearity as long as the underlying relationship is monotonic (as one gets larger, the other variable either keeps getting larger, or keeps getting smaller), and the data can be ordinal as well. Thus, it's a non-parametric test.

There are two methods to calculate Spearman's correlation.

(1) If the data does not have tied ranks:

$$r = 1 - \frac{6\Sigma difference^2}{n(n^2 - 1)}$$

Where

d_i = the difference between paired ranks

d_i^2 = sum of squared differences between ranks

n = number of paired ranks

(2) If the data has tied ranks:

$$r = \frac{\left(\dfrac{n^3 - n}{6}\right) - \Sigma d_i^2 - \Sigma T_x - \Sigma T_y}{\sqrt{\left[\left(\dfrac{n^3 - n}{6}\right) - 2\Sigma T_x\right]\left[\left(\dfrac{n^3 - n}{6}\right) - 2\Sigma T_y\right]}}$$

Where

$$\Sigma T_x = \frac{\Sigma\left(t_i^3 - t_i\right)}{12} \quad \text{where } t_i = \textbf{number of ties} \text{ of group X ties}$$

and

$$\Sigma T_y = \frac{\Sigma\left(t_i^3 - t_i\right)}{12} \quad \text{where } t_i = \textbf{number of ties} \text{ of group Y ties}$$

10.2.2.1 Example

Richard, a professor at the University of California, set two exams in his statistics class and now wants to find the correlation between the two exam scores. He collects the data of ten students and records each of their exam scores.

Student	Exam score 1	Exam score 2
1	73	65
2	56	58
3	81	72
4	84	81
5	85	88
6	91	92
7	94	95
8	78	82
9	75	72
10	66	60

Now since the only criterion we need to fulfill in this test is monotonicity, let's draw the scatterplot of this data.

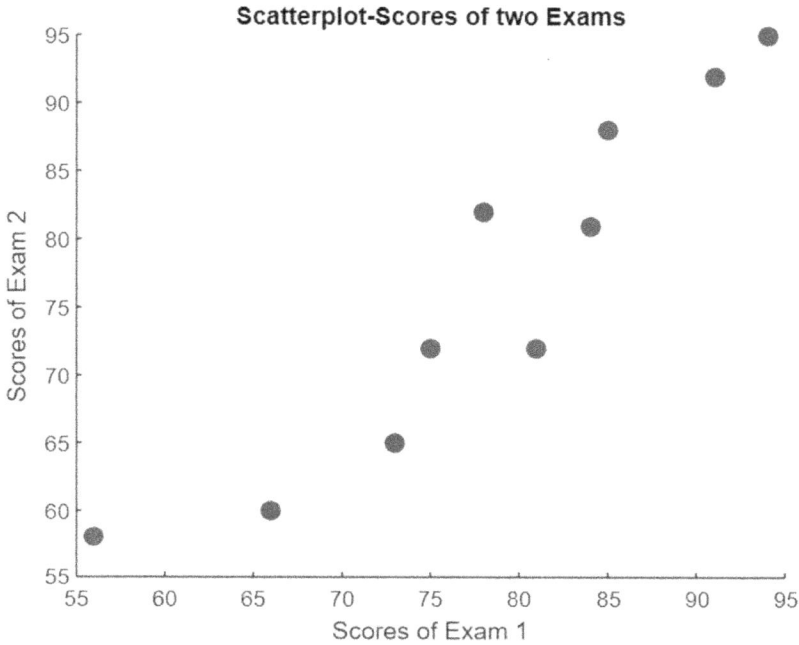

Figure 10.6

In Figure 10.6, the graph shows data points with a positive slope, indicating a positive monotonic relationship between the two variables

Step 1: rank the data of exam score 1 and exam score 2 separately.

Student	Exam score 1	Rank of exam 1	Exam score 2	Rank of exam 2
1	73	3	65	3
2	56	1	58	1
3	81	6	72	4.5
4	84	7	81	6
5	85	8	88	8
6	91	9	92	9
7	94	10	95	10
8	78	5	82	7
9	75	4	72	4.5
10	66	2	60	2

Step 2: in the next column calculate the difference between the two ranks of each student. Then use this value to find the square.

Student	Exam 1	Rank of exam 1	Exam 2	Rank of exam 2	Difference	Difference²
1	73	3	65	3	3 − 3 = 0	$0^2 = 0$
2	56	1	58	1	1 − 1 = 0	$0^2 = 0$
3	81	6	72	4.5	6 − 4.5 = 1.50	$1.50^2 = 2.25$
4	84	7	81	6	7 − 6 = 1	$1^2 = 1$
5	85	8	88	8	8 − 8 = 0	$0^2 = 0$
6	91	9	92	9	9 − 9 = 0	$0^2 = 0$
7	94	10	95	10	10 − 10 = 0	$0^2 = 0$
8	78	5	82	7	7 − 5 = 2	$2^2 = 4$
9	75	4	72	4.5	4.5 − 4 = 0.50	$0.50^2 = 0.25$
10	66	2	60	2	2 − 2 = 0	$0^2 = 0$
						$\Sigma 7.50$

In exam score 2, which is considered group Y, 81 is repeated two times, corresponding to ranks 6 and 7. Therefore

$$r = \frac{\left(\dfrac{10^3 - 10}{6}\right) - 7.50 - 0 - \dfrac{\left(2^3 - 2\right)}{12}}{\sqrt{\left[\left(\dfrac{10^3 - 10}{6}\right) - 2(0)\right]\left[\left(\dfrac{10^3 - 10}{6}\right) - 2 \times \dfrac{\left(2^3 - 2\right)}{12}\right]}} = \frac{157}{\sqrt{165 * 164}} = 0.95441$$

Such a high positive correlation coefficient implies that there is a strong correlation between the two exam scores.

10.2.2.2 Example

An engineer is working on improving the performance of a machine learning model that predicts the energy consumption of a building based on various factors such as the weather, the time of day, and the building's design features. She has collected data on the energy consumption of several buildings and their corresponding weather conditions, and she wants to determine which weather variables have the strongest relationship with energy consumption.

Temperature (°F)	Energy consumption (kWh)
−4	1600
5	1460
14	1450
23	1418
25	1373
30	1399
32	1300
36	1289
41	1278
50	1181
55	1147
59	1050

Temperature (°F)	Energy consumption (kWh)
60	926
63	900
68	894
75	760
77	700
80	750

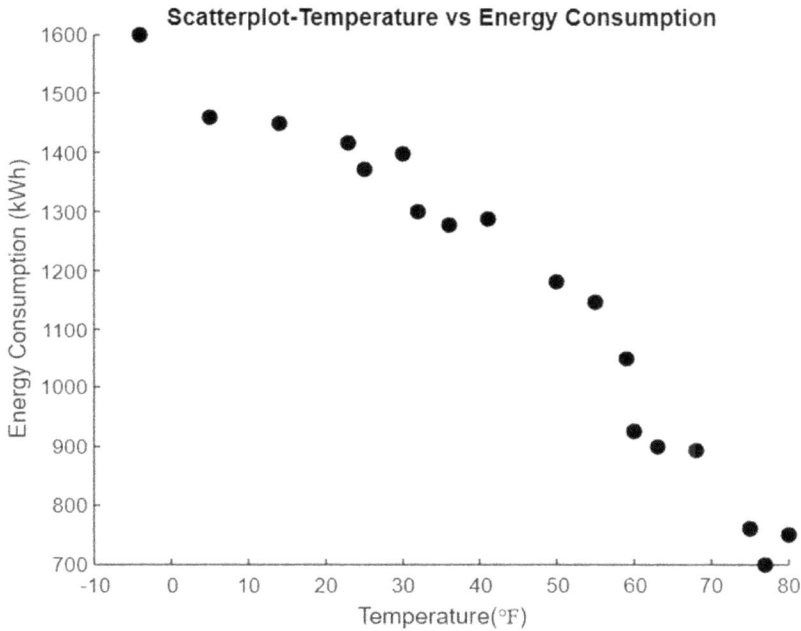

Figure 10.7

In Figure 10.7, we observe that as the temperature rises, the building's energy consumption tends to decrease. The trend line provides a linear fit to the data points and has a negative slope, indicating a negative monotonic relationship between the two variables

Therefore we now can use the Spearman correlation.

Temperature	Rank	Energy consumption	Rank	Difference	Difference^2
-4	1	1600	18	18 − 1 = 17	$17^2 = 289$
5	2	1460	17	17 − 2 = 15	$15^2 = 225$
14	3	1450	16	16 − 3 = 13	$13^2 = 169$
23	4	1418	15	15 − 4 = 11	$11^2 = 121$
25	5	1373	13	13 − 5 = 8	$8^2 = 64$
30	6	1399	14	14 − 6 = 8	$8^2 = 64$
32	7	1300	12	12 − 7 = 5	$5^2 = 25$

(Continued)

36	8	1289	11	11 − 8 = 3	$3^2 = 9$
41	9	1278	10	10 − 9 = 1	$1^2 = 1$
50	10	1181	9	10 − 9 = 1	$1^2 = 1$
55	11	1147	8	11 − 8 = 3	$3^2 = 9$
59	12	1050	7	12 − 7 = 5	$5^2 = 25$
60	13	926	6	13 − 6 = 7	$7^2 = 49$
63	14	900	5	14 − 5 = 9	$9^2 = 81$
68	15	894	4	15 − 4 = 11	$11^2 = 121$
75	16	760	3	16 − 3 = 13	$13^2 = 169$
77	17	700	1	17 − 1 = 16	$16^2 = 256$
80	18	750	2	18 − 2 = 16	$16^2 = 256$
					$\Sigma 1934$

$$r = 1 - \frac{6 \times 1934}{18\left(18^2 - 1\right)} = -0.99587$$

In this case, we conclude that temperature is an extremely significant predictor of energy consumption. The negative sign indicates a negative correlation between rising outdoor temperatures and building power consumption. However, while we might expect the opposite, this result could be influenced by various other factors which is essential to consider the broader context. For example, energy consumption patterns may vary significantly between urban and rural areas and between more and less industrialized regions.

10.2.2.3 Example

Tensile strength (MPa)	Density (g/cm³)	Tensile strength (Mpa)	Density (g/cm³)	Tensile strength (Mpa)	Density (g/cm³)
420	7.9	910	6.82	800	6.57
670	8.03	520	8.19	818	6.30
900	5.71	580	7.56	545	6.10
455	6.70	250	8.59	820	7.52
220	8.92	360	8.46	600	7.03
320	7.94	425	8.42	500	7.48
500	7.85	720	7.21	326	7.55
664	7.80	670	6.96	401	7.38
621	7.50	720	6.60	609	6.59

Using the Spearman formula we get $r = -0.65659$. The value verifies that there is a moderate negative correlation between tensile strength and the density of the material based on the sample.

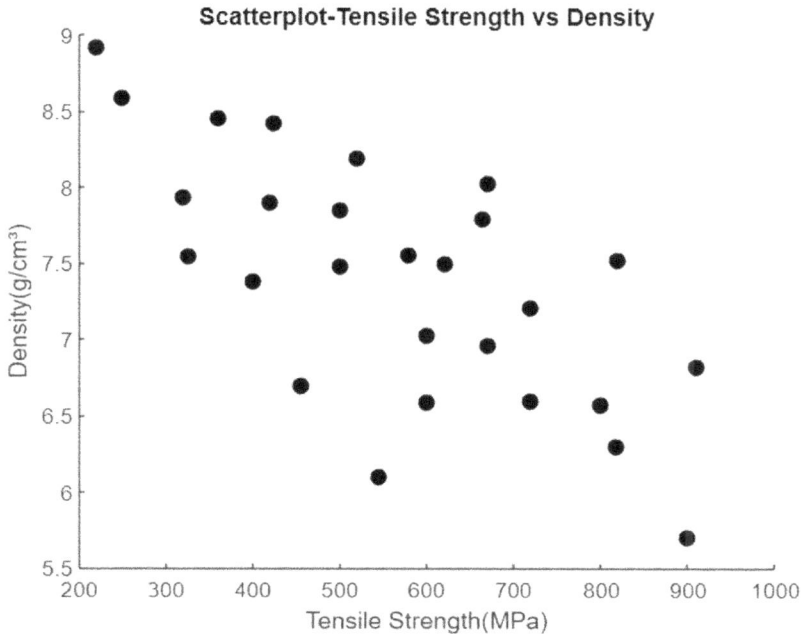

Figure 10.8

10.2.3 Kendall rank correlation coefficient

Another correlation technique for non-parametric data is the Kendall tau. Just like the Spearman, the Kendall is also rank-based. These correlations are preferred when the data is in rank-order or does not follow the assumptions of Pearson's product-moment correlation coefficient.

To calculate the Kendall correlation coefficient we follow some steps:

First step: we first rank the observations on the X and Y variables from 1 to N.

Second step: then we arrange the list of N subjects so that the rank of the subjects on variable X are in their natural order, that is, 1, 2, 3,...N.

Third step: observe the Y ranks in the order in which they occur when the X ranks are in natural order. We call this second column of the ranks of the Y variable the arranged rank.

Fourth step: now the next step is quite interesting; **to make the calculation easier we construct a table containing the arranged rank of the Y variable, placing it horizontally from left to right.** Starting from left we count the number of ranks to its right that are larger than the present rank we are dealing with; these are agreements in order. We subtract from this the number of ranks to its right that are smaller—these are the disagreements in order. If we do this for all the ranks and then sum the results we obtain S.

If there are no ties among either the X or the Y observations then we use the formula:

$$T = \frac{S}{\frac{N(N-1)}{2}}$$

Where:
S = (score of agreement – score of disagreement on X and Y)
N = number of objects or individuals ranked on both X and Y

10.2.3.1 Example

Viscosity	Pressure	Viscosity	Pressure
50.0	19.18	66.8	24.27
63.2	20.92	74.4	26.82
81.4	27.64	47.6	18.10
72.6	25.81	73.5	24.91
70.8	23.10	75.7	25.36
77.1	26.17	60.0	21.00
61.2	22.55	67.8	22.45
76.8	27.56	55.0	20.83
53.6	21.33	70.5	24.61
56.4	19.00	81.9	26.23

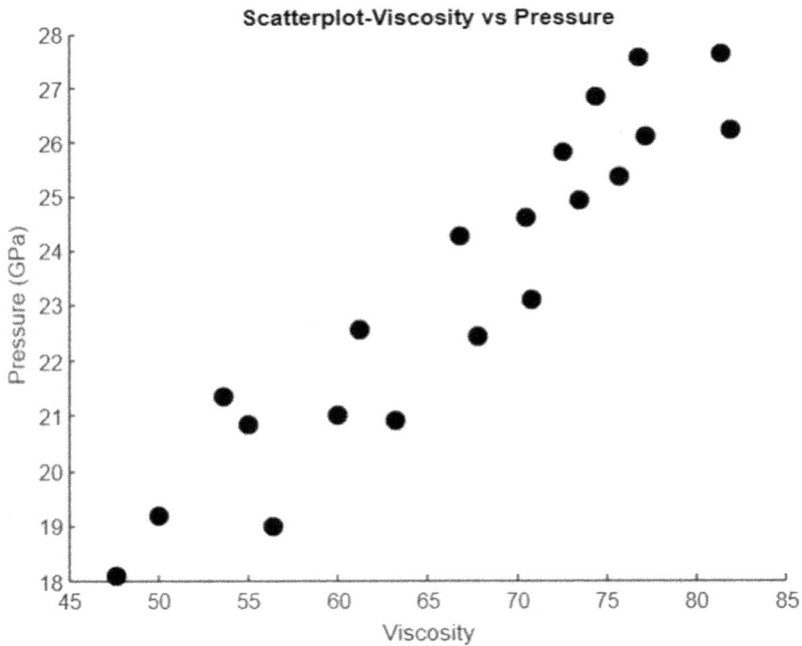

Figure 10.9

Material	Viscosity	Rank	Arranged rank	Pressure	Rank	Arranged rank
A	50.0	2	1(M)	19.18	3	1
B	63.2	8	2(A)	20.92	5	3
C	81.4	19	3(I)	27.64	20	7
D	72.6	13	4(R)	25.81	15	4
E	70.8	12	5(J)	23.10	10	2
F	77.1	18	6(P)	26.17	16	6
G	61.2	7	7(G)	22.55	9	9
H	76.8	17	8(B)	27.56	19	5
I	53.6	3	9(K)	21.33	7	11
J	56.4	5	10(Q)	19.00	2	8
K	66.8	9	11(S)	24.27	11	12
L	74.4	15	12(E)	26.82	18	10
M	47.6	1	13(D)	18.10	1	15
N	73.5	14	14(N)	24.91	13	13
O	75.7	16	15(L)	25.36	14	18
P	60.0	6	16(O)	21.00	6	14
Q	67.8	10	17(H)	22.45	8	19
R	55.0	4	18(F)	20.83	4	16
S	70.5	11	19(C)	24.61	12	20
T	81.9	20	20(T)	26.23	17	17

1	3	7	4	2	6	9	5	11	8	12	10	15	13	18	14	19	16	20	17	
1	+	+	+	+	+	+	+	+	+	+	+	+	+	+	+	+	+	+	+	+19
	3	+	-	+	+	+	+	+	+	+	+	+	+	+	+	+	+	+	+	+16
		7	-	-	-	+	-	+	+	+	+	+	+	+	+	+	+	+	+	+9
			4	-	+	+	+	+	+	+	+	+	+	+	+	+	+	+	+	+14
				2	+	+	+	+	+	+	+	+	+	+	+	+	+	+	+	+15
					6	+	-	+	+	+	+	+	+	+	+	+	+	+	+	+12
						9	-	+	-	+	+	+	+	+	+	+	+	+	+	+9
							5	+	+	+	+	+	+	+	+	+	+	+	+	+12
								11	-	+	-	+	+	+	+	+	+	+	+	+7
									8	+	+	+	+	+	+	+	+	+	+	+10
										12	-	+	+	+	+	+	+	+	+	+7
											10	+	+	+	+	+	+	+	+	+8
												15	-	+	-	+	+	+	+	+3
													13	+	+	+	+	+	+	+6
														18	-	+	-	+	-	-1
															14	+	+	+	+	+4
																19	-	+	-	-1
																	16	+	+	+2
																		20	-	-1
																			17	0
																				150

$$\frac{150}{\dfrac{20(20-1)}{2}} = 0.789473$$

Now the other way is to have two columns where the first column counts the number of ranks below the current rank that are greater than the rank. In the second column we will do the opposite, that is, we count the number of ranks that are below the current rank. Then we sum the values of each column and then subtract the second column from the first:

$$\frac{\sum concordant - \sum discordant}{\dfrac{n(n-1)}{2}}$$

10.2.3.2 Example

A researcher wants to study the relationship between the temperature$(°C)$ of a metal plate and the stress (MPa) it can withstand before breaking.

Metal	Strength	Temp	Metal	Strength	Temp	Metal	Strength	Temp
A	275	20	F	222	40	K	201	80
B	264	30	G	231	32	L	228	75
C	240	50	H	193	90	M	256	55
D	255	24	I	210	53	N	212	86
E	213	60	J	242	35	O	203	70

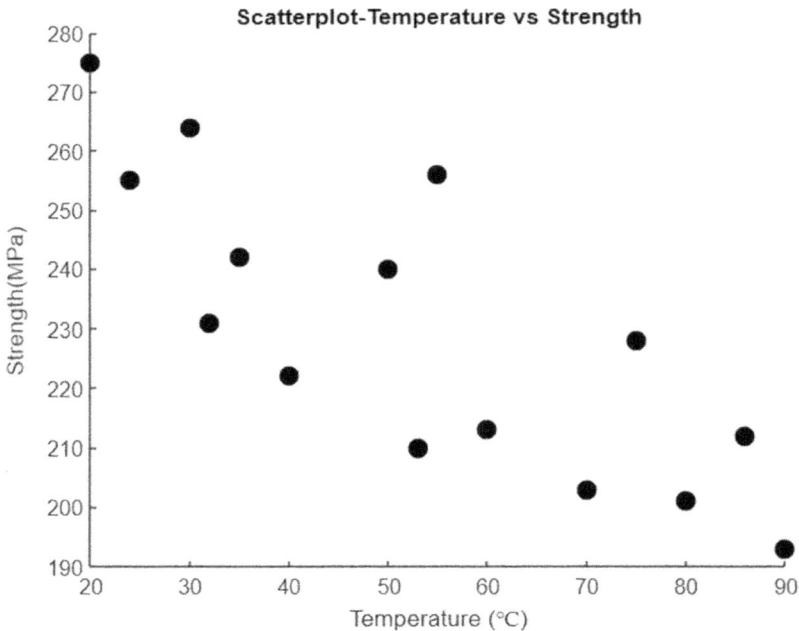

Figure 10.10

Material	Yield strength	Rank	Arranged rank	Temperature	Rank	Arranged rank
A	275	15	1(H)	20	1	15
B	264	14	2(K)	30	3	13
C	240	10	3(O)	50	7	11
D	255	12	4(I)	24	2	8
E	213	6	5(N)	60	10	14
F	222	7	6(E)	40	6	10
G	231	9	7(F)	32	4	6
H	193	1	8(L)	90	15	12
I	210	4	9(G)	53	8	4
J	242	11	10(C)	35	5	7
K	201	2	11(J)	80	13	5
L	228	8	12(D)	75	12	2
M	256	13	13(M)	55	9	9
N	212	5	14(B)	86	14	3
O	203	3	15(A)	70	11	1

Arranged rank	Concordant	Discordant
15	0	14
13	1	12
11	2	10
8	4	7
14	0	10
10	1	8
6	3	5
12	0	7
4	3	3
7	1	4
5	1	3
2	2	1
9	0	2
3	0	1
1	0	0
	$\sum 18$	$\sum 87$

$$\frac{18-87}{\dfrac{15(15-1)}{2}} = -0.657$$

A moderate negative value suggests that the yield strength of a material is affected by temperature, with higher temperatures leading to a decrease in yield strength.

If there are tied (same value) observations:

$$\frac{S}{\sqrt{\left[\dfrac{n(n-1)}{2} - \sum_{i=1}^{t} \dfrac{t_i(t_i-1)}{2}\right]\left[\dfrac{n(n-1)}{2} - \sum_{i=1}^{u} \dfrac{u_i(u_i-1)}{2}\right]}}$$

Where t_i is the number of observations tied at a particular rank of x and
u_i is the number tied at a rank of y.

10.2.3.3 Example

The reactance of a capacitor with frequency of AC source:

Reactance	Frequency	Reactance	Frequency	Reactance	Frequency
3	42	23	15	34	5
4	38	24	13	36	2
5	30	24	14	36	4
5	23	24	6	38	6
10	25	25	8	40	3
10	14	26	8	45	8
10	20	26	10	46	5
13	12	30	2	49	2
15	18	30	3	50	5

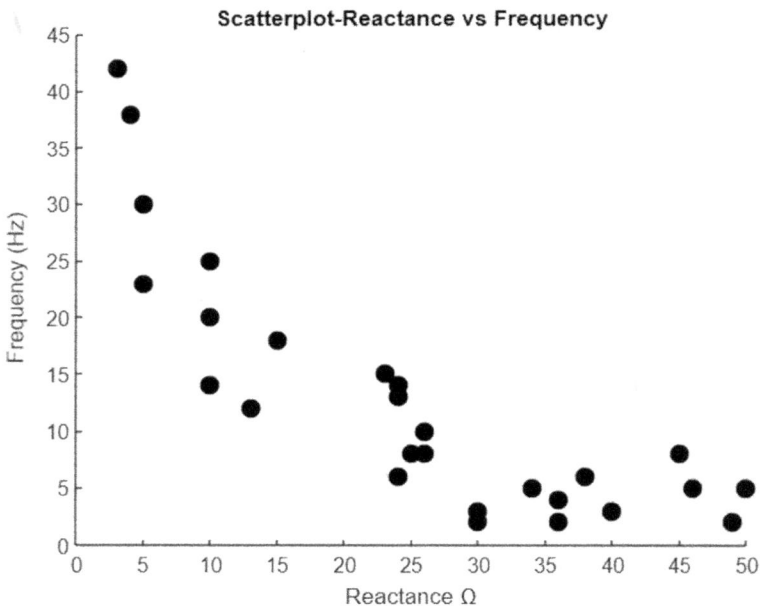

Figure 10.11

Capacitor	Reactance	Rank	Arranged rank	Frequency	Rank	Arranged rank
A	3	1	1(A)	42	27	27
B	4	2	2(B)	38	26	26
C	5	3.5	3.5(C)	30	25	25
D	5	3.5	3.5(D)	23	23	23
E	10	6	6(E)	25	24	24
F	10	6	6(F)	14	18.5	18.5

(Continued)

Capacitor	Reactance	Rank	Arranged rank	Frequency	Rank	Arranged rank
G	10	6	6(G)	20	22	22
H	13	8	8(H)	12	16	16
I	15	9	9(I)	18	21	21
J	23	10	10(J)	15	20	20
K	24	12	12(K)	13	17	17
L	24	12	12(L)	14	18.5	18.5
M	24	12	12(M)	6	10.5	10.5
N	25	14	14(N)	8	13	13
O	26	15.5	15.5(O)	8	13	13
P	26	15.5	15.5(P)	10	15	15
Q	30	17.5	17.5(Q)	2	3	3
R	30	17.5	17.5(R)	3	4.5	4.5
S	34	19	19(S)	5	8	8
T	36	20.5	20.5(T)	2	3	3
U	36	20.5	20.5(U)	4	6	6
V	38	22	22(V)	6	10.5	10.5
W	40	23	23(W)	3	4.5	4.5
X	45	24	24(X)	8	13	13
Y	46	25	25(Y)	5	8	8
Z	49	26	26(Z)	2	3	3
α	50	27	27(α)	5	8	8

Arranged rank	Greater	Smaller	Arranged rank	Greater	Smaller	Arranged rank	Greater	Smaller
27	0	26	20	0	17	8	2	4
26	0	25	17	1	15	3	6	0
25	0	24	18.5	0	15	6	4	2
23	1	22	10.5	4	9	10.5	1	4
24	0	22	13	1	10	4.5	3	1
18.5	3	17	13	1	10	13	0	3
22	0	20	15	0	11	8	0	1
16	4	15	3	8	0	3	1	0
21	0	18	4.5	6	2	8	0	0

$$Concordant = 46$$

$$Discordant = 293$$

$$T_x = 2(2-1)+3(3-1)+3(3-1)+2(2-1)+2(2-1)+2(2-1) = 20$$

$$T_x = 2(2-1)+2(2-1)+3(3-1)+3(3-1)+2(2-1)+3(3-1) = 24$$

$$\frac{46-293}{\frac{1}{2}\sqrt{[27(27-1)-20][27(27-1)-24]}} = -0.7264737$$

10.3 HYPOTHESIS TEST FOR CORRELATION COEFFICIENTS

Just like the hypothesis tests we have discussed so far, this test too takes sample data and draws conclusions about the population from which the sample was drawn.

The hypothesis test lets us decide whether the value of the population correlation coefficient ρ is "close to zero" or "significantly different from zero". We decide this based on the sample correlation coefficient r and the sample size n.

However in this test, instead of making assumptions about the mean or proportions (Chapter), this hypothesis test formally tests if there is a correlation/association between two variables in a population. *Remember that just like any other hypothesis test our sample is our estimate* of the unknown population correlation coefficient. Here we use the symbols:

ρ for the population correlation coefficient (unknown)
r for the sample correlation coefficient (known; calculated from sample data)

Now let's discuss the steps for the hypothesis testing of the coefficient correlations.

10.3.1 First step: Stating the hypotheses

Again:

- **Null hypothesis:** H_0: $\rho = 0$. This means the population correlation coefficient *is not* significantly different from zero. There *is not* a significant linear relationship (correlation) between x and y in the population.
- **Alternate hypothesis:** Ha: $\rho \neq 0$. The population correlation coefficient *is* significantly *different from* zero. There *is a significant linear relationship* (correlation) between x and y in the population.

10.3.2 Calculating the correlation coefficient

(This is equivalent to test statistics when we perform the tests for mean or median.)

Just like we have seen previously, here too we will discuss the two methods for testing the significance of the correlation coefficient.

10.3.2.1 Method 1: Using the t-test

After calculating the appropriate correlation coefficient using any one of the three correlation coefficients we discussed in the previous section, we calculate the test statistic.

10.3.2.1.1 Calculating the test statistic

$$Test\ statistic\ t^* = \frac{r\sqrt{n-2}}{\sqrt{1-r^2}}$$

Like all the other tests we have discussed so far, we can use either of the two approaches to test the significance of the correlation coefficient.

10.3.2.1.2 Using either of the approaches

10.3.2.1.2.1 USING A TABLE OF CRITICAL VALUES OR THE P-VALUE

Let's use the p-value method to test for our first Pearson example. Now generally there is a direct relationship between engine speed and fuel consumption rate. When the engine speed increases, the fuel consumption rate typically increases as well. To test the claim a researcher performs the test at $\alpha = 0.05$.

10.3.2.1.3 Stating the hypotheses:

H_0: $\rho \leq 0$

H_A: $\rho > 0$

The calculated correlation coefficient is 0.967.

$$Calculating\ the\ test\ statistic:$$

$$t^* = \frac{0.967\sqrt{20-2}}{\sqrt{1-(0.967)^2}} = 16.103$$

Now we pull up the t table of critical values for this statistic, to see whether the value we have is significant. We look at row $df = 20 - 2 = 18$ and search for the value closest to 16.103. We cannot find any value close to it; however we can approximate that it is much less than $\alpha = 0.05$.

10.3.2.1.4 Decision

Since the p-value $< \alpha$ we reject the null hypothesis in favor of the alternative and have sufficient evidence at the α level to conclude that there is a significant linear relationship between engine speed and fuel consumption rate because the correlation coefficient is significantly larger than zero.

10.3.2.2 Method 2: Using Pearson

The Pearson approach also offers the same two approaches as any other test we have seen previously.

If we want to use the critical value approach, we use the Pearson table to find the critical value based on the sample size and the significance level and then compare our r value to the critical value.

So let's say we have our $r = 0.967$ given sample size $n = 20$ and $\alpha = 0.05$.

			Proportion in ONE Tail			
	.25	.10	.05	.025	.01	.005
			Proportion in TWO Tails			
DF	.50	.20	.10	.05	.02	.01
1	.7071	.9511	.9877	.9969	.9995	.9999
2	.5000	.8000	.9000	.9500	.9800	.9900
3	.4040	.6870	.8054	.8783	.9343	.9587
4	.3473	.6084	.7293	.8114	.8822	.9172
5	.3091	.5509	.6694	.7545	.8329	.8745
6	.2811	.5067	.6215	.7067	.7887	.8343
7	.2596	.4716	.5822	.6664	.7498	.7977
8	.2423	.4428	.5494	.6319	.7155	.7646
9	.2281	.4187	.5214	.6021	.6851	.7348
10	.2161	.3981	.4973	.5760	.6581	.7079
11	.2058	.3802	.4762	.5529	.6339	.6835
12	.1968	.3646	.4575	.5324	.6120	.6614
13	.1890	.3507	.4409	.5140	.5923	.6411
14	.1820	.3383	.4259	.4973	.5742	.6226
15	.1757	.3271	.4124	.4821	.5577	.6055
16	.1700	.3170	.4000	.4683	.5425	.5897
17	.1649	.3077	.3887	.4555	.5285	.5751
18	.1602	.2992	.3783	.4438	.5155	.5614
19	.1558	.2914	.3687	.4329	.5034	.5487
20	.1518	.2841	.3598	.4227	.4921	.5368
21	.1481	.2774	.3515	.4132	.4815	.5256
22	.1447	.2711	.3438	.4044	.4716	.5151
23	.1415	.2653	.3365	.3961	.4622	.5052

From the table, for a one-tailed test at the 5% level where $n = 20$, the critical value is 0.3783. Since $r = 0.967 > 0.3783 = critical\ value$ we reject the null hypothesis and hence there is sufficient evidence to suggest that there is a strong positive correlation between the variables.

If we use the p-value approach, we compute the probability value for the coefficient and then compare it with the α.

Let's perform the same test using the method.

Since our sample size = 20, our $df = 20 - 2 = 18$, so at row 18 we look for the value closest to 0.967 and we can approximate our p-value < 0.005.

Since the p-value < 0.005 < α = 0.05, we reject the null hypothesis and conclude that the relationship is significantly larger than zero.

10.3.2.3 Example 2

Like the previous test, this is also right-tailed since we expect the thermal conductivity to increase with temperature, and to test the claim we shall perform the test at $\alpha = 0.01$.

Stating the hypotheses:

H_0: $\rho \leq 0$

H_A: $\rho > 0$

The calculated correlation coefficient is 0.4992.

Now we look at row $df = 17 - 2 = 15$ and search for the value closest to 0.4992.

DF	Proportion in ONE Tail					
	.25	.10	.05	.025	.01	.005
	Proportion in TWO Tails					
DF	.50	.20	.10	.05	.02	.01
1	.7071	.9511	.9877	.9969	.9995	.9999
2	.5000	.8000	.9000	.9500	.9800	.9900
3	.4040	.6870	.8054	.8783	.9343	.9587
4	.3473	.6084	.7293	.8114	.8822	.9172
5	.3091	.5509	.6694	.7545	.8329	.8745
6	.2811	.5067	.6215	.7067	.7887	.8343
7	.2596	.4716	.5822	.6664	.7498	.7977
8	.2423	.4428	.5494	.6319	.7155	.7646
9	.2281	.4187	.5214	.6021	.6851	.7348
10	.2161	.3981	.4973	.5760	.6581	.7079
11	.2058	.3802	.4762	.5529	.6339	.6835
12	.1968	.3646	.45	.5324	.6120	.6614
13	.1890	.3507		.5140	.5923	.6411
14	.1820	.3383	.425	.4973	.5742	.6226
15	.1757	.3271	.4124	.4821	.5577	.6055
16	.1700	.3170	.4000	.4683	.5425	.5897
17	.1649	.3077	.3887	.4555	.5285	.5751
18	.1602	.2992	.3783	.4438	.5155	.5614
19	.1558	.2914	.3687	.4329	.5034	.5487
20	.1518	.2841	.3598	.4227	.4921	.5368
21	.1481	.2774	.3515	.4132	.4815	.5256
22	.1447	.2711	.3438	.4044	.4716	.5151
23	.1415	.2653	.3365	.3961	.4622	.5052

Since our coefficient is a little bigger than 0.4821, we can approximate our p-value to be slightly smaller than 0.025 so around 0.020.

10.3.2.3.1 Decision

Since the p-value = 0.020 > 0.01 = *critical value* we fail to reject the null hypothesis and conclude that the relationship is *not* significantly different from zero.

10.3.2.4 Example 3

The student analyzing the relationship between the current density and voltage also wants to analyze if there is a negative correlation between the variables at $\alpha = 0.10$.

Stating the hypotheses:

H_0: $\rho \geq 0$

H_A: $\rho < 0$

The correlation coefficient we calculate is −0.19759.

Now we look at row $df = 20 - 2 = 18$ and search for the value closest to $|r| = 0.19759$.

		Proportion in ONE Tail				
	.25	.10	.05	.025	.01	.005
			Proportion in TWO Tails			
DF	.50	.20	.10	.05	.02	.01
1	.7071	.9511	.9877	.9969	.9995	.9999
2	.5000	.8000	.9000	.9500	.9800	.9900
3	.4040	.6870	.8054	.8783	.9343	.9587
4	.3473	.6084	.7293	.8114	.8822	.9172
5	.3091	.5509	.6694	.7545	.8329	.8745
6	.2811	.5067	.6215	.7067	.7887	.8343
7	.2596	.4716	.5822	.6664	.7498	.7977
8	.2423	.4428	.5494	.6319	.7155	.7646
9	.2281	.4187	.5214	.6021	.6851	.7348
10	.2161	.3981	.4973	.5760	.6581	.7079
11	.2058	.3802	.4762	.5529	.6339	.6835
12	.1968	.3646	.4575	.5324	.6120	.6614
13	.1890	.3507	.4409	.5140	.5923	.6411
14	.1820	.3383	.4259	.4973	.5742	.6226
15	.1757	.3271	.4124	.4821	.5577	.6055
16	.1700	.3170	.4000	.4683	.5425	.5897
17	.1649	.3077	.3887	.4555	.5285	.5751
18	.1602	.2992	.3783	.4438	.5155	.5614
19	.1558	.2914	.3687	.4329	.5034	.5487
20	.1518	.2841	.3598	.4227	.4921	.5368
21	.1481	.2774	.3515	.4132	.4815	.5256
22	.1447	.2711	.3438	.4044	.4716	.5151
23	.1415	.2653	.3365	.3961	.4622	.5052

Our coefficient falls between 0.1602 and 0.2992 which is between the values 0.25 and 0.10. So now if we find the average of the two critical values we get 0.23; therefore our potential p-value could be around 0.18. Again this 0.23 is still not close to the coefficient we got but we know the p-value could now be between 0.25 and 0.18. So we have narrowed the range of our potential p-value. Again if we find the average of 0.1602 and 0.23 we get 0.1951 which is close to 0.19759; therefore we also can find the p-value by finding the average of the corresponding significant levels 0.25 and 0.18 which is 0.215. Now since our coefficient is slightly larger than 0.1951, our p-value is slightly less than 0.215 so approximately 0.20.

Since the p-$value = 0.20 > 0.10 = critical value$, we fail to reject the null hypothesis and conclude that the relationship is *not* significantly different from zero.

A hypothesis test using Spearman's rank correlation coefficient is very similar to one conducting using Pearson. Here we will be using the Spearman table to make conclusions.

10.3.2.5 Example 1

After setting the two exams, Richard decides to determine if the scores of one group of students are significantly different from the scores of another group of students. (Use $\alpha = 0.1$.)

Stating the hypotheses:

H_0: $\rho = 0$

H_A: $\rho \neq 0$

The calculated correlation coefficient is 0.95441.

Now we look at row $df = 10 - 2 = 8$ and search for the value closest to 0.95441.

	One-tailed	Level of significance (H$_0$: $\rho_s \leq 0$ or H$_0$: $\rho_s \geq 0$)					
	$\alpha = 0.1$	$\alpha = 0.05$	$\alpha = 0.025$	$\alpha = 0.01$	$\alpha = 0.005$	$\alpha = 0.0005$	
	Two-tailed	Level of significance (H$_0$: $P_s = 0$ & H$_1$: $P_s \neq 0$)					
df	$\alpha = 0.2$	$\alpha = 0.1$	$\alpha = 0.05$	$\alpha = 0.02$	$\alpha = 0.01$	$\alpha = 0.001$
2	1.000	1.000				
3	0.800	0.900	1.000	1.000		
4	0.657	0.829	0.886	0.943	1.000	
5	0.571	0.714	0.786	0.893	0.929	1.000
6	0.524	0.643	0.738	0.833	0.881	0.976
7	0.483	0.600	0.700	0.783	0.833	0.933
8	0.455	0.564	0.648	0.745	0.794	0.903
9	0.427	0.536	0.618	0.709	0.755	0.873
10	0.406	0.503	0.587	0.678	0.727	0.846
11	0.385	0.484	0.560	0.648	0.703	0.824
12	0.367	0.464	0.538	0.626	0.679	0.802

Since our coefficient is a little slightly larger than 0.903 we can approximate our p-value to be smaller than 0.001.

10.3.2.5.1 Decision

Since the p-value $< 0.001 < 0.01 = \alpha$, we reject the null hypothesis and conclude that the relationship between the two exam scores is significantly different from zero.

10.3.2.6 Example

Let's say the engineer working on improving the performance of a machine learning model predicting the energy consumption of a building and temperature also wants to determine if there is a negative correlation between these two variables at $\pm = 0.10$.

Stating the hypotheses:

H_0: $\rho = 0$

H_A: $\rho < 0$

The calculated correlation coefficient is $= -0.99587$.

Look at row $df = 18 - 2 = 16$ and search for the value closest to the absolute value of -0.99587.

	One-tailed	Level of significance (H$_0$: $\rho_s \leq 0$ or H$_0$: $\rho_s \geq 0$)				
	$\alpha = 0.1$	$\alpha = 0.05$	$\alpha = 0.025$	$\alpha = 0.01$	$\alpha = 0.005$	$\alpha = 0.0005$
	Two-tailed	Level of significance (H$_0$: $P_s = 0$ & H$_1$: $P_s \neq 0$)				
df	$\alpha = 0.2$	$\alpha = 0.1$	$\alpha = 0.05$	$\alpha = 0.02$	$\alpha = 0.01$	$\alpha = 0.001$
2	1.000	1.000				
3	0.800	0.900	1.000	1.000		
4	0.657	0.829	0.886	0.943	1.000	
5	0.571	0.714	0.786	0.893	0.929	1.000
6	0.524	0.643	0.738	0.833	0.881	0.976
7	0.483	0.600	0.700	0.783	0.833	0.933
8	0.455	0.564	0.648	0.745	0.794	0.903
9	0.427	0.536	0.618	0.709	0.755	0.873
10	0.406	0.503	0.587	0.678	0.727	0.846
11	0.385	0.484	0.560	0.648	0.703	0.824
12	0.367	0.464	0.538	0.626	0.679	0.802
13	0.354	0.446	0.521	0.604	0.654	0.779
14	0.341	0.429	0.503	0.582	0.635	0.762
15	0.328	0.414	0.485	0.566	0.615	0.748
16	0.317	0.401	0.472	0.550	0.600	0.728
17	0.309	0.391	0.460	0.535	0.584	0.712

Since our coefficient is a lot larger than 0.728 we can approximate our p-value to be smaller than 0.0005.

10.3.2.6.1 Decision

Since the p-value $< 0.0005 < 0.10 = \alpha$, we reject the null hypothesis and conclude that the relationship between energy consumption **and temperature** is significantly different from zero.

10.3.2.7 Example 3

We want to analyze if there is a negative correlation between tensile strength and density at $\pm = 0.05$.

Stating the hypotheses:

H_0: $\rho = 0$

H_A: $\rho < 0$

The calculated correlation coefficient is $= -0.65659 \sim -0.6566$.

Look at row $df = 27 - 2 = 25$ and search for the value closest to 0.6566.

df	One-tailed	Level of significance (H₀: $\rho_s \leq 0$ or H₀: $\rho_s \geq 0$)					
	$\alpha = 0.1$	$\alpha = 0.05$	$\alpha = 0.025$	$\alpha = 0.01$	$\alpha = 0.005$	$\alpha = 0.0005$	
	Two-tailed	Level of significance (H₀: $P_s = 0$ & H₁: $P_s \neq 0$)					
	$\alpha = 0.2$	$\alpha = 0.1$	$\alpha = 0.05$	$\alpha = 0.02$	$\alpha = 0.01$	$\alpha = 0.001$	
2	1.000	1.000					
3	0.800	0.900	1.000	1.000			
4	0.657	0.829	0.886	0.943	1.000		
5	0.571	0.714	0.786	0.893	0.929	1.000	
6	0.524	0.643	0.738	0.833	0.881	0.976	
7	0.483	0.600	0.700	0.783	0.833	0.933	
8	0.455	0.564	0.648	0.745	0.794	0.903	
9	0.427	0.536	0.618	0.709	0.755	0.873	
10	0.406	0.503	0.587	0.678	0.727	0.846	
11	0.385	0.484	0.560	0.648	0.703	0.824	
12	0.367	0.464	0.538	0.626	0.679	0.802	
13	0.354	0.446	0.521	0.604	0.654	0.779	
14	0.341	0.429	0.503	0.582	0.635	0.762	
15	0.328	0.414	0.485	0.566	0.615	0.748	
16	0.317	0.401	0.472	0.550	0.600	0.728	
17	0.309	0.391	0.460	0.535	0.584	0.712	
18	0.299	0.380	0.447	0.520	0.570	0.696	
19	0.292	0.370	0.435	0.508	0.556	0.681	
20	0.284	0.361	0.425	0.496	0.544	0.667	
21	0.278	0.353	0.415	0.486	0.532	0.654	
22	0.271	0.344	0.406	0.476		0.642	
23	0.265	0.337	0.398	0.466		0.630	
24	0.259	0.331	0.390	0.457		0.619	
25	0.255	0.324	0.382	0.448	0.4	0.608	
26	0.250	0.317	0.375	0.440	0.483	0.598	
27	0.245	0.312	0.368	0.433	0.475	0.589	
28	0.240	0.306	0.362	0.425	0.467	0.580	
29	0.236	0.301	0.356	0.418	0.459	0.571	

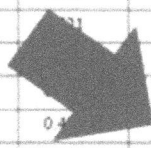

Since our coefficient is slightly larger than 0.6566 we can approximate our p-value to be smaller than 0.0005.

10.3.2.7.1 Decision

Since the p-value $< 0.0005 < 0.05 = \alpha$, we reject the null hypothesis and conclude that the relationship between the tensile strength and density is significantly different from zero.

Now let's perform the hypothesis tests for the Kendal coefficient.

10.3.2.8 Example 1

Stating the hypotheses:

H_0: $\rho \le 0$

H_A: $\rho > 0$

The calculated correlation coefficient is 0.7895.

Now we look at row $df = 27 - 2 = 25$ and search for the value closest to 0.7895.

	Nominal α					
n	0.10	0.05	0.025	0.01	0.005	0.001
4	1.000	1.000	-	-	-	-
5	0.800	0.800	1.000	1.000	-	-
6	0.600	0.733	0.867	0.867	1.000	-
7	0.524	0.619	0.714	0.810	0.905	1.000
8	0.429	0.571	0.643	0.714	0.786	0.857
9	0.389	0.500	0.556	0.667	0.722	0.833
10	0.378	0.467	0.511	0.600	0.644	0.778
11	0.345	0.418	0.491	0.564	0.600	0.709
12	0.303	0.394	0.455	0.545	0.576	0.667
13	0.308	0.359	0.436	0.513	0.564	0.641
14	0.275	0.363	0.407	0.473	0.516	0.604
15	0.276	0.333	0.390	0.467	0.505	0.581
16	0.250	0.317	0.383	0.433	0.483	0.567
17	0.250	0.309	0.368	0.426	0.471	0.544
18	0.242	0.294	0.346	0.412	0.451	0.529
19	0.228	0.287	0.333	0.392	0.439	0.509
20	0.221	0.274	0.326	0.379	0.421	0.495
21	0.210	0.267	0.314	0.371	0.410	0.486
22	0.203	0.264	0.307	0.359	0.394	0.472
23	0.202	0.257	0.296	0.352		0.455
24	0.196	0.246	0.290	0.341		0.449
25	0.193	0.240	0.287	0.333	0.	0.440
26	0.188	0.237	0.280	0.329	0.360	0.428
27	0.179	0.231	0.271	0.322	0.356	0.419
28	0.180	0.228	0.265	0.312	0.344	0.413
29	0.172	0.222	0.261	0.310	0.340	0.404

Since our coefficient is a lot larger than 0.440 we can approximate our p-value to be a lot smaller than 0.001.

10.3.2.8.1 Decision

Since the p-value $< 0.001 < 0.05 = \alpha$, we reject the null hypothesis and conclude that there is a positive correlation between viscosity and pressure.

10.3.2.9 Example 2

Stating the hypotheses:

H_0: $\rho \geq 0$

H_A: $\rho < 0$

The calculated correlation coefficient is −0.657.

Now we look at row $df = 15 - 2 = 13$ and search for the value closest to 0.657.

	Nominal α					
n	0.10	0.05	0.025	0.01	0.005	0.001
4	1.000	1.000	-	-	-	-
5	0.800	0.800	1.000	1.000	-	-
6	0.600	0.733	0.867	0.867	1.000	-
7	0.524	0.619	0.714	0.810	0.905	1.000
8	0.429	0.571	0.643	0.714	0.786	0.857
9	0.389	0.500	0.556	0.667	0.722	0.833
10	0.378	0.467	0.511	0.600	44	0.778
11	0.345	0.418	0.491	0.564		0.709
12	0.303	0.394	0.455	0.545		0.667
13	0.308	0.359	0.436	0.513	0	0.641
14	0.275	0.363	0.407	0.473	0.516	0.604
15	0.276	0.333	0.390	0.467	0.505	0.581
16	0.250	0.317	0.383	0.433	0.483	0.567
17	0.250	0.309	0.368	0.426	0.471	0.544
18	0.242	0.294	0.346	0.412	0.451	0.529
19	0.228	0.287	0.333	0.392	0.439	0.509
20	0.221	0.274	0.326	0.379	0.421	0.495
21	0.210	0.267	0.314	0.371	0.410	0.486
22	0.203	0.264	0.307	0.359	0.394	0.472
23	0.202	0.257	0.296	0.352	0.391	0.455
24	0.196	0.246	0.290	0.341	0.377	0.449
25	0.193	0.240	0.287	0.333	0.367	0.440
26	0.188	0.237	0.280	0.329	0.360	0.428
27	0.179	0.231	0.271	0.322	0.356	0.419
28	0.180	0.228	0.265	0.312	0.344	0.413
29	0.172	0.222	0.261	0.310	0.340	0.404

Since our coefficient is slightly larger than 0.641 we can approximate our p-value to be smaller than 0.001.

10.3.2.9.1 Decision

Since the p-value $< 0.001 < 0.05 = \alpha$, we reject the null hypothesis and conclude that there is a negative correlation between the temperature $(°C)$ of a metal plate and the stress (MPa).

10.3.2.10 Example 3

Stating the hypotheses:

H_0: $\rho \geq 0$

H_A: $\rho < 0$

The calculated correlation coefficient is −0.7265.

Now we look at row $df = 27 - 2 = 25$ and search for the value closest to 0.7265.

				Nominal α		
n	0.10	0.05	0.025	0.01	0.005	0.001
4	1.000	1.000	-	-	-	-
5	0.800	0.800	1.000	1.000	-	-
6	0.600	0.733	0.867	0.867	1.000	-
7	0.524	0.619	0.714	0.810	0.905	1.000
8	0.429	0.571	0.643	0.714	0.786	0.857
9	0.389	0.500	0.556	0.667	0.722	0.833
10	0.378	0.467	0.511	0.600	0.644	0.778
11	0.345	0.418	0.491	0.564	0.600	0.709
12	0.303	0.394	0.455	0.545	0.576	0.667
13	0.308	0.359	0.436	0.513	0.564	0.641
14	0.275	0.363	0.407	0.473	0.516	0.604
15	0.276	0.333	0.390	0.467	0.505	0.581
16	0.250	0.317	0.383	0.433	0.483	0.567
17	0.250	0.309	0.368	0.426	0.471	0.544
18	0.242	0.294	0.346	0.412	0.451	0.529
19	0.228	0.287	0.333	0.392	0.439	0.509
20	0.221	0.274	0.326	0.379	0.421	0.495
21	0.210	0.267	0.314	0.371	0.410	0.486
22	0.203	0.264	0.307	0.359		0.472
23	0.202	0.257	0.296	0.352		0.455
24	0.196	0.246	0.290	0.341		0.449
25	0.193	0.240	0.287	0.333		0.440
26	0.188	0.237	0.280	0.329	0.360	0.428
27	0.179	0.231	0.271	0.322	0.356	0.419
28	0.180	0.228	0.265	0.312	0.344	0.413
29	0.172	0.222	0.261	0.310	0.340	0.404

10.3.2.10.1 Decision

Since the p-*value* $< 0.001 < 0.05 = \alpha$, we reject the null hypothesis and conclude that there is a negative correlation between the reactance of a capacitor and the frequency of the AC source.

10.3.2.11 Confidence interval for correlation coefficient

The sampling distribution of Pearson's r is not normally distributed. Fisher developed a transformation called "Fisher's z-transformation" that converts Pearson's r to the normally distributed variable z. Therefore to calculate a confidence interval r must be transformed to give a normal distribution using Fisher's transformation.

First step: convert a correlation to a z-score or z using the Fisher transformation:

$$z_r = tanh^{-1}(r) = \frac{1}{2} ln \frac{(1+r)}{(1-r)}$$

The standard error of z_r is $\dfrac{1}{\sqrt{n-3}}$

$$Lower\ bound\ L = z_r - \left(z^* \times \frac{1}{\sqrt{n-3}} \right)$$

$$Upper\ bound\ U = z_r + \left(z^* \times \frac{1}{\sqrt{n-3}} \right)$$

For two sided limits $z^* = z_{1-\alpha/2}$ *and for one sided* $z^* = z_{1-\alpha}$

The final confidence interval can be found using the following formula:

$$Confidence\ level = \left[\frac{e^{2L}-1}{e^{2L}+1}, \frac{e^{2U}-1}{e^{2U}+1} \right]$$

10.3.2.12 Example I

Say we want to determine a 95% confidence interval for the correlation coefficient between the engine speed and the fuel consumption rate.

$$z_r = \frac{1}{2} ln \frac{(1+0.967)}{(1-0.967)} = 2.04388$$

Since the value of z^* for a confidence level of 95% is 1.96,

$$Lower\ bound\ L = 2.04388 - \left(1.96 \times \frac{1}{\sqrt{20-3}} \right) = 1.56851$$

$$Upper\ bound\ U = 2.04388 + \left(1.96 \times \frac{1}{\sqrt{20-3}} \right) = 2.51925$$

$$Confidence\ level = \left[\frac{e^{2(1.56851)}-1}{e^{2(1.56851)}+1}, \frac{e^{2(2.51925)}-1}{e^{2(2.51925)}+1} \right] = \left[0.9168, 0.9871 \right]$$

10.3.2.13 Example 2

Let's determine a 90% confidence interval for the correlation coefficient between temperature and conductivity.

$$z_r = \frac{1}{2} ln \frac{(1 + 0.4992)}{(1 - 0.4992)} = 0.54824$$

Since the value of z^* for a confidence level of 95% is 1.645,

$$Lower\ bound\ L = 0.54824 - \left(1.645 \times \frac{1}{\sqrt{17-3}}\right) = 0.10859$$

$$Upper\ bound\ U = 0.54824 + \left(1.645 \times \frac{1}{\sqrt{17-3}}\right) = 0.98788$$

$$Confidence\ level = \left[\frac{e^{2(0.10859)} - 1}{e^{2(0.10859)} + 1}, \frac{e^{2(0.98788)} - 1}{e^{2(0.98788)} + 1}\right] = [0.10816, 0.75645]$$

10.3.2.14 Example 3

Now we determine at 99% confidence interval for the correlation coefficient between voltage and current density.

$$z_r = \frac{1}{2} ln \frac{(1 + (-0.19759))}{(1 - (-0.19759))} = -0.20$$

Since the value of z^* for a confidence level of 99% is 2.576,

$$Lower\ bound\ L = -\left(2.576 \times \frac{1}{\sqrt{20-3}}\right) = -0.8211146$$

$$Upper\ bound\ U = +\left(2.576 \times \frac{1}{\sqrt{20-3}}\right) = 0.424772$$

$$Confidence\ level = \left[\frac{e^{2(-0.8211146)} - 1}{e^{2(-0.8211146)} + 1}, \frac{e^{2(0.424772)} - 1}{e^{2(0.424772)} + 1}\right] = [-0.68, 0.40]$$

10.4 CONFIDENCE INTERVAL OF SPEARMAN CORRELATION COEFFICIENT

The standard error of z_r is $\sqrt{\dfrac{1 + \dfrac{r^2}{2}}{n-3}}$

$$Lower\ bound\ L = z_r - \left(z^* \times \sqrt{\frac{1+\frac{r^2}{2}}{n-3}} \right)$$

$$Upper\ bound\ U = z_r + \left(z^* \times \sqrt{\frac{1+\frac{r^2}{2}}{n-3}} \right)$$

For two sided limits $z^* = z_{1-\alpha/2}$ and for one sided $z^* = z_1$

The final confidence interval can be found using the following formula:

$$Confidence\ level = \left[\frac{e^{2L}-1}{e^{2L}+1}, \frac{e^{2U}-1}{e^{2U}+1} \right]$$

10.4.1 Example

Let's determine a 95% confidence interval for this correlation coefficient between two exam scores.

$$z_r = \frac{1}{2} ln \frac{(1+0.95441)}{(1-0.95441)} = 1.88$$

$$Lower\ bound\ L = 1.88 - \left(1.96 \times \sqrt{\frac{1+\frac{0.95441^2}{2}}{10-3}} \right) = 0.98627$$

$$Upper\ bound\ U = 1.88 + \left(1.96 \times \sqrt{\frac{1+\frac{0.95441^2}{2}}{10-3}} \right) = 2.77372$$

$$Confidence\ level = \left[\frac{e^{2(0.98627)}-1}{e^{2(0.98627)}+1}, \frac{e^{2(2.77372)}-1}{e^{2(2.77372)}+1} \right] = [0.75576, 0.99223]$$

10.4.2 Example

Let's determine a 99% confidence interval for this correlation coefficient between temperature and energy consumption.

$$z_r = \frac{1}{2} \ln \frac{\left(1+\left(-0.99587\right)\right)}{\left(1-\left(-0.99587\right)\right)} = -3.09252$$

$$\text{Lower bound } L = -3.09252 - \left(2.576 \times \sqrt{\frac{1+\dfrac{\left(-0.99587\right)^2}{2}}{18-3}} \right) = -3.906$$

$$\text{Upper bound } U = -3.09252 + \left(2.576 \times \sqrt{\frac{1+\dfrac{\left(-0.99587\right)^2}{2}}{18-3}} \right) = -2.279$$

$$\text{Confidence level} = \left[\frac{e^{2(-3.906)}-1}{e^{2(-3.906)}+1}, \frac{e^{2(-2.279)}-1}{e^{2(-2.279)}+1} \right] = \left[-0.99919, -0.97925 \right]$$

10.4.3 Example

Let's determine a 90% confidence interval for this correlation coefficient between tensile strength and density.

$$z_r = \frac{1}{2} \ln \frac{\left(1+\left(-0.65659\right)\right)}{\left(1-\right)} = -0.786815$$

$$\text{Lower bound } L = -0.786815 - \left(1.645 \times \sqrt{\frac{1+\dfrac{\left(-0.65659\right)^2}{2}}{27-3}} \right) = --.157015$$

$$\text{Upper bound } U = -0.786815 + \left(1.645 \times \sqrt{\frac{1+\dfrac{\left(-0.65659\right)^2}{2}}{27-3}} \right) = --.416615$$

$$\text{Confidence level} = \left[\frac{e^{2(-1.157015)}-1}{e^{-(-1.157015)}+1}, \frac{e^{2(-0.416615)}-1}{e^{-(-0.416615)}+1} \right] = \left[-0.82006, -0.39407 \right]$$

10.5 CONFIDENCE INTERVAL OF THE KENDALL CORRELATION COEFFICIENT

$$\text{The standard error of } z_r \text{ is } \sqrt{\frac{0.437}{n-4}}$$

$$Lower\,bound\,L = z_r - \left(z^* \times \sqrt{\frac{0.437}{n-4}} \right)$$

$$Upper\,bound\,U = z_r + \left(z^* \times \sqrt{\frac{0.437}{n-4}} \right)$$

$$\text{For two sided limits } z^* = z_{1-\alpha/2} \text{ and for one sided } z^* = z_{1-\alpha}$$

The final confidence interval can be found using the following formula:

$$Confidence\,level = \left[\frac{e^{2L}-1}{e^{2L}+1}, \frac{e^{2U}-1}{e^{2U}+1} \right]$$

10.5.1 Example

Let's determine a 99% confidence interval for this correlation coefficient between viscosity and **pressure**.

$$z_r = \frac{1}{2} ln \frac{(1+0.789473)}{(1-0.789473)} = 1.07003$$

$$Lower\,bound\,L = 1.07003 - \left(2.576 \times \sqrt{\frac{0.437}{20-4}} \right) = 0.64430$$

$$Upper\,bound\,U = 1.07003 + \left(2.576 \times \sqrt{\frac{0.437}{20-4}} \right) = 1.49575$$

$$Confidence\,level = \left[\frac{e^{2(0.64430)}-1}{e^{2(0.64430)}+1}, \frac{e^{2(1.49575)}-1}{e^{2(1.49575)}+1} \right] = [0.56782, 0.90437]$$

10.5.2 Example

Let's determine a 95% confidence interval for this correlation coefficient between yield strength and temperature.

$$z_r = \frac{1}{2} ln \frac{(1+(-0.657))}{(1-(-0.657))} = -0.787515$$

$$Lower\ bound\ L = -0.787515 - \left(1.96 \times \sqrt{\frac{0.437}{15-4}}\right) = -1.17817$$

$$Upper\ bound\ U = -0.787515 + \left(1.96 \times \sqrt{\frac{0.437}{15-4}}\right) = -0.39685$$

$$Confidence\ level = \left[\frac{e^{2(-1.17817)}-1}{e^{2(-1.17817)}+1}, \frac{e^{2(-0.39685)}-1}{e^{2(-0.39685)}+1}\right] = [-0.82687, -0.37725]$$

10.5.3 Example

Let's determine a 90% confidence interval for this correlation coefficient between the reactance of a capacitor and the frequency of an AC source.

$$z_r = \frac{1}{2} ln \frac{\left(1+(-0.7264737)\right)}{\left(1-(-0.7264737)\right)} = -0.92122$$

$$Lower\ bound\ L = -0.92122 - \left(1.645 \times \sqrt{\frac{0.437}{27-4}}\right) = -1.14796$$

$$Upper\ bound\ U = -0.92122 + \left(1.645 \times \sqrt{\frac{0.437}{27-4}}\right) = -0.69447$$

$$Confidence\ level = \left[\frac{e^{2(-1.14796)}-1}{e^{2(-1.14796)}+1}, \frac{e^{2(-0.69447)}-1}{e^{2(-0.69447)}+1}\right] = [-0.81707, -0.60084]$$

Now that we know how to perform the hypothesis tests, let's discuss the tests in MATLAB. When we are performing the Pearson test we have two methods we can use.

10.5.4 Method 1

```
[r, pval] = corr(x', y', 'type', '','Tail','')
```

The above function only calculates the r with the input argument 'type' which specifies the type of correlation followed by one of below:

The input argument 'Tail' specifies the alternative hypothesis against which to compute p-values for testing the hypothesis of no correlation.

The above function does not return the upper limit and the lower limit of the confidence interval.

Value	Description
'Pearson'	Pearson's linear correlation coefficient
'Kendall'	Kendall's tau coefficient
'Spearman'	Spearman's rho

10.5.5 Method 2

The other function we can use that performs the task of calculating the Pearson correlation coefficient and p-value is:

```
[r_matrix, pval_matrix, lower_matrix, upper_matrix] = corrcoef(x, y,
                                                    'Alpha', );
```

This syntax not only returns the correlation coefficient and p-value, but also the lower and upper bounds for the correlation coefficient. However all these values are returned as a matrix. So to return the variables as a single value we can use:

```
r = r_matrix(1, 2)
P = pval_matrix(1, 2)
lower = Lower_matrix(1,2)
upper = Upper_matrix(1,2)
```

Notice when we use the above function we do not specify the type of test we will be conducting. Therefore, the P always returns the p-value for a two-tailed test, so in the case of a one-tailed test all we need to do is divide the P by 2.

The only input argument 'Alpha' defines the percent confidence level, 100*(1-Alpha)%, for the correlation coefficients, which determines the upper and lower bounds.

Let's now perform the worked examples using the two different syntaxes and check if they align with our results and with each other below.

10.5.6 Let's do Example 1 using both methods

```
[r,pval] = corr(Speed', ConsumptionRate','type','Pearson','Tail','right')
```

```
r =

    0.9670


pval =

    1.9589e-12
```

It turns out the coefficient correlation we calculated is in fact accurate and our p-value is a lot smaller than our α.

```
[r_matrix, pval_matrix, Lower_matrix, Upper_matrix] = corrcoef(Speed,
                                                    ConsumptionRate);
r = r_matrix(1, 2)
Since the test is one-tailed:
P = pval_matrix(1, 2) / 2
lower = Lower_matrix(1,2)
upper = Upper_matrix(1,2)
```

```
r =

    0.9670


P =

    1.9589e-12


lower =

    0.9168


upper =

    0.9871
```

10.5.7 Example 2

```
[r_matrix, pval_matrix, Lower_matrix, Upper_matrix] = corrcoef
(Temperature', Conductivity', 'Alpha', 0.10)
r = r_matrix(1, 2)
P = pval_matrix(1, 2) / 2
lower = Lower_matrix(1,2)
upper = Upper_matrix(1,2)
```

```
r =

    0.4992


P =

    0.0207


lower =

    0.1083


upper =

    0.7565
```

10.5.8 Example 3

```
[r,pval] = corr(Voltage',CurrentDensity','type','Pearson','Tail','left')
```

```
r =

   -0.1976


pval =

   0.2018
```

```
[r_matrix, pval_matrix, Lower_matrix, Upper_matrix] = corrcoef(Voltage',
                            CurrentDensity', 'Alpha', 0.01)
r = r_matrix(1, 2)
P = pval_matrix(1, 2) / 2
lower = Lower_matrix(1,2)
upper = Upper_matrix(1,2)
```

```
r =

   -0.1976


P =

   0.2018


lower =

   -0.6778


upper =

   0.4007
```

For Spearman and Kendall we can only use Method 1.

10.5.9 Example

```
[r,pval] = corr(Score1', Score2', 'type', 'Spearman')
```

```
r =

   0.9544


pval =

   1.7883e-05
```

10.5.10 Example

```
[r,pval] = corr(Temperature',Consumption','type','Spearman','Tail','left')
```

```
r =

   -0.9959


pval =

   5.1831e-06
```

10.5.11 Example

```
r =

   -0.6566


pval =

   9.9755e-05
```

10.5.12 Example

```
r =

   0.7895


pval =

   1.6097e-08
```

10.5.13 Example

```
[r,pval] = corr(Strength',Temperature','type','Kendall','Tail','left')
```

```
r =

   -0.6571


pval =

   1.6519e-04
```

10.5.14 Example

```
[r,pval] = corr(Reactance', Frequency','type','Kendall','Tail','left')
```

```
r =

   -0.7265


pval =

   1.2586e-07
```

10.5.15 Example

One of the most classic examples of correlation is the height and weight of individuals. Even though there is generally a positive correlation between height and weight, the relationship is not absolute, since weight alone is not a reliable indicator of overall health. We take 26 samples and record their heights and weights

To test whether there is a significant relationship between the two variables.

10.5.15.1 Test for linearity

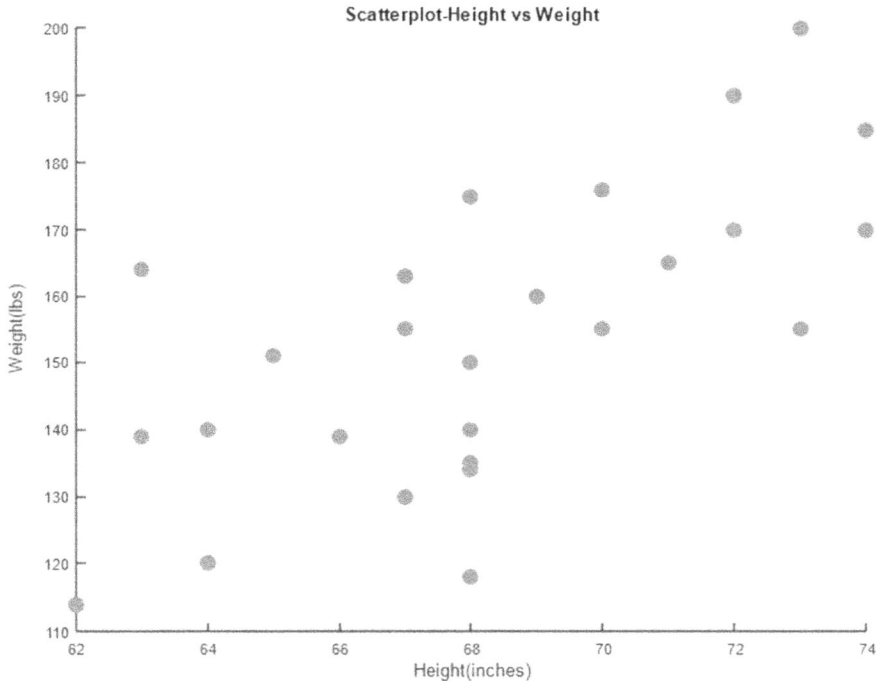

Figure 10.12

Our scatterplot in **Figure 10.12,** displaying heights and weights, meets the linearity assumption as the data points largely align along a straight line. This pattern indicates a strong linear relationship between height and weight

10.5.15.2 Test for normality

10.5.15.2.1 Histogram

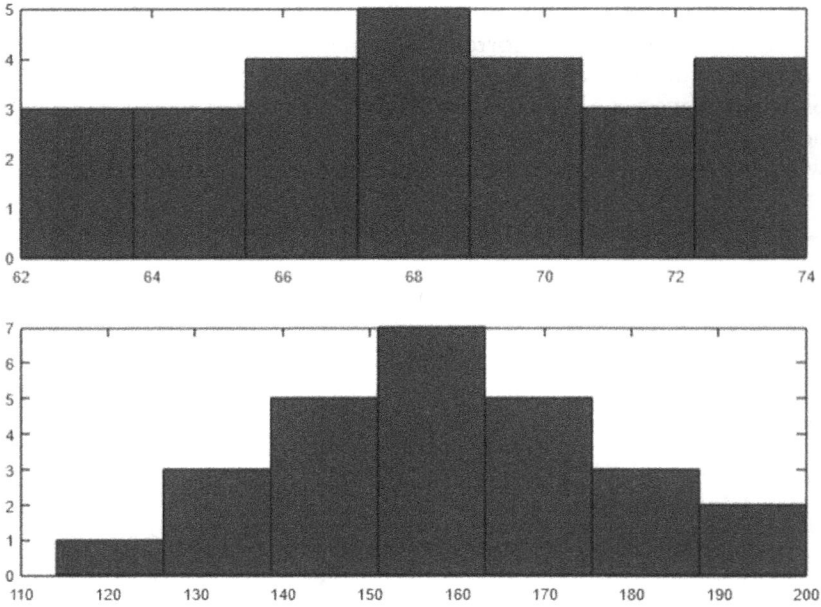

Figure 10.13

To test for normality, we created histograms in Figure 10.13 for both weight and height separately. The histograms show that both variables are approximately symmetrical, with the weight distribution appearing more perfectly symmetrical than the height distribution.

10.5.15.2.2 Boxplot

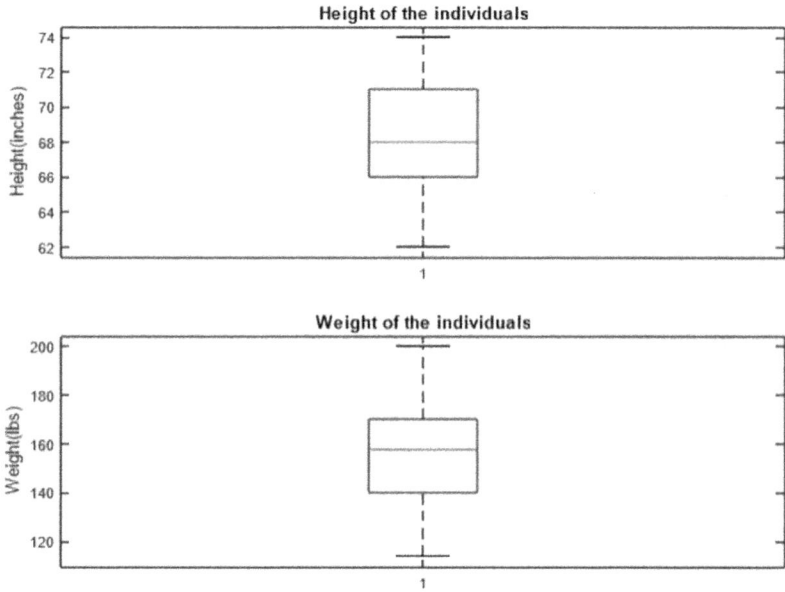

Figure 10.14

Since it passes the assumptions of the Pearson test, we can now perform the test.

To test for normality, we also created boxplots in Figure 10.14 for both weight and height separately. The boxplots show that both variables are approximately symmetrical. The weight distribution, in particular, has the median line closer to the center of the box, appearing more perfectly symmetrical than the height distribution. Both boxplots feature symmetric whiskers, suggesting that the data may come from a normal distribution.

10.5.15.3 Stating the hypotheses

The null hypothesis: there is no correlation between height and weight.

The alternative hypothesis: there is a significant correlation between the two variables.

```
[r, pval] = corr(Height', Weight', 'type', 'Pearson')

                                r =

                              0.6875


                             pval =

                           1.0410e-04
```

Or

```
[r_matrix, pval_matrix, lower_matrix, upper_matrix] = corrcoef(Height,
                                  Weight, 'Alpha', 0.05);
r = r_matrix(1, 2)
P = pval_matrix(1, 2)
lower = Lower_matrix(1,2)
upper = Upper_matrix(1,2)

                                r =

                              0.6875


                                P =

                           1.0410e-04


                             lower =

                              0.6281


                             upper =

                              0.8102
```

The p-*value* $< \alpha = 0.05$, indicating that the correlation is statistically significant, and we reject the null hypothesis. This means that we have evidence to support the hypothesis that there is a correlation between height and weight.

A correlation coefficient of 0.5 to 0.7 is typically considered a moderate correlation. The correlation of 0.6875 implies that there may be other factors that influence an individual's weight besides their height.

10.5.16 Example

https://www.worldometers.info/co2-emissions/co2-emissions-per-capita/

For years there have been studies that have concluded that there is a relationship between per capita GDP and per capita emissions of the greenhouse gas carbon dioxide. Test this claim at the 0.05 level of significance, clearly stating your hypotheses.

10.5.16.1 Stating the hypotheses

The null hypothesis: there is no correlation between GDP and emission of CO_2.

The alternative hypothesis: there is a significant correlation between the two variables.

In this file there are three columns, of which the column GDP has characters like '$' and ',' indicating the amount and currency. Clearly it needs to be converted into numeric data to be able to do computations.

```
data = readtable('GDP vs CO2 emissions.csv');

GDP = table2array(data(:, 'GDP'));
% get rid of $
GDP = strrep(GDP, '$', '');
% get rid of ,
GDP = strrep(GDP, ',', '');
GDP = str2double(GDP);

emission = table2array(data(:, 'emission'));
```

10.5.16.2 Test for linearity

```
scatter(GDP, emission, 10, "blue","filled")
xlabel('GDP per capita')
ylabel('Emission')
title('Scatterplot-GDP vs Emission')
```

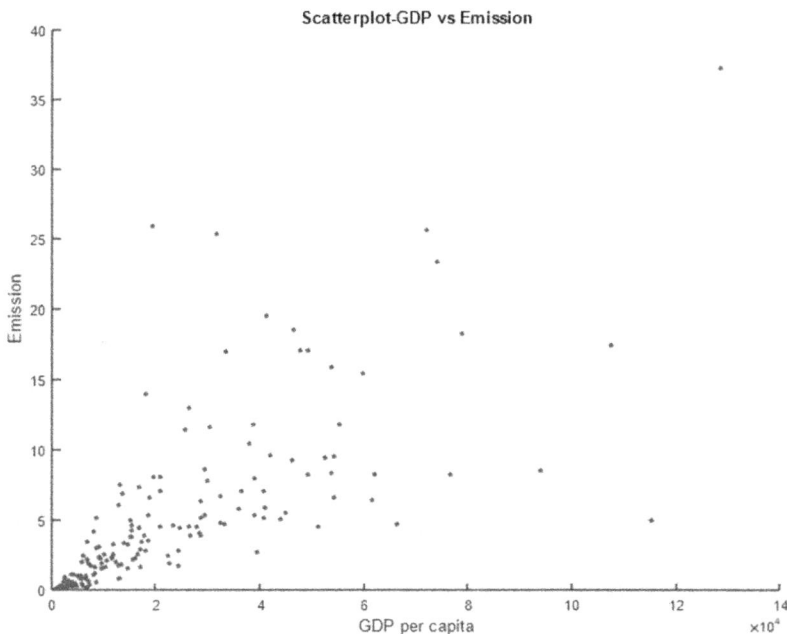

Figure 10.15

In Figure 10.15, we can see that the data points are widely scattered, making it difficult to identify any clear trends. There is no obvious pattern or relationship between the two variables plotted on the scatterplot. The points appear randomly distributed across the plot, with no indication of a linear or monotonic relationship.

10.5.16.3 Test for normality

Let's draw histograms to check if they are normal.

```
tiledlayout(2,1)
% Top axes
ax1 = nexttile;
hist(ax1,GDP,5)
xlabel('GDP')
title('GDP per Capita')
% Bottom axes
ax2 = nexttile;
hist(ax2,emission,5)
xlabel('Emission')
title('Emission of CO2=')
```

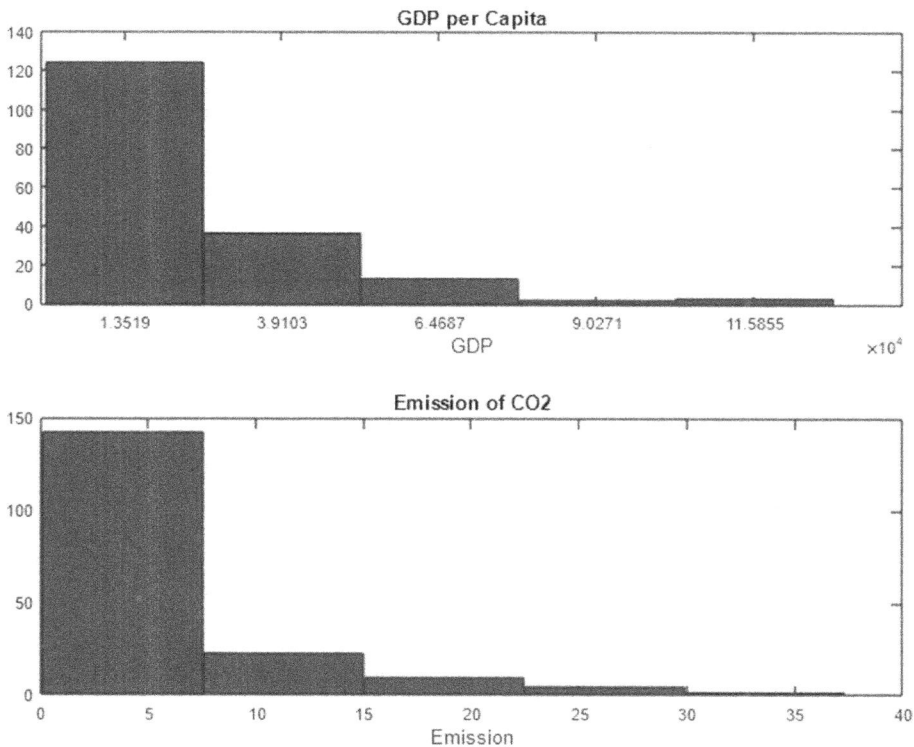

Figure 10.16

Clearly this data does not pass the normality test as it is highly skewed to the right.

So we have a dataset which passes neither the linearity nor the normality test. In fact in practical life most relationships between two variables are never quite linear nor do the variables follow a normal distribution. So what do we do in such a situation?

In this type of situation, we do what we call the mathematical transformation of data which is basically transforming nonlinear, highly skewed, and non-normally distributed data by applying a mathematical function to each participant's data value.

There are several ways to do this. Let's discuss some of them.

10.5.16.3.1 Log transformation

The log transformation is considered the most popular among the different types of transformations used to transform skewed data to approximately conform to normality. If the original data follows a lognormal distribution, the log-transformed data will follow or approximately follow the normal distribution. (You should know by now what the data of a lognormal distribution graph looks like.) In a log transformation each variable of x will be replaced by log(x) with base 10, base 2, or natural log.

```
x = log(data)
```

However one should remember that there is no guarantee that the log transformation will always tranreduce skewness and make the data a better approximation of the normal distribution.

In fact, in some cases, applying the transformation can make the distribution more skewed than the original data.

10.5.16.3.2 Square-root transformation:

Another type of transformation that makes the data less skewed and the variation
more uniform is the square-root transformation. Like the log transformation, the square-root transformation is also used for right-skewed data. The larger values are compressed while the smaller ones become more spread out. However, like any other transformation, the square-root transformation will not fix all skewed data, especially data with a left skew. In fact it will become worse after the square-root transformation.

```
x = sqrt(data)
```

10.5.16.3.3 Cube-root transformation:

The cube root transformation involves converting x to $x^{(1/3)}$. This transformation has a significant impact on the shape of the distribution, though it is not as strong as the logarithmic transformation. One key advantage of the cube root transformation is that it can be applied to negative values and zero, making it useful for datasets with negatively skewed data.

10.5.16.3.4 Reciprocal transformation:

```
x = 1/(data)
```

Reciprocal transformation maps non-zero values of x to 1/x (or -1/x for negative values).

10.5.16.3.5 Power transformation:

This transformation is very useful for heteroscedasticity data (non-constant variance). It supports two methods:
- Box-Cox transformation: Box-Cox can only be applied to strictly positive data.
  - [transdat,lambda] = boxcox(data)
  - transdat = boxcox(lambda,data)

Below are some common values for lambda.

Lambda value	Description
1.0	No transformation
0.0	Log transform.
-1.0	Reciprocal transform.
0.5	Square-root transform.
-0.5	Reciprocal square-root transform

- Yeo–Johnson: the purpose of transforming data is to make the data follow assumptions of linearity, normality, and even homogeneity. Since our data is right skewed, the first choice should be using the log transformation.

```
% Transforming the GDP data
G = log(GDP)

% Transforming the emission data
E = log(emission)
```

Now that we've transformed both the predictor and explanatory values, let's see if it helped correct the nonlinear trend in the data.

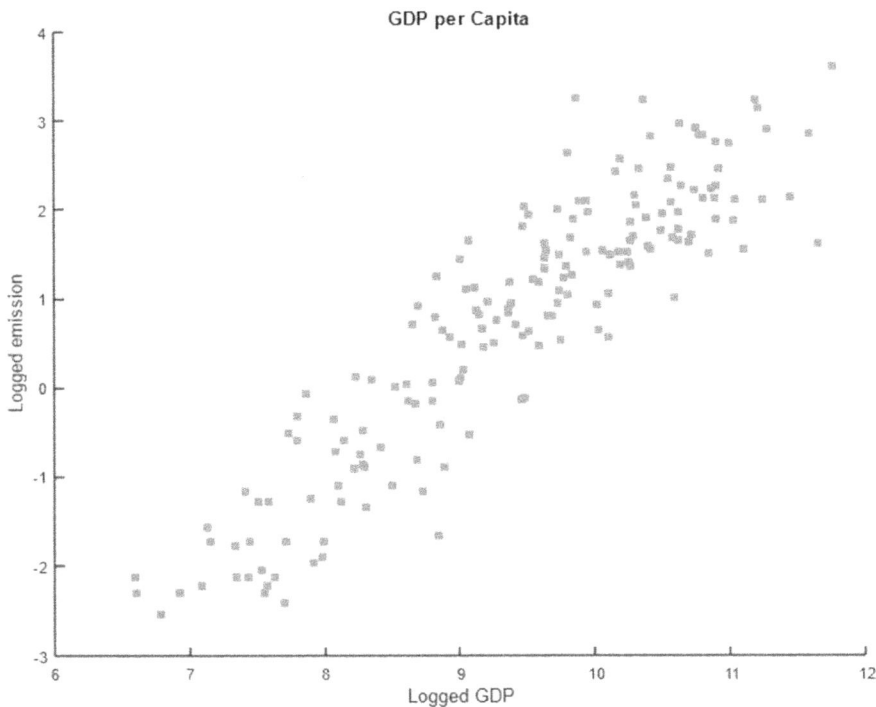

Figure 10.17

In Figure 10.17, we can see that the new scatterplot indicates that taking the natural logarithm of both the x and y variables has been helpful. The transformation has made the relationship between the variables clearer, potentially revealing a more linear pattern.

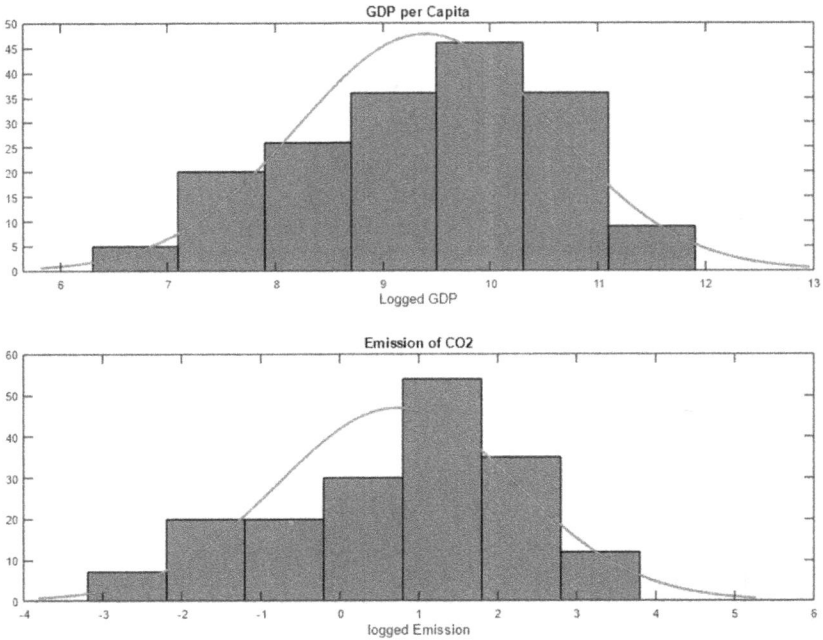

Figure 10.18

From Figure 10.18, we can observe that the log transformation has effectively reduced the skewness of the data. Prior to the transformation, the data was right-skewed, but after applying the logarithm, the distribution has improved significantly with slight right skew, compared to the strong right skew observed before the transformation.

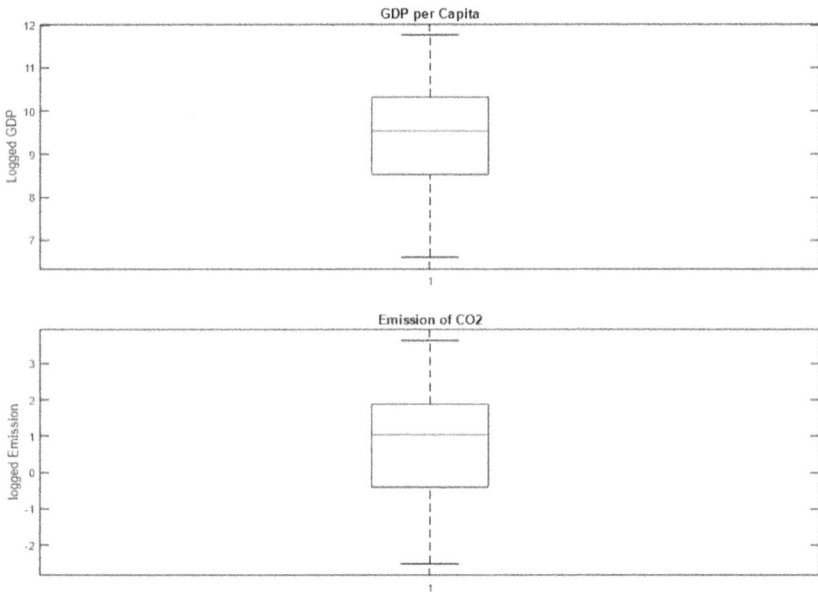

Figure 10.19

From Figure 10.19, we can observe that the median line is nearly centered in the box for both variables, and the lengths of the whiskers are approximately equal. This symmetry suggests that the transformed dataset closely follows a normal distribution, making it reasonable to conclude that the transformation has effectively normalized the data.

Now visually we can conclude that the transformed dataset is normally distributed; however at times this approach proves to be unreliable and does not guarantee that the distribution is normal. So, it's generally recommended to use hypothesis tests as well to check for normality.

10.5.16.4 Using the one-sample Kolmogorov–Smirnov test

```
%testing the GDP variable
G = log(GDP);

Mean_GDP = mean(G);
Std_GDP = std(G);

Data_GDP = (G - Mean_GDP)/Std_GDP;


[h, p] = kstest(Data_GDP)
```

```
h =

  logical

   0


p =

  0.4309
```

Since h = 0 this indicates a failure to reject the null hypothesis at the 0.05 level and we can conclude that the GDP data comes from a standard normal distribution.

Let's do the same for emissions:

```
E = log(emission);

Mean_emission = mean(E);
Std_emission = std(E);

Data_Emission = (E - Mean_emission)/Std_emission;


[h, p = kstest(Data_Emission)
```

```
h =

  logical

   0


p =

  0.0694
```

This too implies that the emission data comes from a standard normal distribution.

Both visual tests and hypothesis tests confirms that the transformation of *both variables* does a great job at correcting the nonlinearity and non-normality as well.

Since it fulfills both the linearity and normality criteria, we can use Pearson.

```
[r, pval] = corr(G, E, 'type', 'Pearson')

                            r =

                          0.9132


                          pval =

                        1.6251e-70
```

Since the p-*value* < $\alpha = 0.05$, the correlation is statistically significant, and we reject the null hypothesis. This means that we have evidence to support the hypothesis that there is a correlation between GDP and emissions per capita. The strength of the correlation implies that as countries become wealthier, they tend to produce more greenhouse gas emissions per person, primarily due to increased energy consumption, industrialization, and transportation.

As we can see, transforming both x and y variables helped correct both the nonlinear and non-normal trends in the data. But do we always transform both variables?

Transforming the *x* values is appropriate **when nonlinearity is the only problem** (i.e., the independence, normality, and equal variance conditions are met).

Let's see some more examples.

10.5.17 Example

https://www.itl.nist.gov/div898/handbook/datasets/BERGER1.DAT
Alaska Pipeline Ultrasonic Defect Calibration Curve

10.5.17.1 Test for linearity

```
data = readtable('field vs lab.csv');

Defect_Field = table2array(data(:, 'field'));
Defect_Lab = table2array(data(:, 'lab'));

scatter(Defect_Field, Defect_Lab, 70, "magenta","filled");
xlabel('Defect size in field')
ylabel('Defect size in lab')
title('Scatterplot-Defect size in field vs Defect size in lab')
```

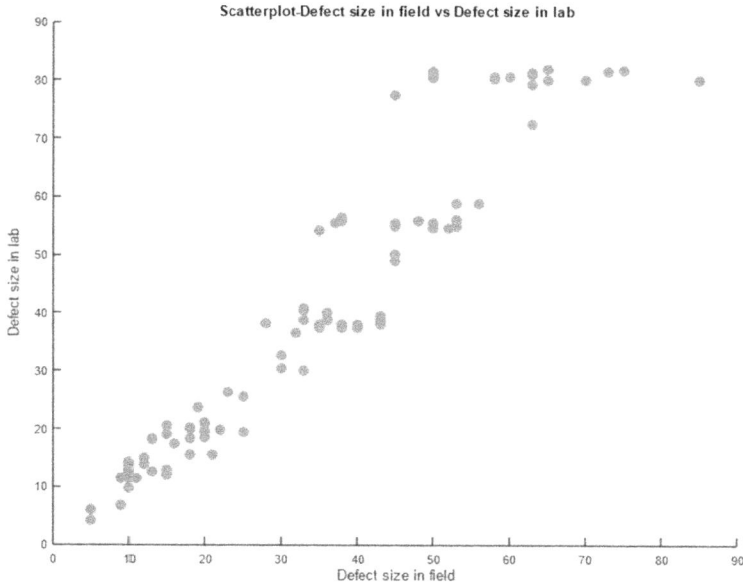

Figure 10.20

The scatterplot has somewhat a linear pattern; therefore it would be safe to assume that it passes the linearity test. However, some might think it does not. Notice that, as x increases, y increases up to a point and then levels off after reaching the maximum point. This makes it a nonlinear relationship. So there might be a slight confusion there. Moreover the variances for larger values are larger compared to the smaller values.

Let's check if it passes the normality test.

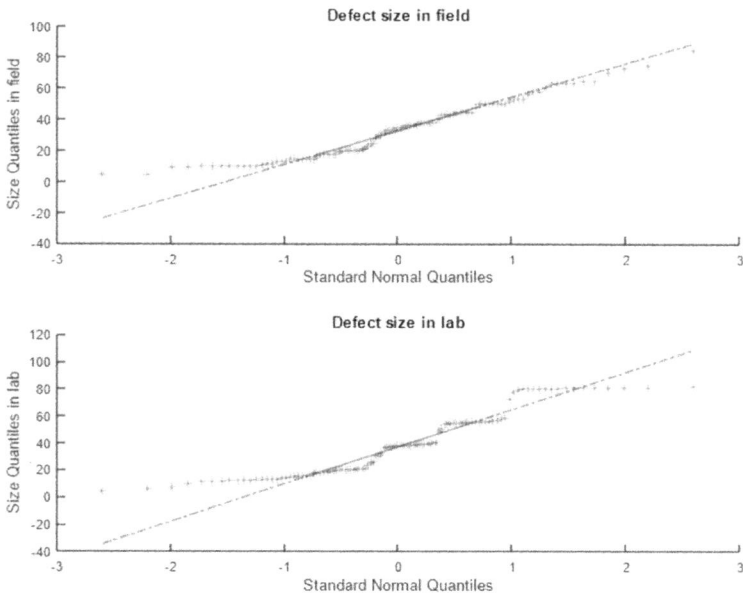

Figure 10.21

The Q-Q plot illustrates the non-normality of the data so it would be safe to conclude that we have a dataset in which both heterogeneity and non-normality are a problem.

Then comes the transformation. So let's try transforming both variables.

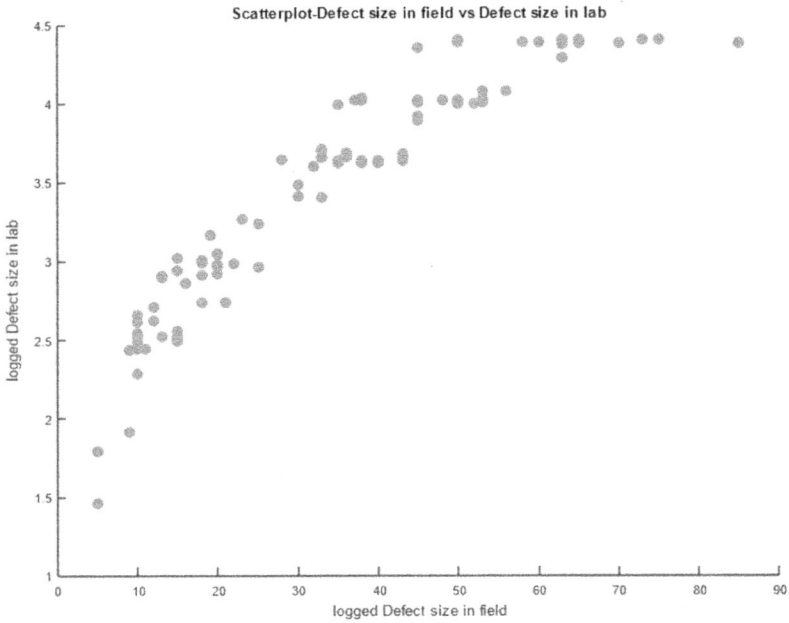

Figure 10.22

As we can see the variance does decrease. But it neither seems linear nor monotonic. So what do we do now?

Let's try something else. Instead of transforming both variables, let's transform one individually and see what happens.

%transforming the x variable using boxcox $\lambda = 0 (log)$

```
field = boxcox(0, Defect_Field)
scatter(field, Defect_Lab, 60, "black", 'filled')
xlabel('Logged size in field')
ylabel('Size in lab')
title('Defect size in field vs lab')
```

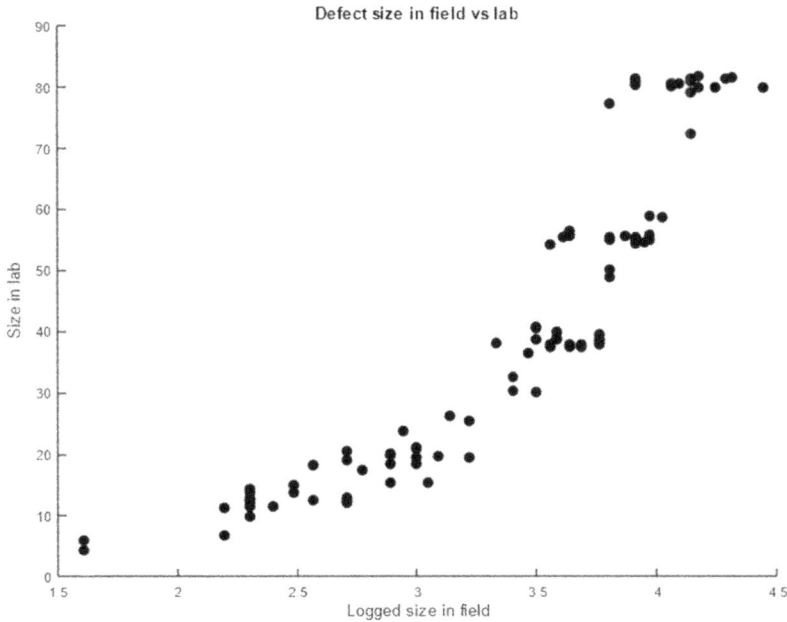

Figure 10.23

After transforming the predictor variable we see that the graph increases rapidly as x increases, but not in a straight line. The graph of this function appears as a curve that starts from the origin and increases rapidly as x increases. So we can conclude from visual inspection of the scatterplot that there is a nonlinear but a monotonic increasing relationship between the two variables.

Let's now try transforming the explanatory variable y and see what the graph looks like.

%transforming the y variable using boxcox λ.

```
lab = boxcox(0, Defect_Lab)

scatter(Defect_Field, lab, 60, "black", 'filled')
xlabel('size in field')
ylabel('logged Size in lab')
title('Defect size in field vs lab')
```

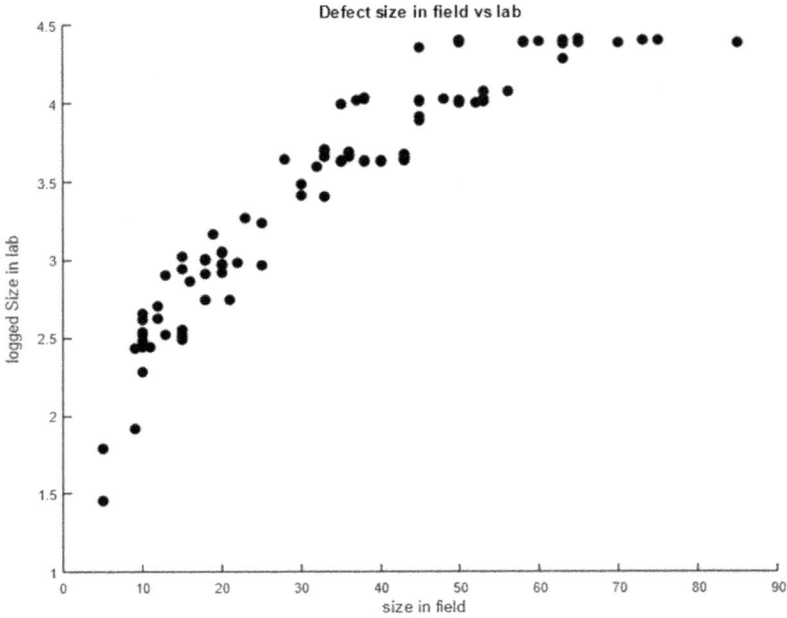

Figure 10.24

Clearly this graph does not fulfil the criteria of either linearity or monotonicity. So evidently the graph of transforming only the predictor variable seems the most appropriate. So even though the graph is not linear it is monotonic which makes it suitable for Spearman.

Let's now test for normality after the transformation.

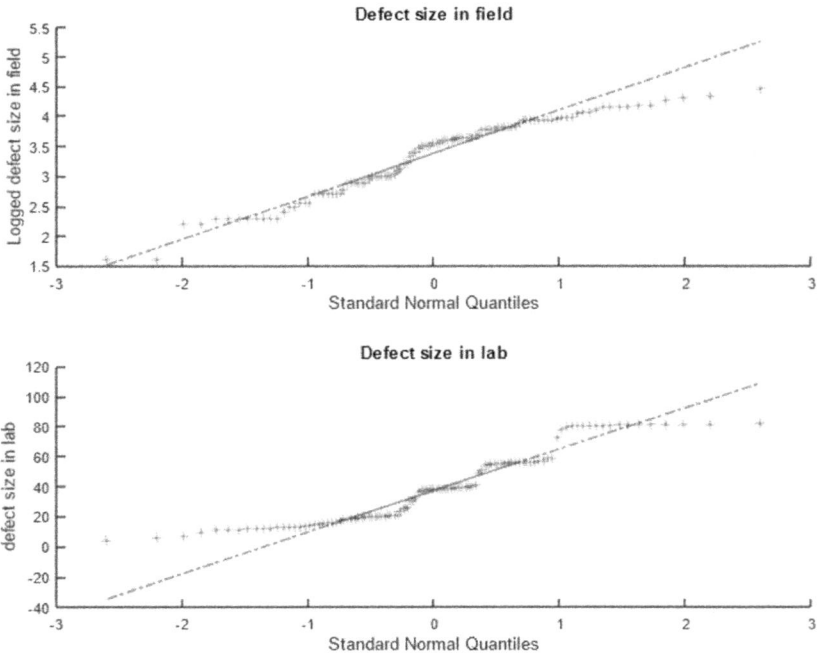

Figure 10.25

The log transformation of the response did not adversely affect the normality; in fact any transformation of any of the variables has no effect on it. So, we are bound to choose the Spearman correlation test.

10.5.17.2 Stating the hypotheses

The null hypothesis: there is no correlation between the defect size in the field and the lab.

The alternative hypothesis: there is a significant correlation between the two.

```
[r,pval] = corr(field, Defect_Lab, 'type', 'Spearman')
```

```
                              r =

                                0.9486


                           pval =

                              2.8443e-54
```

Since the p-*value* < α = 0.05, the correlation is statistically significant, and we reject the null hypothesis. This means that we have evidence to support the hypothesis that there is a correlation between the defect sizes.

10.5.18 Example

Miles per gallon (MPG) and horsepower are two different measures used to describe the performance and efficiency of a vehicle.

https://www.kaggle.com/datasets/uciml/autompg-dataset

```
data = readtable('mpg vs horsepower.csv');

mpg = table2array(data(:, 'mpg'));
horsepower = table2array(data(:, 'horsepower'));

scatter(mpg, horsepower, 30, "red", "filled")
xlabel('mpg')
ylabel('horsepower')
title('Scatterplot-Mileage per gallon vs Horsepower')
```

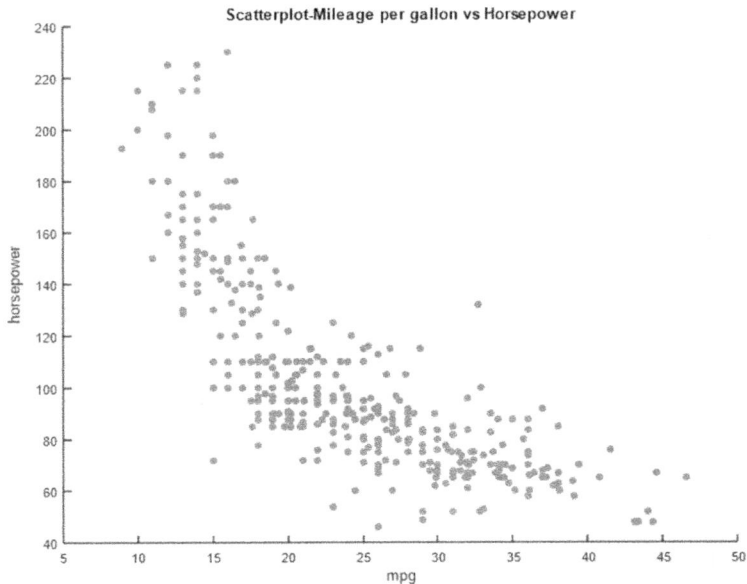

Figure 10.26

This graph in Figure 10.26 displays a downward sloping pattern that shows a negative trend or relationship between the variables being plotted. This means that as MPG increases, the horsepower tends to decrease. Now notice that the data points are arranged in a more curved shape than a straight line.

But let's try transforming it and see if any changes are seen.

```
trans_mpg = boxcox(0, mpg)
trans_horsepower = boxcox(0, horsepower)

scatter(trans_mpg, trans_horsepower, 30, "red", "filled")
xlabel('mpg')
ylabel('horsepower')
title('Scatterplot-Mileage per gallon vs Horsepower')
```

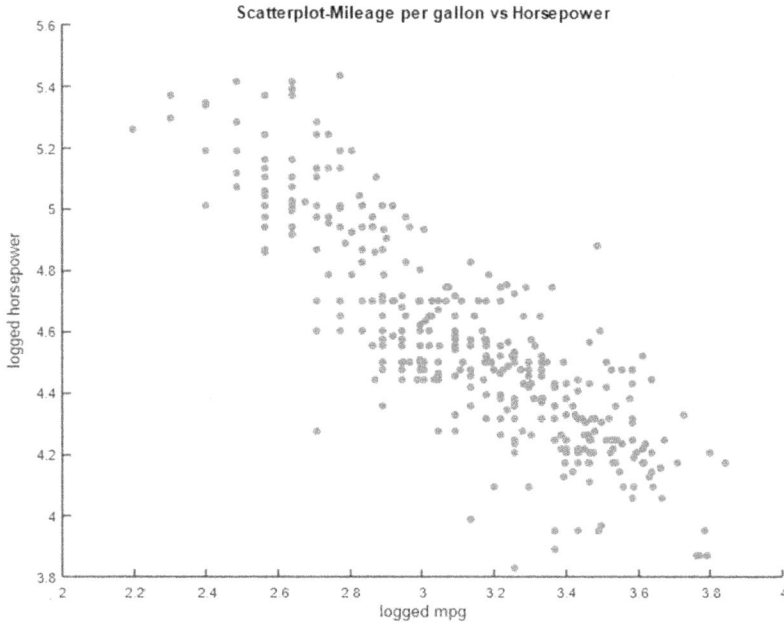

Figure 10.27

Transforming both variables has drastically changed the pattern of the data making it almost perfectly linear.

Let's now look at the histograms before and after transformation.

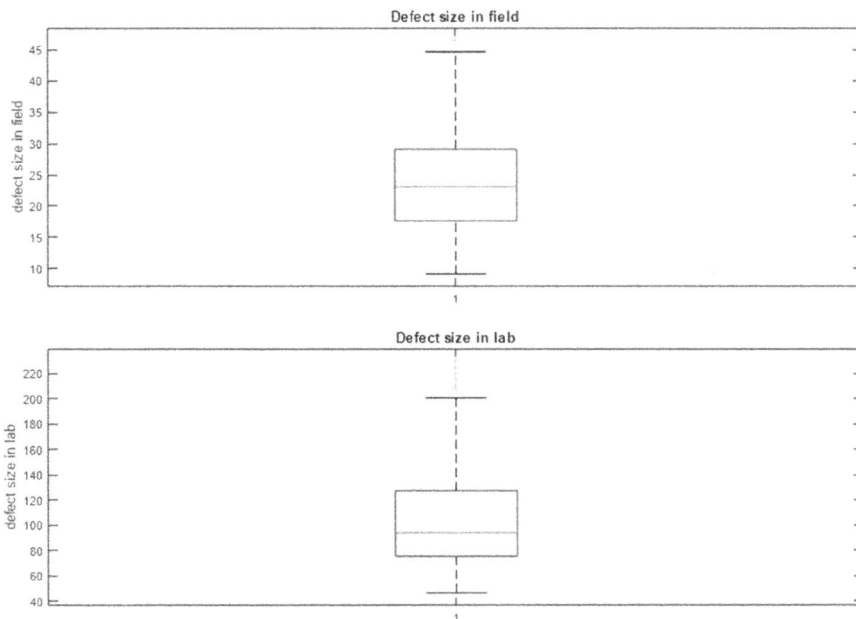

Figure 10.28

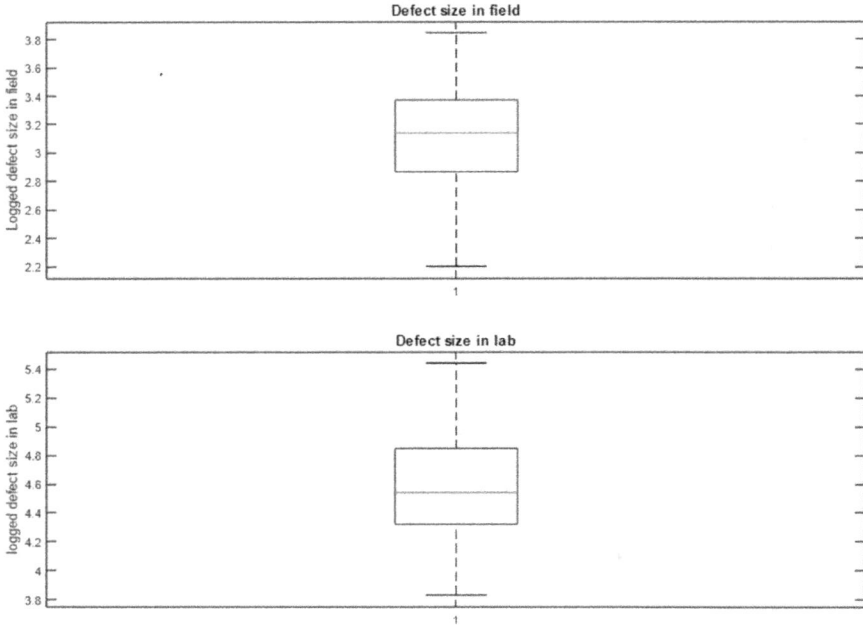

Figure 10.29

While the boxplot of the first variable looks roughly normally distributed, the second boxplot looks a little in doubt. The median line seems to be slightly below the middle line on the plot. So let's use the one-sample Kolmogorov–Smirnov test to overcome these doubts.

```
Mean_mpg = mean(trans_mpg);
Std_mpg = std(trans_mpg);

data = (trans_mpg - Mean_mpg)/Std_mpg;

[h, p] = kstest(data, 'Alpha', 0.01)
```

$$h =$$

$$\underline{logical}$$

$$0$$

$$p =$$

$$0.0816$$

```
Mean_horsepower = mean(trans_horsepower);
Std_horsepower = std(trans_horsepower);

data = (trans_horsepower - Mean_horsepower)/Std_horsepower;

[h, p] = kstest(data, 'Alpha', 0.01)
```

```
h =

  logical

  1


p =

  0.0011
```

The results indicate that the second variable is not normally distributed. Therefore, **Spearman correlation** is most appropriate since one of our two variables is not normally distributed.

10.5.18.1 Stating the hypotheses

The null hypothesis: $r \geq 0$.
the alternative hypothesis: $r < 0$.

```
[r,pval] = corr(trans_mpg, trans_horsepowe', 't'pe', 'Spear'an', 'T'il',
'l'ft')
```
$$r =$$
$$-0.8536$$

```
pval =

  8.0969e-113
```

Since the p-*value* $< \alpha = 0.05$, the correlation is statistically significant, and we reject the null hypothesis. This means that we have evidence to support the hypothesis that there is a negative correlation between miles per gallon and horsepower.

10.5.19 Example

	A	B			A	B				A	B
	Hardness	Strength			Hardness	Strength				Hardness	Strength
	Number	Number			Number	Number				Number	Number
1	Hardness	Strength		11	125	400		21		175	560
2	80	255		12	130	415		22		180	575
3	85	270		13	135	430		23		185	595
4	90	285		14	140	450		24		190	610
5	95	305		15	145	465		25		195	625
6	100	320		16	150	480		26		200	640
7	105	335		17	155	495		27		205	660
8	110	350		18	160	510		28		210	675
9	115	370		19	165	530		29		215	690
10	120	385		20	170	545		30		220	705

	A	B		A	B
	Hardness	**Strength**		**Hardness**	**Strength**
	Number	Number		Number	Number
31	225	720	61	370	1190
32	230	740	62	380	1220
33	235	755	63	390	1255
34	240	770	64	400	1290
35	245	785	65	410	1320
36	250	800	66	420	1350
37	255	820	67	430	1385
38	260	835	68	440	1420
39	265	850	69	450	1455
40	270	865	70	460	1485
41	275	880	71	470	1520
42	280	900	72	480	1555
43	285	915	73	490	1595
44	290	930	74	500	1630
45	255	820	75	510	1665
46	260	835	76	520	1700
47	265	850	77	530	1740
48	270	865	78	540	1775
49	275	880	79	550	1810
50	280	900	80	560	1845
51	285	915	81	570	1880
52	290	930	82	580	1920
53	295	950	83	590	1955
54	300	965	84	600	1995
55	310	995	85	610	2030
56	320	1030	86	620	2070
57	330	1060	87	630	2105
58	340	1095	88	640	2145
59	350	1125	89	650	2180
60	360	1155			

```
data = readtable('Hardness vs Tensile Strength.csv')

Hardness = table2array(data(:, 'Hardness'));
Strength = table2array(data(:, 'Strength'));

scatter(Hardness, Strength, 50)
xlabel('Hardness')
ylabel('Tensile Strength(MPa)')
title("Scatterplot-Hardness vs Tensile Strength")
```

The resulting scatterplot in Figure 10.3 shows a straight line where all the data points are aligned perfectly. Based on our visual observations we can conclude that the slope of the line is positive with a correlation coefficient of +1. Now, a perfectly correlated scatterplot is not commonly observed in real-world data, as most relationships between variables are not perfectly linear.

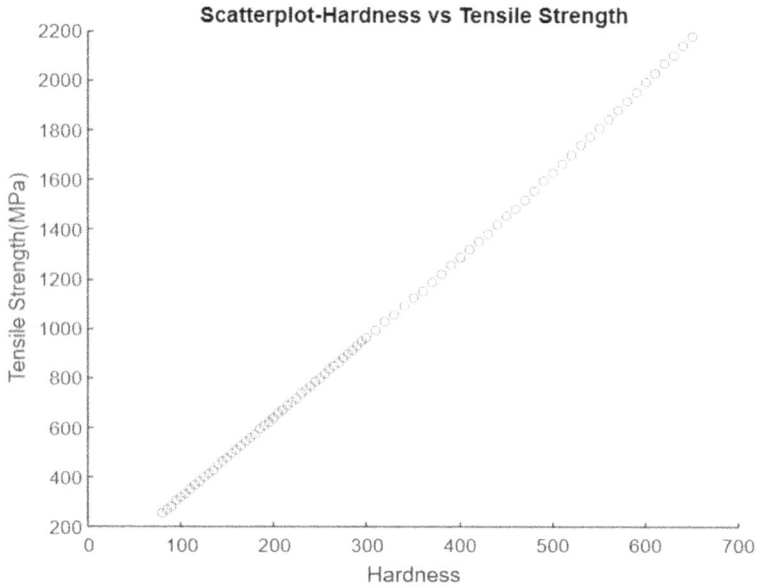

Figure 10.30

10.5.19.1 Let's now test for normality

```
tiledlayout(2,1)
% Top axes
ax1 = nexttile;
histfit(ax1,Hardness,7)
xlabel('Hardness')
title('Hardness of Materials')
% Bottom axes
ax2 = nexttile;
histfit(ax2,Strength,7)
xlabel('Tensile Strength')
title('Tensile Strength(MPa)')
```

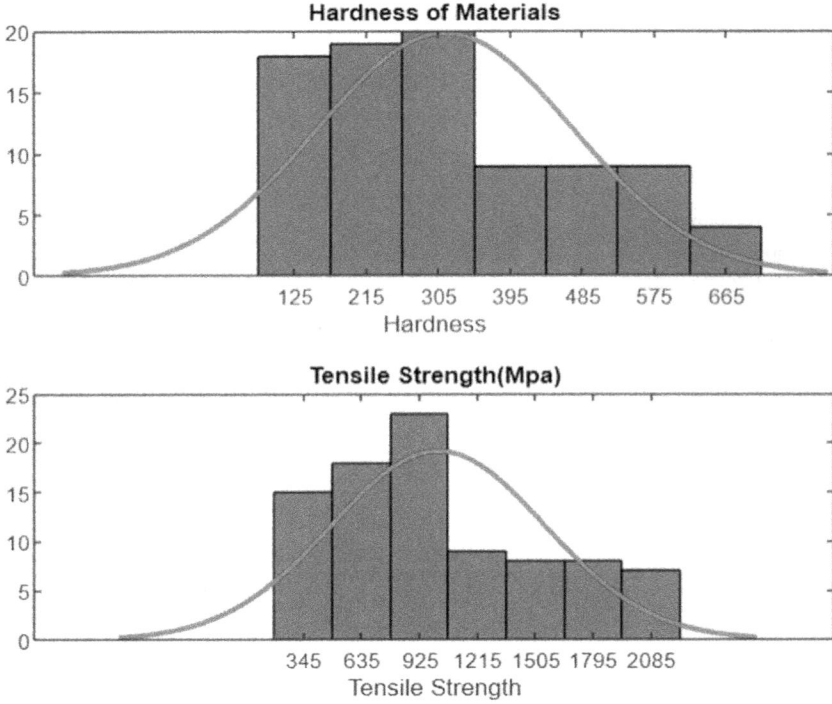

Figure 10.31

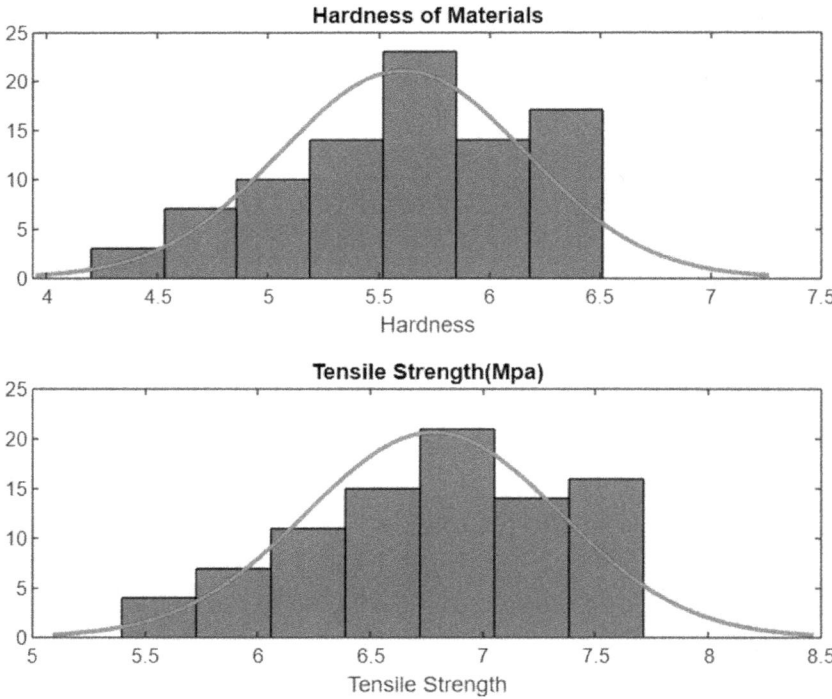

Figure 10.32

Visually we can tell that this histogram is right skewed since the majority of the data points are clustered toward the left side of the graph. Since the normality assumption is not met, our first instinct is to use a non-parametric correlation coefficient such as Spearman's rank correlation coefficient or Kendall's tau. (Remember these non-parametric correlation coefficients do not assume a normal distribution for the variables being correlated and are therefore more robust to violations of normality.)

Since Spearman's rank correlation appears to be the most suitable method for this example, let's use it to determine our correlation

```
[r,pval] = corr(Hardness, Strength, 'type', 'Spearman', 'Tail','right')
                                r =

                             1.0000


                           pval =

                              0
```

Now let's now transform the variables and see if we get the same results.

```
%transforming both variables

trans_hardness = log(Hardness)
trans_strength = log(Strength)
tiledlayout(2,1)

% Top axes
ax1 = nexttile;
histfit(ax1, trans_hardness,7)
xlabel('logged Hardness')
title('Hardness of Materials')

% Bottom axes
ax2 = nexttile;
histfit(ax2, trans_strength,7)
xlabel('logged Tensile Strength')
title('Tensile Strength(MPa)')

%Testing the first variable
h = lillietest(trans_hardness)

%Testing the second variable
h = lillietest(trans_strength)
```

```
h =

        0



h =

        0
```

Given that the normality assumption is satisfied, Pearson's correlation is suitable for this example. Let's apply Pearson and examine if the results differ significantly.

```
[r,pval] = corr(trans_hardness, trans_strength, 'type', 'Pearson',
                                                 'Tail','right')

        r =

           0.9999


        pval =

           4.7299e-160
```

Notice that our results are almost the same, so why go through the process of transformation and not simply use the non-parametric alternative?

One should understand that both approaches are correct. Overall, whether to use original data or transformed data in correlation analysis depends on the nature of the data, the research question, and the specific analysis being performed. It is important to carefully consider the pros and cons of each approach and choose the one that is most appropriate for the research context.

It's important to note, however, that if we choose to use the transformed data, the interpretation of the correlation coefficient will change and we should make sure to interpret the results in the context of the transformed data.

Chapter 11

Regression

In the previous chapter we introduced the concept of correlation. If we recall, correlation analysis allows us to know the association or the absence of a relationship between two variables x and y. We use a scatter diagram to visually see if these variables are correlated, and if they actually are then we use the correlation coefficient to measure the strength of their association.

When data points are concentrated tightly along the line, it is a sign that the correlation coefficient is large. If we look back at our correlation examples we will see that, as the magnitude of coefficient value increases, the scatterplot gets concentrated along the line.

If a linear relationship appears to be reasonable from the scatterplot, we will take the next step of finding a model (an equation of a line) to summarize the relationship. The resulting equation may be used for predicting the response for various values of the explanatory variable.

The objective of correlation analysis is to find the numerical value that measures the degree of the relationship between the two variables. Even though correlation quantifies the relationship, it does not reveal whether x causes y or vice versa, or whether the association is the result of a third component. This is done by regression analysis which quantifies how one variable will change with respect to another variable.

Regression analysis is a method to quantify the relationship between a variable of interest and explanatory variables. In regression analysis, we focus on one measurement of interest: the dependent variable. Other measurements are used as explanatory variables. The goal is to compare differences in the dependent variable in terms of differences in the explanatory variables.

If you were able to make predictions about something important to you, you'd probably love that, right? It's even better if you know that your predictions are sound.

In this chapter, we are going to use regression analysis to make predictions and determine whether they are both unbiased and precise.

However, regression is a parameter to determine how one variable affects another. Basically it predicts the value of the dependent variable based on one or more known independent variables. It plays an important role in many human activities since it is a powerful and flexible tool that is used to forecast and model the future relationship between variables.

So, the million-dollar question is: how does it do it? How does regression analysis predict things?

In one sentence, it provides you with an equation for a graph so that you can make predictions about your data.

To do that we follow certain steps:

1. We load the data into the workspace and create a scatterplot or another suitable plot.
2. Based on the data, we pick a suitable model or regression equation.

DOI: 10.1201/9781003399582-11

We will typically encounter either a linear graph or a nonlinear one. It has multiple variants like linear regression, multi-linear regression, and nonlinear regression, where linear and multi-linear regression are the most common and the only ones we will be dealing with in this textbook.

3. From the equation we will estimate the parameters like slopes and intercepts using the least-squares method.

11.1 LINEAR REGRESSION

The most extensively used modeling technique is linear regression, which assumes a linear connection between a dependent variable (Y) and an independent variable (X). It employs a regression line, also known as a best-fit line. Now what is the best-fit line? To understand the term, let's look at Figure 11.1.

In this graph, the association is visibly positively linear. Since this graph is almost perfectly linear, we can easily draw a straight line that passes through most of the plotted points and form an equation of the graph. Now from algebra, we already know the equation of the graph.

$$y = mx + c$$

Where

m = slope of the graph

$c = y$ intercept

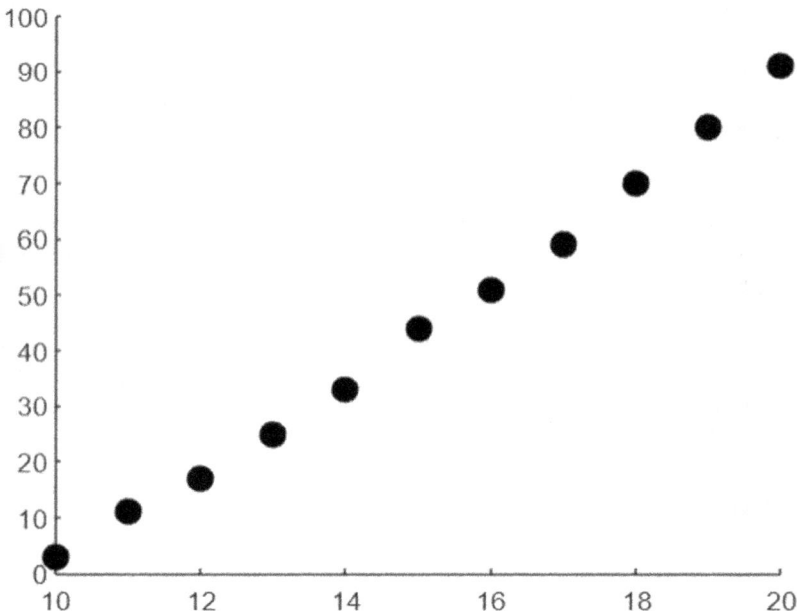

Figure 11.1

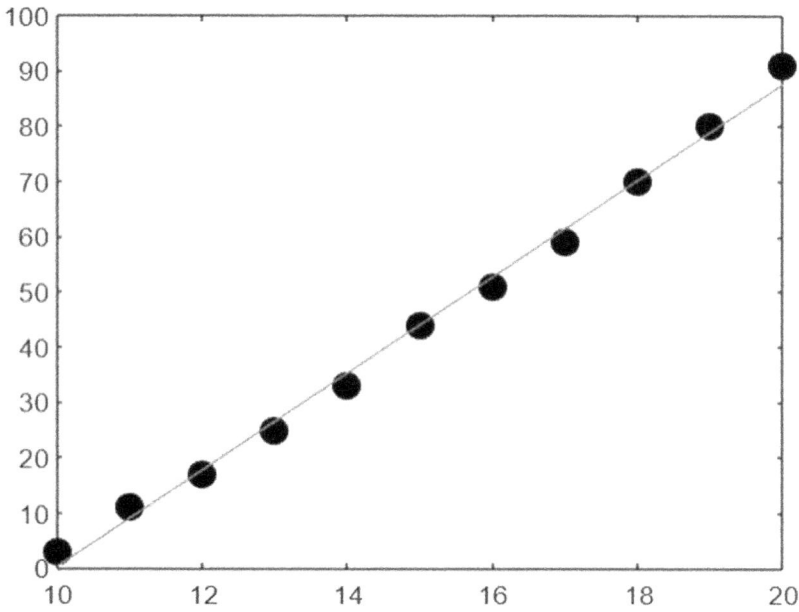

Figure 11.2

Now let's look at a second Figure 11.3.

Clearly, using the existing points to plot a line won't work in this situation, since all the data points do not fall in a straight line. The most we can do in this situation is draw a line that "best fits" all the data points, meaning the number of points above the line and below the line is about equal (and the line passes through as many points as possible).

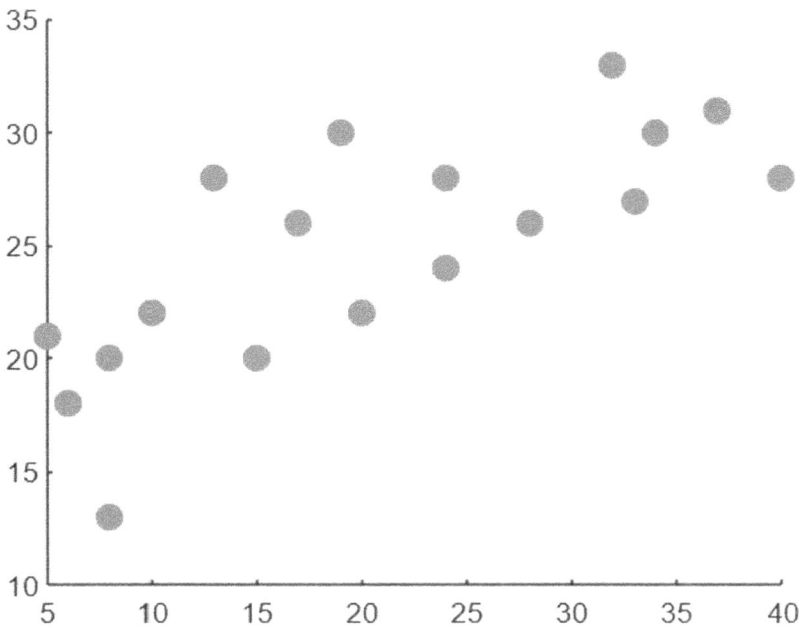

Figure 11.3

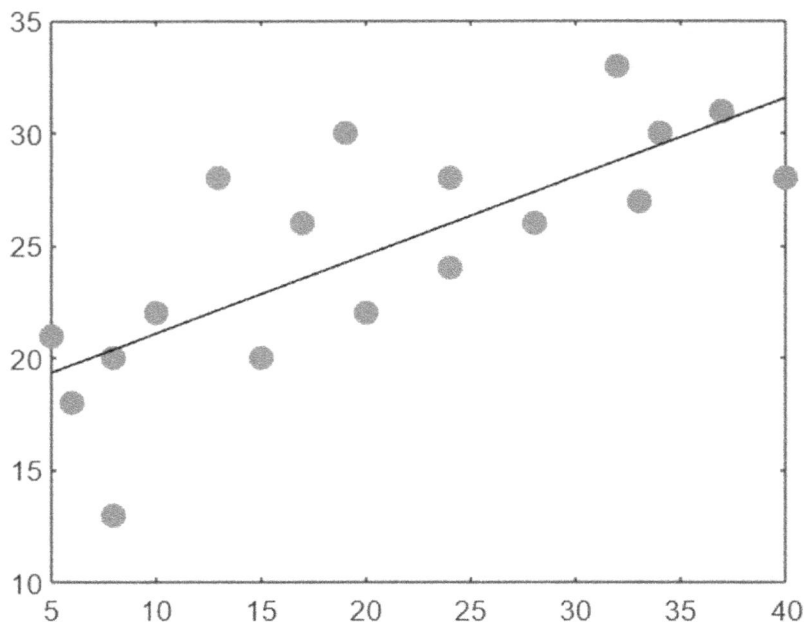

This line which tries to be close as possible to all the data points, with a similar number of points above and below the line, is known as the best-fit line. In regression analysis, we use a similar equation for the best-fit line with different notation,

$$\hat{y} = \beta_0 + \beta_1 x$$

In the model

$\hat{y}$ = the predicted value of the dependent variable y for any given value of the

independent variable x.

x = independent variable.

The scalars β_0 and β_1 are the parameters of the model and are described as regression coefficients or parameters of the linear model. In particular, β_0 is called the intercept term and β_1 is called the slope parameter.

Let's discuss the notion $\hat{y}$. Now if we look at the graph and the line of best fit we will notice that all the data points do not fall exactly on this best-fit line; they are scattered around. So clearly, they are not always equal to the **estimated value of** y, $\hat{y}$. Simply stated, we have two values of y: the **actual values** of the dependent variable, the sample data denoted by y_i, and the predicted value of the dependent variable obtained using the regression line denoted by $\widehat{y_i}$.

As a result of this incomplete relationship, there will always be a difference between the values predicted by a regression model and the actual data. This difference is known as the residual and is represented in models by ε. As residuals are the difference between data points and the regression line, each data point has one residual. To calculate the residual for each data point we find the difference between the predicted values denoted by $\hat{y}$ and the observed values of y.

Residual = observed y value – predicted y value,

$$\varepsilon_i = y_i - \hat{y}_i$$

- This is positive if the data point is above the regression line.
- It is negative if the data point is below the regression line.
- It is zero if the regression line actually passes through the point.

11.1.1 Assumptions of linear regression regarding residuals

a) **Linearity**: the relationship between x and y must be linear. To check this assumption, we use the scatterplot (already discussed in Chapter 11).

b) **Normality assumption**: it is assumed that the error terms, ε, are normally distributed. The normality of errors can be assessed using visual methods, such as a histogram or a Q-Q plot, or through statistical tests, like the Shapiro–Wilk test or the Kolmogorov–Smirnov test.

c) **Constant variance assumption**: it is assumed that the residual terms have the same (but unknown) variance, σ^2. This assumption is also known as the assumption of **homogeneity** or **homoscedasticity**.

d) **Independent error assumption**: it is assumed that the residual terms are independent of each other, i.e. their pair-wise co-variance is zero. This means that there is no correlation between the residuals and the predicted values, or among the residuals themselves.

Even though these might seem like a lot to deal with, what if we can check for all these model assumptions with a single graph? What? Yes, we can actually test if a regression model fits the data properly by creating a **fitted value vs. residuals plot**.

11.1.1.1 Let's first talk about the plot

This is a type of plot that displays the **predicted values** ("fitted values") of the regression model along the x-axis and the residuals of those fitted values along the y-axis. Now how we do we use this plot to check if our data is suitable for linear regression?

Let's use an example. We want to determine if there is a linear relationship between the pressure and temperature of a gas in a sealed container. After collecting the data we calculated the residuals and the predicted value of the temperature and then plotted them in Figure 11.5.

Notice how the residuals are scattered randomly around the horizontal axis in no particular pattern. This random scatter of points indicates that the linear regression model is a good fit for the data. The residuals are normally distributed for each value of x, and they have constant variance.

What are the other possible pattern of residuals that we encounter in practical life?

11.1.1.1.1 Curved-shaped pattern

Calculating the residual provides a valuable clue as to how well our line of best fit fits the dataset. We can examine the quality of our estimated model in terms of the magnitude of

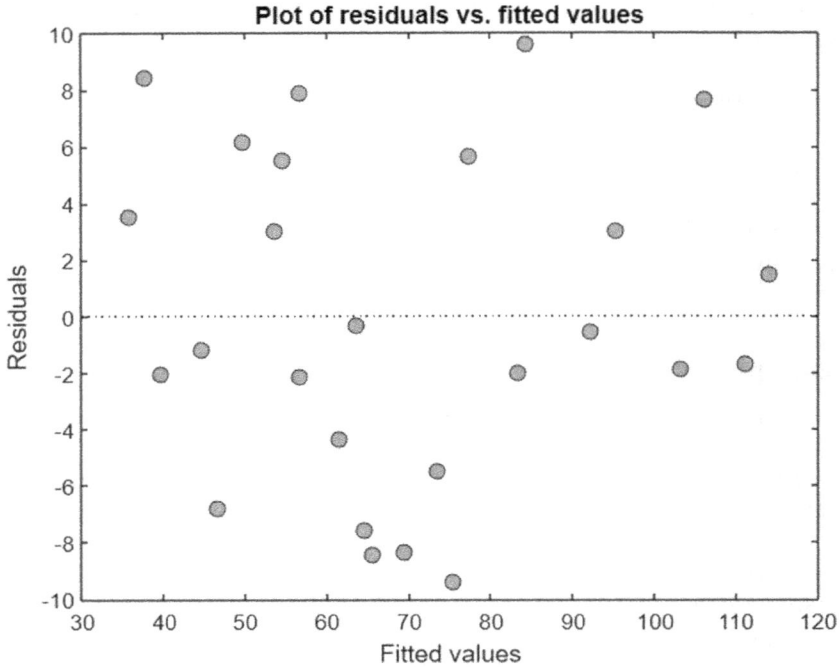

Figure 11.5

the residual. A poorly fit regression model will yield large residuals indicating that the model is not capturing a reliable trend in the dataset while a well-fit regression model will yield small residuals for all data points.

11.1.1.1.2 Residual sum of squares (RSS)

This is also referred to as the sum of squared errors (SSE) and is the variance not explained by the regression model. RSS is obtained by adding the square of residuals. Now we know that

$$\varepsilon_i = y_i - \widehat{y}_i$$

Since we are squaring the residuals

$$\varepsilon_i^2 = (y_i - \widehat{y}_i)^2 = \left(y_i - \left(\beta_0 + \beta_1 x_i\right)\right)^2$$

Adding the square of residuals

$$\sum_{i=1}^{n} \varepsilon_i^2 = \sum_{i=1}^{n}(y_i - \widehat{y}_i)^2 = \sum_{i=1}^{n}\left(y_i - \left(\beta_0 + \beta_1 x_i\right)\right)^2$$

11.2 ESTIMATING THE REGRESSION COEFFICIENTS

There are several methods to estimate the values of the regression parameters. In the remainder of this chapter, we describe the methods of least squares and maximum likelihood estimation.

11.2.1 Least-squares method

The least-squares method is the most widely used procedure for developing estimates of the model parameters. It entails finding the (regression) line that is the best fit for the data points by minimizing the sum of the squared distance from each point to the line. We begin by taking the partial derivative of SSE with respect to b_0 and b_1 and then set them equal to zero.

$$\frac{\partial}{\partial \beta_0} SSE = -2 \cdot \sum_{i=1}^{n} \left(y_i - \beta_0 - \beta_1 x_i \right)$$

$$= -2 \sum_{i=1}^{n} y_i + 2 \sum_{i=1}^{n} \beta_0 + 2 \sum_{i=1}^{n} \beta_1 x_i = -2n\bar{y} + 2n\beta_0 + 2\beta_1 n\bar{x}$$

Setting the derivative equal to 0 and solving for the intercept gives us:

$$-2n\bar{y} + 2n\beta_0 + 2\beta_1 n\bar{x} = 0$$

$$-\bar{y} + \beta_0 + \beta_1 \bar{x} = 0$$

$$\beta_0 = \bar{y} - \beta_1 \bar{x}$$

Now to find the slope we take the partial derivative with respect to the slope:

$$\frac{\partial}{\partial \beta_1} SSE = -2 x_i \cdot \sum_{i=1}^{n} \left(y_i - \beta_0 - \beta_1 x_i \right)$$

$$= -2 \sum_{i=1}^{n} x_i y + 2\beta_0 \sum_{i=1}^{n} x_i + 2\beta_1 \sum_{i=1}^{n} x_i^2$$

$$= -2 \sum_{i=1}^{n} x_i y_i + 2\left(\bar{y} - \beta_1 \bar{x} \right) \sum_{i=1}^{n} x_i + 2\beta_1 \sum_{i=1}^{n} x_i^2$$

$$= -2 \sum_{i=1}^{n} x_i y_i + 2\bar{y} \sum_{i=1}^{n} x_i - 2\beta_1 \bar{x} \sum_{i=1}^{n} x_i + 2\beta_1 \sum_{i=1}^{n} x_i^2$$

$$= 2\beta_1 \left[\sum_{i=1}^{n} x_i^2 - \bar{x} \sum_{i=1}^{n} x_i \right] + 2\bar{y} \sum_{i=1}^{n} x_i - 2 \sum_{i=1}^{n} x_i y_i$$

Setting the derivative equal to 0 and solving for the intercept gives us:

$$2\beta_1 \left[\sum_{i=1}^{n} x_i^2 - \bar{x} \sum_{i=1}^{n} x_i \right] + 2\bar{y} \sum_{i=1}^{n} x_i - 2 \sum_{i=1}^{n} x_i y_i = 0$$

$$\beta_1 = \frac{\sum_{i=1}^{n} x_i\, y_i - \bar{y} \sum_{i=1}^{n} x_i}{\sum_{i=1}^{n} x_i^2 - \bar{x} \sum_{i=1}^{n} x_i} = \frac{\sum_{i=1}^{n} (x_i - \bar{x})\,(y_i - \bar{y})}{\sum_{i=1}^{n} (x_i - \bar{x})^2}$$

We use the slope β_1 and the intercept β to form the equation of the line.

11.2.2 Maximum likelihood estimation method

Maximum likelihood estimation (MLE) is a technique used for estimating the parameters of a given distribution, using some observed data.

This means that, to find the parameters of the linear regression model, we need to know its distribution.

Now from the assumptions of residuals we know that our residuals have a Gaussian distribution with a mean of 0 and some unknown variance σ^2.

$$\varepsilon \sim N\left(0, \tilde{A}^2\right)$$

We have our linear model $\widehat{y}_i = \beta_0 + \beta_1 x_i$ and $\varepsilon_i = y_i - \widehat{y}_i$.

Consequently we get $y_i = \widehat{y}_i + \varepsilon_i = \beta_0 + \beta_1 x_i + \varepsilon_i \left(i = 1, 2, \ldots\ldots, n\right)$.

Remember, in regression analysis, we focus on one measurement of interest: the dependent variable y. Therefore, we need to find the distribution of y_i. From the previous chapter we have already verified that y is normally distributed. We just need to estimate its parameters—mean and variance.

We know the mean is termed as expectation. Let's get the expectation of $E[y_i]$.

$$E[y_i] = E[\beta_0 + \beta_1 x_i + \varepsilon_i] = \beta_0 + \beta_1 x_i + E[\varepsilon_i] = \beta_0 + \beta_1 x_i$$

and

$$\mathrm{Var}[y_i] = \mathrm{Var}[\beta_0 + \beta_1 x_i + \varepsilon_i] = Var[\varepsilon_i] = \sigma^2$$

So now we can say that y_i is a normal distribution with mean $\beta_0 + \beta_1 x_i$ and variance σ^2.

$$y_i \sim N\left(\beta_0 + \beta_1 x_i, \sigma^2\right)$$

Formally, the normal curve is defined by the function

$$f(x|\mu, \sigma^2) = \frac{1}{\sigma\sqrt{2\pi}}\, e^{-\frac{1}{2}\left(\frac{x-\mu}{\sigma}\right)^2}$$

So the PDF of y_i is

$$f(y_i|\beta_0, \beta_1, \sigma^2) = \frac{1}{\sigma\sqrt{2\pi}}\, e^{-\frac{1}{2}\left(\frac{y_i - \beta_0 - \beta_1 x_i}{\sigma}\right)^2}$$

Note that the above equation is the PDF for one y_i. For every observed random sample $y_1, y_2, \ldots, y_n$ we define

$$f(y_1, y_2, \ldots, y_n | \beta_0, \beta_1, \sigma^2) = f(y_1 | \beta_0, \beta_1, \sigma^2) \times f(y_2 | \beta_0, \beta_1, \sigma^2) \times \ldots \times f(y_1 | \beta_0, \beta_1, \sigma^2)$$

$$= \frac{1}{\sigma\sqrt{2\pi}} e^{-\frac{1}{2}\left(\frac{y_1 - \beta_0 - \beta_1 x_1}{\sigma}\right)^2} \times \frac{1}{\sigma\sqrt{2\pi}} e^{-\frac{1}{2}\left(\frac{y_2 - \beta_0 - \beta_1 x_2}{\sigma}\right)^2} \times \ldots \times \frac{1}{\sigma\sqrt{2\pi}} e^{-\frac{1}{2}\left(\frac{y_n - \beta_0 - \beta_1 x_n}{\sigma}\right)^2}$$

$$= \left(\frac{1}{\sigma\sqrt{2\pi}}\right)^n e^{-\frac{1}{2\sigma^2}\sum_{i=1}^{n}(y_i - \beta_0 - \beta_1 x_i)^2}$$

The likelihood function of all the given observations and unknown parameters, $\beta_0, \beta_1,$ and σ^2:

$$L\left(\beta_0, \beta_1, \sigma^2 | y_1, y_2, \ldots, y_n\right) = f(y_1, y_2, \ldots, y_n | \beta_0, \beta_1, \sigma^2) = \left(\frac{1}{\sigma\sqrt{2\pi}}\right)^n e^{-\frac{1}{2\sigma^2}\sum_{i=1}^{n}(y_i - \beta_0 - \beta_1 x_i)^2}$$

Let's get natural logarithms on both sides so that we get the log likelihood:

$$\ln\left(L\left(\beta_0, \beta_1, \sigma^2 | y_1, y_2, \ldots, y_n\right)\right) = \ln\left(\left(\frac{1}{\sigma\sqrt{2\pi}}\right)^n e^{-\frac{1}{2\sigma^2}\sum_{i=1}^{n}(y_i - \beta_0 - \beta_1 x_i)^2}\right)$$

By logarithm properties $= \ln\left(\frac{1}{\sigma\sqrt{2\pi}}\right)^n + \ln\left(e^{-\frac{1}{2\sigma^2}\sum_{i=1}^{n}(y_i - \beta_0 - \beta_1 x_i)^2}\right)$

$$= n\ln\left(\frac{1}{\sigma\sqrt{2\pi}}\right) - \frac{1}{2\sigma^2}\sum_{i=1}^{n}\left(y_i - \beta_0 - \beta_1 x_i\right)^2$$

$$= n\left[\ln 1 - \ln \sigma\sqrt{2\pi}\right] - \frac{1}{2\sigma^2}\sum_{i=1}^{n}\left(y_i - \beta_0 - \beta_1 x_i\right)^2$$

$$= -n[\ln \sigma + \ln\sqrt{2\pi}] - \frac{1}{2\sigma^2}\sum_{i=1}^{n}\left(y_i - \beta_0 - \beta_1 x_i\right)^2$$

$$= -n\ln\sigma - \frac{n}{2}\ln(2\pi) - \frac{1}{2\sigma^2}\sum_{i=1}^{n}\left(y_i - \beta_0 - \beta_1 x_i\right)^2$$

Let's remove the negative signs,

$$-\ln\left(L\left(\beta_0, \beta_1, \sigma^2 | y_1, y_2, \ldots, y_n\right)\right) = n\ln\sigma + \frac{n}{2}\ln(2\pi) + \frac{1}{2\sigma^2}\sum_{i=1}^{n}\left(y_i - \beta_0 - \beta_1 x_i\right)^2$$

Remember our ultimate goal is to find the parameters of our line—β_0 and β_1—so therefore we can remove any constants which don't include our parameters and end up with

$$\sum_{i=1}^{n}\left(y_i - \beta_0 - \beta_1 x_i\right)^2 = \sum_{i=1}^{n}\left(y_i - \hat{y}_i\right)^2 = SSE$$

Consequently, maximizing the likelihood function for the parameters α and β is equivalent to minimizing the above equation, that is, finding the partial derivative of SSE with respect to b_0 and b_1 and then setting them equal to zero.

This looks very familiar, doesn't it? Recall that the least-squares method involves finding the line that minimizes the sum of the squared differences between the observed values and the predicted values of the dependent variable. The regression coefficients are then estimated by finding the slope and the intercept of the line that minimize the sum of squared errors.

The maximum likelihood method, on the other hand, involves finding the line that maximizes the likelihood of the observed data given the model. The likelihood function is the probability of observing the data given the values of the parameters in the model. This method assumes that the errors between the predicted values and the observed values are normally distributed with a mean of zero and constant variance. The regression coefficients are then estimated by finding the values that maximize the likelihood function.

Therefore two common approaches to find the regression coefficients are minimizing the sum of squared errors and maximizing the likelihood function.

11.2.2.1 Using an example

Let's use the engine speed and fuel consumption rate example from Chapter 11 to theoretically demonstrate the method.

Engine speed	Consumption rate	Engine speed	Consumption rate	Engine speed	Consumption rate	Engine speed	Consumption rate
8000	10.5	6800	8.0	5800	8.2	4400	6.5
7800	9.8	6500	8.8	5500	7.1	4200	5.2
7500	8.8	6400	8.3	5200	7.5	3800	5.3
7300	10.1	6200	7.4	5000	6.3	3500	4.6
7000	9.1	5900	7.7	4600	5.9	3000	4.2

Step 1: calculate the mean of the x values and the mean of the y values.

$$x = 8000 + 7800 + \ldots + 6200 + 5900 + \ldots 5000 + 4600 + 3500 + 3000 = 114400$$

$$\bar{x} = \frac{114400}{20} = 5720$$

$$y = 10.5 + 9.8 + \ldots 8.3 + 7.4 + 7.7 + \ldots 7.5 + 6.3 + 5.9 + \ldots 5.3 + 4.6 + 4.2 = 149.30$$

$$\bar{y} = \frac{149.30}{20} = 7.465$$

Step 2: find the slope.

x	y	$x_i - \bar{x}$	$y_i - \bar{y}$	$(x_i - \bar{x})(y_i - \bar{y})$	$(x_i - \bar{x})^2$
8000	10.5	2280	3.035	6919.8	5,198,400
7800	9.8	2080	2.335	4856.8	4,326,400

7500	8.8	1780	1.335	2376.3	3,168,400
7300	10.1	1580	2.635	4163.3	2,496,400
7000	9.1	1280	1.635	2092.8	1,638,400
6800	8.0	1080	0.535	577.8	1,166,400
6500	8.8	780	1.335	1041.3	608,400
6400	8.3	680	0.835	567.8	462,400
6200	7.4	480	−0.065	−31.2	230,400
5900	7.7	180	0.235	42.3	32,400
5800	8.2	80	0.735	58.8	6400
5500	7.1	−220	−0.365	80.3	48,400
5200	7.5	−520	0.035	−18.2	270,400
5000	6.3	−720	−1.165	838.8	518,400
4600	5.9	−1120	−1.565	1752.8	1,254,400
4400	6.5	−1320	−0.965	1273.8	1,742,400
4200	5.2	−1520	−2.265	3442.8	2,310,400
3800	5.3	−1920	−2.165	4156.8	3,686,400
3500	4.6	−2220	−2.865	6360.3	4,928,400
3000	4.2	−2720	−3.265	8880.8	7,398,400
				£49,434	£41,492,000

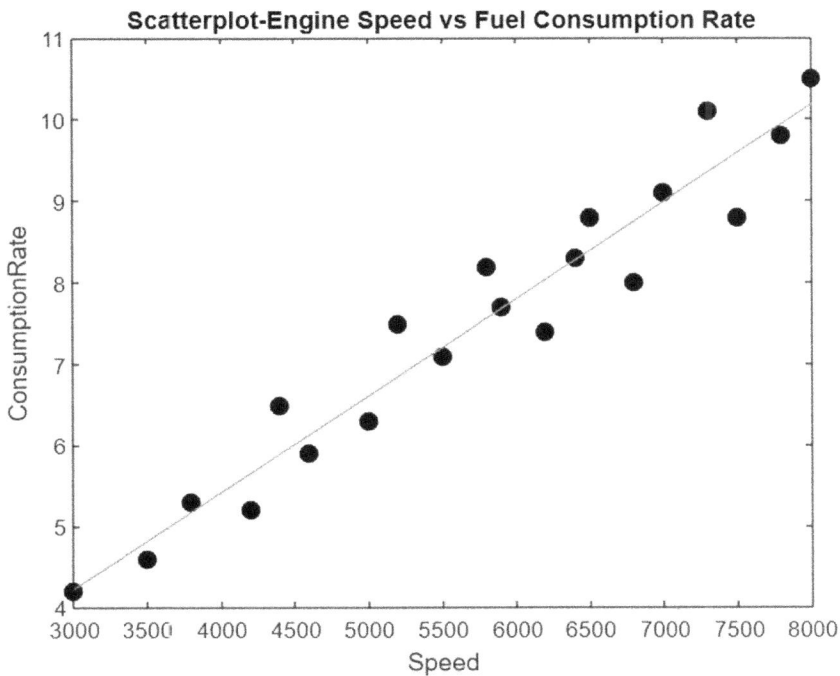

Figure 11.6

$$slope\ \beta_1 = \frac{49434}{41492000} = 0.0011914104$$

Step 3: compute the y-intercept of the line by using the formula:

$$intercept\ \beta_0 = 7.465 - 0.0011914(5720) = 0.6501325557$$

Step 4: equation of the line

$$y = 0.6501 + 0.0012\ x$$

Now that we have the slope and the y-intercept, we can draw the line of best fit.

So what's the next step? How do we use it to make predictions? As we can see from the graph and the equation, x is a predictor of y. So, we can plug any given value of x into the equation and get the value of y. For example, let's say we want to predict the consumption rate based on speed. Therefore, our independent variable x is the speed and y is the consumption rate. So now for any speed we can approximate the fuel consumption rate of the vehicles.

11.3 SUM OF SQUARES

When we plot our scatterplot with the best-fit line, we'll notice that not all data points lie perfectly on the line. This indicates that residuals exist for some points, meaning our predictions are not entirely precise

So let's now talk about how to calculate these residuals.

Observed y_i	x_i	Predicted $\widehat{y_i} = 0.6501 + 0.0012 x_i$	Residual $= y_i - \overline{y_i}$
10.5	8000	10.18141576	0.318584
9.8	7800	9.943133676	−0.14313
8.8	7500	9.585710556	−0.78571
10.1	7300	9.347428476	0.752572
9.1	7000	8.990005356	0.109995
8.0	6800	8.751723276	−0.75172
8.8	6500	8.394300156	0.4057
8.3	6400	8.275159116	0.024841
7.4	6200	8.036877036	−0.63688
7.7	5900	7.679453916	0.020546
8.2	5800	7.560312876	0.639687
7.1	5500	7.202889756	−0.10289
7.5	5200	6.845466636	0.654533
6.3	5000	6.607184556	−0.30718
5.9	4600	6.130620396	−0.23062
6.5	4400	5.892338316	0.607662

5.2		4200	5.654056236		−0.45406
5.3		3800	5.177492076		0.122508
4.6		3500	4.820068956		−0.22007
4.2		3000	4.224363756		−0.02436

Since there is a residual for every data point, we can conclude that our model will always be an approximation.

To calculate the SSE/RSS:

$$\sum_{i=1}^{n} \varepsilon_i^2 = \left(y_i - \widehat{y}_i \right)^2$$

$y_i - \widehat{y}_i$	$\left(y_i - \widehat{y}_i\right)^2$	$y_i - \widehat{y}_i$	$\left(y_i - \widehat{y}_i\right)^2$	$y_i - \widehat{y}_i$	$\left(y_i - \widehat{y}_i\right)^2$	$y_i - \widehat{y}_i$	$\left(y_i - \widehat{y}_i\right)^2$
0.318584	0.101496	−0.75172	0.565083	0.639687	0.409199	0.607662	0.369253
−0.14313	0.020486	0.4057	0.164592	−0.10289	0.010586	−0.45406	0.20617
−0.78571	0.61734	0.024841	0.000617	0.654533	0.428413	0.122508	0.015008
0.752572	0.566365	−0.63688	0.405616	−0.30718	0.09436	−0.22007	0.048431
0.109995	0.012099	0.020546	0.000422	−0.23062	0.053186	−0.02436	0.000593

SSE / RSS = 0.101496 + 0.020486.. + 0.565083 + 0.164592 + ... + 0.409199 + 0.000593

= 4.089317

11.3.1 Other Sums of Squares

Other than the residual there are two more types of sums of squares that are commonly used to measure the amount of variability in the data:

1. **Regression sum of squares:** this sum of squares shows the relationship between the given data and the regression model, where the regression model denotes whether there is a relationship between one or multiple variables. This sums the squared difference between the predicted value and the mean.

$$SSR = \pounds \left(\widehat{y}_i - \bar{y} \right)^2$$

While the sum of squares for residuals represents the variance not explained by the regression model, the regression sum of squares describes the variances that is explained by the regression model.

2. **Total sum of squares (SST)**: the total sum of squares measures the distance between the response variable y and the mean value. It is calculated as the sum of the squared deviations of each observation from the mean of y.

$$\text{SST} / \text{TSS} = \pounds\left(y_i - \bar{y}\right)^2$$

Let's now calculate these sums of squares.

Predicted $\hat{y}_i$	Observed y_i	$\hat{y}_i - \bar{y}$	$\left(\hat{y}_i - \bar{y}\right)^2$	$y_i - \bar{y}$	$\left(y_i - \bar{y}\right)^2$
10.18141576	10.5	2.716416	7.378915	3.035	9.211225
9.943133676	9.8	2.478134	6.141147	2.335	5.452225
9.585710556	8.8	2.120711	4.497413	1.335	1.782225
9.347428476	10.1	1.882428	3.543537	2.635	6.943225
8.990005356	9.1	1.525005	2.325641	1.635	2.673225
8.751723276	8.0	1.286723	1.655657	0.535	0.286225
8.394300156	8.8	0.9293	0.863599	1.335	1.782225
8.275159116	8.3	0.810159	0.656358	0.835	0.697225
8.036877036	7.4	0.571877	0.327043	−0.065	0.004225
7.679453916	7.7	0.214454	0.04599	0.235	0.055225
7.560312876	8.2	0.095313	0.009085	0.735	0.540225
7.202889756	7.1	−0.26211	0.068702	−0.365	0.133225
6.845466636	7.5	−0.61953	0.383822	0.035	0.001225
6.607184556	6.3	−0.85782	0.735847	−1.165	1.357225
6.130620396	5.9	−1.33438	1.780569	−1.565	2.449225
5.892338316	6.5	−1.57266	2.473265	−0.965	0.931225
5.654056236	5.2	−1.81094	3.279517	−2.265	5.130225
5.177492076	5.3	−2.28751	5.232693	−2.165	4.687225
4.820068956	4.6	−2.64493	6.99566	−2.865	8.208225
4.224363756	4.2	−3.24064	10.50172	−3.265	10.66023
			$\Sigma\,58.89618$		$\Sigma\,62.9855$

SSR = 58.89618

RSS / SSE = 4.089317

SST = 62.9855

11.4 COEFFICIENT OF DETERMINATION

You must be wondering why we are discussing the sum of squares. The answer is to determine the goodness of fit. It can be determined using the coefficient of determination denoted by R^2. Now this R is often confused with "r" from Chapter 11, where r is the coefficient correlation. R^2 on the other hand is the coefficient of determination. It is interpreted as the proportion of observed y variation that can be explained by the simple linear regression model. The higher the value of R^2, the more successful is the simple linear regression model in explaining y variation.

The coefficient of determination, denoted by R^2, has two variations:

- Ordinary

$$R^2 = 1 - \frac{SSE}{SST} = \frac{SSR}{SST}$$

- Adjusted for the number of coefficients

$$R^2 = 1 - \left(\frac{n-1}{n-p-1}\right)\frac{SSE}{SST}$$

Where
n = number of observations
p = number of predictors

Let's calculate the ordinary and adjusted R-squared for our example.

Ordinary $R^2 = 1 - \dfrac{4.089317}{62.9855} = \dfrac{58.89618}{62.9855} = 0.93507$

For adjusted R^2

- $n = 20$
- In linear regression $p = 1$

$$R^2 = 1 - \left(\frac{20-1}{20-1-1}\right)\frac{4.089317}{62.9855} = 0.93147$$

The model explains around 93% of the variability in the response variable.

If R^2 is the variance in y explained by x, $1 - R^2$ is the variance in y not explained by x.

An interesting fact about linear regression is that it is made up of two statistical concepts: ANOVA and correlation.

$$\text{Linear regression = correlation + ANOVA}$$

11.5 ERRORS IN THE PARAMETERS OF THE REGRESSION LINE

From previous sections we already learned to find the regression line that best fits a set of points. In the above example we estimated the linear model using a sample size n of 20. Now instead of taking the whole sample, let's take the first ten. When we compute the coefficients of the best fit we get:

$$y = -0.067057 + 0.0012849x$$

What do we notice? A completely different regression line. What does that imply? It implies that there is **variability** in the regression line from sample to sample—the estimated slopes and intercepts vary from sample to sample. In other words, the slope b_1 and intercept b_0 are **sample statistics**. This means that both are estimated on a sample of data so there will always be some variability due to other factors, so the true regression line doesn't describe the population perfectly, making them both incorrect if we compare them to the hypothetical true model.

Now remember our regression line is approximated such that each y_i observed value is estimated at its least squared distance from the line. Even with this predicted $\widehat{y}_i$, there is always some uncertainty (probability) about just where each y is in relation to its x. In this case, you will want to determine the error associated with the slope or intercept. To do this we need the residual standard error (RSE).

Earlier we learned to calculate the uncertainty in the regression in terms of residuals:

$$\varepsilon_i = y_i - \widehat{y}_i$$

Then we obtained the RSS by adding the square of residuals.

$$\sum_{i=1}^{n} \varepsilon_i^2 = \sum_{i=1}^{n} (y_i - \widehat{y}_i)^2$$

The RSE is calculated as the square root of the RSS divided by the degrees of freedom $(n - p - 1)$, where n is the sample size and p is the number of predictors in the model.

$$\text{RSSE} = \sqrt{\frac{\sum_{i=1}^{n} (y_i - \widehat{y}_i)^2}{(n - p - 1)}}$$

Since we have a single predictor in linear regression, our RSS would be

$$\text{RSE for linear regression} = \sqrt{\frac{\sum_{i=1}^{n} (y_i - \widehat{y}_i)^2}{(n - 1 - 1)}} = \sqrt{\frac{\sum_{i=1}^{n} (y_i - \widehat{y}_i)^2}{n - 2}}$$

Now that we have our RSS we can use it to calculate the standard error of the slope and intercept:

$$\text{SE of slope} = \text{RSE} \times \frac{1}{\sqrt{\Sigma(x_i - \bar{x})^2}}$$

$$\text{SE of intercept} = \text{RSE} \times \sqrt{\frac{\Sigma x_i^2}{n\Sigma(x_i - \bar{x})^2}}$$

Let's now calculate these for our example:

$$\text{SE of slope} = \sqrt{\frac{4.089317}{20 - 2}} \times \frac{1}{\sqrt{41492000}} = 7.3996 \times 10^{-5}$$

$$\text{SE of intercept} = \sqrt{\frac{4.089317}{20-2}} \times \sqrt{\frac{695860000}{20 \times 41492000}} = 0.43647$$

11.6 CONFIDENCE INTERVAL FOR REGRESSION PARAMETERS

By now we have already figured out that we will never exactly estimate the true value of these parameters from sample data in an empirical application. This is where the confidence interval comes in.

To construct a confidence interval for the coefficients of the regression line, we need to know the coefficient, the critical value, and the standard error of those coefficients.

The critical value will be of a t-distribution and depends on the confidence level and the degrees of freedom $df = n-2$. We can create a confidence interval using any confidence level between 0 and 100%. However, the confidence levels that you will see most often are 90%, 95%, and 99%. This value is important since our critical value depends on it.

The confidence level equals 100%*(1 – alpha).

11.6.1 For one-sided

The confidence interval for the slope β_1 is

$$\beta_1 \pm t_{1-\alpha, n-2}(\text{SE of slope})$$

Similarly, the confidence interval for the intercept β_0 is

$$\beta_0 \pm t_{1-\alpha, n-2}(\text{SE of intercept})$$

11.6.2 For two-sided

$\pm$ is replaced by $\pm/2$; hence

$$\beta_1 \pm t_{1-\pm/2, n-2}(\text{SE of slope})$$

$$\beta_0 \pm t_{1-\pm/2, n-2}(\text{SE of intercept})$$

So for a confidence level of 90%,

$$90\% = 100\%*(1-\text{alpha})$$

$$0.90 = 1-\text{alpha}$$

$$\text{alpha} = 1-0.90 = 0.10$$

If our test is one-tailed then the confidence interval for the slope β_1 is

$$\beta_1 \pm t_{1-0.10, n-2}(\text{SE of slope})$$

In the case of two-tailed tests our confidence interval then becomes

$$\beta_1 \pm t_{1-0.10/2, n-2} \left(\text{SE of slope} \right)$$

Let's now calculate the confidence interval for our example at the 95% confidence interval. Since our example is right-tailed, our alpha = 0.05 and $df = 20 - 2 = 18$,

$$0.0011914104 \pm 1.734 \left(7.3996 \times 10^{-5} \right) = [0.00106, 0.00131]$$

Let's now do the same for the intercept:

$$0.6501325557 \pm 1.734 \left(0.43647 \right) = [-0.1067064243, 1.40697]$$

The 95% confidence level is commonly interpreted as meaning there is a 95% probability that the true value of the coefficients lies within this interval.

11.7 INFERENCES ABOUT THE SLOPE: THE REGRESSION T TEST

Whenever we perform linear regression, we want to know if there is a statistically significant relationship between the predictor variable and the response variable. To assess the suitability of the linear regression model we use two types of statistics—**t-statistics** and **f-statistics**. In this section we shall discuss the hypothesis testing related to the two statistics.

Both tests are applied to test if the slope, β_1, of the population regression line equals 0. Based on that test we may decide whether x is a useful (linear) predictor of y. The regression t and f test follow the same step-wise procedure as discussed in the previous sections.

11.7.1 State the hypotheses

$$H_0 : \beta_1 = 0$$

There is no relationship between y (response variable) and x (predictor variable).

In other words there is no linear relationship between the explanatory and the dependent variable.

$$H_A : \beta_1 \neq 0$$

11.7.2 Compute the value of the test statistic

The t-statistic is used not just in the calculation of confidence intervals, but also for performing hypothesis testing because the standard deviation of the sampling distribution is unknown. As the standard deviation of a sampling distribution is unknown, the standard error is used to estimate the coefficients. So we use the formula

$$t = \frac{\beta_1}{\text{SE of slope}}$$

11.7.3 Finding the p-value

After calculating the value of the t-statistic for any of the coefficients, it is now time to make a decision about whether to accept or reject the null hypothesis. In order for this decision to be made, we need our p-value to compare it with our alpha level.

Let's perform the hypothesis test for our speed vs. consumption rate example

Null hypothesis H_0: there is no significant linear relationship between speed and consumption rate.

Alternative hypothesis H_A: there is a significant linear relationship between speed and consumption rate.

11.7.3.1 Our test statistic

$$t = \frac{0.0011914104}{7.3996 \times 10^{-5}} = 16.101$$

11.7.3.2 p-value

The p-value that corresponds to $t = 16.101$ with $df = 20 - 2 = 18$ is < 0.0005.

Since the p-value < 0.05 we reject H_0 and conclude that speed and consumption rate are correlated.

Now we can also test for intercept using the same procedure. Then our test statistic for the intercept is

$$t = \frac{0.6501325557}{0.43647} = 1.48952$$

So the p-value that approximately corresponds to the test statistic = 1.48952 at $df = 18$ is 0.15.

Since here we have only one predictor a **t test** should be enough. However, in reality, our model is going to include a **number of independent variables**. This is where the **F-statistic** comes into play. The F test is a statistical test that measures whether a linear regression model fits the data better than a null model. Now what is a null model? It is basically a model that includes only the intercept and no predictor variables.

Null hypothesis H_0: predictor x is not able to explain the variance of the independent variable y.

Alternative hypothesis H_A: x is significant in predicting the value of y.

11.7.3.3 Our test statistic

We already discussed the regression and residual sum of squares. The f-statistic is defined as a function of SSR and SSE in the following manner:

$$f = \frac{\dfrac{SSR}{p}}{\dfrac{SSE}{n-p-1}}$$

Where

SSR = regression sum of squares
SSE = residual sum of squares
p = number of predictors
n = sample size

Now the other variance of the formula is

$$f = \frac{MSR}{MSE}$$

In other words, the F-statistic is a ratio of the mean squared error of the full model to the mean squared error of the null model.

Where

MSR = the mean sum of squares regression calculated by $\dfrac{SSR}{p}$

MSE = the mean sum of squares error estimated by $\dfrac{SSE}{n-p-1}$

The next step will be to find the p-value of the F-statistic at the given level of significance with the degrees of freedom as 3, 196.

Now let's find the f statistic of our example:

$$f = \frac{\dfrac{58.89618}{1}}{\dfrac{4.089317}{20-2}} = 259.244$$

Here $MSR = \dfrac{58.89618}{1}$

and $MSE = \dfrac{4.089317}{20-2}$

Now if we look back, notice we have already discussed degrees of freedom, sum of squares, mean squares, F ratio, and p-value. What does this remind you of?

Your first instinct would probably be ANOVA and you are right.

We know that ANOVA **provides the analysis of the variance in the model**, while in Chapter 9 we learned to use ANOVA to **test for variance between the means of multiple groups**. In regression we use it (ANOVA) to collect information about levels of variability.

Here the ANOVA table breaks down into three sources of variation:

- The total variation
- Variation due to the model or regression
- Variation due to residuals or error

11.7.3.4 The ANOVA table

Source	Sum of squares	df	Mean squares	F	P
Total	TSS / SST	$n-1$	$MST = \dfrac{SST}{n-1}$		
Model	SSR	p	$MSR = \dfrac{SSR}{p}$	$\dfrac{MSR}{MSE}$	
Residual	SSE / RSS	$n-p-1$	$MSE = \dfrac{SSE}{n-p-1}$		

So the ANOVA table for our example would be:

Source	Sum of squares	df	Mean squares	F	P
Total	62.9855	$20 - 1 = 19$	$\dfrac{62.9855}{19} = 3.315$		
Model	58.8962	1	$\dfrac{58.8962}{1} = 58.8962$	$\dfrac{58.8962}{0.22718} = 259.24$	
Residual	4.089317	$20 - 1 - 1 = 18$	$\dfrac{4.089317}{18} = 0.22718$		

11.8 REGRESSION IN MATLAB

Now if you are wondering if finding all these in MATLAB is as exhausting as it is by hand, then get ready to be thrilled. Because there is absolutely no work that needs to be done to get the table. What if I told you all this was achieved with just a few lines of code? Sounds unrealistic? Let's start from the very first step of regression analysis.

First step: check for linearity.
We have already discussed the test of linearity using scatterplots.

Second step: check assumptions of linear regression regarding residuals.
Plot the residuals vs. fitted values. If they are scattered, this implies that all the assumptions are met. Now before we can draw the plot, we first need to fit a regression model to our data. We do this by creating a LinearModel object by using fitlm or stepwiselm. We shall use fitlm for this example. Therefore our syntax is:

```
model = fitlm(x, y)
```

This returns a linear regression model with explanatory variable x and response variable y.

In our example we are performing regression analysis to estimate the vehicle fuel consumption rate based on average travel speed; therefore the speed is the explanatory variable x and the consumption rate the response or 'y variable.

```
model = fitlm(Speed, ConsumptionRate)
```

Then what does the output look like?

```
model =


Linear regression model:
    y ~ 1 + x1

Estimated Coefficients:
                   Estimate        SE         tStat       pValue

    (Intercept)     0.65013     0.43647      1.4895      0.15366
    x1              0.0011914   7.3996e-05   16.101      3.9179e-12


Number of observations: 20, Error degrees of freedom: 18
Root Mean Squared Error: 0.477
R-squared: 0.935,  Adjusted R-Squared: 0.931
F-statistic vs. constant model: 259, p-value = 3.92e-12
```

Wow! All those draining math calculations we had to do by hand are displayed by just one line of code. The magic of MATLAB!

However even though these calculations are accurate, they are not reliable as they may or may not fit the observed data.

After we fit a regression model to our data, it becomes crucial to check the residual plots because, if by any chance our plots display any sort of unwanted patterns, certain parts of the model are violated and, as a consequence, results may become invalid. Therefore, unless we prove that the data follows the assumptions, we can't trust the regression coefficients and other numeric results. So without getting stunned by the output, let's now see if a linear model fits the data. So to plot the residual we use the syntax:

```
plotResiduals(model, plottype, Name, Value)
```

Let's discuss the input arguments:

- Since we want to plot the residuals of the linear regression model, the first argument needs to be the model.
- plottype refers to the type of plot we want. MATLAB gives a bunch of other options to check the validity of these assumptions.

'histogram' (default)	Histogram of residuals using probability density function scaling. The area of each bar is the relative number of observations. The sum of the bar areas is equal to 1.
'fitted'	Residuals vs. fitted (predicted) values.
'caseorder'	Residuals vs. case order (row number).
'lagged'	Residuals vs. lagged residuals ($r(t)$ vs. $r(t-1)$).
'probability'	Normal probability plot of residuals. For details, see probplot.
'symmetry'	Symmetry plot of residuals around their median (residuals in upper tail − median vs. median − residuals in lower tail). This plot includes a dotted reference line of $y=x$ to examine the symmetry of residuals.

Let's discuss each of these and see how it tests the assumptions of residuals.

One of the conditions of the linear regression analysis was the normality of residuals. To test this, we can use the histogram residual plot. A histogram plot of the residuals should exhibit a symmetric bell-shaped distribution, indicating that the normality assumption is likely to be true. Let's see if our example fulfills the normality criteria.

```
plotResiduals(model, 'histogram')
```

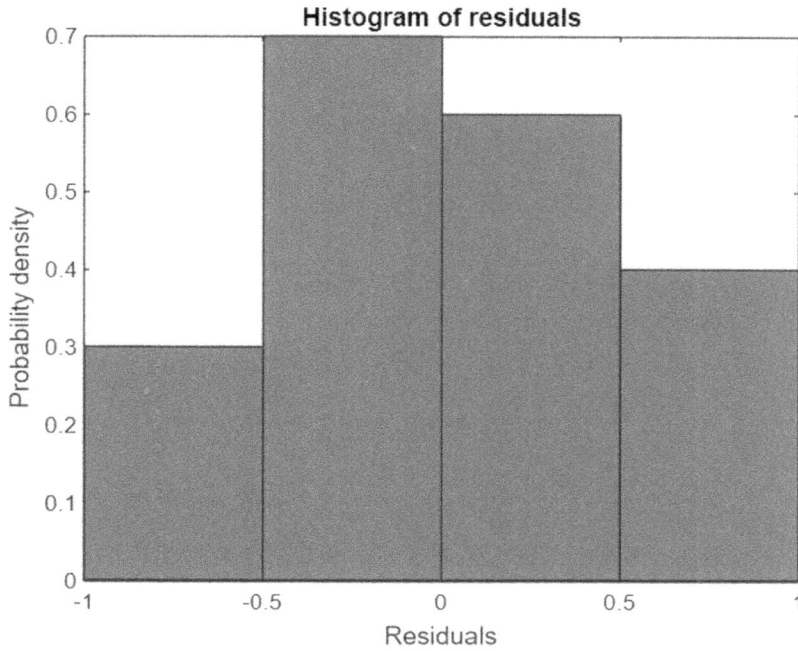

Figure 11.7

Evidently the residuals show a symmetrical curve implying that the normality condition has been met.

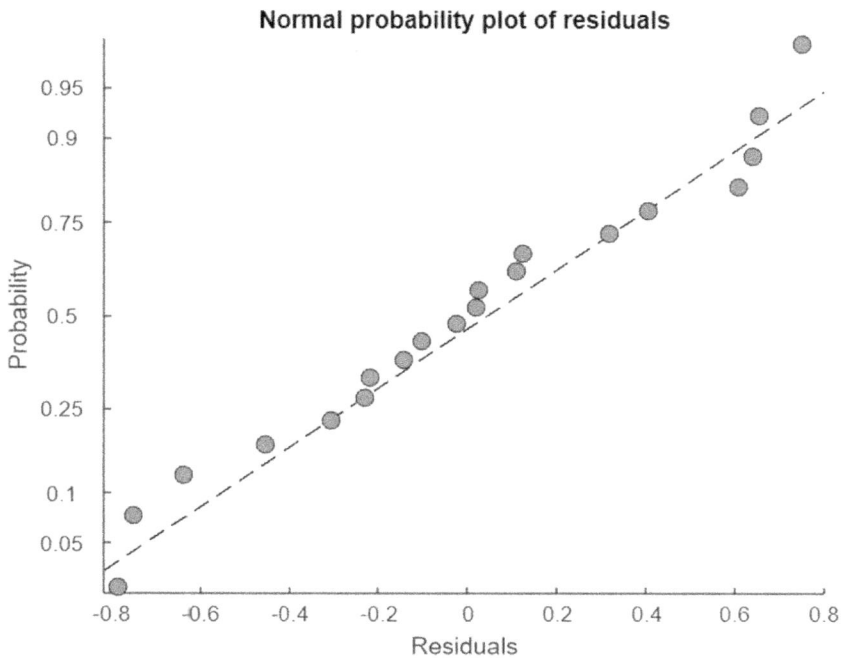

Figure 11.8

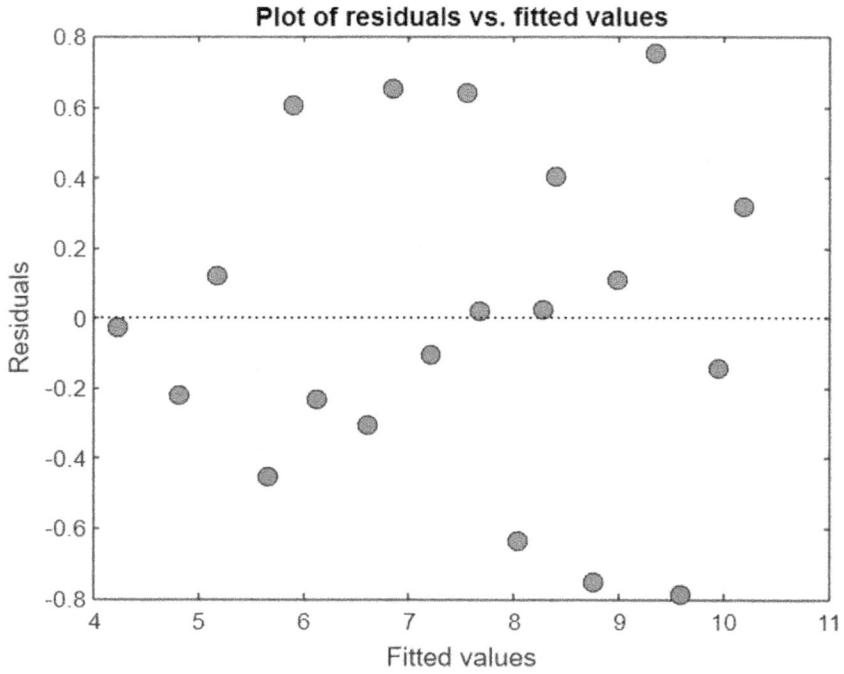

Figure 11.9

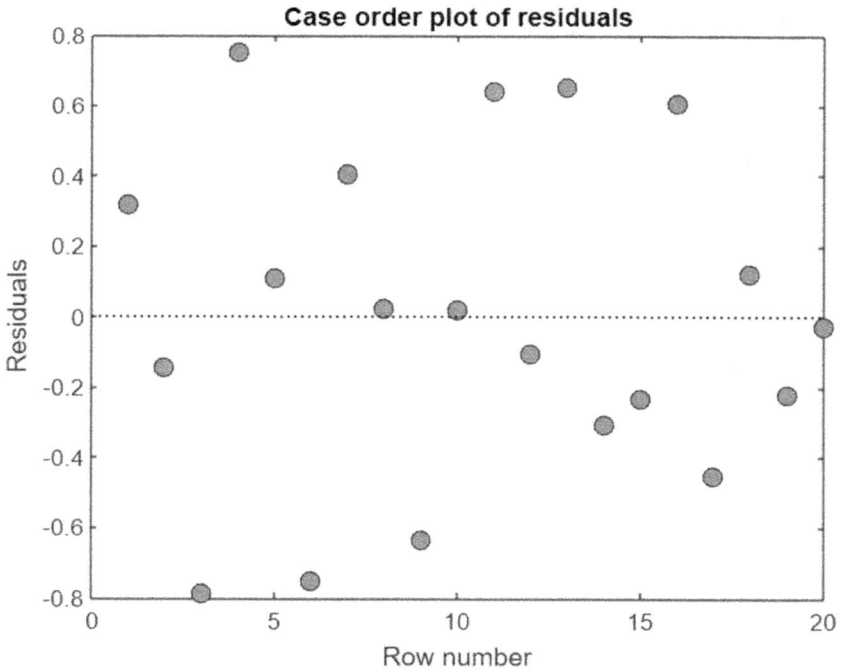

Figure 11.10

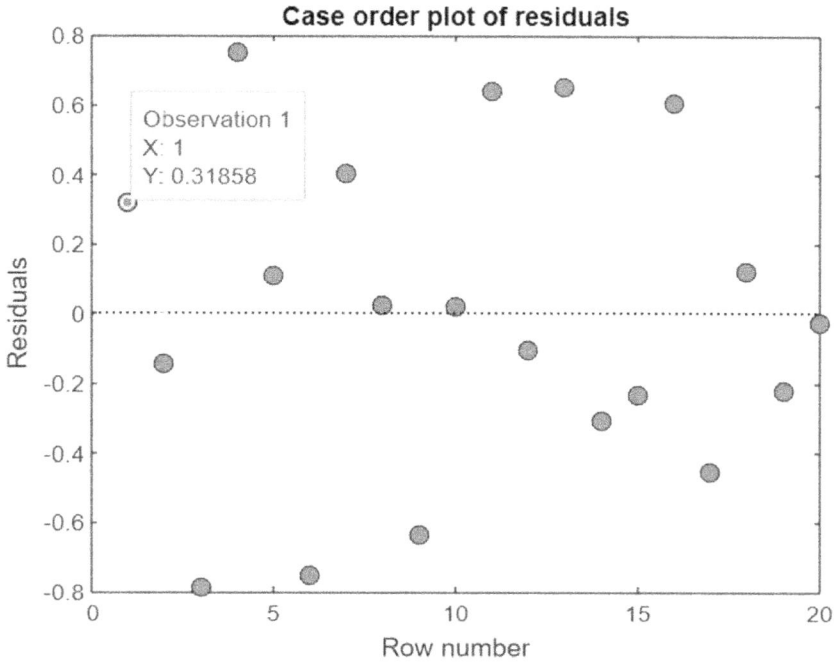

Figure 11.11

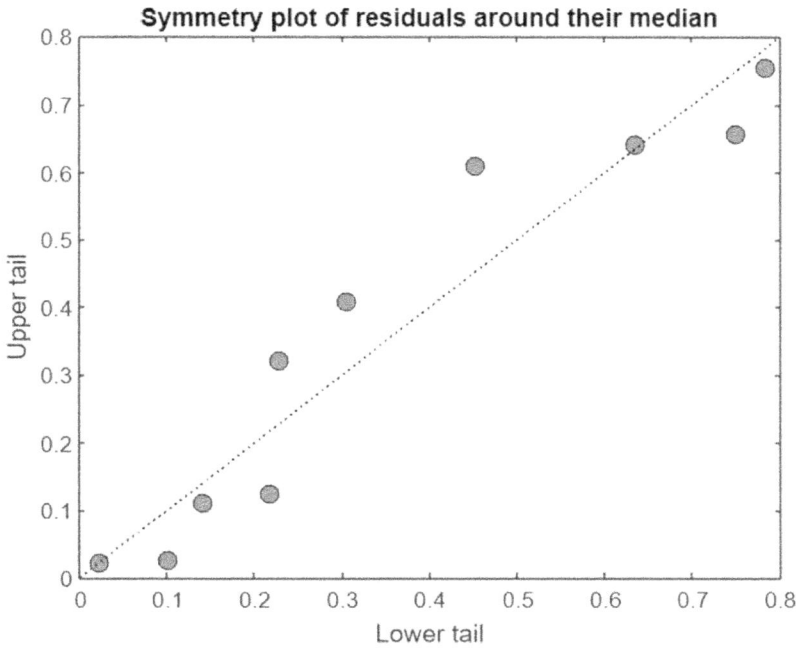

Figure 11.12

Now it is quite unlikely that we do not mention the normal probability plot and, in this case, the normal probability plot of the residuals when we are testing for normality. As we all already know the linearity of the plot confirms its normality. Let's now test for our example (Figure 11.8).

Note that the relationship between the probability and the residuals is approximately linear. Therefore, the normal probability plot of the residuals suggests that the error terms are indeed normally distributed for this example.

Let's now look at the residual plot vs. fitted values in Figure 11.9.

In this plot, the residuals sit on the y-axis and the fitted values (i.e., predicted values) on the x-axis. Each point on the plot represents an observation in our dataset. As we can see, the plots have no discernible pattern. They are all centered around zero with no special clustering pattern. Such random scatter of points indicates a good fit for the model.

If there was a pattern, it could indicate nonlinearity, heteroscedasticity (unequal variances) and the presence of outliers in the model and thus that the model is not a good fit for the data or that there is some underlying relationship that the model is not capturing.

Even though we only discussed the plot of residual vs. fitted all the other plots follow the same rule. In order to be appropriate for the linear model the plot should not follow any trend. A trend would indicate that the residual conditions were not fulfilled making the data unfit for linear regression analysis.

In Figure 11.10, we see another type of plot, called the case order plot of residuals, which can help us assess the quality of our regression model. As the name suggests, this plot is a scatterplot with residuals on the y-axis and the order in which the data was collected on the x-axis.

For example if we look back at our calculations of residuals we will find that the first dataset had the residual 0.31858.

The residuals bounce randomly around the residual = 0 line as we expected. In general, residuals exhibiting normal random noise around the residual = 0 line suggest that there is no serial correlation.

While the reference line of the residual vs. fitted plot was $y = 0$ the symmetry plot, in Figure 11.12, has a dotted reference line of $y = x$ to examine the symmetry of residuals.

The function orders the data (or residuals) and takes the median minus the lower half and plots it against the upper half minus the median.

This plot suggests that the residuals are roughly distributed equally around their median, as would be expected for normal distribution.

In addition to the discussed arguments we have other optional pairs of arguments.

11.8.1 ResidualType

Even though we have learned to calculate the residuals using the formula Observed − Predicted, it turns out there are multiple other ways to calculate residuals. The one we learned is known as the raw method in MATLAB. This is basically the default value of the argument. Other types of residual are:

'pearson'	Raw residuals divided by the root mean squared error
'standardized'	Raw residuals divided by their estimated standard deviation
'studentized'	Raw residuals divided by an independent (delete-1) estimate of their standard deviation

Let's discuss each of these residuals and see how they are different from raw residuals (which we calculated by hand).

11.8.1.1 Pearson

Another type of residual is the Pearson residual. It is calculated by dividing the raw residuals with the root mean squared error, that is,

$$\textbf{Pearson}\_\varepsilon_i = \frac{y_i - \widehat{y_i}}{\sqrt{MSE}} = \frac{\textbf{Raw}}{\sqrt{MSE}}$$

In this example we calculated **MSE** = 0.22718.

Raw	Pearson = $\dfrac{\text{Raw}}{\sqrt{MSE}}$	Raw	Pearson = $\dfrac{\text{Raw}}{\sqrt{MSE}}$
0.318584	0.668404	0.639687	1.342093
−0.14313	−0.30029	−0.10289	−0.21587
−0.78571	−1.64846	0.654533	1.37324
0.752572	1.57893	−0.30718	−0.64448
0.109995	0.230775	−0.23062	−0.48385
−0.75172	−1.57714	0.607662	1.274903
0.4057	0.851177	−0.45406	−0.95264
0.024841	0.052118	0.122508	0.257027
−0.63688	−1.3362	−0.22007	−0.46172
0.020546	0.043106	−0.02436	−0.05111

11.8.1.2 Standardized

Standardized residuals are very similar to the kind of standardization we discussed earlier on. As you'll probably guess, "standardized residuals" normalize the data in regression analysis and are obtained by transforming each of the residuals as follows:

$$\textbf{standardized}\_\varepsilon_i = \frac{y_i - \widehat{y_i}}{\sqrt{\textbf{MSE}(1 - h_i)}} = \frac{\textbf{Raw}}{\sqrt{\textbf{MSE}(1 - h_i)}}$$

Where h_i is the leverage value for observation i.

$$h_i = \frac{1}{n} + \frac{(x_i - \bar{x})^2}{\displaystyle\sum_{i=1}^{n}(x_i - \bar{x})^2}$$

In this example

$n = 20$

$\bar{x} = 5720$

MSE = 0.22718

x	$x_i - \bar{x}$	$(x_i - \bar{x})^2$	$\dfrac{(x_i - \bar{x})^2}{\sum_{i=1}^{n}(x_i - \bar{x})^2}$	$h_i = \dfrac{1}{n} + \dfrac{(x_i - \bar{x})^2}{\sum_{i=1}^{n}(x_i - \bar{x})^2}$	$\sqrt{\mathrm{MSE}(1-h_i)}$
8000	2280	5,198,400	0.125287	0.175287	0.432772882
7800	2080	4,326,400	0.104271	0.154271	0.438252351
7500	1780	3,168,400	0.076362	0.126362	0.445424804
7300	1580	2,496,400	0.060166	0.110166	0.449534585
7000	1280	1,638,400	0.039487	0.089487	0.454727911
6800	1080	1,166,400	0.028111	0.078111	0.457559713
6500	780	608,400	0.014663	0.064663	0.460885037
6400	680	462,400	0.011144	0.061144	0.461751151
6200	480	230,400	0.005553	0.055553	0.46312411
5900	180	32,400	0.000781	0.050781	0.464292649
5800	80	6400	0.000154	0.050154	0.464445875
5500	−220	48,400	0.001166	0.051166	0.464198331
5200	−520	270,400	0.006517	0.056517	0.462887684
5000	−720	518,400	0.012494	0.062494	0.461419135
4600	−1120	1,254,400	0.030232	0.080232	0.457033081
4400	−1320	1,742,400	0.041994	0.091994	0.45410158
4200	−1520	2,310,400	0.055683	0.105683	0.450665492
3800	−1920	3,686,400	0.088846	0.138846	0.442230781
3500	−2220	4,928,400	0.11878	0.16878	0.434476893
3000	−2720	7,398,400	0.178309	0.228309	0.41862992
		£ 41,492,000			

Raw	Standardized $= \dfrac{\text{Raw}}{\sqrt{\mathrm{MSE}(1-h_i)}}$	Raw	Standardized $= \dfrac{\text{Raw}}{\sqrt{\mathrm{MSE}(1-h_i)}}$
0.318584	0.736016314	0.639687	1.377069738
−0.14313	−0.326535149	−0.10289	−0.221611918
−0.78571	−1.763645938	0.654533	1.413772213
0.752572	1.673818892	−0.30718	−0.665611558
0.109995	0.24184931	−0.23062	−0.504513569
−0.75172	−1.642600102	0.607662	1.337927515
0.4057	0.880107898	−0.45406	−1.007354799
0.024841	0.053787902	0.122508	0.276973999
−0.63688	−1.374939977	−0.22007	−0.506427965
0.020546	0.04424447	−0.02436	−0.058179574

From the results we can see that none of the standardized residuals exceed an absolute value of 3. Thus, none of the observations appear to be outliers.

Here, we see that the standardized residual for a given data point depends on the raw residual, the mean square error (MSE), and the leverage h_{ii}.

Now the question is: why should we use these? How are they better than the raw residuals? In one sentence, they help detect outliers when there is inconsistent variance in raw residuals. When there is heteroscedasticity in the raw residuals, the spread of the residuals will be different for different levels of the independent variable(s). In particular, if the residuals have a larger spread for some values of the independent variable(s), it can make it harder to detect outliers in those regions.

The advantage of standardized residuals is that they have variance equal to 1. This makes them simple to compare across model fits.

Typically any observation in a dataset that has a standardized residual greater than an absolute value of 3 is considered an outlier; however in some textbooks the threshold is 2. Regardless of the threshold our data seems to be outlier-free, confirming that the data fits the linear model.

11.8.1.3 Studentized

Studentized residuals are the raw residuals divided by an independent estimate of the residual standard deviation. The residual for observation i is divided by an estimate of the error standard deviation based on all observations except for observation i. For this reason, studentized residuals are sometimes referred to as *externally* studentized residuals.

$$\textbf{studentized}\_\varepsilon_i = \frac{y_i - \widehat{y}_i}{\sqrt{MSE_i(1-h_i)}} = \frac{\textbf{Raw}}{\sqrt{MSE_i(1-h_i)}}$$

As we can see, the above formula is almost equivalent to the standardized residual except that standardized residuals use the MSE for the model based on all observations, while studentized residuals use the mean square error based on the estimated model with the *i*th observation deleted.

Just like the standardized residual, if an observation has a studentized residual that is larger than 3 or 2 we can call it an outlier. However, studentized residuals are going to be more effective for detecting outlying y observations than standardized residuals.

Calculating these residuals can be extremely tedious and laborious; fortunately MATLAB finds all of these for us using:

```
r = model.Residualse.Raw
r = model.Residuals.Pearson
r = model.Residuals.Standardized
r = model.Residuals.Studentized
```

Now that the assumptions of the linear regression model have been met and the residuals have been checked for normality, constant variance, and independence, we can proceed to find the equation of the line.

```
p = polyfit(x, y, n)
```

This returns the coefficients for the polynomial $p(x)$ of degree n that is a best fit (in a least-squares sense) for the data in y. The output displays the coefficients in descending powers, which means the last coefficient is the coefficient of the intercept.

Since we have only one predictor variable, our $n=1$; hence

```
p = polyfit(Speed, ConsumptionRate, 1)
```

$$p =$$

$$0.0012 \qquad 0.6501$$

Using the equation of line now we shall find the predicted values using

```
y = polyval(p, x)
```

This evaluates the value of y at each point in x.

```
f = polyval(p, Speed)
```

```
Columns 1 through 15

10.1814   9.9431   9.5857   9.3474   8.9900   8.7517   8.3943   8.2752   8.0369   7.6795   7.5603   7.2029   6.8455   6.6072   6.1306

Columns 16 through 20

5.8923   5.6541   5.1775   4.8201   4.2244
```

Now we are ready to finally **plot the best-fit line using plot.**
%circular markers represented by o and k specifies the color black then the best fit line represented by - and the color is red which is default

```
h = plot(Speed, ConsumptionRate,'ok', Speed, f, '--', 'MarkerSize', 8)

%filled markers
set(h, {'MarkerFaceColor'}, get(h,'Color'));

%adding the labels
xlabel('Speed')
ylabel('ConsumptionRate')
title('Scatterplot-Engine Speed vs Fuel Consumption Rate')
```

The output has already been shown in Figure 11.6.
The next step is to actually analyze the results. We now bring back the output we initially got from **fitlm.**

```
model =

Linear regression model:
    y ~ 1 + x1

Estimated Coefficients:
                   Estimate          SE          tStat         pValue

    (Intercept)    0.65013        0.43647        1.4895         0.15366
    x1             0.0011914      7.3996e-05     16.101         3.9179e-12


Number of observations: 20, Error degrees of freedom: 18
Root Mean Squared Error: 0.477
R-squared: 0.935,  Adjusted R-Squared: 0.931
F-statistic vs. constant model: 259, p-value = 3.92e-12
```

The output above begins by stating the model of our example. The intercept is displayed as 1. The model formula in the display, $y \sim 1 + x_1$ corresponds to $y = {}^2{}_0 + {}^2{}_1 x_1 + \mu$.

Next is a table of the fit parameters with estimated coefficients, along with the standard error for the estimate and the t-statistic for each coefficient to test the null hypothesis that the corresponding coefficient is zero against the alternative that it is different from zero and then the p-value that represents the probability of observing the calculated t test statistic of the two-sided hypothesis test, assuming the null hypothesis that the true coefficient is zero.

In the table, we usually focus more on the p-values and coefficients of our variables as they provide information about their relationships in the model.

11.8.2 Slope coefficient

Slope coefficients represent the mean change in the response or explanatory variable for one unit of change in the variable while holding other predictors in the model constant.

A predictor that has a low p-value indicates that it is statistically significant and is likely to be a meaningful addition to your model because changes in the predictor's value are related to changes in the response variable.

Conversely, a larger (insignificant) p-value suggests that it is not statistically significant and changes in the predictor are not associated with changes in the response.

In the output above, we can see that the predictor variable of speed is significant because its p-value is extremely small; therefore changes in the consumption rate are related to changes in the speed of the vehicle.

Typically, we use the p-values to determine which terms to keep in the regression model.

11.8.3 Intercept coefficient

The intercept term in a regression table tells us the average expected value for the response variable when the predictor variable is equal to zero.

In this example, the regression coefficient for the intercept is equal to 0.65013. This means that on average, a typical vehicle consumes around 0.65013 liters of fuel per hour while idling. The p-value for the intercept coefficient helps determine whether the intercept significantly differs from zero.

If the p-value is less than our chosen significance level we can conclude that the intercept is statistically significant. It implies that even when the explanatory variable is zero, the expected value of the response variable is significantly different from zero. In contrast, if the p-value is greater than our chosen significance level, we conclude that the intercept is not statistically significant.

In practice, we don't usually care about the p-value for the intercept term. Even if the p-value isn't less than some significance level (e.g. 0.05), we would still keep the intercept term in the model.

Then finally the third output presents the summary statistics on the model as a whole. We can see the number of observations of our dataset = 20; then we are given the degrees of freedom $= n - p - 1 = 20 - 1 - 1 = 18$.

Then it provides the root mean squared error (RMSE). This measures the accuracy of the model by estimating how close the model matches the observed responses. The basic idea is to measure how bad/erroneous the model's predictions are when compared to actual observed values. The gives us an average distance between the predicted values from the model and the actual values in the dataset. This can be interpreted as meaning the lower the RMSE, the better a given model is able to "fit" a dataset, and therefore the more accurate our model is.

In our example our RMSE = 0.477.

Next we have the value of the coefficient of determination. In this example, the R-squared is 0.935, which indicates that 93.50% of the variance in the consumption rate of a vehicle can be explained by the speed it travels. Next we have the adjusted R-squared lower than the R-squared.

Then we have the F-statistic along with the p-value to test if the model performs better than a null model. If the p-value is less than the significance level, our sample data provide sufficient evidence to conclude that our regression model fits the data better than the model with no independent variables.

In our example since our p-value is so small, we can conclude that our regression model fits the data better than the intercept-only model.

Next, we perform ANOVA for the model.

```
Anova(model,'summary')
```

	SumSq	DF	MeanSq	F	pValue
Total	62.986	19	3.315		
Model	58.896	1	58.896	259.24	3.9179e-12
Residual	4.0893	18	0.22718		

Here's how to interpret every value in the output:

DF total: the total degrees of freedom is calculated as #total observations − 1. For this example our value is: 20 − 1 = **19**.

DF model: the degrees of freedom formula for the model is p − the number of predictors. In our example that is 1.

DF residuals: the degrees of freedom for the residuals. This is calculated as #total observations − p − 1. In this case, there were 20 observations and 1 predictor, so this value is: 20 − 1 − 1 = 18.

We are given the F value and the p-value in the linear regression table. The p-value is associated with the F-statistic with numerator df = 1 and denominator df = 18. In this case, the p-value is an extremely tiny number.

Now, we calculate the **coefficient confidence intervals.**

```
coefCI(model, alpha)
```

Remember alpha depends on the confidence level as confidence % = $100(1-\alpha)$%. However, this function only produces the t-critical values for a two-sided test, that is

$$\beta_1 \pm t_{1-\pm/2,\, n-2}(\text{SE of slope})$$

So if our confidence level is 99%, our syntax would be:

```
coefCI(model, 0.01)
```

However we are calculating the confidence interval for our example at 95% for a one-sided test. Hence our alpha would be 0.05, but since MATLAB only produces confidence intervals for two-sided tests, we multiply the alpha by 2; therefore our alpha would be $0.05 \times 2 = 0.10$

```
coefCI(model, 0.10)

                    ans =

                    -0.1067     1.4070
                     0.0011     0.0013
```

The output will always have two columns with the first and second columns representing the lower and the upper confidence limits. The first row will always display the limits of the intercept β_0. Since we have a single predictor, the last row shows the limits for the explanatory variable β_1.

This can be interpreted as: there is a 95% probability that the true value of the

- Intercept coefficient lies within −0.1067 and 1.4070
- Slope coefficient lies within 0.0011 and 0.0013

Since this confidence interval doesn't contain the value 0, we can conclude that there is a statistically significant association between the speed of a vehicle and its consumption rate.

11.8.4 Hypothesis test on coefficients

We perform an F test for the hypothesis where the function returns the p-value, the F-statistic, and the numerator degrees of freedom, that is, of the mean sum of squares regression.

```
[pvalue, Fstat, df] = coefTest(model)

                    pvalue =

                    3.9179e-12


                    Fstat =

                    259.2440


                    df =

                    1
```

From the output above, we can see the F-statistic and p-value. The number of degrees of freedom is 1 since there is a single predictor (speed) in the model.

This is how we perform regression analysis in MATLAB. Even though it might seem like a lot of work, remember how much it offers in return. Not only it can be used to make predictions about the behavior of a system or process but it can also be used as a powerful tool by engineers to help them make data-driven decisions, optimize the performance of systems and processes, and better understand the behavior of complex systems; thus all of that effort pays off in the long run.

Let's now do more examples to deepen our understanding of the concepts, techniques, and their benefits in the engineering world.

11.8.5 Other examples

Let's do the second example we worked on in the correlation chapter, where the
electrical engineer now wants to perform regression analysis to model the relationship between temperature and conductivity, and to predict the conductivity values for different temperatures based on the regression model. Here is the data he collected:

Temperature (°C)	Thermal conductivity (W/m K)
20	2.10
25	2.25
30	2.40
35	2.64
40	2.30
45	2.88
50	2.64
55	2.37
60	2.90
65	3.20
70	2.66
75	3.00
80	2.50
5	2.75
90	2.56
95	2.82
100	2.68

Following the steps:

First we check for the assumptions of residuals.

Instead of one, let's use four types of residual plots to test the:

1) Linearity assumption

We test the linearity using a scatterplot.

```
scatter(Temperature, Conductivity, 90, "blue", '*')
xlabel('Temperature ()')
ylabel('Conductivity (W/m K)')
title('Scatterplot-Temperature vs Conductivity')
```

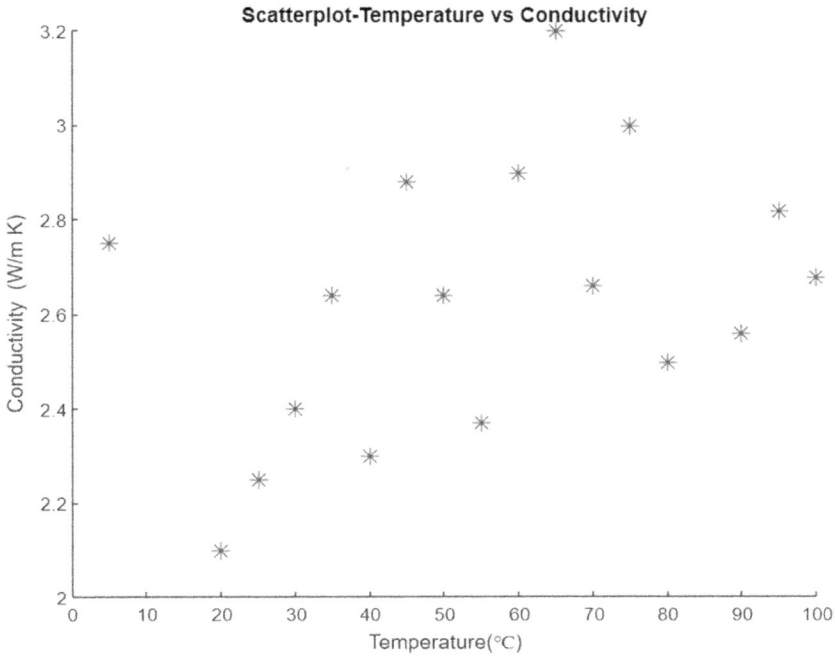

Figure 11.13

3) **Normality assumption**

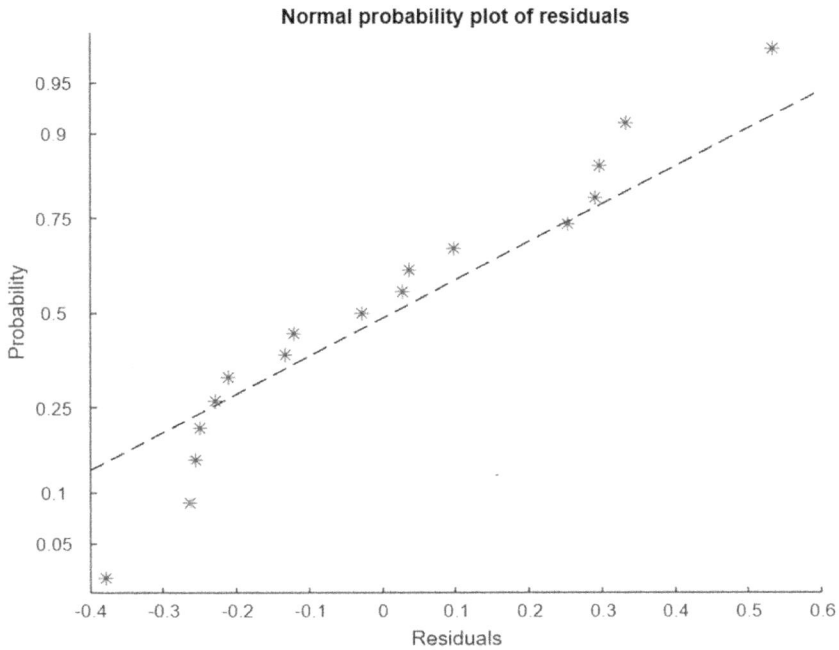

Figure 11.14

As discussed earlier, this assumption can best be checked with a histogram or a Q-Q plot. To check for normality let's use the probability plot in Figure 11.14.

The relationship seems approximately linear; however we can see two data points that seem like outliers.

Now before we come to any conclusion about these two data points, we need to analyze them and verify that those are indeed outliers.

Now how do we analyze them? Remember standardized residuals? Let's now see if we can find any value greater than the cutoff value 3 or 2.

model.Residuals.Standardized	model.Residuals.Studentized
-1.5186	-1.5948
-0.9859	-0.9849
-0.4707	-0.4582
0.3779	0.3668
-1.0034	-1.0037
1.1260	1.1369
0.1348	0.1303
-0.9650	-0.9626
0.9608	0.9582
2.0236	2.2928
-0.1061	-0.1025
1.1206	1.1310
-0.8916	-0.8851
1.4264	1.4822
-0.8438	-0.8352
0.1138	0.1100
-0.5526	-0.5394

When testing twice (standardized and studentized) we can see there is one value that is around 2. However, as mentioned earlier, most textbooks consider 3 as the threshold of the outlier range while some consider 2 as the cutoff, so we still are not 100% confident that this particular data point should be considered an outlier.

So what else can we do?

Let's look at some other statistical techniques offered by MATLAB to test for outliers:

1) Cook's distance

Cook's distance is a summary of how much a regression model changes when the ith observation is removed. When looking to see which observations may be outliers, we count any observation that is larger than 3 times the mean Cook's distance. So let's see how it can be done in MATLAB.

First we plot the Cook's distances of observations and find the outliers visually.

```
plotDiagnostics(model,'cookd','Marker',"*",'Color','blue','MarkerSize', 8)
```

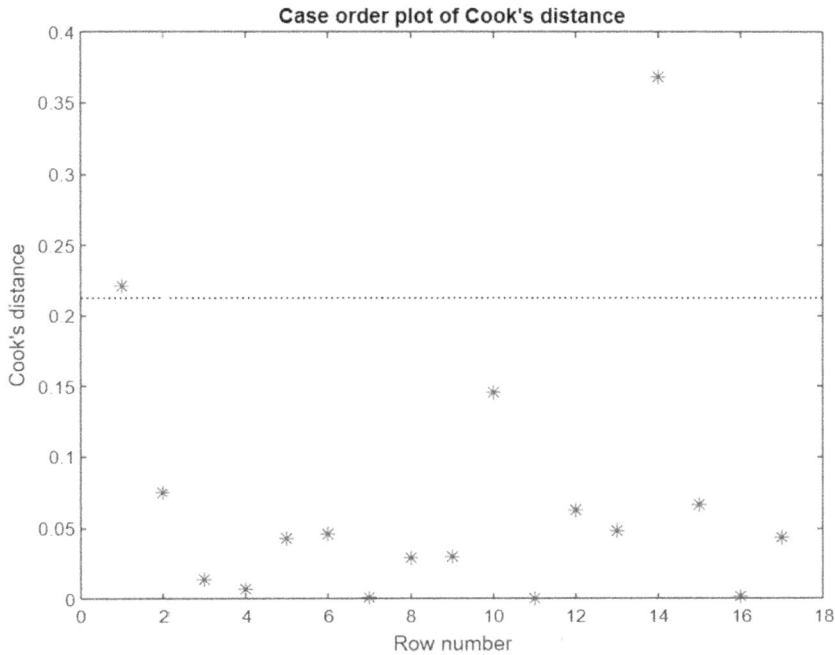

Case order plot of Cook's distance

Figure 11.15

The dotted line represents the standard threshold value (3× the mean of all the distances). As we can see the plot has two observations above the dotted line. Let's find them.

```
find((model.Diagnostics.CooksDistance) > 3*mean(model.Diagnostics.
CooksDistance))
```

```
                              ans =

                               1
                              14
```

Observation 1 and observation 14 exceed the threshold value. However if we look more closely, we will see that observation 14 has a much greater Cook's distance than observation 1 so we can consider this to be an outlier.

2) Dfbetas

This is a measure of how much the slope coefficient changes if the ith observation is deleted. Such change is measured in terms of standard deviation units.

Like in the previous technique this method too uses a plot to find the outliers visually.

```
plotDiagnostics(model,'dfbetas','Marker',"*",'Color','blue','MarkerSize', 8)
```

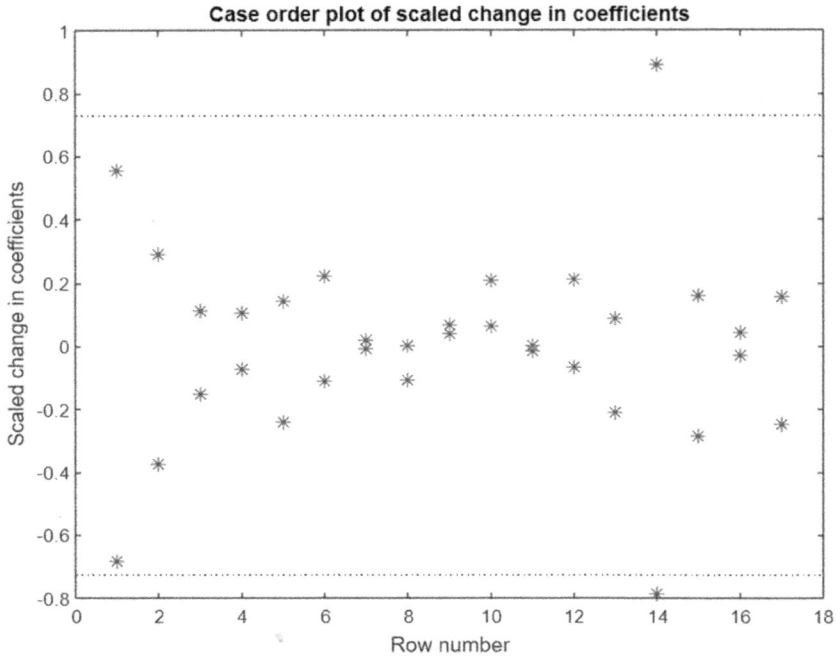

Figure 11.16

Instead of one, we have two threshold values to determine outliers. So Dfbetas larger than $3/\mathrm{sqrt}(n)$ or smaller $3/\mathrm{sqrt}(n)$ indicate outliers. Again we can see only one observation that goes beyond the cutoff value $\pm3/\mathrm{sqrt}(n)$ indicating it to be an outlier.

3) Dffits

This method shows the influence of each observation on the fitted response values. Dffits values larger than $2*\mathrm{sqrt}(p/n)$ or smaller $2*\mathrm{sqrt}(p/n)$ indicate outliers.

```
plotDiagnostics(model,'Dffits','Marker',"*",'Color','blue','MarkerSize', 8)
```

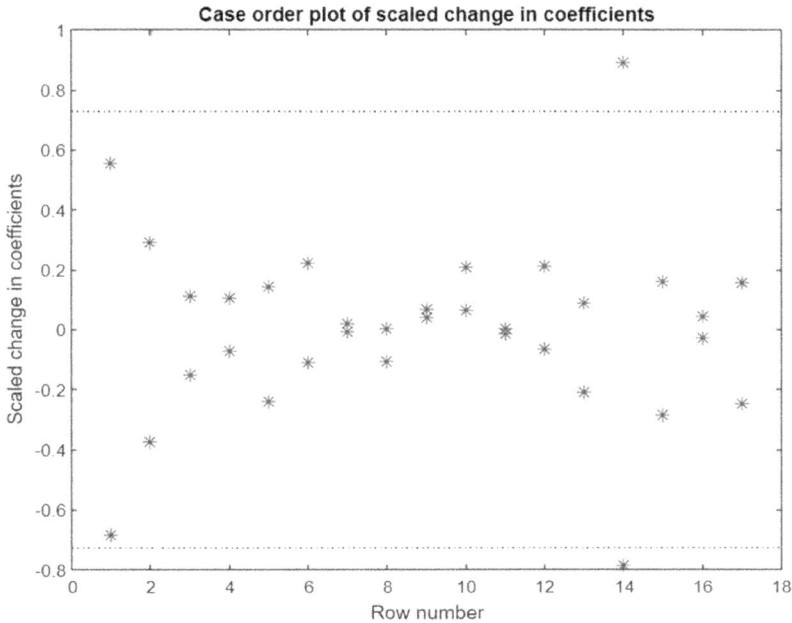

Case order plot of scaled change in coefficients

Figure 11.17

Not surprisingly observation 14 still stands as the only one beyond the dffit threshold.

4) R^2

Another way which might not be effective at all times is measuring the coefficient of determination, R^2.

Remember R-squared is the percentage of the dependent variable variation that a linear model explains. A higher R^2 implies that the data points are closer to the regression line so the regression model accounts for more of the variance. Now that we know observation 14 is the most likely to be an outlier, let's compare the R-squared value with and without this observation and see how much it changes.

```
model =


Linear regression model:
    y ~ 1 + x1

Estimated Coefficients:
                   Estimate         SE        tStat       pValue
                   _____      _____    _____    _____

    (Intercept)     2.3959        0.15155     15.81     9.2222e-11
    x1            0.0041696      0.0024661    1.6908        0.11154


Number of observations: 17, Error degrees of freedom: 15
Root Mean Squared Error: 0.273
R-squared: 0.16,  Adjusted R-Squared: 0.104
F-statistic vs. constant model: 2.86, p-value = 0.112
```

An R-squared value of 0.16 means that 16% of the variation in the conductivity variable is explained by the temperature. In other words, the model is not able to explain a large proportion of the variation in the conductivity, and there may be other factors that are affecting it that are not accounted for in the model.

Now let's look at the R-squared if we exclude this point.

```
model =


Linear regression model:
    y ~ 1 + x1

Estimated Coefficients:
                    Estimate        SE          tStat       pValue

                    _____    _____    _____    _____

    (Intercept)       2.266       0.17017      13.316     2.433e-09
    x1             0.006037     0.0026869      2.2468      0.041301


Number of observations: 16, Error degrees of freedom: 14
Root Mean Squared Error: 0.262
R-squared: 0.265,  Adjusted R-Squared: 0.213
F-statistic vs. constant model: 5.05, p-value = 0.0413
```

It seems like the value has changed a bit, but not much. So in that case it is up to you if you want to take those outliers into a regression model or simply drop them to make a better regression model. For this situation let's drop them and continue the analysis excluding the observation 14.

4) **Constant variance assumption**

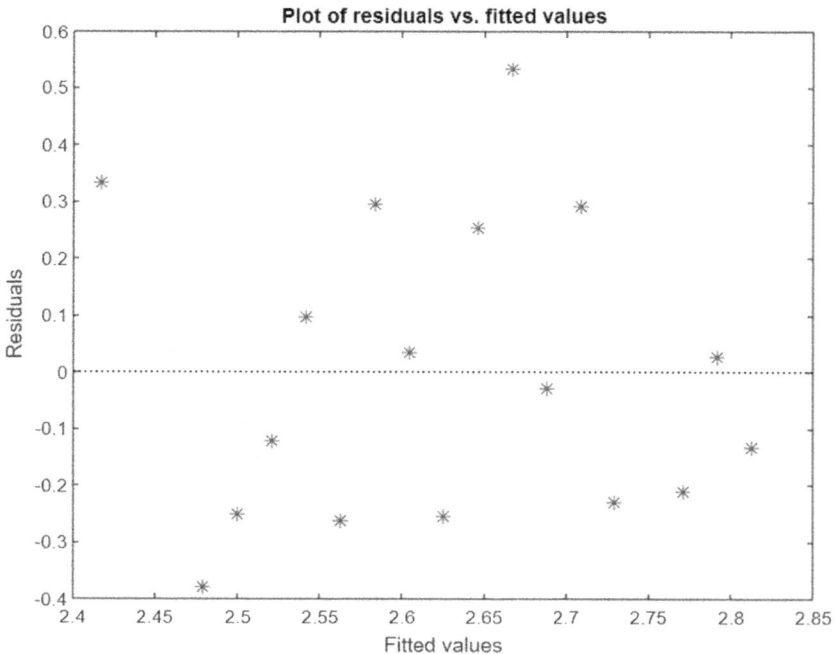

Figure 11.18

As one would expect, the points are randomly scattered around the horizontal line at zero; hence the assumption of constant variance is met.

5) **Independent error assumption**

To check independence, we can use the plot residuals against case order. A common source of non-independence is that observations are close together in time, so now the residuals vs. case order plot helps us to see if there is any correlation between the error terms that are near each other in the sequence.

```
plotResiduals(model,'caseorder','Marker',"*",'Color','blue','MarkerSize',8)
```

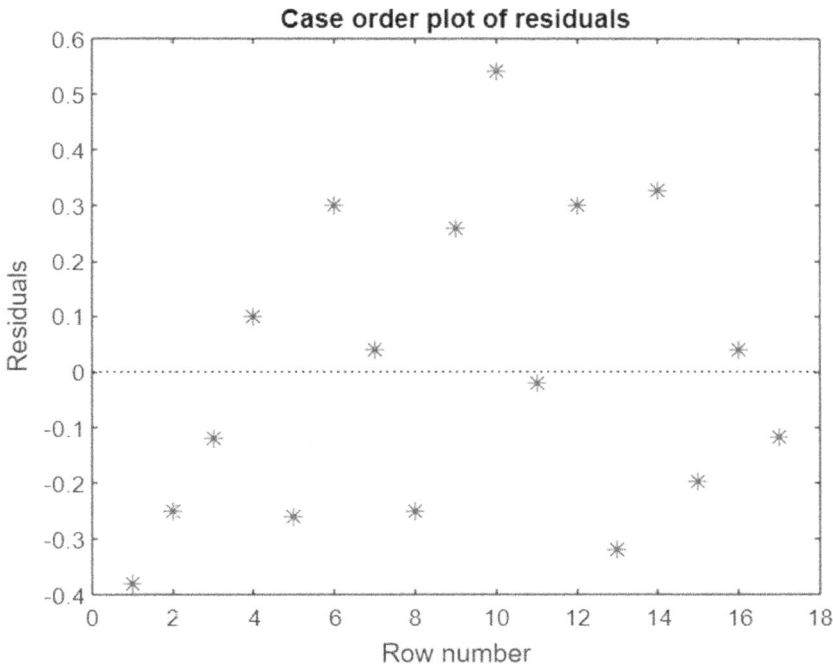

Figure 11.19

Apparently, the residuals bounce randomly around the residual = 0 line as we would hope, suggesting that there is no serial correlation. This implies that the independence of the residuals assumptions has been met.

Next, we proceed to use MATLAB to estimate the coefficients of the regression model

```
p = polyfit(Temperature, Conductivity, 1)
```

$$0.0060 \quad 2.2660$$

As discussed earlier, the last coefficient is the estimated coefficient of the intercept.

Here, with a slope coefficient of 0.006, we can interpret that for every 1 unit increase in temperature, the thermal conductivity increases by 0.006 units.

The intercept coefficient, on the other hand, represents the predicted value of y when x is equal to zero. In this case, with an intercept coefficient of 2.266, we can conclude that the minimum conductivity at the lowest temperature is equal to 2.266 units.

Now we **plot the best-fit line.**

```
%circular markers represented by * and b specifies the color blue then
the best fit line represented by - and the color is red which is default
f = polyval(p, Temperature)
h = plot(Temperature,Conductivity,'*b',Temperature,f,'-','MarkerSize',10)

%adding the labels
xlabel('Temperature(°C)')
ylabel('Conductivity (W/m K)')
title('Scatterplot-Temperature vs Conductivity')
```

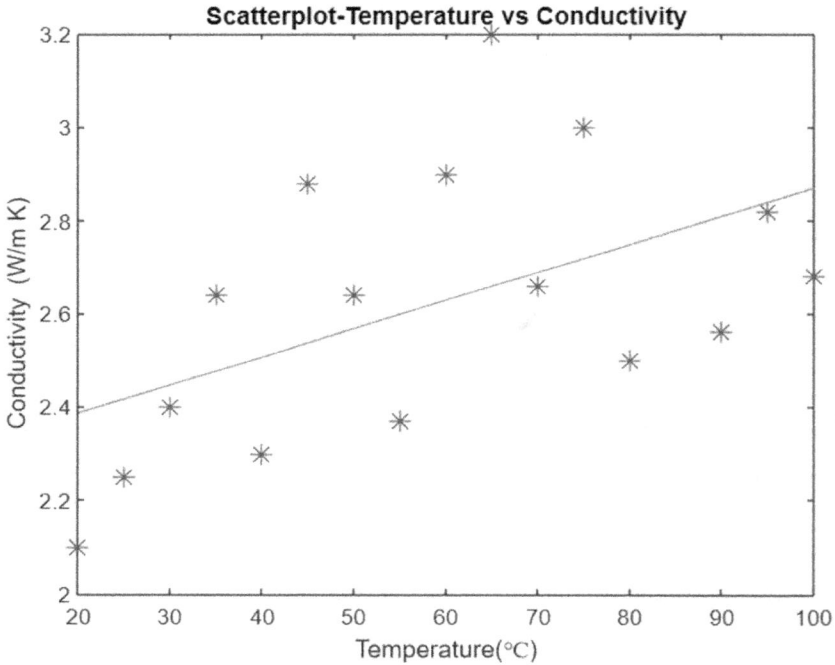

Figure 11.20

The best-fit line indicates a positive correlation between the temperature and conductivity. This means that as the temperature increases, the conductivity also tends to increase.

11.8.6 Assessing the model

Let's say we are performing the analysis at $\alpha = 0.01$. Hence, we see a relatively larger p-value for the slope coefficient indicating that the coefficient is not statistically significant which indicates that the temperature is not quite associated with the conductivity. Since our p-value > 0.01, there is insufficient evidence to conclude that temperature has a significant impact on conductivity and therefore there must be other factors influencing the relationship.

On the other hand we have an extremely small p-value for the intercept coefficient indicating stronger evidence against the null hypothesis, suggesting that the intercept coefficient is significantly different from zero.

The model also reports an RMSE (Root Mean Squared Error) of 0.262, a low value indicating that our estimated model provides a good fit to the data.

Next we have the value of the coefficient of determination. In this example, the R-squared is 0.265, which indicates that 26.50% of the variance in the consumption rate of a vehicle can be explained by the speed at which it travels. Such a low value implies that the temperature and conductivity are unrelated.

Next, we have the F-statistic along with the p-value of 0.0413. Since our p-value is greater than the significance level (0.01), we fail to conclude that our model fits the data better than a model with no predictors.

Now, we calculate the **coefficient confidence intervals**.

```
coefCI(model, 0.01)
            ans =

              1.7594     2.7725
             -0.0020     0.0140
```

This can be interpreted as there is a 99% probability that the true value of the

- Intercept coefficient lies within 1.7594 and 2.7725
- Slope coefficient lies within −0.0020 and 0.0140

Since this confidence interval contains the value 0, we can conclude that there is no statistically significant evidence of a linear relationship between the temperature and the conductivity.

Remember the example of per capita GDP and per capita emissions of the greenhouse gas carbon dioxide. Since the original data could not pass the linearity test, we had to transform it. Let's perform the regression analysis on the transformed data.

11.8.6.1 Test for linearity

As seen earlier, the natural logarithms of both explanatory and response variables transform a scatterplot into a more linear relationship.

11.8.6.2 Test for normality

Using the transformed variables and the residuals we see that the histogram is quite symmetric therefore confirming the normality.

Histogram of residuals

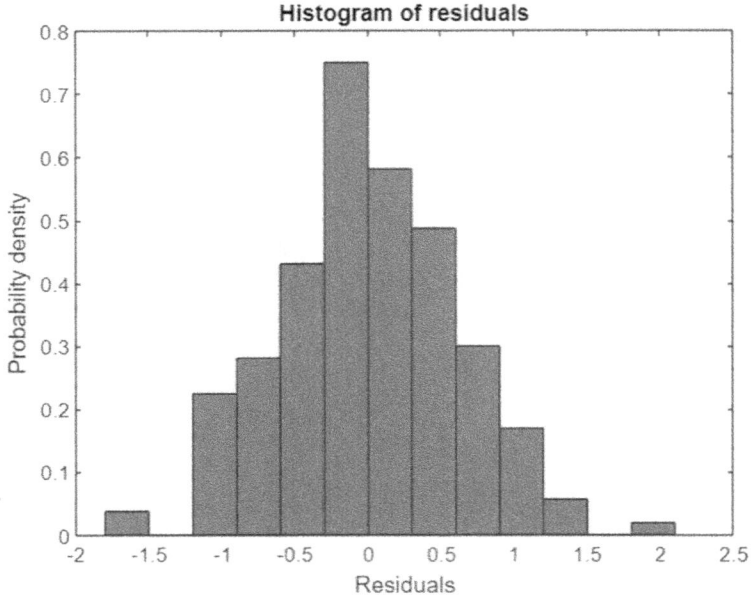

Figure 11.21

11.8.6.3 Test for homoscedasticity

```
h = plotResiduals(model,'fitted','Color','magenta','Marker','square',
                                             'MarkerSize',6)
set(h, {'MarkerFaceColor'}, get(h,'Color'))
```

Plot of residuals vs. fitted values

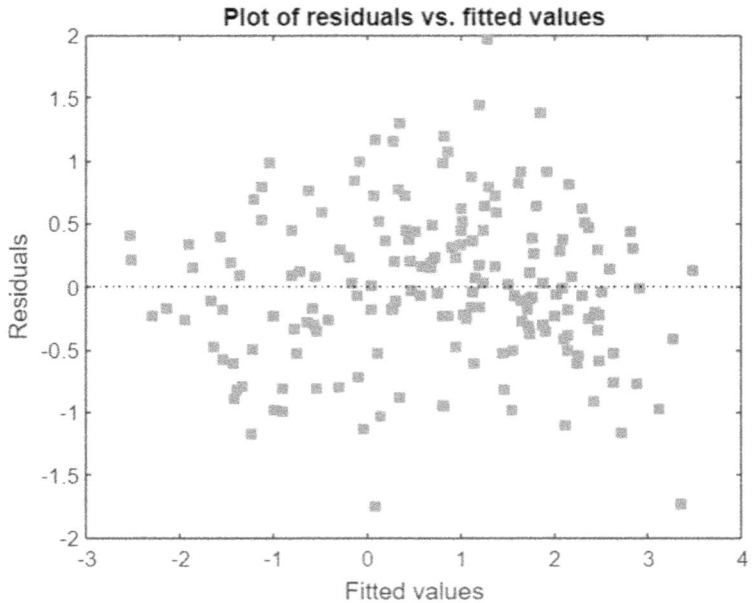

Figure 11.22

The scattered pattern above verifies the homoscedasticity of the residuals.

11.8.6.4 Test for independence

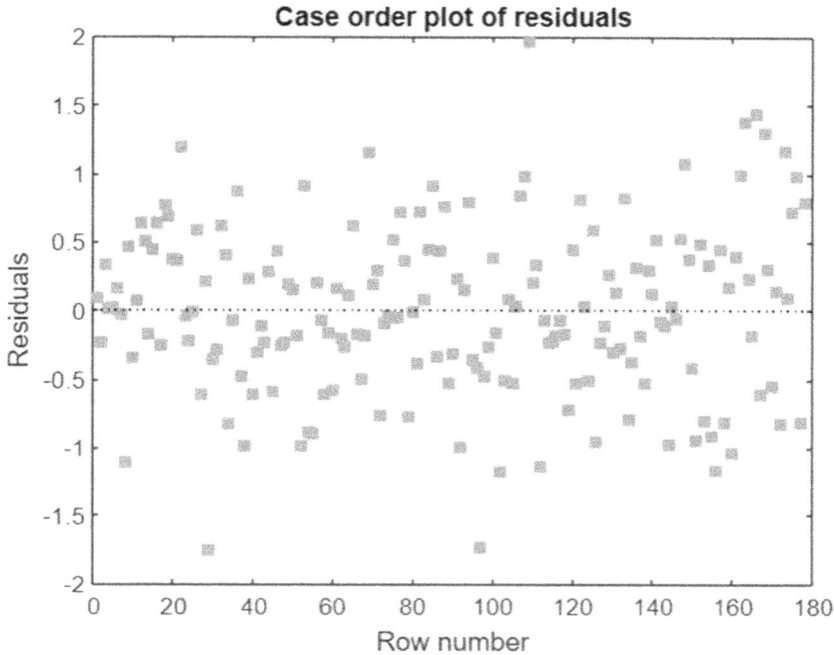

Figure 11.23

Evidently the data seems to meet the independence criteria of residuals as well.

Moving a few steps ahead, let's plot the best-fit line.

```
p = polyfit(G, E, 1)
f = polyval(p, G)
h = plot(G, E,'squaremagenta', G, f,'-k', 'LineWidth', 2, 'MarkerSize', 4)
set(h, {'MarkerFaceColor'}, get(h,'Color'))
xlabel("Logged GDP")
ylabel("Logged Emission")
title("GDP per capita")
```

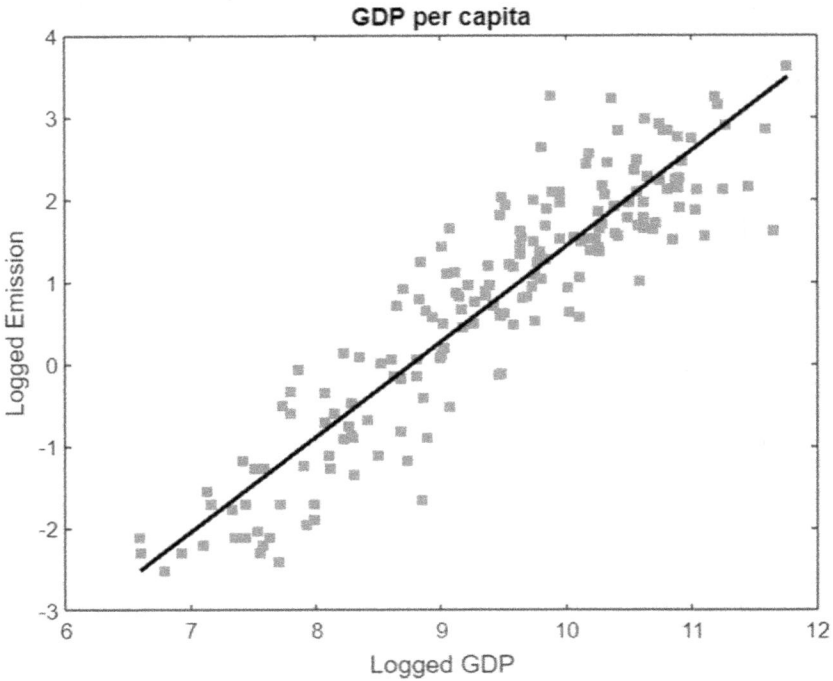

Figure 11.24

The graph implies a positive correlation between GDP and emissions.

11.8.7 Assessing the model

```
model = fitlm(G, E)
            model =

            Linear regression model:
                y ~ 1 + x1

            Estimated Coefficients:
                                Estimate       SE         tStat        pValue
                                _____    _____    _____    _____

                (Intercept)     -10.183     0.36974      -27.54     1.0442e-65
                x1               1.1616     0.039078      29.725     1.6251e-70

            Number of observations: 178, Error degrees of freedom: 176
            Root Mean Squared Error: 0.619
            R-squared: 0.834,   Adjusted R-Squared: 0.833
            F-statistic vs. constant model: 884, p-value = 1.63e-70
```

Looking at the table above, we can conclude that the emissions are predicted to increase by 1.1616 when the GDP goes up by 1 and are predicted to be –10.183 when the GDP is zero. Now looking at their corresponding **p-values**, we see both p-values are extremely small, almost zero.

This implies that the GDP has a statistically significant relationship with the response variable emissions of CO_2. The extremely small p-value for the intercept coefficient also indicates stronger evidence against the null hypothesis, suggesting that the intercept coefficient is significantly different from zero.

Then we see the RMSE = 0.834, indicating that our estimated model fits the data well.

Next, we have a very high value of the coefficient of determination, 83.4%, which indicates that 83.4% of the variance in the emissions can be explained by the GDP per capita. Such a high value implies that the GDP and emissions are related.

Next, we have the F-statistic along with the p-value. Since our p-value is significantly small, we can conclude that our regression model fits the data better than the no predictors model.

Now, we calculate the coefficient confidence intervals. Let's say we take our significance level = 5%.

```
coefCI(model, 0.05)
```

```
          -10.9125   -9.4531
            1.0845    1.2387
```

This can be interpreted as meaning there is a 95% probability that the true value of the

- Intercept coefficient lies within –10.9125 and –9.4531
- Slope coefficient lies within 1.0845 and 1.2387

Since this confidence interval does not contain the value 0, we can conclude that there is statistically significant evidence of a linear relationship between the GDP per capita and the emissions.

Now realize that because we transformed our data, we were able to get good results. Recall why we had to transform it in the first place. In this particular example we had to transform it since the regression function was not linear for the original data. In fact, if any of the linear regression assumptions are violated, transforming the variables becomes valid as they can help address the issue and improve the model's performance.

11.9 MULTIPLE LINEAR REGRESSION

In simple linear regression, our equation was

$$\hat{y} = \beta_0 + \beta_1 x$$

Since we had a single explanatory variable we used just x to represent it.

However if the model involves more than one predictor variable then that linear regression is called a multiple regression model. Many see the multiple linear regression model as just an extension of the simple linear regression model **involving** multiple explanatory variables, and therefore we need multiple "x"s in the equation. So according to the definition our equation of our model looks like:

$$\hat{y} = \beta_0 + \beta_1 x_1 + \beta_2 x_2 + \beta_3 x_3 + \dots \beta_k x_k$$

Where:

- y: dependent variable
- $x_1, x_2, x_3, \dots x_k$ – independent (explanatory) variables
- β_0 – intercept
- $\beta_1, \beta_2, \beta_3, \dots \beta_k$ – slopes

As we can see, multiple linear regression analysis is essentially similar to the simple linear model with one intercept β_0 but multiple independent variables and a slope for each.

As the name suggests, multiple regression is a statistical technique applied on datasets dedicated to drawing out a relationship between one response or dependent variable and multiple independent variables.

Multiple linear regression follows the same conditions as the simple linear model. However, since there are several independent variables in multiple linear analysis, there is another mandatory condition for the model:

- **Non-collinearity:**

    The key purpose of a regression equation is to tell us the individual impact of each of the explanatory variables on the dependent/target variable from the regression coefficients. A regression coefficient captures the average change in the response variable for 1 unit change in the explanatory variable, keeping all the other explanatory variables constant. Therefore, it is necessary that the explanatory variables are not correlated to determine the true significance of individual predictors and to identify which predictors are truly important in explaining the variation in the response variable.

    Now, multicollinearity occurs when there is a high correlation between independent variables. This reduces the precision of the estimated coefficients, making it challenging to rely on p-values to determine which predictors are statistically significant. Multicollinearity complicates assessing true relationships with the dependent variable, impacting the overall interpretation and reliability of the regression model's results.

11.9.1 How to test multicollinearity?

The primary techniques for detecting the multicollinearity are as follows.

11.9.1.1 Correlation matrix/correlation plot

The easiest way to check for multicollinearity is to make a correlation matrix of all predictors. The correlation matrix is a square matrix that displays the correlation coefficients between pairs of variables and has the same number of rows and columns as the number of variables in the dataset. To calculate the correlation, we use the Pearson correlation coefficient, which measures the linear relationship between two variables. That's how we can calculate the correlation coefficient of all the possible pairs of values and make a matrix from it.

In the matrix, The diagonal elements of the matrix will be 1, as they represent the correlation of each variable with itself. The off-diagonal elements will represent the correlations between pairs of variables. To detect multicollinearity using a correlation matrix, we need to examine the correlation coefficients between the predictor variables. The correlation coefficient is a value between -1 and 1 that measures the degree of correlation between two predictors. A high value of the correlation between two variables may indicate that the variables are collinear. For example, A correlation coefficient with an absolute value > 0.7 typically indicates a strong correlation between predictor therefore there is a strong correlation

This method is easy, but it cannot produce a clear estimate of the degree of multicollinearity. High correlation coefficients do not necessarily imply multicollinearity and does not provide information about the severity of multicollinearity.

$$Corr(x,y) = \frac{\sum_{i=1}^{n}(x_i - \bar{x})(y_i - \bar{y})}{\sqrt{\sum(x_i - \bar{x})^2(y_i - \bar{y})^2}}$$

Let's say we have a data of the energy output from a combined cycle power plant (CCPP) varying with the operating thermal parameters like ambient pressure, vacuum, relative humidity, and relative temperature

Temperature (°C)	Vacuum (cmHg)	Pressure (milibar)	Humidity (%)	Energy Output (MW)
8.34	40.77	1010.84	90.01	480.48
23.64	58.49	1011.4	74.2	445.75
29.74	56.9	1007.15	41.91	438.76
19.07	49.69	1007.22	76.79	453.09
11.8	40.66	1017.13	97.2	464.43
13.97	39.16	1016.05	84.6	470.96
22.1	71.29	1008.2	75.38	442.35
14.47	41.76	1021.98	78.41	464
31.25	69.51	1010.25	36.83	428.77
6.77	38.18	1017.8	81.13	484.31
28.28	68.67	1006.36	69.9	435.29
22.99	46.93	1014.15	49.42	451.41
29.3	70.04	1010.95	61.23	426.25
8.14	37.49	1009.04	80.33	480.66
16.92	44.6	1017.34	58.75	460.17
22.72	64.15	1021.14	60.34	453.13
18.14	43.56	1012.83	47.1	461.71
11.49	44.63	1020.44	86.04	471.08
9.94	40.46	1018.9	68.51	473.74
23.54	41.1	1002.05	38.05	448.56
14.9	52.05	1015.11	77.33	464.82
33.8	64.96	1004.88	49.37	427.28
25.37	68.31	1011.12	70.99	441.76
7.29	41.04	1024.06	89.19	474.71
13.55	40.71	1019.13	75.44	467.21
6.39	35.57	1025.53	77.23	487.69
26.64	62.44	1011.81	72.46	438.67
7.84	41.39	1018.21	91.92	485.66
21.82	58.66	1011.71	64.37	452.16
27.17	67.45	1015.67	49.03	429.87
13.42	41.23	994.17	95.79	468.82
20.77	56.85	1012.4	83.63	442.85
8.29	36.08	1020.38	81.53	483.26
30.98	67.45	1015.18	45.4	433.59
31.96	71.29	1008.39	47.51	433.04

Temperature (°C)	Vacuum (cmHg)	Pressure (milibar)	Humidity (%)	Energy Output (MW)
15.83	52.75	1024.3	58.34	458.6
22.56	70.79	1005.85	93.09	435.14
25.91	75.6	1018.23	62.65	443.2
8.24	39.61	1017.99	78.42	477.9
24.66	60.29	1018	59.56	445.26
29.31	68.67	1006.18	63.38	435.57
21.48	66.91	1008.58	84.49	447.42
18.28	44.71	1016.99	33.71	462.28
26.96	65.34	1015.05	46.93	441.81
16.01	65.46	1014	87.68	454.16
27.37	63.73	1009.79	65.25	437.24
16.3	39.63	1004.64	85.61	464.11
23.8	48.6	1002.43	67.32	440.89
8.19	41.66	1016.57	75.38	485.2

Let's calculate the correlation between Temperature and Vacuum. Consider

x_1 = Temperature
x_2 = Vacuum
x_3 = Pressure
x_4 = Humidity

Therefore
$\sum x_1 = 947.7$
$\sum x_2 = 2617.27$
$\sum x_3 = 49647.57$
$\sum x_4 = 3389.13$

x_1^2	x_2^2	x_3^2	x_4^2	$x_1 x_2$	$x_1 x_3$	$x_1 x_4$
69.5556	1662.1929	1021797.506	8101.8001	340.0218	8430.4056	t750.6834
558.8496	3421.0801	1022929.96	5505.64	1382.7036	23909.496	1754.088
884.4676	3237.61	1014351.123	1756.4481	1692.206	29952.641	1246.4034
363.6649	2469.0961	1014492.128	5896.7041	947.5883	19207.6854	1464.3853
139.24	1653.2356	1034553.437	9447.84	479.788	12002.134	1146.96
195.1609	1533.5056	1032357.603	7157.16	547.0652	14194.2185	1181.862
488.41	5082.2641	1016467.24	5682.1444	1575.509	22281.22	1665.898
209.3809	1743.8976	1044443.12	6148.1281	604.2672	14788.0506	1134.5927
976.5625	4831.6401	1020605.063	1356.4489	2172.1875	31570.3125	1150.9375
45.8329	1457.7124	1035916.84	6582.0769	258.4786	6890.506	549.2501
799.7584	4715.5689	1012760.45	4886.01	1941.9876	28459.8608	1976.772
528.5401	2202.4249	1028500.223	2442.3364	1078.9207	23315.3085	1136.1658
858.49	4905.6016	1022019.903	3749.1129	2052.172	29620.835	1794.039
66.2596	1405.5001	1018161.722	6452.9089	305.1686	8213.5856	653.8862
286.2864	1989.16	1034980.676	3451.5625	754.632	17213.3928	994.05
516.1984	4115.2225	1042726.9	3640.9156	1457.488	23200.3008	1370.9248
329.0596	1897.4736	1025824.609	2218.41	790.1784	18372.7362	854.394
132.0201	1991.8369	1041297.794	7402.8816	512.7987	11724.8556	988.5996
98.8036	1637.0116	1038157.21	4693.6201	402.1724	10127.866	680.9894

554.1316	1689.21	1004104.203	1447.8025	967.494	23588.257	895.697
222.01	2709.2025	1030448.312	5979.9289	775.545	15125.139	1152.217
1142.44	4219.8016	1009783.814	2437.3969	2195.648	33964.944	1668.706
643.6369	4666.2561	1022363.654	5039.5801	1733.0247	25652.1144	1801.0163
53.1441	1684.2816	1048698.884	7954.8561	299.1816	7465.3974	650.1951
183.6025	1657.3041	1038625.957	5691.1936	551.6205	13809.2115	1022.212
40.8321	1265.2249	1051711.781	5964.4729	227.2923	6553.1367	493.4997
709.6896	3898.7536	1023759.476	5250.4516	1663.4016	26954.6184	1930.3344
61.4656	1713.1321	1036751.604	8449.2864	324.4976	7982.7664	720.6528
476.1124	3440.9956	1023557.124	4143.4969	1279.9612	22075.5122	1404.5534
738.2089	4549.5025	1031585.549	2403.9409	1832.6165	27595.7539	1332.1451
180.0964	1699.9129	988373.9889	9175.7241	553.3066	13341.7614	1285.5018
431.3929	3231.9225	1024953.76	6993.9769	1180.7745	21027.548	1736.9951
68.7241	1301.7664	1041175.344	6647.1409	299.1032	8458.9502	675.8837
959.7604	4549.5025	1030590.432	2061.16	2089.601	31450.2764	1406.492
1021.4416	5082.2641	1016850.392	2257.2001	2278.4284	32228.1444	1518.4196
250.5889	2782.5625	1049190.49	3403.5556	835.0325	16214.669	923.5222
508.9536	5011.2241	1011734.223	8665.7481	1597.0224	22691.976	2100.1104
671.3281	5715.36	1036792.333	3925.0225	1958.796	26382.3393	1623.2615
67.8976	1568.9521	1036303.64	6149.6964	326.3864	8388.2376	646.1808
608.1156	3634.8841	1036324	3547.3936	1486.7514	25103.88	1468.7496
859.0761	4715.5689	1012398.192	4017.0244	2012.7177	29491.1358	1857.6678
461.3904	4476.9481	1017233.616	7138.5601	1437.2268	21664.2984	1814.8452
334.1584	1998.9841	1034268.66	1136.3641	817.2988	18590.5772	616.2188
726.8416	4269.3156	1030326.503	2202.4249	1761.5664	27365.748	1265.2328
256.3201	4285.0116	1028196	7687.7824	1048.0146	16234.14	1403.7568
749.1169	4061.5129	1019675.844	4257.5625	1744.2901	27637.9523	1785.8925
265.69	1570.5369	1009301.53	7329.0721	645.969	16375.632	1395.443
566.44	2361.96	1004865.905	4531.9824	1156.68	23857.834	1602.216
67.0761	1735.5556	1033414.565	5682.1444	341.1954	8325.7083	617.3622

$\sum x_1^2 = 21426.2236$

$\sum x_2^2 = 147499.4441$

$\sum x_3^2 = 50305703.28$

$\sum x_4^2 = 248144.0909$

$\sum x_1 x_2 = 54715.7778$

$\sum x_1 x_3 = 959073.0701$

$\sum x_1 x_4 = 61309.8618$

$$corr(x_1, x_2) = \frac{(49 \times 54715.7778) - (947.7 \times 2617.27)}{\sqrt{\left[49 \times 21426.2236 - (947.7)^2\right]\left[49 \times 147499.4441 - (2617.27)^2\right]}} = 0.83862$$

$$corr(x_1, x_3) = \frac{(49 \times 959073.0701) - (947.7 \times 49647.57)}{\sqrt{\left[49 \times 21426.2236 - (947.7)^2\right]\left[49 \times 50305703.28 - (49647.57)^2\right]}} = -0.46206$$

$$corr(x_1, x_4) = \frac{(49 \times 61309.8618) - (947.7 \times 3389.13)}{\sqrt{\left[49 \times 21426.2236 - (947.7)^2\right]\left[49 \times 248144.0909 - (3389.13)^2\right]}} = -0.64998$$

Similarly we can calculate the correlation of the rest of the pairs and our correlation matrix becomes

$$Correlation\ Matrix = \begin{bmatrix} 1 & 0.83863 & -0.46207 & -0.6500 \\ 0.83863 & 1 & -0.2796 & -0.3557 \\ -0.46207 & -0.2796 & 1 & 0.06494 \\ -0.6500 & -0.3557 & 0.06494 & 1 \end{bmatrix}$$

	Temperature	Vacuum	Pressure	Humidity
Temperature				
Vacuum	0.83863			
Pressure	-0.46207	-0.2796		
Humidity	-0.6500	-0.3557	0.06494	

The correlation matrix above suggests that Temperature and Vacuum are strongly correlated (r=0.83863) and Temperature and Humidity are fairly strongly negatively correlated (r=-0.6500).

11.9.1.2 Variation inflation factor (VIF)

The second method to check multicollinearity is to use the variance inflation factor (VIF) for each independent variable.

VIF measures how much of the variation in one variable is explained by the other variables. This is done by running a regression using one of the correlated x variables as the dependent variable against the other x variables as the predictor variables.

The VIF is calculated as one divided by the tolerance, which is defined as one minus R-squared when one predictor variable x_1 is the dependent variable in a multiple regression with the other predictor variables as the independent variables. So our formula now is:

$$VIF_i = \frac{1}{1 - R_i}$$

11.9.1.2.1 Now how do we use this to detect multicollinearity?

The higher the value of VIF the higher correlation between this variable and the rest.
$VIF = 1$: not correlated.
$1 < VIF < 5$: moderately correlated.
$VIF > 5$: highly correlated.

In our above example we can calculate the VIF for Temperature by performing a multiple linear regression using it as the response variable and Vacuum, Pressure and Humidity as the explanatory variables. Therefore

$$VIF_{Temp} = \frac{1}{1 - R_{Temp}} = \frac{1}{1 - 0.908256} = 10.8999$$

Similarly we can calculate the VIF for Vacuum using it as the response variable while using Temperature, Pressure and Humidity as the explanatory variables.

$$VIF_{Vacuum} = \frac{1}{1 - R_{Vacuum}} = \frac{1}{1 - 0.81504} = 5.40657$$

This process is repeated for Pressure, using Temperature, Vacuum, and Humidity as predictors, and finally for Humidity, using Temperature, Vacuum, and Pressure as predictors.

$$VIF_{Pressure} = \frac{1}{1 - R_{Pressure}} = \frac{1}{1 - 0.455543} = 1.83669$$

$$VIF_{Humidity} = \frac{1}{1 - R_{Humidity}} = \frac{1}{1 - 0.667293} = 3.00564$$

VIFs are also the diagonal elements of the inverse of the correlation matrix

As we can see, two of the variance inflation factors —10.8999, and 5.40657 —are fairly large. The VIF for the predictor Temperature, for example, tells us that the variance of the estimated coefficient of Temperature is inflated by a factor of 10.8999 because Temperature is highly correlated with at least one of the other predictors in the model.

Detecting Multicollinearity in MATLAB

To identify multicollinearity in MATLAB we use the functions corrcoef. However, MATLAB does not include a built-in function for calculating VIF. Instead, we must manually compute VIF using the formula, which involves taking the diagonal elements of the inverse of the correlation matrix.

```
Data = readtable("4 Independent Variables.xlsx", VariableNamingRule = "preserve");
Temperature = Data.Temperature;
Vacuum = Data.Vacuum;
Pressure = Data.Pressure;
Humidity = Data.Humidity;
Output = Data.Output;

Predictors = [Temperature, Vacuum, Pressure, Humidity];
variableNames = {'Temperature', 'Vacuum', 'Pressure', 'Humidity'};

%correlation matrix
correlation_matrix = corrcoef(Predictors);
correlation_table = array2table(correlation_matrix, 'RowNames', variableNames, 'VariableNames', variableNames);

%VIF
VIF = diag(inv(correlation_matrix));
VIF_table = array2table(VIF, 'VariableNames', {'VIF'}, 'RowNames', variableNames);
```

```
>> Multicollinearitytest

correlation_matrix =

    1.0000    0.8386   -0.4621   -0.6500
    0.8386    1.0000   -0.2796   -0.3557
   -0.4621   -0.2796    1.0000    0.0649
   -0.6500   -0.3557    0.0649    1.0000


correlation_table =

  4x4 table
```

	Temperature	Vacuum	Pressure	Humidity
Temperature	1	0.83863	-0.46207	-0.64998
Vacuum	0.83863	1	-0.27958	-0.35569
Pressure	-0.46207	-0.27958	1	0.064937
Humidity	-0.64998	-0.35569	0.064937	1

```
VIF =

   10.8999
    5.4066
    1.8367
    3.0056


         VIF_table =

           4x1 table
```

	VIF
Temperature	10.9
Vacuum	5.4066
Pressure	1.8367
Humidity	3.0056

Let's start with a very simple classic multiple linear regression example. We want to predict the income of an individual based on their age and total years of experience. This dataset has only two independent variables, "age" and "experience", to predict the "income".

$$\hat{y} = \beta_0 + \beta_1 (\text{Age}) + \beta_2 (\text{Experience})$$

We are given the data:

Age	Experience	Income	Age	Experience	Income
25	2	31,450	29	3	29,840
30	3	35,670	47	9	46,110
47	12	51,580	54	7	46,720
32	5	40,130	51	11	54,800
43	10	47,830	44	12	51,300
51	7	41,630	41	6	38,900
28	5	41,340	58	13	60,600
33	4	37,650	23	1	26,870
37	5	40,250	44	9	44,190
39	8	45,150	37	10	48,700

As we have done for simple linear regression, we first need to verify if the dataset fulfills all the criteria.

11.9.3.1 Test for linearity

Instead, we draw individual plots that display the relationship between the response variable and one predictor variable, while keeping the other predictor variable constant.

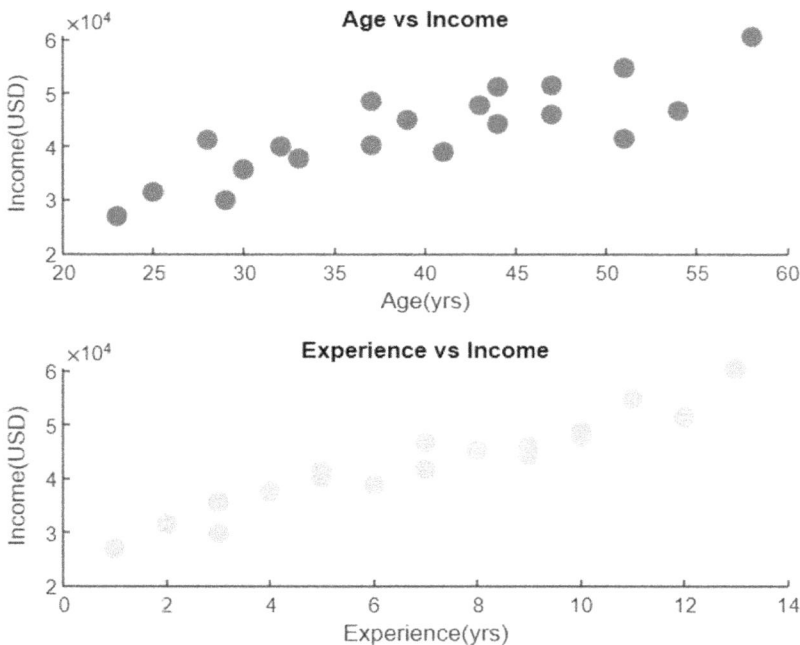

Figure 11.25

```
tiledlayout(2,1)

% Top axes
ax1 = nexttile;
scatter(ax1, Age, Income, 80, "o", "red", 'filled')
xlabel("Age(yrs)")
```

```
ylabel("Income(USD)")
title("Age vs Income")

ax2 = nexttile;
scatter(ax2, Experience, Income, 80, "o", "magenta", 'filled')
xlabel("Experience(yrs)")
ylabel("Income(USD)")
title("Experience vs Income")
```

11.9.3.2 Test for normality

So far we have only used raw/original residuals to test for normality. So let's use a different type of residual and compare it with the raw one and check if we see any differences.

```
tiledlayout(2,1)

% Top axes
ax1 = nexttile;
h = plotResiduals(ax1, model,'probability','Marker',"o",'MarkerSize',
               10,'MarkerEdgeColor',"black", 'MarkerFaceColor',"cyan")

ax2 = nexttile;
h = plotResiduals(ax2, model,'probability','ResidualType','Standardized'
,'Marker',"o", 'MarkerSize',10, 'MarkerEdgeColor',"black",
'MarkerFaceColor',       "#7E2F8E")
```

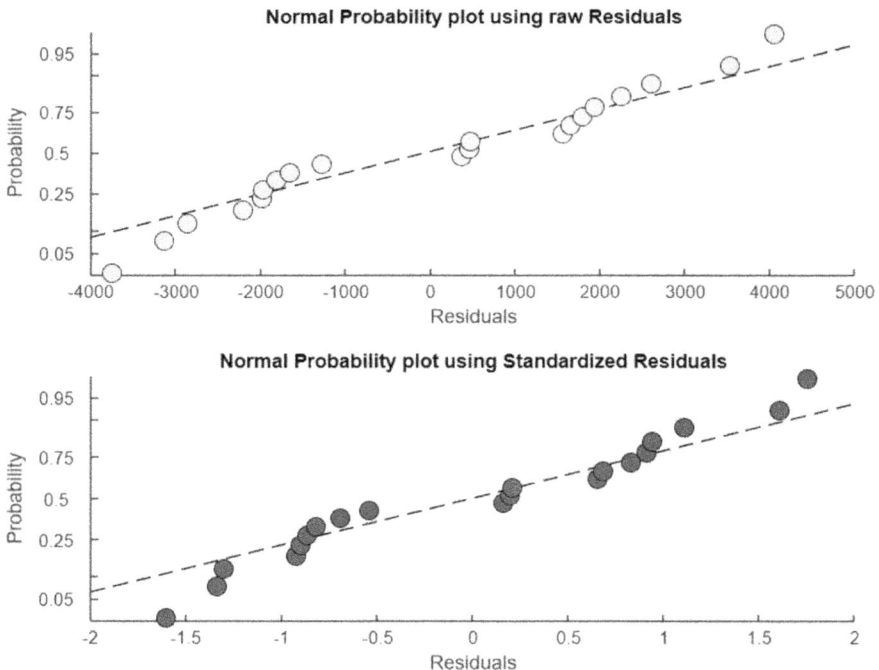

Figure 11.26

Notice that both plots in Figure 11.26 look the same using residuals or standardized residuals. It doesn't matter which ones you use. They both look somewhat straight even though there is one point at the lower end that is slightly off, which needs some further investigation.

11.9.3.3 Test for homoscedasticity

```
tiledlayout(2,1)
% Top axes
ax1 = nexttile;
h = plotResiduals(ax1, model,'fitted','Marker',"o",'MarkerSize', 10,
                'MarkerEdgeColor',"black", 'MarkerFaceColor',"cyan")
title(ax1, "Plot of raw Residuals vs fitted values")

ax2 = nexttile;
h = plotResiduals(ax2, model,'fitted','ResidualType','Standardized',
            'Marker','o', 'MarkerSize',10, 'MarkerEdgeColor',"black",
                'MarkerFaceColor',"#7E2F8E")
title(ax2, "Plot of Standardized Residuals vs fitted values")
```

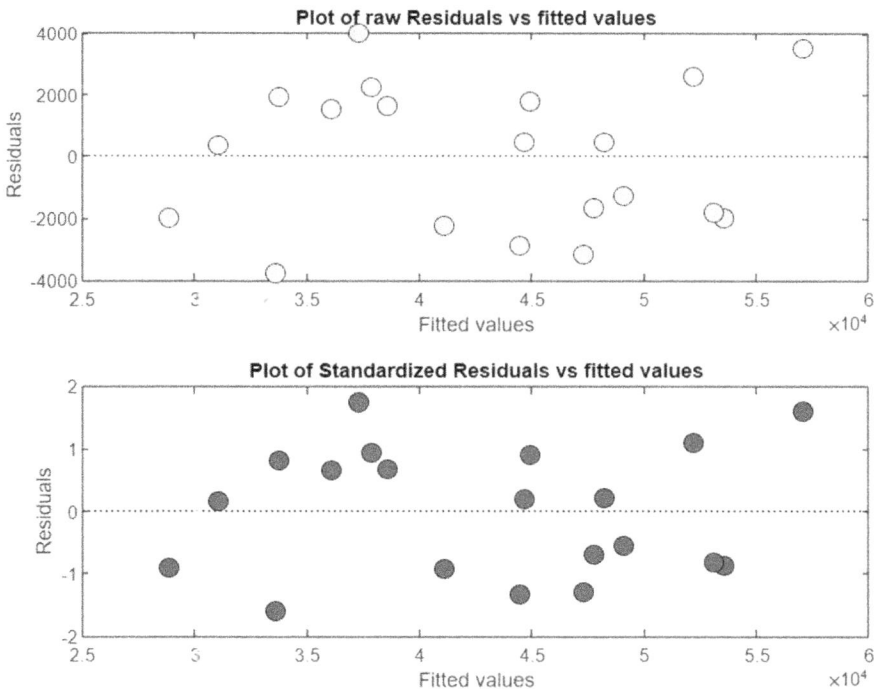

Figure 11.27

11.9.3.4 Test for independence

```
tiledlayout(2,1)
% Top axes
ax1 = nexttile;
```

```
h = plotResiduals(ax1,model,'caseorder', 'Marker',"o", 'MarkerSize',10,
                  'MarkerEdgeColor',"black", 'MarkerFaceColor',"cyan")
title(ax1, "Plot of raw Residuals vs fitted values")

ax2 = nexttile;
h = plotResiduals(ax2,model,'caseorder','ResidualType','Standardized',
            'Marker',"o", 'MarkerSize',10, 'MarkerEdgeColor',"black",
                'MarkerFaceColor','#7E2F8E")
title(ax2, "Plot of Standardized Residuals vs fitted values")
```

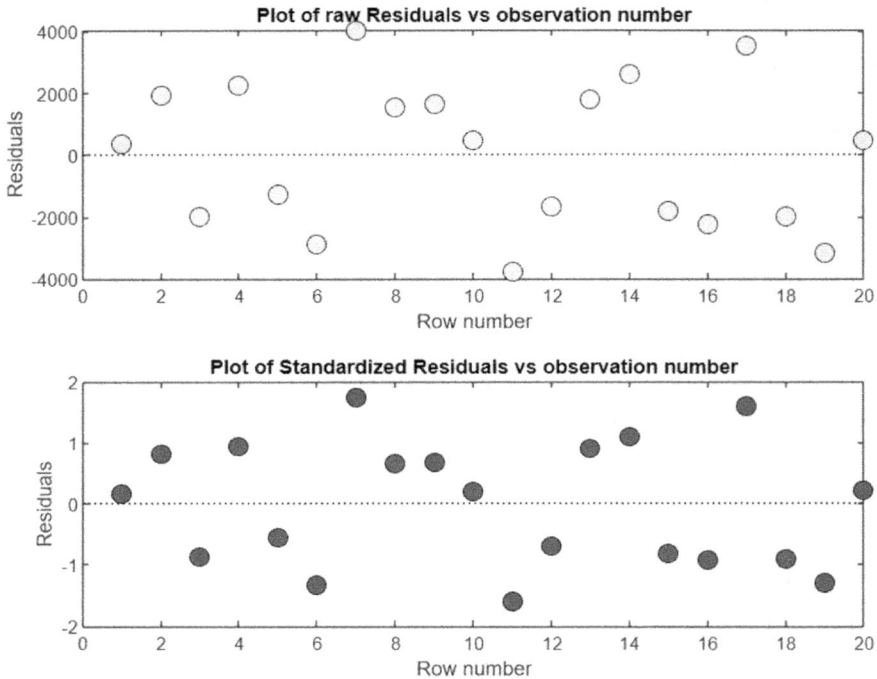

Figure 11.28

11.9.3.5 Test for multicollinearity

11.9.3.5.1 Correlation matrix/correlation plot

```
data = [Age', Experience'];
corr_matrix = corrcoef(data);

% Add row and column labels
variableNames = {'Age', 'Experience'};
corr_table = array2table(corr_matrix, 'RowNames', variableNames,
                            'VariableNames', variableNames)
```

	Age	Experience
Age	1	0.81422
Experience	0.81422	1

A correlation matrix showing a correlation coefficient of 81.42% between the age and experience variables indicates a high degree of correlation between the variables. However, as discussed earlier, it is not sufficient to fully assess multicollinearity in a regression model.

11.9.3.5.2 VIF

```
data = [Age', Experience'];
R = corrcoef(data);
VIF = diag(inv(R));

variableNames = {'Age', 'Experience'};
VIF_table = array2table(VIF, 'VariableNames', {'VIF'}, 'RowNames',
                                                    variableNames)
```

	VIF
Age	2.967
Experience	2.967

Such a low value suggests that there is no severe multicollinearity present in the regression model.

11.9.3.5.3 Collin test

```
Variance Decomposition

 sValue  condIdx   Var1     Var2
---------------------------------
 1.4069    1     0.0104   0.0104
 0.1441  9.7657  0.9896   0.9896

sValue =

    1.4069
    0.1441


condIdx =

    1.0000
    9.7657


VarDecomp =

    0.0104    0.0104
    0.9896    0.9896
```

Now we shall calculate the slopes of our two predictors. As we have two instead of one we will be calculating them in a different way. The formula therefore is:

$$\text{slope}\,\beta_1 = \frac{\left(£\,x_2^2\right)\left(£\,x_1\,y\right)-\left(£\,x_1\,x_2\right)\left(£\,x_2\,y\right)}{\left(£\,x_1^2\right)\left(£\,x_2^2\right)-\left(£\,x_1\,x_2\right)^2}$$

$$\text{slope}\,\beta_2 = \frac{\left(£\,x_1^2\right)\left(£\,x_2\,y\right)-\left(£\,x_1\,x_2\right)\left(£\,x_1\,y\right)}{\left(£\,x_1^2\right)\left(£\,x_2^2\right)-\left(£\,x_1\,x_2\right)^2}$$

Where

$$£\,x_1^2 = £\,x_1^2 - \frac{\left(£\,x_1\right)^2}{n}$$

$$£\,x_2^2 = £\,x_2^2 - \frac{\left(£\,x_2\right)^2}{n}$$

$$£\,x_1\,x_2 = £\,x_1\,x_2 - \frac{£\,x_1\,£\,x_2}{n}$$

$$£\,x_1\,y = £\,x_1\,y - \frac{£\,x_1\,£\,y}{n}$$

$$£\,x_2\,y = £\,x_2\,y - \frac{£\,x_2\,£\,y}{n}$$

However the intercept is calculated in the same way as the simple linear regression with just an additional explanatory variable.

$$\text{Intercept}\,\beta_0 = \bar{y} - \beta_1\left(\overline{x_1}\right) - \beta_2\left(\overline{x_2}\right)$$

Calculating the slopes:

x_1 = age	x_2 = experience	y = income	x_1^2	x_2^2	x_1x_2	x_1y	x_2y
25	2	31,450	625	4	50	7,86,250	62,900
30	3	35,670	900	9	90	1,070,100	107,010
47	12	51,580	2209	144	564	2,424,260	618,960
32	5	40,130	1024	25	160	1,284,160	200,650
43	10	47,830	1849	100	430	2,056,690	478,300
51	7	41,630	2601	49	357	2,123,130	291,410
28	5	41,340	784	25	140	1,157,520	206,700
33	4	37,650	1089	16	132	1,242,450	150,600
37	5	40,250	1369	25	185	1,489,250	201,250

39	8	45,150	1521	64	312	1,760,850	361,200
29	3	29,340	841	9	87	865,360	89,520
47	9	46,110	2209	81	423	2,167,170	414,990
54	7	46,720	2916	49	378	2,522,880	327,040
51	11	54,300	2601	121	561	2,794,800	602,800
44	12	51,300	1936	144	528	2,257,200	615,600
41	6	38,900	1681	36	246	1,594,900	233,400
58	13	60,600	3364	169	754	3,514,800	787,800
23	1	26,870	529	1	23	618,010	26,870
44	9	44,190	1936	81	396	1,944,360	397,710
37	10	48,700	1369	100	370	1,801,900	487,000
£793	£142	£860,710	£33,353	£1252	£6186	£35,476,040	£6,661,710

$$\bar{x_1} = \frac{793}{20} \qquad \bar{x_2} = \frac{142}{20} \qquad \bar{y} = \frac{860710}{20}$$

$$\pounds\, x_1^2 = 33353 - \frac{(793)^2}{20} = 1910.55 \triangleleft$$

$$\pounds\, x_2^2 = 1252 - \frac{(142)^2}{20} = 243.80$$

$$\pounds\, x_1 x_2 = 6186 - \frac{793 \times 142}{20} = 555.7$$

$$\pounds\, x_1 y = 35476040 - \frac{793 \times 860710}{20} = 1348888.5$$

$$\pounds\, x_2 y = 6661710 - \frac{142 \times 860710}{20} = 550669$$

Calculate:

$$\text{slope}\,\beta_1 = \frac{(243.80)(1348888.5) - (555.7)(550669)}{(1910.55)(243.80) - (555.7)^2} = 145.56539$$

$$\text{slope}\,\beta_2 = \frac{(1910.55)(550669) - (555.7)(1348888.5)}{(1910.55)(243.80) - (555.7)^2} = 1926.9$$

$$\beta_0 = \frac{860710}{20} - 145.56539\left(\frac{793}{20}\right) - 1926.9\left(\frac{142}{20}\right) = 23582.8$$

The regression line $\hat{y} = 23582.8 + 145.56539(\text{age}) + 1926.9(\text{experience})$

11.9.3.6 Now how do we interpret this?

The coefficients indicate the expected change in income for a one-unit change in age and experience, respectively. The coefficient for age suggests that, on average, for every one-unit increase in age, income is expected to increase by 145.56539 units, all else being equal. Similarly, for every one-unit increase in experience, income is expected to increase by 1926.9 units, all else being equal.

Like in simple linear regression, we now calculate the standard error of our coefficients.

As seen earlier, we need **the residual standard error to calculate the errors of these coefficients.**

First step: **the residual standard error** RSE $= \dfrac{\sum_{i=1}^{n}(y_i - \widehat{y}_i)^2}{(n - p - 1)}$.

$$SE_{\beta_0} = \sqrt{RSE\left[\frac{1}{n} + \frac{\overline{x_1}^2 \pounds\, x_2^2 + \overline{x_2}^2 \pounds\, x_1^2 - 2\overline{x_1}\overline{x_2}\pounds\, x_1 x_2}{\pounds\, x_1^2 \pounds\, x_2^2 - \left(\pounds\, x_1 x_2\right)^2}\right]}$$

$$SE_{\beta_1} = \sqrt{RSE\,\frac{\pounds\, x_2^2}{\pounds\, x_1^2 \pounds\, x_2^2 - \left(\pounds\, x_1 x_2\right)^2}}$$

$$SE_{\beta_2} = \sqrt{RSE\,\frac{\pounds\, x_1^2}{\pounds\, x_1^2 \pounds\, x_2^2 - \left(\pounds\, x_1 x_2\right)^2}}$$

So let's go ahead and find them.

Let's first calculate our residuals.

x_1 = age	x_2 = experience	Observed y	Predicted $\widehat{y}$	Residual = $y_i - \widehat{y}_i$	$\left(y_i - \widehat{y}_i\right)^2$
25	2	31,450	31,075.73	374.2653	140,074.5
30	3	35,670	33,730.46	1939.538	3,761,809
47	12	51,580	53,547.17	-1967.17	3,869,771
32	5	40,130	37,875.39	2254.608	5,083,255
43	10	47,830	49,111.11	-1281.11	1,641,247
51	7	41,630	44,494.93	-2864.93	8,207,852
28	5	41,340	37,293.13	4046.869	16,377,149
33	4	37,650	36,094.06	1555.942	2,420,956
37	5	40,250	38,603.22	1646.781	2,711,886
39	8	45,150	44,675.05	474.9498	225,577.3
29	3	29,840	33,584.9	-3744.9	14,024,248
47	9	46,110	47,766.47	-1656.47	2,743,904
54	7	46,720	44,931.63	1788.369	3,198,263
51	11	54,800	52,202.53	2597.465	6,746,825
44	12	51,300	53,110.48	-1810.48	3,277,828
41	6	38,900	41,112.38	-2212.38	4,894,630
58	13	60,600	57,075.29	3524.707	12,423,562
23	1	26,870	28,857.7	-1987.7	3,950,967

44	9	44,190	47,329.78	-3139.78	9,858,201
37	10	48,700	48,237.72	462.2806	213,703.3
					105,771,708.4

$$\text{SSE/RSS} = \sum_{i=1}^{n} \left(y_i - \widehat{y}_i \right)^2 = 105771708.4$$

First step: Calculate the RSE $= \dfrac{105771708.4}{20-2-1} = 6221865.2.$

Second step: Calculate the standard error.

$$SE_{\beta_0} = \sqrt{6221865.2 \left[\frac{1}{20} + \frac{\left(\frac{793}{20}\right)^2 (243.80) + \left(\frac{142}{20}\right)^2 (1910.55) - 2\left(\frac{793}{20}\right)\left(\frac{142}{20}\right)(555.7)}{(1910.55)(243.80) - (555.7)^2} \right]} = 2630.3$$

$$SE_{\beta_1} = \sqrt{6221865.2 \frac{243.80}{(1910.55)(243.80) - (555.7)^2}} = 98.297$$

$$SE_{\beta_2} = \sqrt{6221865.2 \frac{1910.55}{(1910.55)(243.80) - (555.7)^2}} = 275.172$$

Recall that we calculated the t-values of the regression slope and intercept for hypothesis testing and determined the statistical significance of the estimated coefficients.

$$\text{T value of } \beta_0 = \frac{\beta_0}{SE_{\beta_0}} = \frac{23582.8}{2630.3} = 8.9658$$

$$\text{t value of } \beta_1 = \frac{\beta_1}{SE_{\beta_1}} = \frac{145.56539}{98.297} = 1.48087$$

$$\text{t value of } \beta_2 = \frac{\beta_2}{SE_{\beta_2}} = \frac{1926.9}{275.172} = 7.0025$$

Now let's find the corresponding p-values of the slope coefficients.

The p-value that corresponds to t= 8.9658 with $df = 20-2 = 18$ is < 0.001.

Similarly, the p-value corresponding to t= 1.48087 with $df = 20-2 = 18$ is around 0.15.

Then finally the p-value for t= 7.0025 is < 0.001.

Therefore our regression table would be:

	Estimate	SE	tStat	pValue
Intercept	23,582.8	2630.3	8.9658	< 0.001
x_1 (age)	145.56539	98.297	1.48087	~0.15
x_2 (experience)	1926.9	275.172	7.0025	< 0.001

Then according to our output we need to estimate the regression statistics to assess the quality of our model and the significance of our two predictors.

Root mean squared error: we already calculated our **RSE**= 6221865.2. Therefore our RMSE = $\sqrt{6221865.2}$ = 2494.367.

R-squared: the coefficient of determination. Remember our $r^2 = 1 - \dfrac{SSE}{SST} = \dfrac{SSR}{SST}$. So now let's calculate our regression and total sum of squares.

Observed y	Predicted y	yi-y	(yi-y)2	yi-y	(yi-y)2
31,450	31,075.73	−11959.8	143,035,984.8	−11,585.5	134,223,810.3
35,670	33,730.46	−9305.04	86,583,737.76	−7365.5	54,250,590.25
51,580	53,547.17	10511.67	110,495,276.2	8544.5	73,008,480.25
40,130	37,875.39	−5160.11	26,626,709.62	−2905.5	8,441,930.25
47,830	49,111.11	6075.612	36,913,058.38	4794.5	22,987,230.25
41,630	44,494.93	1459.435	2,129,950.198	−1405.5	1,975,430.25
41,340	37,293.13	−5742.37	32,974,802.65	−1695.5	2,874,720.25
37,650	36,094.06	−6941.44	48,183,618.84	−5385.5	29,003,610.25
40,250	38,603.22	−4432.28	19,645,111.05	−2785.5	7,759,010.25
45,150	44,675.05	1639.55	2,688,124.891	2114.5	4,471,110.25
29,840	33,584.9	−9450.6	89,313,910.11	−13,195.5	174,121,220.3
46,110	47,766.47	4730.973	22,382,108.65	3074.5	9,452,550.25
46,720	44,931.63	1896.131	3,595,312.997	3684.5	13,575,540.25
54,800	52,202.53	9167.035	84,034,528.67	11,764.5	138,403,460.3
51,300	53,110.48	10,074.98	101,505,164.8	8264.5	68,301,960.25
38,900	41,112.38	−1923.12	3,698,386.727	−4135.5	17,102,360.25
60,600	57,075.29	14,039.79	197,115,776.8	17,564.5	308,511,660.3
26,870	28,857.7	−14,177.8	201,009,900.3	−16,165.5	261,323,390.3
44,190	47,329.78	4294.277	18,440,816.33	1154.5	1,332,870.25
48,700	48,237.72	5202.219	27,063,087	5664.5	32,086,560.25
			$\Sigma 1,257,435,367$		$\Sigma 1,363,207,495$

SSR = 1257435367

SSE = 105771708.4

SST = 1363207495

$$r^2 = 1 - \frac{105771708.4}{1363207495} = \frac{1257435367}{1363207495} = 0.922.$$

In this example, the R-squared is 0.922, which indicates that 92.2% of the variance in the income can be explained by the age and the years of experiences of an individual.

$$\text{Adjusted } R\text{-squared} = 1 - \left(\frac{n-1}{n-p-1}\right)\frac{SSE}{SST} = 1 - \left(\frac{20-1}{20-2-1}\right)\frac{105771708.4}{1363207495} = 0.913.$$

$$\text{F-statistic} = \frac{1257435367}{2} \Big/ \frac{105771708.4}{20-2-1} = 101.$$

p-value: to calculate the p-value corresponding to the f test our degrees of freedom in the numerator are $df_1 = p$ and in the denominator $df_1 = n - p - 1$ where p is the number of predictors. In this example our $F(df_1, df_2) = F(2, 17)$ and is extremely small.

To interpret the results we need to compare the p-value with alpha = 0.025 (two-tailed test). Since the p-value is significantly smaller than the significance level, our sample data provides sufficient evidence to conclude that our regression model fits the data better than the model with no independent variables.

Next, we perform ANOVA for the model.

1) We have the three sums of squares:

- **Total sum of squares** = 1363207495
- Regression sum of squares (model) = 1257435367
- Residual sum of squares = 105771708.4

2) Degrees of freedom

- Total = $\#n - 1 = 20 - 1 = 19$
- Model = $\# p = 2$
- Residual = $\# n - p - 1 = 20 - 2 - 1 = 17$

3) Mean squares

- Total $= \dfrac{1363207495}{19} = 71747762.89473$

- Model $= \dfrac{1257435367}{2} = 628717683.5$

- Residual $= \dfrac{105771708.4}{17} = 6221865.2$

Hence our table looks like:

Source	Sum of squares	df	Mean squares	F	P
Total	1,363,207,495	$20 - 1 = 19$	71,747,762.89473	101	
Model	1,257,435,367	2	628,717,683.5		
Residual	105,771,708.4	$20 - 2 - 1 = 17$	6,221,865.2		

Now, we calculate the **coefficient confidence intervals.**

A 95% confidence interval of $\beta_0 = 23582.8 \pm t_{1-0.05/2,\, 20-2} * 2630.3$

$$= 23582.8 \pm 2.101 * 2630.3$$

$$= [18056.5397, 29109.0603]$$

A 95% confidence interval of $\beta_1 = 145.56539 \pm t_{1-0.05/2,\, 20-2} * 98.297$

$$= 145.56539 \pm 2.101 * 98.297$$

$$= [-60.956607, 352.087387]$$

A 95% confidence interval of $\beta_2 = 1926.9 \pm t_{1-0.05/2,\,20-2} * 275.172$

$$= 1926.9 \pm 2.101 * 275.172$$

$$= [1348.763628, 2505.036372]$$

Matrix Approach to Multiple Linear Regression

The article "How to Optimize and Control the Wire Bonding Process: Part II" (Solid State Technology, Jan 1991: 67 72) described an experiment carried out to asses the impact of force (gm), power (mW), temperature (C) and time (msec) on ball bond shear strength (gm).

Now notice that we have four predictors in this example. Finding the coefficients is tricky for $p > 2$ independent variables, and we really need matrix algebra to see the computations.

Example

Force	Power	Temperature	Time	Strength
30	60	175	15	26.2
40	60	175	15	26.3
30	90	175	15	39.8
40	90	175	15	39.7
30	60	225	15	38.6
40	60	225	15	35.5
30	90	225	15	48.8
40	90	225	15	37.8
30	60	175	25	26.6
40	60	175	25	23.4
30	90	175	25	38.6
40	90	175	25	52.1
30	60	225	25	39.5
40	60	225	25	32.3
30	90	225	25	43.0
40	90	225	25	56.0
25	75	200	20	35.2
45	75	200	20	46.9
35	45	200	20	22.7
35	105	200	20	58.7
35	75	150	20	34.5
35	75	250	20	44.0
35	75	200	10	35.7
35	75	200	30	41.8
35	75	200	20	36.5
35	75	200	20	37.6
35	75	200	20	40.3
35	75	200	20	46.0
35	75	200	20	27.8
35	75	200	20	40.3

$$\bar{y} = \beta_0 + \beta_1\left(\overline{x_1}\right) + \beta_2\left(\overline{x_2}\right) + \beta_3\left(\overline{x_3}\right) + \beta_4\left(\overline{x_4}\right)$$

Let

$x_1 = $ Force

$x_2 = $ Power

$x_3 = $ Temperature

$x_4 = $ Time

$y = $. Strength

In the multiple regression setting, because of the potentially large number of predictors, it is more efficient to use matrices to define the regression model and the subsequent coefficients. As always, let's start with the simple case first. Consider the following simple linear regression function:

$$y_i = \beta_0 + \beta_1 x_{1i} + \beta_2 x_{2i} + \beta_3 x_{3i} + \beta_4 x_{4i} + \ldots \beta_k x_{ki} + \varepsilon_i \quad \text{for } i = 1, 2, 3, 4, .. n$$

r purposes of analysis it is convenient to express the model in matrix form. Let

$$26.2 = \beta_0 + \beta_1(30) + \beta_2(60) + \beta_3(175) + \beta_4(15) + \varepsilon_1$$

$$26.3 = \beta_0 + \beta_1(40) + \beta_2(60) + \beta_3(175) + \beta_4(15) + \varepsilon_2$$

$$39.8 = \beta_0 + \beta_1(30) + \beta_2(90) + \beta_3(175) + \beta_4(15) + \varepsilon_3$$

$$39.7 = \beta_0 + \beta_1(40) + \beta_2(90) + \beta_3(175) + \beta_4(15) + \varepsilon_4$$

$$38.6 = \beta_0 + \beta_1(30) + \beta_2(60) + \beta_3(225) + \beta_4(15) + \varepsilon_5$$

.

.

.

$$27.8 = \beta_0 + \beta_1(35) + \beta_2(75) + \beta_3(200) + \beta_4(20) + \varepsilon_{48}$$

$$40.3 = \beta_0 + \beta_1(35) + \beta_2(75) + \beta_3(200) + \beta_4(20) + \varepsilon_{49}$$

$$
\begin{bmatrix} 26.2 \\ 26.3 \\ 39.8 \\ 39.7 \\ \vdots \\ \vdots \\ \vdots \\ 27.8 \\ 40.3 \end{bmatrix} =
\begin{bmatrix}
1 & 30 & 60 & 175 & 15 \\
1 & 40 & 60 & 175 & 15 \\
1 & 30 & 90 & 175 & 15 \\
1 & 40 & 90 & 175 & 15 \\
\cdot & \cdot & \cdot & \cdot & \cdot \\
\vdots & \vdots & \vdots & \vdots & \vdots \\
\vdots & \vdots & \vdots & \vdots & \vdots \\
\cdot & \cdot & \cdot & \cdot & \cdot \\
1 & 35 & 75 & 200 & 20 \\
1 & 35 & 75 & 200 & 20
\end{bmatrix}
\begin{bmatrix} \beta_0 \\ \beta_1 \\ \beta_2 \\ \beta_3 \\ \beta_4 \end{bmatrix} +
\begin{bmatrix} \varepsilon_1 \\ \varepsilon_2 \\ \varepsilon_3 \\ \varepsilon_4 \\ \vdots \\ \vdots \\ \vdots \\ \varepsilon_{48} \\ \varepsilon_{49} \end{bmatrix}
$$

$$y = X\beta + \varepsilon$$

Therefore,

$$
y = \begin{bmatrix} 26.2 \\ 26.3 \\ 39.3 \\ 39.7 \\ \vdots \\ \vdots \\ \vdots \\ 27.8 \\ 40.3 \end{bmatrix}, \quad
X = \begin{bmatrix} 1 & 30 & 60 & 175 & 15 \\ 1 & 40 & 60 & 175 & 15 \\ 1 & 30 & 90 & 175 & 15 \\ 1 & 40 & 90 & 175 & 15 \\ \cdot & \cdot & \cdot & \cdot & \cdot \\ \vdots & \vdots & \vdots & \vdots & \vdots \\ \vdots & \vdots & \vdots & \vdots & \vdots \\ \cdot & \cdot & \cdot & \cdot & \cdot \\ 1 & 35 & 75 & 200 & 20 \\ 1 & 35 & 75 & 200 & 20 \end{bmatrix}, \quad
\beta = \begin{bmatrix} \beta_0 \\ \beta_1 \\ \beta_2 \\ \beta_3 \\ \beta_4 \end{bmatrix}, \quad
\varepsilon = \begin{bmatrix} \varepsilon_1 \\ \varepsilon_2 \\ \varepsilon_3 \\ \varepsilon_4 \\ \vdots \\ \vdots \\ \vdots \\ \varepsilon_{29} \\ \varepsilon_{30} \end{bmatrix}
$$

$$Y = X\beta$$

Multiplying both sides by X'

$$X'Y = X'X\beta$$

$$\beta = \left(X'X\right)^{-1} X'y$$

$$
X' = \begin{bmatrix}
1 & 1 & 1 & 1 & . & . & . & 1 & 1 \\
30 & 40 & 30 & 40 & . & . & . & 35 & 35 \\
60 & 60 & 90 & 90 & . & . & . & 75 & 75 \\
175 & 175 & 175 & 175 & . & . & . & 200 & 200 \\
15 & 15 & 15 & 15 & . & . & . & 20 & 20
\end{bmatrix}
$$

$$
X'X = \begin{bmatrix}
30 & 1050 & 2250 & 6000 & 600 \\
1050 & 37350 & 78750 & 210000 & 21000 \\
2250 & 78750 & 174150 & 450000 & 45000 \\
6000 & 210000 & 450000 & 1215000 & 120000 \\
600 & 21000 & 45000 & 120000 & 12600
\end{bmatrix}
$$

If you notice this above matrix is in the form:

$$
X'X = \begin{bmatrix}
n & \sum x_1 & \sum x_2 & \sum x_3 & \sum x_4 \\
\sum x_1 & \sum x_1^2 & \sum x_1 x_2 & \sum x_1 x_3 & \sum x_1 x_4 \\
\sum x_2 & \sum x_1 x_2 & \sum x_2^2 & \sum x_2 x_3 & \sum x_2 x_4 \\
\sum x_3 & \sum x_1 x_3 & \sum x_2 x_3 & \sum x_3^2 & \sum x_3 x_4 \\
\sum x_4 & \sum x_1 x_4 & \sum x_2 x_4 & \sum x_3 x_4 & \sum x_4^2
\end{bmatrix}
$$

$$(X'X)^{-1} = \begin{bmatrix} 6.45 & -0.0583333 & -0.0138889 & -0.0133333 & -0.033333 \\ -0.0583333 & 0.0016667 & 0 & 0 & 0 \\ -0.0138889 & 0 & 0.000185185 & 0 & 0 \\ -0.0133333 & 0 & 0 & 0.000066666667 & 0 \\ -0.033333 & 0 & 0 & 0 & 0.00166667 \end{bmatrix}$$

$$\beta = \begin{bmatrix} -37.4767 \\ 0.21167 \\ 0.49833 \\ 0.12967 \\ 0.25833 \end{bmatrix},$$

The regression line

$$\hat{y} = -37.4767 + 0.21167\,(\text{Force}) + 0.49833\,(\text{Power}) + 0.12967\,(\text{Temp}) + 0.25833\,(\text{Time})$$

Using the model we calculate the predicted values of our y i.e the strength.

Force	Power	Temperature	Time	$\hat{y}_i$
30	60	175	15	25.34
40	60	175	15	27.45667
30	90	175	15	40.29
40	90	175	15	42.40667
30	60	225	15	31.82333
40	60	225	15	33.94
30	90	225	15	46.77333
40	90	225	15	48.89
30	60	175	25	27.92333
40	60	175	25	30.04
30	90	175	25	42.87333
40	90	175	25	44.99
30	60	225	25	34.40667
40	60	225	25	36.52333
30	90	225	25	49.35667
40	90	225	25	51.47333
25	75	200	20	36.29
45	75	200	20	40.52333
35	45	200	20	23.45667
35	105	200	20	53.35667
35	75	150	20	31.92333
35	75	250	20	44.89
35	75	200	10	35.82333
35	75	200	30	40.99
35	75	200	20	38.40667
35	75	200	20	38.40667
35	75	200	20	38.40667
35	75	200	20	38.40667
35	75	200	20	38.40667
35	75	200	20	38.40667

y_i	$\widehat{y}_i$	$y_i - \widehat{y}_i$	$\left(y_i - \widehat{y}_i\right)^2$	$y_i - \overline{y}$	$\left(y_i - \overline{y}\right)^2$	$\widehat{y}_i - \overline{y}$	$\left(\widehat{y}_i - \overline{y}\right)^2$
26.2	25.34000	0.86000	0.73960	-12.20667	149.00271	-13.06667	170.73778
26.3	27.45667	-1.15667	1.33788	-12.10667	146.57138	-10.95000	119.90250
39.8	40.29000	-0.49000	0.24010	1.39333	1.94138	1.88333	3.54694
39.7	42.40667	-2.70667	7.32604	1.29333	1.67271	4.00000	16.00000
38.6	31.82333	6.77667	45.92321	0.19333	0.03738	-6.58333	43.34028
35.5	33.94000	1.56000	2.43360	-2.90667	8.44871	-4.46667	19.95111
48.8	46.77333	2.02667	4.10738	10.39333	108.02138	8.36667	70.00111
37.8	48.89000	-11.09000	122.98810	-0.60667	0.36804	10.48333	109.90028
26.6	27.92333	-1.32333	1.75121	-11.80667	139.39738	-10.48333	109.90028
23.4	30.04000	-6.64000	44.08960	-15.00667	225.20004	-8.36667	70.00111
38.6	42.87333	-4.27333	18.26138	0.19333	0.03738	4.46667	19.95111
52.1	44.99000	7.11000	50.55210	13.69333	187.50738	6.58333	43.34028
39.5	34.40667	5.09333	25.94204	1.09333	1.19538	-4.00000	16.00000
32.3	36.52333	-4.22333	17.83654	-6.10667	37.29138	-1.88333	3.54694
43.0	49.35667	-6.35667	40.40721	4.59333	21.09871	10.95000	119.90250
56.0	51.47333	4.52667	20.49071	17.59333	309.52538	13.06667	170.73778
35.2	36.29000	-1.09000	1.18810	-3.20667	10.28271	-2.11667	4.48028
46.9	40.52333	6.37667	40.66188	8.49333	72.13671	2.11667	4.48028
22.7	23.45667	-0.75667	0.57254	-15.70667	246.69938	-14.95000	223.50250
58.7	53.35667	5.34333	28.55121	20.29333	411.81938	14.95000	223.50250
34.5	31.92333	2.57667	6.63921	-3.90667	15.26204	-6.48333	42.03361
44.0	44.89000	-0.89000	0.79210	5.59333	31.28538	6.48333	42.03361
35.7	35.82333	-0.12333	0.01521	-2.70667	7.32604	-2.58333	6.67361
41.8	40.99000	0.81000	0.65610	3.39333	11.51471	2.58333	6.67361
36.5	38.40667	-1.90667	3.63538	-1.90667	3.63538	0.00000	0.00000
37.6	38.40667	-0.80667	0.65071	-0.80667	0.65071	0.00000	0.00000
40.3	38.40667	1.89333	3.58471	1.89333	3.58471	0.00000	0.00000
46.0	38.40667	7.59333	57.65871	7.59333	57.65871	0.00000	0.00000
27.8	38.40667	-10.60667	112.50138	-10.60667	112.50138	0.00000	0.00000
40.3	38.40667	1.89333	3.58471	1.89333	3.58471	0.00000	0.00000
			$\sum 665.1187$		$\sum 2325.26$		$\sum 1660.14$

$$SSR = 1660.14$$

$$SSE = 665.1\text{:}$$

$$SST = 2325.26$$

Errors in the Parameters of the Regression Line:

$$RSSE = \sqrt{\frac{665.11867}{30 - 4 - 1}} = 5.158$$

$$(X'X)^{-1} = \begin{bmatrix} 6.45 & -0.0583333 & -0.0138889 & -0.0133333 & -0.033333 \\ -0.0583333 & 0.0016667 & 0 & 0 & 0 \\ -0.0138889 & 0 & 0.000185185 & 0 & 0 \\ -0.0133333 & 0 & 0 & 0.000066666667 & 0 \\ -0.033333 & 0 & 0 & 0 & 0.00166667 \end{bmatrix}$$

SE of $\beta_0 = 5.158 \times \sqrt{6.45} = 13.0996$

SE of $\beta_1 = 5.158 \times \sqrt{0.0016667} = 0.21057$

SE of $\beta_2 = 5.158 \times \sqrt{0.000185185} = 0.070191$

SE of $\beta_3 = 5.158 \times \sqrt{0.000066666667} = 0.042115$

SE of $\beta_4 = 5.158 \times \sqrt{0.00166667} = 0.21057$

Hypothesis Testing of the Significance of Regression Coefficients

Null hypothesis $H_0 : \beta_1 = \beta_2 = \ldots = \beta_k$

Alternative hypothesis H_A: at least one of the β_k is non-zero

In this example we want to test whether the parameters for Force, Power, Temperature and Time are all zero, against the alternative that at least one of the parameters is not zero in the following model,

H_0: $\beta_1 = \beta_2 = \beta_3 = \beta_4$

H_A: $\beta_1 \neq 0, \beta_2 \neq 0, \beta_3 \neq 0, \beta_4 \neq 0$ or

If this null hypothesis is true, none of the explanatory variables influence y, and thus our model is of little or no value.

However, if the alternative hypothesis is true, then at least one of the parameters is not zero. The alternative hypothesis does not indicate, however, which variables those might be.

t – test

t value of $\beta_0 = \dfrac{\beta_0}{SE_{\beta_0}} = \dfrac{-37.47667}{13.0996} = -2.8609$

t value of $\beta_1 = \dfrac{\beta_1}{SE_{\beta_1}} = \dfrac{0.211667}{0.21057} = 1.0052$

t value of $\beta_2 = \dfrac{\beta_2}{SE_{\beta_2}} = \dfrac{0.498333}{0.070191} = 7.09967$

t value of $\beta_3 = \dfrac{\beta_3}{SE_{\beta_3}} = \dfrac{0.129667}{0.042115} = 3.07887$

t value of $\beta_4 = \dfrac{\beta_4}{SE_{\beta_4}} = \dfrac{0.258333}{0.21057} = 1.2268$

f – test

$$MSE = \frac{665.11867}{30-4-1} = 26.605$$

$$MSR = \frac{1660.14}{4} = 415.035$$

$$MST = \frac{2325.26}{30-1} = 80.181$$

$$f = \frac{415.035}{26.605} = 15.6$$

The p value corresponding to f test our degrees of freedom in the numerator are $df_1 = 4$ and in the denominator $df_2 = 25$. Hence,

$$F(4,25) = \text{extremely small}$$

To interpret the results we need to compare it with **alpha = 0.025 (two tailed test)** Since the p-value is significantly smaller than the significance level, hence we reject the null hypothesis since our sample data provide sufficient evidence to conclude that our regression model fits the data better than the model with no independent variables.

THE ANOVA TABLE

Source	Sum of Squares	df	Mean Squares	F	P
Total	2325.26	29	80.181		
Model	1660.14	4	418.035	15.6	Extremely small
Residual	665.11867	25	26.605		

coefficient of determination

$$-\text{Ordinary } R^2 = 1 - \frac{665.11867}{2325.26} = 0.714$$

$$-\text{adjusted } R^2 = 1 - \left(\frac{30-1}{30-4-1}\right)\frac{665.11867}{2325.26} = 0.668$$

Root Mean Squared Error

$$RMSE = \sqrt{26.605} = 5.158$$

MATLAB

```
x = [Age', Experience'];
```

```
y = Income';
model = fitlm(x, y)
```

```
                model =

        Linear regression model:
            y ~ 1 + x1 + x2

        Estimated Coefficients:
                          Estimate      SE       tStat      pValue
                          _____   _____   _____   _____

            (Intercept)     23583     2630.3    8.9658   7.4797e-08
            x1             145.57     98.297    1.4809      0.15694
            x2             1926.9     275.17    7.0025   2.1284e-06

        Number of observations: 20, Error degrees of freedom: 17
        Root Mean Squared Error: 2.49e+03
        R-squared: 0.922,  Adjusted R-Squared: 0.913
        F-statistic vs. constant model: 101, p-value = 3.66e-10
```

```
anova(model,'summary')
```

```
            ans =

        3×5 table

                         SumSq        DF     MeanSq        F       pValue
                       _____    ___   _____   _____   _____

            Total      1.3632e+09     19   7.1748e+07
            Model      1.2574e+09      2   6.2872e+08   101.05   3.659e-10
            Residual   1.0577e+08     17   6.2219e+06
```

```
coefCI(model, 0.05);
```

```
                    ans =

                1.0e+04 *

                1.8033     2.9132
               -0.0062     0.0353
                0.1346     0.2507
```

As we can see it passes all the tests for non collinearity – low VIF values for each predictor variable. Assuming that it passes all the test of regression assumptions (after testing for linearity, normality, independence and homoscedasticity) lets proceed on the estimation of the linear model

11.10 LINEAR REGRESSION WITH CATEGORICAL VARIABLES

In the context of multiple regression analysis, we often think of predictor variables as being quantitative – e.g., Engine Speed, age of the study participants, number of years of education, temperature, pressure and so on. In many occasions, however, there exists the need of incorporating categorical predictor variables into a multiple regression model. For example, after controlling for differences attributable to, let's say, Age, we may want to investigate whether there is a difference in salary between men and women. In this case, the variable gender is of interest and needs to be included in the regression model. Because the variable gender is not numerical, it cannot be entered directly into a regression model. Therefore, it needs to be given a numerical coding of some type so that it can be included in the model and be meaningfully interpreted,

Recall that in simple linear regression, we had the basic equation:

$$\hat{y} = \beta_0 + \beta_1 x$$

11.10.1 Example with Dichotomous Predictor Variable

Suppose that instead of a continuous X variable such as age, we have a dichotomous variable, such as gender, or a multiple categorical variable, such as country of birth, or type of employment (e.g., student, office worker, or miner, etc.). How does the nature of the regression changes when such variables are used?

To model regression of a categorical predictor with only two levels / dichotomous

$$\hat{y} = \beta_0 + \beta_1 X_i$$

Where

β_0 = mean of the control/referred group lets call it group 1
β_1 = mean difference between the control group or group 1 and group 2

Say we want to predict miles per gallon but instead of using weight or horsepower we use categorical variables, like am (automatic or manual transmission) as well. In the case of a simple linear regression with a binary predictor, x will hold a dummy variable instead.

- when the predictor is automatic,

$$\hat{y} = \beta_0 + \beta_1 \times 0 = \beta_0$$

The mean value miles per gallon for automatic transmission is the intercept β_0.

- when the predictor is manual,

$$\hat{y} = \beta_0 + \beta_1 \times 1 = \beta_0 + \beta_1$$

The mean value miles per gallon for manual transmission is the intercept β_0 plus the difference between automatic and manual transmission.

Put in this manner, the interpretation of the coefficients changes slightly,
When automatic:

- β_0 is the mean miles per gallon for automatic cars.

When manual:

- β_1 is equal to the mean difference in the gas mileage between automatic and manual cars.

Gender Pay Gap Comparisons with Regression Analysis

Gender	y = Salary
Male	90,000
Female	85,000
Female	140,000
Malec	125,000
Male	55,000
Female	62,000
Male	129,000
Female	120,000
Female	117,000
Male	98,000

To visualize the difference in salary between male and female instructors lets look at the boxplot.

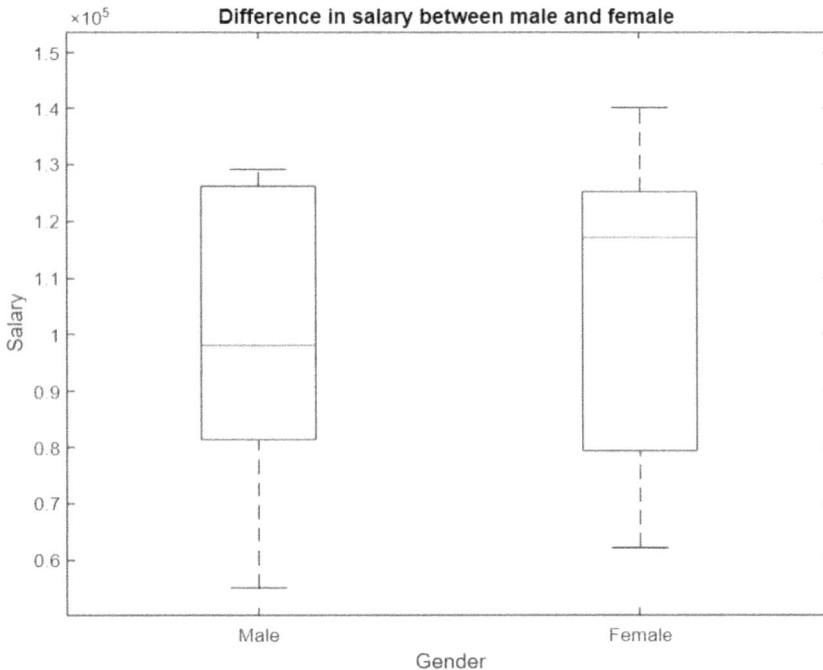

Now notice that Gender is categorical and clearly it cannot be entered directly into a regression model and be meaningfully interpreted, therefore some other method of dealing with information of this type must be developed.

Such variables can be brought within the scope of regression analysis using the method of dummy variables. So the first thing we need to do is to express our categorical variables as one or more dummy variables. Once a categorical variable has been recoded as a dummy variable, the dummy variable can be used in regression analysis just like any other quantitative variable.

Since the explanation sounded a little vague we start with the simplest case, where the qualitative variable is a binary variable, having only two possible value (dichotomous) which in this case is the gender.

The standard approach is to code the binary variable with the values 0 and 1. As a practical matter, regression results are easiest to interpret when dummy variables are limited to two specific values, 1 or 0.

Theoretically we can assign any one of the variables as 1 but in MATLAB the default order is ascending alphabetical which means 1 will be assigned to the female since f comes before m in male. Now typically 1 represents the presence of a qualitative attribute, and 0 represents the absence. Therefore our category now becomes

$x_2 = 1$ for female students.

$x_2 = 0$ for non-female students.

Now, we can replace Gender with 0 and 1in our data table.

x_1 = Gender		x_1 = Gender	
Male	0	**Female**	1
Female	1	Male	0
Female	1	**Female**	1
Male	0	**Female**	1
Male	0	Male	0

Gender		y = Salary
Male	0	90,000
Female	1	85,000
Female	1	140,000
Male	0	125,000
Male	0	55,000
Female	1	62,000
Male	0	129,000
Female	1	120,000
Female	1	117,000
Male	0	98,000

From the boxplots, we can see that the median salary for female is much higher than that of the male. Since least-squares linear regressions are defined by the mean of the outcome variable, it would be useful to calculate and compare the mean salaries according to sex. Lets calculate the mean salaries of both genders.

mean salary of the male group

$$\frac{90,000 + 125,000 + 55,000 + 129,000 + 98,000}{5} = 99400$$

mean salary of the female group

$$\frac{85000 + 140,000 + 62,000 + 120,000 + 117,000}{5} = 104800$$

The mean values indicate that the mean salary for female is higher than the mean salary for male by $5,400$. Thus, the boxplots and the difference in means indicate, overall, that females have a higher salary than male– but is this difference in salary statistically significant?

Since,

β_0 = mean salary of the control/referred group which is male= 99400

β_1 = mean difference of salary between the male and female= $104800 - 99400 = 5400$

Therefore Equation of the line

$$y = 99400 + 5400\left(Gender_{Female}\right)$$

From the model,

the average salary for male is estimated to be $\beta_0 = 99400 + 5400 \times 0 = \$99400$
while the average salary for female is $\beta_0 + \beta_1 = \$99400 + 5400 \times 1 = \$104800$

Sum of Squares

Using the model lets now calculate the predicted values

x_i	Observed y_i	Predicted $\widehat{y}_i = 5400\,x + 99400$
0	90,000	99400
1	85,000	104800
1	140,000	104800
0	125,000	99400
0	55,000	99400
1	62,000	104800
0	129,000	99400
1	120,000	104800
1	117,000	104800
0	98,000	99400

Lets now calculate the three sum of squares

y_i	$\widehat{y}_i$	$\left(y_i - \widehat{y}_i\right)$	$\left(y_i - \widehat{y}_i\right)^2$	$y_i - \bar{y}$	$\left(y_i - \bar{y}\right)^2$	$\widehat{y}_i - \bar{y}$	$\left(\widehat{y}_i - \bar{y}\right)^2$
90000	99400	-9400	88360000	-12100	146410000	-2700	7290000
85000	104800	-19800	392040000	-17100	292410000	2700	7290000
140000	104800	35200	1239040000	37900	1436410000	2700	7290000
125000	99400	25600	655360000	22900	524410000	-2700	7290000
55000	99400	-44400	1971360000	-47100	2218410000	-2700	7290000
62000	104800	-42800	1831840000	-40100	1608010000	2700	7290000
129000	99400	29600	876160000	26900	723610000	-2700	7290000
120000	104800	15200	231040000	17900	320410000	2700	7290000
117000	104800	12200	148840000	14900	222010000	2700	7290000
98000	99400	-1400	1960000	-4100	16810000	-2700	7290000
1021000			7436000000		7508900000		7290000

SSR = 72900000

SSE = 7436000000

SST = 7508900000

Errors in the Parameters of the Regression Line:

$$RSSE = \sqrt{\frac{7436000000}{10-1-1}} = 30487.70243$$

$$\sum x_i^2 = 5 \qquad \sum(x_i - \bar{x})^2 = 2.50$$

$$\text{SE of slope} = 30487.70243 \times \frac{1}{\sqrt{2.50}} = 19282.11606$$

$$\text{SE of intercept} = 30487.70243 \times \sqrt{\frac{5}{10 \times 2.50}} = 13634.51502$$

Hypothesis Testing of the Significance of Regression Coefficients

Null hypothesis H_0 : There is no significant linear relationship between gender and their salary

Alternative hypothesis H_A: There is a significant linear relationship between gender and their salary

t - test

$$t \text{ statistics for intercept} = \frac{99400}{13634.51502} = 7.2903$$

$$t \text{ statistics for slope} = \frac{5400}{19282.11606} = 0.28005$$

The p-value that corresponds to t = 7.2903 with $df = 10 - 2 = 8$ is < 0.001.
Similarly t = 0.28005 with $df = 10 - 2 = 8$ is ≈ 0.75

	Estimate	SE	tStat	pValue
Intercept	99400	13634.51502	7.2903	<0.001
Gender_Female	5400	19282.11606	0.28005	≈0.75

The p-value corresponding to the dummy variable *sex* (Gender_Female in the table) is highly significant, which suggests that there exists statistical evidence of a difference in average salary between female and male.

f – test

$$MSE = \frac{7436000000}{10-1-1} = 929500000$$

$$MSR = \frac{72900000}{1} = 72900000$$

$$MST = \frac{7508900000}{10-1} = 834322222.22222$$

$$f = \frac{72900000}{929500000} = 0.078429$$

THE ANOVA TABLE

Source	Sum of Squares	Df	Mean Squares	F	P
Total	7508900000	9	834322222.22222		
Model	72900000	1	72900000	0.078429	
Residual	7436000000	8	929500000		

coefficient of determination

$$-\text{Ordinary } R^2 = 1 - \frac{7436000000}{7508900000} = \frac{72900000}{7508900000} = 0.009708$$

$$-\text{adjusted } R^2 = 1 - \left(\frac{10-1}{10-1-1}\right)\frac{7436000000}{7508900000} = -0.11407$$

Root Mean Squared Error

$$RMSE = \sqrt{929500000} = 30487.70243$$

Example 2

MPG	Transmssion	MPG	Transmssion
21	Automatic	14.7	Manual
21	Automatic	32.4	Automatic
22.8	Automatic	30.4	Automatic
21.4	Manual	33.9	Automatic
18.7	Manual	21.5	Manual
18.1	Manual	15.5	Manual
14.3	Manual	15.2	Manual
24.4	Manual	13.3	Manual
22.8	Manual	19.2	Manual
19.2	Manual	27.3	Automatic
17.8	Manual	26	Automatic
16.4	Manual	30.4	Automatic
17.3	Manual	15.8	Automatic
15.2	Manual	19.7	Automatic
10.4	Manual	15	Automatic
10.4	Manual	21.4	Automatic

Where

$$n_{Auto} = 13, \ n_{Manual} = 19, \ \sum MPG_{Auto} = 317.1, \ \sum MPG_{Manual} = 325.8,$$

β_0 = mean of the MPG of the control/referred group which is Automatic

$$\frac{\sum MPG_{Manual}}{n_{Auto}} = \frac{317.1}{13} = 24.3923$$

β_1 = mean difference of MPG between the Automatic and Manual Transmission Manual

$$\frac{\sum MPG_{Manual}}{n_{Manual}} = \frac{325.8}{19} = 17.14737$$

$$\beta_1 = 17.14737 - 24.3923 = -7.2449$$

Therefore Equation of the line

$$y = -7.2449 \, x + 24.3923$$

Sum of Squares

Using the model lets now calculate the predicted values

x_i	Observed y_i	Predicted $\hat{y}_i = -7.2449\,x + 24.3923$
0	21	24.3923
0	21	24.3923
0	22.8	24.3923
1	21.4	17.1474
1	18.7	17.1474
1	18.1	17.1474
1	14.3	17.1474
1	24.4	17.1474
1	22.8	17.1474
1	19.2	17.1474
1	17.8	17.1474
1	16.4	17.1474
1	17.3	17.1474
1	15.2	17.1474
1	10.4	17.1474
1	10.4	17.1474
1	14.7	17.1474
0	32.4	24.3923
0	30.4	24.3923
0	33.9	24.3923
1	21.5	17.1474
1	15.5	17.1474
1	15.2	17.1474
1	13.3	17.1474
1	19.2	17.1474
0	27.3	24.3923
0	26	24.3923
0	30.4	24.3923
0	15.8	24.3923
0	19.7	24.3923
0	15	24.3923
0	21.4	24.3923

Lets now calculate the three sum of squares

y_i	$\hat{y}_i$	$\left(y_i - \hat{y}_i\right)$	$\left(y_i - \hat{y}_i\right)^2$	$y_i - \bar{y}$	$\left(y_i - \bar{y}\right)^2$	$\hat{y}_i - \bar{y}$	$\left(\hat{y}_i - \bar{y}\right)^2$
21	24.3923	-3.392	11.508	0.9094	0.827	4.301675	18.50441
21	24.3923	-3.392	11.508	0.9094	0.827	4.301675	18.50441
22.8	24.3923	-1.592	2.535	2.7094	7.3407	4.301675	18.50441
21.4	17.1474	4.253	18.085	1.3094	1.7145	-2.94323	8.662573
18.7	17.1474	1.553	2.411	-1.3906	1.9338	-2.94323	8.662573
18.1	17.1474	0.953	0.907	-1.9906	3.9626	-2.94323	8.662573

y_i	$\widehat{y}_i$	$\left(y_i - \widehat{y}_i\right)$	$\left(y_i - \widehat{y}_i\right)^2$	$y_i - \overline{y}$	$\left(y_i - \overline{y}\right)^2$	$\widehat{y}_i - \overline{y}$	$\left(\widehat{y}_i - \overline{y}\right)^2$
14.3	17.1474	-2.847	8.108	-5.7906	33.5313	-2.94323	8.662573
24.4	17.1474	7.253	52.6	4.3094	18.5707	-2.94323	8.662573
22.8	17.1474	5.653	31.952	2.7094	7.3407	-2.94323	8.662573
19.2	17.1474	2.053	4.213	-0.8906	0.7932	-2.94323	8.662573
17.8	17.1474	0.653	0.426	-2.2906	5.247	-2.94323	8.662573
16.4	17.1474	-0.747	0.559	-3.6906	13.6207	-2.94323	8.662573
17.3	17.1474	0.153	0.023	-2.7906	7.7876	-2.94323	8.662573
15.2	17.1474	-1.947	3.792	-4.8906	23.9182	-2.94323	8.662573
10.4	17.1474	-6.747	45.527	-9.6906	93.9082	-2.94323	8.662573
10.4	17.1474	-6.747	45.527	-9.6906	93.9082	-2.94323	8.662573
14.7	17.1474	-2.447	5.99	-5.3906	29.0588	-2.94323	8.662573
32.4	24.3923	8.008	64.123	12.3094	151.5207	4.301675	18.50441
30.4	24.3923	6.008	36.092	10.3094	106.2832	4.301675	18.50441
33.9	24.3923	9.508	90.396	13.8094	190.6988	4.301675	18.50441
21.5	17.1474	4.353	18.945	1.4094	1.9863	-2.94323	8.662573
15.5	17.1474	-1.647	2.714	-4.5906	21.0738	-2.94323	8.662573
15.2	17.1474	-1.947	3.792	-4.8906	23.9182	-2.94323	8.662573
13.3	17.1474	-3.847	14.802	-6.7906	46.1126	-2.94323	8.662573
19.2	17.1474	2.053	4.213	-0.8906	0.7932	-2.94323	8.662573
27.3	24.3923	2.908	8.455	7.2094	51.9751	4.301675	18.50441
26	24.3923	1.608	2.585	5.9094	34.9207	4.301675	18.50441
30.4	24.3923	6.008	36.092	10.3094	106.2832	4.301675	18.50441
15.8	24.3923	-8.592	73.828	-4.2906	18.4095	4.301675	18.50441
19.7	24.3923	-4.692	22.018	-0.3906	0.1526	4.301675	18.50441
15	24.3923	-9.392	88.215	-5.0906	25.9145	4.301675	18.50441
21.4	24.3923	-2.992	8.954	1.3094	1.7145	4.301675	18.50441
			$\sum 720.897$		$\sum 1126.047$		$\sum 405.15$

SSR = 405.15

SSE = 720.897

SST = 1126.047

Errors in the Parameters of the Regression Line:

$$RSSE = \sqrt{\frac{720.897}{32 - 1 - 1}} = 4.902$$

x_i	x_i^2	$x_i - \bar{x}$	$(x_i - \bar{x})^2$
0	0	-0.59375	0.352539
0	0	-0.59375	0.352539
0	0	-0.59375	0.352539
1	1	0.40625	0.165039
1	1	0.40625	0.165039
1	1	0.40625	0.165039
1	1	0.40625	0.165039
1	1	0.40625	0.165039
1	1	0.40625	0.165039
1	1	0.40625	0.165039
1	1	0.40625	0.165039
1	1	0.40625	0.165039
1	1	0.40625	0.165039
1	1	0.40625	0.165039
1	1	0.40625	0.165039
1	1	0.40625	0.165039
1	1	0.40625	0.165039
0	0	-0.59375	0.352539
0	0	-0.59375	0.352539
0	0	-0.59375	0.352539
1	1	0.40625	0.165039
1	1	0.40625	0.165039
1	1	0.40625	0.165039
1	1	0.40625	0.165039
1	1	0.40625	0.165039
0	0	-0.59375	0.352539
0	0	-0.59375	0.352539
0	0	-0.59375	0.352539
0	0	-0.59375	0.352539
0	0	-0.59375	0.352539
0	0	-0.59375	0.352539
0	0	-0.59375	0.352539
$\sum x_i = 19$	$\sum x_i^2 = 19$		$\sum(x_i - \bar{x})^2 = 7.71875$
$\bar{x} = \dfrac{19}{32}$			

$$\sum x_i^2 = 19 \qquad \sum(x_i - \bar{x})^2 = 7.71875$$

$$\text{SE of slope} = 4.902 \times \frac{1}{\sqrt{7.71875}} = 1.76441$$

$$\text{SE of intercept} = 4.902 \times \sqrt{\frac{19}{32 \times 7.71875}} = 1.3596$$

Hypothesis Testing of the Significance of Regression Coefficients

Null hypothesis H_0 : There is no significant linear relationship in MPG between automatic and manual transmission

Alternative hypothesis H_A: There is a significant linear relationship in MPG between automatic and manual transmission

t – test

$$t \text{ statistics for intercept} = \frac{24.3923}{1.3596} = 17.9408$$

$$t \text{ statistics for slope} = \frac{-7.2449}{1.76441} = -4.1061$$

	Estimate	SE	tStat	pValue
Intercept	24.3923	1.3596	17.9408	<0.05
Manual	-7.2449	1.76441	-4.1061	<0.05

f – test

$$MSE = \frac{720.897}{32-1-1} = 24.03$$

$$MSR = \frac{405.15}{1} = 405.15$$

$$MST = \frac{1126.047}{32-1} = 36.324$$

$$f = \frac{405.15}{24.03} = 16.86$$

THE ANOVA TABLE

Source	Sum of Squares	df	Mean Squares	F	P
Total	1126.047	31	36.324		
Model	405.15	1	405.15	16.86	<0.05
Residual	720.897	30	24.03		

coefficient of determination

$$-\text{Ordinary } R^2 = 1 - \frac{720.897}{1126.047} = \frac{405.15}{1126.047} = 0.36$$

$$-\text{adjusted } R^2 = 1 - \left(\frac{32-1}{32-1-1} \right) \frac{720.897}{1126.047} = 0.338$$

Root Mean Squared Error

$$RMSE = \sqrt{24.03} = 4.9$$

For 3 categories we can expand the model

$$\hat{y} = \beta_0 + \beta_1 X_{1i} + \beta_2 X_{2i}$$

Where
β_0 = mean of the control/referred group lets call it group 1
β_1 = mean difference between the control group or group 1 and group 2
β_2 = mean difference between the control group or group 1 and group 3
Now, because dummy variables were used to compare
experimental 1 $\left(\text{Mean} = 5.00, \text{SD} = 3.16 \right)$
experimental 2 $\left(\text{Mean} = 12.00, SD = 3.16 \right)$
control $(\text{Mean} = 26.00, SD = 3.16)$,

β_0 is the mean of the reference category $\left(\text{i.e. the control group} \right)$ so 26.0

$$\beta_1 = 5 - 26 = -21$$

$$\beta_2 = 12 - 26 = -14$$

$$\hat{y} = 26 - 21 D_1 - 14 D_2$$

11.10.2 Example with 3-level Categorical Variable

In studying the relationship between student's exposure to different types of teaching methods and their performance, the students were divided into three groups consisting of five each which were taught by discovery method, observational method and traditional method.

Group	Score
Discovery	90
	88
	91
	95
	93
Observational	78
	74
	74
	74
	70
Traditional	56
	59
	54
	55
	58

Lets calculate the average scores of the three groups

$$\text{Average Scores (Discovery)} = \frac{90 + 88 + 91 + 95 + 93}{5} = 91.40$$

$$\text{Average Scores (Observational)} = \frac{78 + 74 + 74 + 74 + 70}{5} = 74$$

$$\text{Average Scores (Traditional)} = \frac{56 + 59 + 54 + 55 + 58}{5} = 56.40$$

$$\beta_0 = \bar{y}_{\text{Discovery}} = 91.40$$

$$\beta_1 = \bar{y}_{\text{Observational}} - \bar{y}_{\text{Discovery}} = 74 - 91.40 = -17.4$$

$$\beta_2 = \bar{y}_{\text{Traditional}} - \bar{y}_{\text{Discovery}} = 56.40 - 91.40 = -35$$

$$\hat{y} = 91.40 - 17.4D_1 - 35D_2$$

Sum of Squares

The group variable is the extension of the first variable Gender we learned. The rule is that to code k categories we need $k - 1$ dummy variables, so in this case we need 2 dummy variables, D_1 and D_2. The reason why we take $k - 1$ variables and not k for k number of categories is because the kth dummy variable is redundant; it carries no new information. And it creates a severe multicollinearity problem for the analysis.

We have to choose one of the categories as the "control"; members of this group will be assigned a 0 on all the dummy variables. Beyond that, we need to arrange for each category to be given a unique pattern of 0s and 1s on the set of dummy variables. One way of doing this is shown in the following table, which defines the two variables. Note here we are taking "Discovery" as the control variable

This Group variable has three values assigned to : "Discovery", "Observational" and "Traditional". This variable could be dummy coded into two variables, one called "Observational" and another "Traditional"(ascending alphabetical).

- If Group = **Observational**, then the column D_1 would be coded with a 1 and D_2 with a 0.
- If Group =**Traditional**, then the column D_1 would be coded with a 0 and D_2 with a 1.
- If Group =Discovery then both columns D_1 and D_2 would be coded with a 0.

Using the model lets now calculate the predicted values

Group	D_1	D_2	y_i	$\hat{y} = 91.40 - 17.4D_1 - 35D_2$
Discovery	0	0	90	91.4
Discovery	0	0	88	91.4
Discovery	0	0	91	91.4
Discovery	0	0	95	91.4
Discovery	0	0	93	91.4
Observational	1	0	78	74
Observational	1	0	74	74
Observational	1	0	74	74
Observational	1	0	74	74
Observational	1	0	70	74
Traditional	0	1	56	56.4
Traditional	0	1	59	56.4
Traditional	0	1	54	56.4
Traditional	0	1	55	56.4
Traditional	0	1	58	56.4

Lets now calculate the three sum of squares

y_i	$\hat{y}_i$	$y_i - \hat{y}_i$	$\left(y_i - \hat{y}_i\right)^2$	$y_i - \bar{y}$	$\left(y_i - \bar{y}\right)^2$	$\hat{y}_i - \bar{y}$	$\left(\hat{y}_i - \bar{y}\right)^2$
90	91.4	-1.4	1.96	16.0667	258.1378	17.46667	305.0844
88	91.4	-3.4	11.56	14.0667	197.8711	17.46667	305.0844
91	91.4	-0.4	0.16	17.0667	291.2711	17.46667	305.0844
95	91.4	3 6	12.96	21.0667	443.8044	17.46667	305.0844
93	91.4	1.6	2.56	19.0667	363.5378	17.46667	305.0844
78	74	4	16	4.06667	16.5378	0.066667	0.004444
74	74	0	0	0.06667	0.00444	0.066667	0.004444
74	74	0	0	0.06667	0.00444	0.066667	0.004444
74	74	0	0	0.06667	0.00444	0.066667	0.004444
70	74	-4	16	-3.9333	15.47111	0.066667	0.004444

56	56.4	-0.4	0.16	-17.9333	321.6044	-17.5333	307.4178
59	56.4	2.6	6.76	-14.9333	223.0044	-17.5333	307.4178
54	56.4	-2.4	5.76	-19.9333	397.3378	-17.5333	307.4178
55	56.4	-1.4	1.96	-18.9333	358.4711	-17.5333	307.4178
58	56.4	1.6	2.56	-15.9333	253.8711	-17.5333	307.4178
			$\sum 78.4$		$\sum 3140.93$		$\sum 3062.53$

$SSR = 3062.53$

$SSE = 78.4$

$SST = 3140.93$

Errors in the Parameters of the Regression Line:

$$RSSE = \sqrt{\frac{78.4}{15-2-1}} = 2.556$$

x_1	x_2	y	x_1^2	x_2^2	$x_1 x_2$
0	0	90	0	0	0
0	0	88	0	0	0
0	0	91	0	0	0
0	0	95	0	0	0
0	0	93	0	0	0
1	0	78	1	0	0
1	0	74	1	0	0
1	0	74	1	0	0
1	0	74	1	0	0
1	0	70	1	0	0
0	1	56	0	1	0
0	1	59	0	1	0
0	1	54	0	1	0
0	1	55	0	1	0
0	1	58	0	1	0
$\sum x_1 = 5$	$\sum x_2 = 5$	$\sum y = 1109$	$\sum x_1^2 = 5$	$\sum x_2^2 = 5$	$\sum x_1 x_2 = 0$
$\overline{x_1} = \dfrac{5}{15}$	$\overline{x_2} = \dfrac{5}{15}$	$\overline{y} = \dfrac{1109}{15}$			

Where

$$\pounds x_1^2 = 5 - \frac{(5)^2}{15} = \frac{10}{3}$$

$$£x_2^{\ 2} = 5 - \frac{(5)^2}{15} = \frac{10}{3}$$

$$£x_1 x_2 = 0 - \frac{(5)^2}{15} = -\frac{5}{3}$$

$$SE_{\beta_0} = \sqrt{6.5333 \times \left[\frac{1}{15} + \frac{\left(\left(\frac{5}{15}\right)^2 \times \frac{10}{3}\right) + \left(\left(\frac{5}{15}\right)^2 \times \frac{10}{3}\right) - \left(2 \times \frac{5}{15} \times \frac{5}{15} \times -\frac{5}{3}\right)}{\left(\frac{10}{3} \times \frac{10}{3}\right) - \left(-\frac{5}{3}\right)^2} \right]}$$

$$= 1.1431$$

$$SE_{\beta_1} = \sqrt{6.5333 \times \frac{\frac{10}{3}}{\left(\frac{10}{3} \times \frac{10}{3}\right) - \left(-\frac{5}{3}\right)^2}} = 1.6166$$

$$SE_{\beta_2} = \sqrt{6.5333 \times \frac{\frac{10}{3}}{\left(\frac{10}{3} \times \frac{10}{3}\right) - \left(-\frac{5}{3}\right)^2}} = 1.6166$$

Hypothesis Testing of the Significance of Regression Coefficients

Null hypothesis H_0 : There is no significant linear relationship between gender and Maximum percentage of heart rate

Alternative hypothesis H_A: There is a significant linear relationship between gender and Maximum percentage of heart rate

t - test

$$t \text{ statistics for } \beta_0 = \frac{91.40}{1.1430} = 79.958$$

$$t \text{ statistics for } \beta_1 = \frac{-17.4}{1.6166} = -10.763$$

$$t \text{ statistics for } \beta_2 = \frac{-35}{1.6166} = -21.65$$

	Estimate	SE	tStat	pValue
Intercept	91.40	1.1430	79.958	< 0.05
Observational	−17.4	1.6166	−10.763	< 0.05
Traditional	−35	1.6166	−21.65	< 0.05

f – test

$$MSE = \frac{78.4}{15-2-1} = 6.533$$

$$MSR = \frac{3062.53}{2} = 1531.265$$

$$MST = \frac{3140.93}{15-1} = 224.35$$

$$f = \frac{1531.265}{6.5333} = 234.38$$

THE ANOVA TABLE

Source	Sum of Squares	df	Mean Squares	F	P
Total	3140.93	14	224.35		
Model	3062.53	2	1531.265	234.38	<0.05
Residual	78.4	12	6.533		

coefficient of determination

$$-\text{Ordinary } R^2 = 1 - \frac{78.4}{3140.93} = \frac{3062.53}{3140.93} = 0.975$$

$$-\text{adjusted } R^2 = 1 - \left(\frac{15-1}{15-2-1}\right)\frac{78.4}{3140.93} = 0.971$$

Root Mean Squared Error

$$RMSE = \sqrt{6.533} = 2.56$$

Calculations in MATLAB

To perform regression with categorical predictors in MATLAB, we use the same function fitlm function, which fits linear regression models, including models with categorical predictors. Below is a step-by-step guide on how to do this for the three examples we did previously.

Example 1

```
                data = readtable("Salary.xlsx");
                Gender = data.Gender;
                Salary = data.Salary;
                DATA = table(Gender, Salary);

                model = fitlm(DATA)
                anova(model,'summary')
```

model1 =

Linear regression model:
    Salary ~ 1 + Gender

Estimated Coefficients:

	Estimate	SE	tStat	pValue
(Intercept)	99400	13635	7.2903	8.4677e-05
Gender_Female	5400	19282	0.28005	0.78654

Number of observations: 10, Error degrees of freedom: 8
Root Mean Squared Error: 3.05e+04
R-squared: 0.00971, Adjusted R-Squared: -0.114
F-statistic vs. constant model: 0.0784, p-value = 0.787

ans =

  3x5 table

	SumSq	DF	MeanSq	F	pValue
Total	7.5089e+09	9	8.3432e+08		
Model	7.29e+07	1	7.29e+07	0.078429	0.78654
Residual	7.436e+09	8	9.295e+08		

Example 2

```
            data = readtable("MPGvsTransmission.xlsx");
            MPG = data.MPG;
            TRANSMISSION = data.Transmssion;

            DATA = table(TRANSMISSION, MPG);
            model = fitlm(DATA)
            anova(model,'summary')
```

```
Linear regression model:
    MPG ~ 1 + TRANSMISSION

Estimated Coefficients:
```

	Estimate	SE	tStat	pValue
(Intercept)	24.392	1.3596	17.941	1.3763e-17
TRANSMISSION_Manual	-7.2449	1.7644	-4.1061	0.00028502

```
Number of observations: 32, Error degrees of freedom: 30
Root Mean Squared Error: 4.9
R-squared: 0.36,  Adjusted R-Squared: 0.338
F-statistic vs. constant model: 16.9, p-value = 0.000285

    ans =

    3x5 table
```

	SumSq	DF	MeanSq	F	pValue
Total	1126	31	36.324		
Model	405.15	1	405.15	16.86	0.00028502
Residual	720.9	30	24.03		

Example 3

```
            data = readtable("ScoresvsGroups.xlsx");
            GROUP = data.Group;
            SCORE = data.Score;

            DATA = table(GROUP, SCORE);
            model = fitlm(DATA)
            anova(model,'summary')

    model =


Linear regression model:
    SCORE ~ 1 + GROUP

Estimated Coefficients:
```

	Estimate	SE	tStat	pValue
(Intercept)	91.4	1.1431	79.958	9.7611e-18
GROUP_Observational	-17.4	1.6166	-10.763	1.6084e-07
GROUP_Traditional	-35	1.6166	-21.651	5.5155e-11

```
Number of observations: 15, Error degrees of freedom: 12
Root Mean Squared Error: 2.56
R-squared: 0.975,  Adjusted R-Squared: 0.971
F-statistic vs. constant model: 234, p-value = 2.42e-10
```

```
ans =

  3x5 table
```

	SumSq	DF	MeanSq	F	pValue
Total	3140.9	14	224.35		
Model	3062.5	2	1531.3	234.38	2.4185e-10
Residual	78.4	12	6.5333		

11.11 MULTIPLE LINEAR REGRESSION WITH CATEGORICAL PREDICTORS

In the previous section we performed simple linear regression model where the response variable is predicted using only one categorical variable. In this section we expand the model to predict the response variable using multiple predictors, which may include both categorical and continuous variables.

11.11.1 one continuous and one categorical

We will learn to fit a linear regression model when we have two explanatory variables: one continuous and one categorical.

Let's examine the relationship between Gender and Income further. So let's use our Incomes data to determine whether, after controlling for the age, there is a difference in salary between male and female.

Gender	Age	y = Salary
Male	32	90,000
Female	28	85,000
Female	48	1,40,000
Male	45	1,25,000
Male	29	55,000
Female	36	62,000
Male	43	1,29,000
Female	38	1,20,000
Female	40	1,17,000
Male	41	98,000

Therefore, the first-order regression model is as follows:

$$y = \beta_0 + \beta_1 \left(Gender_{Female} \right) + \beta_2 \left(Age \right)$$

y = Salary	x_1 = Gder	x_2 = Age
90,000	0	32
85,000	1	28
1,40,000	1	48
1,25,000	0	45
55,000	0	29
62,000	1	36

y = Salary	x_1 = Gder	x_2 = Age
1,29,000	0	43
1,20,000	1	38
1,17,000	1	40
98,000	0	41

$$90,000 = \beta_0 + \beta_1(0) + \beta_2(32) + \varepsilon_1 \qquad 62,000 = \beta_0 + \beta_1(1) + \beta_2(36) + \varepsilon_6$$

$$85,000 = \beta_0 + \beta_1(1) + \beta_2(28) + \varepsilon_2 \qquad 129,000 = \beta_0 + \beta_1(0) + \beta_2(43) + \varepsilon_7$$

$$140,000 = \beta_0 + \beta_1(1) + \beta_2(48) + \varepsilon_3 \qquad 120,000 = \beta_0 + \beta_1(1) + \beta_2(38) + \varepsilon_8$$

$$125,000 = \beta_0 + \beta_1(0) + \beta_2(45) + \varepsilon_4 \qquad 117,000 = \beta_0 + \beta_1(1) + \beta_2(40) + \varepsilon_9$$

$$55,000 = \beta_0 + \beta_1(0) + \beta_2(29) + \varepsilon_5 \qquad 98,000 = \beta_0 + \beta_1(0) + \beta_2(41) + \varepsilon_{10}$$

$$
\begin{bmatrix} 90000 \\ 85000 \\ 140000 \\ 125000 \\ 55000 \\ 62000 \\ 129000 \\ 120000 \\ 117000 \\ 98000 \end{bmatrix}
=
\begin{bmatrix} 1 & 0 & 32 \\ 1 & 1 & 28 \\ 1 & 1 & 48 \\ 1 & 0 & 45 \\ 1 & 0 & 29 \\ 1 & 1 & 36 \\ 1 & 0 & 43 \\ 1 & 1 & 38 \\ 1 & 1 & 40 \\ 1 & 0 & 41 \end{bmatrix}
\begin{bmatrix} \beta_0 \\ \beta_1 \\ \beta_2 \end{bmatrix}
+
\begin{bmatrix} \varepsilon_1 \\ \varepsilon_2 \\ \varepsilon_3 \\ \varepsilon_4 \\ \varepsilon_5 \\ \varepsilon_6 \\ \varepsilon_7 \\ \varepsilon_8 \\ \varepsilon_9 \\ \varepsilon_{10} \end{bmatrix}
$$

$$y = X\beta + \varepsilon$$

$$
y = \begin{bmatrix} 90000 \\ 85000 \\ 140000 \\ 125000 \\ 55000 \\ 62000 \\ 129000 \\ 120000 \\ 117000 \\ 98000 \end{bmatrix},\
X = \begin{bmatrix} 1 & 0 & 32 \\ 1 & 1 & 28 \\ 1 & 1 & 48 \\ 1 & 0 & 45 \\ 1 & 0 & 29 \\ 1 & 1 & 36 \\ 1 & 0 & 43 \\ 1 & 1 & 38 \\ 1 & 1 & 40 \\ 1 & 0 & 41 \end{bmatrix},\
\beta = \begin{bmatrix} \beta_0 \\ \beta_1 \\ \beta_2 \end{bmatrix},\
\varepsilon = \begin{bmatrix} \varepsilon_1 \\ \varepsilon_2 \\ \varepsilon_3 \\ \varepsilon_4 \\ \varepsilon_5 \\ \varepsilon_6 \\ \varepsilon_7 \\ \varepsilon_8 \\ \varepsilon_9 \\ \varepsilon_{10} \end{bmatrix}
$$

$$\beta = (X'X)^{-1} X'y$$

$$X'X = \begin{bmatrix} 10 & 5 & 380 \\ 5 & 5 & 190 \\ 380 & 190 & 14848 \end{bmatrix}$$

$$(X'X)^{-1} = \begin{bmatrix} 3.7392157 & -0.20 & -0.093137255 \\ -0.20 & 0.40 & 0 \\ -0.093137255 & 0 & 0.002451 \end{bmatrix}$$

$$\beta = \begin{bmatrix} -34625 \\ 5400 \\ 3526.961 \end{bmatrix}$$

$$\hat{y} = -34625 + 5400 X_1 + 3526\,961 X_2$$

Sum of Squares

Using the model lets now calculate the predicted values

y_i	x_1	x_2	$\hat{y}_i = -34625 + 5400\ X_1 + 3526.961\ X_2$
90,000	0	32	78237.65
85,000	1	28	69529.81
1,40,000	1	48	140069
1,25,000	0	45	124088.1
55,000	0	29	67656.77
62,000	1	36	97745.5
1,29,000	0	43	117034.2
1,20,000	1	38	104799.4
1,17,000	1	40	111853.3
98,000	0	41	109980.3

Lets now calculate the sum of squares

y_i	$\hat{y}_i$	$y_i - \hat{y}_i$	$(y_i - \hat{y}_i)^2$	$y_i - \bar{y}$	$(y_i - \bar{y})^2 \times 10^4$	$\hat{y}_i - \bar{y}$	$(\hat{y}_i - \bar{y})^2$
90,000	78237.65	11,762	1.38E+08	-12100	14641	-23862.3	569411652.1
85,000	69529.81	15,470	2.39E+08	-17100	29241	-32570.2	1060817407
1,40,000	140069	-69	4764.865	37900	143641	37969.03	1441647087
1,25,000	124088.1	912	831479.5	22900	52441	21988.15	483478520.5
55,000	67656.77	-12,657	1.6E+08	-47100	221841	-34443.2	1186336162
62,000	97745.5	-35,745	1.28E+09	-40100	160801	-4354.5	18961705.09
1,29,000	117034.2	11,966	1.43E+08	26900	72361	14934.22	223031016.6
1,20,000	104799.4	15,201	2.31E+08	17900	32041	2699.418	7286857.539
1,17,000	111853.3	5,147	26488109	14900	22201	9753.34	95127641.16
98,000	109980.3	-11,980	1.44E+08	-4100	1681	7880.301	62099143.85
			$\sum 2360703435$		$\sum 750890$		$\sum 5148197193$

The three sum of squares

SSR/ESS=5148197193

RSS/SSE =2360703435

SST/TSS=7508900000

Errors in the Parameters of the Regression Line:

$$RSSE = \sqrt{\frac{2360703435}{10-2-1}} = 18364.19$$

$$(X'X)^{-1} = \begin{bmatrix} 3.7392157 & -0.20 & -0.093137255 \\ -0.20 & 0.40 & 0 \\ -0.093137255 & 0 & 0.002451 \end{bmatrix}$$

SE of $\beta_0 = 18364.19 \times \sqrt{3.7392157} = 35510.93$

SE of $\beta_1 = 18364.19 \times \sqrt{0.40} = 11614.5335568$

SE of $\beta_2 = 18364.19 \times \sqrt{0.002451} = 909.16$

Hypothesis Testing of the Significance of Regression Coefficients

Let us study β_1, which represents the effect of gender on the salary

$$t \text{ statistics for } \beta_1 = \frac{5400}{11614.5335568} = 0.46493$$

Lets now study the effect of Age on the salary

$$t \text{ statistics for } \beta_2 = \frac{3526.961}{909.16} = 3.8794$$

Similarly,

$$t \text{ statistics for } \beta_0 = \frac{-34625}{35510.93} = -0.97504$$

Since there are 10 observations and 3 parameters in the model, the residual sum of squares has 7 degrees of freedom. The probability value of this test statistic for a two-sided alternative

$$P(|T| > 9.7504) = 2P(T > 0.97504) \approx 0.36$$

$$P(|T| > 0.46493) = 2P(T > 0.46493) \approx 0.65$$

$$P(|T| > 3.8794) = 2P(T > 3.8794) \approx 0.006$$

	Estimate	SE	tStat	pValue
Intercept	-346245	35,511	-9.7504	≈ 0.36
Gender_Female	5400	11,615	0.46493	≈ 0.65
Age	3526.961	909	3.8794	≈ 0.006

As discussed earlier, the **intercept** term in a regression table tells us the average expected value for the response variable when all of the predictor variables are equal to zero.

Upon examining the p-values of the coefficient estimates for both independent variables, it is found that the p-value for **Gender _ Female** is 0.65, and for Age, it is 0.006. Based on these results, it can be concluded that at the statistical hypothesis testing yields the rejection of the null hypothesis for Age and the acceptance of the null hypothesis for **Gender _ Female**

f – test

$$MSE = \frac{2360703435}{10-2-1} = 337243347.857$$

$$MSR = \frac{5148197193}{2} = 2574098596.5$$

$$MST = \frac{7508900000}{10-1} = 834322222.22$$

$$f = \frac{2574098596.5}{337243347.857} = 7.63$$

THE ANOVA TABLE

Source	Sum of Squares	df	Mean Squares	F	P
Total	7.51E+09	9	834322222.2		
Model	5.15E+09	2	2574098597	7.63	
Residual	2.36E+09	7	337243347.9		

coefficient of determination

$$-\text{Ordinary } R^2 = 1 - \frac{2360703435}{7508900000} = 0.686$$

$$-\text{adjusted } R^2 = 1 - \left(\frac{10-1}{10-2-1}\right)\frac{2360703435}{7508900000} = 0.596$$

Model explains around 68.6 % of the variability in the response variable.

Root Mean Squared Error

$$RMSE = \sqrt{337243347.857} = 18364.18$$

Example 2

Data were collected to check whether the presence of urea formaldehyde foam insulation (UFFI) has an effect on the ambient formaldehyde concentration (CH_2O) inside the house. Twelve homes with and 12 homes without UFFI were studied, and the average weekly CH_2O concentration (in parts per billion) was measured. It was thought that the CH_2O concentration was also influenced by the amount of air that can move through the house via windows, cracks, chimneys, etc. A measure of "air tightness," on a scale of 0 to 10, was determined for each home. The data are shown in Table 1.2. CH_2O concentration is the response variable (y) that we try to explain through two explanatory variables: the air tightness. of the home (x_1) and the absence/presence of UFFI (x_2).

$$y = \beta_0 + \beta_1 x_1 + \beta_2 x_2$$

where

- y is the CH_2O concentration,
- x_1 is the air tightness of the house,
- x_2 is 1 or 0, depending on whether or not UFFI is present

$y = CH_2O$ concentration	$x_1 =$ Air Tightness	$x_2 =$ Presence of UFFI
31	0	0
29	1	0
40	1	0
45	4	0
40	4	0
38	5	0
51	7	0
49	7	0
52	8	0
63	8	0
60.79	8	0
56.67	9	0
43.58	1	1
43.3	2	1
46.16	2	1
47.66	4	1
55.31	4	1
63.32	5	1
59.65	5	1
62.74	6	1
60.33	6	1
53.13	7	1
56.83	9	1
70.34	10	1

$$31.33 = \beta_0 + \beta_1(0) + \beta_2(0) + \varepsilon_1$$

$$28.57 = \beta_0 + \beta_1(1) + \beta_2(0) + \varepsilon_2$$

$$39.95 = \beta_0 + \beta_1(1) + \beta_2(0) + \varepsilon_3$$

$$44.98 = \beta_0 + \beta_1(4) + \beta_2(0) + \varepsilon_4$$

$$39.55 = \beta_0 + \beta_1(4) + \beta_2(0) + \varepsilon_5$$

$$38.29 = \beta_0 + \beta_1(5) + \beta_2(0) + \varepsilon_6$$

$$50.58 = \beta_0 + \beta_1(7) + \beta_2(0) + \varepsilon_7$$

$$48.71 = \beta_0 + \beta_1(7) + \beta_2(0) + \varepsilon_8$$

$$51.52 = \beta_0 + \beta_1(8) + \beta_2(0) + \varepsilon_9$$

$$62.52 = \beta_0 + \beta_1(8) + \beta_2(0) + \varepsilon_{10}$$

$$60.79 = \beta_0 + \beta_1(8) + \beta_2(0) + \varepsilon_{11}$$

$$56.67 = \beta_0 + \beta_1(9) + \beta_2(0) + \varepsilon_{12}$$

$$43.58 = \beta_0 + \beta_1(1) + \beta_2(1) + \varepsilon_{13}$$

$$43.30 = \beta_0 + \beta_1(2) + \beta_2(1) + \varepsilon_{14}$$

$$46.16 = \beta_0 + \beta_1(2) + \beta_2(1) + \varepsilon_{15}$$

$$47.66 = \beta_0 + \beta_1(4) + \beta_2(1) + \varepsilon_{16}$$

$$55.31 = \beta_0 + \beta_1(4) + \beta_2(1) + \varepsilon_{17}$$

$$63.32 = \beta_0 + \beta_1(5) + \beta_2(1) + \varepsilon_{18}$$

$$59.65 = \beta_0 + \beta_1(5) + \beta_2(1) + \varepsilon_{19}$$

$$62.74 = \beta_0 + \beta_1(6) + \beta_2(1) + \varepsilon_{20}$$

$$60.33 = \beta_0 + \beta_1(6) + \beta_2(1) + \varepsilon_{21}$$

$$53.13 = \beta_0 + \beta_1(7) + \beta_2(1) + \varepsilon_{22}$$

$$56.83 = \beta_0 + \beta_1(9) + \beta_2(1) + \varepsilon_{23}$$

$$70.34 = \beta_0 + \beta_1(10) + \beta_2(1) + \varepsilon_{24}$$

$$
\begin{bmatrix}
31.33 \\
28.57 \\
\cdot \\
\cdot \\
60.79 \\
56.67 \\
43.58 \\
43.30 \\
\cdot \\
\cdot \\
56.83 \\
70.34
\end{bmatrix}
=
\begin{bmatrix}
1 & 0 & 0 \\
1 & 1 & 0 \\
\cdot & \cdot & \cdot \\
\cdot & \cdot & \cdot \\
1 & 8 & 0 \\
1 & 9 & 0 \\
1 & 1 & 1 \\
1 & 2 & 1 \\
\cdot & \cdot & \cdot \\
\cdot & \cdot & \cdot \\
1 & 9 & 1 \\
1 & 10 & 1
\end{bmatrix}
\begin{bmatrix}
\beta_0 \\
\beta_1 \\
\beta_2
\end{bmatrix}
+
\begin{bmatrix}
\varepsilon_1 \\
\varepsilon_2 \\
\cdot \\
\cdot \\
\varepsilon_{11} \\
\varepsilon_{12} \\
\varepsilon_{13} \\
\varepsilon_{14} \\
\cdot \\
\cdot \\
\varepsilon_{23} \\
\varepsilon_{24}
\end{bmatrix}
$$

$$y = X\beta + \varepsilon$$

$$
y = \begin{bmatrix} 31.33 \\ 28.57 \\ \cdot \\ \cdot \\ 60.79 \\ 56.67 \\ 43.58 \\ 43.30 \\ \cdot \\ \cdot \\ 56.83 \\ 70.34 \end{bmatrix}, \quad
X = \begin{bmatrix} 1 & 0 & 0 \\ 1 & 1 & 0 \\ \cdot & \cdot & \cdot \\ \cdot & \cdot & \cdot \\ 1 & 8 & 0 \\ 1 & 9 & 0 \\ 1 & 1 & 1 \\ 1 & 2 & 1 \\ \cdot & \cdot & \cdot \\ \cdot & \cdot & \cdot \\ 1 & 9 & 1 \\ 1 & 10 & 1 \end{bmatrix}, \quad
\beta = \begin{bmatrix} \beta_0 \\ \beta_1 \\ \beta_2 \end{bmatrix}, \quad
\varepsilon = \begin{bmatrix} \varepsilon_1 \\ \varepsilon_2 \\ \cdot \\ \cdot \\ \varepsilon_{11} \\ \varepsilon_{12} \\ \varepsilon_{13} \\ \varepsilon_{14} \\ \cdot \\ \cdot \\ \varepsilon_{23} \\ \varepsilon_{24} \end{bmatrix}
$$

$$ y = X\beta + \varepsilon $$

Brilliant Leadership

$$
X'X = \begin{bmatrix} 1 & 1 & \cdot & \cdot & 1 & 1 & 1 & 1 & \cdot & \cdot & 1 & 1 \\ 0 & 1 & \cdot & \cdot & 8 & 9 & 1 & 2 & \cdot & \cdot & 9 & 10 \\ 0 & 0 & \cdot & \cdot & 0 & 0 & 1 & 1 & \cdot & \cdot & 1 & 1 \end{bmatrix}
\begin{bmatrix} 1 & 0 & 0 \\ 1 & 1 & 0 \\ \cdot & \cdot & \cdot \\ \cdot & \cdot & \cdot \\ 1 & 8 & 0 \\ 1 & 9 & 0 \\ 1 & 1 & 1 \\ 1 & 2 & 1 \\ \cdot & \cdot & \cdot \\ \cdot & \cdot & \cdot \\ 1 & 9 & 1 \\ 1 & 10 & 1 \end{bmatrix}
= \begin{bmatrix} 24 & 123 & 12 \\ 123 & 823 & 61 \\ 12 & 61 & 12 \end{bmatrix}
$$

$$
(X'X)^{-1} = \begin{bmatrix} 0.22194577 & -0.026828213 & -0.08557 \\ -0.026828213 & 0.0052 & 0.000433 \\ -0.08557 & 0.000433 & 0.1667 \end{bmatrix}
$$

$$\beta = \begin{bmatrix} 0.222 & -0.027 & -0.08557 \\ -0.027 & 0.0052 & 0.000433 \\ -0.08557 & 0.000433 & 0.1667 \end{bmatrix} \begin{bmatrix} 1 & 1 & . & . & 1 & 1 & 1 & 1 & . & . & 1 & 1 \\ 0 & 1 & . & . & 8 & 9 & 1 & 2 & . & . & 9 & 10 \\ 0 & 0 & . & . & 0 & 0 & 1 & 1 & . & . & 1 & 1 \end{bmatrix} \begin{bmatrix} 31.33 \\ 28.57 \\ . \\ . \\ 60.79 \\ 56.67 \\ 43.58 \\ 43.30 \\ . \\ . \\ 56.83 \\ 70.34 \end{bmatrix}$$

$$\beta = \begin{bmatrix} 31.3734 \\ 2.8545 \\ 9.312 \end{bmatrix}$$

$$\hat{y} = 31\,3734 + 2\,8545 X_1 + 9\,312 X_2$$

Sum of Squares

Using the model lets now calculate the predicted values

y_i	x_1	x_2	$\hat{y}_i = 31.3734 + 2.8545x_{1i} + 9.312x_{2i}$
31	0	0	31.3734
29	1	0	34.2279
40	1	0	34.2279
45	4	0	42.7914
40	4	0	42.7914
38	5	0	45.6459
51	7	0	51.3549
49	7	0	51.3549
52	8	0	54.2094
63	8	0	54.2094
60.79	8	0	54.2094
56.67	9	0	57.0639
43.58	1	1	43.5399
43.3	2	1	46.3944
46.16	2	1	46.3944
47.56	4	1	52.1034
55.31	4	1	52.1034
63.32	5	1	54.9579
59.55	5	1	54.9579
62.74	6	1	57.8124
60.33	6	1	57.8124
53.13	7	1	60.6669
56.33	9	1	66.3759
70.34	10	1	69.2304

Lets now calculate the sum of squares

y_i	$\hat{y}_i$	$y_i - \hat{y}_i$	$\left(y_i - \hat{y}_i\right)^2$	$y_i - \bar{y}$	$\left(y_i - \bar{y}\right)^2$	$\hat{y}_i - \bar{y}$	$\left(\hat{y}_i - \bar{y}\right)^2$
31.33	31.3734	-0.0434	0.00188356	-19.3288	373.6006	-19.2854	371.9247
29	34.2279	-6	32.01183241	-22.0888	487.9129	-16.4309	269.9728
40	34.2279	6	32.74242841	-10.7088	114.6773	-16.4309	269.9728
45	42.7914	2	4.78996996	-5.67875	32.2482	-7.86735	61.8952
40	42.7914	-3	10.50667396	-11.1088	123.4043	-7.86735	61.8952
38	45.6459	-7	54.10926481	-12.3688	152.986	-5.01285	25.12867
51	51.3549	-1	0.60047001	-0.07875	0.006202	0.69615	0.484625
49	51.3549	-3	6.99549601	-1.94875	3.797627	0.69615	0.484625
52	54.2094	-3	7.23287236	0.86125	0.741752	3.55065	12.60712
63	54.2094	8	69.06607236	11.86125	140.6893	3.55065	12.60712
61	54.2094	7	43.30429636	10.13125	102.6422	3.55065	12.60712
56.67	57.0639	-0.3939	0.15515721	6.01125	36.13513	6.40515	41.02595
43.58	43.5399	0.0401	0.00160801	-7.07875	50.1087	-7.11885	50.67803
43.3	46.3944	-3.0944	9.57531136	-7.35875	54.1512	-4.26435	18.18468
46.16	46.3944	-0.2344	0.05494336	-4.49875	20.23875	-4.26435	18.18468
47.66	52.1034	-4.4434	19.74380356	-2.99875	8.992502	1.44465	2.087014
55.31	52.1034	3.2066	10.28228356	4.65125	21.63413	1.44465	2.087014
63.32	54.9579	8.3621	69.92471641	12.66125	160.3073	4.29915	18.48269
59.65	54.9579	4.6921	22.01580241	8.99125	80.84258	4.29915	18.48269
62.74	57.8124	4.9276	24.28124176	12.08125	145.9566	7.15365	51.17471
60.33	57.8124	2.5176	6.33830976	9.67125	93.53308	7.15365	51.17471
53.13	60.6669	-7.5369	56.80486161	2.47125	6.107077	10.00815	100.1631
56.83	66.3759	-9.5459	91.12420681	6.17125	38.08433	15.71715	247.0288
70.34	69.2304	1.1096	1.23121216	19.68125	387.3516	18.57165	344.9062
			$\sum 572.8947$		$\sum 2636.15$		$\sum 2063.24$

The three sum of squares
SSR/ESS=2063.24
RSS/SSE =572.8947
SST/TSS=2636.15

Errors in the Parameters of the Regression Line:

$$RSSE = \sqrt{\frac{572.8947}{24-2-1}} = 5.2231$$

$$\left(X'X\right)^{-1} = \begin{bmatrix} 0.22194577 & -0.02683 & -0.08557 \\ -0.02683 & 0.0052 & 0.000433 \\ -0.08557 & 0.000433 & 0.1667 \end{bmatrix}$$

SE of $\beta_0 = 5.2231 \times \sqrt{0.22194577} = 2.46066$

SE of $\beta_1 = 5.2231 \times \sqrt{0.0052} = 0.376$

SE of $\beta_2 = 5.2231 \times \sqrt{0.1667} = 2.1325$

Hypothesis Testing of the Significance of Regression Coefficients

Let us study β_1, which represents the effect of airtightness on the average ambient formaldehyde concentration

$$t \text{ statistics for } \beta_1 = \frac{2.8545}{0.376} = 7.584$$

Clearly the value is very large, and the probability of obtaining such an extreme value from a t distribution with 21 degrees of freedom is negligible; the probability value for a two-sided alternative, $2P(T > 7.58)$, is essentially zero. Hence, there is little doubt that airtightness of a home increases the formaldehyde concentration in the home

Lets now repeat these calculations for the other parameter β_2 which represents the effect of formaldehyde insulation on the ambient formaldehyde concentration. Is there a difference in the average concentration between homes of equal airtightness but different UFFI insulation? If insulation does not matter, then $\beta_1 = 0$. To answer this question, we test the hypothesis $\beta_1 = 0$.

$$t \text{ statistics for } \beta_2 = \frac{9.312}{2.1325} = 4.3667$$

We could repeat the calculations for the intercept β_0, which mathematically is the average concentration for homes without UFFI and with airtightness zero. Usually the intercept does not have much physical meaning, and we often skip the calculation.

$$t \text{ statistics for } \beta_0 = \frac{31.3734}{2.46066} = 12.75$$

	Estimate	SE	tStat	pValue
Intercept	31.3734	2.46066	12.75	~0
Air Tightness	2.8545	0	7.584	~0
UFFI_Yes	9.312	2	4.3667	<0.05

f - test

$$MSE = \frac{572.8947}{24 - 2 - 1} = 27.281$$

$$MSR = \frac{2063.24}{2} = 1031.62$$

$$MST = \frac{2636.15}{24 - 1} = 114.62$$

$$f = \frac{1031.62}{27.281} = 37.815$$

THE ANOVA TABLE

Source	Sum of Squares	df	Mean Squares	F	P
Total	2636.15	23	114.62		
Model	2063.24	2	1031.62	37.815	
Residual	572.8947	21	27.281		

coefficient of determination

$$-\text{Ordinary } R^2 = 1 - \frac{572.8947}{2636.15} = 0.783$$

$$-\text{adjusted } R^2 = 1 - \left(\frac{24-1}{24-2-1}\right)\frac{572.8947}{2636.15} = 0.762$$

Model explains around 78.3 % of the variability in the response variable.

Root Mean Squared Error

$$RMSE = \sqrt{27.281} = 5.22$$

Calculations in MATLAB

Example I

```
data = readtable("Salary.xlsx");
Gender = data.Gender;
Age = data.Age;
Salary = data.Salary;
DATA = table(Gender, Age, Salary);

model = fitlm(DATA)
anova(model,'summary')
```

```
model =


Linear regression model:
    Salary ~ 1 + Gender + Age

Estimated Coefficients:
                      Estimate      SE        tStat       pValue

                      _____   _____   _____   _____

    (Intercept)        -34625      35511     -0.97504     0.36202
    Gender_Female        5400      11615      0.46493     0.65609
    Age                  3527     909.16      3.8794     0.0060581


Number of observations: 10, Error degrees of freedom: 7
Root Mean Squared Error: 1.84e+04
R-squared: 0.686,  Adjusted R-Squared: 0.596
F-statistic vs. constant model: 7.63, p-value = 0.0174

ans =

  3x5 table

                 SumSq        DF      MeanSq        F        pValue

              _____    ____   _____   _____   _____

    Total     7.5089e+09      9    8.3432e+08
    Model     5.1482e+09      2    2.5741e+09    7.6328    0.017423
    Residual  2.3607e+09      7    3.3724e+08
```

Example 2

```
        data = readtable("urea formaldehyde foam.xlsx");

        CH2O = data.CH2O;
        Air_Tightness = data.Air_Tightness;
        UFFI_Present = data.UFFI_Present;

        DATA = table(Air_Tightness, UFFI_Present, CH2O);
        model = fitlm(DATA)
        anova(model,'summary')
```

```
model =


Linear regression model:
    CH2O ~ 1 + Air_Tightness + UFFI_Present

Estimated Coefficients:
                        Estimate      SE       tStat      pValue

                        _____    _____    _____    _____

    (Intercept)          31.373      2.4607     12.75     2.3616e-11
    Air_Tightness        2.8545      0.37637    7.5843    1.9166e-07
    UFFI_Present_Yes     9.312       2.1325     4.3666    0.0002704


Number of observations: 24, Error degrees of freedom: 21
Root Mean Squared Error: 5.22
R-squared: 0.783,  Adjusted R-Squared: 0.762
F-statistic vs. constant model: 37.8, p-value = 1.1e-07

ans =

  5x5 table

                SumSq     DF    MeanSq      F         pValue

                _____    __    _____    _____    _____

    Total       2636.1    23    114.62
    Model       2063.3     2    1031.6    37.815    1.0955e-07
    Residual    572.89    21    27.281
```

11.11.2 Two categorical predictors

Now what if in another case we have two categorical predictors?

Gender	Degree	y = Salary
Male	Bachelors	90,000
Female	Masters	85,000
Female	PhD	1,40,000
Male	Masters	1,25,000
Male	Bachelors	55,000
Female	Bachelors	62,000
Male	PhD	1,29,000
Female	Bachelors	1,20,000
Female	Masters	1,17,000
Male	Bachelors	98,000

y = Salary	x_1 = Gender	x_2 = Degree	D_1	D_2
90,000	0	Bachelors	0	0
55,000	0	Bachelors	0	0
62,000	1	Bachelors	0	0
1,20,000	1	Bachelors	0	0

y = Salary	x_1 = Gender	x_2 = Degree	D_1	D_2
98,000	0	Bachelors	0	0
85,000	I	Masters	I	0
1,25,000	0	Masters	I	0
1,17,000	I	Masters	I	0
1,40,000	I	PhD	0	I
1,29,000	0	PhD	0	I

$$90,000 = \beta_0 + \beta_1(0) + \beta_2(0) + \beta_3(0) + \varepsilon_1 \qquad 85,000 = \beta_0 + \beta_1(1) + \beta_2(1) + \beta_3(0) + \varepsilon_6$$

$$55,000 = \beta_0 + \beta_1(0) + \beta_2(0) + \beta_3(0) + \varepsilon_2 \qquad 125,000 = \beta_0 + \beta_1(0) + \beta_2(1) + \beta_3(0) + \varepsilon_7$$

$$62,000 = \beta_0 + \beta_1(1) + \beta_2(0) + \beta_3(0) + \varepsilon_3 \qquad 117,000 = \beta_0 + \beta_1(1) + \beta_2(1) + \beta_3(0) + \varepsilon_8$$

$$120,000 = \beta_0 + \beta_1(1) + \beta_2(0) + \beta_3(0) + \varepsilon_4 \qquad 140,000 = \beta_0 + \beta_1(1) + \beta_2(0) + \beta_3(1) + \varepsilon_9$$

$$98,000 = \beta_0 + \beta_1(0) + \beta_2(0) + \beta_3(0) + \varepsilon_5 \qquad 129,000 = \beta_0 + \beta_1(0) + \beta_2(0) + \beta_3(1) + \varepsilon_{10}$$

$$
\begin{bmatrix}
90000 \\ 55000 \\ 62000 \\ 120000 \\ 98000 \\ 85000 \\ 125000 \\ 117000 \\ 140000 \\ 129000
\end{bmatrix}
=
\begin{bmatrix}
1 & 0 & 0 & 0 \\
1 & 0 & 0 & 0 \\
1 & 1 & 0 & 0 \\
1 & 1 & 0 & 0 \\
1 & 0 & 0 & 0 \\
1 & 1 & 1 & 0 \\
1 & 0 & 1 & 0 \\
1 & 1 & 1 & 0 \\
1 & 1 & 0 & 1 \\
1 & 0 & 0 & 1
\end{bmatrix}
\begin{bmatrix}
\beta_0 \\ \beta_1 \\ \beta_2 \\ \beta_3
\end{bmatrix}
+
\begin{bmatrix}
\varepsilon_1 \\ \varepsilon_2 \\ \varepsilon_3 \\ \varepsilon_4 \\ \varepsilon_5 \\ \varepsilon_6 \\ \varepsilon_7 \\ \varepsilon_8 \\ \varepsilon_9 \\ \varepsilon_{10}
\end{bmatrix}
$$

$$y = X\beta + \varepsilon$$

$$
y = \begin{bmatrix}
90000 \\ 55000 \\ 62000 \\ 120000 \\ 98000 \\ 85000 \\ 125000 \\ 117000 \\ 140000 \\ 129000
\end{bmatrix}, \quad
X = \begin{bmatrix}
1 & 0 & 0 & 0 \\
1 & 0 & 0 & 0 \\
1 & 1 & 0 & 0 \\
1 & 1 & 0 & 0 \\
1 & 0 & 0 & 0 \\
1 & 1 & 1 & 0 \\
1 & 0 & 1 & 0 \\
1 & 1 & 1 & 0 \\
1 & 1 & 0 & 1 \\
1 & 0 & 0 & 1
\end{bmatrix}, \quad
\beta = \begin{bmatrix}
\beta_0 \\ \beta_1 \\ \beta_2 \\ \beta_3
\end{bmatrix}, \quad
\varepsilon = \begin{bmatrix}
\varepsilon_1 \\ \varepsilon_2 \\ \varepsilon_3 \\ \varepsilon_4 \\ \varepsilon_5 \\ \varepsilon_6 \\ \varepsilon_7 \\ \varepsilon_8 \\ \varepsilon_9 \\ \varepsilon_{10}
\end{bmatrix}
$$

$$\beta = (X'X)^{-1} X'y$$

$$X'X = \begin{bmatrix} 10 & 5 & 3 & 2 \\ 5 & 5 & 2 & 1 \\ 3 & 2 & 3 & 0 \\ 2 & 1 & 0 & 2 \end{bmatrix}$$

$$\beta = \begin{bmatrix} 84746.48 \\ 633.803 \\ 23830.986 \\ 49436.62 \end{bmatrix}$$

$$\hat{y} = 84746.48 + 633.803\left(\text{Gender}_{\text{Female}}\right) + 23830.986\left(\text{Degree}_{\text{Masters}}\right) + 49436.62\left(\text{Degree}_{\text{PhD}}\right)$$

Sum of Squares

Using the model lets now calculate the predicted values

$x_1 = $ Gender	$x_2 = $ Degree		y_i	$\hat{y}_i$
	D_1	D_2		
0	0	0	90,000	84746.48
0	0	0	55,000	109211.3
1	0	0	62,000	134816.9
1	0	0	1,20,000	108577.5
0	0	0	98,000	84746.48
1	1	0	85,000	85380.28
0	1	0	1,25,000	134183.1
1	1	0	1,17,000	85380.28
1	0	1	1,40,000	109211.3
	D_1	D_2		

Lets now calculate the sum of squares

y_i	$\hat{y}_i$	$y_i - \hat{y}_i$	$\left(y_i - \hat{y}_i\right)^2$	$y_i - \bar{y}$	$\left(y_i - \bar{y}\right)^2 \times 10^4$	$\hat{y}_i - \bar{y}$	$\left(\hat{y}_i - \bar{y}\right)^2$
90,000	84746.48	5,254	27599472.39	-12100	14641	-17353.5	301144656.4
55,000	109211.3	-24,211	58,61,85,547	-17100	29241	7111.269	50570146.79
62,000	134816.9	5,183	2,68,64,495	37900	143641	32716.9	1070395742
1,20,000	108577.5	16,423	26,96,99,623	22900	52441	6477.466	41957565.78
98,000	84746.48	-29,746	88,48,53,072	-47100	221841	-17353.5	301144656.4
85,000	85380.28	-23,380	54,66,37,633	-40100	160801	-16719.7	279548936.6
1,25,000	134183.1	-5,183	2,68,64,526	26900	72361	32083.1	1029325306
1,17,000	85380.28	34,620	1,19,85,24,805	17900	32041	-16719.7	279548936.6
1,40,000	109211.3	7,789	6,06,64,331	14900	22201	7111.269	50570146.79
1,29,000	84746.48	13,254	17,56,55,792	-4100	1681	-17353.5	301144656.4
			$\Sigma 3803549296$		$\Sigma 750890$		$\Sigma 3705350749$

The three sum of squares
SSR/ESS=3705350749
RSS/SSE =3803549296
SST/TSS=7508900000

Errors in the Parameters of the Regression Line:

$$RSSE = \sqrt{\frac{3803549296}{10-3-1}} = 25177.865$$

$$(X'X)^{-1} = \begin{bmatrix} 0.2676 & -0.169 & -0.155 & -0.1831 \\ -0.169 & 0.422535 & -0.112676 & -0.0422535 \\ -0.155 & -0.112676 & 0.5634 & 0.2112676 \\ -0.1831 & -0.0422535 & 0.2112676 & 0.70422535 \end{bmatrix}$$

SE of $\beta_0 = 25177.865 \times \sqrt{0.2676} = 13024.5268429$

SE of $\beta_1 = 25177.865 \times \sqrt{0.422535} = 16366.29$

SE of $\beta_2 = 25177.865 \times \sqrt{0.5634} = 18898.50$

SE of $\beta_3 = 25177.865 \times \sqrt{0.70422535} = 21128.795$

Let us study β_1, which represents the effect of gender on the salary

$$t \text{ statistics for } \beta_1 = \frac{633.803}{16366.29} = 0.038726$$

Lets now study the effect of Degree on the salary

$$t \text{ statistics for } \beta_2 = \frac{23830.986}{18398.50} = 1.261$$

$$t \text{ statistics for } \beta_3 = \frac{49436.62}{21128.795} = 2.3398$$

Similarly,

$$t \text{ statistics for } \beta_0 = \frac{84746.48}{13024.5268429} = 6.5066$$

Since there are 10 observations and 4 parameters in the model, the residual sum of squares has 6 degrees of freedom. The probability value of this test statistic for a two-sided alternative

$$P(|T| > 6.5066) = 2P(T > 6.5066) = < 0.001$$

$$P\big(|T| > 0.038726\big) = 2P\big(T > 0.038726\big) \approx \leq 1.00$$

$$P\big(|T| > 1.261\big) = 2P\big(T > 1.261\big) \approx 0.250$$

$$P\big(|T| > 2.3398\big) = 2P\big(T > 2.3398\big) \approx 0.06$$

	Estimate	SE	tStat	pValue
Intercept	**84746.48**	**13,025**	**6.5066**	**<0.001**
Gender_Female	**633.803**	**16,366**	**0**	**≈1.00**
°Degree_Masters	**23830.99**	**18,899**	**1**	**≈0.250**
Degree_PhD	49436.62	21,129	2	≈0.06

f – test

$$MSE = \frac{3803549296}{6} = 633924882.66666$$

$$MSR = \frac{3705350749}{3} = 1235116916.33333$$

$$MST = \frac{7508900000}{9} = 834322222.22222$$

$$f = \frac{1235116916.33333}{633924882.66666} = 1.94836$$

THE ANOVA TABLE

Source	Sum of Squares	df	Mean Squares	F	P
Total	7508900000	9	834322222.2		
Model	3705350749	3	1,23,51,16,916	1.94836	
Residual	3803549296	6	63,39,24,883		

coefficient of determination

$$-\text{Ordinary } R^2 = 1 - \frac{3803549296}{7508900000} = 0.493$$

$$-\text{adjusted } R^2 = 1 - \left(\frac{9}{6}\right)\frac{3803549296}{7508900000} = 0.24$$

Model explains around 49.3% of the variability in the response variable.

Root Mean Squared Error

$$RMSE = \sqrt{633924882.66666} = 25177.865$$

Example

x_1 = Gender	x_2 = Education Level	x_3 =Age	y = Salary
Male	Bachelors	32	90,000
Female	Masters	28	85,000
Female	PhD	48	1,40,000
Male	Masters	45	1,25,000
Male	Bachelors	29	55,000
Female	Bachelors	36	62,000
Male	PhD	43	1,29,000
Female	Bachelors	38	1,20,000
Female	Masters	40	1,17,000
Male	Bachelors	41	98,000

$$90,000 = \beta_0 + \beta_1(0) + \beta_2(0) + \beta_3(0) + \beta_4(32) + \varepsilon_1$$

$$55,000 = \beta_0 + \beta_1(0) + \beta_2(0) + \beta_3(0) + \beta_4(29) + \varepsilon_2$$

$$62,000 = \beta_0 + \beta_1(1) + \beta_2(0) + \beta_3(0) + \beta_4(36) + \varepsilon_3$$

$$120,000 = \beta_0 + \beta_1(1) + \beta_2(0) + \beta_3(0) + \beta_4(38) + \varepsilon_4$$

$$98,000 = \beta_0 + \beta_1(0) + \beta_2(0) + \beta_3(0) + \beta_4(41) + \varepsilon_5$$

$$85,000 = \beta_0 + \beta_1(1) + \beta_2(1) + \beta_3(0) + \beta_4(28) + \varepsilon_6$$

$$125,000 = \beta_0 + \beta_1(0) + \beta_2(1) + \beta_3(0) + \beta_4(45) + \varepsilon_7$$

$$117,000 = \beta_0 + \beta_1(1) + \beta_2(1) + \beta_3(0) + \beta_4(40) + \varepsilon_8$$

$$140,000 = \beta_0 + \beta_1(1) + \beta_2(0) + \beta_3(1) + \beta_4(48) + \varepsilon_9$$

$$129,000 = \beta_0 + \beta_1(0) + \beta_2(0) + \beta_3(1) + \beta_4(43) + \varepsilon_{10}$$

$$\begin{bmatrix} 90000 \\ 55000 \\ 62000 \\ 120000 \\ 98000 \\ 85000 \\ 125000 \\ 117000 \\ 140000 \\ 129000 \end{bmatrix} = \begin{bmatrix} 1 & 0 & 0 & 32 \\ 1 & 0 & 0 & 39 \\ 1 & 1 & 0 & 36 \\ 1 & 1 & 0 & 38 \\ 1 & 0 & 0 & 41 \\ 1 & 1 & 1 & 28 \\ 1 & 0 & 1 & 45 \\ 1 & 1 & 1 & 40 \\ 1 & 1 & 0 & 48 \\ 1 & 0 & 0 & 43 \end{bmatrix} \begin{bmatrix} \beta_0 \\ \beta_1 \\ \beta_2 \\ \beta_3 \end{bmatrix} + \begin{bmatrix} \varepsilon_1 \\ \varepsilon_2 \\ \varepsilon_3 \\ \varepsilon_4 \\ \varepsilon_5 \\ \varepsilon_6 \\ \varepsilon_7 \\ \varepsilon_8 \\ \varepsilon_9 \\ \varepsilon_{10} \end{bmatrix}$$

$$y = X\beta + \varepsilon$$

$$y = \begin{bmatrix} 90000 \\ 55000 \\ 62000 \\ 120000 \\ 98000 \\ 85000 \\ 125000 \\ 117000 \\ 140000 \\ 129000 \end{bmatrix}, \quad X = \begin{bmatrix} 1 & 0 & 0 & 32 \\ 1 & 0 & 0 & 39 \\ 1 & 1 & 0 & 36 \\ 1 & 1 & 0 & 38 \\ 1 & 0 & 0 & 41 \\ 1 & 1 & 1 & 28 \\ 1 & 0 & 1 & 45 \\ 1 & 1 & 1 & 40 \\ 1 & 1 & 0 & 48 \\ 1 & 0 & 0 & 43 \end{bmatrix}, \quad \beta = \begin{bmatrix} \beta_0 \\ \beta_1 \\ \beta_2 \\ \beta_3 \end{bmatrix}, \quad \varepsilon = \begin{bmatrix} \varepsilon_1 \\ \varepsilon_2 \\ \varepsilon_3 \\ \varepsilon_4 \\ \varepsilon_5 \\ \varepsilon_6 \\ \varepsilon_7 \\ \varepsilon_8 \\ \varepsilon_9 \\ \varepsilon_{10} \end{bmatrix}$$

$$\beta = \left(X'X\right)^{-1} X'y$$

$$X'X = \begin{bmatrix} 10 & 5 & 3 & 2 & 380 \\ 5 & 5 & 2 & 1 & 190 \\ 3 & 2 & 3 & 0 & 113 \\ 2 & 1 & 0 & 2 & 91 \\ 380 & 190 & 113 & 91 & 14848 \end{bmatrix}$$

$$\beta = \begin{bmatrix} -15281.22242 \\ 2105.97 \\ 16470.16 \\ 20192.244 \\ 2824.967 \end{bmatrix}$$

$$\hat{y} = -15281.22242 + 2105.97\left(\text{Gender}_{\text{Female}}\right) + 16470.16\left(\text{Degree}_{\text{Masters}}\right)$$

$$+20192.244\left(\text{Degree}_{\text{PhD}}\right) + 2824.967\left(\text{Age}\right)$$

Sum of Squares

Using the model lets now calculate the predicted values

x_1 = Gender	x_2 = Degree		x_3 = Age	y_i	$\hat{y}_i$
	D_1	D_2			
0	0	0	32	90,000	75117.72
0	0	0	29	55,000	66642.82
1	0	0	36	62,000	88523.56
1	0	0	38	1,20,000	94173.49
0	0	0	41	98,000	100542.4
1	1	0	28	85,000	82393.98
0	1	0	45	1,25,000	128312.5
1	1	0	40	1,17,000	116293.6
1	0	1	48	1,40,000	142615.4
0	0	1	43	1,29,000	126384.6

Lets now calculate the sum of squares

$y_i - \hat{y}_i$	$\left(y_i - \hat{y}_i\right)^2$	$y_i - \bar{y}$	$\left(y_i - \bar{y}\right)^2$	$\hat{y}_i - \bar{y}$	$\left(\hat{y}_i - \bar{y}\right)^2$
14,882	221482211	-12,100	14,64,10,000	-26982.3	728043348.7
-11,643	135555271	-47,100	2,21,84,10,000	-35,457	1257211572
-26,524	703499213	-40,100	1,60,80,10,000	-13,576	184319734.5
25,827	667008434	17,900	32,04,10,000	-7,927	62829504.03
-2,542	6463922.75	-4,100	1,68,10,000	-1,558	2426041.19
2,606	6791321.58	-17,100	29,24,10,000	-19,706	388327083.1
-3,312	10972342.1	22,900	52,44,10,000	26,212	687092670.3
706	499018.507	14,900	22,20,10,000	14,194	201457928.4
-2,615	6840356.81	37,900	1,43,64,10,000	40,515	1641498251
2,615	6840303.67	26,900	72,36,10,000	24,285	589741922.5
	$\Sigma 1765952394$		$\Sigma 7508900000$		$\Sigma 5742948056$

The three sum of squares
SSR/ESS=5742948056
RSS/SSE =1765952394
SST/TSS=7508900000

Errors in the Parameters of the Regression Line:

$$RSSE = \sqrt{\frac{1765952394}{10-3-1}} = 18793.36262$$

$$\left(X'X\right)^{-1} = \begin{bmatrix} 5.1781 & -0.2413 & 0.2064 & 1.2525 & -0.1387 \\ -0.2413 & 0.4236 & -0.1180 & -0.0634 & 0.0020 \\ 0.2064 & -0.1180 & 0.5900 & 0.3169 & -0.0102 \\ 1.2525 & -0.0634 & 0.3169 & 1.1240 & -0.0405 \\ -0.1387 & 0.0020 & -0.0102 & -0.0405 & 0.0039 \end{bmatrix}$$

$$SE \text{ of } \beta_0 = 18793.36262 \times \sqrt{5.1781} = 42765.1215018$$

$$SE \text{ of } \beta_1 = 18793.36262 \times \sqrt{0.4236} = 12231.5774421$$

$$SE \text{ of } \beta_2 = 18793.36262 \times \sqrt{0.5900} = 14435.4557377$$

$$SE \text{ of } \beta_3 = 18793.36262 \times \sqrt{1.1240} = 19924.5099791$$

$$SE \text{ of } \beta_4 = 18793.36262 \times \sqrt{0.0039} = 1176.13919111$$

Hypothesis Testing of the Significance of Regression Coefficients

Let us study β_1, which represents the effect of gender on the salary

$$t \text{ statistics for } \beta_1 = \frac{2105.97}{12231.5774421} = 0.17217484907$$

Lets now study the effect of Degree on the salary

$$t \text{ statistics for } \beta_2 = \frac{16470.16}{14435.4557377} = 1.14095185488$$

$$t \text{ statistics for } \beta_3 = \frac{20192.244}{19924.5099791} = 1.0134374206$$

Finally, we study the effect of Age on the salary

$$t \text{ statistics for } \beta_4 = \frac{2824.967}{1176.13919111} = 2.40189853493$$

Similarly,

$$t \text{ statistics for } \beta_0 = \frac{-15281.22242}{42765.1215018} = -0.35732910099$$

Since there are 10 observations and 5 parameters in the model, the residual sum of squares has 5 degrees of freedom. The probability value of this test statistic for a two-sided alternative

$$P\left(|T| > -0.35733\right) = 2P\left(T > 0.35733\right) \approx 0.75$$

$$P\left(|T| > 0.172175\right) = 2P\left(T > 0.172175\right) \approx 0.90$$

$$P(|T| > 1.141) = 2P(T > 1.141) \approx 0.3$$

$$P\left(|T| > 1.01344\right) = 2P\left(T > 1.01344\right) \approx 0.35$$

$$P\left(|T| > 2.402\right) = 2P\left(T > 2.402\right) \approx 0.06$$

	Estimate	SE	tStat	pValue
Intercept	-15281.222	42,765	0	≈0.75
Gender_Female	2105.97	12,232	0	≈0.90
°Degree_Masters	16470.16	14,435	I	≈0.30
Degree_PhD	20192.244	19,925	I	≈0.35
Age	2824.967	1,176	2	≈0.06

f − test

$$MSE = \frac{1765952394}{5} = 353190478.8$$

$$MSR = \frac{5742948056}{4} = 1435737014$$

$$MST = \frac{7508900000}{9} = 834322222.22222$$

$$f = \frac{1435737014}{353190478.8} = 4.065$$

THE ANOVA TABLE

Source	Sum of Squares	df	Mean Squares	F	P
Total	7508900000	9	83,43,22,222		
Model	5742948056	4	1,43,57,37,014	4	
Residual	1765952394	5	35,31,90,479		

coefficient of determination

$$-\text{Ordinary } R^2 = 1 - \frac{1765952394}{7508900000} = 0.76482$$

$$-\text{adjusted } R^2 = 1 - \left(\frac{9}{4}\right)\frac{1765952394}{7508900000} = 0.576674$$

Model explains around 76.482 % of the variability in the response variable.

Root Mean Squared Error

$$RMSE = \sqrt{353190478.8} = 18793.3626262$$

Calculations in MATLAB

```
data = readtable("Salary.xlsx");
Gender = data.Gender;
Degree = data.Degree;
Age = data.Age;
Salary = data.Salary;
DATA = table(Gender, Degree, Age, Salary);

model = fitlm(DATA)
anova(model,'summary')
```

model =

Linear regression model:
    Salary ~ 1 + Gender + Degree + Age

Estimated Coefficients:

	Estimate	SE	tStat	pValue
(Intercept)	-15281	42765	-0.35733	0.73543
Gender_Female	2106	12232	0.17217	0.87005
Degree_Masters	16470	14435	1.141	0.30556
Degree_PhD	20192	19924	1.0135	0.35734
Age	2825	1176.1	2.4019	0.061476

Number of observations: 10, Error degrees of freedom: 5
Root Mean Squared Error: 1.88e+04
R-squared: 0.765,   Adjusted R-Squared: 0.577
F-statistic vs. constant model: 4.07, p-value = 0.0781

ans =

3x5 table

	SumSq	DF	MeanSq	F	pValue
Total	7.5089e+09	9	8.3432e+08		
Model	5.7429e+09	4	1.4357e+09	4.065	0.07811
Residual	1.766e+09	5	3.5319e+08		

11.12 INTERACTION IN LINEAR REGRESSION

All the Linear Regression models we have encountered so far, each regression coefficient is interpreted as the effect of a predictor while holding other predictors at fixed values and the effect does not depend on at which specific values you hold the other predictors.

Consider the first multiple linear regression model where we fit a linear model for the Salary data with only two continuous predictors Age and Experience.

$$\hat{y} = 23582.8 + 145.56539(\text{Age}) + 1926.9(\text{Experience})$$

According to this model,

- every one-unit increase in **Age**, income is expected to increase by 145.56539 units, regardless of the **Experience i.e.** holding Experience constant. The presence of **Experience** in the model does not change this interpretation.
- Similarly, for every one-unit increase in **Experience**, income is expected to increase by 1926.9 units, regardless of the Age **i.e.** holding Experience constant. The presence of **Age** in the model does not change this interpretation.
- At **Age** = 0 and **Experience** = 0, we expect Income to be 23582.8

Understand that often its possible that the effect of one of the predictors on **Salary** is dependent on the values of another predictor i.e the effect of **Age** on **Salary** might be dependent on Experience. This dependency is known in statistics as an interaction effect. An interaction between two predictors is referred to as a **two-way interaction**. It is possible to have three-way (or more way) interactions, but this section will only discuss two-way interactions.

Without interaction our model looks like:

$$\hat{y} = \beta_0 + \beta_1(\text{Age}) + \beta_2(\text{Experience})$$

An interaction term is effectively a multiplication of Age and Experience. The following equation presents the model's new specification:

$$\hat{y} = \beta_0 + \beta_1(\text{Age}) + \beta_2(\text{Experience}) + \beta_3 \times Age \times Experience$$

Here the interaction term enables you to examine whether the relationship between the response variable y and the independent variable changes depending on the value of another independent variable.

Interaction terms are a crucial component of regression analysis, and understanding how they work can help practitioners better train models and interpret their data. Despite their importance, however, interaction terms can be difficult to understand.

11.12.1 Calculating the Coefficient of Interaction Term

Lets use our matrix to calculate the new coefficients when an interaction term is added to the model.

$$
X = \begin{bmatrix}
& x_1 & x_2 & x_1x_2 \\
1 & 25 & 2 & 50 \\
1 & 30 & 3 & 90 \\
1 & 47 & 12 & 564 \\
. & . & . & . \\
. & . & . & . \\
. & . & . & . \\
1 & 37 & 5 & 185 \\
1 & 39 & 8 & 312 \\
1 & 29 & 3 & 87 \\
1 & 47 & 9 & 423 \\
. & . & . & . \\
. & . & . & . \\
. & . & . & . \\
1 & 23 & 1 & 23 \\
1 & 44 & 9 & 396 \\
1 & 37 & 10 & 370
\end{bmatrix}, y = \begin{bmatrix}
31450 \\
35670 \\
51580 \\
. \\
. \\
40250 \\
45150 \\
29840 \\
46110 \\
. \\
. \\
26870 \\
44190 \\
48700
\end{bmatrix}
$$

$$\beta = \left(X'X\right)^{-1} X'y$$

$$
X'X = \begin{bmatrix}
20 & 793 & 142 & 6186 \\
793 & 33353 & 6186 & 279684 \\
142 & 6186 & 1252 & 57172 \\
6186 & 279684 & 57172 & 2681402
\end{bmatrix}
$$

$$
\left(X'X\right)^{-1} = \begin{bmatrix}
3.7075211 & -0.110182663532 & -0.3917669396 & 0.0112924888 \\
-0.110182663532 & 0.003647906 & 0.00912290203 & -0.0003208193 \\
-0.3917669396 & 0.00912290203013 & 0.0887078446705 & -0.00193916 \\
0.0112924888 & -0.0003208193 & -0.00193916 & 0.0000491303713
\end{bmatrix}
$$

$$
\beta = \begin{bmatrix}
26309.92154 \\
68.0891 \\
1458.6023 \\
11.86475
\end{bmatrix}
$$

The regression line

$$\hat{y} = 26309.92154 + 68.0891(\text{Age}) + 1458.6023(\text{Experience}) + 11.86475 \times Age \times Experience$$

11.12.2 Interpreting interactions

An interaction occurs when an independent variable has a different effect on the outcome depending on the values of another independent variable. Adding interaction terms to a model drastically changes the interpretation of all the coefficients. Without an interaction term for two predictors, the interpretation of coefficients:

- The intercept is the expected value of the response variable when two predictors X and Z equal zero.
- The slope coefficient on each predictor is the expected difference in the outcome variable for a one-unit increase of the predictor, holding all other predictors constant.

When the model contains an interaction, the magnitude and direction of the relation between one continuous predictor and the outcome depends on the level of the other predictor in the model spanning from very low to very high values.

To understand this, lets observe the above the model without interaction term.

$$\text{Income} = \beta_0 + \beta_1 (\text{Age}) + \beta_2 (\text{Experience}) + \varepsilon$$

The above equation, also known as additive model, investigates only the main effects of predictors. It assumes that the relationship between a given predictor variable and the outcome is independent of the other predictor variables (James et al. 2014,P. Bruce and Bruce (2017)). Considering our example, the additive model assumes that, the effect of Age on Income is independent of the effect of Experience. Without an interaction term, this regression equation shows that

- when age and experience are 0, the expected income is β_0.
- One additional year increase in age will increase salary by β_1,dollars on average, holding experience constant.
- One additional year increase in experience will increase salary by β_2 dollars, on average, holding age constant

Once we add the interaction term, our model looks like:

$$\text{Income} = \beta_0 + \beta_1 (\text{Age}) + \beta_2 (\text{Experience}) + \beta_3 \times \text{Age} \times \text{Experience} + \varepsilon$$

The above equation can be written as:

$$\text{Income} = \beta_0 + (\beta_1 + \beta_3 \times \text{Experience}) \times \text{Age} + \beta_2 (\text{Experience})$$

This interaction now means that the effect of Age on Income is different for different years of Experience. So, the unique effect of Age on Income is not limited to β_1. It also depends on the values of β_3 and Experience. This implies that the unique effect of Age is represented by everything that is multiplied by Age in the model. For any fixed value of Experience, say Experience$_0$, notice that the effect for Age is given by

$$\beta_1 + \beta_3 \times \text{Experience}$$

This means that the effect of Age changes depending on the Experience of the individual, so that there is really no "unique" effect of Age, it is different for each possible Experience value.

For example, for someone with 2 years of Experience, the effect of Age is

$$\beta_1 + \beta_3 \times 2$$

and for someone with Experience of 12 years it is:

$$\beta_1 + \beta_3 \times 12$$

and so on.

Similarly our equation can be rewritten as

$$\text{Income} = \beta_0 + \beta_1 \left(\text{Age}\right) + \left(\beta_2 + \beta_3 \times \text{Age}\right) \times \text{Experience}$$

The effect of Experience is similarly affected by Age. If the Age is, say, Age$_0$, then the effect of Experience becomes:

$$\beta_2 + \beta_3 \times \text{Age}$$

After adding our interaction term, from our calculations we found:

$$\hat{y} = 26309.92154 + 68.0891\left(\text{Age}\right) + 1458.6023\left(\text{Experience}\right) + 11.86475 \times \text{Age} \times \text{Experience}$$

So how do we interpret these new coefficients? We say:

- For every 1 unit increase in Age, Income increases by 68.0891 (holding Experience constant)
- For every 1 unit increase in Experience, Income increases by 1458.6 (holding Age constant)
- At 0 Age and 0 Experience, we expect Income to be 26309.92154
- For every 1 unit increase in Age, Income changes by $68.0891 + 11.86475 \times \text{Experience}$
- For every 1 unit increase in Experience, Income changes by $1458.6023 + 11.86475 \times \text{Age}$

To summarize the slope on Age:Experience means that the effect of Age on Income increases by 11.86475 for every unit increase in Experience. This also works the other way round.

11.12.3 Visualizing interactions

If our moderator is a categorical variable, plotInteraction will calculate the conditional effect for all categories of the moderator. If the moderator is numeric, plotInteraction will compute the conditional effect for three specific values: the minimum, maximum, and the average of the minimum and maximum values.

For instance, let's take Age as the independent variable plotted on the x-axis and designate Experience as the moderator variable. Considering that the minimum and maximum values of Experience are 1 and 13 respectively, the average would be calculated as $(1 + 13) / 2 = 14 / 2 = 7$. Consequently, we have three levels of Experience: 1, 7, and 13.

Lets compute the slope for **Income** on **Age** while holding the value of the moderator variable, **Experience**, constant at values running 1, 7 and 13.

If Experience= 1, $\text{Salary} = 23582.8 + 145.56539(\text{Age}) + 1926.9 \times 1$

$$= 25509.7 + 145.56539(\text{Age})$$

For Experience= 7, $\text{Salary} = 37071.1 + 145.56539(\text{Age})$

Finally for Experience= 13, $\text{Salary} = 48632.5 + 145.56539(\text{Age})$

So, depending on different levels of **Experience**, we get different straight lines with the same slope (parallel lines), which suggests that regardless of the level of **Experience**, a one unit increase in **Age** is, on average, associated with 145.56539 units increase in **Salary**. In other words, this model assumes that the association between **Age** and **Salary** does not depend on the levels of **Experience therefore** the slope of Age on the Income does not change as the values of Experience changes.

Similarly, lets now take Age as the moderator variable. Considering that the minimum and maximum values of Age are 23 and 58, the average would be calculated as $(23 + 58) / 2 = 40.5$. Consequently, we have three levels of Age: 23, 58, and 40.5.

Lets compute the slope for **Income** on **Experience** while holding the value of the moderator variable, **Age**, constant at values running 23, 40.5 and 58.

If Age= 23, $\text{Salary} = 23582.8 + 145.56539 \times 23 + 1926.9(\text{Experience})$

$$= 26930.80397 + 1926.9(\text{Experience})$$

For Age= 40.5, $\text{Salary} = 29478.1983 + 1926.9(\text{Experience})$

Finally for Age= 58, $\text{Salary} = 32025.59262 + 1926.9(\text{Experience})$

So, depending on three levels of **Age**, we get three straight lines with the same slope (parallel lines), which suggests that regardless of the level of **Age**, a one unit increase in **Experience** is, on average, associated with 1926.9 units increase in **Salary**. In other words, this model assumes that the association between **Experience** and **Salary** does not depend on the levels of **Age therefore** the slope of Experience on the Income does not change as the values of Age changes.

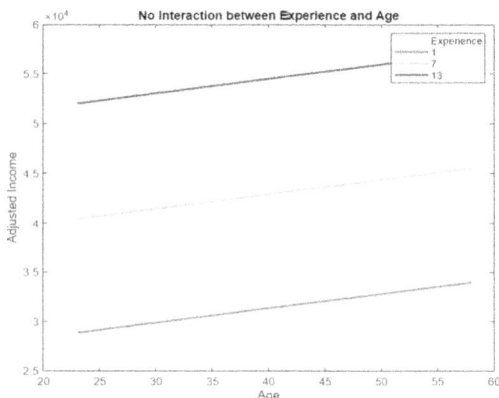

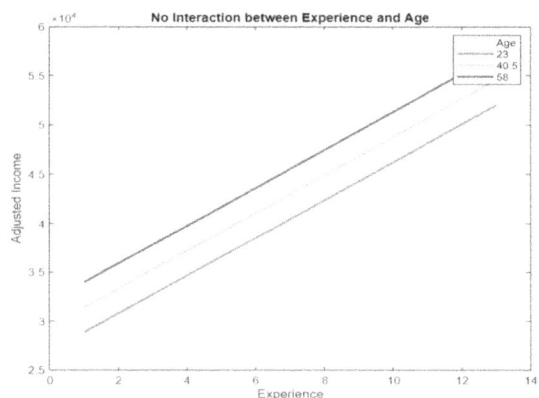

Lets repeat the same calculations for the interaction term
Experience = 1,

$$Salary = 26309.92154 + 68.0891(Age) + 1458.6023 \times 1 + 11.86475 \times Age \times 1$$

$$= 27768.52384 + 79.95385(Age)$$

So when Experience= 1 a one unit increase in Age will produce 79.95385 unit increase in Income

For Experience = 7, $Salary = 36520.13764 + 151.14235(Age)$

Finally for Experience = 13, $Salary = 45271.75144 + 222.33085(Age)$

For the model in above equations at different levels of **Experience**, we find different straight lines with different slopes, which reveals that the lines are not parallel. This model assumes that the positive association between **Age** and **Income** depends on the level of **Experience**. In other words, the slope of Age on the Income changes as the values of Experience changes and vice versa

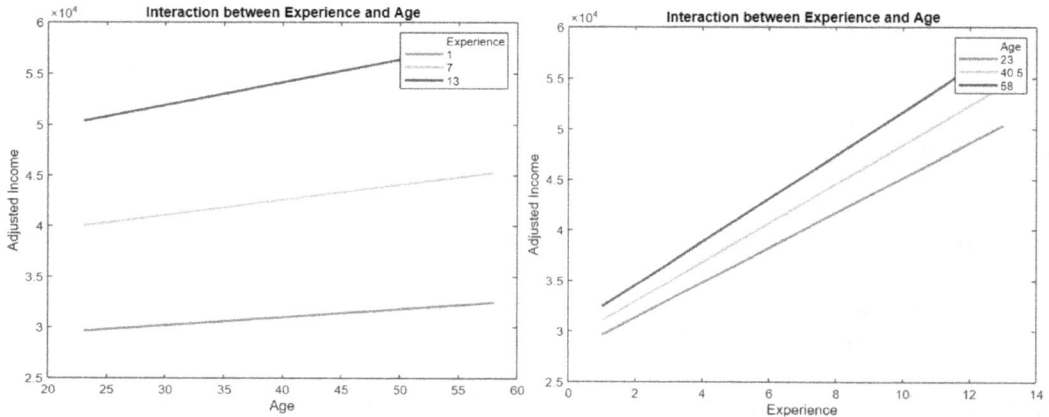

The interaction plot doesn't show significant differences when variables don't interact compared to when they do. Generally, older individuals tend to have more experience, leading to higher income.

11.12.4 Interaction between a continuous predictor and a categorical predictor

Lets now look at the interaction between one continuous and one dichotomous variable i.e Age and Gender. Previously, without interaction terms our model looked like

$$Income = \beta_0 + \beta_1(Gender_{Female}) + \beta_2(Age) + \varepsilon$$

For males i.e. $Gender_{Female} = 0$,

$$Income = \beta_0 + \beta_2(Age)$$

For females i.e. $Gender_{Female} = 1$,

$$\text{Income} = \beta_0 + \beta_1 + \beta_2(\text{Age})$$

Increasing gender by one unit simply moves us from male to female. Therefore

$$\beta_1 = (\beta_0 + \beta_1 + \beta_2(\text{Age})) - (\beta_0 + \beta_2(\text{Age}))$$

Thus β_1 is the difference in Income for females and males of the same age. In other words it is the difference in the intercepts of the two parallel lines which have slope β_2. On the other hand, β_2 represents the amount by which Income changes for each unit change in Age
    We saw our model to be

$$\text{Income} = -34625 + 5400(\textbf{Gender}_{\textbf{Female}}) + 3526.961(\text{Age})$$

- Keeping the Gender constant, a one unit increase in **Age** is, on average, associated with 3526.961 units increase in **Income**
- Keeping the Age constant, the average value of **Income** is 5400 units higher for females $Gender_{Female} = 1$ than males $Gender_{Female} = 0$.

Additionally, from the model, we observe that,
    If $Gender_{Female} = 0$
    the estimated regression equation for males is:

$$\text{Income} = -34625 + 3526.961(\text{Age})$$

And if $Gender_{Female} = 1$
    the estimated regression equation for females is:

$$\text{Income} = -29225 + 3526.961(\text{Age})$$

So, depending on the two Gender, we get two different straight lines with the same slope (parallel lines), which suggests no interaction effect, based on gender. In other words, regardless of Gender, a one unit increase in **Age** is, on average, associated with 3526.961 units increase in **Income**. This model assumes that the association between **Age** and **Income** does not depend on the Gender
    After adding the interaction term

$$\text{Income} = \beta_0 + \beta_1(Gender_{Female}) + \beta_2(\text{Age}) + \beta_3 \times Gender_{Female} \times \text{Age} + \varepsilon$$

There is no longer any unique effect of Age, because it depends upon whether you are talking about the effect of Age in males or females. In other words, the slope of the Age is different for males and females. Similarly, the difference between males and females depends on the Age.
    The interpretation of these coefficients is as follows:

β_0 = The mean income of the males when Age = 0

- females with Age = 0 will have salary have an average Income that is β_1 higher (if the coefficient is positive) than males.

- β_2 is the slope of the regression line for the male group. This is because the interaction term becomes 0 if $Gender_{Female} = 0$ is coded as 0. Therefore a year of increase in age is followed by β_2 increase in salary while a females should expect to earn an additional $\beta_2 + \beta_3$. In other words, the increase in salary due to age differs for men and women
- β_3 is the difference in slope between the male and female group. Therefore the slope for the female group would be $\beta_2 - \beta_3$.

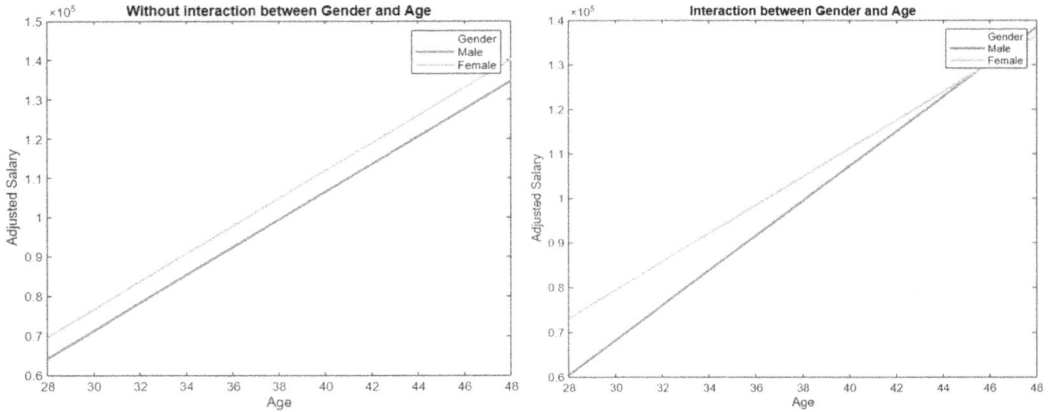

Examining the two graphs, the first depicts parallel lines, suggesting a model without an interaction term. Meanwhile, the second graph displays a model with an interaction term, where the regression lines are not parallel where the line for females falls above the line for males.

How could we tell that females are higher than males? If the coefficient for **female** is positive which tells us that the level for females is higher than for males so lets calculate the coefficients

Insert Equations

$$\beta = (X'X)^{-1} X'y$$

$$\beta = (X'X)^{-1} X'y$$

$$X'X = \begin{bmatrix} 10 & 5 & 380 & 190 \\ 5 & 5 & 190 & 190 \\ 380 & 190 & 14848 & 7428 \\ 190 & 190 & 7428 & 7428 \end{bmatrix}$$

$$(X'X)^{-1} = \begin{bmatrix} 7.42 & -7.42 & -0.19 & 0.19 \\ -7.42 & 14.5623077 & 0.19 & -0.3726923077 \\ -0.19 & 0.19 & 0.005 & -0.005 \\ 0.19 & -0.3716923077 & -0.005 & 0.0098076923077 \end{bmatrix}$$

$$\beta = \begin{bmatrix} -48610 \\ 32833.077 \\ 3895 \\ -721.923077 \end{bmatrix}$$

The regression line

$$\hat{y} = -48610 + 32833.077(Gender_{Female}) + 3895(Age) - 721.923077 \times Gender_{Female} \times Age$$

- $\beta_0 = -48610$ the expected value for **Income** when both $Gender_{Female}$ and **Age** = 0.
- The coefficient for **Age** is 3895 which is the slope of the regression line for the male group (i.e., female=0).
- The value for the **female by Age** interaction is -721.923077 which is the difference in slope between the male and female group, i.e., the slope for the female group would be about $3895 - 721.923077 = 3173.076923$

Therefore, the estimated regression equation for males ($Gender_{Female} = 0$) is:

$$Income = -48610 + 3895(Age)$$

and the estimated regression equation for females ($Gender_{Female} = 1$) is:

$$Income = -48610 + 32833.077 + 3895(Age) - 721.923077 \times Age$$

$$= -15776.923 + 3173.076923(Age)$$

For two genders, we find two straight lines with different slopes (3895 and 3173.076923), which reveals that the lines are not parallel. This model assumes that the positive association between **Age** and **Income** depends on the Gender. Therefore, we interpret the model coefficients as:

- Provided that $Gender_{Female} = 0$, a one unit increase in **Age** is, on average, associated with 3895 units increase in **Income**.
- Provided that $Gender_{Female} = 1$, a one unit increase in **Age** is, on average, associated with 3173.076923 units increase in **Income**.

The coefficient of the interaction term displays the difference in slope between the two lines (i.e., $3895 - 3173.076923 = 721.923077$

Calculations in MATLAB

When performing a regression in MATLAB using the fitlm function, we can easily include interaction terms between predictors by specifying the "interaction" term in the model formula.

```
data = readtable("Salary.xlsx");
Gender = data.Gender;
Age = data.Age;
Salary = data.Salary;
DATA = table(Gender, Age, Salary);

model = fitlm(DATA, "interactions")
plotInteraction(model,'Gender','Age','predictions')
title("Interaction between Gender and Degree")
```

model =

Linear regression model:
    Salary ~ 1 + Gender*Age

Estimated Coefficients:

	Estimate	SE	tStat	pValue
(Intercept)	-48610	53420	-0.90996	0.39792
Gender_Female	32833	74837	0.43873	0.67622
Age	3895	1386.7	2.8088	0.030805
Gender_Female:Age	-721.92	1942.2	-0.37171	0.72288

Number of observations: 10, Error degrees of freedom: 6
Root Mean Squared Error: 1.96e+04
R-squared: 0.693, Adjusted R-Squared: 0.539
F-statistic vs. constant model: 4.51, p-value = 0.0557

We use the plotInteraction function to visualize the interaction between predictors in a regression model. This function helps you explore how the relationship between the response variable and one predictor changes depending on the value of another predictor given above.

11.12.5 Interaction with two binary variables

Another situation when there can be an interaction between two categorical variables.
    Previously without interaction term we had the model

$$\text{Income} = \beta_0 + \beta_1(\text{Gender}_{\text{Female}}) + \beta_2\left(\text{Degree}_{\text{Masters}}\right) + \beta_3\left(\text{Degree}_{\text{PhD}}\right) + \varepsilon$$

Remember **Gender** is a two-level categorical variable and **Degree** is a three-level categorical variable,
    For males i.e. $\text{Gender}_{\text{Female}} = 0$,

$$\text{Income} = \beta_0 + \beta_1 \times 0 + \beta_2\left(\text{Degree}_{\text{Masters}}\right) + \beta_3\left(\text{Degree}_{\text{PhD}}\right) + \varepsilon$$

$$= \beta_0 + \beta_2\left(\text{Degree}_{\text{Masters}}\right) + \beta_3\left(\text{Degree}_{\text{PhD}}\right) + \varepsilon$$

- For Bachelors degree i.e $\text{Degree}_{\text{Masters}} = 0 = \text{Degree}_{\text{PhD}}$

$$\text{Income} = \beta_0 + \beta_2 \times 0 + 3_3 \times 0 = \beta_0$$

the predicted income for males who only have a Bachelor's degree $= \beta_0$

- If $\text{Degree}_{\text{Masters}} = 1$

$$\text{Income} = \beta_0 + \beta_2 \times 1 + \beta_3 \times 0 = \beta_0 + \beta_2$$

Therefore the predicted income for males with a Master's degree$= \beta_0 + \beta_2$

- If $\text{Degree}_{\text{PhD}} = 1$

$$\text{Income} = \beta_0 + \beta_2 \times 0 + \beta_3 \times 1 = \beta_0 + \beta_3$$

Therefore the predicted income for males with a PhD$= \beta_0 + \beta_3$
For females i.e. $\text{Gender}_{\text{Female}} = 1$,

$$\text{Income} = \beta_0 + \beta_1 \times 1 + \beta_2 \left(\text{Degree}_{\text{Masters}}\right) + \beta_3 \left(\text{Degree}_{\text{PhD}}\right) + \varepsilon$$

$$= \beta_0 + \beta_1 + \beta_2 \left(\text{Degree}_{\text{Masters}}\right) + \beta_3 \left(\text{Degree}_{\text{PhD}}\right) + \varepsilon$$

- For Bachelors degree i.e $\text{Degree}_{\text{Masters}} = 0 = \text{Degree}_{\text{PhD}}$

$$\text{Income} = \beta_0 + \beta_1 + \beta_2 \times 0 + \beta_3 \times 0 = \beta_0 + \beta_1$$

the predicted income for females who only have a Bachelor's degree $= \beta_0 + \beta_1$

- If $\text{Degree}_{\text{Masters}} = 1$

$$\text{Income} = \beta_0 + \beta_1 + \beta_2 \times 1 + \beta_3 \times 0 = \beta_0 + \beta_1 + \beta_2$$

So the predicted income for females with a Master's degree$= \beta_0 + \beta_1 + \beta_2$

- If $\text{Degree}_{\text{PhD}} = 1$

$$\text{Income} = \beta_0 + \beta_1 + \beta_2 \times 0 + \beta_3 \times 1 = \beta_0 + \beta_1 + \beta_3$$

Therefore the predicted income for females with a PhD$= \beta_0 + \beta_1 + \beta_3$
Notice,
$\beta_1 = \left(\beta_0 + \beta_1\right) - \beta_0$ which is the difference in Income between females and males with only Bachelors degree.
At the same time,
$\beta_1 = \left(\beta_0 + \beta_1 + \beta_2\right) - \left(\beta_0 + \beta_2\right)$ which is also the difference in income among females and males with Master's degree.
Also,
$\beta_1 = \left(\beta_0 + \beta_1 + \beta_3\right) - \left(\beta_0 + \beta_2\right)$ which is also the difference in income among females and males with PhDs

This means that the difference in predicted **Income** between the two Gender stays the same regardless of the Degree level

$\beta_2 = (\beta_0 + \beta_2) - \beta_0$ The difference in income among males between those with Bachelor's degrees and those with Master's degrees.

Also,

$\beta_2 = (\beta_0 + \beta_1 + \beta_2) - (\beta_0 + \beta_1)$ The difference in income among females between those with Bachelor's degrees and those with Master's degrees.

$\beta_3 = (\beta_0 + \beta_3) - \beta_0$ The difference in income among males between those with Bachelor's degrees and those with PhD degrees.

Likewise,

$\beta_3 = (\beta_0 + \beta_1 + \beta_3) - (\beta_0 + \beta_1)$ The difference in income among females between those with Bachelor's degrees and those with PhDs.

So in conclusion,

- β_1 = the difference in income among females and males holding the Degree constant
- β_2 = the difference in predicted **Income** between the Bachelors and Master's holders
- β_3 = the difference in predicted **Income** between the Bachelors and PhD holders,

Where β_2 and β_3 without accounting for the Gender

$$\hat{y} = 84746.48 + 633.803(\text{Gender}_{\text{Female}}) + 23830.986(\text{Degree}_{\text{Masters}}) + 49436.62(\text{Degr}$$

- The income of a male with a Bachelor's degree is $84746.48
- β_1 = Females earn $633.803 more than males in general.
- β_2 = Individuals with a Masters Degree earn $23830.986 more than Individuals with a Bachelors
- β_3 = Individuals with a PhD earn $49436.62 more than Individuals with a Bachelors

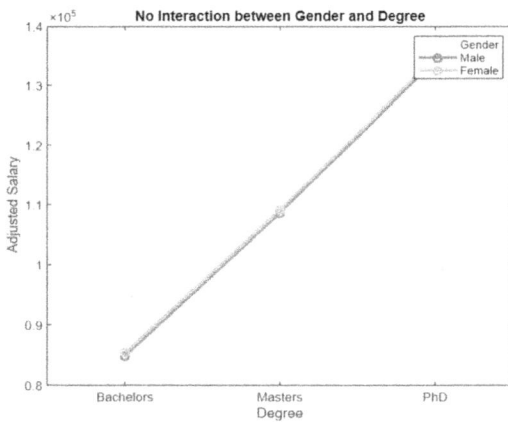

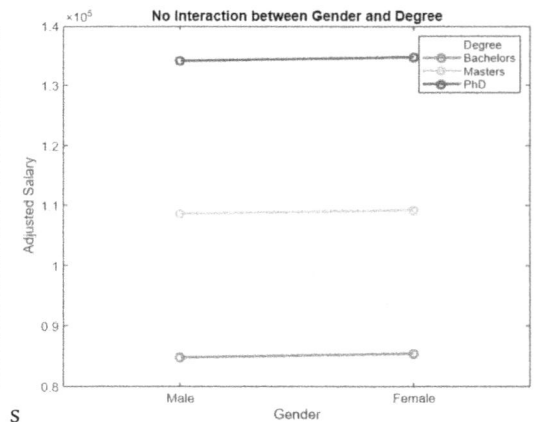

With interaction term we have the model

$$\text{Income} = \beta_0 + \beta_1(\text{Gender}_{\text{Female}}) + \beta_2(\text{Degree}_{\text{Masters}}) + \beta_3(\text{Degree}_{\text{PhD}}) +$$

$$\beta_4 \times \text{Gender}_{\text{Female}} \times \text{Degree}_{\text{Masters}} + \beta_5 \times \text{Gender}_{\text{Female}} \times \text{Degree}_{\text{PhD}} + \varepsilon$$

For males i.e. $Gender_{Female} = 0,$

$$Income = \beta_0 + \beta_1 \times 0 + \beta_2 \left(Degree_{Masters}\right) + \beta_3 \left(Degree_{PhD}\right) + \beta_4 \times 0 \times Degree_{Masters}$$

$$+\beta_5 \times 0 \times Degree_{PhD} + \varepsilon$$

$$= \beta_0 + \beta_2 \left(Degree_{Masters}\right) + \beta_3 \left(Degree_{PhD}\right) + \varepsilon$$

- For Bachelors degree i.e $Degree_{Masters} = 0 = Degree_{PhD}$

$$Income = \beta_0 + \beta_2 \times 0 + \beta_3 \times 0 = \beta_0$$

- the predicted income for males who only have a Bachelor's degree $= \beta_0$
- If $Degree_{Masters} = 1$

$$Income = \beta_0 + \beta_2 \times 1 + \beta_3 \times 0 = \beta_0 + \beta_2$$

Therefore the predicted income for males with a Master's degree$= \beta_0 + \beta_2$
If $Degree_{PhD} = 1$

$$Income = \beta_0 + \beta_2 \times 0 + \beta_3 \times 1 = \beta_0 + \beta_3$$

- Therefore the predicted income for males with a PhD$= \beta_0 + \beta_3$

For females i.e. $Gender_{Female} = 1,$

$$Income = \beta_0 + \beta_1 \times 1 + \beta_2 \left(Degree_{Masters}\right) + \beta_3 \left(Degree_{PhD}\right) + \beta_4 \times 1 \times Degree_{Masters}$$

$$\beta_5 \times 1 \times Degree_{PhD} + \varepsilon$$

$$= \beta_0 + \beta_1 + \beta_2 \left(Degree_{Masters}\right) + \beta_3 \left(Degree_{PhD}\right) + \beta_4 \left(Degree_{Masters}\right) + \beta_5 \left(Degree_{PhD}\right) + \varepsilon$$

- For Bachelors degree i.e $Degree_{Masters} = 0 = Degree_{PhD}$

$$Income = \beta_0 + \beta_1 + \beta_2 \times 0 + \beta_3 \times 0 + \beta_4 \times 0 + \beta_5 \times 0 = \beta_0 + \beta_1$$

- the predicted income for females who only have a Bachelor's degree $= \beta_0 + \beta_1$
- If $Degree_{Masters} = 1$

$$Income = \beta_0 + \beta_1 + \beta_2 \times 1 + \beta_3 \times 0 + \beta_4 \times 1 + \beta_5 \times 0 = \beta_0 + \beta_1 + \beta_2 + \beta_4$$

- So the predicted income for females with a Master's degree$= \beta_0 + \beta_1 + \beta_2 + \beta_4$
- If $Degree_{PhD} = 1$

$$Income = \beta_0 + \beta_1 + \beta_2 \times 0 + \beta_3 \times 1 + \beta_4 \times 0 + \beta_5 \times 1 = \beta_0 + \beta_1 + \beta_3 + \beta_5$$

- Therefore the predicted income for females with a PhD$= \beta_0 + \beta_1 + \beta_3 + \beta_5$

From the above equations we can also find,

β_1 = the difference in Income between females and males with only Bachelors degree.

β_2 = The difference in income among males between those with Bachelor's degrees and those with Master's degrees.

β_3 = The difference in income among males between those with Bachelor's degrees and those with PhD degrees.

- The difference in income among females between those with Bachelor's degrees and those with Master's degrees.

$$(\beta_0 + \beta_1 + \beta_2 + \beta_4) - (\beta_0 + \beta_1) = \beta_2 + \beta_4$$

- The difference in income among females between those with Bachelor's degrees and those with PhD

$$(\beta_0 + \beta_1 + \beta_3 + \beta_5) - (\beta_0 + \beta_1) = \beta_3 + \beta_5$$

- the difference in Income between females and males with Master's degree

$$(\beta_0 + \beta_1 + \beta_2 + \beta_4) - (\beta_0 + \beta_2) = \beta_1 + \beta_4$$

- the difference in Income between females and males with PhD degree

$$(\beta_0 + \beta_1 + \beta_3 + \beta_5) - (\beta_0 + \beta_3) = \beta_1 + \beta_5$$

- The Income gender gap among the Bachelor's degree and the Master's degree

$$(\beta_1 + \beta_4) - \beta_1 = \beta_4$$

- The Income gender gap among the Bachelor's degree and the PhD degree

$$(\beta_1 + \beta_5) - \beta_1 = \beta_5$$

Lets now calculate these coefficients:

$$y = \begin{bmatrix} 90000 \\ 55000 \\ 62000 \\ 120000 \\ 98000 \\ 85000 \\ 125000 \\ 117000 \\ 140000 \\ 129000 \end{bmatrix}, \quad X = \begin{bmatrix} & x_1 & D_1 & D_2 & x_1 D_1 & x_1 D_2 \\ 1 & 0 & 0 & 0 & 0 & 0 \\ 1 & 0 & 0 & 0 & 0 & 0 \\ 1 & 1 & 0 & 0 & 0 & 0 \\ 1 & 1 & 0 & 0 & 0 & 0 \\ 1 & 0 & 0 & 0 & 0 & 0 \\ 1 & 1 & 1 & 0 & 1 & 0 \\ 1 & 0 & 1 & 0 & 0 & 0 \\ 1 & 1 & 1 & 0 & 1 & 0 \\ 1 & 1 & 0 & 1 & 0 & 1 \\ 1 & 0 & 0 & 1 & 0 & 0 \end{bmatrix}$$

$$\beta = (X'X)^{-1} X'y$$

$$X'X = \begin{bmatrix} 10 & 5 & 3 & 2 & 2 & 1 \\ 5 & 5 & 2 & 1 & 2 & 1 \\ 3 & 2 & 3 & 0 & 2 & 0 \\ 2 & 1 & 0 & 2 & 0 & 1 \\ 2 & 2 & 2 & 0 & 2 & 0 \\ 1 & 1 & 0 & 1 & 0 & 1 \end{bmatrix}$$

$$(X'X)^{-1} = \begin{bmatrix} \dfrac{1}{3} & -\dfrac{1}{3} & -\dfrac{1}{3} & -\dfrac{1}{3} & \dfrac{1}{3} & \dfrac{1}{3} \\ -\dfrac{1}{3} & \dfrac{5}{6} & \dfrac{1}{3} & \dfrac{1}{3} & -\dfrac{5}{6} & -\dfrac{5}{6} \\ -\dfrac{1}{3} & \dfrac{1}{3} & \dfrac{4}{3} & \dfrac{1}{3} & -\dfrac{4}{3} & -\dfrac{1}{3} \\ -\dfrac{1}{3} & \dfrac{1}{3} & \dfrac{1}{3} & \dfrac{4}{3} & -\dfrac{1}{3} & -\dfrac{4}{3} \\ \dfrac{1}{3} & -\dfrac{5}{6} & -\dfrac{4}{3} & -\dfrac{1}{3} & \dfrac{7}{3} & \dfrac{5}{6} \\ \dfrac{1}{3} & -\dfrac{5}{6} & -\dfrac{1}{3} & -\dfrac{4}{3} & \dfrac{5}{6} & \dfrac{17}{6} \end{bmatrix}$$

$$\beta = \begin{bmatrix} 81000 \\ 10000 \\ 44000 \\ 48000 \\ -34000 \\ 1000 \end{bmatrix}$$

$$\text{Income} = 81000 + 10000(\text{Gender}_{\text{Female}}) + 44000(\text{Degree}_{\text{Masters}}) + 48000(\text{Degree}_{\text{PhD}})$$

$$-34000 \times \text{Gender}_{\text{Female}} \times \text{Degree}_{\text{Masters}} + 1000 \times \text{Gender}_{\text{Female}} \times \text{Degree}_{\text{PhD}} + \varepsilon$$

- the income for males with only a Bachelor's degree is $81000
- The income gap between male Bachelor's and Master's degree holders is $44000
- The income gap between male Bachelor's and PhD degree holders is $48000
- The income for males with a Master's degree is $(81000 + 44000) = \$125000$
- the income for males with a PhD degree is $(81000 + 48000) = \$129000$
- The first Income gender gap is $10000
- the income for females with a Bachelor's degree is $(10000 + 81000) = \$91000$
- The Income gender gap among the Master's degree holders is $(10000 - 34000) = -\$24000$
- The Income gender gap among the PhD holders is $(10000 + 1000) = \$11000$
- the income for females with a Master's degree is $(81000 + 10000 + 44000 - 34000)$

$$= \$101000$$

- the income for females with a PhD degree is $(81000 + 10000 + 48000 + 1000)$

$$= \$140000$$

- The income gap between female Bachelor's and Master's degree holders is $\$(44000 - 34000)$

  $= \$10000$

- The income gap between female Bachelor's and PhD degree holders is $\$(48000 + 1000)$

  $= \$49000$

The influence of a Master's degree on income is more significant for males than for females, with respective increases of \$44,000 and \$10,000. In contrast, the impact of a PhD degree on income appears to be relatively smaller, with increases of \$48,000 for males and \$49,000 for females.

Another observation that the Income gender gap is somewhat similar for individuals with Bachelors and PhD. (the gap is \$10000 for Bachelors and \$11000 for PhD) but is negative for Masters (−\$24000).

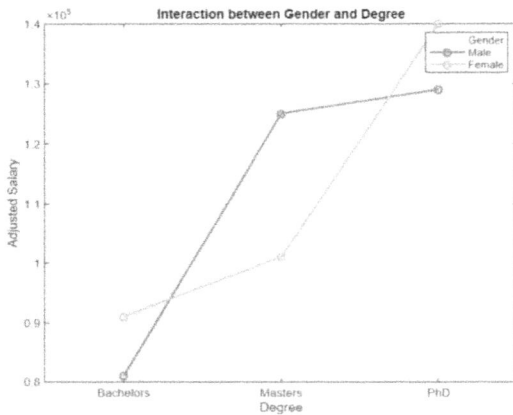

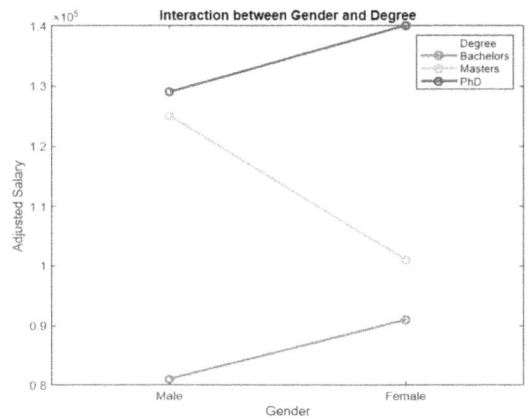

11.12.6 How do we decide whether to include the interaction term or not?

Adding an interaction term to a linear model — estimated using regression — becomes necessary when the statistical association between a predictor and an outcome depends on the value/level of another predictor However, the question remains: How do we ascertain this dependency?

1) Interaction plots:
1. In the three interactions we discussed previously, we utilized interaction plots to understand the potential interaction effects.
   - We saw when the lines are parallel it suggests that there is no interaction effect between the two predictors.
   - On the other hand, the evidence of non-parallel lines at different levels of a categorical predictor suggests that we consider adding an interaction term.
2) P-value of the interaction term:
2. We are going to use the hypothesis testing to conclude if there is a relationship between a predictor and the response variable.

3. After we add the interaction term to the model, if the p-value of the coefficient of the interaction term turns out to be lower than the significance level (usually 0.05), that suggests the interaction term is significantly different from 0 which justifies the inclusion of the term in the model. If the interaction term is statistically significant, the interaction term is probably important.

Finally if the coefficient of determination is also much bigger with the interaction term, it is definitely important. If neither of these outcomes are observed, the interaction term can be removed from the regression equation.

Let's look at the three intersections examples we worked on previously and compare the regression analysis with and without interaction term

```
model =


Linear regression model:
    Income ~ 1 + Age*Experience

Estimated Coefficients:
                    Estimate      SE        tStat       pValue
                    _____    _____    _____    _____

    (Intercept)       26310      4883.2      5.3879     6.0388e-05
    Age              68.089      153.17     0.44452        0.66262
    Experience       1458.6      755.34      1.9311       0.071391
    Age:Experience   11.865      17.776     0.66746        0.51399


Number of observations: 20, Error degrees of freedom: 16
Root Mean Squared Error: 2.54e+03
R-squared: 0.925,  Adjusted R-Squared: 0.91
F-statistic vs. constant model: 65.3, p-value = 3.4e-09
```

We see that the interaction between Age and Experience is not statistically significant at the 0.05 level. When we examine the main effects, we see that both predictors are not statistically significant either.

```
model =


Linear regression model:
    Income ~ 1 + Age + Experience

Estimated Coefficients:
                    Estimate      SE       tStat       pValue
                    _____    _____    _____    _____

    (Intercept)      23583      2630.3     8.9658     7.4797e-08
    Age             145.57      98.297     1.4809        0.15694
    Experience      1926.9      275.17     7.0025     2.1284e-06


Number of observations: 20, Error degrees of freedom: 17
Root Mean Squared Error: 2.49e+03
R-squared: 0.922,  Adjusted R-Squared: 0.913
F-statistic vs. constant model: 101, p-value = 3.66e-10
```

The regression output above shows the p-value= $2.1284 \times 10^{-6} < 0.05$ therefore at the 5% significance level there is enough evidence to suggest that there is a relationship between the dependent variable "Income" and the independent variable "Experience"

On the other hand, because p-value= $0.15694 > 0.05$, we do not reject the null hypothesis. At the 5% significance level there is not enough evidence to suggest that there is a relationship between the dependent variable "Income" and the independent variable "Age."

Results from our sample problem are summarized in the table below:

Analytical output	Without interaction	With interaction
Age, p-value	0	0.66262
Experience, p-value	2.1284×10^{-6}	0.071391
Age:Experience, p-value	NA	0.51399
R^2	1	0.925

The interaction plot doesn't show significant differences when variables don't interact compared to when they do. Generally, older individuals tend to have more experience, leading to higher income

Example 2

With no interaction term

```
model1 =


Linear regression model:
    Salary ~ 1 + Gender*Age

Estimated Coefficients:
                          Estimate        SE        tStat        pValue

                          _____     _____     _____     _____

    (Intercept)            -48610        53420      -0.90996      0.39792
    Gender_Female           32833        74837       0.43873      0.67622
    Age                      3895        1386.7       2.8088      0.030805
    Gender_Female:Age      -721.92       1942.2      -0.37171      0.72288


Number of observations: 10, Error degrees of freedom: 6
Root Mean Squared Error: 1.96e+04
R-squared: 0.693,  Adjusted R-Squared: 0.539
F-statistic vs. constant model: 4.51, p-value = 0.0557
```

Here the interaction term is non-significant, so the two predictors do not have an interactive effect. The p-values for age main effects become more significant compared to the non-interaction only model.

```
model1 =


Linear regression model: .
    Salary ~ 1 + Gender + Age

Estimated Coefficients:
                    Estimate      SE       tStat       pValue
                    _____    _____    _____    _____

    (Intercept)      -34625      35511    -0.97504     0.36202
    Gender_Female      5400      11615     0.46493     0.65609
    Age                3527     909.16      3.8794    0.0060581


Number of observations: 10, Error degrees of freedom: 7
Root Mean Squared Error: 1.84e+04
R-squared: 0.686,  Adjusted R-Squared: 0.596
F-statistic vs. constant model: 7.63, p-value = 0.0174
```

The corresponding p-value of Age variable is **0.0060581**, which is statistically significant at an alpha level of 0.05. This tells us that that the average change in Income for each additional Age is **statistically significantly different than zero.**

Another way to put this: **Age** has a statistically significant relationship with the response variable **Income. In other words,** an individual's age influences their Income.

The corresponding p-value of Gender is **0.65609,** which is not statistically significant at an alpha level of 0.05. Being a female does not significantly increase her Income.

Example 3

```
model1 =


Linear regression model:
    Salary ~ 1 + Gender + Degree

Estimated Coefficients:
                    Estimate      SE      tStat       pValue
                    _____    _____    _____    _____

    (Intercept)       84746      13025     6.5066    0.0006278
    Gender_Female      633.8      16366    0.038726    0.97037
    Degree_Masters     23831      18898     1.261      0.25411
    Degree_PhD         49437      21129     2.3398     0.057862


Number of observations: 10, Error degrees of freedom: 6
Root Mean Squared Error: 2.52e+04
R-squared: 0.493,  Adjusted R-Squared: 0.24
F-statistic vs. constant model: 1.95, p-value = 0.223
```

```
model1 =


Linear regression model:
    Salary ~ 1 + Gender*Degree

Estimated Coefficients:
                                   Estimate      SE       tStat      pValue
                                   _____    _____    _____    _____

    (Intercept)                     81000      16432     4.9295     0.0078756
    Gender_Female                   10000      25981     0.3849     0.7199
    Degree_Masters                  44000      32863     1.3389     0.25164
    Degree_PhD                      48000      32863     1.4606     0.21792
    Gender_Female:Degree_Masters   -34000      43474    -0.78207    0.47788
    Gender_Female:Degree_PhD         1000      47906     0.020874   0.98435


Number of observations: 10, Error degrees of freedom: 4
Root Mean Squared Error: 2.85e+04
R-squared: 0.569,  Adjusted R-Squared: 0.0292
F-statistic vs. constant model: 1.05, p-value = 0.493
```

Although adding an interaction term to a model can make it a better fit with the data, it simultaneously complicates the interpretation of the coefficients of the predictors.

11.13 CENTERING PREDICTORS

In all the models we have discussed so far, each continuous predictor was left uncentered.

11.13.1 Why is it important?

Before we jump into how lets understand its significance. Previously we looked at a model which had

$$\text{Income} = \beta_0 + \beta_1 (\text{Age}) + \beta_2 (\text{Experience}) + \mu$$

2) Interpretation
1. Let's say that now we want to interpret the intercept β_0. We know that the interpretation of intercept is "when all predictors are zero". In our case the intercept would represent the average income of an individual, of zero Experience (Experience = 0) and of zero years (Age = 0).
2. You understand that in this way the value of the intercept is "not interpretable" because at zero years it is impossible to have an income.Therefore, to have an interpretable value, you can analyze the data using a "centered" variable of age, obtained by subtracting the average age from each age value instead of direct use of the age variable. Since when all two predictors are at their average values, the centered variables are 0. Therefore, the intercept can be interpreted as the predicted Income when the predictors are at their average values. So centering completely changes what "zero" represents!
3) Decreases multicollinearity
3. While centering can be done to help interpretation, its real benefits emerge when there are multiplicative terms in a model with interactions or higher order terms(quadratic, cubic etc). Consider our example where two predictor variables are on a positive scale,

Age and Experience. When we multiplied these two variables to create the interaction variable, the numbers near 0 stay near 0 and the high numbers get really high. In other words, if Age is high no matter what value Experience takes (unless it is really close to zero), the product Age×Experience will be high. The interaction term is then highly correlated with original variables. Why does this happen? When all the X values are positive, higher values produce high products and lower values produce low products. So, the product variable is highly correlated with each component variable.

```
Corr_Age_Experience = corrcoef(Age, Experience)
Corr_Interaction_Age = corrcoef(Interaction, Age)
Corr_Interaction_Experience = corrcoef(Interaction, Experience)
                        Corr_Age_Experience =

                            1.0000    0.8142
                            0.8142    1.0000


                        Corr_Interaction_Age =

                            1.0000    0.8982
                            0.8982    1.0000


                        Corr_Interaction_Experience =

                            1.0000    0.9684
                            0.9684    1.0000
```

So

- the correlation between the two predictors is 0.8142
- the correlation between the interaction and the Age is 0.8982
- the correlation between the interaction and the Experience is 0.9684

We observe from the results that the Product-interaction terms are generally highly correlated with their main effects (0.8982 and 0.9684). This problem can obscure the statistical significance of model terms, produce imprecise coefficients, and make it more difficult to choose the correct model.

So what centering does is it reduces the correlation between the individual variables and the product term.

Hence centering becomes necessary both to simplify the interpretation of the coefficients and to reduce the problem of multicollinearity.

11.13.2 How do we do it?

So far, I have been mostly referring centering to as **grand-mean centering**. To grand-mean center a variable, we simply subtract the mean of a predictor variable from each value of that variable, thereby creating a new variable in which the mean is zero and the standard deviation is the same as it was before centering. In our example we have the variable Age for a sample of individuals, where Age is measured in years. In it's raw format, the mean of Age is $793 / 20 = 39.65$ years. So when we center the variable we take the Age of each individual and subtract the mean Age of 39.65.

Our first individual has an Age of 25, then their centered Age would be $25 - 39.65 = -14.65$. By centering each individual's Age relative to the grand-mean, we end up with a variable that has a mean of 0.0, but with the same standard deviation equal to the standard deviation of the original Age variable. Why is this the case? Well, we have just performed a linear shift of Age, which affects only the mean and not the standard deviation.

Similarly we can grand mean center the Experience variable.

x_1 = Age	Centered Age	x_2 = Experience	Centered Experience
25	25-39.65=-14.65	2	2 - 7.1 = -5.1
30	30-39.65=-9.65	3	3 - 7.1 = -4.1
47	47-39.65=7.35	12	12 - 7.1 = 4.9
32	32-39.65=-7.65	5	5 - 7.1 = -2.1
43	43-39.65=3.35	10	10 - 7.1 = 2.9
51	51-39.65=11.35	7	7 - 7.1 = -0.1
28	28-39.65=-11.65	5	5 - 7.1 = -2.1
33	33-39.65=-6.65	4	4 - 7.1 = -3.1
37	37-39.65=-2.65	5	5 - 7.1 = -2.1
39	39-39.65=-0.65	8	8 - 7.1 = 0.9
29	29-39.65=-10.65	3	3 - 7.1 = -4.1
47	47-39.65=7.35	9	9 - 7.1 = 1.9
54	54-39.65=14.35	7	7 - 7.1 = -0.1
51	$51 - \dfrac{793}{20} = 11.35$	11	11 - 7.1 = 3.9
44	$44 - \dfrac{793}{20} = 4.35$	12	12 - 7.1 = 4.9
41	$41 - \dfrac{793}{20} = 1.35$	6	6 - 7.1 = -1.1
58	$58 - \dfrac{793}{20} = 18.35$	13	13 - 7.1 = 5.9
23	$23 - \dfrac{793}{20} = -16.65$	1	1 - 7.1 = -6.1
44	$44 - \dfrac{793}{20} = 4.35$	9	9 - 7.1 = 1.9
37	$37 - \dfrac{793}{20} = -2.65$	10	10 - 7.1 = 2.9
$\sum 793$		$\sum 142$	
$\overline{x_1} = \dfrac{793}{20} = 39.65$		$\overline{x_2} = \dfrac{142}{20} = 7.1$	

Now that we centered our two predictors we can use them to find our new regression line.

$$X = \begin{bmatrix} & x_1 & x_2 \\ 1 & -14.65 & -5.1 \\ 1 & -9.65 & -4.1 \\ 1 & 7.35 & 4.9 \\ \cdot & \cdot & \cdot \\ \cdot & \cdot & \cdot \\ \cdot & \cdot & \cdot \\ 1 & -2.65 & -2.1 \\ 1 & -0.65 & 0.9 \\ 1 & -10.65 & -4.1 \\ 1 & 7.35 & 1.9 \\ \cdot & \cdot & \cdot \\ \cdot & \cdot & \cdot \\ \cdot & \cdot & \cdot \\ 1 & -16.65 & -6.1 \\ 1 & 4.35 & 1.9 \\ 1 & -265 & 2.9 \end{bmatrix}$$

$$\beta = (X'X)^{-1} X'y$$

$$X'X = \begin{bmatrix} 20 & 0 & 0 \\ 0 & 1910.55 & 555.7 \\ 0 & 555.7 & 243.8 \end{bmatrix}$$

$$(X'X)^{-1} = \begin{bmatrix} 0.05 & 0 & 0 \\ 0 & 0.00155297 & -0.003539725 \\ 0 & -0.003539725 & 0.01217 \end{bmatrix}$$

$$\beta = \begin{bmatrix} 43035.5 \\ 145.5654 \\ 1926.90037 \end{bmatrix}$$

The regression line

$$\hat{y} = 43035.5 + 145.5654(\text{Age}) + 1926.90037(\text{Experience})$$

Using the centered variable, the interpretation of β_0 is now the mean Income of an individual with an average Age and Experience.

Lets now find the new regression line with interaction term

With interaction:

C.Age	C. Exp	C. Age × C. Exp
-14.65	-5.1	74.715
-9.65	-4.1	39.565
7.35	4.9	36.015
-7.65	-2.1	16.065
3.35	2.9	9.715
11.35	-0.1	-1.135
-11.65	-2.1	24.465
-6.65	-3.1	20.615
-2.65	-2.1	5.565
-0.65	0.9	-0.585
-10.65	-4.1	43.665
7.35	1.9	13.965
14.35	-0.1	-1.435
11.35	3.9	44.265
4.35	4.9	21.315
1.35	-1.1	-1.485
18.35	5.9	108.265
-16.65	-6.1	101.565
4.35	1.9	8.265
-2.65	2.9	-7.685

$$X = \begin{bmatrix} & x_1 & x_2 & x_1x_2 \\ 1 & -14.65 & -5.1 & 74.715 \\ 1 & -9.65 & -4.1 & 39.565 \\ 1 & 7.35 & 4.9 & 36.015 \\ \cdot & \cdot & \cdot & \cdot \\ \cdot & \cdot & \cdot & \cdot \\ \cdot & \cdot & \cdot & \cdot \\ 1 & -2.65 & -2.1 & 5.565 \\ 1 & -0.65 & 0.9 & -0.585 \\ 1 & -10.65 & -4.1 & 43.665 \\ 1 & 7.35 & 1.9 & 13.965 \\ \cdot & \cdot & \cdot & \cdot \\ \cdot & \cdot & \cdot & \cdot \\ \cdot & \cdot & \cdot & \cdot \\ 1 & -16.65 & -6.1 & 101.565 \\ 1 & 4.35 & 1.9 & 101.565 \\ 1 & -2.65 & 2.9 & -7.685 \end{bmatrix}, y = \begin{bmatrix} 31450 \\ 35670 \\ 51580 \\ \cdot \\ \cdot \\ \cdot \\ 40250 \\ 45150 \\ 29840 \\ 46110 \\ \cdot \\ \cdot \\ \cdot \\ 26870 \\ 44190 \\ 48700 \end{bmatrix}$$

$$\beta = (X'X)^{-1} X'y$$

$$X'X = \begin{bmatrix} 20 & 0 & 0 & 555.7 \\ 0 & 1910.55 & 555.7 & -1189.31 \\ 0 & 555.7 & 243.8 & -360.74 \\ 555.7 & -1189.31 & -360.74 & 36537.1465 \end{bmatrix}$$

$$\beta = \begin{bmatrix} 42705.838 \\ 152.3288 \\ 1929.040116 \\ 11.86475 \end{bmatrix}$$

The regression line

$$\hat{y} = 42705.838 + 152.3238(\text{Age}) + 1929.040116(\text{Experience}) + 11.86475 \times \text{Age} \times \text{Experience}$$

11.13.3 Centering in MATLAB

In MATLAB to center the variables we use the function normalize and then specify which type of centering method we want to use. The syntax of our function would be:

$$\text{Centered\_Predictor} = \text{normalize}(\_\_\_, \text{method}, \text{methodtype})$$

Method	Method Type Options	Description
"zscore"	"std" (default)	Compute the z-score. Center data to have mean 0, and scale data to have standard deviation 1.
	"robust"	Compute the z-score. Center data to have mean 0, and scale data to have median absolute deviation 1.
"norm"	Positive numeric scalar (default is 2)	Scale data by the p-norm, where p is a positive numeric scalar.
	Inf	Scale data by the p-norm, where p is Inf. The infinity norm, or maximum norm, is the same as the largest magnitude of the elements in the data.
"scale"	"std" (default)	Scale data to have standard deviation 1.
	"mad"	Scale data to have median absolute deviation 1.
	"first"	Scale data by the first element of the data.
	"iqr"	Scale data to have interquartile range 1.
	Numeric array	Scale data by an array of numeric values. The array must have a compatible size with input A.
	Table	Scale data by variables in a table. Each table variable in the input data A is scaled using the value in the similarly named variable in the scaling table.
"range"	2-element row vector (default is [0 1])	Rescale range of data to [a b], where a < b.
"center"	"mean" (default)	Center data to have mean 0.
	"median"	Center data to have median 0.
	Numeric array	Shift center by an array of numeric values. The array must have a compatible size with input A.
	Table	Shift center by variables in a table. Each table variable in the input data A is centered using the value in the similarly named variable in the centering table

Here we shall use the method "center" and the default type "mean" to center the data to have mean zero

Now that we have seen how we can get our centered interaction terms lets now check if it affects the correlation.

```
Age = normalize(Age,"center");
Experience = normalize(Experience, "center");
Interaction = Age.*Experience;
Corr_Age_Experience = corrcoef(Age, Experience)
Corr_Interaction_Age = corrcoef(Interaction, Age)
Corr_Interaction_Experience = corrcoef(Interaction, Experience)
```

```
Corr_Age_Experience =

       1.0000    0.8142
       0.8142    1.0000


Corr_Interaction_Age =

       1.0000   -0.1873
      -0.1873    1.0000


Corr_Interaction_Experience =

       1.0000   -0.1591
      -0.1591    1.0000
```

The output illustrate how centering variables can reduce the correlation between the variables and their interaction terms. Note that centering two variables does NOT change the correlation between them.(Before and after centering the correlation between the two predictors Age and Experience stays the same).

Now that there is no issues of multicollinearity we can use our new centered predictors to estimate the regression line.

11.13.4 How Centering affects the coefficients

We will now investigate how centering our continuous variables will affect the estimates from our model

As mentioned earlier, Centering is the rescaling of predictors by subtracting the mean. So before centering the predictors the linear regression followed a model of the form:

$$y = \beta_0 + \beta_1 x_{1i} + \beta_2 x_{2i} + \beta_3 x_3 + \ldots + \beta_{ki} x_{ki} + \varepsilon$$

After centering them the predictors in the equation are

$$y = \beta_0 + \beta_1 \left(x_{1i} - \overline{x_1} \right) + \beta_2 \left(x_{2i} - \overline{x_2} \right) + \beta_3 \left(x_{3i} - \overline{x_3} \right) + \ldots \beta_{ki} (x_{ki} - \overline{x_k}) + \varepsilon \qquad \text{(eq 1)}$$

$$= \left(\beta_0 - \beta_1 \overline{x_1} - \beta_2 \overline{x_2} - \beta_3 \overline{x_3} - \ldots \beta_k \overline{x_k} \right) + \beta_1 x_{1i} + \beta_2 x_{2i} + \beta_3 x_3 + \ldots + \beta_{ki} x_{ki} + \varepsilon \qquad \text{(eq 2)}$$

Comparing the two models, only the intercepts are different.

Therefore Centering predictors in a regression model with only main effects (no interaction) has no influence on the main effects but as we have seen earlier, it *does* change what we get out of it. If we re-center a predictor in our linear model, the only thing that changes is our intercept.

So for our example

$$\text{Income} = \beta_0 + \beta_1 \left(\text{Age} - \overline{Age}\right) + \beta_2 \left(\text{Experience} - \overline{Experience}\right) + \varepsilon$$

β_1 can be interpreted as the average change in Income when $(Age - \overline{Age})$ increases by 1 unit, which is equivalent to the average change in Income when the Age increases by 1 unit. This implies that when we center the independent variable, the interpretation of β_1 remains the same.

β_2 can be interpreted as the average change in Income when $(\text{Experience} - \overline{Experience})$ increases by 1 unit, which is equivalent to the average change in Income when the Experience increases by 1 unit. This implies that when we center the independent variable, the interpretation of β_2 remains the same.

$$\text{Income} = \left(\beta_0 - \beta_1 \overline{Age} - \beta_2 \overline{Experience}\right) + \beta_1 \left(\text{Age}\right) + \beta_2 \left(\text{Experience}\right) + \varepsilon$$

However, from eq 2 we observe that the intercept parameter has changed. Using the centered variable, the interpretation of β_0 is now the mean Income of someone whose Age and Experience is equal to the average in the sample. This is of course much more interpretable.

To verify this lets look at the linear regression with only main effects with uncentered predictors and then with centered predictors

```
model =


Linear regression model:
    Income ~ 1 + Age + Experience

Estimated Coefficients:
                   Estimate      SE       tStat      pValue

                   _____    _____    _____    _____

    (Intercept)      23583     2630.3    8.9658    7.4797e-08
    Age             145.57     98.297    1.4809       0.15694
    Experience      1926.9     275.17    7.0025    2.1284e-06


Number of observations: 20, Error degrees of freedom: 17
Root Mean Squared Error: 2.49e+03
R-squared: 0.922,  Adjusted R-Squared: 0.913
F-statistic vs. constant model: 101, p-value = 3.66e-10
```

Now with centering the predictor data

```
Age = normalize(Age, "center");
Experience = normalize(Experience, "center");
```

```
model =


Linear regression model:
    Income ~ 1 + Age + Experience

Estimated Coefficients:
                   Estimate      SE       tStat       pValue

                   _____    _____   _____    _____

    (Intercept)      43035     557.76    77.158     4.4024e-23
    Age             145.57     98.297    1.4809       0.15694
    Experience      1926.9     275.17    7.0025     2.1284e-06


Number of observations: 20, Error degrees of freedom: 17
Root Mean Squared Error: 2.49e+03
R-squared: 0.922,  Adjusted R-Squared: 0.913
F-statistic vs. constant model: 101, p-value = 3.66e-10
```

The parameter estimates of the regression with uncentered predictors are

$\beta_1 = \beta_1^c = 145.57$ and $\beta_2 = \beta_2^c = 1926.9$. One can clearly see that centering the predictors only changes the intercept which confirms that centering the predictors in a regression model has no influence on the main effects.

Lets now include the interaction term and show that centering the predictors now does affect the main effects. We first fit the regression model without centering

```
model =


Linear regression model:
    Income ~ 1 + Age*Experience

Estimated Coefficients:
                     Estimate     SE       tStat       pValue

                     _____   _____   _____    _____

    (Intercept)        26310    4883.2    5.3879     6.0388e-05
    Age               68.089    153.17    0.44452      0.66262
    Experience        1458.6    755.34    1.9311       0.071391
    Age:Experience    11.865    17.776    0.66746      0.51399


Number of observations: 20, Error degrees of freedom: 16
Root Mean Squared Error: 2.54e+03
R-squared: 0.925,  Adjusted R-Squared: 0.91
F-statistic vs. constant model: 65.3, p-value = 3.4e-09
```

With centering

```
model =


Linear regression model:
    Income ~ 1 + Age*Experience

Estimated Coefficients:
                    Estimate      SE       tStat       pValue
                    _____    _____    _____    _____

    (Intercept)       42706     752.02     56.788     6.9448e-20
    Age              152.33     100.45     1.5164        0.14892
    Experience         1929     279.79     6.8946     3.6069e-06
    Age:Experience   11.865     17.776     0.66746       0.51399


Number of observations: 20, Error degrees of freedom: 16
Root Mean Squared Error: 2.54e+03
R-squared: 0.925,  Adjusted R-Squared: 0.91
F-statistic vs. constant model: 65.3, p-value = 3.4e-09
```

Ostensibly, that's a big difference! It's worth mentioning that the fit didn't change, we still have a reported $R^2 = 0.925$. The big differences are reflected in the standard errors between the two models. When data are mean centered the standard errors are much smaller. Notice that the interaction term did not change.

Therefore If we only have main effects in your model then centering does not influence results, BUT if the predictors are involved in an interaction, centering DOES influence results!

11.13.5 Standardizing variables

If we go back and look at the table, we have multiple methods of normalizing the predictors. In regression besides using the center mean method another common method of transforming our predictors is scaling. Scaling changes the *units* of the variable, and the most common transformation that involves scaling is 'standardisation'. During standardization, we remove the mean from each value, and then scale them by dividing them by their standard deviation. The result is data with a mean of zero and the standard deviation of one.

Lets now standardize our variable Age. In its raw format, the mean of Age is 39.65 years with a standard deviation of 10.0277. For the individual whose Age is 25 their scaled Age would be $\frac{25 - 39.65}{10.0277} = -1.4609$. Similarly, the mean of Experience is 7.1 years with a standard deviation of 3.5821. For the individual with Experience of 2 years their scaled Experience would be $\frac{2 - 7.1}{3.5821} = -1.4237$

So how does scaling/standardizing differ from centering?

While they are similar in the way that both of them change the interpretation of the intercept, they differ in their impact on the coefficients of our model.

In centering, we are changing the values but not the scale. As seen earlier, without the interaction term, the intercept will change, but the regression coefficient for that variable will not. Since the regression coefficient is interpreted as the effect on the mean of Y for each one unit difference in X, it doesn't change when X is centered.

When we scale a predictor, this will change both the intercept and the slope. However, note that the significance of the slopes remains exactly the same, we are only changing the units that we are using to expressing that slope. A one-unit difference now means a one-standard deviation difference. This is usually done so you can compare coefficients for predictors that were measured on different scales. In other words when one regressor is measured on a very small order, while another on a very large order. For example, take a Linear Regression model that uses 3 variables, high school GPA, SAT score and letters of recommendation (X variables), to predict the college GPA (y variable). While GPA range from 0 to 4 in 0.1 increments, SAT scores range from 400 to 1600 in 10-point increments, and number of letter of recommendation is simple discrete values. Standardization allows the units of regression coefficients to be expressed in the same units.

```
Predictors = [Age', Experience'];
model = fitlm(Predictors, Income)
%centered the variables
AgeC = normalize(Age, "center");
ExperienceC = normalize(Experience, "center");
PredictorsC = [AgeC', ExperienceC'];
Centered_model = fitlm(PredictorsC, Income)
%scaled the variables
AgeZ = normalize(Age, "zscore");
ExperienceZ = normalize(Experience, "zscore");
PredictorsZ = [AgeZ', ExperienceZ'];
Scaled_model = fitlm(PredictorsZ, Income)
```

model =

Linear regression model:
    y ~ 1 + x1 + x2

Estimated Coefficients:

	Estimate	SE	tStat	pValue
(Intercept)	23583	2630.3	8.9658	7.4797e-08
x1	145.57	98.297	1.4809	0.15694
x2	1926.9	275.17	7.0025	2.1284e-06

Number of observations: 20, Error degrees of freedom: 17
Root Mean Squared Error: 2.49e+03
R-squared: 0.922, Adjusted R-Squared: 0.913
F-statistic vs. constant model: 101, p-value = 3.66e-10

Centered_model =

Linear regression model:
    y ~ 1 + x1 + x2

Estimated Coefficients:

	Estimate	SE	tStat	pValue
(Intercept)	43035	557.76	77.158	4.4024e-23
x1	145.57	98.297	1.4809	0.15694
x2	1926.9	275.17	7.0025	2.1284e-06

Number of observations: 20, Error degrees of freedom: 17
Root Mean Squared Error: 2.49e+03
R-squared: 0.922, Adjusted R-Squared: 0.913
F-statistic vs. constant model: 101, p-value = 3.66e-10

Scaled_model =

Linear regression model:
    y ~ 1 + x1 + x2

Estimated Coefficients:

	Estimate	SE	tStat	pValue
(Intercept)	43035	557.76	77.158	4.4024e-23
x1	1459.7	985.7	1.4809	0.15694
x2	6902.4	985.7	7.0025	2.1284e-06

Number of observations: 20, Error degrees of freedom: 17
Root Mean Squared Error: 2.49e+03
R-squared: 0.922, Adjusted R-Squared: 0.913
F-statistic vs. constant model: 101, p-value = 3.66e-10

In this chapter, we have explored the fundamental concepts of regression analysis, focusing on how to work with both continuous and categorical predictors. We studied the importance of interaction effects and how to identify and interpret them within a regression model. Additionally, we covered the process of standardizing predictors to ensure consistency and comparability across variables. These foundational techniques are essential for building robust regression models and will serve as a strong basis for more advanced topics in the field of statistics.

Index

For Product Safety Concerns and Information please contact our EU
representative GPSR@taylorandfrancis.com
Taylor & Francis Verlag GmbH, Kaufingerstraße 24, 80331 München, Germany